VOLUME FIFTY FIVE

ADVANCES IN
ECOLOGICAL RESEARCH
Large-Scale Ecology: Model Systems to Global Perspectives

ADVANCES IN ECOLOGICAL RESEARCH

Series Editors

GUY WOODWARD
Professor of Ecology
Imperial College London
Silwood Park Campus
Buckhurst Road
Ascot Berkshire
SL5 7PY United Kingdom

DAVID A. BOHAN
Directeur de Recherche
UMR 1347 Agroécologie
AgroSup/UB/INRA
Pôle GESTAD, Dijon, France

ALEX J. DUMBRELL
School of Biological Sciences
University of Essex
Wivenhoe Park, Colchester
Essex, United Kingdom

VOLUME FIFTY FIVE

Advances in
ECOLOGICAL RESEARCH

Large-Scale Ecology: Model Systems to Global Perspectives

Edited by

ALEX J. DUMBRELL
School of Biological Sciences
University of Essex
Wivenhoe Park, Colchester, Essex,
United Kingdom

REBECCA L. KORDAS
Department of Life Sciences
Imperial College London, Silwood Park Campus,
Ascot, Berkshire, United Kingdom

GUY WOODWARD
Department of Life Sciences
Imperial College London
Silwood Park Campus, Ascot, Berkshire,
United Kingdom

AMSTERDAM • BOSTON • HEIDELBERG • LONDON
NEW YORK • OXFORD • PARIS • SAN DIEGO
SAN FRANCISCO • SINGAPORE • SYDNEY • TOKYO
Academic Press is an imprint of Elsevier

Academic Press is an imprint of Elsevier
125 London Wall, London, EC2Y 5AS, United Kingdom
The Boulevard, Langford Lane, Kidlington, Oxford OX5 1GB, United Kingdom
525 B Street, Suite 1800, San Diego, CA 92101-4495, United States
50 Hampshire Street, 5th Floor, Cambridge, MA 02139, United States

First edition 2016

© 2016 Elsevier Ltd. All rights reserved.

No part of this publication may be reproduced or transmitted in any form or by any means, electronic or mechanical, including photocopying, recording, or any information storage and retrieval system, without permission in writing from the publisher. Details on how to seek permission, further information about the Publisher's permissions policies and our arrangements with organizations such as the Copyright Clearance Center and the Copyright Licensing Agency, can be found at our website: www.elsevier.com/permissions.

This book and the individual contributions contained in it are protected under copyright by the Publisher (other than as may be noted herein).

Notices
Knowledge and best practice in this field are constantly changing. As new research and experience broaden our understanding, changes in research methods, professional practices, or medical treatment may become necessary.

Practitioners and researchers must always rely on their own experience and knowledge in evaluating and using any information, methods, compounds, or experiments described herein. In using such information or methods they should be mindful of their own safety and the safety of others, including parties for whom they have a professional responsibility.

To the fullest extent of the law, neither the Publisher nor the authors, contributors, or editors, assume any liability for any injury and/or damage to persons or property as a matter of products liability, negligence or otherwise, or from any use or operation of any methods, products, instructions, or ideas contained in the material herein.

ISBN: 978-0-08-100935-2
ISSN: 0065-2504

For information on all Academic Press publications
visit our website at https://www.elsevier.com/

Publisher: Zoe Kruze
Acquisition Editor: Alex White
Editorial Project Manager: Helene Kabes
Production Project Manager: Radhakrishnan Lakshmanan
Cover Designer: Greg Harris

Typeset by SPi Global, India

CONTENTS

Contributors xi
Preface xix

Part I
Large Spatial Scale Ecology

1. The Unique Contribution of Rothamsted to Ecological Research at Large Temporal Scales 3
J. Storkey, A.J. Macdonald, J.R. Bell, I.M. Clark, A.S. Gregory, N.J. Hawkins, P.R. Hirsch, L.C. Todman, and A.P. Whitmore

  1. Introduction to Long-Term Ecological Research at Rothamsted 4
  2. Monitoring the Impact of Environmental Change 13
  3. Community Ecology 17
  4. Ecosystem Stability and Resilience 24
  5. Evolutionary Ecology 28
  6. Soil Microbial Ecology 32
  7. Conclusion 35
  Acknowledgements 36
  References 36

2. How Agricultural Intensification Affects Biodiversity and Ecosystem Services 43
M. Emmerson, M.B. Morales, J.J. Oñate, P. Batáry, F. Berendse, J. Liira, T. Aavik, I. Guerrero, R. Bommarco, S. Eggers, T. Pärt, T. Tscharntke, W. Weisser, L. Clement, and J. Bengtsson

  1. Introduction 44
  2. The CAP and AI 49
  3. Local-Level and Landscape-Level Effects of AI 60
  4. Organic-Conventional Comparisons 79
  5. Linking AI to Biodiversity and Ecosystem Services 83
  6. Conclusions 87
  Acknowledgements 88
  References 89

3. Litter Decomposition as an Indicator of Stream Ecosystem Functioning at Local-to-Continental Scales: Insights from the European *RivFunction* Project 99

E. Chauvet, V. Ferreira, P.S. Giller, B.G. McKie, S.D. Tiegs, G. Woodward, A. Elosegi, M. Dobson, T. Fleituch, M.A.S. Graça, V. Gulis, S. Hladyz, J.O. Lacoursière, A. Lecerf, J. Pozo, E. Preda, M. Riipinen, G. Rîşnoveanu, A. Vadineanu, L.B.-M. Vought, and M.O. Gessner

1. Introduction 101
2. Nutrient Enrichment Effects on Leaf Litter Decomposition 107
3. Effects of Riparian Forest Modifications on Leaf Litter Decomposition 118
4. Biodiversity-Related Mechanisms Underlying Altered Litter Decomposition 137
5. Accomodating Natural Variability When Using Litter Decomposition in Stream Assessment 153
6. Towards the Integration of Ecosystem Functioning into Stream Management 161
Acknowledgements 168
References 168

4. Unravelling the Impacts of Micropollutants in Aquatic Ecosystems: Interdisciplinary Studies at the Interface of Large-Scale Ecology 183

C. Stamm, K. Räsänen, F.J. Burdon, F. Altermatt, J. Jokela, A. Joss, M. Ackermann, and R.I.L. Eggen

1. Large-Scale Ecology and Human Impacts on Ecosystems 184
2. Water Management as a *Real-World Experiment* 193
3. Outlook: Potential of Combining *Real-World* and *Research-Led Experiments* 211
Acknowledgements 215
References 216

Part II
Large/Long Temporal Scale Ecology and Model Systems

5. The Colne Estuary: A Long-Term Microbial Ecology Observatory 227

D.B. Nedwell, G.J.C. Underwood, T.J. McGenity, C. Whitby, and A.J. Dumbrell

1. Introduction 228
2. Study Site Description 231
3. Functional Ecology of Estuarine Microbes 238
4. Estuarine Saltmarshes 263
5. Estuaries and Climatically Important Trace Gases 265

6. Stressors and Pollution	268
7. Future Directions	270
Acknowledgements	273
References	273

6. Locally Extreme Environments as Natural Long-Term Experiments in Ecology   283
I. Maček, D. Vodnik, H. Pfanz, E. Low-Décarie, and A.J. Dumbrell

1. Introduction	284
2. Locally Extreme Environments as Long-Term Experiments	285
3. Case Study: Mofettes	290
4. Conclusions	314
Acknowledgements	316
References	316

7. Climate-Driven Range Shifts Within Benthic Habitats Across a Marine Biogeographic Transition Zone   325
N. Mieszkowska and H.E. Sugden

1. Introduction	326
2. The Rise of Natural History and Species Recording	328
3. History and Development of Biogeographic Research in the Northeast Atlantic	330
4. Patterns of Change Across the Boreal–Lusitanian Biogeographic Breakpoint in the Northeast Atlantic	332
5. Factors Setting Biogeographic Range Limits	336
6. Long-Term Time-Series for Benthic Ecosystems in the Northeast Atlantic and Regional Seas	341
7. Observed Changes in the Physical Environment	344
8. Impacts of Climate Change on Intertidal Benthic Species	347
9. Future Advances in Quantifying and Modelling Distributional Responses to Climate Change	352
References	356

8. Cross-Scale Approaches to Forecasting Biogeographic Responses to Climate Change   371
J.L. Torossian, R.L. Kordas, and B. Helmuth

1. Introduction	372
2. Common Pitfalls and Their Unintended Consequences	383

3. Moving Forward: How Do We Make Useful Forecasts While Recognizing Limitations?	407
4. Conclusions	413
Acknowledgements	414
References	414

Part III
Large SpatioTemporal Scale Ecology

9. Shifting Impacts of Climate Change: Long-Term Patterns of Plant Response to Elevated CO_2, Drought, and Warming Across Ecosystems — 437

L.C. Andresen, C. Müller, G. de Dato, J.S. Dukes, B.A. Emmett, M. Estiarte, A. Jentsch, G. Kröel-Dulay, A. Lüscher, S. Niu, J. Peñuelas, P.B. Reich, S. Reinsch, R. Ogaya, I.K. Schmidt, M.K. Schneider, M. Sternberg, A. Tietema, K. Zhu, and M.C. Bilton

1. Introduction	439
2. Methods for Data Analysis	444
3. Results	450
4. Discussion	456
5. Conclusions	462
Acknowledgements	462
Appendix A. Details of the Database I	463
Appendix B. Details of the Database II	464
Appendix C. Site Details	465
Appendix D. Site Groupings	467
References	469

10. Recovery and Nonrecovery of Freshwater Food Webs from the Effects of Acidification — 475

C. Gray, A.G. Hildrew, X. Lu, A. Ma, D. McElroy, D. Monteith, E. O'Gorman, E. Shilland, and G. Woodward

1. Introduction	476
2. Methods	487
3. Results	491
4. Discussion	502
5. Conclusion	507
Appendix	508
Acknowledgements	528
References	528

11. Effective River Restoration in the 21st Century: From Trial and Error to Novel Evidence-Based Approaches — 535

N. Friberg, N.V. Angelopoulos, A.D. Buijse, I.G. Cowx, J. Kail, T.F. Moe, H. Moir, M.T. O'Hare, P.F.M. Verdonschot, and C. Wolter

1. Introduction — 537
2. Responses of River Biota to Hydrology and Physical Habitats — 545
3. The Current Restoration Paradigm — 560
4. Effects of Restoration — 569
5. Future Directions — 578
6. Conclusions — 593
Acknowledgements — 600
References — 600

Part IV
A Look To the Future

12. Recommendations for the Next Generation of Global Freshwater Biological Monitoring Tools — 615

M.C. Jackson, O.L.F. Weyl, F. Altermatt, I. Durance, N. Friberg, A.J. Dumbrell, J.J. Piggott, S.D. Tiegs, K. Tockner, C.B. Krug, P.W. Leadley, and G. Woodward

1. Introduction — 616
2. Invertebrates as Indicators of Ecosystem State — 618
3. Decomposition-Based Indicators — 623
4. Fishery Indicators: Learning from the Marine Realm — 624
5. Molecular-Based Indicators — 625
6. Indicators of Change Across Space and Time — 628
7. Conclusions and Future Directions — 630
Acknowledgments — 631
References — 631

Index — *637*
Cumulative List of Titles — *649*

CONTRIBUTORS

T. Aavik
Institute of Ecology and Earth Sciences, University of Tartu, Tartu, Estonia

M. Ackermann
Eawag, Swiss Federal Institute of Aquatic Science and Technology, Dübendorf; ETH Zürich, Environmental Systems Science, Zürich, Switzerland

F. Altermatt
Eawag, Swiss Federal Institute of Aquatic Science and Technology, Dübendorf; University of Zurich, Zürich, Switzerland

L.C. Andresen
University of Gothenburg, Gothenburg, Sweden; Justus-Liebig-University Giessen, Gießen, Germany

N.V. Angelopoulos
University of Hull, Hull, United Kingdom

P. Batáry
Georg-August-University, Göttingen, Germany

J.R. Bell
Rothamsted Research, Harpenden, United Kingdom

J. Bengtsson
Swedish University of Agricultural Sciences, Uppsala, Sweden

F. Berendse
Wageningen University, Wageningen, The Netherlands

M.C. Bilton
University of Tübingen, Tübingen, Germany

R. Bommarco
Swedish University of Agricultural Sciences, Uppsala, Sweden

A.D. Buijse
Deltares, Delft, The Netherlands

F.J. Burdon
Eawag, Swiss Federal Institute of Aquatic Science and Technology, Dübendorf, Switzerland

E. Chauvet
EcoLab, University of Toulouse, CNRS, INPT, UPS, Toulouse, France

I.M. Clark
Rothamsted Research, Harpenden, United Kingdom

L. Clement
School of Life Sciences Weihenstephan, Technische Universität München, Freising, Germany

I.G. Cowx
University of Hull, Hull, United Kingdom

G. de Dato
Council for Agricultural Research and Economics–Forestry Research Centre (CREA-SEL), Arezzo, Italy

M. Dobson
APEM Limited, Edinburgh Technopole, Penicuik, United Kingdom

J.S. Dukes
Purdue University, West Lafayette, IN, United States

A.J. Dumbrell
School of Biological Sciences, University of Essex, Colchester, United Kingdom

I. Durance
Cardiff Water Research Institute, School of Biosciences, Cardiff University, Cardiff, United Kingdom

R.I.L. Eggen
Eawag, Swiss Federal Institute of Aquatic Science and Technology, Dübendorf; ETH Zürich, Institute of Biogeochemistry and Pollutant Dynamics, Zürich, Switzerland

S. Eggers
Swedish University of Agricultural Sciences, Uppsala, Sweden

A. Elosegi
Faculty of Science and Technology, University of the Basque Country, Bilbao, Spain

M. Emmerson
Institute of Global Food Security, School of Biological Sciences, Queen's University Belfast, Belfast, United Kingdom

B.A. Emmett
Center for Ecology and Hydrology (CEH), Bangor, United Kingdom

M. Estiarte
CSIC, Global Ecology Unit CREAF-CSIC-UAB; CREAF, Cerdanyola del Vallès, Barcelona, Catalonia, Spain

V. Ferreira
MARE, University of Coimbra, Coimbra, Portugal

T. Fleituch
Institute of Nature Conservation, Polish Academy of Sciences, Krakow, Poland

N. Friberg
Norwegian Institute for Water Research (NIVA), Oslo, Norway; water@leeds, School of Geography, University of Leeds, Leeds, United Kingdom

M.O. Gessner
Leibniz Institute of Freshwater Ecology and Inland Fisheries; Berlin Institute of Technology, Berlin, Germany

P.S. Giller
School of Biological, Earth and Environmental Sciences, University College Cork, Cork, Ireland

M.A.S. Graça
MARE, University of Coimbra, Coimbra, Portugal

C. Gray
Imperial College London, Ascot, Berkshire; School of Biological and Chemical Sciences, Queen Mary University of London, London, United Kingdom

A.S. Gregory
Rothamsted Research, Harpenden, United Kingdom

I. Guerrero
Terrestrial Ecology Group (TEG), Universidad Autónoma de Madrid, Madrid, Spain

V. Gulis
Coastal Carolina University, Conway, SC, United States

N.J. Hawkins
Rothamsted Research, Harpenden, United Kingdom

B. Helmuth
Northeastern University, Marine Science Center, Nahant, MA, United States

A.G. Hildrew
School of Biological and Chemical Sciences, Queen Mary University of London, London; Freshwater Biological Association, The Ferry Landing, Ambleside, Cumbria, United Kingdom

P.R. Hirsch
Rothamsted Research, Harpenden, United Kingdom

S. Hladyz
School of Biological Sciences, Monash University, Melbourne, Australia

M.C. Jackson
University of Pretoria, Hatfield, Gauteng, South Africa; Imperial College London, Ascot, Berkshire, United Kingdom

A. Jentsch
Bayreuth Center of Ecology and Environmental Research (BayCEER), Bayreuth, Germany

J. Jokela
Eawag, Swiss Federal Institute of Aquatic Science and Technology, Dübendorf; ETH Zürich, Institute of Integrative Biology, Zürich, Switzerland

A. Joss
Eawag, Swiss Federal Institute of Aquatic Science and Technology, Dübendorf, Switzerland

J. Kail
University of Duisburg-Essen, Essen, Germany

R.L. Kordas
Imperial College London, Ascot, Berkshire, United Kingdom

G. Kröel-Dulay
MTA Centre for Ecological Research, Institute of Ecology and Botany, Budapest, Hungary

C.B. Krug
Laboratoire ESE, Université Paris-Sud, UMR 8079 CNRS, UOS, AgroParisTech, Orsay, France

J.O. Lacoursière
Kristianstad University, Kristianstad, Sweden

P.W. Leadley
Laboratoire ESE, Université Paris-Sud, UMR 8079 CNRS, UOS, AgroParisTech, Orsay, France

A. Lecerf
EcoLab, University of Toulouse, CNRS, INPT, UPS, Toulouse, France

J. Liira
Terrestrial Ecology Group (TEG), Universidad Autónoma de Madrid, Madrid, Spain; Institute of Ecology and Earth Sciences, University of Tartu, Tartu, Estonia

E. Low-Décarie
School of Biological Sciences, University of Essex, Colchester, United Kingdom

A. Lüscher
ETH Zürich, Institute of Agricultural Sciences; Institute for Sustainability Sciences, Agroscope, Zürich, Switzerland

X. Lu
School of Electronic Engineering and Computer Science, Queen Mary University of London, London, United Kingdom

A. Ma
School of Electronic Engineering and Computer Science, Queen Mary University of London, London, United Kingdom

A.J. Macdonald
Rothamsted Research, Harpenden, United Kingdom

I. Maček
Biotechnical Faculty, University of Ljubljana, Ljubljana; Faculty of Mathematics, Natural Sciences and Information Technologies (FAMNIT), University of Primorska, Koper, Slovenia

D. McElroy
Centre for Research on Ecological Impacts of Coastal Cities, University of Sydney, Sydney, NSW, Australia

T.J. McGenity
School of Biological Sciences, University of Essex, Colchester, United Kingdom

B.G. McKie
Aquatic Sciences & Assessment, Swedish University of Agricultural Sciences, Uppsala, Sweden

N. Mieszkowska
School of Environmental Sciences, University of Liverpool, Liverpool; The Marine Biological Association of the UK, Plymouth, United Kingdom

H. Moir
cbec eco-engineering Ltd., Inverness, Scotland, United Kingdom

C. Müller
Justus-Liebig-University Giessen, Gießen, Germany; School of Biology and Environmental Science, University College Dublin, Dublin, Ireland

T.F. Moe
Norwegian Institute for Water Research, Oslo, Norway

D. Monteith
Centre for Ecology & Hydrology, Lancaster Environment Centre, Lancaster, United Kingdom

M.B. Morales
Terrestrial Ecology Group (TEG), Universidad Autónoma de Madrid, Madrid, Spain

D.B. Nedwell
School of Biological Sciences, University of Essex, Colchester, United Kingdom

S. Niu
Key Laboratory of Ecosystem Network Observation and Modelling, Institute of Geographic Sciences and Natural Resources Research, Chinese Academy of Sciences, Beijing, China

J.J. Oñate
Terrestrial Ecology Group (TEG), Universidad Autónoma de Madrid, Madrid, Spain

R. Ogaya
CSIC, Global Ecology Unit CREAF-CSIC-UAB; CREAF, Cerdanyola del Vallès, Barcelona, Catalonia, Spain

E. O'Gorman
Imperial College London, Ascot, Berkshire, United Kingdom

M.T. O'Hare
Centre for Ecology & Hydrology, Edinburgh, Scotland, United Kingdom

J. Peñuelas
CSIC, Global Ecology Unit CREAF-CSIC-UAB; CREAF, Cerdanyola del Vallès, Barcelona, Catalonia, Spain

H. Pfanz
Lehrstuhl für Angewandte Botanik, University Duisburg-Essen, Essen, Germany

J.J. Piggott
University of Otago, Dunedin, New Zealand; Center for Ecological Research, Kyoto University, Otsu, Japan

J. Pozo
Faculty of Science and Technology, University of the Basque Country, Bilbao, Spain

E. Preda
Research Center in Systems Ecology and Sustainability, University of Bucharest, Bucharest, Romania

T. Pärt
Swedish University of Agricultural Sciences, Uppsala, Sweden

P.B. Reich
University of Minnesota, Minneapolis, MN, United States; Hawkesbury Institute for the Environment, Western Sydney University, Richmond, Australia

S. Reinsch
Center for Ecology and Hydrology (CEH), Bangor, United Kingdom

M. Riipinen
Plymouth University, Plymouth, United Kingdom

K. Räsänen
Eawag, Swiss Federal Institute of Aquatic Science and Technology, Dübendorf, Switzerland

G. Rîşnoveanu
Research Center in Systems Ecology and Sustainability, University of Bucharest, Bucharest, Romania

I.K. Schmidt
University of Copenhagen, København, Denmark

M.K. Schneider
Institute for Sustainability Sciences, Agroscope, Zürich, Switzerland

E. Shilland
Environmental Change Research Centre, University College London, London, United Kingdom

C. Stamm
Eawag, Swiss Federal Institute of Aquatic Science and Technology, Dübendorf, Switzerland

M. Sternberg
Tel Aviv University, Tel Aviv, Israel

J. Storkey
Rothamsted Research, Harpenden, United Kingdom

H.E. Sugden
School of Marine Science and Technology, The Dove Marine Laboratory, Newcastle University, Cullercoats, United Kingdom

S.D. Tiegs
Oakland University, Rochester, MI, United States

A. Tietema
University of Amsterdam, ESS, Amsterdam, The Netherlands

K. Tockner
Leibniz-Institute of Freshwater Ecology and Inland Fisheries; Freie Universität Berlin, Berlin, Germany

L.C. Todman
Rothamsted Research, Harpenden, United Kingdom

J.L. Torossian
Northeastern University, Marine Science Center, Nahant, MA, United States

T. Tscharntke
Georg-August-University, Göttingen, Germany

G.J.C. Underwood
School of Biological Sciences, University of Essex, Colchester, United Kingdom

A. Vadineanu
Research Center in Systems Ecology and Sustainability, University of Bucharest, Bucharest, Romania

P.F.M. Verdonschot
Alterra, Wageningen, The Netherlands

D. Vodnik
Biotechnical Faculty, University of Ljubljana, Ljubljana, Slovenia

L.B.-M. Vought
Kristianstad University, Kristianstad, Sweden

W. Weisser
School of Life Sciences Weihenstephan, Technische Universität München, Freising, Germany

O.L.F. Weyl
South African Institute for Aquatic Biodiversity, Grahamstown, Eastern Cape, South Africa

C. Whitby
School of Biological Sciences, University of Essex, Colchester, United Kingdom

A.P. Whitmore
Rothamsted Research, Harpenden, United Kingdom

C. Wolter
IGB, Berlin, Germany

G. Woodward
Imperial College London, Ascot, Berkshire, United Kingdom

K. Zhu
Rice University, Houston, TX, United States

PREFACE

This thematic volume is focused on recognising the role of model systems and comparative approaches over large scales in space and/or over time. Until recently, ecological research has generally been conducted on a relatively contingent and ad hoc basis, with little scope for true long-term, strategic studies or those that operate at larger spatial scales. This has often reflected the realpolitik of research funding, and also a lack of a strong scientific underpinning behind the collection of biomonitoring data that could otherwise have addressed these gaps, but which tends to be lost to the grey literature.

Sustained research over many decades in a model system (or systems) is still exceedingly rare, and is dwindling in many parts of the world due to a perceived lack of novelty, and the prevailing view that we should focus on providing answers to immediate and obvious questions, rather than building the capacity to address unforeseen problems that may arise in the future. Paradoxically, however, the very rarity of large-scale studies is that it confers uniqueness and value upon them—especially as many long-term initiatives are being wound down and the thread of continued observation is broken. Research at larger spatial scales is far more common as it is logistically more feasible within the time-limited remit of funding frameworks; it is far easier to secure funding for a study that spans 10 sites for 1 year than it is for one that spans 10 years at one site. This has increased reliance on space-for-time substitutions in correlational survey data in particular, which can be problematic when there are time lags or hystereses in biological responses to environmental change, as opposed to the instantaneous mapping of cause onto effect that is otherwise assumed.

There are some ways that such biases might be minimised, however, such as using microbial systems to shrink spatial and temporal dimensions into more tractable studies, which can address issues such as metacommunity dynamics and evolution that would not be possible with larger or longer-lived organisms. Defining what we mean by "large scale" in absolute terms is thus somewhat arbitrary. As a rule of thumb, we have used large scale to refer temporally to timescales that span at least a decade, and spatially to scales of several square kilometres and upwards within a given system or group of systems. Most cases presented in this volume represent a study that is large scale on just one of these axes, but also a few apply to both space and time.

Several of the papers, although not necessarily rooted in a model system that occupies particular physical space per se, may be considered large scale in their philosophical framework rather than simply in terms of the empirical boundaries within which they operate. In these cases the authors are searching for globally recurrent phenomena or ways to standardise data gathering at continental to global scales.

The papers in this volume are laid out in order such that they provide a conceptual thread that runs through them sequentially. The first paper by Storkey et al. (2016) opens with the longest ecological and agricultural field experiments ever conducted, which was set up in Rothamsted in the UK over a century (>170 years) ago to understand how land-use practices can alter terrestrial ecosystems. In addition, shorter but still exceptionally long-term ecological surveys have been running since the mid-1900s, expanding the focus across a range of taxonomic groups (e.g. Rothamsted Insect Survey). Arguably, long-term ecology started with the original Rothamsted studies, and the insights provided are invaluable to our understanding of agroecology. The second paper by Emmerson et al. (2016) switches the temporal and spatial scales, but stays focused on terrestrial (agro)ecosystems, by taking a snapshot of multiple sites across Europe in the EU-funded Agripopes project. This research tackles a timely and highly topical issue—contemporary increases in agricultural intensification across Europe, the subsequent decline in biodiversity across trophic levels, and the consequences of this for important ecosystem services. Ultimately this work showcases the need to consider the large-scale ecological impacts of anthropogenic landscape-use change across geopolitical boundaries.

It is followed by another large-scale EU-funded project—RivFunction—described by Chauvet et al. (2016), which sought to unravel the responses of a key ecosystem process—terrestrial leaf-litter decomposition—to environmental change via a range of stressors, but primarily focused on nutrient enrichment and riparian alteration, which are both characteristic by-products of the continental scales of agricultural intensification described in the preceding paper. The research presented in Chauvet et al. (2016) has pioneered the development of novel functional approaches to biomonitoring that have emerged over the past decade, as well as revolutionising much of the more "pure" research into Biodiversity–Ecosystem Functioning Relationships by introducing multiple new aspects of both drivers and responses that were previously unknown, but which we now know are essential for understanding how these ecosystems operate under stress at local-to-continental scales. The fourth paper, by Stamm et al. (2016), extends beyond several of the methods pioneered in the RivFunction project to take a more holistic view of how stream ecosystems respond to a complex array of micropollutants, including agricultural pesticides as well as urban pharmaceutical pollution, in multiple sites across Switzerland. At present there are many thousands of these chemicals in play in natural systems, and the task of disentangling their impacts and identifying cause-and-effect relationships is daunting to say the least, as it requires sophisticated multiscale and combined experimental and observational approaches, as clearly advocated by Stamm et al. (2016) and that moves far beyond the traditional small-scale ecotoxicological studies of dose responses of individual species in the laboratory.

The fifth paper, by Nedwell et al. (2016), moves downstream from the headwaters covered in the preceding two chapters to focus on long-term and large-scale (particularly in relation to the microbial taxa investigated) studies on a single model system, the Colne Estuary in southern England, which has pioneered much of our modern understanding of in situ microbial ecology and biogeochemistry. This system, which is a useful general model for many estuaries in the temperate zone, has been key for unlocking novel pathways within the major macronutrient cycles that fuel the planet's ecosystems, and the role microbes play in regulating these. The importance of microbes as model organisms is pursued further in the subsequent chapter by Maček et al. (2016), which makes the case for studying their ecology in extreme environments—natural CO_2 vents known as "mofettes"—across the globe as an ideal test bed for understanding how environmental and biogeographical filters in general, and proxies of climate change in particular, shape natural communities and their associated ecosystem processes. Moreover, terrestrial mofette systems are also used as proxies to assess if Carbon Capture Storage (CCS) solutions are a viable method for reducing global CO_2 levels. However, the microbial ecology of mofette systems demonstrates that any escape of CO_2 from CCS may promote local soil hypoxia and shift the microbial communities to dominance by anaerobic methane producing Archaea, highlighting the role long-term extreme environments may play in providing evidence for landscape management.

Climate change is the main theme of the next collection of chapters, which starts with a paper by Mieszkowska and Sugden (2016), moving into the marine realm to consider how large-scale range shifts have been triggered across a transitional zone. Species abundance data have been gathered since the 1950s, across 120 intertidal and subtidal sites from northern Africa to northern UK, making this one of the longest-term and largest-scale marine datasets in existence. The theme is extended further still in the subsequent chapter by Torossian et al. (2016), which aims to unify differing predictive approaches that forecast ecological changes to climate change. A better understanding of the mechanistic drivers of change can be achieved by combining physiological (small-scale) and biogeographical (large-scale) modelling approaches, which will improve our ability to develop a more accurate global perspective on the ecology of climate change. The paper by Andresen et al. (2016) then explores the impacts of the main driver of climate change—elevated atmospheric CO_2—on long-term terrestrial plant experiments by examining a large-scale network of Free Air Carbon Dioxide Exchange projects. Indeed, the research presented by Andresen et al. (2016) clearly highlights the importance of considering space and time in the analysis of a key response variable—biomass—when considering how plants respond to elevated CO_2, demonstrating clear spatiotemporal scale dependencies on how we perceive these responses.

The next group of three chapters returns to the aquatic realm, and each has a distinct conservation and biomonitoring flavour. The first of these, by Gray et al. (2016), like the preceding paper of Andresen et al. (2016), covers large scales in both space and time, but here the focus is on very different drivers and responses, namely tracking food-web responses to reversals in acidification following several decades of strong regulation on emissions. This set of >20 streams and lakes from across the UK, which have been tracked since the 1980s, generated >400 food webs in total—by far the largest standardised suite of trophic networks published to date for any ecosystem. This unique dataset unearths some unexpected results and also compelling evidence that space-for-time proxies are far from perfect matches for true chronosequences, with many systems showing marked lags or hysteresis in their ecological recovery that do not match the much faster and clearer rates of chemical recovery. The penultimate chapter remains in the freshwater realm but moves beyond simply anticipating and predicting stressor impacts, to consider the scope for active restoration across a large number of European rivers. Here, Friberg et al. (2016) demonstrate clearly how carrying out manipulations at appropriate temporal and spatial scales is one of the major determinants of the success (or failure) of such interventions. The final paper in the volume, by Jackson et al. (2016), looks forward to gauge future prospects for meeting the urgent need for a new generation of freshwater biomonitoring approaches that can track changes in biodiversity and also ecosystem processes and services at global scales. At present these tools are poorly developed for application at large scales, but this paper highlights some existing approaches that could be adapted and harnessed relatively quickly, as well as novel ones that could potentially be better placed to address the needs of the emerging monitoring or regulatory frameworks (e.g. IPBES) which are driving much of the biomonitoring research agenda around the world.

In conclusion, we hope the diverse set of studies highlighted here and drawn from both the aquatic and terrestrial realms provide some stimulating and novel insights into the challenges—and potential rewards—of carrying out large-scale ecological research. One of the strongest themes to emerge from this volume is the invaluable, and often unexpected, new insights that can be gleaned from supporting long-term research in model systems—just as developmental biologists have *Caenorhabditis elegans*, geneticists have *Drosophila melanogaster*, and plant biologists have *Arabidopsis* spp. as core study systems, ecologists need comparable long-term observational systems that can complement experimental and comparative space-for-time approaches. Contrary to the perception of many Governmental agencies that fund the majority of these programmes, the very fact that they do the same thing repeatedly over long periods of time is their great strength and not a weakness reflecting a supposed lack of novelty, which means that funding is constantly under threat. Once the data series is interrupted, however, its value is hugely diminished and we would be well advised to protect these unique observational platforms for both current and future generations.

ALEX J. DUMBRELL
REBECCA L. KORDAS
GUY WOODWARD

REFERENCES

Andresen, L.C., Müller, C., de Dato, G., Dukes, J.S., Emmett, B.A., Estiarte, M., et al., 2016. Shifting impacts of climate change: long-term patterns of plant response to elevated CO_2, drought and warming across ecosystems. Adv. Ecol. Res. 55, 437–473.

Chauvet, E., Ferreira, V., Giller, P.S., Mckie, B.G., Tiegs, S.D., Woodward, G., et al., 2016. Litter decomposition as an indicator of stream ecosystem functioning at local-to-continental scales: insights from the European RivFunction project. Adv. Ecol. Res. 55, 99–182.

Emmerson, M., Morales, M.B., Oñate, J.J., Batáry, P., Berendse, F., Liira, J., et al., 2016. How agricultural intensification affects biodiversity and ecosystem services. Adv. Ecol. Res. 55, 43–97.

Friberg, N., Angelopoulos, N.V., Buijse, A.D., Cowx, I.G., Kail, J., Moe, T.F., et al., 2016. Effective river restoration in the 21st century: from trial and error to novel evidence-based approaches. Adv. Ecol. Res. 55, 535–611.

Gray, C., Hildrew, A., Lu, X., Ma, A., McElroy, D., Monteith, D., et al., 2016. Recovery and nonrecovery of freshwater food webs from the effects of acidification. Adv. Ecol. Res. 55, 475–534.

Jackson, M.C., Weyl, O.L.F., Altermatt, F., Durance, I., Friberg, N., Dumbrell, A.J., et al., 2016. Recommendations for the next generation of global freshwater biological monitoring tools. Adv. Ecol. Res. 55, 615–636.

Maček, I., Vodnik, D., Pfanz, H., Low-Décarie, E., Dumbrell, A.J., 2016. Locally extreme environments as natural long-term experiments in ecology. Adv. Ecol. Res. 55, 283–323.

Mieszkowska, N., Sugden, H.E., 2016. Climate-driven range shifts within benthic habitats across a marine biogeographic transition zone. Adv. Ecol. Res. 55, 325–369.

Nedwell, D.B., Underwood, G.J.C., McGenity, T.J., Whitby, C., Dumbrell, A.J., 2016. The Colne Estuary: a long-term microbial ecology observatory. Adv. Ecol. Res. 55, 227–281.

Stamm, C., Räsänen, K., Burdon, F.J., Altermatt, F., Jokela, J., Joss, A., et al., 2016. Unravelling the impacts of micropollutants in aquatic ecosystems: interdisciplinary studies at the interface of large-scale ecology. Adv. Ecol. Res. 55, 183–223.

Storkey, J., Macdonald, A.J., Bell, J.R., Clark, I.M., Gregory, A.S., Hawkins, N.J., et al., 2016. The unique contribution of Rothamsted to ecological research at large temporal scales. Adv. Ecol. Res. 55, 3–42.

Torossian, J.L., Kordas, R.L., Helmuth, B., 2016. Cross-scale approaches to forecasting biogeographic responses to climate change. Adv. Ecol. Res. 55, 371–433.

PART I

Large Spatial Scale Ecology

CHAPTER ONE

The Unique Contribution of Rothamsted to Ecological Research at Large Temporal Scales

J. Storkey[1], A.J. Macdonald, J.R. Bell, I.M. Clark, A.S. Gregory, N.J. Hawkins, P.R. Hirsch, L.C. Todman, A.P. Whitmore

Rothamsted Research, Harpenden, United Kingdom
[1]Corresponding author: e-mail address: jonathan.storkey@rothamsted.ac.uk

Contents

1. Introduction to Long-Term Ecological Research at Rothamsted — 4
   1.1 The Rothamsted Classical Experiments — 5
   1.2 Highfield, Fosters and Woburn Ley-Arable Experiments — 9
   1.3 Broadbalk and Geescroft Wildernesses — 10
   1.4 The Rothamsted Insect Survey — 11
2. Monitoring the Impact of Environmental Change — 13
   2.1 Response of Plant Communities to Environmental Change — 13
   2.2 Response of Plant Pathogens to Environmental Change — 15
   2.3 Phenological Change and Trophic Asynchrony — 17
3. Community Ecology — 17
   3.1 Value of the Rothamsted Experiments to Plant Community Ecology — 18
   3.2 Trophic Interactions — 21
   3.3 Environmental Drivers of Insect Abundance and Distribution — 23
4. Ecosystem Stability and Resilience — 24
   4.1 Plant Community Stability — 24
   4.2 Resilience of Ecosystem Function — 26
5. Evolutionary Ecology — 28
   5.1 Beyond Snaydon and Davies — 29
   5.2 Evolutionary Ecology of Pathogens and Weeds — 30
6. Soil Microbial Ecology — 32
   6.1 Microbial Communities on the PGE — 33
   6.2 Microbial Communities on Broadbalk — 34
7. Conclusion — 35
Acknowledgements — 36
References — 36

Abstract

The Rothamsted Estate in Hertfordshire, United Kingdom, is home to the longest running ecological and agricultural experiments in the world that have generated unique data sets on the assembly and functioning of ecosystems that stretch back more than 170 years. In addition, the Rothamsted Sample Archive contains over 300,000 samples of dried soil, herbage, straw and grain dating back to the start of the first experiments. Additional long-term experiments were set up in the mid-1900s and the systematic sampling of invertebrates at Rothamsted started in 1964, which continues to this day in the form of the Rothamsted Insect Survey. Here, we introduce the resources available at Rothamsted for research on ecological processes that can only be understood using data over long time periods. Rather than cataloguing all the work that has been done using the data, we focus on new advances made in the last decade in areas of environmental monitoring, community ecology, evolutionary biology, ecosystem stability and resilience and microbial ecology. The combination of long-term data sets with archived plant and soil samples together with new analytical techniques mean that the Rothamsted long-term experiments and insect collections continue to be as relevant and valuable to scientists today as when they were originally set up.

1. INTRODUCTION TO LONG-TERM ECOLOGICAL RESEARCH AT ROTHAMSTED

The Rothamsted Estate can be found on the edge of the town of Harpenden in the South East of England approximately 50 km north of London. It comprises a 330 ha experimental farm in an arable rotation and includes Rothamsted Manor, a Grade 1 listed building dating back to the early 1200s. This was the home of Sir John Bennet Lawes who inherited the Estate in 1822 and established Rothamsted as a site for experimental research. As well as being a respected scientist of the Victorian age, Lawes was also a successful business man, running a profitable fertiliser business, largely based on the manufacture of superphosphate. Using monies from the sale of this business, he established the Lawes Agricultural Trust, which continues to this day as a registered charity, owning the freehold on the Estate and supporting the maintenance of the 'Classical' field experiments described later. Now the longest running agricultural research station in the world, in its current form, Rothamsted Research is an independent research institute employing approximately 550 staff working on agroecology, plant biology, crop science, biological chemistry, computational and systems biology and sustainable soils and grassland systems.

1.1 The Rothamsted Classical Experiments

The year 1843 is taken as the start of agroecological research at Rothamsted, as it is when the first of the 'Classical' experiments was established by Sir John Bennet Lawes and his colleague, the chemist Sir Joseph Henry Gilbert (Fig. 1). Some of these experiments continue today and are now widely recognised to be the oldest continuing agricultural field experiments in the world. Most of the early experiments were located at Rothamsted Experimental Farm (Hertfordshire, United Kingdom), on a moderately well-drained flinty silty clay loam, but later, other long-term experiments (LTEs) were established on the sandy loam soil at Woburn Experimental Farm (78 ha), 40 km north of Rothamsted. Today there are more than 15 LTEs (including the Classicals); two of the best known are the Broadbalk Wheat Experiment and the Park Grass Continuous Hay Experiment (Anonymous, 2006). These two experiments have been the focus of the majority of ecological work using data from the Classical Experiments. Initially started by Lawes and Gilbert in 1843 and 1856, respectively, they examined the importance of different mineral nutrients (N, P, K, Na and Mg) and organic manures for crop growth and

Fig. 1 Sir John Bennet Lawes (1814–1900), *left*, and Sir Joseph Henry Gilbert (1817–1901), *right*. John Lawes was the owner of the Rothamsted Estate and Henry Gilbert was his long-term colleague and scientific collaborator. Together, they were responsible for establishing Rothamsted as the longest running agricultural research station in the world and setting up the 'Classical' field experiments, some of which continue today.

development. The agronomic questions they were originally designed to address were soon answered, but the experiments continued and are now a unique resource for quantifying long-term effects of management and environmental change on the assembly and functioning of biological communities. Over time, the experiments were modified to include the use of lime, herbicides, fungicides, pesticides and new crop varieties to ensure their management was relevant to modern farming systems, whilst maintaining their long-term integrity.

About 300,000 samples of dried plant and soil material, collected from the LTEs since the 1840s, are stored in the Rothamsted Sample Archive (RSA). Data collected from the experiments are stored in the electronic Rothamsted Archive (e-RA). The LTEs, together with the RSA and e-RA, form the Rothamsted LTEs National Capability (Macdonald et al., 2015). Collaborative research based on these resources, both national and international, is actively encouraged and data are available on request (http://www.era.rothamsted.ac.uk).

1.1.1 The Park Grass Experiment

The Park Grass experiment (PGE) is the oldest experiment on permanent grassland in the world. It was established in 1856 to examine the effects of different mineral fertilisers and organic manures on hay production, but in recent decades has become of more interest as an ecological experiment (Silvertown et al., 2006). The experiment is located on a moderately well-drained silty clay loam overlying clay-with-flints. The soil pH was slightly acidic when the experiment began (5.4–5.6) and the nutrient status was poor. The field was in permanent pasture for at least 100 years prior to the start of the experiment and the original vegetation was classified by Dodd et al. (1994) as dicotyledon-rich *Cynosurus cristatus*–*Centaurea nigra* grassland. Treatments imposed in 1856 included controls (nil, no fertiliser or manure), and various combinations of P, K, Mg and Na, with N applied as either sodium nitrate or ammonium salts. Farmyard manure (FYM) was applied to two plots but was discontinued after 8 years because, when applied annually to the surface in large amounts, it had adverse effects on the sward. FYM, applied every 4 years, was reintroduced on three plots in 1905.

The experiment consists of 20 main plots (Fig. 2). The plots are cut in mid-June and made into hay. For 19 years the regrowth was grazed by sheep penned to individual plots but since 1875 a second cut, usually carted green, has been taken. The plots were originally cut by scythe, then by horse-drawn and then tractor-drawn mowers. Yields were originally estimated by

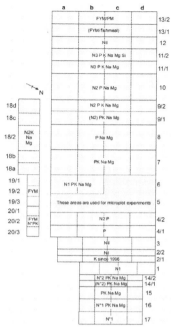

Fig. 2 The Park Grass experiment began in 1856 on permanent pasture. The experiment is managed as a hay meadow, cut twice a year. The main plots (1–20) receive fertiliser treatments with different combinations of N, P, K, Na, Si and Mg; some receive farmyard manure (FYM), poultry manure (PM) or fishmeal. Nitrogen is applied at three rates (N1, N2 and N3; 48, 96 and 144 kg N ha^{-1}, respectively) and in two forms, ammonium sulphate or sodium nitrate (N*). Subplots a, b and c have different amounts of lime added every 3 years, when necessary, to achieve a target pH of 7, 6 and 5, respectively. The d subplots are unlimed. Treatment shown on the plan in *parenthesis* has been withheld since the 1980s or 1990s.

weighing the produce, either of hay (first harvest) or green crop (second harvest), and dry matter determined from the whole plot. Since 1960 yields of dry matter have been estimated from strips cut with a forage harvester. However, for the first cut the remainder of the plot is still mown and made into hay, continuing earlier management and ensuring return of seed. For the second cut the whole plot is cut with a forage harvester. Most of the plots were divided into two in 1903 to introduce a test of liming on one half. In 1965 they were further divided into four subplots (a, b, c and d) to establish a wider range of soil pH within each plot, including the optimum (pH 6) for long-term grass (Anonymous, 2010). The a, b and c subplots receive lime every 3 years, if necessary, to maintain a target soil pH of 7, 6 and 5, respectively. The d subplots are unlimed.

1.1.2 The Broadbalk Wheat Experiment

The Broadbalk experiment is the oldest continuous agronomic experiment in the world. Winter wheat has been sown and harvested on all or part of the field every year since 1843. The original aim of the experiment was to test the effects of various combinations of mineral fertilisers (supplying N, P, K, Na and Mg) and organic manures (FYM, rape cake or castor bean meal) on the yield of winter wheat compared with an unfertilised control plot. The experiment was divided into different Strips or 'Plots' (Fig. 3) receiving different fertiliser and manure treatments each year. The experiment has had three main phases. In 1843–1925, winter wheat was grown continuously, apart from occasional fallowing to control weeds. Most treatment strips were established by 1852, except for strip 2A (now 2.1), which began in 1885, and strip 20, which began in 1906. Initially, the plots were long (c. 300 m).

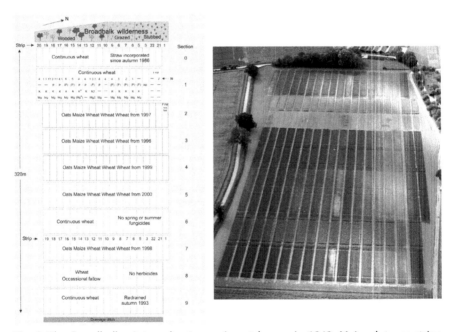

Fig. 3 The Broadbalk winter wheat experiment begun in 1843. Main plots, or strips, numbered 1–20 have different combinations of P (35 kg ha^{-1}), K (90 kg ha^{-1}) and Mg (12 kg ha^{-1}) with six rates of N (N1, N2, N3, N4, N5 and N6; 48, 96, 144, 192, 240 and 288 kg N ha^{-1}, respectively). Plots 2.1 and 2.2 receive farmyard manure (FYM) with or without additional mineral N, respectively; plot 2.1 received 96 kg N ha^{-1} from 1968 and 144 kg N ha^{-1} since 2005. The experiment was split into 10 sections in 1968 (see text for details of treatments in different sections). P and Mg have been withheld from some plots since 2001; (P) and (Mg), respectively.

Consequently, between 1894 and 1925 many plots were harvested in two halves, Top (T) and Bottom (B), equivalent to the Western and Eastern parts of the experiment.

In 1926 the strips were divided into five 'Sections' (I–V) to allow the introduction of a rotational fallow to help improve weed control. One section was bare fallowed every 5 years as part of 5-year rotation with four successive crops of wheat. In 1968, Sections I–V were further divided to create 10 sections (0–9, Fig. 3) and rotational cropping was introduced on some sections so that the yield of wheat grown continuously could be compared with that of wheat grown after a 2-year break. In addition, short-strawed wheat cultivars were introduced, which lead to an increase in grain yields and a decrease in straw yields. After 1968, Sections 0, 1, 8 and 9 continued to grow winter wheat only, whilst Sections 2, 4, 7 and 3, 5, 6 went into two different three-course rotations; wheat: potatoes: beans and wheat: wheat; fallow, respectively. In 1978, Section 6 reverted to continuous wheat and the other five sections went into a 5-year rotation, currently oats: forage maize: wheat: wheat: wheat (Fig. 3). Pesticides are applied where necessary, except on Section 6, which does not receive spring or summer fungicides, and Section 8 which has never received herbicides (applied to the rest of the experiment since 1964). Section 8 has proved to be a valuable resource for studying the assembly and functioning of weed communities adapted to different soil properties. On Section 0 the straw on each plot has been chopped after harvest and incorporated in the soil since autumn 1986; on all other sections the straw is baled and removed.

In 1849 Lawes and Gilbert installed tile drains at a depth of about 60 cm in the centre of each treatment strip. In 1866 small pits were dug to allow the collection of water samples from the drains so that the nutrients lost in the drainage could be measured. After 150 years many of the drains had collapsed and in 1993 Section 9 was redrained so that water leaching through the soil could again be collected and analysed. Lime has been applied as required since the 1950s to maintain soil pH at a level at which crop yield is not limited (Anonymous, 2010). From 2001 P was withheld from some plots until levels of plant available P have decreased to more appropriate agronomic levels (Index 2–3; Anonymous, 2010). This is reviewed each year.

1.2 Highfield, Fosters and Woburn Ley-Arable Experiments

The Rothamsted Highfield and Fosters Ley-Arable experiments started in 1949. Their aims were to examine the effects of crop rotations with and without grass leys on soil organic matter (SOM; Johnston et al., 2009) and crop yields. The two experiments were located on similar soil types (Batcombe and Batcombe–Carstens series) but differed in their cropping histories. Highfield had been in permanent grass (since 1838) before the arable rotations were established in 1949; some plots stayed in permanent grass, whilst others alternated between leys and arable crops. In contrast, Fosters was in arable cropping for several centuries; on this site, some plots stayed in continuous arable, whilst others went into permanent grass; others alternated between leys and arable crops. Yields are no longer measured but SOM continues to be monitored and the experiment has provided a valuable resource to examine the effects of contrasting management practices on soil microbial diversity and SOM.

In addition, in 1959 an area of permanent grass (since 1838) adjacent to the Highfield Ley-Arable experiment was ploughed and has remained bare fallow ever since. This is the Highfield Bare Fallow (Barre et al., 2010; Hirsch et al., 2009). It is kept free of weeds by frequent cultivation, but herbicides are used occasionally to ensure that carbon inputs to the soil in vegetation are negligible. SOM is measured periodically, and it has declined substantially since it was first ploughed out of grass (Barre et al., 2010). Finally, the Woburn Ley-Arable experiment started in 1938 to compare the effects of rotations with or without grass or grass-clover leys on SOM and the yield of two arable test crops (Anonymous, 2006).

1.3 Broadbalk and Geescroft Wildernesses

Two Rothamsted sites that had previously been arable for centuries were fenced off in the 1880s and left unmanaged to allow natural vegetation to regenerate. They subsequently became the Broadbalk and Geescroft Wildernesses. Whilst they are not experiments in the usual sense, they have been of value for looking at long-term changes in soil and woodland vegetation (Poulton et al., 2003). Broadbalk Wilderness was originally part of the Broadbalk wheat experiment, and had grown unmanured winter wheat since 1843. The last wheat crop was sown in autumn 1881, but not harvested. The site was then fenced off to allow natural vegetation to regenerate. In around 1900 it was divided into two halves, one remained as regenerating woodland whilst on the other woody species were removed (stubbed) each year, to allow open ground vegetation to develop. In 1957 the stubbed section was divided into two, one half remains 'stubbed', the other half was mown for 3 years, grazed by sheep each year from 1960 to 2000, and has been mown each year from 2001 (herbage not removed). Geescroft Wilderness is a larger site about 1 km away from Broadbalk. It was originally part of an experimental field that grew field beans (*Vicia faba*) from 1847 to 1878, with frequent breaks due to crop failure. After bare fallowing for 4 years, clover was grown from 1883 to 1885, and the wilderness area was fenced off in January 1886. It has been uncultivated since spring 1883. The surface soil is now acidic (pH 4.4 in 1999).

1.4 The Rothamsted Insect Survey

The systematic, frequent sampling of invertebrates at Rothamsted started in the second half of the 20th century. Now called the 'Rothamsted Insect Survey' (RIS) two very different trap networks monitoring migrating insects are operated. In both cases, the networks grew from original traps set up at Rothamsted which continue to operate, providing continuity of data. Firstly, the suction-trap network began operation on the 29th April 1964 and continues to this day to monitor the aerial fauna migrating at a height of 12.2 m (calculated as the logarithmic mean of aphid flight; Bell et al., 2015). Essentially, these suction traps can be thought of as upside-down hoovers which indiscriminately catch small- to medium-sized insects (≤ 5 mg), particularly aphids (Fig. 4). Currently, 16 traps are in operation in the United Kingdom, with two still based at Rothamsted, although over the last 50 years a total of 37 traps with at least 2 years' worth of data have been in service (Bell et al., 2015). Suction-trap catches are monitored daily in the aphid season for the agricultural industry, transferring knowledge of pest incidence to farmers, consultants and levy boards in a bid to reduce the prophylactic use of insecticides and therefore resistance. It is perhaps important to note that as our food chains have become global, so too as the suction-trap network. Currently, there are 129 traps in 17 countries serving a number of purposes with aims to monitor rice, potato and cereal pests as well as changes the nonaphid fraction (NJF, 2013).

The second network is made up of light traps. Currently, 78 light traps are in service in a wide range of habitats, although 569 traps have provided data to the network since the first was turned on in 1933. Although the network was formally established in 1964 it was not until 1968 that a fair coverage of the habitats and regions of the United Kingdom was achieved. Currently three traps continue to operate at Rothamsted in woodland, farmland and grassland, including at the site of the original 1933 trap. Each trap uses a 200-W tungsten bulb that produces a wide wavelength

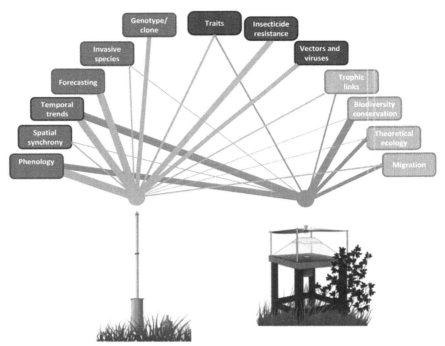

Fig. 4 Linkage of scientific topics covered by the Rothamsted Insect Survey's two networks. The suction- and light-trap networks have the potential to cover four main areas of science that include climate change (*blue, grey* in the print version), population dynamics and monitoring (*red, grey* in the print version), selection and evolution (*purple, black* in the print version) and ecology (*green, light grey* in the print version). The degree to which a network covers any topic is indicated by an *arc* or *line*. *Grey arcs* for suction traps, *red* (*grey* in the print version) for light traps. The thickness of an arc is equivalent to an estimate of the volume of research published in that area.

spectrum (400–700 nm) that attracts various night-flying insects. The light traps were originally set up to monitor moths and in this activity the network is world leading; recently Fox et al. (2013) demonstrated that over the last 40 years two-thirds of our moth fauna are in decline but new arrivals to the United Kingdom were also apparent. Interestingly, both nocturnal wasps (*Ichneumoidae*) and winter gnats (*Trichoceridae*) are also attracted to these traps and recent studies of these two groups have revealed new species to science for the United Kingdom from trap catches (C. Shortall, personal communication). Other groups have also been studied including biting midges, craneflies, St Mark's flies, black flies, beetles, caddisflies, true bugs, and lacewings. However, these groups have not been well studied from light trap catches in the recent past but remain a curiosity of the 20th century.

Combined, the light- and suction-trap networks comprise the most comprehensive, standardised data set on terrestrial invertebrates in the world that have generated high impact research from the outset. In lieu of a detailed historical analysis, an overview of research activities by trap network is provided in this paper which summarises the topics that have dominated over the last 50 years (Fig. 4) and recent findings are presented in subsequent sections. The main emphasis of the light-trap network has been associated with the themes of ecology and latterly climate change, whereas the suction-trap network's activities have been more wide ranging.

2. MONITORING THE IMPACT OF ENVIRONMENTAL CHANGE

Environmental stochasticity often makes isolating treatment effects from background environmental trends difficult in short-term manipulative experiments and hampers the development of robust predictive models. In this respect, the long-term nature of the Rothamsted experiments and RIS is of particular value to ecologists—assuming that sufficient data are available on environmental variables. Fortunately, John Lawes understood the importance of the weather in driving the responses observed in his experiments and comprehensive weather data stretching back to the mid-1800s are held in the Rothamsted Electronic Archive. Rainfall has been measured at Rothamsted since 1853 and air temperatures have been recorded since 1873 (Fig. 5). As new equipment became available for measuring additional weather variables, it was periodically added to the weather station located at Rothamsted which was fully automated in 2004. Since 1992, Rothamsted has been part of the UK Environmental Change Network (ECN) that monitors a range of environmental and biological variables across multiple sites and habitats using standard protocols (http://www.ecn.ac.uk). Together with the weather data, these measurements provide a detailed environmental context for understanding the local scale ecological dynamics observed at the site. Some recent examples are presented in the next section.

2.1 Response of Plant Communities to Environmental Change

C_3 plants generally increase photosynthetic carbon assimilation (A) and decrease stomatal conductance (g_s) under elevated atmospheric CO_2

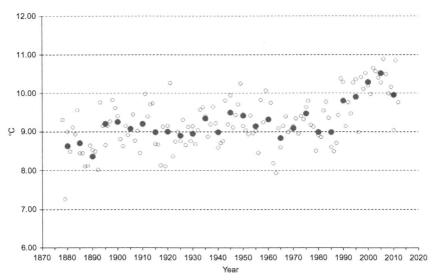

Fig. 5 Annual mean air temperature at Rothamsted between 1878 and 2012. Also shown is the mean over each 5-year period, 1878–82, 1883–87, etc. The mean air temperature at Rothamsted is now 10°C, 1°C higher than the 1878–1987 average. The 10 warmest years on record occurred in the last 17 years.

(Kohler et al., 2010). Thus, intrinsic water-use efficiency W_i ($W_i = A/g_s$) would also be expected to increase in response to rising global CO_2 levels. However, past studies found the photosynthetic response to elevated CO_2 was dependent on nitrogen nutritional status (Ainsworth and Rogers, 2007; Stitt and Krapp, 1999). Consequently, W_i of grasslands with limited N availability may be constrained despite the observed increases in atmospheric CO_2 that have been observed over the last century. The PGE combines a long-term data series of archived herbage samples with a gradient of N treatments making it the ideal system to challenge this hypothesis.

W_i can be derived from estimates of carbon isotope discrimination ($^{13}\Delta$) using measurements of atmospheric CO_2 concentration (c_a) and $\delta^{13}C$ enrichment in herbage samples ($\delta^{13}C_p$) and free air ($\delta^{13}C_a$). An estimate of the ratio between the CO_2 concentration in the internal leaf space (c_i) and c_a can be obtained from the known relationship with carbon isotope discrimination ($^{13}\Delta$) (Farquhar et al., 1989) allowing W_i to be estimated. By measuring the $\delta^{13}C$ enrichment of archived herbage, samples from selected plots of the PGE from 1915 to 2009 Kohler et al. (2012) were able to examine the effects of annual applications of different fertilisers on changes in the intrinsic water-use efficiency (W_i) of the plant communities over a period of nearly 100 years under conditions of increasing atmospheric CO_2. Results from the limed subplots indicated that carbon isotope discrimination ($^{13}\Delta$) increased significantly ($P<0.001$) on the unfertilised control (0.9‰ per 100 ppm CO_2 increase), but this trend differed significantly ($P<0.01$) from those observed on the fertilised treatments (PK, N and NPK). The $^{13}\Delta$ trends on fertilised treatments did not differ significantly from each other, but N status, assessed as N fertiliser supply plus an estimate of biologically fixed N, was negatively related ($r^2=0.88$; $P<0.02$) to the trend for $^{13}\Delta$ against CO_2. Consequently, the increase of W_i at high N+PK (96 kg N ha^{-1} + PK) was twice that of the control (+28% resp., +13% relative to 1915). In addition, the CO_2 responsiveness of $^{13}\Delta$ was related to the grass content of the plant community. This may have been due to the greater CO_2 responsiveness of stomatal conductance in grasses relative to forbs, indicating that the greater CO_2 response of fertilised swards may be related to effects of N supply on botanical composition (see Section 3.1).

Vegetation surveys have been carried out on Park Grass on more than 30 occasions since the experiment began. The most thorough assessments of botanical composition were done by Crawley et al. (2005) who between 1991 and 2012 collected herbage annually from six randomly located quadrats (each 50 × 25 cm) within each plot. The herbage was cut with scissors to ground level in early June, immediately before harvesting the first hay crop, and separated to species quantified as relative dry weight. From these data, an empirical model of the drivers of grassland diversity was derived; the main ones being pH and levels of soil nitrogen and phosphorus. More recently, the same protocol was used to sample a subset of plots between 2010 and 2012 with a focus on plots that stopped receiving nitrogen fertiliser in 1989. These 'transition' plots were found to have recovered rapidly from the negative effects of eutrophication (possibly facilitated by the frequent mowing and removal of biomass) but, more interestingly, an increase in plant diversity was also observed across the whole experiment since the samples taken during the 1990s. This recovery was attributed to a decrease in atmospheric nitrogen deposition resulting in an observed increase in legumes and decrease in grasses as well as greater overall plant diversity, Fig. 6 (Storkey et al., 2015).

2.2 Response of Plant Pathogens to Environmental Change

The response of plant communities to long-term changes in atmospheric chemistry could only have been detected using data sets of the large

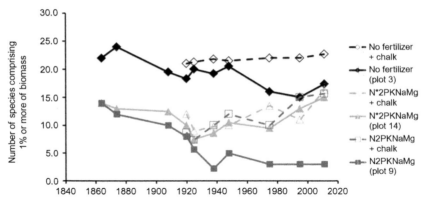

Fig. 6 Changes in species richness on the Park Grass experiment over time showing the effect of treatments and atmospheric deposition of N and S. Adding N and P reduced species richness in contrast to the control plots with no fertiliser inputs. This effect was exacerbated on plots receiving ammonium sulphate without lime owing to the lowering of soil pH to 3.5. Between 1960 and 1990, species richness declined on the control plot in response to eutrophication and acidification form atmospheric pollution. Emissions of N and S have declined since the 1990s and this is reflected in a recovery of diversity on Park Grass (Storkey et al., 2015).

temporal scale covered by the Rothamsted experiments. Another example, using the model system of plant pathogens, reinforces this point and also shows how new technologies can be combined with archived samples from the experiments to reveal new insights. In this case, quantitative PCR assays were developed to detect overall population levels of different plant pathogens in wheat straw and grain from the Broadbalk experiment using archived samples. Bearchell et al. (2005) used quantitative polymerase chain reaction (qPCR) analysis to measure the relative abundance of two foliar pathogens, *Zymospetoria tritici* and *Stagonospora nodorum*, in the samples. For those years in which disease surveys had been carried out, pathogen abundance as measured by qPCR correlated well with survey data. Relative abundance of the two pathogens was correlated with air pollution as measured by sulphur dioxide (SO_2) emissions. *S. nodorum* became the more abundant pathogen in the early 20th century as SO_2 emissions increased, then *Z. tritici* took over as the dominant pathogen in the late 1980s as SO_2 emissions decreased again. Further analysis of pathogen abundance and long-term weather data revealed spring rainfall and summer temperature as further factors correlated with plant pathogen abundance (Shaw et al., 2008).

2.3 Phenological Change and Trophic Asynchrony

Probably the work with the largest impact in the last 5 years using data from the RIS documents the impact of environmental change on phenology (Thackeray et al., 2010, 2016), which has since been cited in the IPCC fifth assessment in both terrestrial systems and ocean chapters. Thackeray et al. (2010) was able to show that in all major UK systems including aquatic and terrestrial habitats where time series were available, rapid phenological change was well underway indicating that UK flora and fauna were being driven by a shared large-scale process. The RIS contributed aphid and moth trends to these analyses and when combined with other time series showed that rates of change differed significantly among trophic levels. As a consequence, Thackeray et al.'s (2010) results inferred that there was a risk that phenological mismatch could disrupt these habitats.

These phenological changes were further explored by Bell et al. (2015) who went beyond looking at first flights to also consider the duration of the flight season, last flights and annual abundances for 55 species of aphid. Phenologies were statistically linked to a large-scale weather pattern, the North Atlantic Oscillation (NAO), which determined what types of winter the United Kingdom was likely to experience given the atmospheric pressure difference between the Azores and Iceland. The minimum threshold for flight in aphids is 11°C, but Bell et al. (2015) hypothesised that 16°C would be the average threshold for the range of aphids studied. Using the NAO and the accumulated day degrees above 16°C the authors were able to explain patterns across the United Kingdom over the last five decades. Under increasing temperatures, particularly milder winters, communities are unlikely to remain stable over time. This is a challenging area that requires new statistical techniques to demonstrate empirically. Sheppard et al. (2016) also used RIS data to show how phenological synchrony of aphid first flights was changing, driven by synchrony in winter climate. Periods of long-term synchrony in the fluctuations of aphid first flights (>4 years) were shown to be in decline but short-term synchrony (<4 years) was on the increase after 1993, due to changeable winters.

3. COMMUNITY ECOLOGY

The Rothamsted experiments constitute a unique resource for studying community ecology for two important reasons. Firstly, the age of the experiments means that plant communities are in dynamic equilibrium

(Silvertown, 1980). Therefore, when analysing differences between the plots, we know we are not observing transient or successional dynamics. In the case of PGE, the plots are generally recognised to have reached equilibrium with the respective fertiliser treatments by the early 1900s, after which the overall species list and relative abundance of plant functional groups have remained relatively stable (although temporal trends of individual species continue to be more dynamic—see Section 4.1). Secondly, the communities on Park Grass and weed floras on Section 8 of Broadbalk are naturally assembled from the local species pool; Broadbalk is thought to have been in arable cropping for many centuries before 1843 and Park Grass had been in permanent pasture for at least 100 years before 1856. This avoids the common criticism of short-term diversity manipulation experiments that artificial selection of component species can lead to a 'sampling' or 'positive selection' effect whereby there is a greater chance of randomly selecting a more productive species in more diverse treatments (Jiang et al., 2009).

3.1 Value of the Rothamsted Experiments to Plant Community Ecology

The PGE and, to a lesser extent, Broadbalk, have been used extensively as model systems to challenge ecological theories that seek to explain the structuring of plant communities and quantify relationships between biodiversity and ecosystem function. When taken together, Park Grass and Broadbalk include a range of plots that differ in the two main drivers of plant community assembly, soil fertility (or resource availability) and the level of disturbance. Early analyses of the plant communities on Park Grass established a generic, additive effect of decreasing pH and a negative relationship between biomass and plant diversity (Silvertown, 1980; Wilson et al., 1996). The plots have also been assessed in terms of Ellenberg N indicator values (Hill and Carey, 1997) which were found to be more useful for predicting variation in productivity as opposed to nitrogen inputs. The classic humpbacked relationship of species richness with soil fertility has been found on Section 8 of Broadbalk (Moss et al., 2004) but on Park Grass, species richness increases monotonically with decreasing fertiliser inputs and biomass (Crawley et al., 2005).

These studies on plant community ecology across the Rothamsted experiments have tended to be empirical in nature, describing variance in vegetation patterns in contrasting environments without necessarily increasing our understanding of the mechanisms that explain why different plots have reached different end points. In this regard, the Rothamsted experiments still have enormous potential to further elucidate our understanding of plant ecological strategy and community assembly. In particular, the fact that Park Grass and Broadbalk both have plots positioned along a fertility gradient creates an opportunity to challenge conflicting theories about the importance of interspecific resource competition in structuring plant communities in environments with low fertility (Craine, 2005)—discussed in the following section.

3.1.1 Reconciling Resource Ratio and C–S–R Theories of Plant Community Assembly

Data from the PGE have been used in the past in support of the resource ratio hypothesis of plant community structure (Harpole and Tilman, 2007; Tilman, 1982) which postulates that where a single resource is limiting, the species able to tolerate the lowest concentration of that resource will dominate. Further, species will be able to coexist where multiple resources are limiting and species differ in their respective resource-specific carrying capacities resulting in increased niche dimensionality. When applied to the Rothamsted experiments, this theory assumes that: (a) interspecific competition has been an important process structuring communities along the fertility gradient found on both Park Grass and Broadbalk, (b) inherent trade-offs in ecophysiological traits mean that a species that is more competitive for light will be less competitive for nutrients and (c) competitive hierarchies should vary with the experimental treatments.

Circumstantial evidence for a trade-off between the allocation of resource to shoot vs root expressed in contrasting plant communities either limited by light or nutrients has been found on Section 8 of Broadbalk (Storkey et al., 2010). Weighted averages for plant traits of communities sampled across the nitrogen gradient were positively correlated with soil fertility for maximum canopy height and negatively correlated for seed weight—taller, small-seeded weeds were associated with the more fertile plots. Previous studies had also found a strong negative relationship between seed weight and root weight ratio among groups of annual weeds (Seibert and Pearce, 1993; Storkey, 2006), implying that larger seeded species, found on less-fertile plots, were allocating more assimilate to below-ground resource capture. From this work, Storkey et al. (2010) proposed that species with a combination of large seed, late flowering and short stature (that tended to be found on plots with low to medium nitrogen inputs) represented a functional group that had largely disappeared from the weed flora in the wider landscape and could be interpreted as a 'rare weed traits syndrome'.

However, as well as providing evidence for the resource ratio hypothesis, it is also possible to find empirical support from the Rothamsted experiments for a competing hypothesis; that interspecific competition is not an important process driving community assembly in low-fertility environment but rather contrasts between communities can be explained by a resource acquisition vs conservation trade-off (Grime, 1974). In this case, species either have high relative growth rates and tissue turn over (palatable, short-lived leaves), adapted for 'competitive' (C), fertile environments or slow growth rates with increased investment in herbivore defence leading to longer plant residence time of nutrients and a more 'stress-tolerant' (S) community. In environments with frequent disturbance, a third plant ecological strategy characterised by short lifespan, early flowering and rapid growth is defined as 'ruderality' (R). It has been proposed that the C–S–R plant ecological strategy scheme can be reduced to two primary axes of specialisation reflecting trade-offs in functional traits along gradients of fertility and disturbance (Grime et al., 1997).

For the purposes of this discussion, we present a new analysis of Park Grass and Broadbalk data that shows evidence for these two functional axes from a comparison of contrasting plant communities (Figs. 7 and 8). Biomass samples separated to species from the PGE (0.75 m^{-2} from each plot averaged over 10 years of sampling, 1991–2000) and Broadbalk (1.5 m^{-2} from Plots 5, 6, 7, 8, 9, 15, 16 averaged over three sampling years, 2004, 2010 and 2014—the only years for which biomass data were available) were used to calculate the community-weighted mean (CWM) for a range of functional traits. The plot CWM trait values were used as response variables in redundancy analyses. When comparing the two experiments (Fig. 7A), the primary ordination axis corresponds to the trade-off between competitive and ruderal strategies, with Broadbalk dominated by ruderal species adapted to frequent disturbance. The second axis reflects the trade-off between stress-tolerant and competitive species driven by soil fertility, driven largely by specific leaf area. When the data from each experiment are analysed separately, this emerges as the primary ordination axis (Fig. 7B and C) that is positively correlated with biomass production and negatively with species richness (Fig. 8).

Having shown that data from the Rothamsted experiments can be used to support either the resource ratio or C–S–R theories for plant community assembly in low-fertility environments, we suggest that additional work needs to be done to determine the relative importance of interspecific competition and selective herbivory across the nutrient gradients present on the experiments.

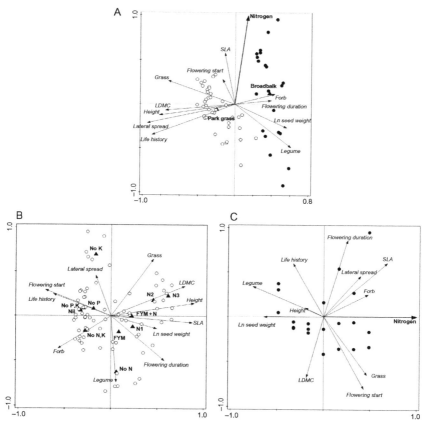

Fig. 7 Redundancy analysis (RDA) of vegetation data on the PGE and Section 8 of Broadbalk using community-weighted mean (by biomass) of plant functional traits as response variables and experimental treatment as explanatory variables. (A) Data from the PGE 'b' subplots and Broadbalk Section 8 along the comparable nitrogen gradient on each experiment, (B) Partial RDA for Park Grass Data (10-year means of species abundance on each plot) using pH as a covariate and combinations of nutrients as categorical explanatory variables. (C) Equivalent analysis for the nitrogen gradient on Section 8 of Broadbalk using year as covariate and nitrogen as a continuous explanatory variable.

3.2 Trophic Interactions

An area where the RIS is increasingly collaborating and expanding its interest with those who hold long time series or spatially extensive data is trophic ecology. For example, in a study of high altitude migration of ladybirds which also used radar data to understand the flight behaviour of our native (*Coccinella septempunctata*) and invasive (*Harmonia axyridis*) species, aphid abundance from the suction traps together with independent temperature

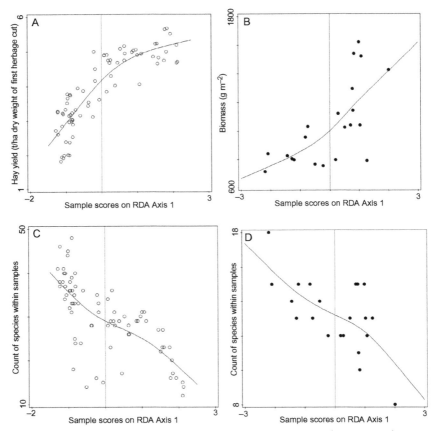

Fig. 8 Relationships between functional trait composition, productivity and species richness on PGE and Section 8 of Broadbalk. Sample scores on Axis 1 of the partial RDA analyses are plotted against (A) hay yield on the PGE experiment, (B) total above ground dry weight (gm^2) including crop on Broadbalk and species richness on (C) PGE and (D) Broadbalk.

measures were successful terms that predicted peaks in ladybird flight (Jeffries et al., 2013). There is considerable interest in the effect of climate change on trophic levels and whether this yields a mismatch between predator and prey. Our collaboration with the British Trust for Ornithology (BTO) has since revealed that although insect phenology has advanced more rapidly than swallow (*Hirundo rustica*) phenology, there was no evidence of increasing phenological mismatch over 30 years. This was in part because the number of second broods was increasing over time. Another type of top-down relationship with insects is parasitism which functions differently to predation, with arguably tighter statistical associations. Perez-Rodriguez et al.

(2015) was interested in the parasitoid wasps that are associated with the grain aphid (*Sitobion avenae*) and the level of synchrony between their migrations. The co-occurrence of the summer peaks is between 51% and 64% and unlike the leading edge of the aphid population (Bell et al., 2015), this mid-season peak appears somewhat stationary in time (Perez-Rodriguez et al., 2015).

A final top-down relationship is herbivory for which aphids are infamous. Zust et al. (2012) investigated whether aphid herbivory can lead to the maintenance of genetic variation in host plants across large geographic scales. They showed that aphids select for different 'chemotypes' which drive the variation in plant defences across Europe. Indeed at the level of the clone, there is evidence of host plant specialisation. Fenton et al. (2010) studied the peach–potato aphid (*Myzus persicae*), arguably the most serious aphid pest in Europe, and showed that Scottish populations feeding on either downy ground-cherry, radish, spring oilseed rape or potato, produced clonal lineages with different host performances, indicating specialisation. The link between aphids and their host plants was further explored using the NBN gateway, a digital storage facility for the UK's biological records. Using this, Bell et al. (2012) was able to show that for many of the 170 species of aphid studied, their abundances over the United Kingdom were resource controlled which is suggestive of a strong bottom-up effect that is particularly prevalent for species that have an alternative winter host plant. In these cases, aphids were shown to be constrained to low abundances if their winter host had a low area of occupancy over the United Kingdom.

3.3 Environmental Drivers of Insect Abundance and Distribution

The RIS is not all about aphids but instead has a very broad taxonomic remit. For example, *Culicoides* biting midges have been studied in both light- and suction-trap catches. These midges transmit arboviruses, including bluetongue virus, and thus understanding their phenology is important for the prediction of transmission and spread. Sanders et al. (2011) showed that total abundance of *Culicoides* increased significantly with the density of cattle per square km around suction traps. Midge migration is also driven by weather and Sanders et al. (2011) showed that larger catches were associated with higher temperatures, lower wind speeds and days on which it was not raining at sunset.

The conservation of species is yet another area in which weather, or more precisely climate, has played a key part in the research associated with the light-trap network. Palmer et al. (2016) studied 155 butterfly and moth species quantifying the sensitivity of each species according to the amount of variation in year-to-year abundance or range size changes that could be attributed to seasonal mean temperature and rainfall. Perhaps unexpectedly given often strong latitudinal effects in many species, the changes in range size (22% variance explained) were much less associated with climate than abundance (55% variance explained) changes. These results also support previous moth work done by Mutshinda et al. (2011) who established that winter rainfall and temperature were the most important drivers of moth population dynamics that accounted for 45–75% of the variation. During this period, a much smaller study was completed by O'Neill et al. (2012) which highlighted both the geographical reach of the RIS through its volunteer network and also how moths need to adapt to changing weather patterns in Ireland. Finally, the RIS data were exploited by Chapman et al. (2012, 2013) who used the light-trap network to confirm suspected migration of the silver Y moth (*Autographa gamma*) as determined by radar, showing the likely path that the moths would take in and out of the United Kingdom and their annual patterns of abundance on the ground.

4. ECOSYSTEM STABILITY AND RESILIENCE

The study of the stability of ecosystems in the context of a variable environment and their resilience to environmental or management perturbations necessitates data on a large temporal scale encompassing gradients of environmental variables. It is unsurprising, therefore, that the Rothamsted experiments have been used extensively to address ecological theory in this area including: (i) understanding the importance of environmental variability in maintaining community stability, (ii) the relationship between species diversity and variability of biomass production and (iii) the resilience of ecosystem functions to disturbance.

4.1 Plant Community Stability

Initial analyses of long time series of data on biomass and species richness on the PGE established that the plots were in a botanical equilibrium with a negative relationship between biomass and species richness. The slope of this relationship has changed little with time, although the splitting of the plots in 1903 had the effect of lowering the intercept (Silvertown, 1980). This equilibrium between species richness and biomass is maintained in the context of year-to-year variability in biomass mainly resulting from differences in rainfall patterns (Silvertown et al., 1994). The underlying community dynamics that explain the dynamic botanical equilibrium observed on Park Grass are still not clearly understood. However, recent papers around this topic have helpfully clarified the conceptual framework in which current understanding can be formulated and new hypotheses proposed. Micheli et al. (1999) emphasised the importance of quantifying the dual nature of ecosystem variability—both the compositional (variability in components of the ecosystem) and aggregate (variability in summary properties of the ecosystem such as productivity). In terms of the former on the Rothamsted experiments, this can be studied at the level of functional group, species richness or genetic differentiation between species (Silvertown, 1987). More recently, mathematical models have been developed that separate the variability in an ecosystem property due to longer term, seasonal (structured) variability in rainfall and temperature from stochastic environmental fluctuations (Schulte, 2003).

Studies of the variability in community composition on Park Grass at the level of plant functional group (grass/legume/miscellaneous) have concluded that, although the relative proportion of the functional groups can vary at the scale of 1–2 years in response to differences in annual rainfall, the absence of negative feedback between functional groups at the decadal scale means that the equilibrium dynamics cannot be explained at this level (Silvertown, 1980). Rather, it is likely that community dynamics operating at the scale of individual species (or even genotypes) in response to within-year environmental variability are responsible for maintaining stability in botanical composition and ecosystem function. Interestingly, evidence supporting this hypothesis has been found using data from Section 8 on the Broadbalk experiment. Using data from 20 years of surveys on the composition of weed communities found on contrasting fertiliser plots, the authors attempted to identify the causes of within-plot temporal variability in the relative proportion of component species (De Leon et al., 2014). Support was found for the 'storage effect hypothesis' of plant coexistence (Chesson, 2000); species that responded similarly to fertiliser treatment had subtly different responses to within-year variability in temperature and rainfall. The resulting community dynamics in a given year were further regulated by differences in intra- and interspecific competition. It is likely that similar processes are operating on Park Grass to maintain long-term community stability, for example, the functionally similar grasses *Arrhenatherum elatius* and *Alopecurus pratensis* tend to be found on the same plots but one tends to dominate in any given year (unpublished data).

Finally, the potential importance of interactions of pathogens in maintaining taxonomic and genetic diversity of plant communities should not be forgotten and has been the subject of a number of recent papers using Park Grass as a model system (Jeger et al., 2014; Salama et al., 2010, 2012).

4.2 Resilience of Ecosystem Function

As well as being valuable for studying the stability of ecosystems over time, the Rothamsted experiments are also an important resource for studying the resilience of ecosystems in response to a disturbance and have been used to investigate resilience of soil function. Resilience is a difficult concept to capture and define; however, it usually encompasses two important aspects: *resistance* to and *recovery* from change. In a widely cited paper, Orwin and Wardle (2004) derive metrics from a time course of the substrate-induced respiration in soils following a disturbance. They focus attention on particular parts of the response only, so: resistance is the change in a soil function immediately after the disturbance (relative to a control sample), whilst recovery is how nearly the disturbed soil returns from a perturbed level to the same level of function as the control. Such metrics are helpful but disregard much of the data that describe the entirety of the response. Alternatively, Todman et al. (2016) identify four characteristics of the whole, dynamic response of a soil function to disturbance that pertain to the resilience, these are (i) the degree of return to the level of function prior to disturbance, (ii) the time taken to return to this level, (iii) the rate at which the response returns and (iv) the area under the response curve (a measure of the efficiency of the response period).

The resilience of ecosystems is a subject that has received much study, partly because the failure of ecosystems has wide-reaching consequences but also because humans are causing enormous disturbance to ecosystems (Steffen et al., 2015). A prime example is the soil. Almost all soil in the United Kingdom is subject to some kind of management. Soil is a complex, but relatively easy ecosystem to study and wider terrestrial ecosystems all depend on it as the source of water and nutrients for primary production. Although it is opaque and irreducible to components that still function, a functioning system can still be found at scales suitable for laboratory manipulation. If soil has properties that make it suitable to study resilience in laboratories and over experimental timescales, it also retains an imprint of the way in which it has been used or managed that will influence any measures made on it—including an assay of resilience. The Rothamsted experiments have many plots that have had consistent or at least known management for periods of up to 172 years. Such self-contained consistency is key to understanding how the capacity for a resilient response to disturbance arises, but is difficult to find in large-scale ecosystems. As a result, research of this nature has often been confined to islands, lakes or reefs. In turn, these suffer that they are difficult to replicate and the fact that they are impossible to reproduce in the laboratory under controlled conditions. In the Rothamsted field plots, differences in long-term management (from decades to almost two centuries) exist side by side in space. This therefore removes the major complication in ecosystem science that treatments cannot be implemented on identical underlying conditions. It is true that soil varies in space and true also that some of the Rothamsted experiments (particularly the 'Classicals' begun in the 1840s and 1850s) were set out before the need for statistical replication and design was understood, but the closeness of parent soil material in most of the LTEs at Rothamsted still allows us to draw sound conclusions about the effect of management on the underlying stability and resilience of soil responses to disturbance and the possible ways in which resilience might be enhanced where it is lacking.

Gregory et al. (2009) compared both the physical and biological resilience of soils from Broadbalk, Highfield and the Woburn Ley-Arable experiments. They first assayed biological resilience by measuring the response in substrate-induced respiration during a 4-week incubation at 16°C following imposition of transient (heating to 40°C for 18 h) or persistent (dosing with 1.6 M solution of copper sulphate) stresses relative to control soils. Physical resilience was then assayed by compressing soils uniaxially on a loading frame to 200 kPa and monitoring the resultant change in void ratio. Unaided recovery of compacted soils by the action of microbes is unlikely and although macroscale biology such as earthworms, moles or tree roots can alleviate density in the field, these were absent from these small-scale assays. Accordingly, recovery was initiated by freeze–thaw and wet–dry cycles which rely on changes in energy potential of water at the microscale (Gregory et al., 2009). Building on a survey of the resilience of soils in Scotland Kuan et al. (2007), Gregory et al. (2009) and Corstanje et al. (2015) worked with a wider range of soils than from the LTEs alone, in order to infer what qualities of soil or land management endow soil with the ability to mount a resilient response. Gregory et al. (2009) found that the initial recovery of the compacted pore space (void ratio) in a soil was related to the management of the land from where the soil was taken. Soils under grassland or woodland tended to recover pore space more or less immediately. In contrast, soils from arable or fallow fields recovered much less pore space following compaction, if at all. Land use is almost certainly confounded with SOM in these studies, but the amount of SOM was not quantitatively related to the extent of rebound in a manner that explained more of the variation than land use. Wet–dry or freeze–thaw cycles had the effect of restoring pore space lost during compaction of soil, but only in soils containing more than 26% clay. Pore space continued to recover in soils with much clay (>60%) with a second wet–dry or freeze–thaw cycle.

On the whole, Gregory et al. (2009) found that respiration in soils from the Rothamsted experiments tended to recover from the imposition of a transient stress (heat) more fully than from a persistent stress (Cu). The extent of recovery was significantly related to the amount of SOM but the relationship was better in the Cu-stressed soils. As with the physical stress, land use also appeared to play a role in that recovery of grassland and woodland sites was better than arable sites. In neither case were there significant or obvious relationships between the resistance (initial susceptibility) to heat or Cu stress. Corstanje et al. (2015) have built on these pioneering observations and systematically compared intrinsic factors such as texture or SOM with extrinsic factors such as management in order to explain the observed variation in resilience of respiration in soils from the Rothamsted experiments and elsewhere. They used both CART (classification and regression tree) models and Bayesian Belief networks to try to attribute causes to the observed resistance and recoveries from stress. They concluded that multiple factors operating at a whole system level were probably responsible for observed differences in resilience.

This conclusion that systems-level understanding is needed to explain differences in resilience amplifies the value of long-term experimental soils for further investigations into the nature of resilience. It is only where the complete systems history of land is known that we will have the necessary background information with which to interpret differences in the resilient behaviour of soil fully. This can then be used to attribute causes to these differences and infer strategies that might increase the resilience of our land in response to stresses to which it is likely to be subjected to in the future.

5. EVOLUTIONARY ECOLOGY

As with the response of communities studied at the level of species or functional group discussed in Section 2, the confidence we can have in data on genetic adaptation to management or environmental change increases the longer a population has had to adapt to a new environment. Consequently, the area of evolutionary ecology represents an exciting avenue for research utilising data (and archived samples) from the Rothamsted experiments, particularly as new analytical technologies become available.

5.1 Beyond Snaydon and Davies

Perhaps one of the most well-known ecological advances to come from the PGE is the early demonstration of local adaptation of plant populations to a changing environment over a relatively short time period. The classic studies of Snaydon and Davies using the polyploid grass species, *Anthoxanthum odoratum* L. (sweet vernal grass), showed that populations grown from seed sampled from plots with contrasting pH were higher yielding when grown in soils more closely resembling their 'home' environment (Davies and Snaydon, 1976; Snaydon, 1970; Snaydon and Davies, 1972a,b, 1976). They thus demonstrated the heritability of fitness traits relating to aluminium (Al) tolerance and local adaptation over distances of less than 30 m and timescales of less than 40 years (measured from when liming began). More recent studies have confirmed that genome-wide population differentiation has occurred in the populations of *A. odoratum* on Park Grass despite the close proximity of the plots (Freeland et al., 2010) which has been related to the concept of reinforcement through differentiation of flowering times between populations (Silvertown et al., 2005).

Further insight into the mechanisms underlying the spatial structuring of the genetics of *A. odoratum* on the experiment has also come from a recent analysis of the relative contribution of pollen and seed dispersal to gene flow (Freeland et al., 2012). This work highlighted the role of environment in facilitating the establishment of immigrant seeds from neighbouring plots, increasing intraspecific genetic diversity. An interesting future avenue of research would be to combine these concepts of spatial dynamics of intraspecific genetic diversity with studies on source sink environments and the mass effects hypothesis that predicts greater species diversity at the interface of plots with moderate environmental differences (Kunin, 1998). This would further elucidate the relationship between fertiliser treatments and genetic and species diversity across the experiment, building on a recent study by Silvertown et al. (2009) that found evidence for a negative correlation between genetic and species diversity across a subset of plots which was explained in terms of the resource ratio hypothesis discussed in Section 3.1.1. The historical data and archived samples from the

Rothamsted experiments make them ideal model systems for further exploring the combined effects of colonisation–competition–extinction ecological dynamics with local adaptation and evolutionary dynamics in the structuring of communities in response to environmental change (Loeuille and Leibold, 2014; Smith et al., 2009).

Continuing on the theme of using modern tools to elucidate the mechanisms underlying the population genetics observed on the Park Grass plots, two recent papers by Gould et al. (2014, 2015) have built on the classic experiments of Snaydon and Davies. In this case, the *A. odoratum* populations studied by Snaydon and Davies were revisited and additional measurements taken on root traits in combination with sequenced RNA expressed in response to high Al low pH conditions. By identifying outlier tests of single-nucleotide polymorphisms (SNPs), the authors were able to identify a number of candidate genes underlying the expression of Al tolerance including a malate transporter, a tonoplast intrinsic transporter and a homologue of the STAR1 cell wall modification gene. Interestingly, from an ecological viewpoint, the results suggested that the traits expressed in low pH environments were more typical of an Al excluding strategy (through root cell wall resistance to the attachment of Al) as opposed to an Al accumulator. This resulted in reduced root growth rate of populations adapted to low pH in the absence of Al and reduced competitive ability which the authors suggest may lead to the exclusion of these biotypes on more alkaline plots restricting the tolerant genotypes to more acid soils. One further study has also increased our understanding of the mechanism by which grasses on the PGE have adapted to low pH and Al stress. In this case populations of *Holcus lanatus* L. (Yorkshire fog) from plots with contrasting pH were found to differ in the promotion of the gene controlling the rate of secretion of malate (Chen et al., 2013).

5.2 Evolutionary Ecology of Pathogens and Weeds

The majority of studies on evolutionary ecology on the Rothamsted experiments have built on the original work on *A. odoratum* on the Park Grass plots. However, there is also the opportunity to employ modern molecular techniques to investigate the genetic differentiation of weed populations along the soil fertility gradient on Section 8 of Broadbalk. Although some work was done in 2000 (Cavan et al., 2000), demonstrating local adaptation of *Stellaria media* L. (common chickweed), advances in sequencing technology now mean that the tools are available to address more thoroughly the question of whether similar processes of local selection and adaptation observed on the PGE are also responsible for structuring weed populations on Broadbalk. The plots may be particularly useful for quantifying gene flow and local adaptation to environments within and between populations of the important agricultural weed, *Alopecurus myosuroides* Huds (black grass) that may provide important insights into the evolution of weeds to changing weed management practices (Neve et al., 2009). The persistence of small isolated populations of weed species that have now become rare in the wider environment (including a population of one species, *Galium tricornutum* L., on Broadbalk that is the last naturally occurring population in the United Kingdom) could also be studied in the context of the impact of fragmentation and population bottlenecks on population genetics (Brutting et al., 2012).

Long-term data from Broadbalk show that wheat yields, and yield responses to higher fertiliser levels, have increased since the introduction of foliar fungicides to control plant diseases in 1979 (Dungait et al., 2012). However, resistance to various classes of single-site-inhibiting fungicides has subsequently evolved in many plant pathogens (Lucas et al., 2015). The Rothamsted LTEs allow researchers to investigate the dynamics of resistant alleles in pathogen populations over time, as the harvested straw and grain also contained plant pathogens, the DNA of which can now be amplified and analysed.

It is now also possible to use quantitative SNP detection techniques such as pyrosequencing to assess the proportion of a pathogen population carrying a specific genotype, such as a mutation conferring fungicide resistance. Hawkins et al. (2014) investigated levels of the *CYP51A* gene, conferring resistance to azole fungicides, in the barley pathogen *Rhynchosporium commune* in archived straw samples from one of the less well-known Classical Experiments at Rothamsted, Hoosfield Barley. This experiment was established in 1852 to test the effects of mineral fertilisers (supplying N, P, K, Mg and sodium silicate) and organic manures (FYM, rape cake and castor bean meal) on the growth and yield of spring barley. Hawkins et al. showed that *CYP51A* was not present at detectable levels until 1983, but from 1985, shortly after the introduction of azole fungicides for use on cereal crops in the United Kingdom, *CYP51A* levels increased rapidly. This corresponds to reports of less-sensitive isolates of *R. commune* from contemporaneous field surveys (Kendall et al., 1993). The molecular mechanism of resistance was not known at that time, but the analysis of archived samples means that the recently discovered resistance mechanism can be correlated with the previously reported reduction in sensitivity.

Alongside the practical implications of fungicide resistance for crop protection, such studies also provide an evolutionary time course, which can be used to address fundamental questions relating to temporal processes in evolution. The reemergence of the *CYP51A* paralogue under selection by azoles shows the role of temporal processes in the fate of duplicate genes. Previous studies of duplicate gene fate focused on the factors leading to loss or neofunctionalisation immediately after gene duplication, but in the case of *CYP51A*, the Hoosfield data set, combined with phylogenetic reconstructions of the paralogue's origins, showed that a previously declining paralogue may reemerge in a population following a change in selective pressure such as the introduction of fungicides.

Future uses of archive samples include measuring the dynamics of recent and future emerging pathogens, such as *Ramularia collo-cygni* in barley (Havis et al., 2015), or evolving pathotypes able to infect previously resistant crop varieties. Further studies of fungicide resistance will look at resistance in other pathogen species and to other fungicide classes. Resistance to MBCs (Kendall et al., 1994) and QoIs (Fraaije et al., 2005) is conferred by a single mutation in any one isolate, and so these SNPs can easily be detected and quantified by qPCR probes or pyrosequencing. For the azole fungicides, resistance mechanisms are more complex. In *Z. tritici*, multiple mutations within the azole target site-encoding gene *CYP51* contribute to less-sensitive haplotypes (Cools et al., 2011), and while current methods could quantify each individual mutation, new sequencing and bioinformatics approaches would be needed to reconstruct full *CYP51* haplotypes from the shorter DNA fragments recovered from older samples (Yoshida et al., 2015). Target site overexpression also contributes to azoles resistance in some isolates (Cools et al., 2012), and the current growth of next-generation sequencing and transcriptomics is enabling the detection of nontarget site resistance mechanisms, such as enhanced efflux (Omrane et al., 2015). Where overexpression of a target site or efflux pump is due to promoter inserts rather than single-nucleotide polymorphisms, different assay types, such as isothermal DNA amplification (Fraaije, 2014), will need to be developed to detect these resistance mechanisms.

6. SOIL MICROBIAL ECOLOGY

Perhaps the area of research on the Rothamsted experiments that has expanded most rapidly in recent years in response to advances in genome sequencing technology is soil microbial ecology. This emphasis on belowground diversity and function is to be welcomed as relatively little is known about the importance of plant/soil microbe interactions in driving the assembly and temporal dynamics of plant and invertebrate communities on the experiments. Next-generation sequencing applied to community-extracted DNA or RNA has revolutionised molecular microbial ecology by analysing metagenomes or metatranscriptomes rather than attempting to enumerate individual species. Because the unaligned sequences are relatively short (50–250 bp) and many individuals are from poorly described groups, it is not possible to assign all reads to species: rather, they are referred to as operational taxonomic units (Schloss and Westcott, 2011). Depending on the stringency with which sequences are matched to the nucleic acid databases, they may be assigned at different taxonomic levels from kingdom to subspecies. Other methods applied to soil microbial community-extracted DNA or RNA include PCR amplification of taxonomic identifiers, usually the ribosomal RNA genes (16S for Bacteria and Archaea, 18S for Eukarya and ITS for fungi), or known functional genes of interest, and hybridisation to microarrays that contain function and phylogenetic sequences (Hirsch et al., 2010, 2013).

6.1 Microbial Communities on the PGE

High-profile reports concerning management-driven changes in plant communities in the PGE resulted in this site being internationally famous. However, information on the below-ground biota was lacking until the first metagenome was recently derived from the untreated control plot 3d by the international TerraGenome consortium (Vogel et al., 2009). One aim of the work was to establish the potential sources of variability in soil metagenomes due to sampling approaches (spatial, temporal, depth) and the application of different methods to extract the soil community DNA (Delmont et al., 2011a,b, 2012). From 5 Gb sequence, c. 91% could be assigned with 89% belonging to the Bacteria, 1.4% to the Archaea and 1.0% to Eukarya (Delmont et al., 2012). The DNA extraction method was the most important factor in establishing which groups were detected and their relative abundance; the depth, season and spatial separation of samples had relatively little influence (Delmont et al., 2011a, 2012) within this one plot. The work provides an important reminder of the importance of reporting sampling depth and molecular methodology. The relatively low contribution of Eukarya to the metagenome, compared to Bacteria, was surprising because fungal activity is reported to be an important component of grassland soil ecosystems. However, such comparisons are not straightforward as soil fungi differ from bacteria in scale and growth habits, with cytoplasm-depleted hyphae connecting the actively growing tips where cytoplasm and nuclei are located. Furthermore, the metagenome indicates the genetic potential of a community rather than the relative activity of the different component groups.

Following the publication of the Park Grass metagenome from the control plots, different molecular approaches have been applied to study how different treatments influence the soil microbiome. A survey of 16S rRNA amplicons in community DNA collected from across the pH gradient on Park Grass plots with different N and P fertilisation regimes and controls showed soil pH to correlate most strongly with microbial diversity (H') with the soil C/N ratio and concentration of ammonia–N also playing a significant role (Zhalnina et al., 2015). Whilst pH has previously been reported to have a major influence on soil microbial communities (Fierer and Jackson, 2006), responses to changes in ammonia concentration are likely to influence the nitrifiers that depend on it as a substrate. Nitrifier groups are included in the Proteobacteria, Thaumarchaeota and Nitrospirae which were amongst the 12 most abundant phyla detected (Zhalnina et al., 2015). A study using a nested sampling strategy on plot 11/2c that receives mineral fertilisation (NPK) and control plot 12d employed the GeoChip functional microarray to compare plant species–area relationships (SARs) above ground with microbial functional gene–area relationships (GARs) below ground (Liang et al., 2015). Results showed that the long-term fertilisation treatment had decreased both plant and microbial α diversity when compared to the control treatment. In addition, both microbial GARs (fungal and bacterial) and plant SARs increased significantly in the fertilised plot, influenced to a similar extent by fertilisation, plant diversity and spatial distance with a lesser role played by soil geochemical properties. The authors concluded that long-term fertilisation may magnify existing divergent spatial patterns of both plants and microorganisms, and that this is likely to be influenced by long-term agricultural practice.

6.2 Microbial Communities on Broadbalk

Loss of N fertiliser due to bacterial denitrification, where nitrate is successively reduced to nitrite and nitrous oxide, represents a major route for greenhouse gas emissions from soil. However, many bacteria also contain a gene to reduce nitrous oxide to nitrogen gas, *nosZ* and some denitrifiers carry, but do not express it. Measurement of bacterial genes involved in denitrification in Broadbalk soil indicated that, in general, the dissimilatory nitrite reductase genes *nirK* and *nirS* increased in abundance with increasing N fertiliser, consistent with the increased N_2O emissions from soils receiving high N rates although it is assumed that *nosZ* converts a proportion to N_2. However, the woodland soil which does not receive fertiliser N had much higher emissions when fertiliser was applied. It also had a relatively lower abundance of *nosZ* compared to *nirK* which implies that the woodland soil harbours a distinctly different microbiome compared to the plots remaining under arable management (Clark et al., 2012).

A survey of 16S rRNA amplicons in soil sampled monthly over the growing season in plots with a range of N fertiliser inputs as well as the grassland and woodland sections of Broadbalk confirmed the difference in community structure (Zhalnina et al., 2013). The communities in plots receiving no fertiliser, or different rates of mineral N, were not distinctly different, although addition of FYM had an effect. However, major differences were seen in the communities in the 'nonagricultural' grassland and woodland that are not tilled and receive no fertiliser. This supports the previous suggestion that the woodland soil community was substantially different from that in plots under arable management. The microbial groups having the largest single effect were the Proteobacteria *Bradyrhizobium* which was significantly more abundant in untilled soil, and the Thaumarchaeota *Nitrososphaera* which showed a converse preference for tilled soil (Zhalnina et al., 2013). Both of these groups are involved in the N cycle as denitrifiers and nitrifiers, respectively.

7. CONCLUSION

The wealth of new insights gained over recent years across a range of disciplines that have resulted from new analyses of Rothamsted data and samples demonstrates the continuing relevance and importance of the Rothamsted experiments and insect collections. The advance of technology and ecological theory means that future uses for these resources cannot always be anticipated and the work presented here emphasises again the need to maintain experimental platforms that are not constrained by short-term, hypothesis-driven milestones that can lead to scientific myopia (Silvertown et al., 2010). Having said this, it may be that additional value can only be derived from some LTEs by introducing novel treatments or manipulation. This highlights the difficulties that exist when trying to maintain the integrity of the experiments whilst maximising their value to the research community. One recent example of how this has been done successfully is the change to the design of the Highfield Ley-Arable experiment that has had plots of permanent fallow, grass or arable since 1949. Some of these plots were split in 2008 to introduce alternative treatment, for example, old arable → fallow or grassland, such that all combinations were represented. These 'reversion' plots are now being used to study the recovery and transition of ecosystem properties including soil microbial diversity and functioning in response to land-use change. Realising the potential of this and other long-term data and experiments at Rothamsted into the future will depend on each generation of ecologists identifying similar novel and exciting avenues of research.

ACKNOWLEDGEMENTS

The Rothamsted Classical and long-term experiments and Rothamsed Insect Survey are UK National Capabilities supported by the UK Biotechnology and Biological Sciences Research Council (BBSRC; BBS/E/C/00005189) and the Lawes Agricultural Trust (LAT). Rothamsted Research is a National Institute of Bioscience strategically funded by the BBSRC. We thank the e-RA curators for their contribution of data and figures and Iris Köhler for helpful comments. We also thank Tony Scott for the Rothamsted meteorological data.

REFERENCES

Ainsworth, E.A., Rogers, A., 2007. The response of photosynthesis and stomatal conductance to rising CO2: mechanisms and environmental interactions. Plant Cell Environ. 30, 258–270.

Anonymous, 2006. Rothamsted Long-term Experiments: Guide to the Classicals and Other Long-term Experiments, Datasets and Sample Archive. Lawes Agricultural Trust Co. Ltd., Harpenden, Hertfordshire, UK

Anonymous, 2010. Fertiliser Manual (RB209), eighth ed. TSO Press, London, UK.

Barre, P., Eglin, T., Christensen, B.T., Ciais, P., Houot, S., Katterer, T., Van Oort, F., Peylin, P., Poulton, P.R., Romanenkov, V., Chenu, C., 2010. Quantifying and isolating stable soil organic carbon using long-term bare fallow experiments. Biogeosciences 7, 3839–3850.

Bearchell, S.J., Fraaije, B.A., Shaw, M.W., Fitt, B.D.L., 2005. Wheat archive links long-term fungal pathogen population dynamics to air pollution. Proc. Natl. Acad. Sci. U.S.A. 102, 5438–5442.

Bell, J.R., Taylor, M.S., Shortall, C.R., Welham, S.J., Harrington, R., 2012. The trait and host plant ecology of aphids and their distribution and abundance in the United Kingdom. Glob. Ecol. Biogeogr. 21, 405–415.

Bell, J.R., Alderson, L., Izera, D., Kruger, T., Parker, S., Pickup, J., Shortall, C.R., Taylor, M.S., Verrier, P., Harrington, R., 2015. Long-term phenological trends, species accumulation rates, aphid traits and climate: five decades of change in migrating aphids. J. Anim. Ecol. 84, 21–34.

Brutting, C., Meyer, S., Kuhne, P., Hensen, I., Wesche, K., 2012. Spatial genetic structure and low diversity of the rare arable plant Bupleurum rotundifolium L. indicate fragmentation in Central Europe. Agric. Ecosyst. Environ. 161, 70–77.

Cavan, G., Potier, V., Moss, S.R., 2000. Genetic diversity of weeds growing in continuous wheat. Weed Res. 40, 301–310.

Chapman, J.W., Bell, J.R., Burgin, L.E., Reynolds, D.R., Pettersson, L.B., Hill, J.K., Bonsall, M.B., Thomas, J.A., 2012. Seasonal migration to high latitudes results in major reproductive benefits in an insect. Proc. Natl. Acad. Sci. U.S.A. 109, 14924–14929.

Chapman, J.W., Lim, K.S., Reynolds, D.R., 2013. The significance of midsummer movements of Autographa gamma: implications for a mechanistic understanding of orientation behavior in a migrant moth. Curr. Zool. 59, 360–370.

Chen, Z.C., Yokosho, K., Kashino, M., Zhao, F.J., Yamaji, N., Ma, J.F., 2013. Adaptation to acidic soil is achieved by increased numbers of cis-acting elements regulating ALMT1 expression in Holcus lanatus. Plant J. 76, 10–23.

Chesson, P., 2000. Mechanisms of maintenance of species diversity. Annu. Rev. Ecol. Syst. 31, 343–366.

Clark, I.M., Buchkina, N., Jhurreea, D., Goulding, K.W.T., Hirsch, P.R., 2012. Impacts of nitrogen application rates on the activity and diversity of denitrifying bacteria in the Broadbalk Wheat Experiment. Philos. Trans. R. Soc. B 367, 1235–1244.

Cools, H.J., Mullins, J.G.L., Fraaije, B.A., Parker, J.E., Kelly, D.E., Lucas, J.A., Kelly, S.L., 2011. Impact of recently emerged sterol 14{alpha}-demethylase (CYP51) variants of Mycosphaerella graminicola on azole fungicide sensitivity. Appl. Environ. Microbiol. 77, 3830–3837.

Cools, H.J., Bayon, C., Atkins, S., Lucas, J.A., Fraaije, B.A., 2012. Overexpression of the sterol 14alpha-demethylase gene (*MgCYP51*) in *Mycosphaerella graminicola* isolates confers a novel azole fungicide sensitivity phenotype. Pest Manag. Sci. 68, 1034–1040.

Corstanje, R., Deeks, L.R., Whitmore, A.P., Gregory, A.S., Ritz, K., 2015. Probing the basis of soil resilience. Soil Use Manag. 31, 72–81.

Craine, J.M., 2005. Reconciling plant strategy theories of Grime and Tilman. J. Ecol. 93, 1041–1052.

Crawley, M.J., Johnston, A.E., Silvertown, J., Dodd, M., De Mazancourt, C., Heard, M.S., Henman, D.F., Edwards, G.R., 2005. Determinants of species richness in the Park Grass experiment. Am. Nat. 165, 179–192.

Davies, M.S., Snaydon, R.W., 1976. Rapid population differentiation in a mosaic environment. 3. Measures of selection pressures. Heredity 36, 59–66.

De Leon, D.G., Storkey, J., Moss, S.R., Gonzalez-Andujar, J.L., 2014. Can the storage effect hypothesis explain weed co-existence on the Broadbalk long-term fertiliser experiment? Weed Res. 54, 445–456.

Delmont, T., Robe, P., Cecillon, S., Clark, I., Constancias, F., Simonet, P., Hirsch, P., Vogel, T., 2011a. Accessing the soil metagenome for studies of microbial diversity. Appl. Environ. Microbiol. 77, 1315–1324.

Delmont, T.O., Robe, P., Clark, I., Simonet, P., Vogel, T.M., 2011b. Metagenomic comparison of direct and indirect soil DNA extraction approaches. J. Microbiol. Methods 86, 397–400.

Delmont, T.O., Prestat, E., Keegan, K.P., Faubladier, M., Robe, P., Clark, I.M., Pelletier, E., Hirsch, P.R., Meyer, F., Gilbert, J.A., Le Paslier, D., Simonet, P., Vogel, T.M., 2012. Structure, fluctuation and magnitude of a natural grassland soil metagenome. ISME J. 6, 1677–1687.

Dodd, M.E., Silvertown, J., Mcconway, K., Potts, J., Crawley, M., 1994. Application of the British National Vegetation Classification to the communities of the Park Grass experiment through time. Folia Geobot. Phytotx. 29, 321–334.

Dungait, J.A.J., Cardenas, L.M., Blackwell, M.S.A., Wu, L., Withers, P.J.A., Chadwick, D.R., Bol, R., Murray, P.J., MacDonald, A.J., Whitmore, A.P., Goulding, K.W.T., 2012. Advances in the understanding of nutrient dynamics and management in UK agriculture. Sci. Total Environ. 434, 39–50.

Farquhar, G.D., Ehleringer, J.R., Hubick, K.T., 1989. Carbon isotope discrimination and photosynthesis. Annu. Rev. Plant Physiol. Plant Mol. Biol. 40, 503–537.

Fenton, B., Kasprowicz, L., Malloch, G., Pickup, J., 2010. Reproductive performance of asexual clones of the peach-potato aphid (Myzus persicae, Homoptera: Aphididae), colonising Scotland in relation to host plant and field ecology. Bull. Entomol. Res. 100, 451–460.

Fierer, N., Jackson, R.B., 2006. The diversity and biogeography of soil bacterial communities. Proc. Natl. Acad. Sci. U.S.A. 103, 626–631.

Fox, R., Parsons, M.S., Chapman, J.W., Woiwod, I.P., Warren, M.S., Brooks, D.R., 2013. The State of Britain's Larger Moths 2013. Butterfly Conservation, Wareham, Dorset, UK.

Fraaije, B.A., 2014. Use of loop-mediated isothermal amplification assays to detect azole-insensitive CYP51-overexpressing strains of Zymoseptoria tritici. Phytopathology 104, 41–42.

Fraaije, B.A., Cools, H.J., Fountaine, J., Lovell, D.J., Motteram, J., West, J.S., Lucas, J.A., 2005. Role of ascospores in further spread of QoI-resistant cytochrome b alleles (G143A) in field populations of Mycosphaerella graminicola. Phytopathology 95, 933–941.

Freeland, J.R., Biss, P., Conrad, K.F., Silvertown, J., 2010. Selection pressures have caused genome-wide population differentiation of Anthoxanthum odoratum despite the potential for high gene flow. J. Evol. Biol. 23, 776–782.

Freeland, J.R., Biss, P., Silvertown, J., 2012. Contrasting patterns of pollen and seed flow influence the spatial genetic structure of sweet vernal grass (Anthoxanthum odoratum) populations. J. Hered. 103, 28–35.

Gould, B., McCouch, S., Geber, M., 2014. Variation in soil aluminium tolerance genes is associated with local adaptation to soils at the Park Grass experiment. Mol. Ecol. 23, 6058–6072.

Gould, B., McCouch, S., Geber, M., 2015. De novo transcriptome assembly and identification of gene candidates for rapid evolution of soil Al tolerance in Anthoxanthum odoratum at the long-term Park Grass experiment. PLoS One 10, e0124424.

Gregory, A.S., Watts, C.W., Griffiths, B.S., Hallett, P.D., Kuan, H.L., Whitmore, A.P., 2009. The effect of long-term soil management on the physical and biological resilience of a range of arable and grassland soils in England. Geoderma 153, 172–185.

Grime, J.P., 1974. Vegetation classification by reference to strategies. Nature 250, 26–31.

Grime, J.P., Thompson, K., Hunt, R., Hodgson, J.G., Cornelissen, J.H.C., Rorison, I.H., Hendry, G.A.F., Ashenden, T.W., Askew, A.P., Band, S.R., Booth, R.E., Bossard, C.C., Campbell, B.D., Cooper, J.E.L., Davison, A.W., Gupta, P.L., Hall, W., Hand, D.W., Hannah, M.A., Hillier, S.H., Hodkinson, D.J., Jalili, A., Liu, Z., Mackey, J.M.L., Matthews, N., Mowforth, M.A., Neal, A.M., Reader, R.J., Reiling, K., Rossfraser, W., Spencer, R.E., Sutton, F., Tasker, D.E., Thorpe, P.C., Whitehouse, J., 1997. Integrated screening validates primary axes of specialisation in plants. Oikos 79, 259–281.

Harpole, W.S., Tilman, D., 2007. Grassland species loss resulting from reduced niche dimension. Nature 446, 791–793.

Havis, N., Fountaine, J., Gorniak, K., Paterson, L., Taylor, J., 2015. Diagnosis of Ramularia collo-cygni and Rhynchosporium spp. in Barley. Methods Mol. Biol. 1302, 29–36.

Hawkins, N.J., Cools, H.J., Sierotzki, H., Shaw, M.W., Knogge, W., Kelly, S.L., Kelly, D.E., Fraaije, B.A., 2014. Paralog re-emergence: a novel, historically contingent mechanism in the evolution of antimicrobial resistance. Mol. Biol. Evol. 31, 1793–1802.

Hill, M.O., Carey, P.D., 1997. Prediction of yield in the Rothamsted Park Grass experiment by Ellenberg indicator values. J. Veg. Sci. 8, 579–586.

Hirsch, P.R., Gilliam, L.M., Sohi, S.P., Williams, J.K., Clark, I.M., Murray, P.J., 2009. Starving the soil of plant inputs for 50 years reduces abundance but not diversity of soil bacterial communities. Soil Biol. Biochem. 41, 2021–2024.

Hirsch, P.R., Mauchline, T.H., Clark, I.M., 2010. Culture-independent molecular techniques for soil microbial ecology. Soil Biol. Biochem. 42, 878–887.

Hirsch, P.R., Mauchline, T.H., Clark, I.M., 2013. Culture-independent molecular approaches to microbial ecology in soil and the rhizosphere. In: Molecular Microbial Ecology of the Rhizosphere. John Wiley & Sons, Inc., Hoboken, NJ.

Jeffries, D.L., Chapman, J., Roy, H.E., Humphries, S., Harrington, R., Brown, P.M.J., Handley, L.J.L., 2013. Characteristics and drivers of high-altitude ladybird flight: insights from vertical-looking entomological radar. PLoS One 8, e82278.

Jeger, M.J., Salama, N.K.G., Shaw, M.W., Van Den Berg, F., Van Den Bosch, F., 2014. Effects of plant pathogens on population dynamics and community composition in grassland ecosystems: two case studies. Eur. J. Plant Pathol. 138, 513–527.

Jiang, L., Wan, S.Q., Li, L.H., 2009. Species diversity and productivity: why do results of diversity-manipulation experiments differ from natural patterns? J. Ecol. 97, 603–608.

Johnston, A.E., Poulton, P.R., Coleman, K., 2009. Soil organic matter: its importance in sustainable agriculture and carbon dioxide fluxes. In: Sparks, D.L. (Ed.), Advances in Agronomy vol. 101. Elsevier, Inc., San Diego, pp. 1–57.

Kendall, S.J., Hollomon, D.W., Cooke, L.R., Jones, D.R., 1993. Changes in sensitivity to DMI fungicides in *Rhynchosporium secalis*. Crop. Prot. 12, 357–362.

Kendall, S.J., Hollomon, D.W., Ishii, H., Heaney, S.P., 1994. Characterisation of benzimidazole-resistant strains of *Rhynchosprium secalis*. Pestic. Sci. 40, 175–181.

Kohler, I.H., Poulton, P.R., Auerswald, K., Schnyder, H., 2010. Intrinsic water-use efficiency of temperate seminatural grassland has increased since 1857: an analysis of carbon isotope discrimination of herbage from the Park Grass experiment. Glob. Chang. Biol. 16, 1531–1541.

Kohler, I.H., Macdonald, A., Schnyder, H., 2012. Nutrient supply enhanced the increase in intrinsic water-use efficiency of a temperate seminatural grassland in the last century. Glob. Chang. Biol. 18, 3367–3376.

Kuan, H.L., Hallett, P.D., Griffiths, B.S., Gregory, A.S., Watts, C.W., Whitmore, A.P., 2007. The biological and physical stability and resilience of a selection of Scottish soils to stresses. Eur. J. Soil Sci. 58, 811–821.

Kunin, W.E., 1998. Biodiversity at the edge: a test, of the importance of spatial "mass effects" in the Rothamsted Park Grass experiments. Proc. Natl. Acad. Sci. U.S.A. 95, 207–212.

Liang, Y., Wu, L., Clark, I.M., Xue, K., Yang, Y., Van Nostrand, J.D., Deng, Y., He, Z., Mcgrath, S., Storkey, J., Hirsch, P.R., Sun, B., Zhou, J., 2015. Over 150 years of long-term fertilisation alters spatial scaling of microbial biodiversity. mBio 6, e00240-15.

Loeuille, N., Leibold, M.A., 2014. Effects of local negative feedbacks on the evolution of species within metacommunities. Ecol. Lett. 17, 563–573.

Lucas, J.A., Hawkins, N.J., Fraaije, B.A., 2015. The evolution of fungicide resistance. Adv. Appl. Microbiol. 90, 29–92.

Macdonald, A.J., Powlson, D.S., Poulton, P., Watts, C.J., Clark, I.M., Storkey, J., Hawkins, N.J., Glendining, M.J., Goulding, K.W.T., Mcgrath, S.P., 2015. The Rothamsted long-term experiments. Asp. Appl. Biol. 128, 1–10.

Micheli, F., Cottingham, K.L., Bascompte, J., Bjornstad, O.N., Eckert, G.L., Fischer, J.M., Keitt, T.H., Kendall, B.E., Klug, J.L., Rusak, J.A., 1999. The dual nature of community variability. Oikos 85, 161–169.

Moss, S., Storkey, J., Cussans, J.W., Perryman, S.A.M., Hewitt, M.V., 2004. The Broadbalk long-term experiment at Rothamsted: what has it told us about weeds? Weed Sci. 52, 864–873.

Mutshinda, C.M., O'Hara, R.B., Woiwod, I.P., 2011. A multispecies perspective on ecological impacts of climatic forcing. J. Anim. Ecol. 80, 101–107.

Neve, P., Vila-Aiub, M., Roux, F., 2009. Evolutionary-thinking in agricultural weed management. New Phytol. 184, 783–793.

NJF, 2013. Suction traps in studying distribution and occurrence of insects and forecasting pests. In: Seminar 468NJF Report, 9(7).

Omrane, S., Sghyer, H., Audeon, C., Lanen, C., Duplaix, C., Walker, A.S., Fillinger, S., 2015. Fungicide efflux and the MgMFS1 transporter contribute to the multidrug resistance phenotype in Zymoseptoria tritici field isolates. Environ. Microbiol. 17, 2805–2823.

O'Neill, B.F., Bond, K., Tyner, A., Sheppard, R., Bryant, T., Chapman, J., Bell, J.R., Donnelly, A., 2012. Climatic change is advancing the phenology of moth species in Ireland. Entomol. Exp. Appl. 143, 74–88.

Orwin, K.H., Wardle, D.A., 2004. New indices for quantifying the resistance and resilience of soil biota to exogenous disturbances. Soil Biol. Biochem. 36, 1907–1912.

Palmer, G., Hill, J.K., Brereton, T.M., Brooks, D.R., Chapman, J.W., Fox, R., Oliver, T.H., Thomas, C.D., 2016. Individualistic sensitivities and exposure to climate change explain variation in species' responses. Sci. Adv. 1, e1400220.

Perez-Rodriguez, J., Shortall, C.R., Bell, J.R., 2015. Large-scale migration synchrony between parasitoids and their host. Ecol. Entomol. 40, 654–659.

Poulton, P.R., Pye, E., Hargreaves, P.R., Jenkinson, D.S., 2003. Accumulation of carbon and nitrogen by old arable land reverting to woodland. Glob. Chang. Biol. 9, 942–955.

Salama, N.K.G., Edwards, G.R., Heard, M.S., Jeger, M.J., 2010. The suppression of reproduction of Tragopogon pratensis infected by the rust fungus Puccinia hysterium. Fungal Ecol. 3, 406–408.

Salama, N.K.G., Van Den Bosch, F., Edwards, G.R., Heard, M.S., Jeger, M.J., 2012. Population dynamics of a non-cultivated biennial plant Tragopogon pratensis infected by the autoecious demicyclic rust fungus Puccinia hysterium. Fungal Ecol. 5, 530–542.

Sanders, C.J., Shortall, C.R., Gubbins, S., Burgin, L., Gloster, J., Harrington, R., Reynolds, D.R., Mellor, P.S., Carpenter, S., 2011. Influence of season and meteorological parameters on flight activity of Culicoides biting midges. J. Appl. Ecol. 48, 1355–1364.

Schloss, P.D., Westcott, S.L., 2011. Assessing and improving methods used in operational taxonomic unit-based approaches for 16S rRNA gene sequence analysis. Appl. Environ. Microbiol. 77, 3219–3226.

Schulte, R.P.O., 2003. Analysis of the production stability of mixed grasslands II: a mathematical framework for the quantification of production stability in grassland ecosystems. Ecol. Model. 159, 71–99.

Seibert, A.C., Pearce, R.B., 1993. Growth analysis of weed and crop species with reference to seed weight. Weed Sci. 41, 52–56.

Shaw, M.W., Bearchell, S.J., Fitt, B.D.L., Fraaije, B.A., 2008. Long-term relationships between environment and abundance in wheat of Phaeosphaeria nodorum and Mycosphaerella graminicola. New Phytol. 177, 229–238.

Sheppard, L., Bell, J.R., Harrington, R., Reuman, D.C., 2016. Changes in large-scale climate alter spatial synchrony of aphid pests. Nat. Clim. Change 6, 610–613.

Silvertown, J., 1980. The dynamics of a grassland ecosystem—botanical equilibrium in the Park Grass experiment. J. Appl. Ecol. 17, 491–504.

Silvertown, J., 1987. Ecological stability—a test case. Am. Nat. 130, 807–810.

Silvertown, J., Dodd, M.E., Mcconway, K., Potts, J., Crawley, M., 1994. Rainfall, biomass variation, and community composition in the Park Grass experiment. Ecology 75, 2430–2437.
Silvertown, J., Servaes, C., Biss, P., Macleod, D., 2005. Reinforcement of reproductive isolation between adjacent populations in the Park Grass experiment. Heredity 95, 198–205.
Silvertown, J., Poulton, P., Johnston, E., Edwards, G., Heard, M., Biss, P.M., 2006. The Park Grass experiment 1856-2006: its contribution to ecology. J. Ecol. 94, 801–814.
Silvertown, J., Biss, P.M., Freeland, J., 2009. Community genetics: resource addition has opposing effects on genetic and species diversity in a 150-year experiment. Ecol. Lett. 12, 165–170.
Silvertown, J., Tallowin, J., Stevens, C., Power, S.A., Morgan, V., Emmett, B., Hester, A., Grime, P.J., Morecroft, M., Buxton, R., Poulton, P., Jinks, R., Bardgett, R., 2010. Environmental myopia: a diagnosis and a remedy. Trends Ecol. Evol. 25, 556–561.
Smith, M.D., Knapp, A.K., Collins, S.L., 2009. A framework for assessing ecosystem dynamics in response to chronic resource alterations induced by global change. Ecology 90, 3279–3289.
Snaydon, R.W., 1970. Rapid population differentiation in a mosaic environment. 1. Response of Anthoxanthum odoratum populations to soils. Evolution 24, 257–269.
Snaydon, R.W., Davies, M.S., 1972a. Rapid population differentiation in a mosaic environment. 2. Morphological variation in Anthoxanthum odoratum. Evolution 26, 390–405.
Snaydon, R.W., Davies, M.S., 1972b. Rapid population differentiation in a mosaic environment. II. Morphological variation in Anthoxanthum odoratum L. Evolution 26, 390–405.
Snaydon, R.W., Davies, M.S., 1976. Rapid population differentiation in a mosaic environment. 4. Populations of Anthoxanthum odoratum at sharp boundaries. Heredity 37, 9–25.
Steffen, W., Richardson, K., Rockstrom, J., Cornell, S.E., Fetzer, I., Bennett, E.M., Biggs, R., Carpenter, S.R., De Vries, W., De Wit, C.A., Folke, C., Gerten, D., Heinke, J., Mace, G.M., Persson, L.M., Ramanathan, V., Reyers, B., Sorlin, S., 2015. Planetary boundaries: guiding human development on a changing planet. Science 347, 736.
Stitt, M., Krapp, A., 1999. The interaction between elevated carbon dioxide and nitrogen nutrition: the physiological and molecular background. Plant Cell Environ. 22, 583–621.
Storkey, J., 2006. A functional group approach to the management of UK arable weeds to support biological diversity. Weed Res. 46, 513–522.
Storkey, J., Moss, S.R., Cussans, J.W., 2010. Using assembly theory to explain changes in a weed flora in response to agricultural intensification. Weed Sci. 58, 39–46.
Storkey, J., MacDonald, A.J., Poulton, P.R., Scott, T., Kohler, I.H., Schnyder, H., Goulding, K.W.T., Crawley, M.J., 2015. Grassland biodiversity bounces back from long-term nitrogen addition. Nature 528, 401–404.
Thackeray, S.J., Sparks, T.H., Frederiksen, M., Burthe, S., Bacon, P.J., Bell, J.R., Botham, M.S., Brereton, T.M., Bright, P.W., Carvalho, L., Clutton-Brock, T., Dawson, A., Edwards, M., Elliott, J.M., Harrington, R., Johns, D., Jones, I.D., Jones, J.T., Leech, D.I., Roy, D.B., Scott, W.A., Smith, M., Smithers, R.J., Winfield, I.J., Wanless, S., 2010. Trophic level asynchrony in rates of phenological change for marine, freshwater and terrestrial environments. Glob. Chang. Biol. 16, 3304–3313.
Thackeray, S.J., Henrys, P.A., Hemming, D., Bell, J.R., Botham, M.S., Burthe, S., Helaouet, P., Johns, D.G., Jones, I.D., Leech, D.I., MacKay, E.B., Massimino, D., Atkinson, S., Bacon, P.J., Brereton, T.M., Carvalho, L., Clutton-Brock, T.H., Duck, C., Edwards, M., Elliott, J.M., Hall, S.J.G., Harrington, R., Pearce-Higgins,

J.W., Høye, T.T., Kruuk, L.E.B., Pemberton, J.M., Sparks, T.H., Thompson, P.M., White, I., Winfield, I.J., Wanless, S., 2016. Phenological sensitivity to climate across taxa and trophic levels. Nature 535, 241–245.

Tilman, D., 1982. Resource Competition and Community Structure. Princeton University Press, Princeton, NJ.

Todman, L.C., Fraser, F.C., Corstanje, R., Deeks, L.K., Harris, J.A., Pawlett, M., Ritz, K., Whitmore, A.P., 2016. Defining and quantifying the resilience of responses to disturbance: a conceptual and modelling approach from soil science. Sci. Rep. 6, 28426.

Vogel, T., Simonet, P., Jansson, J., Hirsch, P., Tiedje, J., Van Elsas, J., Bailey, M., Nalin, R., Philippot, L., 2009. TerraGenome: a consortium for the sequencing of a soil metagenome. Nat. Rev. Microbiol. 7, 252.

Wilson, J.B., Crawley, M.J., Dodd, M.E., Silvertown, J., 1996. Evidence for constraint on species coexistence in vegetation of the Park Grass experiment. Vegetatio 124, 183–190.

Yoshida, K., Sasaki, E., Kamoun, S., 2015. Computational analyses of ancient pathogen DNA from herbarium samples: challenges and prospects. Front. Plant Sci. 6, 771.

Zhalnina, K., De Quadros, P.D., Gano, K.A., Davis-Richardson, A., Fagen, J.R., Brown, C.T., Giongo, A., Drew, J.C., Sayavedra-Soto, L.A., Arp, D.J., Camargo, F.A.O., Daroub, S.H., Clark, I.M., Mcgrath, S.P., Hirsch, P.R., Triplett, E.W., 2013. Ca. Nitrososphaera and Bradyrhizobium are inversely correlated and related to agricultural practices in long-term field experiments. Front. Microbiol. 4, 104.

Zhalnina, K., Dias, R., De Quadros, P.D., Davis-Richardson, A., Camargo, F.A.O., Clark, I.M., Mcgrath, S.P., Hirsch, P.R., Triplett, E.W., 2015. Soil pH determines microbial diversity and composition in the Park Grass experiment. Microb. Ecol. 69, 395–406.

Zust, T., Heichinger, C., Grossniklaus, U., Harrington, R., Kliebenstein, D.J., Turnbull, L.A., 2012. Natural enemies drive geographic variation in plant defenses. Science 338, 116–119.

CHAPTER TWO

How Agricultural Intensification Affects Biodiversity and Ecosystem Services

M. Emmerson[*,1], M.B. Morales[†], J.J. Oñate[†], P. Batáry[‡], F. Berendse[§], J. Liira[†,¶], T. Aavik[¶], I. Guerrero[†], R. Bommarco[‖], S. Eggers[‖], T. Pärt[‖], T. Tscharntke[‡], W. Weisser[#], L. Clement[#], J. Bengtsson[‖]

[*]Institute of Global Food Security, School of Biological Sciences, Queen's University Belfast, Belfast, United Kingdom
[†]Terrestrial Ecology Group (TEG), Universidad Autónoma de Madrid, Madrid, Spain
[‡]Georg-August-University, Göttingen, Germany
[§]Wageningen University, Wageningen, The Netherlands
[¶]Institute of Ecology and Earth Sciences, University of Tartu, Tartu, Estonia
[‖]Swedish University of Agricultural Sciences, Uppsala, Sweden
[#]School of Life Sciences Weihenstephan, Technische Universität München, Freising, Germany
[1]Corresponding author: e-mail address: m.emmerson@qub.ac.uk

Contents

1. Introduction — 44
   1.1 General Objective and Goals — 48
2. The CAP and AI — 49
   2.1 CAP as a Driver of Agriculture in Europe — 49
   2.2 How Does CAP Affect AI? — 52
   2.3 How Does CAP Affect Biodiversity and Ecosystem Services Through AI? — 53
   2.4 The AGRIPOPES Project—Examining the Multiple Effects of AI on Biodiversity and Ecosystem Services — 54
3. Local-Level and Landscape-Level Effects of AI — 60
   3.1 Local (Field)-Level Components of AI and Their Effect on Biodiversity — 60
   3.2 Landscape-Level Components of AI and Their Effect of Biodiversity — 74
   3.3 Farm-Level Components of AI — 76
   3.4 Comparing the Importance of Local- vs Landscape-Level Components of AI on Biodiversity and Ecosystem Services — 77
4. Organic-Conventional Comparisons — 79
   4.1 Landscape Context — 80
   4.2 Functional and Taxon-Specific Responses of Biodiversity to Farming Practices — 82
5. Linking AI to Biodiversity and Ecosystem Services — 83
   5.1 General Model — 83
   5.2 Study Area, Biodiversity Surveys and Biocontrol Experiment — 84
   5.3 Structural Equation Modelling — 84
   5.4 Results and Discussion of SEM — 86

6. Conclusions 87
Acknowledgements 88
References 89

Abstract

As the world's population continues to grow, the demand for food, fodder, fibre and bioenergy will increase. In Europe, the Common Agricultural Policy (CAP) has driven the intensification of agriculture, promoting the simplification and specialization of agroecosystems through the decline in landscape heterogeneity, the increased use of chemicals per unit area, and the abandonment of less fertile areas. In combination, these processes have eroded the quantity and quality of habitat for many plants and animals, and hence decreased biodiversity and the abundance of species across a hierarchy of trophic levels and spatial scales within Europe. This biodiversity loss has led to profound changes in the functioning of European agroecosystems over the last 50 years. Here, we synthesize the findings from a large-scale pan-European investigation of the combined effects of agricultural intensification on a range of agroecosystem services. These include (1) the persistence of high conservation value species; (2) the level of biological control of agricultural pests and (3) the functional diversity of a number of taxonomic groups, including birds, beetles and arable weeds. The study encompasses a gradient of geography-bioclimate and agricultural intensification that enables the large-scale measurement of ecological impacts of agricultural intensification across European agroecosystems. We provide an overview of the role of the CAP as a driver of agricultural intensification in the European Union, and we demonstrate compelling negative relationships between the application of pesticides and the various components of biodiversity studied on a pan-European scale.

1. INTRODUCTION

The world's population is predicted to grow from 7 to at least 9 billion by 2050, whilst simultaneously the climate is predicted get warmer globally and the frequency of extreme climate events to increase. At the same time, crop production is not increasing and as a consequence of economic and global climatic changes is even declining in major agricultural regions despite technological advances (Ray et al., 2012). This combination has the potential to create a global food crisis (Global Food Security, 2011; Godfray et al., 2010; Lennon, 2015; Poppy et al., 2014).

Farmlands are the most extensive habitat for biodiversity in Europe, harbouring, for example more than one half (250 species) of European bird species, of which 50% are either threatened or have suffered steep population declines (Chamberlain et al., 2000; Donald et al., 2001, 2006;

Krebs et al., 1999; Robinson and Sutherland, 2002). There were 10.8 million farms across the EU-28 in 2013, working 174.4 million ha of land (the utilized agricultural area or UAA), which represents roughly 40% of the total land area of the EU-28 (Eurostat, 2015). Nearly 60% of the UAA was used as arable land (104.2 million ha), including 57.6 million ha being used for cereal production. A large proportion of the production of crops such as cereals is not used for direct human consumption, but rather is fed to livestock (Foley et al., 2011), and estimates show that in 2011 around 6.1 million ha of agricultural land (3.4% of the total UAA) were directly and increasingly devoted to the production of biomass and energy crops (Eurostat, 2015).

Until recently in Europe, the Common Agricultural Policy (CAP) drove the intensification of agriculture in order to meet increased demands for food and fodder starting in the 1950s. European farmed landscapes have traditionally consisted of complex mosaics of extensive crops that sustained high levels of biodiversity (Potter, 1997; Walk and Warner, 2000). Over the last 50 years, however, the farmlands of western European countries have experienced dramatic changes, mainly through the intensification of farming techniques (Björklund et al., 1999; Robinson and Sutherland, 2002; Siriwardena et al., 2000). For example, the yield of cereals has increased steadily (Liira et al., 2008a), although the total application of fertilizers has dropped by 30% since the 1980s (Eurostat, 2015; Liira et al., 2008a; see Fig. 1). The loss of biodiversity driven by agricultural intensification (AI) is judged to be similar in scale to that expected from climate change (Tilman et al., 2001).

AI occurs at multiple spatial scales, with particular focus on local and landscape scales (Benton et al., 2003; Firbank et al., 2008). On the one hand, crop yield and revenue optimization lead to an increased impact by agricultural activities, at the cost of noncultivated components of the field and its immediate surroundings, and thereby, severe losses of wild plant and animal populations (Firbank et al., 2008; Haberl et al., 2004). On the other hand, large-scale field-level intensification leads to landscape simplification and homogenization, which further reduces habitat availability for wild species (Tscharntke et al., 2005). At regional scales that incorporate several landscapes, specialization, focused on particular monocultures, can occur, e.g. on fertile alluvial soils. In contrast, some regions can experience land abandonment, for example in less productive areas such as mountainous regions (Benton et al., 2003; Tivy, 1990) or forest-dominated regions in northern Europe (Wretenberg et al., 2007). In summary, AI operates primarily at

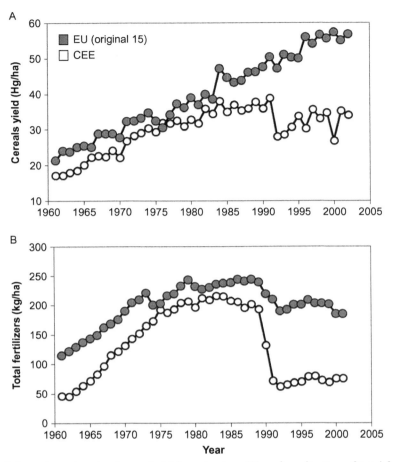

Fig. 1 Long-term trends of cereal yield per hectare (A) and application of total fertilizers (active substances) per hectare (B) in early (western) members of European Union (EU(15)) and in countries of Central- and Eastern Europe (CEE). *Source: FAOSTAT, 2004. Agricultural data; see details in Liira, J., Aavik, T., Parrest, O., Zobel, M., 2008a. Agricultural sector, rural environment and biodiversity in the central and eastern European EU member states. Acta Geograph. Debrecina Landsc. Environ. Ser. 2, 46–64.*

the field level (increased fertilizers, pesticides, employment of machinery, increased sowing density, ploughing depth, etc.), but may also dominate whole landscapes and regions, thereby contributing to biodiversity declines and homogenization of agricultural ecosystems at larger scales (Flohre et al., 2011a).

In contrast, many management decisions made by farmers do not concern particular fields, but rather the entire farm. Such decisions often affect the production mode and sequence of crop rotations, for example the decision to manage conventionally vs organic farming techniques, and may generate a certain degree of spatial aggregation of farming types in the landscape, particularly in regions where the agrarian property is spatially concentrated, e.g. Western Europe. Therefore, farm-level measures of intensification might often be required in some regions, and field-level measures in others.

The increased human demand of food and energy crops is predicted to contribute to the continuing intensification of European agriculture, but it also poses an environmental and sustainability problem (Godfray et al., 2010). The intensive management practices that prevail in agricultural ecosystems have the potential to affect a wide range of plant and animal species, as well as ecosystem processes underpinning agricultural production, at local to very large spatial scales (Oliver et al., 2015). For example, over the last 50 years, AI has led to marked declines of numerous species of European flora and fauna at local, national and regional scales (Donald et al., 2001; Kleijn et al., 2009; Stoate et al., 2001; Tilman et al., 2001). This decline is reflected in the National Biodiversity Indices (SCBD, 2001) of old EU states (15) subject to AI, relative to new EU member states that have significantly higher indices of biodiversity (Liira et al., 2008a).

Agricultural ecosystems harbour part of a wide ecological network of interacting organisms from arable and nonarable habitats that form the ecological context for food production. This ecological network comprises the biodiversity elements that confer a range of beneficial ecosystem services underpinning the production of food and other commodities in agroecosystems. At large spatial scales, there are clines in species richness and biodiversity that reflect underlying spatial variation in the physical environment, e.g. patterns in precipitation, temperature and soil conditions that in combination comprise the bioclimate of a region.

The physical environment, including bioclimate and human management activities that modify the physical environment, creates a range of different environmental contexts that are intimately associated with the presence and relative abundance of species, and therefore drive large and fine scale patterns in biodiversity. In addition, the environmental context can affect the physical attributes or traits of species that in turn can alter the way in which species interact within their ecological networks (Poisot et al., 2012;

Woodward et al., 2005). For example, changes in body mass can alter the relative strength of trophic or competitive interactions among species (Vucic-Pestic et al., 2010). Changing patterns of interactions can affect the flow of energy and nutrients and alter the resilience and functioning of an ecosystem.

1.1 General Objective and Goals

How AI processes drive habitat degradation and the loss of biodiversity and associated ecosystem services is the focus of this chapter. Since the impacts of AI manifest at different spatial scales and can affect biodiversity and ecosystem service delivery in unique ways, we pay particular attention to the relationships between biodiversity and AI factors operating at local and landscape scales. To manage agroecosystems for secure and sustainable food production requires that we understand how the different components of agricultural management practices affect biodiversity and ecosystem functioning. To achieve this, we first provide an overview of the role of the CAP as a driver of AI in the European Union. We then review and synthesize evidence quantifying where and how AI has had impacts on the taxonomic and functional diversity (FD) of agricultural ecosystems, and the provision of biological control of pests, a key ecosystem service for the sustainability of food production.

We draw predominantly, but not exclusively, on the results of AGRIPOPES (AGRIcultural POlicy-Induced landscaPe changes: effects on biodiversity and Ecosystem Services), a large-scale pan-European research project focused on quantifying agricultural policy-induced landscape changes and their effects on taxonomic and FD and the associated delivery of biological pest control. Our goal is to use the results obtained within AGRIPOPES that address the impacts of AI on biodiversity in farmed landscapes, but we highlight and review evidence from the relevant literature that is far broader than the AGRIPOPES project in scope. We focus on the three processes that are understood to drive biodiversity loss through AI: (1) increased use of farm chemicals, like fertilizers, herbicides and pesticides; (2) mechanization and crop and husbandry specialization and (3) simplification of farmed landscapes leading to loss of landscape diversity. Initially, we provide a historical background of the CAP as the main driver of agricultural change and AI over recent decades in Europe, and we present the general methodology of AGRIPOPES. We then review the evidence for impacts of AI on biodiversity throughout Europe and beyond. We have concentrated on agricultural areas dominated by cereals, since they alone comprise more than 13% of EU-28 land area, that is, more than 57 million ha, which amounts to one-third of the total area devoted to agriculture in Europe (Eurostat, 2015).

2. THE CAP AND AI
2.1 CAP as a Driver of Agriculture in Europe

The CAP is frequently considered the main instrument behind the dual process of intensification and abandonment observed in agricultural systems in the European Union over the last few decades. Paradoxically, the CAP is also now expected to play a role in the environmental conservation of these systems (Pe'er et al., 2014). This apparent contradiction is explained by considering the founding principles and evolution of one of the oldest policies of the European Union, which has been strongly rooted in the European integration project since its effective inception in 1962.

The original priority of the CAP was to increase agricultural production in order to stabilize agricultural markets and farmers incomes, this was enshrined in the Treaty of Rome, which came into effect in 1957. The core mechanisms were based on (1) price support, i.e. maintaining prices to producers above world levels through intervention, e.g. buying food; (2) import levies, i.e. raising the price of imports above world prices; and (3) export subsidies, i.e. compensating producers selling at lower world market prices. The economic environment provided by the security of commodity prices offered by the CAP also stimulated scientific research and development aimed at increasing crop and animal yields, much of it financed by public funds (Robson, 1997). In this way the CAP incentivised the intensive use of farm chemicals (e.g. fertilizers, pesticides and herbicides) and machinery, and so increased the yields of all supported commodities. Originally these were cereals, milk and livestock, but the list increased as the European Community expanded in subsequent years. Further, the CAP also promoted capital-intensive operations, such as drainage, land consolidation and irrigation expansion, and was often supported by National governments that reinforced the trend towards greater intensity and scale of agricultural production (Robson, 1997).

The resulting multivariate process of change is embodied in the term 'Agricultural Intensification', which from an agronomic and economic view, is related to the increase of production output, i.e. yield, per unit of area or time (Turner and Doolittle, 1978) or per unit of inputs (FAO, 2004). Agricultural production can be intensified by increasing inputs of capital, such as machinery, energy and biotechnology, or by increasing inputs of manual labour (Börjeson, 2010). The continued decrease of agricultural labour input in Western Europe over the last decades (4.2% of total EU-15 employment in 1999 to 2.9% in 2013; Eurostat, 2015; Eastern Europe ca. 12% in 2013) indicates that AI in the EU has been based on high inputs of entrants and capital investment per unit land area.

Biodiversity declines are more likely to be related to inputs than to yield per se. Therefore, in the ecological domain, AI refers to the increased use per unit area of fertilizers, pesticides, water and machinery for crop production and high-density housing systems for animals. Broadly, the concept of AI also relates to the landscape-scale consequences of these and other changes, such as reduced crop rotations, or loss of noncrop features (e.g. Donald et al., 2001; Kleijn et al., 2009; Stoate et al., 2001). These variables are often used as surrogate measures of intensity (Herzog et al., 2006), an approach that was adopted in the AGRIPOPES project.

The changes observed in agricultural landscapes and the declines in biodiversity during CAP implementation are well documented (Lefebvre et al., 2012; Meeus, 1993; Tscharntke et al., 2005; Wretenberg et al., 2007). Whilst the CAP played a major role in transforming the rural landscape of Europe, other drivers were also important. It is likely that the observed patterns of agricultural change would have taken place even without government interventions, since they are ultimately related to the processes of technological development (supporting advanced use of machinery, chemicals, biology and information-related developments) and social and demographic change, including industrialization, urbanization, migration to cities and the ageing of rural populations, processes that have been ongoing since the beginning of the 20th century (Brouwer and Lowe, 2000; Buckwell, 1990). It is likely, however, that the financial support under the CAP played a major role in determining the profitability of the adoption and diffusion of novel crops, livestock and techniques in the agricultural sector, thus creating the conditions for their general adoption (Potter, 1997). Therefore the confluence between capacity (technical progress) and opportunity (farm support) has in the long run brought about the changes in the structure, pattern and practice of farming now recognized to have had profound environmental effects (Benton et al., 2003).

The CAP has evolved and has undergone numerous reforms throughout its 54-year history. These modifications to the CAP reflect attempts at remediating unforeseen consequences of varied internal (e.g. production surpluses, budget deficits, environmental concerns) and external problems (e.g. trade conflicts) that arose because of legislation embodied in the CAP (Robson, 1997). Initiatives to reform the CAP were developed in the early 1970s, although the major shift came with the 'MacSharry reform', adopted in 1992. This included replacement of price support with direct aid payments per hectare and the introduction of compulsory set aside, and the formal introduction of environmental objectives in the CAP, i.e. promoting the adoption of environmentally friendly farming methods. Since then, Agri-Environmental Schemes (AESs) represent the most prominent instrument available to achieve alignment between agricultural practices and nature conservation policies (Hodge et al., 2015). These measures provide economic incentives to compensate for the additional costs and income foregone resulting from the voluntary adoption by farmers of practices that protect the environment. These practices include the reduction of agro-chemical inputs, adoption of organic farming, and extensive forms of production, based on longer rotations and allowing the presence of fallows and unploughed landscape features (hedgerows, trees, small woods, ponds, wetlands, field borders), as well as reduced livestock density. Expenditure on these schemes and area covered by their prescriptions has steadily increased across the EU, but they have had only partial success in the environmental enhancement of the targeted systems, due to unclear objectives, inadequate design, or low uptake (see Batáry et al., 2015 and references herein). Although several studies have assessed the effectiveness of these schemes in different countries (Concepción et al., 2012; Kleijn et al., 2001; Tuck et al., 2014), the lack of continuous and homogeneously applied monitoring schemes may have limited their efficacy.

The Agenda 2000 reform subsequently recognized the multifunctionality of European agricultural systems, and the 2003 reform of the CAP removed the link between the receipt of a direct payment and the production of a specific commodity (known as 'decoupling'). These reforms also introduced 'cross-compliance', as a series of rules that the farmer had to respect in order to receive direct payments. These rules were related to the environment, the protection of water resources, and the condition in which farmland was maintained. In 2007, AESs were further reinforced with the creation of the European Agricultural Fund for Rural Development as the second pillar of the CAP. Direct payments to farmers and market management measures continued in the so-called first pillar, but with a budget close to four times the size of that of pillar 2 in the last programming period 2007–2013. Decoupling was completed by the 'health check' of the CAP, a range of streamlining measures introduced in 2008 that also included the abolition of arable set aside, new cross-compliance requirements, and a reduction of direct payments to farmers with the money transferred to the Rural Development Fund instead.

With the CAP reform in 2013, covering the period 2014–2020, new so-called 'greening measures' were introduced affecting direct payments to farmers. These measures made 30% of the payment conditional upon the maintenance of permanent pastures, the diversification of crops, and the establishment of 'ecological focus areas'. However, the new environmental prescriptions are so diluted—only applying to roughly 50% of EU farmland, and with most farmers exempted from deploying them—that they are unlikely to be of benefit to biodiversity (Pe'er et al., 2014). For example, in the AGRIPOPES data, only around 12% of the farms grew less than three crops, the majority of those were in Spain.

2.2 How Does CAP Affect AI?

In a European context, AI is therefore a multifactorial process that leads to increased yields (Donald et al., 2006; Herzog et al., 2006; Matson et al., 1997). Crop and livestock specialization, increased synthetic inputs and soil-disrupting operations or removal of semi-natural elements and landscape features are all components acting at field and landscape levels which have interacted jointly to modify the agricultural ecosystems of Europe over the last decades (Chamberlain et al., 2000). Two related processes acting at different scales underlie increased AI: (1) intensification of management practices at the field level through the increased use of farm chemicals (herbicides, pesticides and fertilizers) per unit area (see Fig. 1) and soil-disrupting operations (ploughing, refining) and (2) simplification of landscape diversity through crop specialization, removal of landscape elements, conversion of permanent pastures into arable land, and land abandonment on less fertile areas and expansion of early successional, homogeneous shrub replacing the landscape mosaic landscape typical of extensive farmland. These processes in combination have led to a degradation of habitat quality and an overall decline in the diversity and total biomass of a wide range of species. These species make up the biological matrix within which agricultural ecosystems are nested. Ultimately, the processes underpinning AI have led to similar rates of local species loss for several taxonomic groups across European agricultural landscapes (Robinson and Sutherland, 2002). The massive use of entrants also has additional detrimental effects like soil and water eutrophication and contamination, whilst soil-disrupting operations favour erosion (Stoate et al., 2001).

2.3 How Does CAP Affect Biodiversity and Ecosystem Services Through AI?

AI is considered to be the main process driving the generalized decline of farmland biodiversity observed in Europe over the last decades. Such declines affect organisms from different taxa, including birds (e.g. Donald et al., 2001, 2006), vascular plants (Marshall et al., 2003; Storkey, 2012), invertebrates (e.g. Aebischer, 1990; Östman et al., 2001a; Weibull et al., 2000) and soil organisms (Kladivko, 2001).

Arguably, species loss in European agricultural landscapes is driven by changes in food web, and more generally, ecological network structure, e.g. plant–pollinator networks (Ings et al., 2009). Structural habitat modification and changes in the supply and diversity of the species' resource base have altered the availability of food and shelter, which has in turn driven changes in the abundance and diversity of species. The simplification of agricultural landscapes has also affected ecosystem services, i.e. the benefits for human society provided by different ecological processes, produced by subsets of biodiversity, for example biological control of agricultural pests (Östman et al., 2001b, 2003; Thies and Tscharntke, 1999) and pollination (Garibaldi et al., 2011; Kennedy et al., 2013). However, the documentation of changes in biodiversity and ecosystem services at a European-wide scale is largely lacking.

Plants, insects and especially birds have all declined in European farmland at the community and landscape level (Billeter et al., 2008; Chamberlain et al., 2000; Pain and Dixon, 1997; Wretenberg et al., 2007). Different local and country scale studies have shown that AI and landscape homogenization can induce biodiversity loss (e.g. Benton et al., 2003; Robinson and Sutherland, 2002; Tuck et al., 2014). For example, the study by Wretenberg et al. (2007) showed that changes in farmland bird population trends in Sweden were directly linked to changes in agricultural intensity caused by corresponding changes in agricultural policies between 1970 and 2000. That study also showed that these relationships were partly dependent on landscape heterogeneity. However, until recently, there were limited data available to support the statement that intensification led directly to biodiversity loss at the European scale. The lack of such supporting evidence was the basis of much of the work summarized here in the context of the AGRIPOPES project.

2.4 The AGRIPOPES Project—Examining the Multiple Effects of AI on Biodiversity and Ecosystem Services

The AGRIPOPES project was developed within the EuroDiversity programme of the European Science Foundation during the years 2006–2009. A consortium of nine research teams, representing a latitudinal gradient and two different land intensification histories (east–west) from eight European countries (Sweden, Estonia, Poland, the Netherlands, Germany (two areas: close to Göttingen, reflecting West Germany, and Jena, reflecting East Germany), France, Spain and Ireland), examined multiple aspects of the intensification of agriculture associated with the CAP and its consequences for the biodiversity and the ecosystem services associated with cereal agroecosystems. Three major issues were addressed: the persistence of taxonomic diversity, the responses to AI of FD of a number of taxa and the prevalence of sustained biological control of important agricultural pests. The design of the project used the double gradient of geography or bioclimate, and AI from Northern and Eastern Europe (Sweden, Estonia) through Germany, the Netherlands, Ireland and France to Mediterranean climate in Spain, which made it possible to assess large-scale ecological impacts of AI across European cereal-dominated agroecosystems (Fig. 2).

The project aimed to quantify the effects of AI on landscape composition and the taxonomic and FD of selected vertebrate (birds), insect (carabid beetles) and wild plant groups in European agroecosystems. AGRIPOPES also examined the effects of AI on the potential for biological control of pests on common crops, in relation to local and regional landscape composition. The main working hypotheses formulated at the start of the project were (1) AI leads to a loss of biodiversity and decrease in density in many organisms associated with the agricultural landscape, and this effect across Europe is similar to that found when examining regions within a country and (2) the loss of biodiversity due to AI leads to a simplification of food webs associated with biological control of insect pests, consequently entailing less efficient biological control.

The eight study areas were all agricultural regions where winter cereal was the dominant crop. Hence we chose winter wheat as a standard crop and were careful to standardize measurements of biodiversity and biological

Fig. 2 Location of the 9 AGRIPOPES European study sites. All sites are named after the corresponding country, except for Germany, where two sites were located. 1: Spain; 2: France; 3: Ireland; 4: the Netherlands; 5: West Germany (Göttingen); 6: East Germany (Jenna); 7: Sweden; 8: Poland; 9: Estonia.

control potential across all field sites, whilst allowing landscape composition and AI to vary within, as well as between countries.

2.4.1 General Methodology
2.4.1.1 Selection of Farms and Fields

Given the diversity of agricultural structures in the different countries involved in AGRIPOPES, a farm was considered the ecological unit under study and was recognized as a set of one or more fields, separated by a distance of not more than 1 km, which were cultivated by the same farmer (owned or leased), and occupying an area not exceeding 1 km^2.

In each sampling area (one per country, except in the case of Germany, see Fig. 2), 30 farms separated by at least 1 km were selected, and considered to be representative of a gradient of regional AI. These farms were situated in regions between 30×30 and 50×50 km^2 in area, in order to limit variation in within-region species pools and β diversity, and to avoid an excessive heterogeneity of landscapes and soil types within each study area. Farms were selected so that the range of cereal productivity in the sample was as large as possible, based on information obtained from the farmers on cereal yields in the years preceding the study, and with a representative and even distribution across the gradient of productivity in each area.

On selected farms (which could be conventional or organic agriculture farms) cereal had to be grown during biodiversity sampling, mainly winter wheat (80% of the fields). Only cereal crops were sampled on each farm. Sampled fields were never smaller than 1 ha in size nor irrigated. To assess the gradient of regional AI, the average cereal yield in the three previous years to sampling (2004–2006) was used. Sampling took place during spring and summer 2007 and was synchronized using the phenological stages of winter wheat in each study area. Winter wheat passes through well-recognized growth stages that are used by farmers to assess crop development. We timed the sampling of biodiversity components, i.e. plants, invertebrates and birds to coincide with the different growth stages of winter wheat (see Section 2.4.1.2 for details of each taxonomic group).

2.4.1.2 Selection of Points for Biodiversity Sampling

For each farm unit, five points distributed over no more than five arable fields were selected for sampling wild plants and carabids and estimating the biological control potential. Sampling points were located, whenever possible, in five different fields of the same farm and, when possible, always on winter wheat. When this was not feasible (less than 20% of all studied fields), winter barley was used instead. When there were fewer than five fields available, the points were stratified in proportion to size of sampled fields. Sampling points were placed parallel to an herbaceous (not woody) field edge and at 10 m distance from the edge towards the centre of the field. When more than one sampling point was placed in the same field, they were placed at opposite sides of the field. For the survey of breeding birds, one area of 500×500 m^2 was selected around one of the sampled fields on each farm.

Plants were sampled in three 2×2 m^2 plots parallel to the edge of the field and separated from each other by a distance of 5 m. In each plot, information on presence and abundance of all present species was collected. Sampling was performed once during the flowering to the milk-ripening stage of winter wheat. To further avoid phenological effects of sampling regionally, the sequence of farm surveys was randomized over the AI gradient within each study area.

Carabids were sampled using two pitfall traps per sampling point located in the middle of the two outer vegetation plots. The traps were protected with a covered with a plastic lid, suspended 1 cm above the ground by 100 mm nails, to avoid the effect of precipitation. Each pitfall trap was filled with 150 ml of ethylene glycol at 50%. Traps were opened during two periods of 7 days. The first sampling period occurred 1 week after the appearance of spikes of winter wheat and the second coincided with the milk-ripening stage of winter wheat. Specimens caught in pitfall traps were stored in 70% ethanol, and all the species caught in one trap randomly selected from each pair of traps were identified.

A modified version of the British Trust for Ornithology's Common Bird Census (Bibby et al., 1992) was used for the bird surveys. Given their mobility characteristics, birds were sampled over a 500×500 m^2 square centred on the largest field of each sampled farm, in such a way that each spot within the quadrat was no more than 100 m from the surveyor's route. Surveys were conducted three times, at intervals of 3 weeks during the spring and summer 2007 (March–June) to cover the breeding time of farmland birds. Based on local information on the phenology of breeding birds, this meant that the start and end of the surveys varied from south to north reflecting local information on timing of known breeding seasons, e.g. in Spain surveys began in March, whilst in Sweden they started in April. Individuals of all ground-nesting farmland species showing some activity inside the square sample were counted. Breeding bird territories were determined considering the three visits and applying the following three categories, depending on species detectability and reproductive behaviour, and the number of visits in which they were recorded (see Geiger et al., 2010a for details):
- Category A: Easily detectable species, present throughout the spring, detected in at least two visits as showing territorial behaviour (song, call and defence of territory) in the same location.
- Category B: Species difficult to detect and species that were less likely to be detected during the three visits (e.g. long distance migrant species and strictly summer visitors), detected at least once showing territorial behaviour.
- Category C: Direct evidence was required of breeding activities to confirm a territory of these species.

2.4.1.3 Biological Control Potential

During the emergence of the first inflorescence of winter wheat, biological control potential was estimated by a 2-day experiment, which was repeated once within 8 days (Östman et al., 2001b). In the morning of the first day, three live pea aphids (*Acyrthosiphon pisum*) of the third or fourth instar were glued to plastic labels by at least two of their legs and part of their abdomen using odourless superglue. At noon, three labels were placed on the ground along the diagonal of each plot, at three of the five sampling points per farm. The labels were bent and placed, so the aphids were on the lower surface, protected from rain. Hence, at each farm there were 27 labels, with 81 aphids in total. The labels were checked at the start of the experiment and four more times over a 30 h interval: around 6 p.m. of the first day, at 8 a.m., 1 p.m. and about 6 p.m. on the following day, the exact time varying depending on the study area. The labels with the remaining aphids after the last counts were taken to the lab to check under stereomicroscopes whether remaining aphids could not have been removed by predators because they were covered with glue. The data from one or both of the rounds from each study area were used for the analyses, depending on whether the measurements were reliable or interrupted by, for example heavy rains.

2.4.1.4 Field-Level Intensification Variables

AI variables at the field scale were obtained through questionnaires conducted by personal interviews with all farmers owning the fields or responsible for their management. Spatial measures were obtained using digital maps processed in a Geographic Information System. The farmers' response rate was 98%, and information about yields and farming practices (pesticide and fertilizer use, ploughing and mechanical weed control regime) and farm layout (number of crops, percentage land covered by AES, field size) was collected.

2.4.1.5 Landscape-Level Intensification Variables

Four variables reflecting landscape structure and composition were estimated from aerial photographs with ArcView Patch Analyst 3.12 tool (Rempel et al., 1999) from 500 m radius circles around each sampling point and coinciding with the centre of the bird survey area: mean field size and its standard deviation, the percentage of land planted with arable crops within the area, and the Shannon habitat diversity index. For the latter, the following habitat classes were used (according to the definitions from the European Topic Centre on Land Use and Spatial Information; Büttner et al., 2000): continuous urban fabrics, discontinuous urban fabrics, cultivated arable lands, fallow lands under rotation systems, permanent crops, pastures, forests, transitional woodland–scrub and water (Table 1).

Table 1 Main Response and Explanatory Variables Considered in AGRIPOPES

Variable		Description
Response		
	Vascular plant species richness	Number of plant species detected in five sampling points (three 2×2 m² plots each) distributed over focal fields within 1×1 km² squares
	Carabid richness	Number of carabid species detected in five sampling points (one pitfall traps each) distributed over focal fields within 1×1 km² squares
	Bird species richness	Number of wintering bird species detected in 500×500 m² squares centred on focal fields
Explanatory		
Field level	Focal field size	Size of each surveyed plot's focal field (ha)
	Yield	Cereal grain obtained in focal field (tons/ha)
	Amount of herbicide	Total amount of herbicide active ingredients applied on focal field (g/ha)
	Frequency of herbicide	Number of herbicide applications on focal field during the previous agricultural year
	Amount of insecticide	Total amount of insecticide active ingredients applied on focal field (g/ha)
	Frequency of insecticide	Number of insecticide applications on focal field during the previous agricultural year
	Amount of fungicide	Total amount of fungicide active ingredients applied on focal field (g/ha)
	Frequency of fungicide	Number of fungicide applications on focal field during the previous agricultural year
	N fertilizer	Total amount of nitrogen applied on focal field (kg/ha)
	Frequency of tillage	Number of soil-disrupting operations carried out on focal field during the previous agricultural year
Landscape level	Mean field size	Mean size of fields with arable crops within a 500 m radius circle centred on focal field (ha)
	Substrate diversity	Shannon–Wiener index of agricultural, natural and artificial substrates within a 500 m radius circle centred on focal field
	Percentage cover of arable crops	Percentage area of arable crops within a 500-m radius circle centred on focal field

Continued

Table 1 Main Response and Explanatory Variables Considered in AGRIPOPES—cont'd

Variable		Description
Farm level	Area under AES	Percentage area of a particular farm under Agri-Environmental Schemes
	Number of crops	Number of different crop types in the farm
	Organic vs conventional	Whether a particular farm is under organic farming or not

Explanatory variables are classified as field, landscape or farm-level variables.

3. LOCAL-LEVEL AND LANDSCAPE-LEVEL EFFECTS OF AI

Most studies quantifying AI effects on biodiversity and ecosystem functions have measured variables at several spatial scales. At the field scale, yield has been frequently used as a main output variable summarizing the effects of management practices such as agrochemical applications (Donald et al., 2006; Geiger et al., 2010a; Tilman et al., 2002). At the landscape scale, metrics describing heterogeneity due to configuration and composition are usually employed (Fahrig et al., 2011; Hiron et al., 2015; Teillard et al., 2014). The heterogeneity of configuration can be measured by mean field size or the total length of borders (Benton et al., 2003; Teillard et al., 2014). Landscape composition indices like Shannon–Wiener H' or the proportion of arable land are commonly employed (Chiron et al., 2010; Ekroos et al., 2010; Filippi-Codaccioni et al., 2010). However, configurational and compositional heterogeneity components are usually strongly correlated (Fahrig et al., 2011), which may confound landscape effects on biodiversity. In fact, few studies have attempted to disentangle the effects of the two types of heterogeneity (Hiron et al., 2015; Teillard et al., 2014), and further research would help designing more targeted landscape management prescriptions. As an example, Table 1 summarizes the variables measured to cover both field and landscape-scale components of AI, as well as farm-level intensification measures, in the AGRIPOPES project.

3.1 Local (Field)-Level Components of AI and Their Effect on Biodiversity

Field-level AI is associated with an increase in the amount and frequency of application of pesticides and synthetic fertilizers, which allows for an increase in the density of sown cereal crops. The result is a significant increase in yields and a concomitant loss of spatiotemporal heterogeneity of crops. In turn, higher field productivity allows the abandonment of crop rotation systems, leading to landscape simplification (e.g. Benton et al., 2003).

Among the field-level intensification variables measured in AGRIPOPES (Table 1), yield and those related to pesticide application accounted most for variation in the various measures of biodiversity studied and are discussed later, although results on other important field-level factors like fertilizer input, tillage and sowing density are important and are also summarized.

In all study sites, winter cereal crops were the dominant land use. However there were important differences between the sites in their level of AI, which are reflected in the average values of the different intensification components considered (Figs. 3 and 4). At the field scale, the Irish study site was the most intensively managed one according to both yield and entrants such as nitrogen fertilizer and insecticides, followed by the East German site (Jena), which was the most intensively managed when considered from the perspective of herbicides (Fig. 3). Spain was situated at the opposite end of the continental intensification gradient with the lowest average values of yield and synthetic inputs. Consistently, sites with low entrant levels showed a much higher degree of other intensification practices at the field scale that allows for some compensation. This can be seen in the high frequency of tillage operations (an alternative to weed elimination through herbicide application) and sowing density shown not only by the Spanish area but also by the Swedish and Polish areas, where entrants were also relatively low.

With regard to landscape-scale intensification, Jena and Estonia had the largest field and farm sizes (Fig. 4). Along with the Dutch and Polish sites, they also contained the largest proportions of arable land in the landscape, as well as the greatest diversity of crops. The Polish, Spanish and Irish sites had the smallest field size, which probably suggests lower historical intensification levels and indicates a very strong intensification process per unit area in the case of Ireland. Finally, Spain had the lowest crop diversity, as expected for a continental Mediterranean region where soil and rainfall are particularly limiting for crops other than cereals.

3.1.1 Relationships with Yield

As a fundamental output variable in AGRIPOPES, yield was used as a proxy for AI. Yield had a strong negative influence on wild vascular plant, carabid beetle and ground-nesting farmland bird species richness across all study sites (Wald tests: plants: $\chi^2_1 = 141.42$, $p < 0.001$; carabids $\chi^2_1 = 23.33$, $p < 0.001$;

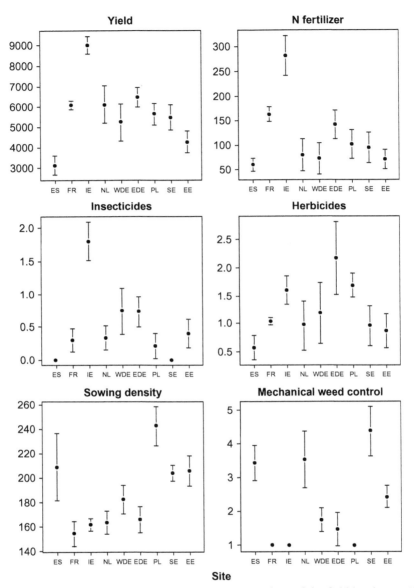

Fig. 3 Differences between study areas in the mean values of the field-level intensification variables measured in the AGRIPOPES project (see Table 1 for descriptions). Error bars indicate 95% confidence intervals. *ES*, Spain; *FR*, France; *IE*, Ireland; *NL*, Netherlands; *WDE*, West Germany (Göttingen); *EDE*, East Germany (Jena); *PL*, Poland; *SE*, Sweden; *EE*, Estonia.

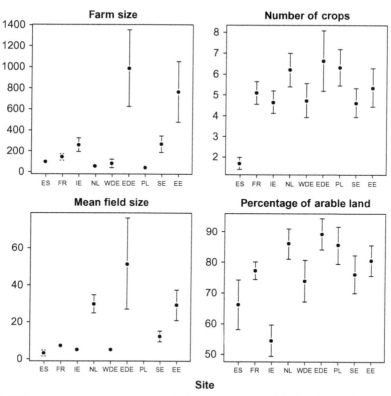

Fig. 4 Differences between study areas in the mean values of the landscape-level intensification variables measured in the AGRIPOPES project (see Table 1 for descriptions). Error bars indicate 95% confidence intervals. *ES*, Spain; *FR*, France; *IE*, Ireland; *NL*, Netherlands; *WDE*, West Germany (Göttingen); *EDE*, East Germany (Jena); *PL*, Poland; *SE*, Sweden; *EE*, Estonia.

birds: $\chi^2_1 = 7.33$, $p = 0.007$; regional landscape differences were controlled for by treating them as random factors using General Linear Mixed Model analyses; see Fig. 4). On average, an increase in cereal yield from 4 to 8 ton/ha resulted in the loss of five of nine plant species, two of seven carabid species and one of three bird species (Fig. 5A–C).

The overall negative relationships between yield and different components of biodiversity were not always consistent and these varied with country and taxa, i.e. wild plants and carabid beetles (yield × study area interaction: $\chi^2_8 = 36.87$, $p < 0.001$; $\chi^2_8 = 24.35$, $p = 0.002$; respectively). Comparison of the yield effects among study areas revealed that in some countries, yield had negative effects, but in other countries there was no

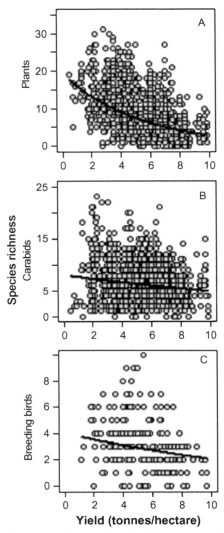

Fig. 5 Effects of cereal yield (tons/ha) on (A) wild plant species richness per sampling point (in three plots of 4 m^2), (B) carabid species richness per sampling point (per pitfall trap during two sampling periods) and (C) ground-nesting bird species richness per farm (one survey plot of 500 × 500 m^2). Trend lines were calculated using GLMM including the two surrounding landscape variables as covariates and field, farm and study area as nested random effects. *Based on Geiger et al. (2010a).*

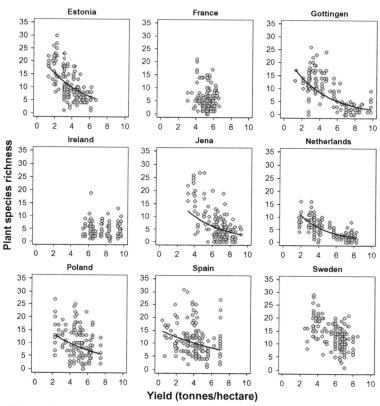

Fig. 6 Effects of cereal yield (tons/ha) on the number of wild plant species per sampling point (in three plots of 4 m^2) in each of the study areas. Trend lines were calculated using GLMM including the two surrounding landscape variables as covariates and field, farm and study area as nested random effects (see Section 2.4.1). Trend lines were only plotted when the relationship was significant ($p < 0.05$). *Based on Geiger, F., Bengtsson, J., Berendse, F., Weisser, W.W., Emmerson, M., Morales, M.B., Ceryngier, P., Liira, J., Tscharntke, T., Winqvist, C., Eggers, S., Bommarco, R., Pärt, T., Bretagnolle, V., Plantegenest, M., Clement, L.W., Dennis, C., Palmer, C., Oñate, J.J., Guerrero, I., Hawro, V., Aavik, T., Thies, C., Flohre, A., Hänke, S., Fischer, C., Goedhart, P.W., Inchausti, P., 2010a. Persistent negative effects of pesticides on biodiversity and biological control potential on European farmland. Basic Appl. Ecol. 11, 97–105.*

relationship (Fig. 6). In two of the three study areas where no relationship was found, the variation in yield among fields and farms was much smaller than in the other countries, which probably explains the lack of significant effects.

Yield also had a negative effect on farmland bird FD (quantified using diet type, nesting behaviour, and foraging and migration strategy, see

Guerrero et al., 2011 for details of methodologies employed), and individual and breeding pair abundance across the continent (Guerrero et al., 2012). In the case of carabid beetles, their total abundance decreased with yield, but species richness and the abundance of some functional groups (small and medium-sized beetles and wingless carabids) did not show any response (Winqvist et al., 2014).

Species richness and abundance of overwintering birds was also negatively associated with yield (Geiger et al., 2010b), which indicates that AI might limit the resources available for birds in this particularly constrained period of the year, with direct impacts on survival and future reproductive performance (Newton, 1998). Notably, when these relationships were examined on a national basis, it was found that species richness in some instances increased with yield, e.g. wintering birds in Spain. This suggests that fields with higher yields sometimes could provide a larger resource base to overwintering birds (Morales et al., 2015). The Spanish study area contained some of the least intensively managed fields measured by the metrics of intensification used in AGRIPOPES (see Figs. 3 and 4), suggesting that higher yielding fields in relatively less intensified areas still provide sufficient and beneficial resources to overwintering birds. For example, the stubble remaining after harvesting of higher yielding fields might provide greater quantities of waste or spilt cereal seeds to species that rely on them during winter, such as skylarks and corn buntings (Morales et al., 2015).

The impacts of increased yield on the delivery of biocontrol were also negative. Here, the survival time of tethered aphids increased with increased yield ($\chi_1^2 = 6.85$; $p=0.009$, see Fig. 7) and was therefore inversely related to predation. However, the effect of yield on aphid survival time was not consistent geographically, and differed among study areas (yield × study area interaction: $\chi_6^2 = 17.84$, $p=0.007$).

Our results support the suitability of yield as a general measure of intensification and demonstrate its overall negative effect on biodiversity components. However, the individual effects of the different intensification factors summarized by yield are complex and warrant further examination. These intensification factors are measured through variables that describe field-scale management activities (see Table 1 and Figs. 3 and 4). To disentangle these effects, the importance of 13 variables considered as relevant components of AI were investigated (see Table 2). Here, we present and explore their main effects and discuss them in a broader context.

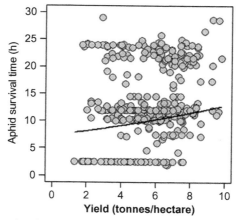

Fig. 7 Effects of cereal yield (tons/ha) on the median survival time of aphids (h). The trend line was calculated using GLMM including the two surrounding landscape variables as covariates and field, farm and study area as nested random effects. *Based on Geiger, F., Bengtsson, J., Berendse, F., Weisser, W.W., Emmerson, M., Morales, M.B., Ceryngier, P., Liira, J., Tscharntke, T., Winqvist, C., Eggers, S., Bommarco, R., Pärt, T., Bretagnolle, V., Plantegenest, M., Clement, L.W., Dennis, C., Palmer, C., Oñate, J.J., Guerrero, I., Hawro, V., Aavik, T., Thies, C., Flohre, A., Hänke, S., Fischer, C., Goedhart, P.W., Inchausti, P., 2010a. Persistent negative effects of pesticides on biodiversity and biological control potential on European farmland. Basic Appl. Ecol. 11, 97–105.*

3.1.2 Effects of Pesticides

Pesticides had consistently negative effects on the studied components of farmland biodiversity (see Table 2; Geiger et al., 2010a). Wild plant species richness declined as the frequency of herbicide and insecticide applications and the amounts of active ingredients of fungicides increased. Carabid species richness was negatively associated with the amounts of active ingredients of insecticide applied, whilst bird species richness declined with increasing frequency of fungicide application. Fungicide application rate is strongly correlated with the frequency of insecticide applications (Pearson's correlation coefficient $r=0.732$; $p<0.001$), and consequently it was difficult to disentangle their relative effects. In terms of ecosystem services, the predation rate of aphids measured in the field significantly declined as the amounts of insecticide applied increased, suggesting reduced activity or abundance of natural predators such as beetles and spiders.

The dominant effect of pesticides on species diversity and biological control potential was one of the key results emerging from the analysis of the field data collected in the AGRIPOPES project (Geiger et al., 2010a).

Table 2 Effects of Different Components of Agricultural Intensification on the Number of Plant, Carabid and Bird Species and Median Aphid Survival Time

Response Variable	Explanatory Variable	Standardized Effect	χ^2	p-Value
Number of plant species	Mean field size	−0.094	6.09	0.014
	% of land under AES	0.149	12.23	<0.001
	Frequency of herbicide applications	−0.106	8.88	0.003
	Frequency of insecticide applications	−0.105	6.15	0.013
	Applied amounts of a.i. of fungicides	−0.262	31.45	<0.001
Number of carabid species	% of land under AES	0.062	6.31	0.012
	Applied amounts of a.i. of insecticides	−0.061	10.87	0.001
Number of breeding bird species	Frequency of fungicide applications	−0.127	5.71	0.017
Median survival time of aphids	% of land under AES	−0.144	9.43	0.002
	Applied amounts of a.i. of insecticides	0.114	11.17	0.001

The models were selected after considering 13 intensification variables using forward selection (backward selection produced identical models). All models included two landscape variables (mean field size and percentage of land planted with arable crops within a radius of 500 m), even if these had no significant effects (nonsignificant effects are not shown). Intensification variables were only included, if they had significant effects using the Wald test ($p < 0.05$). AES, Agri-Environmental Schemes; *amount of a.i.*, amount of active ingredients.

After Geiger, F., Bengtsson, J., Berendse, F., Weisser, W.W., Emmerson, M., Morales, M.B., Ceryngier, P., Liira, J., Tscharntke, T., Winqvist, C., Eggers, S., Bommarco, R., Pärt, T., Bretagnolle, V., Plantegenest, M., Clement, L.W., Dennis, C., Palmer, C., Oñate, J.J., Guerrero, I., Hawro, V., Aavik, T., Thies, C., Flohre, A., Hänke, S., Fischer, C., Goedhart, P.W., Inchausti, P., 2010a. Persistent negative effects of pesticides on biodiversity and biological control potential on European farmland. Basic Appl. Ecol. 11, 97–105.

The result was novel in that whilst it might be unsurprising to find negative effects of pesticides on biodiversity, they were found consistently at a pan-European scale, and despite decades of policy implementation regulating their controlled use.

In the particular case of birds, the application of fungicides was significantly and negatively associated with the total (combined) abundance of all breeding birds surveyed. In four breeding bird species—Yellow Wagtail,

Whinchat, Corn Bunting and Quail—significant negative effects of the application of fungicides were found, and these were also strongly correlated to the application of insecticides.

The numbers of overwintering birds in the surveyed areas were also assessed (Geiger et al., 2010b). Notably, pesticides did not have any associations with the abundance of birds during winter. Similar results were obtained when the Spanish wintering bird data alone were analyzed (Morales et al., 2015). This suggests that the most important effect of insecticides or fungicides on the abundance of breeding birds acts through their impacts on food supply during the period in which most species feed their nestlings with insects. However, a potential effect of pesticide-coated seeds on seed-eating wintering birds cannot be discounted (López-Antia et al., 2015).

Organic farms and AESs had positive effects on plant and carabid diversity, but not on breeding birds. It is unclear why birds were not positively affected. Organic farming, whilst not intensive by measures relevant to AI, is an intensive, yet benign, form of management, and it is possible that breeding birds are disturbed by some organic management practices. Alternatively, a possible explanation is the large spatial scale at which pesticide pollution occurs, which inevitably leads to negative effects—even in areas where the application of these chemicals has ceased locally. Such large-scale impacts are especially relevant for highly mobile organisms such as birds, bees and butterflies (Clough et al., 2007; Rundlöf et al., 2008).

In Western Europe many birds of prey such as Kestrel, Sparrow Hawk and Buzzard showed large-scale declines between 1950 and 1970. After banning most of the responsible pesticides such as DDT and dieldrin, many of these species recovered rather quickly, whilst others with much slower life cycles, such as White-tailed Eagle, needed several decades (Newton, 1998). In the subsequent years the European Union developed policies to restrict the negative effects of pesticides, resulting in the EU's Sustainable Pesticides Directive in 2009. There were therefore good reasons to assume that the pesticide load of European agricultural landscapes had been substantially reduced. However, the results reported here and others recently published (e.g. Hallmann et al., 2014) do not support such an assumption.

3.1.2.1 Pending Questions in Pesticide Research

Although the results from AGRIPOPES are correlational and thus do not necessarily reflect cause–effect relationships, they consistently reveal a negative association, suggesting that the pesticide load on European farmland continues to impact biodiversity and relevant ecosystem services such as the biological control of harmful organisms. Therefore, landscape-wide experiments are needed to stringently test potential cause–effect relationships.

European farmers apply a large variety of chemicals to protect their crops against herbivorous insects, aphids and pathogenic fungi. An important, yet unanswered question concerns the identity of the active ingredients that were responsible for the observed negative relations between pesticides and biodiversity components. The most important groups of compounds are organophosphates, carbamates, pyrethroids and the recently introduced neonicotinoids. The last group of chemicals has recently received much attention and been the subject of intense scientific and societal debate.

The debate regarding the effects of neonicotinoids on biodiversity and ecosystem services is illustrated by the contrasting results of two reviews that appeared in 2014. Godfray et al. (2014) reviewed 259 peer-reviewed papers and concluded that the evidence for negative impacts on honeybee colonies was not yet convincing. In contrast, Van der Sluijs et al. (2015) reviewed 800 scientific studies and concluded that there was sufficient evidence for direct toxic effects on honeybees and for sublethal effects of concentrations that overlap with concentrations measured in nectar and pollen in the field (see also Rundlöf et al., 2015).

In 2014 the European Commission asked the European Academy of Sciences (EASAC) to analyse these contrasting views and to review the most recent scientific evidence on the use of neonicotinoids (EASAC, 2015). The resulting report reached some conclusions relevant in the context of this review. First, the existing debate was focussed almost completely on the survival of honeybee colonies. It is true that honeybees are extremely important pollinators, accounting for about 50% of crop pollination. However, wild solitary bees, bumblebees and hover flies perform the other 50% (Kleijn et al., 2015). The honeybee forms large colonies that provide a resilient buffer against forager losses, which explains the absence of net negative results on this exceptional species. However, bumblebees have much smaller colonies and solitary bees lack any form of buffering capacity to ensure pollination potential. Therefore, honeybee colonies do not provide a useful model system to assess the impacts on the broader group of wild crop pollinators.

Secondly, the EASAC (2015) report noted that the different approaches applied in the assessment (laboratory studies, field correlational studies and field experiments) had their own specific weaknesses, concluding that such weaknesses are inherent to the scientific analysis of complex ecological problems. Reviewing the combined evidence generated by the different studies, the report concluded that an increasing body of evidence supports the severe negative effects of the prophylactic use of neonicotinoids on nontarget organisms (see also Gibbons et al., 2015). Furthermore, there was clear evidence for sublethal effects of very low concentrations over long periods.

These conclusions were supported by recent studies. Rundlöf et al. (2015) showed experimentally that neonicotinoids strongly affected solitary bees and bumblebee colonies, whilst honeybee colonies did not respond (probably due to colony resilience). Williams et al. (2015) demonstrated significant effects on honeybee queen reproductive organs resulting in decreased fertility (e.g. Hallmann et al., 2014). Therefore, the scientific evidence that neonicotinoids can have dramatic effects on nontarget insects and on insectivorous organisms such as birds or bats is increasing rapidly. However, similar information about the impacts of the other pesticides commonly applied in the European agricultural landscapes is still lacking. An assessment of these impacts using large-scale field experiments, longer time scales and including a broad variety of nontarget organisms is urgently required.

3.1.3 Effects of Fertilization

Inorganic fertilizer input is one of the main components of AI (Firbank et al., 2008; Tivy, 1990) with a number of potential direct and indirect effects on biodiversity (Robinson and Sutherland, 2002). Although artificial fertilization clearly contributes to the monopolization of primary production by agriculture, it also increases the global productivity of the system, and thus some mixed responses of biodiversity components might be expected. Geiger et al. (2010a) did not demonstrate any significant effect of inorganic fertilizers on the species richness of vascular plants, carabid beetles or birds. However, Guerrero et al. (2011) focused on the FD of birds using the AGRIPOPES data and showed a positive relationship between the diversity of some functional groups, such as diet type and nesting strategy, and the amount of N fertilizers probably related to the increased productivity of agroecosystems.

Studies at particular sites also provided mixed results. Aavik and Liira (2009, 2010) found in Estonian farmland that higher fertilization rates negatively affected small-scale plant species richness among taxa with a high tolerance to agriculture, but also among less tolerant ones (more associated with natural habitats). Such responses could be explained by the enhanced growth of cultivated cereals due to fertilizers, which in turn could favour their ability to monopolize other resources such as light and water, outcompeting weeds and other wild plants. Guerrero et al. (2010) did not find such effects in the Spanish study area, where no response of arable weeds to the application of N, P and K was observed. These differences might be explained by the high prevalence of agricultural weeds at the Spanish sites, and these species are well adapted to local crop conditions. It is also likely that farmers apply fertilizers more intensively when sowing at higher density, so that the negative effect of sowing density (see in the following section) might mask the influence of fertilizers. Guerrero et al. (2010) found that carabid species richness was negatively influenced by inorganic N input, but positively affected by applied P. Previous studies on organic farming have indicated a negative effect of synthetic fertilizers on this group, possibly mediated by decreased prey abundance (e.g. Bengtsson et al., 2005), although the influence of different inorganic nutrients was not examined. The positive effect of P found by Guerrero et al. (2010) could result from an indirect relationship of P with other factors affecting carabids such as sward structure or microclimate (Holland, 2002). In summary, the effect of inorganic fertilizers on biodiversity components is less consistent across taxa and geography than that of other inputs such pesticides, which probably results from differences in local management and landscape configuration.

3.1.4 Effects of Tillage

Field ploughing and other mechanical operations cause soil disruption, thus becoming a source of disturbance for plants growing in fields. Ploughing was the traditional technique used to eliminate weeds until the use of herbicides was generalized (Tivy, 1990). In regions where 1 year rotation is still frequent, it continues to be the dominant procedure of weed control. For example, in central Spain fallow fields used to be employed in a 3-year rotation cycle, where fallow fields were kept as bare terrain until sown with a legume crop in the second year and then a cereal crop on the third (Suárez et al., 1997). As a result of AI, fallow fields are frequently ploughed and treated with herbicides throughout the agrarian season and then followed directly by the cereal crop. Such treatment not only eliminates plants but reduces resources for other organisms, such as food and cover for insects and birds (Robinson and Sutherland, 1999; Suárez et al., 1997). The retention of unploughed winter stubbles as an agri-environmental measure, is extensively applied in different European countries (Suárez et al., 1997) and has benefited farmland birds in many regions (Wakeham-Dawson and Aebischer, 1998; Wilson et al., 1996). Despite the positive effects of such measures, the frequency of ploughing and other mechanical operations, per se, was not significantly associated to the biodiversity components considered across the areas studied in AGRIPOPES (Geiger et al., 2010a), possibly due to the correlational structure of data. However, when the frequency of soil disruption was integrated with pesticide and fertilizer inputs to define an index of AI, it was found that there was a negative correlation with the species richness of plants and birds (but not carabids) at local, landscape and regional scales (Flohre et al., 2011a).

In the particular case of wintering birds (Geiger et al., 2010b), mechanical weed control did negatively affect species richness, though not abundance, presumably due to food reduction. Guerrero et al. (2010) found no significant relationship between mechanical weed control and plant, bird or carabid species richness in central Spain, which suggests that the frequent ploughing typical of this area, which exerts important deleterious effects on fallow-nesting birds (Morales et al., 2013), might be a spurious control measure of weed abundance.

3.1.5 Effects of Sowing Density

Field-level intensification, through the use of increasingly efficient machinery and the application of inorganic fertilizer, has allowed for an increase in the density of sown grains, and, subsequently, of yields (Firbank et al., 2008; Tivy, 1990). As humans increasingly monopolize the primary production in agroecosystems, increased sowing densities generate denser and more homogeneous sward structures (Benton et al., 2003; Robinson and Sutherland, 2002), sequestering resources and modifying habitats for plants, invertebrates and birds. At the European scale, the influence of sowing density on total bird abundance, the number of nesting territories, skylark abundance and the number of skylark nesting territories was examined along with other field and landscape-scale intensification components (Guerrero et al., 2012). Sowing density did not have a significant influence on the bird response variables, and its importance relative to that of other intensification components was minor. Crop vegetation structure has proved to be a relevant component of habitat suitability for farmland birds (Chamberlain et al., 1999; Donald, 2004; Donald et al., 2001; Eggers et al., 2011; Morales et al., 2008), yet these results were not supported by the AGRIPOPES findings. It should also be noted that not all sward structure features are directly determined by sowing density. For example, crop height, which is key for many ground-nesting birds, may vary with management, the wind and rain, and factors such as date of sowing (e.g. Eggers et al., 2011). Alternatively, the lack of sowing density effects presented by Guerrero et al. (2012) might only indicate that this measure has poor explanatory capacity compared to other intensification factors considered, particularly yield as the main surrogate of the process. At local scales, sowing density negatively affected breeding bird and weed abundances at the Spanish study sites, consistent with expectations and results of previous studies.

3.2 Landscape-Level Components of AI and Their Effect of Biodiversity

Practices that lead to AI include a range of activities that occur in the landscape, that is, farmers specializing on one or few (arable) crops instead of mixed farming, converting perennial habitat (grassland) to arable fields, destroying edge habitats (e.g. hedges, field boundaries, buffer zones along creeks) and reallocating land to increase field size and make farms more compact. These activities further simplify landscapes by limiting the spatial and temporal variety of land-use types, ultimately increasing landscape homogeneity (Tscharntke et al., 2005).

In the AGRIPOPES project, several studies focused on how landscape context affected local biodiversity and community composition, and how those effects might impact upon the ecosystem service of biological control.

Flohre et al. (2011a) analyzed the diversity of vascular plants, carabid beetles and birds in agricultural landscapes in cereal crop fields at the field ($n=1350$), farm ($n=270$) and European region ($n=9$) scales, and partitioned diversity into its additive components, the alpha, beta and gamma diversity at each spatial scale. AI negatively affected the species richness of plants and birds at all spatial scales, but not carabid beetles (see Fig. 8). Local AI was closely correlated to beta diversity at larger scales up to the farm and region level and was hence a good indicator of farm- and region-wide biodiversity losses.

Winqvist et al. (2011) found that landscape simplification from 20% to 100% arable land reduced plant species richness by about 16% and cover by 14% in organic fields, and by 33% and 5.5% in conventional fields. For birds, landscape simplification reduced species richness and abundance by 34% and 32% in organic fields and by 45.5% and 39% in conventional fields. In contrast, ground beetles were more abundant in simple landscapes but were unaffected by farming practice. This Europe-wide study suggested that organic farming enhanced the biodiversity of plants and birds in all landscapes, but only improved the potential for biological control in heterogeneous landscapes.

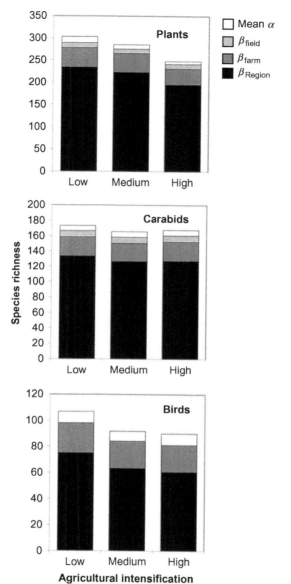

Fig. 8 Effect of agricultural intensification (AI) level (low, medium and high) on species richness of plants, carabids and birds, separating (α) diversity in the field and (β) diversity between fields, farms and regions. *Based on Flohre, A., Hänke, S., Fischer, C., Geiger, F., Bengtsson, J., Berendse, F., Weisser, W.W., Emmerson, M., Ceryngier, P., Liira, J., Tscharntke, T., Winqvist, C., Eggers, S., Bommarco, R., Pärt, T., Bretagnolle, V., Plantegenest, M., Clement, L.W., Dennis, C., Palmer, C., Morales, M.B., Oñate, J.J., Guerrero, I., Hawro, V., Aavik, T., Thies, C., Inchausti, P., 2011a. Agricultural intensification and biodiversity partitioning in European landscapes comparing plants, carabids, and birds. Ecol. Appl. 21, 1772–1781; Flohre, A., Rudnick, M., Traserc, G., Tscharntke, T., Eggers, T., 2011b. Does soil biota benefit from organic farming in complex vs. simple landscapes? Agric. Ecosyst. Environ. 141, 210–214.*

Guerrero et al. (2012) studied the response of ground-nesting farmland birds to AI in six European countries and found that landscape factors accounted for most of the variation of ground-nesting farmland bird breeding pair densities. In the case of breeding Skylarks, field factors were found to be more important. These results suggest that landscape management for farmland bird conservation is most important, but that field-level practices are also relevant, as crop yield and bird densities were negatively related.

Although belowground biodiversity is an important part of the agroecosystem, studies focussed on the interacting effects of local and landscape intensification on the belowground detritivore community, including bacteria, fungi, collembola and earthworms are rare. Flohre et al. (2011b) found that landscape context plays a significant role in shaping the effects of organic vs conventional farming on soil biota. Earthworm species richness in simple landscapes, where predation pressure is reduced, was enhanced by organic farming, whereas in complex landscapes conventional farming enhanced earthworm species richness. As the same pattern has been found for microbial carbon biomass, earthworms might play a role in enhancing microbial biomass.

The impacts of the various components differ considerably among bird species. Many species are affected by mean field size in the area around the sampled field. Mean field size had negative effects on the abundance of Skylark, Yellowhammer, Whinchat and Marsh Warbler (Guerrero et al., 2012), illustrating the important role of field margins and less intensively exploited areas in-between the fields. Field size had positive effects on lapwing densities, probably due to the preference of this species for open areas (Klomp, 1954).

In conclusion, AESs need to expand the view from the local field and farm to the landscape and region level to improve their effectiveness. Taxon-specific responses also need to be considered in conservation efforts.

3.3 Farm-Level Components of AI

As mentioned earlier, the farm scale represents an administrative and decision-making level that translates into specific types of field management as well as landscape features. For example, a farmer's decision to consolidate land, thus enlarging fields, to convert to organic farming with its more diverse crop rotations, or to devote a certain surface area to AES generates changes both at field and landscape scales. The response of biodiversity and ecosystem functioning to the application of organic instead of conventional farming is discussed in Section 4 later. Here, we discuss farm-level responses of biodiversity components to field-scale intensification.

Flohre et al. (2011a) addressed this question by partitioning diversity of plants, birds and carabids across European sites. They found that field-level intensification explained 12.83–20.52% of the variation in farm-level beta diversity of the three groups, which indicated important differences in environmental conditions between farms. The response of beta diversity to local intensification was even stronger when the regional scale (i.e. study area) was considered. In the case of birds and carabids, beta diversity increased at the farm-scale with increased field-level intensification. Therefore, and contrary to initial expectations, field-level intensification did not necessarily homogenize local communities. Rather, these communities retained their compositional differences, and it is likely that this is due to the relative heterogeneity of farm-scale management practices. It is thus necessary to better understand intensification effects on biodiversity patterns of different groups and at multiple spatial scales so that more efficient AES can be designed.

3.4 Comparing the Importance of Local- vs Landscape-Level Components of AI on Biodiversity and Ecosystem Services

Our results clearly support the negative influence of landscape-level intensification factors on biodiversity and ecosystem function, as well as the effects of landscape structure on biodiversity responses shown by other authors (Concepción et al., 2012; Tscharntke et al., 2005; Whitthingham, 2007). However, they also highlight the importance of field-level management. Therefore, the question of how biodiversity and associated ecosystem services change in response to the relative importance of intensification at each scale remains pertinent. Different studies carried out in the AGRIPOPES context have specifically addressed this question.

Guerrero et al. (2012) employed a PCA and variance partitioning analysis, and showed that field-level factors explained a smaller amount of variation when the overall abundance of ground-nesting farmland birds was considered (2.9%) than when the density of breeding territories was examined (13.1%). However, landscape-scale factors always explained more variation (11.3% and 20.1%, respectively). This effect was stronger in the particular case of the skylark (from 12.9% to 18%), for which field-level intensification outweighed landscape-scale factors (11.2%). These results indicate that the influence of field-level intensification on farmland birds is particularly important for open or simplified habitat specialists, like the skylark, which rely on the arable area of fields for both foraging and breeding.

In their study of wintering birds across Europe, Geiger et al. (2010b) also found an important influence of landscape-scale intensification: farmland bird abundance was higher in areas with more stubble pasture and green manure crops, as well as in heterogeneous landscapes comprising arable crops as well as grasslands, whilst species richness was higher in areas with more pasture. This is consistent with later results of Morales et al. (2015) for the Spanish study site, where landscape-scale factors explained 70.79% of the variation in wintering bird community composition vs the 29.21% explained by field-level intensification.

In contrast, field-level intensification was particularly important for plant FD in Spain (Guerrero et al., 2014). Plant species richness decreased linearly with field-level intensification, but showed no response to landscape-level intensification. Community-weighted mean and diversity of the different functional traits considered (plant height, specific leaf area, seed mass and flowering onset) were affected by intensification at the field scale in nonlinear ways, but no influence of landscape-level intensification was found. More specifically, the diversity of all functional traits decreased with AI at the field scale, although specific leaf area and seed mass followed marked nonlinear relationships, showing the strongest decreases at medium to high intensification levels, and an increase at low values. In contrast, the greatest loss of species was not accompanied by similar changes in FD and vice versa (Carmona et al., 2016). In the lowest levels of field-scale intensification, species were lost without a decrease in FD, which implies a reduction in functional redundancy of communities, and thus in their resistance to environmental change (Mouillot et al., 2013). At intermediate levels, FD decreased rapidly with the loss of few species, that is, the community had become more functionally vulnerable due to reduced redundancy. At the highest end of the field-level intensification gradient, no important FD reductions were observed, which suggests that these poorer communities were characterized by intensification-resistant species and traits. Similarly, functional vulnerability of arable plant communities due to species loss showed a nonlinear positive relationship with field-level intensification, whilst no landscape-scale effect was found (Carmona et al., 2016).

Winqvist et al. (2014) also addressed the influence of scale on FD, in this case for carabid beetles. They showed that functional traits (size, diet and type of dispersal) in these ground beetles responded differently to local and landscape management. Field-scale intensification (yield) reduced overall carabid abundance, although it did not affect abundance of small and medium-sized beetles, or that of wingless carabids. Species richness was not affected either, although the increased proportion of arable land in the landscape increased overall carabid abundance, an effect that was driven by an increase of omnivorous beetles. Total carabid species richness did not increase with the proportion of arable land, although richness of wingless beetles did increase with that variable.

These results support the view that landscape-scale factors are the main drivers of biodiversity responses in highly mobile groups like birds and carabids, but that this influence may be relatively reduced by field-level factors, e.g. for birds that rely on the crop field for both foraging and nesting (Butler et al., 2007; Guerrero et al., 2012). However, for sessile organisms like arable plants, field-level management seems to be the key factor influencing their populations, community structure and FD.

4. ORGANIC-CONVENTIONAL COMPARISONS

Comparing conventional farming to organic practices provides an approach to quantify the effects of agricultural management intensity on the diversity of farmland systems, because organic farming is often regarded as less intensive. AESs in Europe differ between countries, but organic farming is a more uniformly understood agri-environmental measure, as insecticides, herbicides and synthetic fertilizers are forbidden. Organic farming combines the best environmental practices expected to preserve a high level of biodiversity and natural resources, thereby also contributing to human welfare (e.g. Council Regulation (EC) No. 834/2007). Consequently, various AES, and organic farming in particular, have been widely advertised and supported (Bengtsson et al., 2005; Gibson et al., 2007; Kleijn et al., 2006, 2009). Some reports have shown, however, that organic farming practices might not be effective (Kleijn et al., 2001, 2006), or that the effects vary among organism groups and landscapes (Hiron et al., 2013; Tuck et al., 2014). For example, organic farming appeared to be more efficient in conserving aboveground than belowground species diversity (Tuck et al., 2014). Other studies contend that the overall benefit of organic farming for biodiversity compared to the costs related to lower yields is ambiguous, and the cost efficiency is not properly linked with the reality of ecological processes (Gabriel et al., 2013; Kremen, 2015). In AGRIPOPES, we examined the biodiversity of 151 farms occurring across a subset of five study regions (Sweden, Estonia, the Netherlands, Western Germany and Eastern

Germany), reflecting the high prevalence of organic farms in these regions (33%). Among the 151 farms studied, 51 were organically managed. We found that environmental conditions and associated biodiversity responses of our focal groups did not always differ between organic vs conventional farms, although there were positive effects of AES, which included organic farming, on diversity of plants, carabids and biocontrol potential across the nine study regions (Geiger et al., 2010a). Winqvist et al. (2011) found that organic farming had higher levels of species richness in plants and birds, whilst carabids showed no response to farming system, and biological control potential only increased with organic farming in heterogeneous landscapes. These results show that a detailed analysis of various organism groups at different trophic levels and delivering a range of ecosystem services in organic farming should include several other factors, such as landscape context and taxon-specific responses, to accurately estimate the potential advantages of organic practices over other types of farming.

4.1 Landscape Context

In addition to the intensive application of agrochemicals, contemporary agricultural landscapes have experienced a severe loss of the area and connectivity of natural and semi-natural habitats, which impose another major pressure on biodiversity (Benton et al., 2003; Fahrig, 2003; Liira et al., 2008a). One of the main challenges in evaluating the effect of organic farming relative to conventional farming practices is the frequent correlation between these two groups of factors of farming practices and landscape characteristics. Conventional farming systems and related land-use intensity are often accompanied by larger fields and homogeneous landscapes, whilst organic fields are smaller and located in remote areas with an increased representation of natural habitats around (e.g. Bengtsson et al., 2005; Hole et al., 2005; Norton et al., 2009). In AGRIPOPES, the aim was to avoid potentially confounding effects of farming practice and landscape characteristics by examining biodiversity patterns and biocontrol potential in organic and conventional farms along a gradient of landscape complexity (Geiger et al., 2010a; Winqvist et al., 2011).

The relative benefits of organic farming on biodiversity have been found to be highest in simple homogeneous landscapes characterized by a high proportion of croplands (Batáry et al., 2011; Tuck et al., 2014). Indeed, such an interactive effect of landscape structure and farming practices has also been observed for plants (Roschewitz et al., 2005), birds (Dänhardt et al., 2010; Hiron et al., 2013) and various insect groups (Holzschuh et al., 2007;

Rundlöf et al., 2008). Nevertheless, the analysis of the pan-European biodiversity dataset within the AGRIPOPES framework revealed that positive effects of organic farming on the diversity of plants and birds did not differ between complex and simple landscapes (Geiger et al., 2010a), whereas the species richness of ground beetles did not depend on landscape characteristics nor land-use practices (Winqvist et al., 2011). Furthermore, biocontrol potential was highest in the organic farms of complex landscapes. In contrast, in homogeneous landscapes biological control potential was higher in conventional fields (Thies et al., 2011; Winqvist et al., 2011). These mixed effects of organic farming in simple and complex landscapes suggest that biodiversity and related ecosystem services, such as biocontrol, may in fact show a differential response to land-use and landscape structure depending on the studied ecosystems. Future AES and related monitoring programmes should therefore place more emphasis on enhancing and evaluating the potential of ecosystem services in addition to biodiversity per se, in order to maximize the outcome of environmentally sound management practices. The relative role of farming practices on biodiversity is also location-specific at the field scale. For example, the positive influence of organic farming on plants has been most evident and widely reported at the field scale (Gabriel et al., 2013; Gibson et al., 2007; Hole et al., 2005), a result consistent with the findings observed in AGRIPOPES (Geiger et al., 2010b; Winqvist et al., 2011). Hiron et al. (2013) also found a positive field-scale response of birds to organic farming. Conventional farming tends to have impacts in habitats adjacent to focal fields, e.g. due to leaching of agrochemicals (de Snoo and Van der Poll, 1999; Kleijn and Snoeijing, 1997), whilst the effects of organic farming on plant species richness in field boundaries and in other habitats adjacent to agricultural land have been less apparent (Clough et al., 2007; Gibson et al., 2007). Indeed, the analysis of field margin vegetation (Aavik and Liira, 2009, 2010; Aavik et al., 2008) showed that large- and local-scale landscape structure and the presence and abundance of source habitats for species were the main determinants of species richness and composition, whilst only a relatively low amount of variation in species patterns was explained by farming type (organic and conventional). Nevertheless, as the results of AGRIPOPES have shown, focusing only on species richness might not provide sufficient detail regarding the effects of land-use intensification on biodiversity in and around agricultural land. In contrast, in-depth analyses within groups of species with different traits and conservation value would significantly advance our understanding of agriculture-related drivers of biodiversity change.

4.2 Functional and Taxon-Specific Responses of Biodiversity to Farming Practices

A recent meta-analysis concluded that organic farming increases overall species richness approximately by 30% (Tuck et al., 2014), but that different organisms vary in their responses to organic practices. In accordance with this, various studies of bird (Birkhofer et al., 2014), insect (Birkhofer et al., 2014), plant (Boutin et al., 2008; Gabriel and Tscharntke, 2007; Petersen et al., 2006) and microbial (Hartmann et al., 2015) communities in agricultural systems show that responses of biodiversity to farming practices are largely taxon-specific and/or varies among functional groups (see also Section 3.1). Indeed, such taxon-specific responses of biodiversity to organic and conventional farming systems were also observed in AGRIPOPES, as described earlier.

Plants, whose habitat conditions are directly influenced by the application of agrochemicals, show a clearly different response to AI depending on the functional group considered. Enhanced nitrogen and phosphorus concentrations facilitate the growth of competitive plants, and herbicides affect large-sized species, so that resulting communities mainly consist of grasses and other generalists, as well as fast-growing ruderal plants (Marshall and Moonen, 2002). Indeed, boundaries of organic fields may support a higher diversity of hemerophobic and habitat specialist species at local scales (Aavik and Liira, 2009, 2010; Manhoudt et al., 2007), whilst species tolerating agricultural management, such as nitrophilous and disturbance-tolerant species persist in the boundaries of all farming types (Aavik and Liira, 2009). In an analysis of landscape-scale vegetation, Liira et al. (2008b) showed that land-use intensity decreased the species richness of two growth forms—sedges and pteridophytes. However, in the same study, Liira et al. (2008b) showed that an increased number of crops caused an increase in the richness of annuals and a decrease in the richness of perennials. Thus, the responses of different plant functional groups to organic and conventional farming are highly complex and depend on the scale (local vs landscape) of the study.

Changes in plant productivity and plant species composition induced by intensive agricultural practices are expected to covary with the composition of species at other trophic levels due to altered interactions among species groups, e.g. plant–pollinator interactions (Gabriel and Tscharntke, 2007). Nevertheless, whilst the cover, species richness and functional group composition of plants were observed to vary between farming systems with contrasting management (Aavik and Liira, 2010; Geiger et al., 2010a; Winqvist et al., 2011), Winqvist et al. (2014) observed no trait-specific response of ground beetles to organic and conventional farming. Instead, continuous variables describing farm-level land-use intensity, such as yield, were significantly better predictors of the trait composition of ground beetles.

The results obtained from a large-scale comparison of organic and conventional farming systems within the framework of AGRIPOPES suggest that whilst organic methods often benefit biodiversity and related ecosystem services, the relative effects of farming system may also depend on landscape context. In addition, the oversimplification of evaluating only species richness may lead to underestimation of the role of land-use intensification on biodiversity. More in-depth analyses of species responses within different functional groups may help to target future AES towards species groups with higher conservation needs and/or to related ecosystem services.

5. LINKING AI TO BIODIVERSITY AND ECOSYSTEM SERVICES

5.1 General Model

Using structural equation modelling (SEM), it was possible to examine the respective contributions of agricultural land use, plant diversity and predator (carabid) diversity on the biological control potential measures presented in Geiger et al. (2010a). The SEM method makes it possible to disentangle the indirect pathways leading to biological control (Fig. 9).

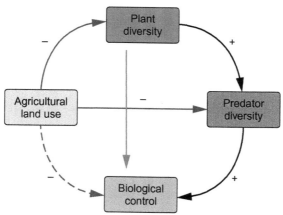

Fig. 9 General model for predicting the influence of agricultural land use on biological pest control. Biological control can be affected directly by predator diversity, which in turn may be affected by agricultural land use (AI), either through changes in plant diversity or through other unknown mechanisms (direct path). There may also be a direct effect of plant diversity (independent of predator diversity).

5.2 Study Area, Biodiversity Surveys and Biocontrol Experiment

Using a subset of AGRIPOPES data from six regions (FR, IR, JE, NL, PL, SW) with 90 sampling points per region, we assessed the relative importance of agricultural land use, plant diversity and predatory carabid diversity on aphid survival rates. We excluded Estonian and Göttingen data due to extremely high predation rates (nearly all aphids were gone after 36 h). Given our study design, the SEM approach requires seven variables per sampling point represented as nodes (Fig. 10A and B). These consisted of four variables reflecting measures of AI: three local intensification variables (the application of mineral fertilizer measured as N kg/ha, the frequency of herbicide use and the frequency of insecticide use) and one landscape-scale variable (the percentage of arable land measured in a 1000 m buffer area). We also included two variables characterizing biodiversity at the field scale (mean plant and carabid species richness per sampling point). Aphid survival was treated as a binomial response variable, i.e. the number of aphids predated as a proportion of the number provided at the start of the experiment. Sampling points where any of the seven required data records were missing were excluded, resulting in a dataset of 436 sampling points.

5.3 Structural Equation Modelling

The initial full SEM contained all possible paths (including correlations among AI variables) with the exception of the effect of herbicide application frequency on biocontrol, which was the least justifiable relationship. Both initial full and the simplified SEMs consisted of three models analyzed together with the piecewise SEM (Lefcheck, 2016) in R (R Development Core Team, 2015):

(1) mean plant species richness per sampling point analyzed with a General Linear Mixed Model based on a normal distribution with the nlme package (Pinheiro et al., 2015);

(2) carabid species richness per sampling point analyzed with a General Linear Mixed Model based on a normal distribution with the nlme package;

(3) aphid survival per sampling point analyzed with a Generalized Linear Mixed Model based on a binomial distribution with the lme4 package (Bates et al., 2014).

In all three models, we used the following hierarchical random structure: sampling points were nested in fields, fields were nested in farms and farms

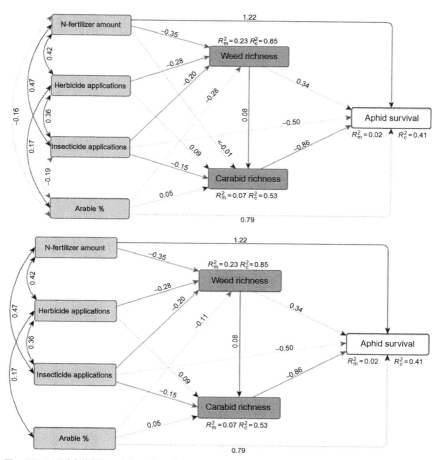

Fig. 10 Initial full (A) and final fitted (B) SEM models for aphid survival (Fisher's C statistic for initial full and final fitted models: $C_2 = 3.04$, $p = 0.218$; $C_6 = 8.86$, $p = 0.181$, respectively). Standardized path coefficients are presented for richness models (*grey*), unstandardized path coefficients for binomial aphid survival model (*white*) and Pearson's *r* correlation coefficients for correlations among agricultural intensification variables (*light grey*). For all three models per SEM marginal (R_m^2) and conditional (R_c^2) R-squared are shown. Negative path *arrows* are in *dark grey*, and *dotted lines* mean non-significant effects.

were nested in regions. The fixed effects and correlations that were tested are shown in Fig. 10A and B. The initial full SEM was simplified with manual backward selection by excluding the least significant variable considering all model coefficients until reaching the minimal SEM based on AIC (Shipley, 2013).

5.4 Results and Discussion of SEM

Aphid survival was negatively affected by predator diversity, which was in turn (weakly) affected by plant species richness. The number of plant species was negatively affected by local AI variables and by AI at the landscape scale (percentage arable land) (Geiger et al., 2010a). The negative effect of insecticide applications on the predators indicates a mechanism by which AI decreased biological control potential, i.e. that insecticide use reduced the diversity of carabid beetles (Fig 10A and B; cf. Geiger et al., 2010a). Among the direct effects of AI variables on aphid survival retained in the reduced model, the amount of N fertilizer was significantly positive, whilst application of insecticides (negative) and percentage of arable land in the surroundings (positive) were nonsignificant. The synthetic inputs (pesticides, herbicides and fertilizer) affected the measure of pest control indirectly via the number of plant species in the field or via the diversity of carabids in the field (insecticide input). Insecticide applications may also have affected other invertebrates that are prey for carabid beetles, additionally decreasing carabid richness and indirectly enhancing aphid survival. Thus the spectrum of prey and also its abundance is reduced in fields managed by using insecticides.

Although the focal variables (percentage of surviving aphids survived and median survival time of aphids) were highly correlated ($R^2 = 0.58$, $p < 0.001$), there were some differences in interpreting the outcomes of our experiment. The number of surviving aphids is a general measure for the effectiveness of the aphidophagous predators in the field and for the attractiveness of the glued aphids. The time when half of the aphids where removed, or the median survival time of aphids, is a measure of the predators' detection rate of the aphids. The shorter this period then the faster the aphids were found and consumed.

Our results show that biological pest control of aphids was more than the simple effects of land use on pest control. It is a multitrophic system where different stages are affected by land use in a range of different ways. The decrease in biological control potential with AI was mediated by decreases in predator diversity related to increased insecticide use, and to lower weed richness because of herbicide and fertilizer applications. In turn, more ground-dwelling predators reduced the number of aphids in our experiment, as indicated by the effect of predator diversity. The number of weed species seemed to be an important part of the system because of its mediating position in our model. Many of our AI measures influenced biological control of cereal aphids indirectly via weeds and their positive effect on carabids. These findings seem to be repeatable at a European scale, which makes them relevant for a general understanding of biological control in agricultural landscapes altered by the agricultural policy of the European Community.

6. CONCLUSIONS

AI has been related to biodiversity declines both globally and notably within the agroecosystems of the European Union, where biodiversity has developed and been shaped by agricultural land-use history. Such biological impoverishment may compromise the delivery of ecosystem services important for human welfare. However, a comprehensive review of the response of different components of biodiversity and ecosystem services to the AI process resulting from the European Union agricultural policy has remained lacking.

In this chapter, we have synthesized the findings from a large-scale pan-European study investigating the combined effects of AI and large-scale climatic gradients on taxonomic and functional biodiversity of key taxa (birds, carabid beetles, arable plants), as well the biological pest control ecosystem service provided by biodiversity in European agroecosystems.

We found that the three-service providing taxa studied (birds, carabid beetles, plants) were negatively related to AI when measured in terms of yield, but when these effects were examined in different regions across Europe the results were variable. Diversity often had negative relationships to yield, but in some regions there was no relationship, probably because intensification has coevolved with developments of varieties and technologies. On the other hand, and for some individual (input) measures such as fertilizers, we even found positive effects. However, diversity was generally distinctly negatively related to the use of pesticides (herbicides, insecticides or fungicides) and fertilizer application, although the relationships varied extensively between organism groups and regions. Notably, we also found clear negative effects of AI on the biological control potential of aphids, an important pest of cereals. Using a SEM model, we linked the decrease in biological control potential to a decrease in diversity of predatory insects (carabid beetles) and to pesticide use.

With regard to landscape-wide impacts, AI clearly had negative relationships with plant and bird diversity from farm to landscape scales, whilst relationships with beetles were more ambiguous. Notably, communities were not homogenized in intensively farmed landscapes, which may be attributed to variation in farming practices among farms in such landscapes. The effects of AI varied not just with region and taxa, but were further complicated when large-scale comparisons of organic and conventional farming systems were taken into consideration. Whilst organic methods and other AESs generally seem to benefit biodiversity and related ecosystem services, such as biological pest control, the relative effects of especially organic farming varied from none to large and depended on, for example the organisms studied and landscape context.

Several policy-relevant results have emerged from the project. Most important is that AESs, which mainly target single sites and are uncoordinated among farmers and landowners, need to expand from the local field and farm to the landscape (several neighbouring farms) and regional levels to be effective conservation tools. Also, the results strongly suggest that AESs need to be more taxon specific, as different components of biodiversity react differently to such measures. Finally, the large-scale results highlight that the CAP-supported AI not only has affected many components of biodiversity negatively but also has had negative effects on a critical ecosystem service like biological pest control, on which sustainable farming relies. Conventionally, pesticides are used to help produce food by controlling a range of pest species that have economic impacts on food production. Prevailing wisdom considers their use to be essential, yet here we have shown pervasive and compelling evidence demonstrating that pesticides have detrimental impacts on a range of biodiversity components at a pan-European scale. Recently, there have been calls to halt the *routine* and *indiscriminate* use of antibiotics in agriculture (Neff et al., 2015); our results indicate that it would be of particular importance to substantially decrease or halt the *routine* and *indiscriminate* use of pesticides in many European agricultural landscapes as well.

ACKNOWLEDGEMENTS

The AGRIPOPES project was funded within the European Science Foundation EuroDiversity programme through each partners' national research councils (Polish Ministry of Science and Higher Education, Spanish Ministry of Science and Education and Madrid Regional Government through project REMEDINAL3 P2013-MAE 2719, Estonian Scientific Foundation, Irish Research Council, German Federal Ministry of Education and Science BMBF, German Research Foundation, Netherlands Organization for Scientific Research, Swedish Research Council, Centre National de la Recherche Scientifique, Departement Ecologie et Developpement Durable). We are indebted to all who participated and worked in the project, too numerous to mention. Of particular importance at the outset of the project were, among others, Pablo Inchausti, William Sutherland, Pavel Kindlmann and Carsten Thies. We also want to especially thank the PhD students in the project: Flavia Geiger, Chris Dennis, Catherine Palmer, Violetta Hawro, Andreas Flohre, Sebastian Hänke, Christina Fischer and Camilla Winqvist.

REFERENCES

Aavik, T., Liira, J., 2009. Agrotolerant and high nature-value species—plant biodiversity indicator groups in agroecosystems. Ecol. Indic. 9, 892–901.

Aavik, T., Liira, J., 2010. Quantifying the effect of organic farming, field boundary type and landscape structure on the vegetation of field boundaries. Agric. Ecosyst. Environ. 135, 178–186.

Aavik, T., Augenstein, I., Bailey, D., Herzog, F., Zobel, M., Liira, J., 2008. What is the role of local landscape structure in the vegetation composition of field boundaries? Appl. Veg. Sci. 11, 375–386.

Aebischer, N.J., 1990. Assessing pesticide effects on nontarget invertebrates using long-term monitoring and time-series modeling. Funct. Ecol. 3, 369–373.

Batáry, P., Báldi, A., Kleijn, D., Tscharntke, T., 2011. Landscape-moderated biodiversity effects of agri-environmental management—a meta-analysis. Proc. R. Soc. Lond. B 278, 1894–1902.

Batáry, P., Dicks, L.V., Kleijn, D., Sutherland, W.J., 2015. The role of agri-environment schemes in conservation and environmental management. Conserv. Biol. 29, 1006–1016.

Bates, D., Maechler, M., Bolker, B., Walker, S., 2014. lme4: Linear mixed-effects models using Eigen and S4. R Package Version 1.1-7. http://CRAN.R-project.org/package=lme4.

Bengtsson, J., Ahnström, J., Weibull, A.-C., 2005. The effects of organic agriculture on biodiversity and abundance: a meta-analysis. J. Appl. Ecol. 42, 261–269.

Benton, T.G., Vickery, J.A., Wilson, J.D., 2003. Farmland biodiversity: is habitat heterogeneity the key? Trends Ecol. Evol. 18, 182–188.

Bibby, C.J., Burgess, N.D., Hill, D.A., 1992. Bird Census Techniques. Academic Press, London.

Billeter, R., Liira, J., Bailey, D., Bugter, R., Arens, P., Augenstein, I., Aviron, S., Baudry, J., Bukacek, R., Burel, F., Cerny, M., De Blust, G., De Cock, R., Diekotter, T., Dietz, H., Dirksen, J., Dormann, C., Durka, W., Frenzel, M., Hamersky, R., Hendrickx, F., Herzog, F., Klotz, S., Koolstra, B., Lausch, A., Le Coeur, D., Maelfait, J.P., Opdam, P., Roubalova, M., Schermann, A., Schermann, N., Schmidt, T., Schweiger, O., Smulders, M.J.M., Speelmans, M., Simova, P., Verboom, J., van Wingerden, W.K.R.E., Zobel, M., Edwards, P.J., 2008. Indicators for biodiversity in agricultural landscapes: a pan-European study. J. Appl. Ecol. 45, 141–150.

Birkhofer, K., Ekroos, J., Corlett, E.B., Smitha, H.G., 2014. Winners and losers of organic cereal farming in animal communities across Central and Northern Europe. Biol. Conserv. 175, 25–33.

Björklund, J., Limburg, K.E., Rydberg, T., 1999. Impact of production intensity on the ability of the agricultural landscape to generate ecosystem services: an example from Sweden. Ecol. Econ. 29, 269–291.

Börjeson, L., 2010. Agricultural intensification. In: Warf, B. (Ed.), Encyclopedia of Geography. Sage Publications, Thousand Oaks.

Boutin, C., Baril, A., Martin, P.A., 2008. Plant diversity in crop fields and woody hedgerows of organic and conventional farms in contrasting landscapes. Agric. Ecosyst. Environ. 123, 185–193.

Brouwer, F., Lowe, P., 2000. CAP and the environment: policy development and the state of research. In: Brouwer, F., Lowe, P. (Eds.), CAP Regimes and the European Countryside. CAB International, The Hague.

Buckwell, A., 1990. Economic signals, farmers' response and environmental change. J. Rural Stud. 5, 149–160.

Butler, S.J., Vickery, J.A., Norris, K., 2007. Farmland biodiversity and the footprint of agriculture. Science 315, 381–384.

Büttner, G., Feranec, J., Jaffrain, G., 2000. Corine land cover update 2000: Technical guidelines. In: European Environmental Agency (Ed.), Technical Report.

Carmona, C.P., Guerrero, I., Morales, M.B., Oñate, J.J., Peco, B., 2016. Assessing vulnerability of functional diversity to species loss: a case in Mediterranean agricultural systems. Funct. Ecol. http://dx.doi.org/10.1111/1365-2435.12709.

Chamberlain, D.E., Wilson, A.M., Browne, S.J., Vickery, J.A., 1999. Effects of habitat type and management on the abundance of skylarks in the breeding season. J. Appl. Ecol. 36, 856–870.

Chamberlain, D.E., Fuller, R.J., Bunce, R.G.H., Duckworth, J.C., Shrubb, M., 2000. Changes in the abundance of farmland birds in relation to the timing of agricultural intensification in England and Wales. J. Appl. Ecol. 37, 771–788.

Chiron, F., Filippi-Codaccioni, O., Jiguet, F., Devictor, V., 2010. Effects of non-cropped landscape diversity on spatial dynamics of farmland birds in intensive farming systems. Biol. Conserv. 143, 2609–2616.

Clough, Y., Kruess, A., Tscharntke, T., 2007. Local and landscape factors in differently managed arable fields affect the insect herbivore community of a non-crop plant species. J. Appl. Ecol. 44, 22–28.

Concepción, E.D., Díaz, M., Kleijn, D., Baldi, A., Batary, P., Clough, Y., 2012. Interactive effects of landscape context constrain the effectiveness of local agri-environmental management. J. Appl. Ecol. 49, 695–705.

Dänhardt, J., Green, M., Lindström, A., Rundlöf, M., Smith, H.G., 2010. Farmland as stopover habitat for migrating birds- effects of organic farming and landscape structure. Oikos 119, 1114–1125.

de Snoo, G.R., van der Poll, R.J., 1999. Effect of herbicide drift on adjacent boundary vegetation. Agric. Ecosyst. Environ. 73, 1–6.

Donald, P.F., 2004. The Skylark. T. & A.D. Poyser, London.

Donald, P.F., Green, R.E., Heath, M.F., 2001. Agricultural intensification and the collapse of Europe's farmland bird populations. Proc. R. Soc. Lond. B 268, 25–29.

Donald, P.F., Sanderson, F.J., Burfield, I.J., van Bommel, F.P.J., 2006. Further evidence of continent-wide impacts of agricultural intensification on European farmland birds, 1990–2000. Agric. Ecosyst. Environ. 116, 189–196.

EASAC, 2015. Ecosystem services, agriculture and neonicotinoids. http://www.easac.eu/fileadmin/Reports/Easac_15_ES_web_complete_01.pdf.

Eggers, S., Unell, M., Pärt, T., 2011. Autumn sowing of cereals reduces breeding bird numbers in a heterogeneous agricultural landscape. Biol. Conserv. 144, 1137–1144.

Ekroos, J., Heliola, J., Kuussaari, M., 2010. Homogenization of lepidopteron communities in intensively cultivated agricultural landscapes. J. Appl. Ecol. 47, 459–467.

Eurostat, 2015. Statistics explained. http://ec.europa.eu/eurostat/statistics-explained/index.php/Main_Page. last accessed 14.07.2016.

Fahrig, L., 2003. Effects of habitat fragmentation on biodiversity. Annu. Rev. Ecol. Syst. 34, 487–515.

Fahrig, L., Baudry, J., Brotons, L., Burel, F.G., Crist, T.O., Fuller, R.J., Sirami, C., Siriwardena, G.M., Martin, J.L., 2011. Functional landscape heterogeneity and animal biodiversity in agricultural landscapes. Ecol. Lett. 14, 101–112.

FAO, 2004. The ethics sustainable agricultural intensification. In: FAO Ethics Series 3, Food and Agriculture Organization of the United Nations, Rome.

FAOSTAT, 2004. Agricultural data. http://faostat.fao.org/faostat/collections?subset=agriculture2004. Visited 1 March 2004.

Filippi-Codaccioni, O., Devictor, V., Bas, Y., Clobert, J., Julliard, R., 2010. Specialist response to proportion of arable land and pesticide input in agricultural landscapes. Biol. Conserv. 143, 883–890.

Firbank, L.G., Petit, S., Smart, S., Blain, A., Fuller, R.J., 2008. Assessing the impacts of agricultural intensification on biodiversity: a British perspective. Philos. Trans. R. Soc. B 363, 777–787.

Flohre, A., Hänke, S., Fischer, C., Geiger, F., Bengtsson, J., Berendse, F., Weisser, W.W., Emmerson, M., Ceryngier, P., Liira, J., Tscharntke, T., Winqvist, C., Eggers, S., Bommarco, R., Pärt, T., Bretagnolle, V., Plantegenest, M., Clement, L.W., Dennis, C., Palmer, C., Morales, M.B., Oñate, J.J., Guerrero, I., Hawro, V., Aavik, T., Thies, C., Inchausti, P., 2011a. Agricultural intensification and biodiversity partitioning in European landscapes comparing plants, carabids, and birds. Ecol. Appl. 21, 1772–1781.

Flohre, A., Rudnick, M., Traserc, G., Tscharntke, T., Eggers, T., 2011b. Does soil biota benefit from organic farming in complex vs. simple landscapes? Agric. Ecosyst. Environ. 141, 210–214.

Foley, J.A., Ramankutty, N., Brauman, K.A., Cassidy, E.S., Gerber, J.S., Johnston, M., Mueller, M.D., O'Connell, C., Ray, D.K., West, P.C., Balzer, C., Bennett, E.M., Carpenter, S.R., Hill, J.S., Monfreda, C., Polasky, S., Rockström, J., Sheehan, J., Siebert, S., Tilman, D., Zaks, D.P.M., 2011. Solutions for a cultivated planet. Nature 478, 337–342.

Gabriel, D., Tscharntke, T., 2007. Insect pollinated plants benefit from organic farming. Agric. Ecosyst. Environ. 118, 43–48.

Gabriel, D., Sait, S.M., Kunin, W.E., Benton, T.G., 2013. Food production vs. biodiversity: comparing organic and conventional agriculture. J. Appl. Ecol. 50, 355–364.

Garibaldi, L.A., Steffan-Dewenter, I., Kremen, C., Morales, J.M., Bommarco, R., Cunningham, S.A., Carvalheiro, L.G., Chacoff, N.P., Dudenhöffer, J.H., Greenleaf, S.S., 2011. Stability of pollination services decreases with isolation from natural areas despite honey bee visits. Ecol. Lett. 14, 1062–1072.

Geiger, F., Bengtsson, J., Berendse, F., Weisser, W.W., Emmerson, M., Morales, M.B., Ceryngier, P., Liira, J., Tscharntke, T., Winqvist, C., Eggers, S., Bommarco, R., Pärt, T., Bretagnolle, V., Plantegenest, M., Clement, L.W., Dennis, C., Palmer, C., Oñate, J.J., Guerrero, I., Hawro, V., Aavik, T., Thies, C., Flohre, A., Hänke, S., Fischer, C., Goedhart, P.W., Inchausti, P., 2010a. Persistent negative effects of pesticides on biodiversity and biological control potential on European farmland. Basic Appl. Ecol. 11, 97–105.

Geiger, F., de Snoo, G.R., Berendse, F., Guerrero, I., Morales, M.B., Oñate, J.C., Eggers, E., Pärt, T., Bommarco, R., Bengtsson, J., Clement, L.W., Weisser, W.W., Olszewski, A., Ceryngier, P., Hawro, V., Inchaustih, P., Fischer, C., Flohre, A., Thies, C., Tscharntke, T., 2010b. Landscape composition influences farm management effects on farmland birds in winter: a pan-European approach. Agric. Ecosyst. Environ. 139, 571–577.

Gibbons, D., Morrissey, C., Mineau, P., 2015. A review of the direct and indirect effects of neonicotinoids and fipronil on vertebrate wildlife. Environ. Sci. Pollut. Res. 22, 103–118.

Gibson, R.H., Pearce, S., Morris, R.J., Symondson, W.O.C., Memmott, J., 2007. Plant diversity and land use under organic and conventional agriculture: a whole-farm approach. J. Appl. Ecol. 44, 792–803.

Global Food Security, 2011. Global Food Security Strategic Plan 2011–2016. Global Food Security, UK.

Godfray, H.C.J., Beddington, J.R., Crute, I.R., Haddad, L., Lawrence, D., Muir, J.F., Pretty, J., Robinson, S., Thomas, S.M., Toulmin, C., 2010. Food security: the challenge of feeding 9 billion people. Science 327, 812–818.

Godfray, H.C.J., Blacquiere, T., Field, L.M., Hails, R.S., Petrokofsky, G., Potts, S.G., Raine, N.E., Vanbergen, A.J., McLean, A.R., 2014. A restatement of the natural science evidence base concerning neonicotinoid insecticides and insect pollinators. Proc. R. Soc. Lond. B 281, 1786. 20140558.

Guerrero, I., Martínez, P., Morales, M.B., Oñate, J.J., 2010. Agricultural factors influencing bird, carabid and weed richness in a high conservation value, low-intensity cereal system. Agric. Ecosyst. Environ. 138, 103–108.

Guerrero, I., Morales, M.B., Oñate, J.J., Aavik, T., Bengtsson, J., Berendse, F., Clement, L.W., Dennis, C., Eggers, S., Emmerson, M., Fischer, C., Flohre, A., Geiger, F., Hawro, V., Inchausti, P., Kalamees, A., Kinks, R., Liira, J., Meléndez, L., Pärt, T., Thies, C., Tscharntke, T., Olszewski, A., Weisser, W.W., 2011. Taxonomic and functional diversity of farmland bird communities across Europe: effects of biogeography and agricultural intensification. Biodivers. Conserv. 20, 3663–3681.

Guerrero, I., Morales, M.B., Oñate, J.J., Geiger, F., Berendse, F., Snoo, G.D., Eggers, S., Pärt, T., Bengtsson, J., Clement, L.W., Weisser, W.W., Olszewski, A., Ceryngier, P., Hawro, V., Liira, J., Aavik, T., Fischer, C., Flohre, A., Thies, C., Tscharntke, T., 2012. Response of ground-nesting farmland birds to agricultural intensification across Europe: landscape and field level management factors. Biol. Conserv. 152, 74–80.

Guerrero, I., Carmona, C.P., Morales, M.B., Oñate, J.J., Peco, B., 2014. Non-linear responses of functional diversity and redundancy to agricultural intensification at the field scale in Mediterranean arable plant communities. Agric. Ecosyst. Environ. 195, 36–43.

Haberl, H., Schulz, N.B., Plutzar, C., Erb, K.H., Krausmann, F., Loibl, W., Moser, D., Sauberer, N., Weisz, H., Zechmeister, H.G., Zulka, P., 2004. Human appropriation of net primary production and species diversity in agricultural landscapes. Agric. Ecosyst. Environ. 102, 213–218.

Hallmann, C.A., Foppen, R.P.B., van Turnhout, C.A.M., de Kroon, H., Jongejans, E., 2014. Declines in insectivorous birds are associated with high neonicotinoid concentrations. Nature 511, 341.

Hartmann, M., Frey, B., Mayer, J., Mäder, P., Widmer, F., 2015. Distinct soil microbial diversity under long-term organic and conventional farming. ISME J. 9, 1177–1194.

Herzog, F., Steiner, B., Bailey, D., Baudry, J., Billeter, R., Bukácek, R., De Blust, G., De Cock, R., Dirksen, J., Dormann, C.F., De Filippi, R., Frossard, E., Liira, J., Schmidt, T., Stöckli, R., Thenail, C., van Wingerden, W., Bugter, R., 2006. Assessing the intensity of temperate European agriculture at the landscape scale. Eur. J. Agron. 24, 165–181.

Hiron, M., Berg, Å., Eggers, S., Josefsson, J., Pärt, T., 2013. Bird diversity relates to agri-environment schemes at local and landscape level in intensive farmland. Agric. Ecosyst. Environ. 176, 9–16.

Hiron, M., Berg, Å., Eggers, S., Berggren, Å., Josefsson, J., Pärt, T., 2015. The relationship of bird diversity to crop and non-crop heterogeneity in agricultural landscapes. Landsc. Ecol. 30, 2001–2013.

Hodge, I., Hauck, J., Bonn, A., 2015. The alignment of agricultural and nature conservation policies in the European Union. Conserv. Biol. 29, 996–1005.

Hole, D.G., Perkins, A.J., Wilson, J.D., Alexander, I.H., Grice, P.V., 2005. Does organic farming benefit biodiversity? Biol. Conserv. 122, 113–130.
Holland, J.M., 2002. Carabid beetles: their ecology, survival, and use in agroecosystems. In: Holland, J.M. (Ed.), The Agroecology of Carabid Beetles. Intercept, Andover.
Holzschuh, A., Steffan-Dewenter, I., Kleijn, D., Tscharntke, T., 2007. Diversity of flower-visiting bees in cereal fields: effects of farming system, landscape composition and regional context. J. Appl. Ecol. 44, 41–49.
Ings, T.C., Montoya, J.M., Bascompte, J., Blüthgen, N., Brown, L., Dormann, C.F., Edwards, F., Figueroa, D., Jacob, U., Jones, J.I., Lauridsen, R.B., Ledger, M.E., Lewis, H.M., Olesen, J.M., Van Veen, F.J.F., Warren, P.H., Woodward, G., 2009. Ecological networks—beyond food webs. J. Anim. Ecol. 78, 253–269.
Kennedy, C.M., Lonsdorf, E., Neel, M.C., Williams, N.M., Ricketts, T.H., Winfree, R., Bommarco, R., Brittain, C., Burley, A.L., Cariveau, D., 2013. A global quantitative synthesis of local and landscape effects on wild bee pollinators in agroecosystems. Ecol. Lett. 16, 584–599.
Kladivko, E.J., 2001. Tillage systems and soil ecology. Soil Tillage Res. 61, 61–76.
Kleijn, D., Snoeijing, G.I.J., 1997. Field boundary vegetation and the effects of agrochemical drift: botanical change caused by low levels of herbicide and fertilizer. J. Appl. Ecol. 34, 1413–1425.
Kleijn, D., Berendse, F., Smit, R., Gilissen, N., 2001. Agri-environment schemes do not effectively protect biodiversity in Dutch agricultural landscapes. Nature 413, 723–725.
Kleijn, D., Baquero, R.A., Clough, Y., Diaz, M., De Esteban, J., Fernandez, F., Gabriel, D., Herzog, F., Holzschuh, A., Jöhl, R., Knop, E., Kruess, A., Marshall, E.J.P., Steffan-Dewenter, I., Tscharntke, T., Verhulst, J., West, T.M., Yela, J.L., 2006. Mixed biodiversity benefits of agri-environment schemes in five European countries. Ecol. Lett. 9, 243–254.
Kleijn, D., Kohler, F., Báldi, A., Batáry, P., Concepción, E.D., Clough, Y., Díaz, M., Gabriel, D., Holzschuh, A., Knop, E., Kovács, A., Marshall, E.J.P., Tscharntke, T., Verhulst, J., 2009. On the relationship between farmland biodiversity and land use intensity in Europe. Proc. R. Soc. Lond. B 276, 903–909.
Kleijn, D., Winfree, R., Bartomeus, I., Carvalheiro, L.G., Henry, M., Isaacs, R., Klein, A., Kremen, C., M'Gonigle, L.K., Rader, R., Ricketts, T.H., Williams, N.M., Lee Adamson, N., Ascher, J.S., Baldi, A., Batáry, P., Benjamin, F., Biesmeijer, J.C., Blitzer, E.J., Bommarco, R., Brand, M.R., Bretagnolle, V., Button, L., Cariveau, D.P., Chifflet, R., Colville, J.F., Danforth, B.N., Elle, E., Garratt, M.P.D., Herzog, F., Holzschuh, A., Howlett, B.G., Jauker, F., Jha, S., Knop, E., Krewenka, K.M., Le Feon, V., Mandelik, Y., May, E.A., Park, M.G., Pisanty, G., Reemer, M., Riedinger, V., Rollin, O., Rundlof, M., Sardinas, H.S., Scheper, J., Sciligo, A.R., Smith, H.G., Steffan-Dewenter, I., Thorp, R., Tscharntke, T., Verhulst, J., Viana, B.F., Vaissiere, B.E., Veldtman, R., Westphal, C., Potts, S.G., 2015. Delivery of crop pollination services is an insufficient argument for wild pollinator conservation. Nat. Commun. 6, 7414.
Klomp, H., 1954. Die terreikus van der Kievit *Vanellus vanellus* (L.). Ardea 42, 1–139.
Krebs, J.R., Wilson, J.D., Bradbury, R.B., Siriwardena, G.M., 1999. The second silent spring? Nature 400, 611–612.
Kremen, C., 2015. Reframing the land-sparing/land-sharing debate for biodiversity conservation. Ann. N.Y. Acad. Sci. 1355, 52–76.
Lefcheck, J.S., 2016. PiecewiseSEM: piecewise structural equation modeling in R for ecology, evolution, and systematics. Methods Ecol. Evol. 7, 573–579. http://dx.doi.org/10.1111/2041-210X.12512.

Lefebvre, M., Espinosa, M., Gómez, S., 2012. The influence of the Common Agricultural Policy on agricultural landscapes: European Commission (Report EUR 25459 EN), Luxembourg. http://www.jrc.ec.europa.eu/.

Lennon, J.J., 2015. Potential impacts of climate change on agriculture and food safety within the island of Ireland. Trends Food Sci. Technol. 44, 1–10.

Liira, J., Aavik, T., Parrest, O., Zobel, M., 2008a. Agricultural sector, rural environment and biodiversity in the central and eastern European EU member states. Acta Geograph. Debrecina Landsc. Environ. Ser. 2, 46–64.

Liira, J., Schmidt, T., Aavik, T., Arens, P., Augenstein, I., Bailey, D., Billeter, R., Bukacek, R., Burel, F., De Blust, G., De Cock, R., Dirksen, J., Edwards, P.J., Hamersky, R., Herzog, F., Klotz, S., Kühn, I., Le Coeur, D., Miklova, P., Roubalova, M., Schweiger, O., Smulders, M.J.M., van Wingerden, W.K.R.E., Bugter, R., Zobel, M., 2008b. Plant functional group composition and large-scale species richness in the agricultural landscapes of Europe. J. Veg. Sci. 19, 3–14.

Lopez-Antia, A., Ortiz-Santaliestra, M.E., Garcia-de Blas, E., Camarero, P.R., Mougeot, F., Mateo, R., 2015. Adverse effects of thiram-treated seed ingestion on the reproductive performance and the offspring immune function of the red-legged partridge. Environ. Toxicol. Chem. 34, 1320–1329.

Manhoudt, A.G.E., Visser, A.J., de Snoo, G.R., 2007. Management regimes and farming practices enhancing plant species richness on ditch banks. Agric. Ecosyst. Environ. 119, 353–358.

Marshall, E.J.P., Moonen, A.C., 2002. Field margins in northern Europe: their functions and interactions with agriculture. Agric. Ecosyst. Environ. 89, 5–21.

Marshall, E.J.P., Brown, V.K., Boatman, N.D., Lutman, P.J.W., Squire, G.R., Ward, L.K., 2003. The role of weeds in supporting biological diversity within crop fields. Weed Res. 43, 77–89.

Matson, P.A., Parton, W.J., Power, A.G., Swift, M.J., 1997. Agricultural intensification and ecosystem properties. Science 277, 504–509.

Meeus, J.H.A., 1993. The transformation of agricultural landscapes in Western Europe. Sci. Total Environ. 129, 171–190.

Morales, M.B., Traba, J., Carriles, E., Delgado, M.P., García de la Morena, E.L., 2008. Sexual differences in microhabitat selection of breeding Little Bustards *Tetrax tetrax*: ecological segregation based on vegetation structure. Acta Oecol. 34, 345–353.

Morales, M.B., Traba, J., Delgado, M.P., García de la Morena, E.L., 2013. The use of fallows by nesting little bustard *Tetrax tetrax* females: implications for conservation in mosaic cereal farmland. Ardeola 60, 85–97.

Morales, M.B., Oñate, J.J., Guerrero, I., Meléndez, L., 2015. Influence of landscape and field-level agricultural management on a Mediterranean farmland winter bird community. Ardeola 62, 49–65.

Mouillot, D., Bellwood, D.R., Baraloto, C., Chave, J., Galzin, R., Harmelin-Vivien, M., Kulbicki, M., Lavergne, S., Lavorel, S., Mouquet, N., Paine, C.E.T., Renaud, J., Thuiller, W., 2013. Rare species support vulnerable functions in high-diversity ecosystems. PLoS Biol. 11, e1001569.

Neff, R.A., Merrigan, K., Wallinga, D., 2015. A food systems approach to healthy food and agriculture policy. Health Aff. 34, 1908–1915.

Newton, I., 1998. Population Limitations in Birds. Academic Press, London.

Norton, L., Johnson, P., Joys, A., Stuart, R., Chamberlain, D., Feber, R., Firbank, L., Manley, W., Wolfe, M., Hart, B., Mathews, F., Macdonald, D., Fuller, R.J., 2009. Consequences of organic and non-organic farming practices for field, farm and landscape complexity. Agric. Ecosyst. Environ. 129, 221–227.

Oliver, T.H., Isaac, N.J.B., August, T.A., Woodcock, B.A., Roy, D.B., Bullock, J.M., 2015. Declining resilience of ecosystem functions under biodiversity loss. Nat. Commun. 6, 10122. http://dx.doi.org/10.1038/ncomms10122.

Östman, Ö., Ekbom, B., Bengtsson, J., Weibull, A.C., 2001a. Landscape complexity and farming practice influence the condition of polyphagous carabid beetles. Ecol. Appl. 11, 480–488.

Östman, Ö., Ekbom, B., Bengtsson, J., 2001b. Landscape heterogeneity and farming practice influence biological control. Basic Appl. Ecol. 2, 365–371.

Östman, Ö., Ekbom, B., Bengtsson, J., 2003. Yield increase attributable to aphid predation by ground-living polyphagous natural enemies in spring barley in Sweden. Ecol. Econ. 45, 149–158.

Pain, D.J., Dixon, J., 1997. Why farming birds in Europe? In: Pain, D.J., Pienkowski, M.W. (Eds.), Farming Birds in Europe: The Common Agricultural Policy and Its Implications for Bird Conservation. Academic Press, London.

Pe'er, G., Dicks, L.V., Visconti, P., Arlettaz, R., Baldi, A., Benton, T.G., Collins, S., Dieterich, M., Gregory, R.D., Hartig, F., Henle, K., Hobson, P.R., Kleijn, D., Neumann, R.K., Robijns, T., Schmidt, J., Shwartz, A., Sutherland, W.J., Turbe, A., Wulf, F., Scott, A.V., 2014. EU agricultural reform fails on biodiversity. Science 344, 1090–1092.

Petersen, S., Axelsen, J.A., Tybirk, K., Aude, E., Vestergaard, P., 2006. Effects of organic farming on field boundary next term vegetation in Denmark. Agric. Ecosyst. Environ. 113, 302–306.

Pinheiro, J., Bates, D., DebRoy, S., Sarkar, D., Core Team, R., 2015. nlme: linear and nonlinear mixed effects models. R Package Version 3.1-120. http://CRAN.R-project.org/package=nlme.

Poisot, T., Canard, E., Mouillot, D., Mouquet, N., Gravel, D., 2012. The dissimilarity of species interaction networks. Ecol. Lett. 15, 1353–1361.

Poppy, G.M., Jepson, P.C., Pickett, J.A., Birkett, M.A., 2014. Achieving food and environmental security: new approaches to close the gap. Philos. Trans. R. Soc. B 369. 20120272.

Potter, C., 1997. Europe's changing farmed landscapes. In: Pain, D.J., Pienkowski, M.W. (Eds.), Farming and Birds in Europe: The Common Agricultural Policy and Its Implications for Bird Conservation. Academic Press, London.

R Development Core Team, 2015. R version 3.2.0: A Language and Environment for Statistical Computing. R Foundation for Statistical Computing, Vienna.

Ray, D.K., Ramankutty, N., Mueller, N.D., West, P.C., Foley, J.A., 2012. Recent patterns of crop yield growth and stagnation. Nat. Commun. 3, 1293. http://dx.doi.org/10.1038/ncomms2296.

Rempel, R.S., Carr, A., Elkie, P., 1999. Patch Analyst and Patch Analyst (Grid) Function Reference. Centre for Northern Forest Ecosystems Research, Ontario Ministry of Natural Resources, Lakehead University, Thunder Bay, Canada.

Robinson, R.A., Sutherland, W.J., 1999. The winter distribution of seed-eating birds: habitat structure, seed density and seasonal depletion. Ecography 22, 447–454.

Robinson, R.A., Sutherland, W.J., 2002. Post-war changes in arable farming and biodiversity in Great Britain. J. Appl. Ecol. 39, 157–176.

Robson, N., 1997. The evolution of Common Agricultural Policy and the incorporation of the environmental considerations. In: Pain, D., Pienkowski, M.W. (Eds.), Farming and Birds in Europe: The Common Agricultural Policy and Its Implications for Bird Conservation. Academic Press, London.

Roschewitz, I., Gabriel, D., Tscharntke, T., Thies, C., 2005. The effects of landscape complexity on arable weed species diversity in organic and conventional farming. J. Appl. Ecol. 42, 873–882.

Rundlöf, M., Nilsson, H., Smith, H.G., 2008. Interacting effects of farming practice and landscape context on bumble bees. Biol. Conserv. 141, 417–426.

Rundlöf, M., Andersson, G.K.S., Bommarco, R., Fries, I., Hederstrom, V., Herbertsson, L., Jonsson, O., Klatt, B.K., Pedersen, T.R., Yourstone, J., Smith, H.G., 2015. Seed coating with a neonicotinoid insecticide negatively affects wild bees. Nature 521, 77–80.

SCBD, 2001. Global Biodiversity Outlook. Secretariat of the Convention on Biological Diversity, Montreal.

Shipley, B., 2013. Structural equation modeling: a confirmatory analysis of computer self-efficacy. Struct. Equ. Model. 10, 214–221.

Siriwardena, G.M., Crick, H.Q.P., Baillie, S.R., Wilson, J.D., 2000. Agricultural land-use and the spatial distribution of granivorous lowland farmland birds. Ecography 23, 702–719.

Stoate, C., Boatman, N.D., Borralho, R., Rio Carvalho, C., de Snoo, G., Eden, P., 2001. Ecological impacts of arable intensification in Europe. J. Environ. Manage. 63, 337–365.

Storkey, J., Meyer, S., Still, K.S., Leuschner, C., 2012. The impact of agricultural intensification and land-use change on the European arable flora. Proc. R. Soc. Lond. B 279, 1421–1429.

Suárez, F., Naveso, M.A., De Juana, E., 1997. Farming in the drylands of Spain: birds of the pseudosteppes. In: Pain, D.J., Pienkowsky, M.W. (Eds.), Farming and Birds in Europe: The Common Agricultural Policy and Its Implications for Bird Conservation. Academic Press, London.

Teillard, F., Antoniucci, D., Jiguet, F., Tichit, M., 2014. Contrasting distributions of grassland and arable birds in heterogenous farmlands: implications for conservation. Biol. Conserv. 176, 243–251.

Thies, C., Tscharntke, T., 1999. Landscape structure and biological control in agroecosystems. Science 285, 893–895.

Thies, C., Haenke, S., Scherber, C., Bengtsson, J., Bommarco, R., Clement, L.W., Ceryngier, P., Dennis, C., Emmerson, M., Gagic, V., Hawro, V., Liira, J., Weisser, W.W., Winqvist, C., Tscharntke, T., 2011. The relationship between agricultural intensification and biological control: experimental tests across Europe. Ecol. Appl. 21, 2187–2196.

Tilman, D., Fargione, J., Wolff, B., D'Antonio, C., Dobson, A., Howarth, R., Schindler, D., Schlesinger, W.H., Simberloff, D., Swackhamer, D., 2001. Forecasting agriculturally driven global environmental change. Science 292, 281–284.

Tilman, D., Cassman, K.G., Matson, P.A., Naylor, R., Polasky, S., 2002. Agricultural sustainability and intensive production practices. Nature 418, 671–677.

Tivy, J., 1990. Agricultural Ecology. Longman, Essex.

Tscharntke, T., Klein, A.M., Kruess, A., Steffan-Dewenter, I., Thies, C., 2005. Landscape perspectives on agricultural intensification and biodiversity—ecosystem service management. Ecol. Lett. 8, 857–874.

Tuck, S.L., Winqvist, C., Mota, F., Ahnstrom, J., Turnbull, L.A., Bengtsson, J., 2014. Land-use intensity and the effects of organic farming on biodiversity: a hierarchical meta-analysis. J. Appl. Ecol. 51, 746–755.

Turner, B.L., Doolittle, W.E., 1978. The concept of agricultural intensity. Prof. Geogr. 30, 297–301.

Van der Sluijs, J.P., Amaral-Rogers, V., Belzunces, L.P., Bonmatin, J.M., Chagnon, M., Downs, C., Furlan, L., Gibbons, D.W., Giorio, C., Girolami, V., Goulson, D., Kreutzweiser, D.P., Krupke, C., Liess, M., Long, E., McField, M., Mineau, P., Mitchell, E.A.D., Morrissey, C.A., Noome, D.A., Pisa, L., Settele, J., Simon-Delso, N., Stark, J.D., Tapparo, A., van Dyck, H., van Praagh, J., Whitehorn, P.R., Wiemers, M., 2015. Conclusions of the Worldwide Integrated Assessment on the risks of neonicotinoids and fipronil to biodiversity and ecosystem functioning. Environ. Sci. Pollut. Res. 22, 148–154.

Vucic-Pestic, O., Rall, B.C., Kalinkat, G., Brose, U., 2010. Allometric functional response model: body masses constrain interaction strengths. J. Anim. Ecol. 79, 249–256.

Wakeham-Dawson, A., Aebischer, N.J., 1998. Factors determining winter densities of birds on environmentally sensitive area arable reversion grassland in southern England, with special reference to skylarks (*Alauda arvensis*). Agric. Ecosyst. Environ. 70, 189–201.

Walk, J.W., Warner, R.E., 2000. Grassland management for the conservation of songbirds in the Midwestern USA. Biol. Conserv. 94, 165–172.

Weibull, A.-C., Bengtsson, J., Nohlgren, E., 2000. Diversity of butterflies in the agricultural landscape: the role of farming system and landscape heterogeneity. Ecography 23, 743–749.

Whitthingham, M.J., 2007. Will agri-environment schemes deliver substantial biodiversity gain and if not why not? J. Appl. Ecol. 44, 1–5.

Williams, G.R., Troxler, A., Retschnig, G., Roth, K., Yanez, O., Shutler, D., Neuman, P., Gauthier, L., 2015. Neonicotinoid pesticides severely affect honey been queens. Nat. Sci. Rep. 5, 14621.

Wilson, J.D., Taylor, R., Muirhead, L.B., 1996. Field use by farmland birds in winter: an analysis of field type preferences using resampling methods. Bird Study 43, 320–332.

Winqvist, C., Bengtsson, J., Aavik, T., Berendse, F., Clement, L.W., Eggers, S., Fischer, C., Flohre, A., Geiger, F., Liira, J., 2011. Mixed effects of organic farming and landscape complexity on farmland biodiversity and biological control potential across Europe. J. Appl. Ecol. 48, 570–579.

Winqvist, C., Bengtsson, J., Öckinger, E., Aavik, T., Berendse, F., Clement, L.W., Fischer, C., Flohre, A., Geiger, F., Liira, J., Thies, C., Tscharntke, T., Weisser, W.W., Bommarco, R., 2014. Species' traits influence ground beetle responses to farm and landscape level agricultural intensification in Europe. J. Insect. Conserv. 18, 837–846.

Woodward, G., Ebenman, B., Emmerson, M.C., Montoya, J.M., Olesen, J.M., Valido, A., Warren, P.H., 2005. Body size in ecological networks. Trends Ecol. Evol. 20, 402–409.

Wretenberg, J., Lindström, Å., Svensson, S., Pärt, T., 2007. Linking agricultural policies to population trends of Swedish farmland birds in different agricultural regions. J. Appl. Ecol. 44, 933–994.

CHAPTER THREE

Litter Decomposition as an Indicator of Stream Ecosystem Functioning at Local-to-Continental Scales: Insights from the European *RivFunction* Project

E. Chauvet[*,1], V. Ferreira[†], P.S. Giller[‡], B.G. McKie[§], S.D. Tiegs[¶],
G. Woodward[||], A. Elosegi[#], M. Dobson[**], T. Fleituch[††], M.A.S. Graça[†],
V. Gulis[‡‡], S. Hladyz[§§], J.O. Lacoursière[¶¶], A. Lecerf[*], J. Pozo[#],
E. Preda[||||], M. Riipinen[##], G. Rîşnoveanu[||||], A. Vadineanu[||||],
L.B.-M. Vought[¶¶], M.O. Gessner[***,†††]

[*]EcoLab, University of Toulouse, CNRS, INPT, UPS, Toulouse, France
[†]MARE, University of Coimbra, Coimbra, Portugal
[‡]School of Biological, Earth and Environmental Sciences, University College Cork, Cork, Ireland
[§]Aquatic Sciences & Assessment, Swedish University of Agricultural Sciences, Uppsala, Sweden
[¶]Oakland University, Rochester, MI, United States
[||]Imperial College London, Ascot, Berkshire, United Kingdom
[#]Faculty of Science and Technology, University of the Basque Country, Bilbao, Spain
[**]APEM Limited, Edinburgh Technopole, Penicuik, United Kingdom
[††]Institute of Nature Conservation, Polish Academy of Sciences, Krakow, Poland
[‡‡]Coastal Carolina University, Conway, SC, United States
[§§]School of Biological Sciences, Monash University, Melbourne, Australia
[¶¶]Kristianstad University, Kristianstad, Sweden
[||||]Research Center in Systems Ecology and Sustainability, University of Bucharest, Bucharest, Romania
[##]Plymouth University, Plymouth, United Kingdom
[***]Leibniz Institute of Freshwater Ecology and Inland Fisheries, Berlin, Germany
[†††]Berlin Institute of Technology, Berlin, Germany
[1]Corresponding author: e-mail address: eric.chauvet@univ-tlse3.fr

Contents

1. Introduction — 101
2. Nutrient Enrichment Effects on Leaf Litter Decomposition — 107
3. Effects of Riparian Forest Modifications on Leaf Litter Decomposition — 118
   3.1 Deciduous Broadleaf Plantations — 122
   3.2 Conifer Plantations — 124
   3.3 Eucalyptus Plantations — 126
   3.4 Invasion of Riparian Areas by Exotic Woody Species — 130
   3.5 Forest Clear Cutting — 131
   3.6 Pasture — 134
4. Biodiversity-Related Mechanisms Underlying Altered Litter Decomposition — 137

4.1 Is There a General Relationship Between Biodiversity and Ecosystem Functioning, and Which Aspects of Biodiversity Are Most Important?	142
4.2 How Does Biodiversity Influence the Stability of Decomposition Rates, Including Under Variable Environmental Conditions?	150
4.3 Implications for the Use and Interpretation of a Litter Decomposition Assay in Bioassessment	152
5. Accomodating Natural Variability When Using Litter Decomposition in Stream Assessment	153
5.1 Extrinsic Factors	153
5.2 Temporal Variability	154
5.3 Intrinsic Factors: Partitioning and Minimising Variability in Leaf Litter Resource Quality and Potential Alternatives	157
6. Towards the Integration of Ecosystem Functioning into Stream Management	161
6.1 Ecosystem Functioning and Stream Management	161
6.2 Rationale and Steps in the Use of Litter Decomposition for Functional Assessment	163
6.3 An Example of National Adoption	164
6.4 Proposed Metrics	166
Acknowledgements	168
References	168

Abstract

RivFunction is a pan-European initiative that started in 2002 and was aimed at establishing a novel functional-based approach to assessing the ecological status of rivers. Litter decomposition was chosen as the focal process because it plays a central role in stream ecosystems and is easy to study in the field. Impacts of two stressors that occur across the continent, nutrient pollution and modified riparian vegetation, were examined at >200 paired sites in nine European ecoregions. In response to the former, decomposition was dramatically slowed at both extremes of a 1000-fold nutrient gradient, indicating nutrient limitation in unpolluted sites, highly variable responses across Europe in moderately impacted streams, and inhibition via associated toxic and additional stressors in highly polluted streams. Riparian forest modification by clear cutting or replacement of natural vegetation by plantations (e.g. conifers, eucalyptus) or pasture produced similarly complex responses. Clear effects caused by specific riparian disturbances were observed in regionally focused studies, but general trends across different types of riparian modifications were not apparent, in part possibly because of important indirect effects. Complementary field and laboratory experiments were undertaken to tease apart the mechanistic drivers of the continental scale field bioassays by addressing the influence of litter, fungal and detritivore diversity. These revealed generally weak and context-dependent effects on decomposition, suggesting high levels of redundancy (and hence potential insurance mechanisms that can mitigate a degree of species loss) within the food web. Reduced species richness consistently increased decomposition variability, if not the absolute rate. Further field studies were aimed at identifying important sources of this variability (e.g. litter quality, temporal variability) to help constrain ranges of predicted decomposition rates in different field situations. Thus, although many details still need to be resolved, litter decomposition holds considerable potential in some circumstances to capture impairment of stream ecosystem functioning. For instance, species traits associated with the body size and metabolic capacity of the consumers were often the main driver at local scales, and these were often translated into important determinants of otherwise apparently contingent effects at larger scales. Key insights gained from conducting continental scale studies included resolving the apparent paradox of inconsistent relationships between nutrients and decomposition rates, as the full complex multidimensional picture emerged from the large-scale dataset, of which only seemingly contradictory fragments had been seen previously.

1. INTRODUCTION

Although its roots can be traced back deep into the ecological literature (e.g. even indirectly in Darwin, 1881), interest in what we now call ecosystem functioning and its relationship with biodiversity gained momentum particularly towards the end of the 20th century (Jax, 2010). This was prompted by the increasing recognition that species can have strong effects on their environments (Lawton, 1994; Wallace and Webster, 1996) and growing concerns that population declines and high rates of species extinctions could eventually lead to the loss of key ecosystem functions, ultimately threatening human life support mechanisms (Ehrlich and Ehrlich, 1981). The argument is based on the recognition that organisms in ecosystems ultimately regulate biogeochemical cycles and provide resources essential to humans, such as clean water, timber or fish (Jackson et al., 2016). Three decades of intensive research have established that biodiversity loss can indeed have important repercussions on ecosystem functioning (Cardinale et al., 2012; Hooper et al., 2005), but also that many idiosyncrasies exist, partly because some species show a larger degree of redundancy than others (Rosenfeld, 2002).

Riverine ecosystems are especially vulnerable to the loss of both biodiversity and ecosystem functioning via a wide range of stressors, on local-to-global scales, yet it is only in the last decade or so that the full implications of these threats have become appreciated. The majority of drainage pathways in river catchments worldwide are small wooded streams (Allan and Castillo, 2007), where the closed riparian vegetation limits instream primary production but supplies large amounts of litter (Vannote et al., 1980), and this is where much of riverine biodiversity is concentrated in the landscape. Consequently, aquatic food webs in these streams obtain most of their energy and carbon from land-derived allochthonous organic matter, which ultimately also fuel the lower reaches into which the headwaters flow (Hladyz et al., 2011a*; Wallace et al., 1997). The decomposition of this litter is mainly a biological process, driven by microbial decomposers (fungi and bacteria) and macroinvertebrate detritivores (Gessner et al., 1999; Hieber and Gessner, 2002; Webster and Benfield, 1986), and it is highly sensitive to changes in environmental conditions (Ferreira et al., 2015a; Rosemond et al., 2015; Webster and Benfield, 1986). The central role of litter decomposition in streams, which represents the major 'brown pathways' in the food web, means that this process needs to be considered in order to capture and assess the broader ecological status of these ecosystems (Gessner and Chauvet, 2002*).

Despite seminal work by Odum (1956), for many years stream ecosystems have been considered mere conduits instead of biologically active ecosystems in their own right (Battin et al., 2009; Raymond et al., 2013). This view is changing and the necessity to incorporate ecosystem functioning into stream assessment and environmental management schemes has become increasingly evident (Christensen et al., 1996; Giller, 2005). The *RivFunction* initiative developed against this backdrop and evolved into a large-scale EU-funded project that aimed at elaborating a novel methodology for assessing the ecological status of European rivers in functional terms by focusing on leaf litter decomposition as a key ecosystem-level process. The goal was to unravel the relationships between environmental drivers, community structure and litter decomposition in streams at unprecedented scale.

This was achieved by conducting coordinated large-scale field experiments and bioassays across Europe (Hladyz et al., 2011a*; Woodward et al., 2012*), as well as smaller-scale field studies and controlled microcosm experiments (e.g. McKie et al., 2008*) within five Workpackages (Fig. 1). Emphasis of the large-scale field studies was on two types of widespread impacts in European streams: pollution by high nutrient inputs and modification of riparian vegetation. Two potential pathways affecting ecosystem functioning as indicated by leaf litter decomposition were examined, where: (i) stressors directly affect the activities of organisms and (ii) shifts in community structure towards species with intrinsically different activity potentials lead to changes in the emergent properties of the community (Fig. 1).

The importance of the second pathway was elucidated in microcosm and field experiments, especially by focusing on the diversity of plant litter, fungi and litter-consuming macroinvertebrates as one important aspect of

---

References marked with an '' are derived from the *RivFunction* project.

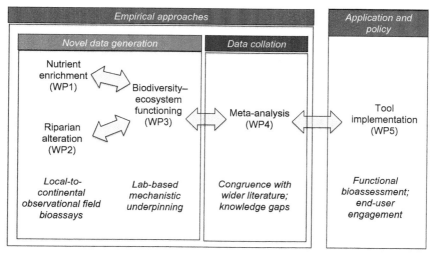

Fig. 1 A schematic highlighting the interconnections among the five original RivFunction Workpackages, from field-based surveys and bioassays to laboratory experiments, meta-analysis and tool development and implementation.

community structure that drives functioning at local scales (e.g. McKie et al., 2008*). This information, together with published data and results from studies addressing important sources of variability in decomposition served as the basis for elaborating a methodology to assess the ecological status of European streams and rivers from a functional point of view at local-to-continental scales (Hladyz et al., 2011a*; Woodward et al., 2012*).

Standardising RivFunction Protocols for Pan-European Comparisons

A key challenge when designing field studies across large and heterogenous areas such as Europe is ensuring that differences among ecoregions are due to the anthropogenic impacts of interest, rather than intrinsic biogeographical or other differences. To avoid bias due to plant litter species and quality, decomposition studies were carried out using leaves from two tree species—alder (*Alnus glutinosa* (L.) Gaertn.) and oak (*Quercus robur* L.)—that both occur nearly throughout the continent (Graça and Poquet, 2014). Alder and oak have contrasting physical and chemical characteristics (greater softness, greater concentrations of nutrients and lower concentrations of structural and secondary compounds in alder than in oak; Gulis et al., 2006*; Hladyz et al., 2009*), differ greatly in their palatability to litter-consuming detritivores (greater for alder than oak; Canhoto and Graça, 1995) and decompose at different rates (faster for alder and slow for oak; Gulis et al., 2006*; Riipinen et al., 2010*), thus allowing assessment of whether the effects of change in nutrient concentration or riparian forest modification on litter decomposition are moderated by litter quality.

The same set of response variables was assessed at each study site (i.e. individual streams nested and replicated across ecoregions) following standardised field and laboratory protocols to investigate impacts along a gradient of nutrient enrichment (Workpackage 1), as well as impacts of various types of riparian forest modification (Workpackage 2), in addition to enabling pairwise comparisons of impacted and corresponding reference streams within individual ecoregions. The coordinated and standardised field experiments were carried out by 10 research teams from 9 European ecoregions and countries (England, France, Ireland, Poland, Portugal, Romania, Spain, Sweden and Switzerland) and included a total of >200 streams across north–south and east–west gradients in Europe (Figs 2 and 3). All streams were <5 m wide, <50 cm deep at winter baseflow, first to fourth order, with a stony substrate and bordered with native deciduous riparian vegetation except where riparian forests were clear-cut or replaced by pasture, exotic invaders or plantations. A range of other measurements to characterise streams were standardised. Water samples were analysed in the laboratory for NH_4^+, NO_3^-, NO_2^- (total dissolved inorganic nitrogen (DIN) = nitrogen in $NH_4^+ + NO_3^- + NO_2^-$) and soluble reactive phosphorus (SRP ≈ phosphorus in PO_4^{3-}).

Detailed protocols describing the preparation, field placement and retrieval of litter bags from streams, as well as subsequent laboratory procedures were implemented to ensure comparable data across the entire study and to enable integrated analyses of data gathered at all sites. Alder and oak leaf litter were collected locally after senescence and used in both large-scale coordinated litter decomposition experiments conducted during autumn and winter 2002/2003. Both litter species were incubated in coarse- and fine-mesh bags, and litter mass remaining and decomposition rates were determined based on six replicate litter bags collected once from each stream. Mesh bags, each containing 5.00 ± 0.25 g of air-dried leaves, were deployed in at least 10 streams per ecoregion for each Workpackage, with apertures of 10 (coarse) or 0.5 mm (fine) to permit or prevent macroinvertebrate colonisation, respectively. This enabled us to quantify total, microbial and macroinvertebrate-driven decomposition rates of oak (slow-decomposing) and alder (fast-decomposing) litter. Litter bags were collected

Fig. 2 Types of reference and impacted streams from Workpackage 1, designed to isolate the effects of nutrient enrichment while standardising for riparian coverage and other physicochemical properties of the system. An example of a pair of sites in the Alps (Switzerland) with reference and nutrient-impacted streams on *left* and *right*, respectively. Average water temperature (15°C) and depth (0.1 m), and substratum composition (*gravel*, *cobble*) were identical. Electrical conductivity (449 and 513 µs/cm, respectively), water discharge (150 and 250 L/s, respectively) and stream width (2.5 and 1.7 m, respectively) were similar. *Photo credit: M.O. Gessner.*

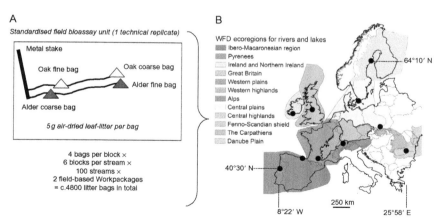

Fig. 3 The standardised field bioassay unit (A) and (B) regions and locations (*dots*) where the litter bioassay units were used in the initial continental-scale surveys in Workpackages 1 and 2. This basic design was adapted in subsequent studies: for instance in the follow-up Workpackage 2 studies, algal processes were also explored via the addition of a colonisation tile adjacent to each litter bag, or different types of litter were also added. The positioning of the fine (0.5 mm aperture mesh) vs coarse (1 cm aperture mesh) litter bags per leaf type per rebar was randomised within blocks.

when additional coarse-mesh bags (sampled on several occasions at reference sites) had lost ∼50% of their initial mass (T_{50}) (see Riipinen et al., 2009*). Thus, we standardised among regions and leaf species for the degree of decomposition rather than for exposure time. The retrieved leaf litter was oven-dried and ash-free dry mass (AFDM) determined after ashing of sub-samples. Correction factors derived in the laboratory for leaching losses and moisture content were applied to the initial air-dry mass and AFDM (Hladyz et al., 2009*).

Litter decomposition rates were expressed as the exponential decay rate coefficient, k, in the model $(m_t/m_0) = e^{-kt}$, where m_0 is the initial AFDM and m_t is AFDM at time t (see, e.g. Riipinen et al., 2009*). Macroinvertebrate-driven decomposition was derived by calculating the difference in the mean percent mass remaining in coarse- and fine-mesh bags in each stream and then subtracting the difference from the initial 100% mass before calculating a k value (k_{inv}) indicating the contribution of litter-consuming macroinvertebrates to mass loss (Woodward et al., 2012*). To correct for potential temperature differences among streams and regions (Fig. 4), t was expressed in terms of thermal sums (degree–days).

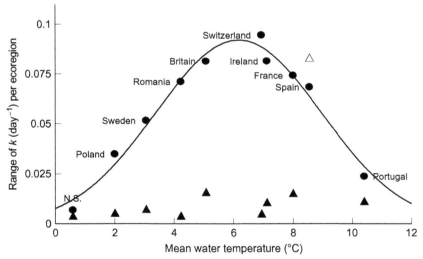

Fig. 4 Among-stream variation in total and microbial-mediated decomposition rates among reference sites within the study regions across Europe as a function of stream temperature. The range was compressed at either extreme, with the greatest apparent potential for biotic responses, especially those driven by macroinvertebrates, evident in the mid-range. The *circles* represent total decomposition in coarse-mesh bags (1 cm mesh aperture) and the *triangles* microbial-only mediated decomposition in fine-mesh bags (0.5 mm mesh aperture).

2. NUTRIENT ENRICHMENT EFFECTS ON LEAF LITTER DECOMPOSITION

The susceptibility of freshwater ecosystems to anthropogenically derived pollutants and nutrient enrichment is now well known and the impacts are likely to intensify in the future (Ferreira et al., 2015a; Friberg et al., 2016; Gessner and Chauvet, 2002*; Jackson et al., 2016; Stamm et al., 2016; Young et al., 2008). Impacts in rivers in particular are exacerbated by their intimate terrestrial linkages in dendritic networks, such that almost any activity within a river catchment has the potential to cause environmental change within the river system, and any significant pollutant entering a river may exert some effect for large distances downstream (Malmqvist and Rundle, 2002; Thompson et al., 2015). The wide range of stressors that can affect freshwater systems can be classified into four major types: ecosystem destruction, physical habitat alteration, water chemistry alteration and direct species additions and removals (Malmqvist and Rundle, 2002) resulting in 14 major threats that interact with the 6 major services provided by freshwater systems (Giller et al., 2004a; Fig. 5). There is a strong regional influence in this context, dependent largely on economic activity and state of development.

In Europe, nutrient enrichment from agricultural run-off, sewage inputs and nitrogen deposition have been occurring in surface waters for centuries, but the dramatic increases evident from the second half of the 20th century have placed the continent's vulnerable freshwaters in a precarious position (Woodward et al., 2012*). Attempts to reverse the damage done to these ecosystems and to the 'goods and services' they provide has triggered the introduction of far-reaching international legislation, such as the EU Water Framework Directive (WFD) (Hering et al., 2010). The US Clean Water Act was implemented for similar reasons (Adler et al., 1993), and ambitious environmental legislation is currently being drawn up in many other parts of the world to protect global water resources (such as New Zealand's National Policy Statement (NPS) for Freshwater Management). These legislative provisions stress the need to consider environmental impacts on processes as well as structural elements and to include assessment of such processes in biomonitoring and restoration.

Biodiversity and ecosystem functioning are intimately linked and both are vulnerable to environmental stressors. Aquatic ecosystems and freshwater ecosystems in particular are impacted by multiple stressors (Giller et al., 2004b) and experimental studies have begun to explore the range

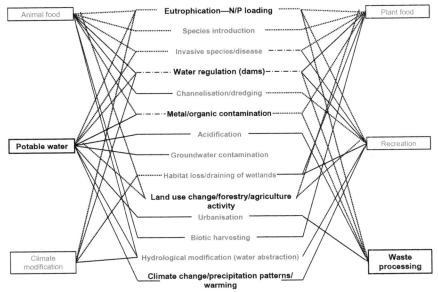

Fig. 5 A schematic diagram illustrating the interaction between 6 major ecosystem services provided by freshwater systems and 14 potential threats and stressors affecting the freshwater domain. Those threats and services related to the research associated with *RivFunction* are highlighted. *Dotted lines* indicate possible mediation through the benthos. Modified from Giller, P.S., Covich, A.P., Ewel, K.C., Hall, R.O., Merritt, D., 2004a. Vulnerability and management of ecological services in freshwater systems: case studies of freshwater ecosystem services. In: Wall, D.H. (Ed.), Sustaining Biodiversity and Functioning in Soils and Sediments. SCOPE Series, vol. 64, © Island Press, Washington, DC, pp. 137–160.

of synergistic and antagonistic impacts (e.g. McKie et al., 2009*; Piggott et al., 2015; Tolkkinen et al., 2013, 2015). Assessing and understanding the consequences of biodiversity change on ecosystem functioning are critical, given the threats to ecosystem processes and, in turn, to the associated ecological goods and services (e.g. drinking water quality, fisheries) (Giller et al., 2004a), especially since many of the 'ecological surprises' that often arise can be linked to the indirect effects of biotic interactions among different species in the food web (Gray et al., 2016). In the present context, we are concerned explicitly with the important ecosystem process of litter decomposition in streams, and both its direct and indirect drivers. As the trophic state of many streams is likely to deteriorate in the future due to the continuing increase in human-induced nutrient availability

(Jackson et al., 2016), it is of fundamental importance to understand how nutrient enrichment affects litter decomposition.

At large biogeographical scales, litter decomposition rates are strongly influenced by temperature and, consequently, they are likely to vary markedly across latitudinal and climatic gradients (Fig. 4). The few studies that have compared tropical and temperate freshwaters, however, have revealed that although microbial activity increases with temperature, macroinvertebrate leaf-shredders, which drive leaf litter decomposition rates at higher latitudes, are often missing from tropical streams (Boyero et al., 2011; Dobson et al., 2002; Graça et al., 2015; Irons et al., 1994; Rosemond et al., 1998). Decomposition rates are also governed indirectly by the nutrient status of the water, which is itself affected by environmental factors, including anthropogenic inputs of domestic sewage and agricultural run-off (e.g. Gulis et al., 2006*; Lecerf et al., 2006*; Niyogi et al., 2003; Pascoal et al., 2003; Stamm et al., 2016). A meta-analysis of 99 studies suggests that the effect of nutrient enrichment might be strongest in cold oligotrophic streams driven by patterns of biogeography of invertebrate decomposers which may be modulating the effect of nutrient enrichment on litter decomposition (Ferreira et al., 2015a). At more local scales, the availability and quality of leaf litter, in addition to the abundance, diversity and activity of aquatic consumers, combine to determine the rate of decomposition and energy flux to the higher trophic levels (Dangles et al., 2004*; Fleituch, 2013*; Gessner and Chauvet, 1994, 2002*; Gulis et al., 2006*; Hladyz et al., 2011a*). Decomposition by purely physical forces, such as sediment abrasion, is generally trivial relative to these biotic drivers (e.g. Ferreira et al., 2006b*; Hieber and Gessner, 2002; Hladyz et al., 2009*). Some evidence of synergistic effects of nitrogen and phosphorus has also been identified from field manipulative experiments (Ferreira et al., 2015a; Kominoski et al., 2015; Rosemond et al., 2015).

Experimental studies have reported elevated decomposition rates in nutrient-enriched systems, reflecting concomitant increases in both litter quality (i.e. lower carbon-to-nutrient ratios that enhance microbial conditioning and/or litter stoichiometry) and consumer abundance and activity (e.g. Bergfur et al., 2007; Ferreira et al., 2006c*; Greenwood et al., 2007; Gulis et al., 2006*; Halvorson et al., 2016; Manning et al., 2015; Rosemond et al., 2002). Nevertheless, there may be no clear effect of nutrient enrichment of stream water on the leaf decomposition rate where levels of eutrophication are relatively low, as in headwater streams (Pérez et al., 2013) or where nitrogen levels may not in fact be limiting (Royer and

Minshall, 2001; Stallcup and Ardón, 2006). In addition, such monotonic responses are unlikely to be ubiquitous, as many streams receive considerable inputs of domestic sewage and agricultural run-off that can induce anoxia, mobilise heavy metals and physically smother the benthos by so-called 'sewage fungus' (in fact a filamentous bacteria in the *Sphaerotilus* genus) (Curtis and Harrington, 1971; Lecerf et al., 2006*; Niyogi et al., 2003; Pascoal and Cássio, 2004). Such direct and indirect, potentially toxic, effects will also lead to community-level changes related to physiological tolerance of species, multistressor impacts (Giller et al., 2004b) and the well documented negative impacts on stream communities (e.g. Wright et al., 2000), including widespread species loss and catastrophic fish kills. Thus, some form of unimodal relationship between nutrient concentrations and decomposition rates in streams and rivers is predicted, with slow decomposition at both extremes, due to nutrient limitation in oligotrophic systems and toxic effects of other pollutants in hypertrophic systems.

Most studies have been conducted at relatively small scales and over a limited range of nutrient concentrations, thus the seemingly monotonic (and occasionally contradictory—positive and negative) responses that have often been reported to date (e.g. Ferreira and Chauvet, 2011; Gulis and Suberkropp, 2003; Ramírez et al., 2003; and those above) might simply reflect a truncated portion of an underlying unimodal response that is only evident at much larger scales and/or across a larger nutrient gradient. The picture is complicated further at large scales, as climatic and biogeographical effects come into play, and if we are to develop the new generation of ecosystem-based approaches to bioassessment, as required under current and emerging legislative provisions, we need to know how eutrophication–functioning relationships are influenced by these factors. The *RivFunction* project was able to present the results of a pan-European experiment that measured this crucial ecosystem process across a gradient that spans three orders of magnitude in nutrient concentrations and several biogeographic areas (Figs 6 and 7). We were also able to quantify the relative importance of microbial and macroinvertebrate-driven decomposition, by measuring both simultaneously. This was the first and to date only study where such comparisons have been made at multiple sites at a truly continental scale.

The objective of Workpackage 1 was to assess the impacts of nutrient loading (eutrophication) on river ecosystem functioning through the quantification of leaf litter decomposition and associated parameters. The underlying hypothesis was that increases in nutrient levels through pollution (eutrophication) fundamentally alters the functioning of the river ecosystem,

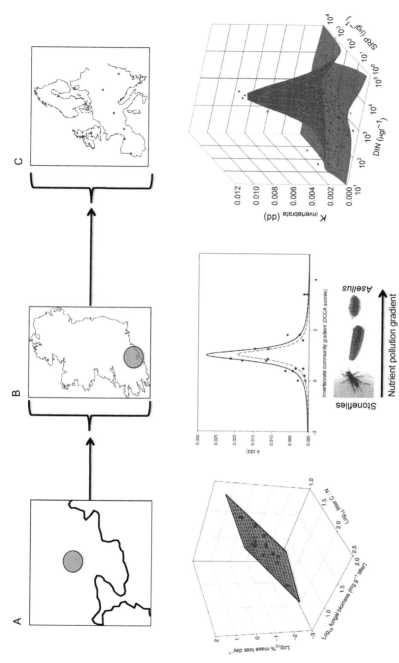

Fig. 6 Workpackage 1: Nutrient enrichment and environmental filters—from local-to-continental scales. (A) Decomposition rates as a function of litter quality (CN content) and fungal conditioning among 15 species of trees' litter in a single site in Ireland; (B) total (*black solid line*) and invertebrate-mediated decomposition (*red (grey in the print version) dashed line*) rates of alder in 10 streams across Ireland, as a function of invertebrate assemblage composition, from pristine to heavily-polluted systems; (C) volume-filling relationship between two major macronutrients and alder litter decomposition in the 100 streams across Europe.

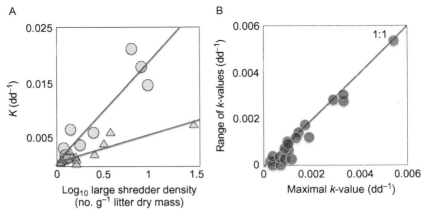

Fig. 7 Workpackage 1: (A) Decomposition rates as a function of consumer density in 10 streams in Ireland, providing mechanistic insight into the potential drivers behind the unimodal curve in Ireland in Fig. 6B. Total decomposition rates in coarse-mesh bags per degree–day plotted as a function of the abundance of large litter-consuming detritivores (Limnephilidae+Gammaridae) per gram of leaf litter ash-free dry mass in oak (*triangles*) and alder (*circles*). Equations for the regression lines are: (1) alder k_{total}: $y=-0.0007+0.019x$; $r^2=0.86$, $F_{1,9}=50.0$, $P<0.001$; (2) oak k_{total}: $y=-0.0009+0.0049x$; $r^2=0.80$, $F_{1,9}=32.2$, $P<0.001$. (B) The range of decomposition rates vs maxima across Europe, showing that within the grid of combinations of SRP and DIN across Europe that describes the volume-filling relationship in Fig. 6C, even at the highest peak rates decomposition can also be extremely slow. The *1:1 line* shows where the range is equal to the maximum. This suggests the potentially key role of local and contingent community-level effects related to the abundance, size and identity of the consumer assemblages (e.g. as per Fig. 7A and B). *Redrawn from Woodward, G., Gessner, M.O., Giller, P.S., Gulis, V., Hladyz, S., Lecerf, A., Malmqvist, B., McKie, B.G., Tiegs, S.D., Cariss, H., Dobson, M., Elosegi, A., Ferreira, V., Graça, M.A.S., Fleituch, T., Lacoursière, J.O., Nistorescu, M., Pozo, J., Risnoveanu, G., Schindler, M., Vadineanu, A., Vought, L.B.-M., Chauvet, E., 2012. Continental-scale effects of nutrient pollution on stream ecosystem functioning. Science, 336, 1438–1440.*

and that this impairment can be quantified and evaluated by following changes in a key ecosystem-level process, litter decomposition, with a simple assay. Suitable indicators of, and critical thresholds for, proper river ecosystem functioning were identified based on litter decomposition experiments conducted under field conditions. We hypothesised that rates of litter decomposition are constrained by microbial nutrient limitation on the rising limb along nutrient pollution gradients and by the effects of environmental degradation and other accompanying pollutants on invertebrates on the falling limb of a unimodal relationship between nutrient levels and leaf litter decomposition.

Key Findings

The effects of nutrient (nitrogen and phosphorus) loading on river functioning were examined through a single, large scale, pan-European experiment, by comparing decomposition rates of leaf litter at impacted and corresponding reference sites in which both point source and nonpoint source pollution were considered. The 100 investigated streams across Europe spanned 1000-fold differences in nutrient concentrations on both axes (SRP: <1–926 µg L^{-1}, DIN: 13–15,700 µg L^{-1}). Elevated dissolved nutrient concentrations relative to regional baselines were due to agricultural run-off and sewage effluents. The validity of the representativeness and scalability of this approach is highlighted by a clear positive relationship between BOD_5 (5-day biochemical oxygen demand) and nutrient concentrations in over 8000 European streams, and the comparable frequency distributions of nutrient concentrations between these and our study sites (Woodward et al., 2012*). In total, over 2400 leaf bags were exposed in the 100 streams (6 replicates × 2 mesh sizes × 2 leaf species × 10 streams × 10 regions), of which 2161 were retrieved at the end of the experiment. Results from these core Workpackage 1 studies have been published in a variety of journals over the past decade (e.g. Gulis et al., 2006*; Lecerf et al., 2006*; Woodward et al., 2012*) and the key findings are summarised and revisited below (see also Figs 6 and 7).

Across the nine ecoregions, total decomposition was considerably faster than microbial-mediated decomposition only, and the higher quality alder litter decomposed considerably faster than oak litter. The ratio of invertebrate-only to microbial-only k-values was close to two for both leaf species: thus, invertebrates were the main drivers of decomposition across Europe as a whole, although their relative contribution varied among ecoregions. Many of the patterns reported for oak were also evident in alder, although variability was higher in the latter, reflecting this species' faster decomposition rate vs, whereas the more refractory oak leaves gave a more integrated signal over a far longer exposure time in each stream.

Decomposition rates varied markedly among ecoregions, being suppressed at both high and low latitudes and in the more eastern sites, as the climate became less maritime and more continental. Similar latitudinal and longitudinal patterns were evident in the ratio of invertebrate: microbially driven decomposition, with invertebrates being less important at the extremes and in the more eastern ecoregions. Simple linear regression models using ecoregions as replicates (i.e. removing the lower levels of nesting) yielded significant relationships between mean temperature and

k-values, when outliers were excluded (northern Sweden for fine-mesh bags, Portugal for coarse-mesh bags). The slope of the line was also steeper for total decomposition than for microbial-only and, consequently, invertebrate-only decomposition also increased with temperature, when the two warmest ecoregions were excluded ($y = 0.0022x - 0.0005$; $R^2 = 0.76$). The fact that the outliers to these fits were always at the extremes of the temperature gradient suggests that other factors, in addition to temperature, were influencing decomposition rates. Multiple regression models on the full dataset revealed significant effects of water quality, temperature and geographical location: i.e. the spatial and environmental context of the ecoregions explained some of the variation not accounted for by temperature. When normalised per degree–day, Irons et al. (1994) found that decomposition rates actually increased with latitude and that shredders become progressively more important as agents of decomposition as latitude increases. However, many of the previous studies have compared a small set of sites clustered together in a temperate setting, with a similar set in the tropics, rather than a true gradient-based approach. Thus, many of the more subtle shifts within temperate regions identified in the present study have been previously obscured: although shredders were the main drivers of decomposition across Europe as a whole (Fig. 4), they did not become progressively more important with latitude, as suggested by other studies (Boyero et al., 2011; Irons et al., 1994; Rosemond et al., 1998). In fact, most of the decomposition in northern Sweden, close to the arctic circle, appeared to be driven almost entirely by microbial activity.

In addition to the responses in mean decomposition rates, there were further temperature-related patterns within the data that were also not evident at smaller spatial scales. For instance, the range of variation in total decomposition among streams within ecoregions followed a unimodal Gaussian curve ($R^2 = 0.97$) in response to temperature (Fig. 4). The range of microbial decomposition, however, was much lower and effectively constant across the temperature gradient, with the possible exception of the more variable Spanish sites.

Intriguingly, microbially driven decomposition did not display any clear or consistent response to nutrient concentrations, whereas, in contrast to the monotonic changes in decomposition rates in response to nutrients that have been described in previous studies (e.g. see references earlier), our overall invertebrate-driven spatial data for decomposition rates were best described by a unimodal surface in two dimensions and a unimodal volume in three dimensions—as we predicted at the outset (Fig. 6C). This applied for both leaf species and as a function of SRP and DIN concentrations. The rising limb of the unimodal curve was likely due to nutrient stimulation of microbes and subsequent increased consumption by invertebrates (Woodward et al., 2012*). In contrast, the falling limb was likely due to deteriorating environmental conditions accompanying excessive nutrient supply (e.g. low oxygen concentrations, presence of other pollutants), which suppressed invertebrate-driven decomposition. Rates were always low at the extremes, but ranged from low-to-high at intermediate nutrient concentrations and were bounded by a hump-shaped envelope (Fig. 6C). Within a given range of nutrient concentrations, the spread of decomposition rates was almost equal to the maximum: i.e. there was an upper, but not a lower, limit to decomposition rates (Fig. 7B). Essentially, there was a bounding envelope that delimited maximal decomposition rates, but within that envelope a broad distribution of process rates was possible, such that at any given nutrient concentration the maximum rate was almost equal to the range. The wide range of decomposition rates at intermediate nutrient concentrations were likely found due to a combination of site-specific differences in environmental parameters, particularly water temperature, and intrinsic ecosystem properties, such as the taxonomic composition, abundance and (functional) activity of the shredder assemblage, in turn related to other environmental gradients. Indeed, within one of the Irish sites we found that >90% of the variance in litter decomposition rates was explained by litter resource quality and the degree of microbial conditioning (Fig. 6A), which in turn determined the density of large shredders that drove overall consumption rates (Fig. 7A). The communities at the extreme ends of the nutrient gradient will fall within a more constrained range of possible configurations than those in more mesotrophic waters, where enrichment effects are more modulated and community structure is determined to a larger extent by additional factors, such as hydrology, pH and the biogeographical filtering of the species pool.

In addition to the responses to nutrients, there was considerable scale-dependent spatial variation in decomposition rates. In particular, larger-scale effects reflected temperature differences among ecoregions but within ecoregions, where temperature differences among streams were minimal, changes in macroinvertebrate community structure were important. This is highlighted by the data from Ireland, which provided a useful subset of sites because they spanned the entire nutrient gradient across Europe (Fig. 6B). Condensing the principal gradient of invertebrate community structure across these sites into a single x-axis, via detrended correspondence analysis, enabled us to plot invertebrate community structure against decomposition rate (Woodward et al., 2012*). The species scores reflected the classic community response to eutrophication, with indicators of 'pristine' conditions (e.g. nemourid stoneflies) at one end of the gradient and indicators of organic pollution (e.g. *Asellus* isopods) at the other end. The moderately enriched sites exhibited the highest decomposition rates and were characterised by the largest consumer taxa (i.e. gammarid shrimps and limnephilid caddisflies). As a result, this integrated, biotic gradient yielded a clearer unimodal relationship than for nutrient concentration alone, with R^2 values of 0.96 and 0.95 for total and invertebrate-mediated decomposition, respectively. The abundance of large detritivores was the most powerful predictor of decomposition rate across the Irish sites (Fig. 7A), whereas no significant relationship emerged for the smaller, but far more abundant taxa such as stoneflies, suggesting that consumer body size was the key functional trait driving the patterns in the field and when multiplied by abundance this gave a measure of the total functional capacity of the assemblage, in ways analogous to what we found in the subsequent laboratory studies (e.g. Perkins et al., 2010; Reiss et al., 2011).

The concentrations of individual nutrients thus represented only part of what is a more complex and multifaceted stressor in the real world. Indeed, gradients in community structure, rather than nutrient concentrations per se, provide a more holistic measure of the trophic status of freshwaters, as evidenced by the widespread use of diversity indices rather than chemical analysis to detect organic pollution (Friberg et al., 2011; Karr, 1999; Wright et al., 2000). The problem with this approach, though, is that the species composition of communities differs across countries and ecoregions, due to underlying biogeographical effects, making it difficult to standardise taxonomic-based bioassessment metrics at large scales (see chapter "Recommendations for the next generation of global freshwater biological monitoring tools" by Jackson et al.; Wright et al., 2000).

It is instructive that the intermediate 'mesotrophic' study sites were characterised by shredders and that the peak decomposition rates corresponded to a subset of these sites with communities dominated by large taxa, particularly Gammaridae shrimps and Limnephilidae and other cased caddis (Figs 6B and 7A). This suggests that body size and abundance of the consumer assemblage is a more important driver of stream ecosystem functioning than other aspects of community structure, including species richness (e.g. Huryn et al., 2002). Indeed, decomposition rates in Ireland were some of the fastest recorded across the entire European range, yet this country has a relatively depauperate invertebrate fauna (Giller et al., 1998), with only about 70% of the species found in Great Britain, which in turn possesses a similar percentage of the species found in continental Europe. The importance of body size within ecological systems has become a major focus of recent food web research (Brose et al., 2005; Mancinelli et al., 2013; Woodward et al., 2005b), in contrast to the field of biodiversity–ecosystem functioning (B-EF), which has traditionally focused mainly on taxonomic diversity. These two previously disparate fields are now converging, and there is increasing interest in the role of functional diversity and species traits, such as body size, in the more recent literature (Woodward et al., 2005a,b, 2010).

A potential bias in the broader literature is that terrestrial grassland ecosystems have dominated much of the B-EF research until recently (e.g. Hector et al., 2002; Spehn et al., 2005; Tilman et al., 2014), whereas freshwaters may be fundamentally different. For example, the importance of vertical consumer–resource interactions, as opposed to horizontal competitive interactions, within aquatic food webs has been strongly emphasised (Kohler and Wiley, 1997; Petchey et al., 2004, Woodward, 2009). The fact that the most marked changes in decomposition rates occurred in the midportion of the nutrient gradient where structural measures of community composition are often least sensitive, suggests that this functional approach is a potentially important complement to the traditional structural-based biomonitoring techniques in current usage (Gessner and Chauvet, 2002*). The next logical step is to develop integrated structural–functional metrics for assessing ecological integrity that combine community descriptors (consumer traits and abundance) with ecosystem data (process rates). The advantage of using a trait-based structural–functional approach is that it is effectively independent of taxonomy, and could therefore be applied across large biogeographical regions (Dolédec et al., 1999; Friberg et al., 2011; Usseglio-Polatera et al., 2000; Woodward and Hildrew, 2002). This would be a major leap forward and would serve to unite the community and ecosystem approaches to ecology that have developed in parallel, but in relative isolation, over the last century.

Our results also raise fundamental questions about how to determine ecosystem health in the context of eutrophication. Firstly, water resource managers normally aspire to naturally low-nutrient conditions, and yet ecosystem functioning in such systems, as assessed by leaf litter decomposition rates, was indistinguishable from that of heavily polluted streams (Fig. 6B and C). This suggests that ensuring both low-nutrient water and effective resource use in stream food webs (from leaf litter to detritivores to fish) coupled with high process rates might be irreconcilable goals in stream management. Second, stream managers currently rely primarily on structural bioassessment measures to assess stream ecosystem health because they provide a reliable time-integrated response to stressors such as organic pollution or acidification. Biogeographical constraints can be overcome using litter decomposition to monitor nutrient loading because biogeography is a minor issue (for example, black alder or similar species of the genus are common throughout most of Europe and the Holarctic), and marked changes in decomposition rate occurred in the rising portion of the pollution gradient, in which established structural measures (such as metrics based on fish, invertebrate or algal communities) are typically least sensitive.

As in many other parts of the world, Europe is a highly industrialised, intensively managed continent, with a large proportion of the landscape characterised by agriculture and other human land-uses leading to the significant pollution of receiving freshwater environments (Hilton et al., 2006; Vörösmarty et al., 2010). Our study reveals that as Europe's freshwaters drift away from their natural conditions, along with rapid biodiversity losses, ecosystem functioning is changed, too—and on a continental scale. The natural controlling factors appear to have been altered so fundamentally that drivers and responses might not continue to operate in easily predictable ways. Given the huge uncertainties surrounding human environmental impacts it will be challenging to manage European surface waters sustainably and meet the demands of biodiversity conservation and environmental legislation. As eutrophication is recognised as one of the major threats to water quality throughout Europe, the outcomes of Workpackage 1 of the *RivFunction* project can play an important role in the context of overall social objectives of understanding and improving quality of the water resources in the future.

3. EFFECTS OF RIPARIAN FOREST MODIFICATIONS ON LEAF LITTER DECOMPOSITION

Small forest streams constitute the majority of water courses in undisturbed catchments (Downing et al., 2012) and are often densely shaded by riparian vegetation, deriving most of their energy and carbon from the decomposition of litter of terrestrial origin, rather than instream primary production (Wallace et al., 1997). Given their strong dependence on litter resources, aquatic communities and processes can be very sensitive to changes in riparian forests driven by forestry practices, invasion by exotic tree species or conversion to pasture. The replacement of native forests by plantations or invasion by exotic tree species has the potential to affect aquatic communities and processes, especially when invasive tree species or those used in plantations introduce novel traits or strongly alter the composition of functional traits (leading to litter of different quality), relative to an undisturbed assemblage (Kominoski et al., 2013). In such cases, even when a closed canopy is maintained, the allochthonous trophic pathway may still be affected (Hladyz et al., 2011b*; Murphy and Giller, 2001). When the native forest is clear-cut or converted into pasture, the increase in solar irradiation can promote instream primary production and autochthonous trophic pathways (Hladyz et al., 2011b*), while the reduction in organic matter input to streams can inhibit allochthonous trophic pathways, although aquatic communities may still retain the capacity to decompose terrestrially derived leaf litter (Hladyz et al., 2011b*)—thus there can be fundamental shifts between the roles of the green vs brown pathways at the base of the food web as a result of land-use change.

In this context, the overarching goal of *RivFunction* Workpackage 2 was to assess the impacts of riparian forest modifications on leaf litter decomposition in streams. Four broad types of forest change were assessed: (i) replacement of native forests by commercial tree plantations (deciduous plantations, conifer plantations and eucalyptus plantations), (ii) invasion of native forests by exotic tree species, (iii) forest clear cutting and (iv) replacement of native forests by pasture (Fig. 8). In the main coordinated litter decomposition experiment, each of 10 research teams selected 5 stream pairs (eight stream pairs for Ireland), each composed of a stream with native riparian forest (reference stream) and a partner with altered riparian vegetation (altered stream), making a total of 53 reference—altered stream pairs across Europe. A total of 2544 bags (10 regions × 10 streams (+6 in Ireland) × 2 leaf litter species × 2 mesh sizes × 6 replicates) were thus deployed in European streams for the main Workpackage 2 experiment. In contrast to the Workpackage 1 sites, all these streams were selected to be minimally impacted by agricultural run-off, sewage effluents or other human disturbances (apart from changes in riparian vegetation in altered streams), and thus variation in water chemistry primarily reflected regional context (e.g. geology, atmospheric nitrogen deposition). Within each pair, streams were as similar as possible regarding environmental characteristics other than riparian vegetation to isolate the effect of forest change. Riparian forests in reference streams were representative of the natural vegetation and riparian vegetation in altered streams was representative of the major anthropogenic alteration to riparian forests in each region (Table 1).

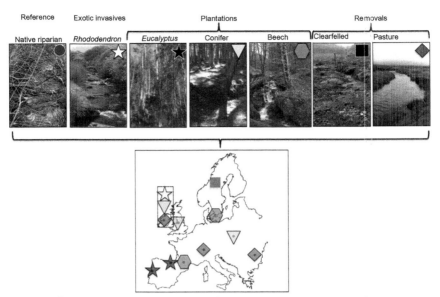

Fig. 8 Workpackage 2: The major riparian alterations investigated across Europe—note, some were unique to particular regions, others were repeated across Europe and in other instances several were studied in the same place (e.g. Ireland).

The experimental design used in Workpackage 2 was otherwise broadly similar to that used in Workpackage 1 (Fig. 3), bar the use of categorical treatments in the former as opposed to the gradient approach in the latter (cf. Fig. 6 vs Fig. 8). In some regions, detailed macroinvertebrate (e.g. taxa richness, density, biomass) and microbial (e.g. fungal biomass and reproductive activity) parameters associated with decomposing litter were also determined (Ferreira et al., 2015b*; Lecerf et al., 2005*; Riipinen et al., 2010*). However, to better understand the dynamics of litter decomposition and of associated parameters (e.g. macroinvertebrate and aquatic hyphomycete colonisation, litter chemical composition) under forest alteration, a complementary coordinated experiment was carried out simultaneously to the main litter decomposition experiment, in one reference—altered stream pair per region (10 reference—altered stream pairs across Europe). In this experiment, additional alder and oak litter bags were deployed in the streams, sampled on up to five occasions over time (i.e. the 'complete time series'), and AFDM remaining and associated variables were determined (Ferreira et al., 2006a*; Lecerf and Chauvet, 2008a*). Additional experiments were also carried out in three European regions to assess

Table 1 European Region and Terrestrial Ecoregion, and Riparian Vegetation in Altered Streams for the Main and Complementary Coordinated Experiments in *RivFunction* Workpackage 2

Region	Ecoregion[a]	Altered Streams	Additional Information
Central England	Celtic broadleaf forests	Conifer plantations	Riipinen et al. (2009*, 2010*) and Hladyz et al. (2011a)*
Southwestern France	Western European broadleaf forests	Beech plantations	Lecerf et al. (2005)*, Lecerf and Chauvet (2008a)* and Hladyz et al. (2011a)*
Western Ireland	Celtic broadleaf forests	Pasture	Hladyz et al. (2010*, 2011a*,b*)
	Celtic broadleaf forests	Conifer plantations	Riipinen et al. (2010)*
Southern Poland	Central European mixed forests	Conifer plantations	Riipinen et al. (2010)* and Hladyz et al. (2011a)*
Central Portugal	Cantabrian mixed forests	Eucalyptus plantations	Ferreira et al. (2006a*, 2015b*) and Hladyz et al. (2011a)*
Romanian Danube plains	Pontic steppe	Pasture	Hladyz et al. (2010*, 2011a*)
Northern Spain	Cantabrian mixed forests	Eucalyptus plantations	Ferreira et al. (2006a*, 2015b*) and Hladyz et al. (2011a)*
Northern Sweden	Scandinavian and Russian taiga	Forest clear cutting	McKie and Malmqvist (2009)* and Hladyz et al. (2011a)*
Southern Sweden	Sarmatic mixed forests	Beech plantations	Hladyz et al. (2011a)*
Swiss plateau	Alps conifer and mixed forests	Pasture	Hladyz et al. (2010*, 2011a*)

[a]Ecoregions defined on the basis of climatic, topographic and geobotanical European data (European Environment Agency; http://www.eea.europa.eu/data-and-maps/figures/dmeer-digital-map-of-european-ecological-regions).

the effects of invasion of riparian forests by exotic woody species of major concern in these regions on litter decomposition in streams (Hladyz et al., 2011b*; Lecerf et al., 2007a*) (Fig. 8).

When the 10 regions were considered together, modifications of riparian forests (i.e. replacement by conifer, beech and eucalyptus plantations, forest clear cutting and replacement by pasture) did not have an overall significant effect on leaf litter decomposition (Hladyz et al., 2011a*). However, strong effects associated with specific disturbances were observed in more regionally focused studies (Hladyz et al., 2011b*; McKie and Malmqvist, 2009*). These contrasting findings at the pan-European and regional scales are not surprising since different types of forest modification have contrasting effects on leaf litter provision and decomposition, which also depends on leaf litter species and type of decomposer community. Nevertheless, at the European scale, total litter decomposition was generally slower in altered than in reference streams, while microbially driven litter decomposition was not significantly affected by forest change (Hladyz et al., 2011a*) suggesting that these alterations affected macroinvertebrates more strongly. Also, the response of litter decomposition to forest change depended on the region (Hladyz et al., 2011a*), indicating that effects on litter decomposition were contingent upon the type of forest change, aquatic communities and/or environmental conditions. Differences between regions were exacerbated further when temperature-corrected litter decomposition rates (k, dd^{-1}) were considered, suggesting that differences other than in temperature are responsible for the observed differences among regions (Hladyz et al., 2011a*). The effects of different types of forest modification on litter decomposition in streams are discussed in more detail below.

3.1 Deciduous Broadleaf Plantations

The effects of the replacement of native deciduous broadleaf forests by deciduous broadleaf plantations on litter decomposition were assessed in south-western France where plantations of beech (*Fagus sylvatica* L.) have replaced large areas of native forests where hazel (*Corylus avellana* L.), chestnut (*Castanea sativa* Mill.), oaks (*Quercus petraea* (Mattuschka) Liebl, *Q. robur*) and beech were previously common (Lecerf et al., 2005*) (Table 2). Total oak leaf litter decomposition was inhibited in streams flowing through beech plantations (altered streams) compared with streams in native forests (reference streams), which was likely due to reduced shredder biomass in litter bags in the former. However, no significant differences were found for total alder leaf litter decomposition or both alder and oak microbially driven leaf litter decomposition between stream types (decomposition rates determined based on the simplified time-series experiment; Lecerf et al., 2005*). Aquatic hyphomycete species richness was also reduced in plantation compared with reference streams, but this did not

Table 2 Summary Table of the Literature Assessing the Effects of the Replacement of Native Deciduous Broadleaf Forests by Deciduous Broadleaf Plantations on Litter Decomposition in Streams

Reference[a]	Region	Litter Substrate	Decomposer Community[b]	No. Reference/ Altered Streams or Sites	Response to Forest Change[c]
*Lecerf et al. (2005)	South-western France	Alder leaves	Microbial	4/4	∼
			Total	4/4	∼
		Oak leaves	Microbial	4/4	∼
			Total	4/4	−
*Lecerf and Chauvet (2008a)	South-western France	Alder leaves	Total	1/1	−
			Microbial	1/1	−
Menéndez et al. (2013)	North-eastern Spain	Alder leaves	Total	3/3	+

[a]References marked with an '*' are derived from the *RivFunction* project.
[b]Total decomposer community: microbes + macroinvertebrates.
[c]Response of litter decomposition to forest change: −, significant inhibition of litter decomposition in altered streams; ∼, no significant effect of forest change on litter decomposition and +, significant stimulation of litter decomposition in altered streams.

translate into slower microbially driven litter decomposition in the plantation streams (Lecerf et al., 2005*), suggesting some degree of functional redundancy among fungal species (Dang et al., 2005*; Ferreira and Chauvet, 2012). However, reduced aquatic hyphomycete species richness may have limited leaf litter palatability to shredders, indirectly contributing to the lower total oak leaf litter decomposition in altered streams. No effect of forest change on alder leaf litter decomposition was found, possibly because alder litter formed 'islands' of high quality resource in a stream bed dominated by nutrient-poor beech litter, which could attract shredders over small microhabitat scales within a given stream (Lecerf et al., 2005*). A similar explanation was invoked by Menéndez et al. (2013) in relation to faster decomposition of alder leaf litter in streams flowing through plane (*Platanus hybrida* (*hispanica*) Brot.) plantations compared with reference streams.

When data from the complete time-series experiment was used, total and microbially driven alder leaf litter decomposition rates were slightly lower in the altered vs reference streams (Lecerf and Chauvet, 2008a*), which may have reflected the greater fungal species richness associated with decomposing litter per date in reference sites (Lecerf et al., 2005*).

Differences in the effect of beech plantations on alder leaf litter decomposition between simplified and complete time-series experiments demonstrated the need to consider multiple sampling dates when using leaf litter decomposition to assess impacts of forest change on stream ecosystem functioning, as results based on a single sampling date are likely conservative (Lecerf et al., 2005* vs Lecerf and Chauvet, 2008a*). Nevertheless, leaf litter decomposition was sensitive to forest change, showing its potential to be used as a bioassessment tool of stream functional integrity, although the small number of studies investigating the replacement of native forests by deciduous plantations on litter decomposition (Table 2) currently limits the development of specific management recommendations.

3.2 Conifer Plantations

The effects of replacing native broadleaf forests with conifer plantations on litter decomposition in streams were assessed in central England, western Ireland and southern Poland (Table 3). In England, conifer plantations are dominated by Sitka spruce (*Picea sitchensis* (Bong.) Carr.), Norway spruce (*Picea abies* (L.) H. Karst.) and Scots pine (*Pinus sylvestris* L.), while native broadleaf forests are dominated by oak, hazel (*C. avellana* L.), lime (*Tilia* spp.) and alder. In Ireland, Sitka spruce and Lodgepole pine (*Pinus contorta* Douglas) are common species in conifer plantations, and oak, alder, hawthorn (*Crataegus monogyna* Jacq.) and holly (*Ilex aquifolium* L.) are common in native forests. In Poland, the most common species in conifer plantations are Norway spruce and silver fir (*Abies alba* Mill.), while native forests are dominated by beech. No overall effect of the replacement of native broadleaf forests by conifer plantations on leaf litter decomposition in the standardised leaf litter bags was found when considering 13 reference—altered stream pairs in the three regions (Riipinen et al., 2010*). However, a significant vegetation × region interaction revealed that the effect of conifer plantations on leaf litter decomposition was contingent on the region (Riipinen et al., 2010*). It is important to note that this interaction subsequently disappeared after fitting pH as a covariable, indicating its overall importance in controlling litter decomposition (Riipinen et al., 2010*). In central England, total and microbially driven litter decomposition rates were faster in streams flowing through conifer plantations (altered streams) than through broadleaf forests (reference streams), which can be partially attributed to the greater shredder abundance in the former (Riipinen et al., 2009*,

Table 3 Summary Table of the Literature Assessing the Effects of the Replacement of Native Broadleaf Forests by Conifer Plantations on Litter Decomposition in Streams

Reference[a]	Region	Litter Substrate	Decomposer Community[b]	No. Reference/ Altered Streams or Sites	Response to Forest Change[c]
Whiles and Wallace (1997)	North Carolina, USA	White pine needles	Total	2/2	+
		Red maple leaves	Total	2/2	+
Riipinen et al. (2009*, 2010*)	Central England	Alder leaves	Microbial	5/5	+
			Total	5/5	+
		Oak leaves	Microbial	5/5	+
			Total	5/5	+
Riipinen et al. (2010)*	Western Ireland	Alder leaves	Microbial	3/3	−
			Total	3/3	∼
		Oak leaves	Microbial	3/3	−
			Total	3/3	∼
	Southern Poland	Alder leaves	Microbial	5/5	∼
			Total	5/5	+
		Oak leaves	Microbial	5/5	∼
			Total	5/5	∼
Hisabae et al. (2011)	Southwestern Japan (winter)	Japanese cedar needles	Total	1/1	+
		Fusazakura leaves	Total	1/1	+
	Southwestern Japan (summer)	Japanese cedar needles	Total	1/1	∼
		Fusazakura leaves	Total	1/1	∼
Martínez et al. (2013)	Northern Spain	Alder leaves	Total	3/3	−
		Pine needles	Total	3/3	∼

[a]References marked with an '*' are derived from the *RivFunction* project.
[b]Total decomposer community: microbes + macroinvertebrates.
[c]Response of litter decomposition to forest change: −, significant inhibition of litter decomposition in altered streams; ∼, no significant effect of forest change on litter decomposition and +, significant stimulation of litter decomposition in altered streams.

2010*). In southern Poland, total alder leaf litter decomposition was faster in altered than in reference streams, despite the greater abundance, species richness and biomass of shredders in reference streams (Riipinen et al., 2010*), which suggests that structure and function are not always closely linked. In western Ireland, there was a tendency for slower microbially driven leaf litter decomposition in altered than in reference streams (Riipinen et al., 2010*).

Differences in the magnitude and direction of the effect of conifer plantation on litter decomposition among regions suggest that the identity of the conifer species, local communities and/or environmental conditions may be of prime importance. Shredder species composition differed between vegetation types, with small stoneflies most strongly associated with conifer streams while broadleaved streams generally had a higher proportion of larger taxa, such as limnephilid caddisflies and gammarid shrimps (echoing the main drivers of breakdown seen in Workpackage 1), although the latter were excluded from sites with low pH. The maintenance of decomposition rates irrespective of shredder community composition suggested a high degree of functional redundancy: indeed, similar decomposition rates were observed between streams with high numbers of nemourids and those with only a few limnephilids or gammarids, suggesting that density compensation among consumers might stabilise process rates (Riipinen et al., 2009*).

Despite conifer plantations being the most common plantations worldwide, only five studies have so far addressed their effects on litter decomposition in streams (Hisabae et al., 2011; Martínez et al., 2013; Riipinen et al., 2009*, 2010*; Whiles and Wallace, 1997; Table 3). The effects vary within and among studies suggesting that they depend on the type of plantation, environmental context, identity of litter and type of decomposer community. Given the potential for effects of conifer plantations on stream processes, more studies are needed to support the development of future management recommendations.

3.3 Eucalyptus Plantations

The effects of the replacement of native deciduous broadleaf forests by eucalyptus (*Eucalyptus globulus* Labill.) plantations on litter decomposition in streams were assessed in central Portugal and northern Spain (Table 4). Eucalyptus plantations cover > 1.5 million ha in the Iberian Peninsula and in many cases these replace native deciduous broadleaf forests dominated

Table 4 Summary Table of the Literature Assessing the Effects of the Replacement of Native Broadleaf Forests by Eucalyptus Plantations on Litter Decomposition in Streams

Reference[a]	Region	Litter Substrate	Decomposer Community[b]	No. Reference/ Altered Streams or Sites	Response to Forest Change[c]
Pozo (1993)	Northern Spain	Alder leaves	Total	2/1	~
		Eucalyptus leaves	Total	2/1	+
Abelho and Graça (1996)	Central Portugal	Chestnut leaves	Total	3/3	−
		Eucalyptus leaves	Total	3/3	−
Molinero et al. (1996)	Northern Spain	Chestnut leaves	Total	2/2	~
		Oak leaves	Total	2/2	~
		Eucalyptus leaves	Total	2/2	~
Pozo et al. (1998)	Northern Spain	Alder leaves	Total	1/1	~
		Eucalyptus leaves	Total	1/1	~
Díez et al. (2002)	Northern Spain	Alder branches	Total	1/1	~
		Alder heartwood	Total	1/1	~
		Oak branches	Total	1/1	~
		Eucalyptus branches	Total	1/1	~
		Pine branches	Total	1/1	~
Bärlocher and Graça (2002)	Central Portugal	Chestnut leaves	Microbial	2/3	~
			Total	2/3	~
		Eucalyptus leaves	Microbial	2/3	~
			Total	2/3	~

Continued

Table 4 Summary Table of the Literature Assessing the Effects of the Replacement of Native Broadleaf Forests by Eucalyptus Plantations on Litter Decomposition in Streams—cont'd

Reference	Region	Litter Substrate	Decomposer Community	No. Reference/ Altered Streams or Sites	Response to Forest Change
*Ferreira et al. (2006a)	Central Portugal	Alder leaves	Total	1/1	∼
		Oak leaves	Total	1/1	∼
	Northern Spain	Alder leaves	Total	1/1	∼
		Oak leaves	Total	1/1	+[d]
Laćan et al. (2010)	California, USA	Native litter mixture	Total	3/3	∼
		Eucalyptus leaves	Total	3/3	∼
Larrañaga et al. (2014)	Northern Spain	Alder leaves	Total	2/2[e]	−
		Eucalyptus leaves	Total	2/2[e]	−
*Ferreira et al. (2015b)	Central Portugal	Alder leaves	Microbial	5/5	∼
			Total	5/5	−
		Oak leaves	Microbial	5/5	∼
			Total	5/5	∼
	Northern Spain	Alder leaves	Microbial	5/5	∼
			Total	5/5	−
		Oak leaves	Microbial	5/5	∼
			Total	5/5	∼

[a]References marked with an '*' are derived from the *RivFunction* project.
[b]Total decomposer community: microbes + macroinvertebrates.
[c]Response of litter decomposition to forest change: −, significant inhibition of litter decomposition in altered streams; ∼, no significant effect of forest change on litter decomposition and +, significant stimulation of litter decomposition in altered streams.
[d]Stimulation of litter decomposition in the altered stream attributed to a flood event (alder litter bags had all been sampled by this time).
[e]The reference and altered stream sites resulted from the experimental addition of native broadleaf and eucalyptus litter, respectively, in each of two streams.

by oak and chestnut. Total alder leaf litter decomposition was slower in streams flowing through eucalyptus plantations (altered streams) than through native forests (reference streams), which was attributed to lower macroinvertebrate and shredder colonisation in altered streams (Ferreira et al., 2015b*). Total oak leaf litter decomposition was not significantly affected by forest change (Ferreira et al., 2015b*), likely due to the lower contribution of macroinvertebrates to the decomposition of nutrient-poor litter (Hieber and Gessner, 2002). Microbially driven alder and oak leaf litter decomposition were generally not affected by forest change (Ferreira et al., 2015b*), despite differences in aquatic hyphomycete community structure between stream types (Bärlocher and Graça, 2002; Ferreira et al., 2006a*), suggesting again a degree of functional redundancy among microbes (Dang et al., 2005*). The replacement of native deciduous broadleaf forests with eucalyptus plantations had stronger negative effects on aquatic communities in central Portugal than in northern Spain (Ferreira et al., 2006a*, 2015b*), likely due to the drier climate in the former promoting summer droughts in eucalyptus streams and limiting the development of an understory of deciduous vegetation. The maintenance of a native riparian buffer may thus partially mitigate the negative effects of eucalyptus plantation on aquatic communities.

The effects of eucalyptus plantations on litter decomposition vary within and among studies (Abelho and Graça, 1996; Bärlocher and Graça, 2002; Díez et al., 2002; Ferreira et al., 2006a*, 2015b*; Laćan et al., 2010; Larrañaga et al., 2014; Molinero et al., 1996; Pozo, 1993; Pozo et al., 1998; Table 4), but a recent meta-analysis has found an overall 20% inhibition of litter decomposition in streams flowing through eucalyptus plantations compared with reference streams (Ferreira et al., 2016a). Eucalyptus plantations cover >20 million ha worldwide (Iglesias-Trabado et al., 2009), but their impacts on litter decomposition in streams have been addressed mostly in the Iberian Peninsula and thus generalisation to other regions is limited due to differences in the type of native forest, eucalyptus species used in plantations, climate, etc. A new collaborative experiment is currently underway to assess the effects of the replacement of native forests by eucalyptus plantations on alder leaf litter decomposition in streams in eight locations distributed across seven countries in the Iberian Peninsula, East Africa and South America to wider our understanding of the effects of eucalyptus plantation on stream functioning.

3.4 Invasion of Riparian Areas by Exotic Woody Species

The effects of the invasion of riparian forests by exotic woody species were assessed in northern England, south-western France and western Ireland, where native riparian vegetation is being invaded in the former two regions by Japanese knotweed (*Fallopia japonica* (Houtt.) Ronse Decr.) and in the latter by Rhododendron (*Rhododendron ponticum* L.) (Table 5). Lecerf et al. (2007a)* found no significant differences in total oak and knotweed leaf litter decomposition between a reference stream and a stream flowing through a riparian forest invaded by knotweed (altered stream) in north-central England, likely due to the low level of invasion, while leaf litter decomposition was stimulated in the altered stream in south-western France likely due to the greater abundance of larger shredders in the invaded stream. Hladyz et al. (2011b)* found slower total alder and oak leaf litter decomposition in streams where riparian forests were invaded by *Rhododendron* (altered streams) than in reference streams, reflecting the overall lower shredder abundance in altered streams (Fig. 9). In contrast, total *Rhododendron* leaf litter decomposition was not significantly affected by forest change, probably because its decomposition was already slow in reference streams as a result of the overriding effect of poor resource quality (Hladyz et al., 2011b*). Thus, even though the canopy cover was maintained in altered streams, the allochthonous trophic pathway was negatively affected by *Rhododendron* invasion of riparian forests (Hladyz et al., 2011b*). As altered streams have closed canopies resulting from dense stands of *Rhododendron*, the autochthonous trophic pathway was also negatively affected (Hladyz et al., 2011b*).

The invasion of riparian forests by exotic plant species is a serious problem in many regions of the world (e.g. Friedman et al., 2005; Lorenzo et al., 2010), which can affect aquatic systems by multiple pathways (Hladyz et al., 2011b*; Roon et al., 2014; Schulze and Walker, 1997; Serra et al., 2013). However, results from the few studies that have addressed the effects of the invasion of riparian forests by exotic tree species on litter decomposition in streams are conflicting (Table 5), which suggests that the effect may depend on multiple factors, e.g. the identity of invasive species and/or of the quality of decomposing litter, type of decomposer community involved in the process, environmental conditions, etc. Thus, there is urgent need to increase our knowledge on the response of aquatic communities and processes to plant species invasions to better manage aquatic resources (Hladyz et al., 2011b*).

Table 5 Summary Table of the Literature Assessing the Effects of the Invasion of Native Forests by Exotic Woody Species on Litter Decomposition in Streams

Reference[a]	Region	Litter Substrate[b]	Decomposer Community[c]	No. Reference/ Altered Streams or Sites	Response to Forest Change[d]
Schulze and Walker (1997)	South Australia	Redgum leaves	Microbial	3/3	∼
			Total	3/3	∼
		*Weeping willow leaves	Microbial	3/3	∼
			Total	3/3	∼
*Lecerf et al. (2007a)	North-central England	Oak leaves	Total	1/1	∼
		*Japanese knotweed leaves	Total	1/1	∼
	South-western France	Oak leaves	Total	1/1	+
		*Japanese knotweed leaves	Total	1/1	+
*Hladyz et al. (2011b)	Western Ireland	Alder leaves	Total	3/3	−
		Oak leaves	Total	3/3	−
		*Rhododendron leaves	Total	3/3	∼
Roon et al. (2014)	Alaska, USA	Thin-leaf alder leaves	Total	1/1	−
		*European bird cherry leaves	Total	1/1	−

[a]References marked with an '*' are derived from the *RivFunction* project.
[b]Substrates marked with an '*' originated from the exotic invader.
[c]Total decomposer community: microbes + macroinvertebrates.
[d]Response of litter decomposition to forest change: −, significant inhibition of litter decomposition in altered streams; ∼, no significant effect of forest change on litter decomposition and +, significant stimulation of litter decomposition in altered streams.

3.5 Forest Clear Cutting

The effects of forest clear cutting on leaf litter decomposition were assessed in northern Sweden (Table 6). Leaf litter decomposition was stimulated in

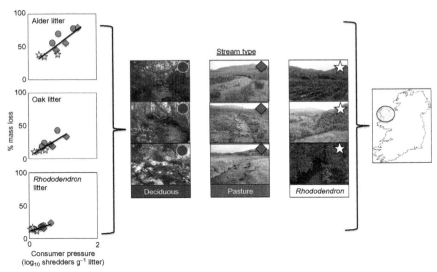

Fig. 9 Decomposition of different litter types in different stream types in Ireland, representing the three major land uses in the locale—native deciduous woodland, land cleared for unimproved pasture, and *Rhododendron* invasion. Note, in each of the nine streams invertebrate consumer abundance was quantified per gram of litter (cf. Fig. 7A), as well as mass loss, which revealed strong differences due to both bottom-up effects of resource quality and top-down effects. Redrawn after *Hladyz, S., Tiegs, S.D., Gessner, M.O., Giller, P.S., Rîşnoveanu, G., Preda, E., Nistorescu, M., Schindler, M., Woodward, G., 2010. Leaf-litter breakdown in pasture and deciduous woodland streams: a comparison among three European regions. Freshw. Biol. 55, 1916–1929.*

streams flowing through forest clear-cuts (altered streams) compared with streams flowing through old mixed boreal forests (reference streams), mostly for alder in coarse-mesh bags (McKie and Malmqvist, 2009*). No significant differences were found for macroinvertebrate abundance, diversity, assemblage composition or functional feeding groups abundances and species densities between reference and altered streams (except for scraper species density that was higher in reference streams), suggesting that macroinvertebrate community structure was not tightly coupled to variability in leaf litter decomposition (McKie and Malmqvist, 2009*). Rather, higher decomposition rates in the clear-cut streams were associated with an increase in decomposition efficiency by microbes and shredders compared with reference streams (McKie and Malmqvist, 2009*). Notably, this increase in decomposition efficiency occurred even though mean temperatures were actually lower in the clear-cut sites during the study period. This can be explained by the joint effects of three variables that differed between clear-cut and forested streams: increased nutrient concentrations, a shift in the

Table 6 Summary Table of the Literature Assessing the Effects of Forest Logging on Litter Decomposition in Streams

Reference[a]	Region	Litter Substrate	Decomposer Community[b]	No. Reference/ Altered Streams or Sites	Response to Forest Change[c]
Benfield et al. (1991)	South Appalachian Mountains, USA	Dogwood leaves	Total	1–3/1	+
		Red maple leaves	Total	1–3/1	+
		White oak leaves	Total	1–3/1	+
		Rhododendron leaves	Total	1–3/1	+
Kreutzweiser et al. (2008)	Canada	Speckled alder leaves	Total	9/12	−
*McKie and Malmqvist (2009)	Northern Sweden	Alder leaves	Microbial	5/5	+
			Total	5/5	+
		Oak leaves	Microbial	5/5	+
			Total	5/5	+
Lecerf and Richardson (2010b)	Canada	Red alder leaves	Total	13/3	−

[a]References marked with an '*' are derived from the *RivFunction* project.
[b]Total decomposer community: microbes + macroinvertebrates.
[c]Response of litter decomposition to forest change: −, significant inhibition of litter decomposition in altered streams and +, significant stimulation of litter decomposition in altered streams.

composition of litter inputs, and increased shredder biomass. Firstly, phosphate concentrations were slightly greater in the clear-cut streams (McKie and Malmqvist, 2009*), which might have stimulated decomposition from the bottom-up by favouring increased microbial activity (Ferreira et al., 2006c*, 2015a; Gulis and Suberkropp, 2003; Robinson and Gessner, 2000). Secondly, benthic litter standing stocks in the clear-cut streams were dominated by broadleaf (*Betula* spp.) litter, while the forested streams were dominated by refractory conifer needles, reflecting the dominance of birch saplings in the recovering riparian vegetation of the clear-cut streams. This greater incidence of broadleaf litter coupled with higher phosphorus concentrations together likely resulted in greater availability of nutrient rich and palatable litter in the clear-cut streams, in turn explaining why shredder biomass was overall higher in these streams (McKie and Malmqvist, 2009*). Higher shredder biomass in turn increased the resource-processing potential of detritivore assemblages, providing a further potential explanation for elevated decomposition rates in the clear-cut streams. Additionally, the potential increase in primary production in clear-cut streams may have stimulated litter decomposition by the release of labile carbon that could have stimulated the use of leaf litter by decomposers in a case of priming effect (Danger et al., 2013). Increased primary production may have also contributed to the increased shredder biomass at clear-cut streams if these were feeding on algal resources associated with decomposing litter (Franken et al., 2005).

Again, there are conflicting results among studies addressing the effects of forest harvest on litter decomposition in streams (Benfield et al., 1991, 2001; Kreutzweiser et al., 2008; Lecerf and Richardson, 2010a; McKie and Malmqvist, 2009*; Table 6) suggesting that effects are context dependent and in particular related to the clear-cut type (deciduous/broadleaf vs coniferous). A recent meta-analysis addressing the effects of forest harvest on several stream parameters also found contradictory results among primary studies, i.e. negative and positive responses of the same parameter to forest harvest among studies, highlighting the 'need to consider site-specific mechanisms by which such changes occur' (Richardson and Béraud, 2014).

3.6 Pasture

The effects of the long-term conversion of native broadleaf forests to grazing pasture on leaf litter decomposition were assessed in western Ireland, the Romanian Danube plains and the Swiss plateau where this type of forest change is widespread (Hladyz et al., 2010*) (Fig. 8; Table 7). When the three regions were considered together, no significant overall effects of forest change on litter decomposition were found (Hladyz et al., 2010*). However, in some regions macroinvertebrate-driven leaf litter decomposition was faster in streams flowing through deciduous forests (reference streams) and microbially driven leaf litter decomposition was faster in streams flowing through pastures (altered streams), although a significant difference in leaf litter decomposition between stream types was found only for total alder leaf litter decomposition—which was slower in altered than in reference streams in the Swiss plateau (Hladyz et al., 2010*). This suggests a shift in the relative contribution of macroinvertebrates and microbes to leaf litter decomposition between reference and altered

Table 7 Summary Table of the Literature Assessing the Effects of the Conversion of Native Forests to Pasture on Litter Decomposition in Streams

Reference[a]	Region	Litter Substrate	Decomposer Community[b]	No. Reference/ Altered Streams or Sites	Response to Forest Change[c]
Bird and Kaushik (1992)	Canada	Maple leaves	Total	1/1	~
Danger and Robson (2004)	South-eastern Australia	Eucalyptus leaves	Total	2/2	~
Encalada et al. (2010)	North-western Ecuador	Alder leaves	Microbial	3/3	~
			Total	3/3	−
		Guaba leaves	Microbial	3/3	~
			Total	3/3	−
*Hladyz et al. (2010)	Western Ireland	Alder leaves	Microbial	5/5	~
			Total	5/5	~
		Oak leaves	Microbial	5/5	~
			Total	5/5	~
	Romanian Danube plains	Alder leaves	Microbial	5/5	~
			Total	5/5	~
		Oak leaves	Microbial	5/5	~
			Total	5/5	~
	Swiss plateau	Alder leaves	Microbial	5/5	~
			Total	5/5	−
		Oak leaves	Microbial	5/5	~
			Total	5/5	~
*Hladyz et al. (2011b)	Western Ireland	Alder leaves	Total	3/3	~
		Oak leaves	Total	3/3	~
		Rhododendron leaves	Total	3/3	~

[a]References marked with an '*' are derived from the *RivFunction* project.
[b]Total decomposer community: microbes + macroinvertebrates.
[c]Response of litter decomposition to forest change: −, significant inhibition of litter decomposition in altered streams; ~, no significant effect of forest change on litter decomposition and +, significant stimulation of litter decomposition in altered streams.

streams. In fact, the ratio between macroinvertebrate- and microbially driven leaf litter decomposition was generally greater in reference than in altered streams, suggesting that reduced performance by macroinvertebrates in altered streams is compensated for by stimulated microbial activity (Hladyz et al., 2010*). These results are consistent with those of some previous studies comparing streams flowing through forest and pasture with comparable water characteristics (Ontario: Bird and Kaushik, 1992; Australia: Danger and Robson, 2004; Ecuador: Encalada et al., 2010; Table 7).

Altogether these results suggest that decomposition rates may, in some cases, not be very useful as simple indicators of stream functional impairment due to forest change, whereas the ratio between macroinvertebrate- and microbially driven leaf litter decomposition may be more sensitive to environmental change (Gulis et al., 2006*; Hladyz et al., 2010*), and there are some clear similarities, as well as notable differences, to the responses to nutrient enrichment in the other large-scale bioassay (e.g. Fig. 10 for comparison of main results from the Irish sites for both Workpackages).

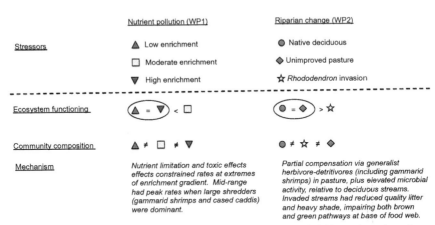

Fig. 10 Summary of contrasts of results from the two major field-based Workpackages (1 and 2) in Ireland, where both consumer assemblages (structure) and decomposition rates (function) were quantified. Note identical rates were observed in highly polluted vs pristine conditions, which were both markedly lower than in moderately enriched sites, with community composition changing progressively along this gradient. Under riparian change, there was similar but less extreme patterns of community turnover, but here pasture and woodland sites had comparable breakdown rates, and both were higher than those invaded by *Rhododendron*.

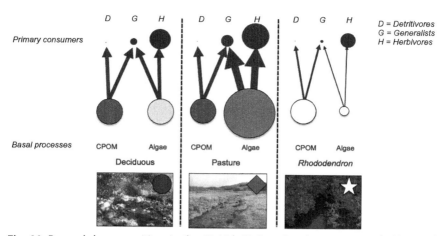

Fig. 11 Beyond decomposition: in the 10 Irish *RivFunction* sites we expanded beyond focusing on decomposition in the 'brown pathways' of the food web to include algal processes in the 'green pathways' and the links to the consumers of both sets of dominant basal resources, as a function of riparian land-use change. Here, we found evidence of compensation for the partial loss of detrital pathways in pasture streams, via elevated algal production and resource quality, whereas in *Rhododendron* streams both pathways were severely compromised. This was due to reduced resource quality and availability, due to the high C:N content of the litter and the heavy shade cast by the canopy: the transfer of energy through the food web therefore collapsed as the compensation of one pathway by the other was not possible. The area of the *circles* are scaled by standing stocks of consumers and resources, the *arrows* by process rates as measured using litter bags and algal colonisation tiles, both in the presence and absence of macroinvertebrates. *Redrawn after Hladyz, S., Tiegs, S.D., Gessner, M.O., Giller, P.S., Rîșnoveanu, G., Preda, E., Nistorescu, M., Schindler, M., Woodward, G., 2010. Leaf-litter breakdown in pasture and deciduous woodland streams: a comparison among three European regions. Freshw. Biol. 55, 1916–1929.*

It is also clear that the brown and green pathways in the food web are often intimately connected and one may partly compensate for impairment of the other, especially if generalist consumers are present that can access both major basal energy inputs (Fig. 11).

4. BIODIVERSITY-RELATED MECHANISMS UNDERLYING ALTERED LITTER DECOMPOSITION

Accurate diagnosis of the underlying causes of changes in functional integrity, whether under anthropogenic or natural disturbance, requires a sound mechanistic understanding of the abiotic and biotic factors driving variation in ecosystem processes (Truchy et al., 2015). A key component of the *RivFunction* framework is the idea that anthropogenic pressures can alter leaf decomposition both by affecting the activities and hence resource-processing potential of decomposing organisms, and by favouring consumer species with intrinsically different resource-processing characteristics. Such trait-mediated and indirect effects could also arise from a loss of diversity, which could impair ecosystem functioning if key species are lost, or if important interactions among species or groups of organisms that underpin ecosystem processes are compromised (Gessner et al., 2010; Truchy et al., 2015).

Much of the variation in leaf litter decomposition along anthropogenic stressor gradients can be attributed to direct influences of abiotic drivers on activities of organisms, such as the initially positive effects of increasing nutrients and temperature on processing rates (Ferreira and Chauvet, 2011; Gulis and Suberkropp, 2003; Salinas et al., 2011). Even so, clear associations between changes in biodiversity and community composition of detrital food webs and rates of leaf decomposition were frequently detected in the *RivFunction* field studies. For example, declines in leaf litter decomposition seen at very high levels of nutrient enrichment appeared partly driven by the loss of (large) detritivore species (Woodward et al., 2012*). Disentangling interactions between abiotic and biotic drivers are often challenging, but may be crucial for interpreting and managing the effects of human disturbances on ecosystem functioning (Frainer and Mckie, 2015; Giller, 2005; McKie et al., 2006*). In a study of the effects of a stream-liming program on acid streams in Sweden, microbially mediated decomposition was stimulated by the addition of calcium, which otherwise limits microbial activity in these systems (McKie et al., 2006*). However, this increase was completely offset by reductions in detritivore-driven litter decomposition in limed stream sections, which was associated with declines in the species richness and evenness of shredders, and an increase in the dominance of less efficiently feeding stoneflies at the expense of more efficient caddisflies (McKie et al., 2006*). The net result was that overall rates of leaf decomposition did not differ between limed and unlimed stream sections, although it would be erroneous to conclude that there were no direct or indirect effects of liming on the decomposition process.

A limitation of field-based studies is that while associations between ecosystem functioning and community composition and diversity can be detected, often with the aid of sophisticated statistical approaches, such associations remain essentially correlative (Frainer and McKie, 2015).

A substantial portion of the research conducted within *RivFunction* focused on the role of biodiversity, with Workpackage 3 dedicated to understanding the roles played by the community composition and diversity of litter-decomposing invertebrates, microbes and of the litter resource itself, in regulating decomposition rates.

The approach taken in *RivFunction* for investigating these relationships drew heavily on the theoretical and empirical framework developed within the wider field of B-EF research. Until *RivFunction* was launched in 2002, B-EF research had focused on two main questions: (1) is there a general relationship between increasing biodiversity of producers or consumers and key ecosystem processes and (2) what is the importance of biodiversity per se relative to the presence of particular species for ecosystem functioning (Loreau et al., 2002)? A consensus on both issues was developing, which would be reflected subsequently in key meta-analyses and review papers (e.g. Balvanera et al., 2006; Cardinale et al., 2006, 2007): increasing biodiversity was often, but not universally, associated with increasing ecosystem process rates (especially the primary productivity of grassland plants), and these relationships were often driven by both nonadditive effects of multiple species (i.e. complementarity) and the presence of particular, highly influential species (i.e. the selection effect). However, more subtle and sophisticated questions were increasingly being posed when *RivFunction* was launched, including those related to the roles of other measures of biodiversity than species richness (e.g. evenness, functional diversity), interactions between biodiversity and other environmental variables, and the role of biodiversity in maintaining the stability of ecosystem functioning, including under environmental stress (Cardinale and Palmer, 2002; Giller et al., 2004b; Loreau et al., 2002; Wilsey and Polley, 2004). These and other questions informed the development of B-EF research in *RivFunction* (Fig. 12).

Each of the three main groups involved in the decomposition of litter—macroinvertebrate detritivores, aquatic hyphomycetes and the litter resource itself—were considered in turn, none of which had been extensively studied previously in B-EF research. The role of aquatic hyphomycetes in particular had been barely investigated and remains limited to this day. The logistical and statistical challenges involved in working with these organism groups were substantial, whether it be in restraining mobile aquatic insects in realistic experimental units or inoculating leaf discs with known spores of different fungal species, all at predefined levels of biodiversity. A further challenge arose from the impossibility of exactly

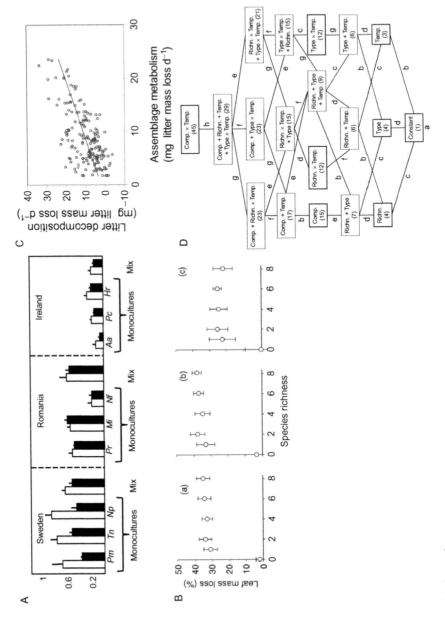

Fig. 12 See legend on next page.

quantifying the contribution of individual consumer species to decomposition rates in species mixture, complicating the assessment of the effects of diversity per se vs individual species (e.g. by using the Loreau and Hector (2001) diversity effect partition). Finally, litter decomposition is an integrative process involving the interplay among organisms at different trophic levels, and capturing this complexity is logistically very challenging in manipulative experiments, not least because of the degree of replication required to vary diversity across two or more trophic groups (Jabiol et al., 2013). To overcome these and other challenges, *RivFunction* employed a number of novel experimental designs and innovative statistical approaches.

Results from the *RivFunction* diversity experiments were a component of the landmark paper by Gessner et al. (2010), which focused on reviews of B-EF relationships separately at leaf litter, microbial, and detritivore levels, and multitrophic relationships. Here, we discuss the *RivFunction* results in the context of two broad research questions: (1) *Is there a general relationship between biodiversity and leaf decomposition, and which aspects of diversity are most important?* (2) *How does biodiversity influence the stability of decomposition rates, including under variable environmental conditions?* This section

Fig. 12 The ever increasing complexity of biodiversity–ecosystem functioning experiments (Workpackage 3), during and after *RivFunction*. Here we show experiments involving a range of invertebrate shredder and microbial species. The early Workpackage 3 studies revealed evidence of both idiosyncracy and redundancy among and within assemblages, with weak or inconsistent richness effects (Panels A and B), though effects of species evenness were stronger, but still inconsistent (McKie et al., 2008*). Subsequent work that followed after *RivFunction* highlighted how body size and the metabolic capacity of the assemblage could account for much (>40%) of this variance, irrespective of richness (Panel C). More recent work suggested richness was more important when multiple processes were considered, especially in a heterogeneous environment, using far more complex experimental designs (Panel D). *Panel A: A subset of data redrawn after McKie, B.G., Woodward, G., Hladyz, S., Nistorescu, M., Preda, E., Popescu, C., Giller, P.S., Malmqvist, B., 2008. Ecosystem functioning in stream assemblages from different regions: contrasting responses to variation in detritivore richness, evenness and density. J. Anim. Ecol. 77, 495–504. Panel B: Dang, C.K., Chauvet, E., Gessner, M.O., 2005. Magnitude and variability of process rates in fungal diversity-litter decomposition relationships. Ecol. Lett. 8, 1129–1137. Panel C: Perkins, D.M., Mckie, B.G., Malmqvist, B., Gilmour, S.G., Reiss, J., Woodward, G., 2010. Environmental warming and biodiversity-ecosystem functioning in freshwater microcosms: partitioning the effects of species identity, richness and metabolism. Adv. Ecol. Res. 43, 177–209. Panel D: Perkins, D.M., Bailey, R.A., Dossena, M., Gamfeldt, L., Reiss, J., Trimmer, M., Woodward, G., 2015. Higher biodiversity is required to sustain multiple ecosystem processes across temperature regimes. Glob. Chang. Biol. 21, 396–406.*

concludes with a discussion of implications of the findings of Workpackage 3 for development of leaf decomposition as a tool in environmental assessment.

4.1 Is There a General Relationship Between Biodiversity and Ecosystem Functioning, and Which Aspects of Biodiversity Are Most Important?

In line with most previous B-EF research, the *RivFunction* biodiversity manipulations focused primarily on whether general relationships between increasing biodiversity and ecosystem functioning could be detected (cf. Gessner et al., 2010). This was motivated by the large number of drivers of diversity loss in streams that could potentially affect organisms within the detrital food web (Gessner et al., 2010). However, many of the *RivFunction* studies emphasised other aspects of biodiversity than species richness within a single functional guild, such as evenness and diversity across multiple trophic levels, and the idea of species traits was also investigated (McKie et al., 2008*; Sanpera-Calbet et al., 2009*; Schindler and Gessner, 2009*).

Broadly, the functional traits of an organism are those that regulate its responses to environmental variability and its effects on ecosystem processes (Violle et al., 2007). The potential for species traits to underpin a predictive framework linking environmental and community change with ecosystem processes is being recognised increasingly (Reiss et al., 2009; Truchy et al., 2015), but research on these linkages was still nascent when *RivFunction* started. Similarly, relationships between species evenness and ecosystem functioning remain poorly investigated for most ecosystem processes, even though human activities affect relative abundances of species far more often than their presence (Chapin et al., 2000). Increases in species dominance can in turn increase the concentration of particular traits associated with those species in functional guilds, which may either increase the importance of those traits for functioning per se, or alter the strength of key species interactions associated with them (Hillebrand et al., 2008; Truchy et al., 2015). Finally, although most B-EF studies focused on variation in diversity within a single trophic level, interactions across levels can shape the effects of species loss on ecosystem functioning (Raffaelli et al., 2002). Biodiversity changes within one trophic level not only affect species richness and composition at other levels, but can also alter ecosystem processes, via both top-down and bottom-up drivers (Jabiol et al., 2013; Mancinelli and Mulder, 2015; Srivastava et al., 2009).

The potential importance of the species evenness of detritivores for litter decomposition was demonstrated by Dangles and Malmqvist (2004) in their analysis of decomposition datasets from field studies conducted in France and Sweden. They found that decomposition rates generally increased with increasing species richness, but also observed contrasting relationships between assemblages of differing evenness (Dangles and Malmqvist, 2004). Leaf litter decomposition rates were always higher in the dominated communities, but also plateaued sooner, with no further increases in species richness effects (i.e. redundancy) beyond four species. In contrast, although decomposition rates were lower in more even communities, rates were still increasing at higher levels of richness (6–7 species). Dangles and Malmqvist (2004) further documented spatiotemporal variation in species dominance patterns, which was associated with substantial variability in decomposition rates, especially across seasons.

McKie et al. (2008)* expanded on this work by conducting a microcosm experiment in Sweden, Romania and Ireland to investigate effects of the species richness, evenness and density of detritivores on leaf decomposition. Assemblage composition at each laboratory was chosen to reflect the main species comprising natural stream communities in the autumn. Across all three experiments, there was no general relationship between increasing diversity and leaf decomposition rates (Fig. 12; McKie et al., 2008*). Rather, effects of both species richness and evenness varied according to the composition of each species pool.

Decomposition in the Romanian study was generally enhanced as richness—but not evenness—increased, whereas in Ireland it was affected by evenness but not richness, with the effects of the former further depending on the identity of the dominant species. There was no relationship between any diversity parameter and decomposition rates in Sweden. The mechanisms underlying these diversity effects contrasted markedly among the regions. The Romanian results were predominantly attributable to the selection effect (i.e. driven by the presence of particular species), whereas there was evidence for positive complementarity among species in Ireland, particularly when the isopod crustacean *Asellus aquaticus* was dominant. Overall, the occurrence of B–EF relationships depended on the degree of taxonomic heterogeneity within each assemblage: the Swedish assemblage consisted of three closely related stoneflies, while the other experiments consisted of detritivores from two orders (McKie et al., 2008*). This points towards the importance of heterogeneity in the functional traits of species in regulating the occurrence and strengths of relationships between biodiversity and ecosystem functioning, a key topic of later B-EF research (e.g. Flynn et al., 2011; Frainer and McKie, 2015; Frainer et al., 2014).

Schindler and Gessner (2009)* focused more explicitly on the role of species trait heterogeneity in regulating relationships between the diversity of the litter resource and decomposition rates. Litter packs of species that contrast strongly in nutrient concentration or structural carbon compounds might decompose more rapidly than single-species packs, due to detritivores optimising nutrient and energy acquisition among multiple species, or microbially mediated transfers of nutrients from nutrient rich to recalcitrant species (Gessner et al., 2010). Both mechanisms could potentially suggest species complementarity enhances ecosystem process rates (e.g. Cardinale, 2011; Tylianakis et al., 2008). However, despite the plausibility of these mechanisms, subsequent studies have generally not found consistent effects of increasing litter diversity per se on decomposition (Gessner et al., 2010). Nevertheless, particular species combinations have been associated with nonadditive outcomes for decomposition (Handa et al., 2014; Lecerf et al., 2007b*, 2011), with higher or lower rates than derived from the component species in isolation. This suggests that there may be specific combinations of litter species traits that might be particularly influential in regulating decomposition, and that more consistent effects of diversity might emerge when it is quantified in its 'functional' rather than taxonomic form.

Schindler and Gessner (2009)* studied nine deciduous tree species, which were categorised into three categories (three species per category) according to their expected decay rate, as the focal functional 'effect' trait: fast, medium and slow-decomposing species. They also quantified differences in litter chemistry traits: phosphorus, nitrogen and lignin concentration. They mixed the 9 species in a total of 40 species combinations, including replication of 'homogenous' (where all species represent the same litter decay class) and 'heterogeneous' (where the species are drawn from different decay classes). Contrary to their hypotheses, they did not find generally elevated rates of litter decomposition between heterogeneous and homogenous litter packs, with decomposition rates largely controlled by a trait–litter lignin concentration. However, they did find that the most recalcitrant and most labile species decomposed slower and faster, respectively, in litter mixtures comprising different decay categories than in homogenous mixtures, or in single-species litter bags. Schindler and Gessner's (2009*; also see Frainer et al., 2015) results thus point towards the value of litter traits in a framework for predicting not only purely additive effects on decomposition, but also nonadditive effects arising from particular litter species combinations (e.g. Handa et al., 2014).

Sanpera-Calbet et al. (2009)* specifically examined the interaction across trophic levels, between diversity of litter resources vs that at microbial and detritivore consumer levels. They found that for two rapidly decomposing species, hazel (*C. avellana*) and ash (*Fraxinus excelsior* L.), rates were strongly elevated when either or both species were mixed with a refractory species, beech (*F. sylvatica*). Intriguingly, this interaction was not driven primarily by the qualities of beech as a food resource, but rather as habitat, and as a material for the construction of protective cases by particular group of highly efficient consumers in the limnephilid caddisfly genus *Potamophylax* (Sanpera-Calbet et al., 2009*). Litter packs including beech supported higher than expected abundances and biomasses of leaf-shredding invertebrates, including higher abundances of *Potamophylax*, and this resulted in accelerated decomposition of other species mixed with beech (but not beech itself). This represents a form of facilitation, whereby beech leaf litter facilitates the activities of a key consumer and thereby enhances the decomposition process (Bruno et al., 2003; Gessner et al., 2010). There was no evidence that interactions between litter species composition and microbial consumers influenced decomposition rates.

The importance of the diversity of aquatic fungi was also investigated in *RivFunction*. There is particularly strong potential for increasing microbial species diversity to elevate decomposition rates, since species possess different litter degrading enzymes which may complement one another, and some types of microbes may facilitate penetration of the litter matrix by others (Gessner et al., 2010). Despite this, studies conducted within *RivFunction* and elsewhere have found scant evidence that increases in fungal diversity have consistent general effects on microbially mediated decomposition, although a laboratory microcosm study (Lecerf et al., 2005*) found increasing aquatic hyphomycete diversity stimulated consumption rates by the crustacean *Gammarus fossarum* Koch. This finding suggests that complementary interactions and chemical or functional differences among multiple fungal species increase the availability of nutrients or otherwise enhance the palatability of the litter to *Gammarus*, and further emphasise the importance of interactions across trophic levels for regulating B-EF relationships (Lecerf et al., 2005*). This was even more thoroughly investigated in an experiment partly conceived during the *RivFunction* project, and ultimately conducted by Jabiol et al. (2013). They found that rates of leaf litter decomposition were maximised when diversity of both fungi and invertebrate decomposers was the highest, and when fish predator cues were present, with the cumulative effects of species loss within and across trophic levels reducing process rates. This partly reflected both bottom-up effects of fungal diversity and top-down effects of fish cues on the performance of detritivores, including increased feeding by a caddisfly species which was both an efficient leaf-shredder and predation-resistant (Jabiol et al., 2013). Results from this and other experiments (e.g. O'Connor and Donohue, 2013; Perkins et al., 2015), suggest that functional ecosystem impairment resulting from widespread biodiversity loss could be more severe than inferred from previous experiments confined to varying diversity within single trophic levels (Lecerf and Richardson, 2010a).

Other aspects of the functional diversity of litter, microbes and detritivores were also assessed within RivFunction. Lecerf and Chauvet (2008b)* focused on intraspecific diversity of a single, key riparian species—*A. glutinosa*—collected from five widely spaced source populations across Europe. They found wide variation in phosphorus and lignin concentrations, which together explained much of the variability in litter decomposition rates. Significantly, intraspecific variation in leaf decomposition rate was within a similar range to that reported for interspecific variation among cooccurring riparian plant species in Europe (Lecerf and Chauvet, 2008b*). Rather than litter nutrient concentrations per se, Hladyz et al. (2009)* focused on the degree of stoichiometric imbalance between the nutrient ratios (e.g. C:N and C:P) of different litter species and the stoichiometric requirements of the main shredder consumers (Fig. 6A). They exposed different single-species litter bags in a stream field experiment, and quantified nutrient ratios of both leaf litter and consumers, and found that litter stoichiometric ratios indeed predicted a significant portion of variability in decomposition, with decomposition rates generally declining as the degree of stoichiometric imbalance with the main consumers in a stream increased (also see Frainer et al., 2016). Finally, McKie et al. (2008)* and McKie et al. (2009)* built on previous research (Jonsson and Malmqvist, 2003; Ruesink and Srivastava, 2001) on the interplay between detritivore density, biomass and biodiversity in regulating B-EF relationships. This ultimately contributed to refinements in the definition of 'density-dependent diversity effects' (Gessner et al., 2010), and approaches for unifying B-EF research with metabolic theory (Perkins et al., 2010). The results of Lecerf and Chauvet (2008b)*, Hladyz et al. (2009)* and McKie et al. (2008, 2009)* point towards additional aspects of biodiversity which can strongly regulate decomposition rates at local and regional scales, and thus need to be incorporated into a more complete framework for understanding variability in functioning, including variability attributable to human activities.

Following on from the suggestions that emerged from the correlative field studies in *RivFunction*, that body size was a key trait for detritivore consumption rates, and that assemblage total abundance and biomass could be a strong predictor of decomposition rates (Hladyz et al., 2011a,b*; Woodward et al., 2012*), the next generation of B-EF experiments (e.g. Perkins et al., 2010) and field studies (e.g. Frainer et al., 2014) explored these drivers in more detail. Many of these post-*RivFunction* studies turned to metabolic frameworks and (supposedly) universal allometric scaling relationships to develop the theoretical underpinning of the novel experimental designs in what became evermore complex and sophisticated laboratory experiments aimed at disentangling these different drivers, and how they responded to environmental change—with a strong focus on climate change and global warming (Table 8). These studies were natural extensions of the pioneering *RivFunction* work and many of them unearthed new insights—for instance, the role of species richness per se appeared to have been overemphasised, and in most cases the majority of variation in the data could be explained by the distribution of biomass among and within species in the assemblage, and the resulting metabolic capacity of the consumers as a whole. This tallied with the observations in the field, in which large detritivores, when abundant, dominated processing rates (e.g. Frainer and McKie, 2015), with cased caddis and *Gammarus* shrimps being keystone species in the regard and important conduits to the higher trophic levels. It also emerged from many of these studies that effects were often simply additive—at least when a single process was measured and diversity quantified as taxonomic richness—but that more complex biodiversity effects were manifested as environmental conditions changed and more processes were quantified (Fig. 12C and D; Perkins et al., 2010, 2015; Reiss et al., 2011), or when diversity was quantified as functional diversity of species traits rather than taxonomic richness (Frainer et al., 2014). For instance, 'multifunctionality' became an important consideration across a thermal gradient, in which species richness became important for delivering a range of process rates closer to their maxima across but not within temperature regimes (Perkins et al., 2015).

Table 8 Beyond *Rivfunction*—Illustrative Examples of the Next Generation of Research Based on the *RivFunction* Approach, from 2005 Onwards

Study	Approach	Drivers	Responses
Water characteristics (extension of WP1)			
Ferreira et al. (2015a)	MA	Nutrients	
Ferreira et al. (2015c)	WEM	Temperature and litter species	
Ferreira and Chauvet (2011)*	LME	Temperature and nutrients	
Ferreira et al. (2016b)	MA	Heavy metals	
Dossena et al. (2012)	FME	Temperature	Algal production Benthic invertebrates
Rosemond et al. (2015)	WEM	Nutrients	Community respiration Carbon cycle
Land-use (extension of WP2)			
Ferreira et al. (2015b)	MA	Changes in forest composition	
Tolkkinen et al. (2015)	FS	Forest drainage and pH	DNA-based assessment of fungal community
Burrows et al. (2014)	WEM	Forest clear-felling	Cotton strip tensile strength loss Bacterial production
Jinggut et al. (2012)	FS	Deforestation	
Arroita et al. (2013)	FS	Agriculture/irrigation	Breakdown rate of wooden stick
Kominoski et al. (2011)	FS	Changes in forest composition	DNA-based assessment of fungal and bacterial communities
Biotic controls (extension of WP3)			
Dangles et al. (2011)	FS, Mod	Detritivore richness	
Majdi et al. (2014)	FME	Invertebrate predators	Meiofauna community Trophic cascade
Frainer et al. (2016)	LME	Stoichiometric traits of leaf litter and shredders	Consumer–resource elemental imbalance
Frainer et al. (2014)	FS	Shredder functional diversity	Constancy of diversity–function relationship
Danger et al. (2013)	LME	Algae	Algal production
Woodward et al. (2008)	FME	Fish predator	Algal production Trophic cascade

Table 8 Beyond *Rivfunction*—Illustrative Examples of the Next Generation of Research Based on the *RivFunction* Approach, from 2005 Onwards—cont'd

Study	Approach	Drivers	Responses
Reiss et al. (2011)	LME	Evenness and body size of detritivores	
Alp et al. (2016) Macroscale patterns (extension of WP4)	FS, Mod	Biological invasion	Ecosystem phenology
Handa et al. (2014)	FME	Biome, ecosystem type, and litter diversity	Carbon and nitrogen cycling
Boyero et al. (2011)	FS	Temperature, latitude	Activation energy of litter breakdown
Boyero et al. (2016)	FS	Specific leaf area, litter phylogenetic diversity, channel width, and pH	
Graça and Poquet (2014) Biomonitoring (extension of WP5)	FS	Climatic and edaphic factors	Litter palatability
Young and Collier (2009)	FS	Land-use, nutrients	Ecosystem metabolism Leaf toughness loss Cotton strip tensile strength loss Breakdown rate of wooden stick Invertebrate-based biotic index
Clapcott et al. (2012)	FS	Land-use	Ecosystem metabolism Cotton strip tensile strength loss ^{15}N natural abundance Invertebrate-based biotic index
Thompson et al. (2015)	FS	Pesticide	Algal production Nutrient cycling Ecosystem metabolism Genes
Feio et al. (2010) FWB	FS	Land-use, pollution	Sediment respiration rate Periphyton biomass
Lepori et al. (2005)	FS	Stream restoration	

The table illustrates work that built on the original sets of drivers examined in the five project Work-packages (WP). The different approaches are: *FME*, field microcosm/mesocosm experiment; *LME*, laboratory microcosm/mesocosm experiment; *EM*, whole ecosystem manipulation; *MA*, meta-analysis; *FS*, field survey; *Mod*, modelling. Displayed responses are other than litter breakdown rate and "standard" metrics related to microbial decomposers and shredders associated to litterbags.

4.2 How Does Biodiversity Influence the Stability of Decomposition Rates, Including Under Variable Environmental Conditions?

Theory suggests that more species-rich communities may better able to buffer environmental variability and maintain ecosystem processes within 'normal' bounds due to (i) statistical averaging, also called the portfolio effect, whereby functioning is inherently more stable for species-rich systems, as the responses of extreme species are diluted over a more diverse assemblage and (ii) insurance effects, or the greater likelihood that a more species-rich assemblage will include species with some level of tolerance to the altered conditions (Doak et al., 1998; Loreau et al., 2002). These scenarios had scarcely been tested when *RivFunction* was launched.

Two key studies from *RivFunction* focused on the potential for the portfolio effect to stabilise ecosystem process rates. In a microcosm experiment, Dang et al. (2005)* found that while increased fungal species diversity was not associated with any systematic increase in leaf decomposition rates (Fig. 12B), it was associated with reduced levels of variability. This effect was weakened by increased levels of species dominance in microbial assemblages. Lecerf et al. (2007b)* observed similar phenomena at the level of the litter resource in a field study, with no overall relationship between increased litter diversity and decomposition rates; however, higher litter diversity was associated with reduced variability of decomposition. Together, these results point towards the potential for losses of not only species richness but also species evenness to increase variability in leaf decomposition rates, and hence reduce stability in the processing of litter at local and regional scales.

Other *RivFunction* studies focused on interactions between community change and abiotic environmental parameters. McKie et al. (2009)* investigated how variation in detritivore species richness and two environmental perturbations interacted to affect litter decomposition and detritivore growth. The assessed environmental perturbations were nutrient enrichment, which was expected to enhance decomposition from the bottom-up by stimulating microbial activity, and stream liming, which also stimulates microbial activity by increasing availability of limiting cations, but which has been associated with reduced decomposition by stonefly (Plecoptera) detritivores in Swedish boreal streams (McKie et al., 2006*). Both treatments constituted perturbations for the naturally acidic and nutrient-poor streams of the Swedish boreal region. McKie et al. (2009)* expected the effects of liming on decomposition by the selected detritivore species to range from positive to negative, reflecting differences in their pH preferences (Lillehammer, 1988; McKie et al., 2006*), implying potential for species mixtures to buffer the effects of liming on ecosystem functioning, in line with the insurance effect hypothesis. Replicate enclosures containing litter and the detritivore assemblages were deployed in the field, with liming manipulated at the whole reach scale, and nutrients varied at the level of the individual enclosure. Surprisingly, increased detritivore richness reduced both leaf decomposition and detritivore growth. These negative effects of richness in our field study were opposite to previous laboratory observations (Jonsson, 2006), further illustrating the importance of environmental context for B-EF relationships (discussed in detail by McKie et al., 2009*, also see Cardinale, 2011; Cardinale et al., 2000). Effects of the abiotic manipulations were similar in magnitude to these diversity effects, but positive, with leaf decomposition increasing by 18% and 8% following liming and nutrient enrichment, respectively. Finally, the effects of liming were reduced in most species mixtures relative to the monocultures, suggesting increased functional stability when multiple species were present under an anthropogenic perturbation.

Temperature is a basic driver of metabolic processes (Brown et al., 2004), with the chemical reactions underpinning respiration, resource assimilation and organismal growth all generally increasing with temperature, in line with Van't Hoff's rule (Myers, 2003). Mean temperatures as well as daily and seasonal variability are increasingly being altered worldwide in aquatic environments as a result of thermal pollution and hydropeaking, and these changes are expected to intensify as a result of global climate change (Céréghino and Lavandier, 1998; Salinas et al., 2011). Dang et al. (2009)* investigated how shifts in both mean temperatures and the degree of diel temperature oscillations affected the community composition of aquatic hyphomycetes and litter decomposition rates. They tested the effect of 5°C warming with and without diel oscillations on litter decomposition by fungal communities in stream-mimicking laboratory microcosms. Five temperature regimes with identical thermal sums (degree–days) were applied: constant 3°C (representing an ambient scenario) and 8°C (representing a warming scenario); diel temperature oscillations of 5°C around each mean and oscillations of 9°C around 8°C. Temperature oscillations around 8°C but not 3°C accelerated decomposition markedly, by 18% (5°C oscillations) and 31% (9°C oscillations), respectively, compared to the constant temperature regime at 8°C. These outcomes for decomposition were regulated by a combination of both direct (the effect of temperature on processing rates of individual species, reflecting their individual temperature response curves) and indirect (the effect of an increase in mean temperature on dominance by a functionally important species) pathways (Dang et al., 2009*).

4.3 Implications for the Use and Interpretation of a Litter Decomposition Assay in Bioassessment

The *RivFunction* Workpackage 3 experiments provided several key insights with important bearings on the inferences that can be drawn from decomposition assay studies in both the lab and the field:

(1) Variation in species composition, reflecting the presence of functionally significant traits, is frequently associated with variation in decomposition at local scales (see also Jonsson and Malmqvist, 2000, 2003).

(2) Variation in biodiversity per se, can also be associated with significant variation in leaf decomposition rates, but these effects are inconsistent and contingent (Fig. 12) on the composition of regional species pools and environmental context.

(3) Strong impacts of species loss on ecosystem functioning can be expected when linkages among food web compartments are weakened or lost, and overall trophic complexity reduced.

(4) Nonadditive effects associated with particular species combinations, rather than increasing diversity per se, may also be common, especially at the level of the litter resource.

(5) Variability in leaf decomposition regulated by biotic factors (biodiversity, species composition, biomass and density) can equal that regulated by abiotic factors. Litter decomposition assays which only consider abiotic drivers risk missing interactions with biotic drivers that can dampen or amplify the effects of human pressures.

(6) Biodiversity per se might assist in maintaining the inherent stability of ecosystem functioning under stress, particularly reflecting the portfolio effect, but these relationships remain poorly assessed empirically.

(7) Other attributes of the functional diversity of detrital food webs also appear to have strong potential predictive power at both local and larger scales, including intraspecific diversity, and stoichiometric characteristics of litter and consumers.

A clear limitation of the *RivFunction* Workpackage 3 manipulations, despite being at the cutting-edge of the field at the time, is that they were often still limited in scope (e.g. number of species or trophic levels considered, spatiotemporal scale) and/or conducted under experimental settings with varying degrees of realism. However, recently, more spatiotemporally extensive field and mesocosm studies have been conducted which draw on insights partly derived from the *RivFunction* experiments to disentangle interactions between abiotic and biotic drivers and their effects on ecosystem functioning (e.g. Dangles et al., 2011; Frainer and Mckie, 2015; Frainer et al., 2014; Tolkkinen et al., 2013; Table 8). Further scope for doing so can only be enhanced as leaf decomposition is more frequently incorporated into environmental assessment schemes and more extensive datasets are generated.

5. ACCOMODATING NATURAL VARIABILITY WHEN USING LITTER DECOMPOSITION IN STREAM ASSESSMENT

Natural or background variation in process rates is a key consideration when using ecosystem processes such as litter decomposition to evaluate human impacts on streams. This variation can be broadly assigned to two sources: extrinsic environmental conditions and consumer activity, and intrinsic litter quality. These factors need to be understood if human impacts are to be evaluated accurately. Important extrinsic factors include temperature, dissolved nutrients (e.g. N, P, Ca) and pH of the stream water. The biomass and composition of litter-consuming detritivore communities respond to these abiotic drivers and form a bridge to the intrinsic drivers of breakdown inherent in the litter itself. Intrinsic factors that relate primarily to intra- and interspecific variation in litter quality, include their stoichiometric and biochemical composition—especially in terms of CNP ratios and lignin and cellulose content (e.g. Fig. 6A). Overlain on these sources of biotic and abiotic variation are those stemming from methodological choices and experimental design of the study itself (cf. Gessner et al., 2007). Many of these factors vary over broad temporal and spatial scales and some of this variation is relevant for using decomposition to evaluate human impacts on ecosystem functioning. While sometimes considerable in aggregate, extrinsic and intrinsic variation in decomposition can be minimised or accommodated to develop a more-sensitive decomposition-based tool (Gessner and Chauvet, 2002*).

5.1 Extrinsic Factors

Extrinsic factors vary widely across a range of spatial scales, from large scale, such as those that span continents and latitudinal gradients (Boyero et al., 2011; Woodward et al., 2012*) to local habitat scale within a stream

(Frainer et al., 2014; Langhans et al., 2008*; Tiegs et al., 2008*). Variation across these scales acts as filters that often need to be partitioned to detect human impacts, depending on the goals of the assessment. At the largest spatial scales, variation stemming from climatic and geological differences matters for streams by influencing water quality (e.g. nutrient concentrations and temperature) and hydrologic conditions. For example, regional variation in water quality can be reflected in the nutrient-limitation status of microbial heterotrophs (Reisinger et al., 2016). Additionally, biogeographic variation in species pools can influence decomposition (e.g. litter-consuming macroinvertebrate taxa). For example, invertebrate shredding is often insignificant in many low-latitude streams, due to the absence or rarity of invertebrate detritivores, but elevated microbial activity due to warmer temperatures may compensate for this effect (Boyero et al., 2011; see also Section 2). Variation at spatial scales smaller than regions, such as drainage networks, may be significant especially in geologically heterogeneous areas (Casas et al., 2011; Pozo et al., 2011), but not necessarily so. For example, litter decomposition rates were highly consistent among drainage networks in a geologically uniform region of the Black Forest (Germany) with minimal human impacts when efforts were made to control for stream size, and within stream habitats (Tiegs et al., 2009*). In the same study, decomposition rates were also highly consistent among reaches within streams and among riffles. Taken together, these results suggest that even though decomposition rates are inherently spatially variable among different habitats (Frainer et al., 2014; Langhans et al., 2008*) or longitudinal temperature gradients (Griffiths and Tiegs, 2016), when study design controls for a few extrinsic factors, highly consistent background values can be obtained (Tiegs et al., 2009*). These values provide a useful backdrop against which changes in decomposition due to human activities can be evaluated.

5.2 Temporal Variability

Like spatial variation in decomposition, unaccounted for temporal variation in decomposition can compromise the detection of human impacts on ecosystem functioning. Leaf litter decomposition in streams is typically measured over weeks to months, thereby integrating variation over (relatively) small temporal scales. However, temporal variability over longer timescales poses a problem as a source of statistical noise. While many decomposition studies have been conducted in temperate latitudes where there is a strong seasonality in stream environmental factors, most have been conducted in autumn to coincide with peak leaf fall. For this reason, we have a more limited understanding of decomposition in other seasons (but see, e.g. Bergfur, 2007; Ferreira et al., 2006b*; McKie et al., 2006*), and temporal variation in general. Studies during other seasons have typically found that organic matter decomposition varies through time, and tracks water quality attributes such as nutrients and temperature, as well as shifts in consumer assemblages (e.g. Griffiths and Tiegs, 2016; Hladyz et al., 2011a,b*; Mora-Gómez et al., 2015). In temperate zones, decomposition appears to be more rapid in warmer seasons, probably due to both higher metabolic activity of consumers, but also particularly to the phenology of large shredders (Dangles and Malmqvist, 2004; McKie et al., 2006*) and a general scarcity of high-quality litter in streams resulting in detritivores aggregating in artificially inserted litter bags (Murphy et al., 1998; Murphy and Giller, 2000). Unless accommodated in experimental designs or accounted for in statistical analyses, such temporal background variation can impair the sensitivity of assessments of ecosystem functioning.

Several solutions are available to address these issues. The first is to assume that effects brought on by human activities will swamp natural temporal background variability, and thus simply ignore the latter. A more satisfactory approach is to use temporally stratified designs whereby sampling is conducted only during specified times of the year to control for seasonal variability, preferably including periods of peak organic matter input. A logistically more involved option is to repeat decomposition experiments on multiple sampling dates to integrate temporal variability and/or to use slow-decomposing litter, such as oak leaves, that integrates decomposer activities over larger timespans. These latter two options may be less critical in ecoregions where there is little seasonality, and rates of decomposition might be relatively consistent throughout the year. However, many tropical areas experience significant seasonality.

An important downside of using recalcitrant organic matter is the greater potential for loss of experimental units during floods, thus fast-decomposing leaf species are preferable in streams with unpredictably variable flow regimes. High flow also removes litter naturally deposited on stream beds: a consequence of this is that the release of fungal spores from decomposing leaves into the water diminishes, thus slowing fungal colonisation of leaves freshly submerged in streams and, by extension, microbial decomposition. Conversely, aggregations of litter-consuming macroinvertebrates on small patches of experimental litter introduced into streams after flow recedes could artificially elevate decomposition rates. Consequently, the timing of leaf litter exposure in streams relative to flow events, including their legacy, is an important consideration for sensitive assessment of stream ecosystem functioning.

Variation in decomposition rates across large time scales (i.e. years–decades), is likely, but poorly understood given the small number of decomposition studies that are repeated on an annual basis, or even resampled many years apart (but see Fig. 13). Long-term variation might be related to periodic events such as El Niño–Southern Oscillation (ENSO), and therefore predictable to some extent, or be independent of known meteorological or other events. This variation might stem from long-term fluctuations in hydrology (e.g. above- or below-average discharges), water quality (e.g.

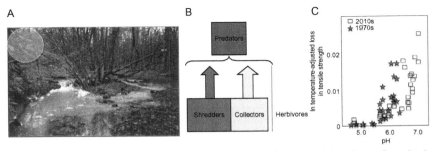

Fig. 13 The next steps beyond *RivFunction*: towards a consideration of multiple stressors, multiple responses and long-term trends. The image (A) shows one of the most intensively studied model stream systems in the world—Broadstone Stream in S. England—highlighting how it, like many other systems, is exposed to a mix of natural and anthropogenic drivers. The image (B) shows a simplified version of the food web, in which the trophic basis of total secondary production via detrital vs algal pathways is drawn to scale. The stream has had little exposure to direct human modification, and its high retentiveness and large amounts of woody debris and leaf litter represent close to the ancestral state for this part of the world, as revealed by the dominance of detrital inputs to the food web. Despite its limited land-use modification, it has been exposed to diffuse acidification for many decades from which it and adjacent sites are now recovering, as revealed using cotton-strip assays (Panel C). In addition, the vegetation is now a mix of native and nonnative species, with *Rhododendron* (cf. Fig. 11) and other exotic evergreens having invaded in recent decades (highlighted in (A) within the *yellow* (*light grey* in the print version) *inset circle*), and it has also been subjected to extreme events, including floods (A) and droughts, over a similar timespan. Combining the use of such model systems with larger-scale comparative approaches (e.g. Figs 3, 6 and 11) will improve future understanding of the drivers of stream ecosystem functioning at larger scales in both space and time, and across multiple stressor gradients. Panel B: Data derived from Woodward, G., Speirs, D.C., Hildrew, A.G., 2005b. Quantification and resolution of a complex, size-structured food web. Adv. Ecol. Res. 36, 85–135. Panel C: Jenkins et al. (2012).

temperature) and be reflected in biotic communities (e.g. invertebrates). Long-term studies in decomposition are very rare, but needed in order to better understand natural temporal variation in decomposition, and to track human impacts to ecosystem functioning that occur gradually at large spatial scales.

Another approach to accommodate temporal variability is to normalise decomposition data with environmental data that have a well-understood influence on decomposition. In particular, if temporal variability stems from variation in temperature, the use of 'temperature-corrected' decomposition rates should minimise noise in the dataset. This is commonly done by assuming a linear relationship between temperature and decomposition rate and replacing elapsed time with degree–days in regression analyses to calculate decomposition rates (e.g. Woodward et al., 2012*). A key drawback with this approach, however, is that there may be instances when the environmental factor that is being controlled for, such as temperature, is also confounded with and altered by human activity, and is of interest in bioassessment.

Lastly, it may be possible to derive a 'disturbance index' to assess the degree to which any human-induced disturbance has impacted a stream over and above the degree of natural variability in decomposition under normal conditions (cf. Johnson et al., 2005). This approach allows for the objective assessment of the occurrence and direction of change as well as the duration of an impact. The disturbance index can be applied at different scales—for a single stream, a catchment or a region.

5.3 Intrinsic Factors: Partitioning and Minimising Variability in Leaf Litter Resource Quality and Potential Alternatives

While leaf litter is a logical choice as a type of organic matter to use in assessment and monitoring programs (Gessner and Chauvet, 2002*), its use in projects that span large temporal or spatial scales presents considerable methodological and logistical challenges (Tiegs et al., 2013). Leaf litter is the largest source of organic matter that enters many stream ecosystems (Abelho, 2001), and it therefore constitutes an environmentally realistic material for use in decomposition assays. Additionally, there is a large body of literature on decomposition, providing considerable background information for interpreting data. In order for leaf litter decomposition to be directly comparable among sites, however, a single homogenous batch of leaf litter is required. In studies that span large spatial scales, a homogenous batch of leaf litter could seemingly be amassed choosing litter from a broadly distributed tree species that grows near each site, therefore ruling out the interspecific variability in litter quality (Webster and Benfield, 1986). However, researchers have recently also documented considerable intraspecific variability in leaf litter quality among biogeographic regions (Graça and Poquet, 2014; Lecerf and Chauvet, 2008b*; LeRoy et al., 2007), confounding such experimental designs if the goal is to compare the decomposition rates among streams using locally collected batches of leaf litter. Even within a region, researchers have documented variability within a species and species–hybrid complexes (LeRoy et al., 2007). This solution of amassing a single batch of litter is additionally limited by several factors, including the fragility of air-dried litter, which tends to become fragmented during transport, and the use of fresh (i.e. not dried) litter presents additional problems (e.g. decomposition prior to incubation in the field). For long-term interannual projects, the need for prolonged storage of the litter presents further logistical difficulties, as when the original batch has been used up subsequent batches collected on other dates are unlikely to be identical matches.

An approach to overcome the problem of variable litter quality is the use of other standardised forms of organic matter instead of natural leaf litter. For example, agar-based pellets containing ground leaf litter and referred to as DecoTabs (Kampfraath et al., 2012) are consumed by litter-consuming macroinvertebrates, and also act as a viable substrate for microbes. Advantages of this approach are that DecoTabs are highly standardised, inexpensive and easy to prepare and deploy. Moreover, chemicals, whether in dissolved or particulate form, can be added, so that to DecoTabs can be used to test for specific effects of nutrients, metals, xenobiotics or other substances in the field. A potential disadvantage is that the texture of DecoTabs greatly differs from that of natural leaf litter, suggesting that the assay can at best be used as a proxy of natural leaf litter decomposition.

A second alternative to measuring decomposition with organic matter other than leaf litter is the cotton-strip assay (Jenkins et al., 2013; Slocum et al., 2009; Tiegs et al., 2007*, 2013). Advantages are that the material is made of cellulose—the most abundant organic polymer on Earth and the primary constituent of leaf litter—in the form of highly standardised woven cotton fabric. In contrast to assays using DecoTabs or natural leaf litter, in which mass loss is the standard response variable, the cotton-strip assay relies on loss of tensile strength, a measure that corresponds to cellulose degradation. Advantages of the cotton-strip assay are that the incubation period is short (i.e. often around 20–30 days) relative to most assays using natural leaf litter, and that the material is durable. Furthermore, being small, light and nonliving, or previously living, cotton strips can be readily shipped across the globe in large numbers (Fig. 14). For example, the assay was recently deployed in over 500 streams as part of a global-scale experiment, CELLDEX (CEllulose Decomposition EXperiment). In this experiment, more than 5000 cotton strips were prepared from cotton in the form of artist's fabric, following a protocol detailed in Tiegs et al. (2013). Each strip was 27 threads in width and 8 cm in length, and loss of tensile strength was compared to cotton strips that were not incubated in the field. Data were expressed in percent tensile strength loss per day. This assay has been shown to be sensitive to several environmental parameters, including those that are impacted by human activities, including concentrations of dissolved nutrients and water temperature (Griffiths and Tiegs, 2016) and geomorphic alterations (Wensink and Tiegs, 2016). A key drawback is that the assay does not explicitly include the impact of litter-consuming macroinvertebrates, which, as discussed earlier, are often more sensitive to perturbations than microbial decomposers. Also, it is essential that details of the cloth's density and thread number per unit area are standardised as this can have strong effects on its loss of tensile strength (Jenkins et al., 2013). Future

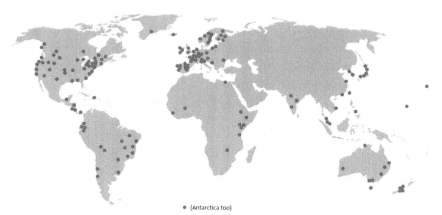

● (Antarctica too)

Fig. 14 The next steps beyond *RivFunction*: towards global-scale biomonitoring. The CELLDEX Project builds on the *RivFunction* Project by measuring decomposition rates using a standardised field bioassay—here this involves cotton strips, rather than leaf litter, as a trade-off between capturing the reality of local litter types and standardisation is unavoidable at these large scales. Each *red* (*dark grey* in the print version) *dot* represents a location where the cotton-strip assay was deployed in four streams.

work is needed to establish the degree to which microbial communities colonising cotton strips resemble those of natural leaf litter.

A third alternative is to incubate commercially available tea bags containing tea leaves of varying nutrient quality in the field for determination of mass loss (Keuskamp et al., 2013). As with the cotton-strip and DecoTab assays, key advantages are ease of use, low cost and high standardisation. Moreover, although fragmented, the material resembles natural leaf litter much more closely than cotton strips and especially DecoTabs. A noteworthy aspect of the tea-bag assay is the simultaneous use of two different types of tea that vary in their carbon-to-nitrogen ratios. The idea is that the different tea types will decompose at different rates, depending on the nutrient status of the ecosystem, as *RivFunction* revealed in its simultaneous use of alder and oak (and other) leaves (e.g. Hladyz et al., 2009*). Major drawbacks are that the assay currently remains untested in aquatic environments, and given the fragile nature of the materials involved (e.g. a delicate rope for attaching the bags to an anchor), may not be suitable for high-energy streams. Additionally, given the very fine mesh used, problems with sedimentation inside the tea bags need consideration, especially under high concentrations of suspended sediments.

A fourth alternative to leaf litter is standardised pieces of wood, which are inexpensive, can be stored for long periods prior to deployment without suffering decay, are readily send to collaborating laboratories and are easily deployed in the field. The most commonly used wood substrata are ice cream, popsicle sticks or medical tongue depressors, and wood veneers have also been used. Although little comparative data are currently available, wood appears to respond to environmental factors in ways that are consistent with responses of leaf litter (Arroita et al., 2012; Díez et al., 2001; Ferreira et al., 2006c*; Gulis et al., 2004). A further advantage of wood is that its decomposition rate varies with the surface-to-volume ratio of individual pieces (Spänhoff and Meyer, 2004) and thus, by using pieces of different size and shape, it is possible to perform decomposition experiments that last and integrate environmental conditions from weeks to years. Aristi et al. (2012) detected a 50-fold variation in the decomposition rate of tongue depressors deployed at 66 sites across the Iberian Peninsula, thus demonstrating large differences in ecosystem functioning. Currently, the approach is used for routine monitoring of the functional status of rivers in New Zealand (Collier and Hamer, 2014). However, it is unclear the extent to which the efficacy of this method is dependent on the presence of xylophagous invertebrates.

In aggregate, these alternatives to the use of leaf litter provide solutions to most of the limitations associated with litter bag assays, and especially those identified during and after the *RivFunction* project. While each has certain advantages that can be exploited to meet the needs of particular research projects, an overarching goal of stream assessment is to rely on a recognised standard approach to facilitate comparisons among studies conducted by different teams at different times and different places. Although the development of numerous alternative assays, irrespective of their individual strengths, could undermine this goal, none of the alternatives is compelling enough at present to abandon further testing of the strengths and limitations of different approaches.

6. TOWARDS THE INTEGRATION OF ECOSYSTEM FUNCTIONING INTO STREAM MANAGEMENT

6.1 Ecosystem Functioning and Stream Management

River management and restoration are a worldwide practice of growing importance as we attempt to redress the problems that have arisen from our longstanding use and misuse of freshwater habitats and resources (Giller, 2005). The interest of managers for ecosystem functioning is at least twofold. On the one hand, they may be interested in a given ecosystem-level process, such as fish production or nutrient retention, and thus, manage the ecosystem to maximise this process, or in more modern terms, the associated ecosystem services (Corvalan et al., 2005; Costanza et al., 1997). This approach is increasingly taken into account, for instance, in river restoration projects, in which desired outcomes are not necessarily determined by the guiding image of an unrealistically pristine situation (Palmer and McDonough, 2013; see chapter "Effective river restoration in the 21th century: from trial and error to novel evidence-based approaches" by Friberg et al.). The full implementation of this type of restoration project would require *measuring* the ecosystem process of interest, or some other highly related proxy, but does not imply determining what the 'natural' or 'reference' conditions should be, nor whether 'more is necessarily better' in relation to the overall efficient functioning of the system as a whole. For instance, river managers in some regions may deliberately elevate nutrient pollution to increase fish yield, even at the cost of decreasing water quality or reducing biodiversity (Stevenson and Esselman, 2013). On the other hand, managers may be interested in maintaining 'natural' or 'healthy' ecosystems, which usually involves *assessing* the status of the ecosystem against a specific benchmark. This is, for instance, the prism of the ecological status in the EU Water Framework Directive (WFD, 2000), which explicitly includes functioning as one goal for ecosystem management (ecological status is an expression of the quality of the structure and functioning of aquatic ecosystems associated with surface waters). This is a more controversial area, since it implies that ecosystems are entities whose limits and status can be clearly delimited, what is often not the case (Jax, 2010). Nevertheless, the implementation of the WFD clearly shows that, although ecosystems can be open systems with undefined boundaries, highly variable in time and space, and affected by humans since prehistoric times, they can be described in typologies consistently enough (but see, Friberg et al., 2011) that their status assessed accurately enough across an entire continent (Hering et al., 2010).

As with a growing number of other regulatory frameworks worldwide, the EU WFD explicitly includes ecosystem functioning in the definition of ecological status, but does not include clear definitions of what the functional properties among the assessment elements might be for streams and rivers. This contradiction led Gessner and Chauvet (2002)* to their initial advocacy for using litter decomposition for functional assessment and has later been expanded by other authors to include additional functional metrics (e.g. Palmer and Febria, 2012). From the outset, the *RivFunction* project involved environmental managers concerned by the challenge of introducing ecosystem functioning into river management. To date, some progress has been made based on individual initiatives, but there have been diverse responses among managers across Europe, ranging from scepticism to approval and, even more than a decade later, we are still far from its general implementation. The causes for this are multiple, and include the difficulties in defining 'reference' conditions for ecosystem functioning, the specialised skills and equipment necessary to determine some ecosystem-level processes, as well as the frequent nonlinearities of ecosystem processes vs environmental stressors (Woodward et al., 2012*). Nevertheless, the general success of structural biomonitoring techniques (Birk et al., 2012), despite the inherent spatial and temporal variability of river communities, suggests that there is a priori circumspection for similar techniques to be developed for functional assessment.

River ecosystem functioning and services are based on multiple processes, which include nutrient retention, secondary production, decomposition of organic matter, pollutant attenuation or whole-stream metabolism (Giller et al., 2004a; Truchy et al., 2015). They offer a potentially powerful and flexible toolbox to gauge functional impairment and recovery (Young et al., 2008). Each one of these integrates different spatiotemporal scales, from minutes to months and from millimetres to entire river sections, and responds to different environmental stressors (Elosegi and Sabater, 2013), thus offering managers the possibility to tailor their monitoring programs according to the functional variables of most relevance or interest.

6.2 Rationale and Steps in the Use of Litter Decomposition for Functional Assessment

An essential prerequisite for any project within the Fourth European Framework Programme (Key Action EESD-1999-1: *Sustainable Management and Quality of Water*)—which was the key driver of *RivFunction* from a legislative and management perspective—was to respond to social expectations in terms of water quality improvement. The RTD Priorities EESD-1999-1.2.1 and 1.2.2 were dedicated to *Ecosystem functioning* and *Ecological quality* targets, respectively. *RivFunction* was thus designed to acquire new knowledge and summarise existing knowledge on decomposition responses to environmental disturbances. Information on the two major sources of alteration, water quality degradation (nutrient enrichment) and land-use change (riparian forest modification), was obtained within the project from 100 European streams for each of both impairment sources. In addition, the compilation of published data provided information on litter decomposition sensitivity to other local environmental stresses and/or global scale environmental changes. The rationale for this was to underpin a robust and widely applicable functional assessment tool, with the potential to be universally applicable, far beyond the borders of the EU and EEA.

In terms of its implementation as a monitoring tool an inherent limitation can be summarised by the following paradox: (i) to be routinely used, authorities need evidence that a tool is reliable, applicable to a wide variety of situations, and fully finalised, whereas (ii) the refined parametrisation of the tool before regional or national implementation would require an adoption by authorities at very large scales, requiring generation of a greatly expanded dataset both in terms of scale and representativeness. As an illustration, compared with the 200 streams sampled in *RivFunction* Workpackages 1 and 2, a target site number for Europe would need to be higher by at least one order of magnitude given the number of sites routinely monitored for biological assessment, i.e. several tens of thousands on the European continent alone. The fact that litter decomposition (like most functional processes) responds to environmental stressors in nonlinear ways (Woodward et al., 2012*) also complicated the implementation of such a tool.

Two developments have, however, strengthened the case for such a functional assessment tool. Firstly, a large body of literature has appeared in the last decade, often from outside the original *RivFunction* consortium and in many cases inspired by the project rationale and outcomes (Table 8): more publications on leaf litter decomposition response to environmental disturbances were produced during the last 10 years than the 30 preceding years. This recent interest is even more pronounced when examining litter decomposition specifically as an assessment tool, including those generated from within *RivFunction* (e.g. Arroita et al., 2012; Castela et al., 2008*; Lecerf et al., 2006*; Riipinen et al., 2009; Tiegs et al., 2007*), but also from elsewhere (Bergfur, 2007; Pascoal et al., 2003), North America (Fritz et al., 2011; Hagen et al., 2006; Johnson et al., 2014), including tropical regions (Lopes et al., 2015; Silva-Junior et al., 2014). Secondly, following these scientific advances, interest has grown among end-user representatives, together with a sufficient investment, in the litter decomposition prototype tool, e.g. as designed in Gessner and Chauvet (2002)* and documented in *RivFunction* and other case studies.

6.3 An Example of National Adoption

An example from France is illuminating within this context, where the functional approach has rapidly gained in popularity over the last 10 years with river basin and national water agencies in charge of water quality assessment/improvement. This enthusiasm may have been caused by the apparent simplicity of its practical implementation, especially at a time when water authorities have had to face new challenges (i.e. all waters reaching 'good ecological status' by 2015, as stipulated in the European WFD). From 2011, the Onema (The French National Agency for Water and Aquatic Environments) initiated an ambitious research programme on the testing and application of alternative functional quality indicators of water bodies, e.g. including isotopic approaches to evaluate food web integrity, cellulose decomposition and associated enzymatic activities, biofilm production and photosynthetic activity, and leaf litter or cotton strip decomposition, together with the comparison of such indicators (F. Guérold, Y. Reyjol and J.-M. Baudoin, pers. comm.). The need for functional bioassessment tools to complement classical WFD-compliant tools is illustrated in two specific situations: (i) streams from overseas European territories, typically in tropical regions, where no or poor biomonitoring information is available and (ii) when attempting to restore ecosystem integrity (Castela et al., 2008). A recent illustration of the Onema support is the application of the litter decomposition assay to 85 headwater stream sites across France, corresponding to various stages and types of hydromorphological disturbance or restoration, i.e. dam removal, remeandering, pond disconnection and sedimentary recharge (Colas et al., 2016). At the regional scale, several programmes have been initiated by river basin agencies, including those for the Loire-Bretagne and Adour-Garonne. The latter recently funded a project aiming at the comparison of structural indicators, as currently used in river quality monitoring, and the newly developed functional indicators based on litter decomposition (Brosed et al., 2016). The 84 regularly monitored stream sites were subjected to various degrees of single to multiple stressors, mostly including point source and diffuse pollution. Even though the functional metrics and thresholds were still tentative, this project identified the complementarity of both types of indicators.

The success of the *RivFunction* project can also be measured by the currently growing interest in applying the litter-bag approach to assess other types of ecosystems beyond streams and rivers. In France, the project communication has been sufficiently broad to reach stakeholders engaged in the management of novel (man-made) ecosystems for which there is an urgent need to develop indicators of 'good ecological potential' and to define relevant ecological thresholds. Litter breakdown metrics have served as benchmark in the Onema-funded project *IsoLac* focusing on gravel-pit lakes to measure the performances of promising food web level indicators based on stable isotope analyses. *RivFunction*'s basic premise that litter breakdown rate can characterise ecosystem health has also been embraced in a research project aiming to develop cost-effective biotic indicators for soil functions in brownfields and old-abandoned mines (*Ifons* project funded by the ADEME, The French National Agency for Environmental Management and Sustainable Development).

In contrast to France, there is no sign yet that the other European countries (at least those involved in the EU *RivFunction* consortium) are fully engaged in formally applying the litter-bag approach to evaluate the functional integrity of streams and rivers, although there is growing interest on a more ad hoc basis at regional scales (e.g. Jackson et al., 2016). Importantly, it must be stressed that the initial arguments for the introduction of litter decomposition in ecological assessment originated from outside Europe: as early as 1986, Webster and Benfield pointed out the potential of litter decomposition to assess the effect of anthropogenic stresses on stream integrity, and in conjunction with complementary functional approaches it has already been implemented to some extent in New Zealand (Young et al., 2004).

6.4 Proposed Metrics

By exploring the various European situations examined within *RivFunction*, the application of the simplest metric proposed in Gessner and Chauvet (2002)*, i.e. the ratio of decomposition rates at impacted and reference site, has provided promising results (Table 9). Although tentatively proposed, the thresholds were sensitive enough to detect responses even to weak environmental changes, as far as the anthropogenic stressor was not complicated by background perturbations and other confounding factors. The situations where reference conditions are unknown or not documented (or even do not exist) are, however, still commonplace. This becomes especially critical when working at the scale of small hydroecoregions, which multiplies the number of cases to consider before carrying out any comparison. In France, for instance, there are 120 level 2-ecohydroregions, with distinct abiotic settings that can induce substantial variability in decomposition even between reference conditions. Ideally, a database covering reference situations from all such hydroecoregions would be needed to standardise stream functional assessment across a heterogeneous territory like France. In the absence of such a database, the ratio of decomposition rates in fine-mesh and coarse-mesh bags can be used as an indicator of the shift in the contributions of microbial and invertebrate decomposers, resulting from differential, possibly compensatory, responses (Gessner and Chauvet, 2002*). Indeed, the ratio of total to microbially driven decomposition rates was useful for detecting the functional impairment of pesticide-affected streams along a gradient of toxicity due to the strongly deleterious effects of pesticides on macroinvertebrates contrasting with the relatively insensitive leaf-associated microflora both in SW France (Brosed et al., 2016) and in the UK (Thompson et al., 2015). No compensatory decomposing activity of microorganisms was observed in the impacted streams in France, and there was only a weak and partial compensation in the UK study. This means that total decomposition rates were also dramatically reduced, thus enabling reliance on either decomposition rate ratios or absolute decomposition rates to quantify the functional impairment of such stream ecosystems. In addition to litter decomposition parameters, several fungal-based metrics have been proposed (e.g. Castela et al., 2008; Colas et al., 2016; Lecerf and Chauvet, 2008a*) and other new molecular microbial-based tools may also serve a complementary

Table 9 Summary Table of *RivFunction* Case Studies Assessing the Functional Impairment of Stream Ecosystems by Various Stressors, Based on Ratio of Decomposition Rates in Coarse-Mesh Bags at Impacted (k_i) and Reference (k_r) Sites

Stressor	Study Reference	k_i:k_r
Nutrient enrichment		
Moderate	Gulis et al. (2006)*	2.3–2.7
Moderate	Elosegi et al. (2006)*	1.2–8.9
Moderate	Castela et al. (2008)*	1.5–4.7
High	Lecerf et al. (2006)*	0.11–0.36
Nitrate addition (0.9 mg L^{-1})	Ferreira et al. (2006c)*	1.3–1.4
Riparian forest modification		
Spruce monoculture	Riipinen et al. (2010)*	0.88–0.92
Beech monoculture	Lecerf et al. (2005)*	0.31–0.77
Eucalypt monoculture, Portugal	Ferreira et al. (2006a)*	0.75–0.79
Eucalypt monoculture, Spain	Ferreira et al. (2006a)*	1.0–2.1
Plant invasion (*Rhododendron ponticum*)	Hladyz et al. (2011b)*	0.44–0.51
Plant invasion (*Fallopia japonica*)	Lecerf et al. (2007a)*	1.0–1.5
Others		
Acidification	Dangles et al. (2004)*	0.09–0.18
Acidification	Baudoin et al. (2008)*	0.16–0.38
Liming of humic streams	McKie et al. (2006)*	0.50–0.75
Mine pollution	Lecerf and Chauvet (2008a)*	0.58
Mine pollution	Medeiros et al. (2008)*	0.36–0.84
Copper contamination (75 µg L^{-1})	Roussel et al. (2008)*	0.28–0.73

Three classes [(0.75–1.33)], [(0.50–0.75), (1.33–2.00)] and [(<0.50), (>2.00)] represent high, moderate and low levels of stream functional resilience to impact, respectively (Gessner and Chauvet, 2002)*, as illustrated by *light grey* to *black shading*. Note that the response to nutrient enrichment has been shown to display bell-shape along the nutrient concentration gradient, in contrast to other stressors, tending to cause monotonic changes in decomposition rate ratios.

role here (see chapter "Recommendations for the next generation of global freshwater biological monitoring tools" by Jackson et al.; Thompson et al., 2015). Maximum leaf-associated spore production was found to be the most sensitive indicator of human impacts on streams, including eutrophication, pollution from mine drainage and alteration of riparian vegetation (Lecerf and Chauvet, 2008a*). Intriguingly, the ratio of microbial to total breakdown was also a clear indicator of the switch from deciduous forest to pastureland (Hladyz et al., 2011a*), especially where grass litter as opposed to leaf litter was used as the resource, suggesting this metric has potential to pick up both instream and riparian impacts.

Finally, further work could also incorporate the idea of providing a more 'robust' reference value through application of the concept of a disturbance index (cf. Johnson et al., 2005). This approach was developed to take on board natural variability of macroinvertebrate community metrics and hydrochemical parameters for assessing the impact of land-use change (e.g. clearfelling) on stream ecosystems. It allows for objective assessment of the occurrence and direction of change as well as the duration of an impact and should be readily applicable to monitor and assess changes in ecosystem processes such as litter decomposition rates.

In conclusion, the *RivFunction* project has generated wide range important and novel insights into the functioning of freshwater ecosystems and also laid the foundations for a stronger coupling or pure and applied research, as well as advancing ecological understanding in each respective discipline. It has opened several exciting new avenues of research, from understanding how biodiversity shapes ecosystem functioning in a changing environment to providing a template for developing the next generation of global biomonitoring tools.

ACKNOWLEDGEMENTS

We are most grateful to Helen Cariss, Christian K. Dang, Aitor Larrañaga, Marius Nistorescu, Markus Schindler and many technicians and field and laboratory assistants for their invaluable contributions to *RivFunction*. We would also like to thank three anonymous reviewers, whose helpful and insightful comments greatly improved an earlier version of the manuscript. This research was carried out with financial support from the EU Commission within the *RivFunction* project (contract EVK1-CT-2001-00088). V.F. also acknowledges the financial support received from the Portuguese Foundation for Science and Technology (FCT; SFRH/BD/11350/2002, program POPH/FSE; IF/00129/2014 and UID/MAR/04292/2013). This paper is dedicated to the memory of our colleague Björn Malmqvist, who sadly passed away in 2010 and whose intellectual contribution to the whole *RivFunction* project was inestimable.

REFERENCES

Abelho, M., 2001. From litterfall to breakdown in streams: a review. ScientificWorldJournal 1, 656–680.

Abelho, M., Graça, M.A.S., 1996. Effects of eucalyptus afforestation on leaf litter dynamics and macroinvertebrate community structure of streams in Central Portugal. Hydrobiologia 324, 195–204.

Adler, R.W., Landman, J.C., Cameron, D.M., 1993. The Clean Water Act: 20 Years Later. Natural Resources Defense Council. Island Press, Washington, DC.
Allan, J.D., Castillo, M.M., 2007. Stream Ecology—Structure and Function of Running Waters, second ed. Springer-Verlag, Dordrecht, The Netherlands.
Alp, M., Cucherousset, J., Buoro, M., Lecerf, A., 2016. Phenological response of a key ecosystem function to biological invasion. Ecol. Lett. 19, 519–527.
Aristi, A., Díez, J., Larrañaga, A., Navarro-Ortega, A., Barceló, D., Elosegi, A., 2012. Assessing the effects of multiple stressors on the functioning of Mediterranean rivers using poplar wood breakdown. Sci. Total Environ. 440, 272–279.
Arroita, M., Aristi, I., Flores, L., Larrañaga, A., Díez, J., Mora, J., Romaní, A.M., Elosegi, A., 2012. The use of wooden sticks to assess stream ecosystem functioning: comparison with leaf breakdown rates. Sci. Total Environ. 440, 115–122.
Arroita, M., Causapé, J., Comín, F.A., Díez, J., Jimenez, J.J., Lacarta, J., Lorente, C., Merchán, D., Muñiz, S., Navarro, E., 2013. Irrigation agriculture affects organic matter decomposition in semi-arid terrestrial and aquatic ecosystems. J. Hazard. Mater. 263, 139–145.
Balvanera, P., Pfisterer, A.B., Buchmann, N., He, J.S., Nakashizuka, T., Raffaelli, D., Schmid, B., 2006. Quantifying the evidence for biodiversity effects on ecosystem functioning and services. Ecol. Lett. 9, 1146–1156.
Bärlocher, F., Graça, M.A.S., 2002. Exotic riparian vegetation lowers fungal diversity but not leaf decomposition in Portuguese streams. Freshwat. Biol. 47, 1123–1135.
Battin, T.J., Luyssaert, S., Kaplan, L.A., Aufdenkampe, A.K., Richter, A., Tranvik, L.J., 2009. The boundless carbon cycle. Nat. Geosci. 2, 598–600.
Baudoin, J.M., Guérold, F., Felten, V., Chauvet, E., Wagner, P., Rousselle, P., 2008. Elevated aluminium concentration in acidified headwater streams lowers aquatic hyphomycete diversity and impairs leaf-litter breakdown. Microb. Ecol. 56, 260–269.
Benfield, E.F., Webster, J.R., Golladay, S.W., Peters, G.T., Stout, B.M., 1991. Effects of forest disturbance on leaf breakdown in southern Appalachian streams. Verh. Int. Ver. Theor. Angew. Limnol. 24, 1687–1690.
Benfield, E.F., Webster, J.R., Tank, J.L., Hutchens, J.J., 2001. Long-term patterns in leaf breakdown in streams in response to watershed logging. Int. Rev. Hydrobiol. 86, 467–474.
Bergfur, J., 2007. Seasonal variation in leaf-litter breakdown in nine boreal streams: implications for assessing functional integrity. Fundam. Appl. Limnol./Arch. Hydrobiol. 169, 319–329.
Bergfur, J., Johnson, R.K., Sandin, L., Goedkoop, W., 2007. Effects of nutrient enrichment on boreal streams: invertebrates, fungi and leaf-litter breakdown. Freshwat. Biol. 52, 1618–1633.
Bird, G.A., Kaushik, N.K., 1992. Invertebrate corobsonlonization and processing of maple leaf litter in a forested and an agricultural reach of a stream. Hydrobiologia 234, 65–77.
Birk, S., Bonne, W., Borja, A., Brucet, S., Courrat, A., Poikane, S., Solimini, A., Van De Bund, W., Zampoukas, N., Hering, D., 2012. Three hundred ways to assess Europe's surface waters: an almost complete overview of biological methods to implement the Water Framework Directive. Ecol. Monogr. 18, 31–41.
Boyero, L., Pearson, R.G., Gessner, M.O., Barmuta, L.A., Ferreira, V., Graça, M.A.S., Dudgeon, D., Boulton, A.J., Callisto, M., Chauvet, E., Helson, J.E., Bruder, A., Albariño, R.J., Yule, C.M., Arunachalam, M., Davies, J.N., Figueroa, R., Flecker, A.S., Ramírez, A., Death, R.G., Iwata, T., Mathooko, J.M., Mathuriau, C., Gonçalves, J.F., Moretti, M.S., Jinggut, T., Lamothe, S., M'Erimba, C., Ratnarajah, L., Schindler, M.H., Castela, J., Buria, L.M., Cornejo, A., Villanueva, V.D., West, D.C., 2011. A global experiment suggests climate warming will not accelerate litter decomposition in streams but might reduce carbon sequestration. Ecol. Lett. 14, 289–294.

Boyero L., Pearson R.G., Hui C., Gessner M.O., Alexandrou M.A., Grac M.A.S., Pérez J., Alexandrou M.A., Graça M.A.S., Cardinale B.J., Albariño R.J., Arunachalam M., Barmuta L.A., Boulton A.J., Bruder A., Callisto M., Chauvet E., Death R.G., Dudgeon D., Encalada A.C., Ferreira V., Figueroa R., Flecker A.S., Gonçalves J.F., Helson J.E., Iwata T., Jinggut T., Mathooko J.M., Mathuriau C., M'Erimba C., Moretti M.S., Pringle C.M., Ramírez A., Ratnarajah L., Rincon J. and Yule C.M., 2016. Biotic and abiotic variables influencing plant litter breakdown in streams: a global study. Proc. R. Soc. B 283, (in press).

Brose, U., Berlow, E.L., Martinez, N.D., 2005. Scaling up keystone effects from simple to complex ecological networks. Ecol. Lett. 8, 1317–1325.

Brosed, M., Lamothe, S., Chauvet, E., 2016. Litter breakdown for ecosystem integrity assessment also applies to streams affected by pesticides. Hydrobiologia 773 (1), 87–102.

Brown, J.H., Gillooly, J.F., Allen, A.P., Savage, V.M., West, G.B., 2004. Toward a metabolic theory of ecology. Ecology 85, 1771–1789.

Bruno, J.F., Stachowicz, J.J., Bertness, M.D., 2003. Inclusion of facilitation into ecological theory. Trends Ecol. Evol. 18, 119–125.

Burrows, R.M., Magierowski, R.H., Fellman, J.B., Clapcott, J.E., Munks, S.A., Roberts, S., Davies, P.E., Barmuta, L.A., 2014. Variation in stream organic matter processing among years and benthic habitats in response to forest clearfelling. For. Ecol. Manag. 327, 136–147.

Canhoto, C., Graça, M.A.S., 1995. Food value of introduced eucalypt leaves for a Mediterranean stream detritivore: Tipula lateralis. Freshwat. Biol. 34, 209–214.

Cardinale, B.J., 2011. Biodiversity improves water quality through niche partitioning. Nature 472, 86–89.

Cardinale, B.J., Palmer, M.A., 2002. Disturbance moderates biodiversity–ecosystem function relationships: evidence from suspension feeding caddisflies in stream mesocosms. Ecology 83, 1915–1927.

Cardinale, B.J., Nelson, K., Palmer, M.A., 2000. Linking species diversity to the functioning of ecosystems: on the importance of environmental context. Oikos 91, 175–183.

Cardinale, B.J., Srivastava, D.S., Duffy, J.E., Wright, J.P., Downing, A.L., Sankaran, M., Jouseau, C., 2006. Effects of biodiversity on the functioning of trophic groups and ecosystems. Nature 443, 989–992.

Cardinale, B.J., Wright, J.P., Cadotte, M.W., Carroll, I.T., Hector, A., Srivastava, D.S., Loreau, M., Weis, J.J., 2007. Impacts of plant diversity on biomass production increase through time because of species complementarity. Proc. Natl. Acad. Sci. U.S.A. 104, 18123–18128.

Cardinale, B.J., Duffy, J.E., Gonzalez, A., Hooper, D.U., Perrings, C., Venail, P., Narwani, A., Mace, G.M., Tilman, D., Wardle, D.A., Kinzig, A.P., Daily, G.C., Loreau, M., Grace, J.B., Larigauderie, A., Srivastava, D.S., Naeem, S., 2012. Biodiversity loss and its impact on humanity. Nature 486, 59–67.

Casas, J.J., Gessner, M.O., López, D., Descals, E., 2011. Leaf-litter colonisation and breakdown in relation to stream typology: insights from Mediterranean low-order streams. Freshwat. Biol. 56, 2594–2608.

Castela, J., Ferreira, V., Graça, M.A.S., 2008. Evaluation of stream ecological integrity using litter decomposition and benthic invertebrates. Environ. Pollut. 153, 440–449.

Céréghino, R., Lavandier, P., 1998. Influence of hypolimnetic hydropeaking on the distribution and population dynamics of ephemeroptera in a mountain stream. Freshwat. Biol. 40, 385–399.

Chapin, F.S.I., Zavaleta, E.S., Eviner, V.T., Naylor, R.L., Vitousek, P.M., Reynolds, H.L., Hooper, D.U., Lavorel, S., Sala, O.E., Hobbie, S.E., Mack, M.C., Díaz, S., 2000. Consequences of changing biodiversity. Nature 405, 234–242.

Christensen, N.L., Bartuska, A.M., Brown, J.H., Antonio, C.D., Francis, R., Franklin, J.F., Macmahon, J.A., Noss, F., Parsons, D.J., Peterson, C.H., Turner, M.G., Robert, G., Brown, J.H., Antonio, C.D., 1996. The report of the Ecological Society of America Committee on the scientific basis for ecosystem management. Ecol. Appl. 6, 665–691.

Clapcott, J.E., Collier, K.J., Death, R.G., Goodwin, E.O., Harding, J.S., Kelly, D., Leathwick, J.R., Young, R.G., 2012. Quantifying relationships between land-use gradients and structural and functional indicators of stream ecological integrity. Freshw. Biol. 57, 74–90.

Colas, F., Baudoin, J., Chauvet, E., Clivot, H., Danger, M., Devin, S., 2016. Dam-associated multiple-stressor impacts on fungal biomass and richness reveal the initial signs of ecosystem functioning impairment. Ecol. Indic. 60, 1077–1090.

Collier, K., Hamer, M., 2014. Aquatic invertebrate communities and functional indicators along the Lower Waikato River. Waikato Regional Council Technical Report 2014/02.

Corvalan, C., Hales, S., McMichael, A., 2005. Ecosystems and Human Well-Being: Health Synthesis: A Report of the Millennium Ecosystem Assessment. Island Press, Washington, DC.

Costanza, R., Arge, R., Groot, R. De, Farber, S., Grasso, M., Hannon, B., Limburg, K., Naeem, S., Neill, R.V.O., Paruelo, J., Raskin, R.G., Sutton, P., 1997. The value of the world's ecosystem services and natural capital. Nature 387, 253–260.

Curtis, E.J.C., Harrington, D.W., 1971. The occurence of sewage fungus in rivers in United Kingdom. Water Res. 5, 281–290.

Dang, C.K., Chauvet, E., Gessner, M.O., 2005. Magnitude and variability of process rates in fungal diversity-litter decomposition relationships. Ecol. Lett. 8, 1129–1137.

Dang, C.K., Schindler, M., Chauvet, E., Gessner, M.O., 2009. Temperature oscillation coupled with fungal community shifts can modulate warming effects on litter decomposition. Ecology 90, 122–131.

Danger, A.R., Robson, B.J., 2004. The effects of land use on leaf-litter processing by macroinvertebrates in an Australian temperate coastal stream. Aquat. Sci. 66, 296–304.

Danger, M., Cornut, J., Chauvet, E., Chavez, P., Elger, A., Lecerf, A., 2013. Benthic algae stimulate leaf litter decomposition in detritus-based headwater streams: a case of aquatic priming effect? Ecology 94, 1604–1613.

Dangles, O., Malmqvist, B., 2004. Species richness–decomposition relationships depend on species dominance. Ecol. Lett. 7, 395–402.

Dangles, O., Gessner, M.O., Guerold, F., Chauvet, E., 2004. Impacts of stream acidification on litter breakdown: implications for assessing ecosystem functioning. J. Appl. Ecol. 41, 365–378.

Dangles, O., Crespo-Pérez, V., Andino, P., Espinosa, R., Calvez, R., Jacobsen, D., 2011. Predicting richness effects on ecosystem function in natural communities: insights from high-elevation streams. Ecology 92, 733–743.

Darwin, C., 1881. The Formation of Vegetable Mould through the Action of Worms, with Observations on Their Habits. John Murray, London, UK.

Díez, J.R., Elosegi, A., Pozo, J., 2001. Woody debris in North Iberian streams: influence of geomorphology, vegetation, and management. Environ. Manag. 28, 687–698.

Díez, J., Elosegi, A., Chauvet, E., Pozo, J., 2002. Breakdown of wood in the Agüera stream. Freshwat. Biol. 47, 2205–2215.

Doak, D.F., Bigger, D., Harding, E.K., Marvier, M.A., O'Malley, R.E., Thomson, D., 1998. The statistical inevitability of stability–diversity relationships in community ecology. Am. Nat. 151, 264–276.

Dobson, M., Magana, A., Mathooko, J., Ndewga, F.K., 2002. Detritivores in Kenyan highland streams: more evidence for the paucity of shredders in the tropics? Freshwat. Biol. 47, 909–919.

Dolédec, S., Statzner, B., Bournaud, M., 1999. Species traits for future biomonitoring across ecoregions: patterns along a human-impacted river. Freshwat. Biol. 42, 737–758.
Dossena, M., Yvon-Durocher, G., Grey, J., Montoya, J.M., Perkins, D.M., Trimmer, M., Woodward, G., 2012. Warming alters community size structure and ecosystem functioning. Proc. R. Soc. B 279, 3011–3019.
Downing, J.A., Cole, J.J., Duarte, C.M., Middelburg, J.J., Melack, J.M., Prairie, Y.T., Kortelainen, P., 2012. Global abundance and size distribution of streams and rivers. Inland Waters 2, 229–236.
Ehrlich, P.R., Ehrlich, A.H., 1981. Extinction: The Causes and Consequences of the Disappearance of Species. Random House, New York.
Elosegi, A., Sabater, S., 2013. Effects of hydromorphological impacts on river ecosystem functioning: a review and suggestions for assessing ecological impacts. Hydrobiologia 712, 129–143.
Elosegi, A., Basaguren, A., Pozo, J., 2006. A functional approach to the ecology of Atlantic Basque streams. Limnetica 25, 123–134.
Encalada, A.C., Calles, J., Ferreira, V., Canhoto, C., Graça, M.A.S., 2010. Riparian land use and the relationship between the benthos and litter decomposition in tropical montane streams. Freshwat. Sci. 55, 1719–1733.
Feio, M.J., Alves, T., Boavida, M., Medeiros, A.O., Graça, M.A.S., 2010. Functional indicators of stream health: a river-basin approach. Freshw. Biol. 55, 1050–1065.
Ferreira, V., Chauvet, E., 2011. Synergistic effects of water temperature and dissolved nutrients on litter decomposition and associated fungi. Glob. Chang. Biol. 17, 551–564.
Ferreira, V., Chauvet, E., 2012. Changes in dominance among species in aquatic hyphomycete assemblages do not affect litter decomposition rates. Aquat. Microb. Ecol. 66, 1–11.
Ferreira, V., Elosegi, A., Gulis, V., Pozo, J., Graça, M.A.S., 2006a. Eucalyptus plantations affect fungal communities associated with leaf-litter decomposition in Iberian streams. Arch. Hydrobiol. 166, 467–490.
Ferreira, V., Graça, M.A.S., de Lima, J.L.M.P., Gomes, R., 2006b. Role of physical fragmentation and invertebrate activity in the breakdown rate of leaves. Arch. Hydrobiol. 165, 493–513.
Ferreira, V., Gulis, V., Graça, M.A.S., 2006c. Whole-stream nitrate addition affects litter decomposition and associated fungi but not invertebrates. Oecologia 149, 718–729.
Ferreira, V., Castagneyrol, B., Koricheva, J.K., Gulis, V., Chauvet, E., Graça, M.A.S., 2015a. A meta-analysis of the effects of nutrient enrichment on litter decomposition in streams. Biol. Rev. 90, 669–688.
Ferreira, V., Larrañaga, A., Gulis, V., Basaguren, A., Elosegi, A., Graça, M.A.S., Pozo, J., 2015b. The effects of eucalypt plantations on plant litter decomposition and macroinvertebrate communities in Iberian streams. For. Ecol. Manage. 335, 129–138.
Ferreira, V., Chauvet, E., Canhoto, C., 2015c. Effects of experimental warming, litter species, and presence of macroinvertebrates on litter decomposition and associated decomposers in a temperate mountain stream. Can. J. Fish. Aquat. Sci. 72, 206–216.
Ferreira, V., Koricheva, J., Pozo, J., Graça, M.A.S., 2016a. A meta-analysis on the effects of changes in the composition of native forests on litter decomposition in streams. For. Ecol. Manage. 364, 27–38.
Ferreira, V., Koricheva, J., Duarte, S., Niyogi, D.K., Guérold, F., 2016b. Effects of anthropogenic heavy metal contamination on litter decomposition in streams—a meta-analysis. Environ. Pollut. 210, 261–270.
Fleituch, T., 2013. Effects of nutrient enrichment and activity of invertebrate shredders on leaf litter breakdown in low order streams. Int. Rev. Hydrobiol. 98, 191–198.

Flynn, D.F.B., Mirotchnick, N., Jain, M., Palmer, M.I., Naeem, S., 2011. Functional and phylogenetic diversity as predictors of biodiversity–ecosystem–function relationships. Ecology 92, 1573–1581.

Frainer, A., Mckie, B.G., 2015. Shifts in the diversity and composition of consumer traits constrain the effects of land use on stream ecosystem functioning. Adv. Ecol. Res. 52, 169–200.

Frainer, A., Mckie, B.G., Malmqvist, B., 2014. When does diversity matter? Species functional diversity and ecosystem functioning across habitats and seasons in a field experiment. J. Anim. Ecol. 83, 460–469.

Frainer, A., Moretti, M.S., Xu, W., Gessner, M.O., 2015. No evidence for leaf-trait dissimilarity effects on litter decomposition, fungal decomposers, and nutrient dynamics. Ecology 96, 550–561.

Frainer, A., Jabiol, J., Gessner, M.O., Bruder, A., Chauvet, E., McKie, B.G., 2016. Stoichiometric imbalances between detritus and detritivores are related to shifts in ecosystem functioning. Oikos 125 (6), 861–871.

Franken, R.J.M., Waluto, B., Peeters, E.T.H.M., Gardeniers, J.J.P., Beiker, J.A.J., Scheffer, M., 2005. Growth of shredders on leaf litter biofilms: the effect of light intensity. Freshwat. Biol. 50, 459–466.

Friberg, N., Bonada, N., Bradley, D.C., Dunbar, M.J., Edwards, F.K., Grey, J., Hayes, R.B., Hildrew, A.G., Lamouroux, N., Trimmer, M., Woodward, G., 2011. Biomonitoring of human impacts in freshwater ecosystems: the good, the bad and the ugly. Adv. Ecol. Res. 44, 1–68.

Friberg, N., Angelopoulos, N.V., Buijse, A.D., Cowx, I.G., Kail, J., Moe, T.F., Moir, H., O'Hare, M.T., Verdonschot, P.F.M., Wolter, C., 2016. Effective river restoration in the 21st century: from trial and error to novel evidence-based approaches. Adv. Ecol. Res. 55, 535–611.

Friedman, J.M., Auble, G.T., Shafroth, P.B., Scott, M.L., Merigliano, M.F., Freehling, M.D., Griffin, E.R., 2005. Dominance of non-native riparian trees in western USA. Biol. Invasions 7, 747–751.

Fritz, K.M., Fulton, S., Johnson, B.R., Barton, C.D., Jack, J.D., Word, D.A., Burke, R.A., 2011. An assessment of cellulose filters as a standardized material for measuring litter breakdown in headwater streams. Ecohydrology 4, 469–476.

Gessner, M.O., Chauvet, E., 1994. Importance of stream microfungi in controlling breakdown rates of leaf litter. Ecology 75, 1807.

Gessner, M.O., Chauvet, E., 2002. A case for using litter breakdown to assess functional stream integrity. Ecol. Appl. 12, 498–510.

Gessner, M.O., Chauvet, E., Dobson, M., 1999. A perspective on leaf litter breakdown in streams. Oikos 85, 377–384.

Gessner, M.O., Gulis, V., Kuehn, K.A., Chauvet, E., Suberkropp, K., 2007. Fungal decomposers of plant litter in aquatic ecosystems. In: Kubikak, C., Druzhinina, I. (Eds.), second ed. The Mycota—Environmental and Microbial Relationships, vol. IV. Springer, Berlin, pp. 301–324.

Gessner, M.O., Swan, C.M., Dang, C.K., McKie, B.G., Bardgett, R.D., Wall, D.H., Hättenschwiler, S., 2010. Diversity meets decomposition. Trends Ecol. Evol. 25, 372–380.

Giller, P.S., 2005. River restoration: seeking ecological standards. Editor's introduction. J. Appl. Ecol. 42, 201–207.

Giller, P.S., O'Connor, J., Kelly-Quinn, M., 1998. Macroinvertebrates. In: Giller, P.S. (Ed.), Studies in Irish Limnology. Marine Institute, Dublin, Ireland, pp. 125–157.

Giller, P.S., Covich, A.P., Ewel, K.C., Hall, R.O., Merritt, D., 2004a. Vulnerability and management of ecological services in freshwater systems: Case studies of freshwater ecosystem services. Wall, D.H. (Ed.), *Sustaining Biodiversity and Functioning in Soils and Sediments.* In: *SCOPE Series,* vol. 64. © Island Press, Washington, DC, pp. 137–160.

Giller, P.S., Hillebrand, H., Berninger, U.-G., Gessner, M.O., Hawkins, S., Inchausti, P., Inglis, C., Leslie, H., Malmqvist, B., Monaghan, M.T., Morin, P.J., O'Mullan, G., 2004b. Biodiversity effects on ecosystem functioning: emerging issues and their experimental test in aquatic environments. Oikos 104, 423–436.

Graça, M.A.S., Poquet, J.M., 2014. Do climate and soil influence phenotypic variability in leaf litter, microbial decomposition and shredder consumption? Oecologia 174, 1021–1032.

Graça, M.A.S., Ferreira, V., Canhoto, C., Encalada, A.C., Guerrero-Bolaño, F., Wantzen, K.M., Boyero, L., 2015. A conceptual model of litter breakdown in low order streams. Int. Rev. Hydrobiol. 100, 1–12.

Gray, C., Hildrew, A., Lu, X., Ma, A., McElroy, D., Monteith, D., et al., 2016. Recovery and nonrecovery of freshwater food webs from the effects of acidification. Adv. Ecol. Res. 55, 475–534.

Greenwood, J.L., Rosemond, A.D., Wallace, J.B., Cross, W.F., Weyers, H.S., Greenwood, J.L., Rosemond, A.D., 2007. Nutrients stimulate leaf breakdown rates and detritivore biomass: bottom-up effects via heterotrophic pathways. Oecologia 151, 637–649.

Griffiths, N.A., Tiegs, S.D., 2016. Organic-matter decomposition along a temperature gradient in a forested headwater stream. Freshwat. Sci. 35 (2), 518–533.

Gulis, V., Suberkropp, K., 2003. Leaf litter decomposition and microbial activity in nutrient-enriched and unaltered reaches of a headwater stream. Freshwat. Biol. 48, 123–134.

Gulis, V., Rosemond, A.D., Suberkropp, K., Weyers, H.S., Benstead, J.P., 2004. Effects of nutrient enrichment on the decomposition of wood and associated microbial activity in streams. Freshwat. Biol. 49, 1437–1447.

Gulis, V., Ferreira, V., Graça, M.A.S., 2006. Stimulation of leaf litter decomposition and associated fungi and invertebrates by moderate eutrophication: implications for stream assessment. Freshwat. Biol. 51, 1655–1669.

Hagen, E.M., Webster, J.R., Benfield, E.F., 2006. Are leaf breakdown rates a useful measure of stream integrity along an agricultural landuse gradient? J. N. Am. Benthol. Soc. 25, 330–343.

Halvorson, H.M., White, G., Scott, J.T., 2016. Dietary and taxonomic controls on incorporation of microbial carbon and phosphorus by detritivorous caddisflies. Oecologia 180, 567–579.

Handa, I.T., Aerts, R., Berendse, F., Berg, M.P., Bruder, A., Butenschoen, O., Chauvet, E., Gessner, M.O., Jabiol, J., Makkonen, M., McKie, B.G., Malmqvist, B., Peeters, E.T.H.M., Scheu, S., Schmid, B., van Ruijven, J., Vos, V.C.A., Hättenschwiler, S., 2014. Consequences of biodiversity loss for litter decomposition across biomes. Nature 509, 218–221.

Hector, A., Schmid, B., Beierkuhnlein, C., Caldeira, M.C., Diemer, M., Dimitrakopoulos, P.G., Finn, J.A., Freitas, H., Giller, P.S., Good, J., Harris, R., Högberg, P., Huss-Danell, K., Joshi, J., Jumpponen, A., Körner, C., Leadley, P.W., Loreau, M., 2002. Biodiversity manipulationexperiments: studies replicated at multiple sites. In: Loreau, M., Naeem, S., Inchausti, P. (Eds.), Biodiversity and Ecosystem Functioning. Oxford University Press, Princeton, pp. 36–46.

Hering, D., Borja, A., Carstensen, J., Carvalho, L., Elliott, M., Feld, C.K., Heiskanen, A.-S., Johnson, R.K., Moe, J., Pont, D., 2010. The European Water Framework Directive at the age of 10: a critical review of the achievements with recommendations for the future. Sci. Total Environ. 408, 4007–4019.

Hieber, M., Gessner, M.O., 2002. Contribution of stream detritivores, fungi, and bacteria to leaf breakdown based on biomass estimates. Ecology 83, 1026–1038.

Hillebrand, H., Bennett, D.M., Cadotte, M.W., 2008. A review of evenness effects on local and regional ecosystem processes. Ecology 89, 1510–1520.

Hilton, J., O'Hare, M., Bowes, M.J., Jones, J.I., 2006. How green is my river? A new paradigm of eutrophication in rivers. Sci. Total Environ. 365, 66–83.

Hisabae, M., Sone, S., Inoue, M., 2011. Breakdown and macroinvertebrate colonization of needle and leaf litter in conifer plantation streams in Shikoku, southwestern Japan. J. For. Res. 16, 108–115.
Hladyz, S., Gessner, M.O., Giller, P.S., Pozo, J., Woodward, G., 2009. Resource quality and stoichiometric constraints on stream ecosystem functioning. Freshwat. Biol. 54, 957–970.
Hladyz, S., Tiegs, S.D., Gessner, M.O., Giller, P.S., Rîşnoveanu, G., Preda, E., Nistorescu, M., Schindler, M., Woodward, G., 2010. Leaf-litter breakdown in pasture and deciduous woodland streams: a comparison among three European regions. Freshwat. Biol. 55, 1916–1929.
Hladyz, S., Abjörnsson, K., Chauvet, E., Dobson, M., Elosegi, A., Ferreira, V., Fleituch, T., Gessner, M.O., Giller, P.S., Gulis, V., Hutton, S.A., Lacoursière, J.O., Lamothe, S., Lecerf, A., Malmqvist, B., Mckie, B.G., Nistorescu, M., Preda, E., Riipinen, M.P., Risneanu, G., Schindler, M., Tiegs, S.D., Vought, L.B., Woodward, G., 2011a. Stream ecosystem functioning in an agricultural landscape: the importance of terrestrial—aquatic linkages. Adv. Ecol. Res. 44, 211–276.
Hladyz, S., Åbjörnsson, K., Giller, P.S., Woodward, G., 2011b. Impacts of an aggressive riparian invader on community structure and ecosystem functioning in stream food webs. J. Appl. Ecol. 48, 443–452.
Hooper, D.U., Chapin, F.S., Ewel, J.J., Hector, A., Inchausti, P., Lavorel, S., Lawton, H., Lodge, D.M., Loreau, M., Naeem, S., Schmid, B., Setälä, H., Symstad, A.J., Wardle, D.A., 2005. Effects of biodiversity on ecosystem functioning: a consensus of current knowledge. Ecol. Monogr. 75, 3–35.
Huryn, A.D., Butz Huryn, V.M., Arbuckle, C.J., Tsomides, L., 2002. Catchment land-use, macroinvertebrates and detritus processing in headwater streams: taxonomic richness versus function. Freshwat. Biol. 47, 401–415.
Iglesias-Trabado, G., Carballeira-Tenreiro, R., Folgueiro-Lozano, J., 2009. Eucalyptus Universalis: Global cultivated Eucalyptus forests. Map version 1.2. GIT Forestry Consulting's EUCALYPTOLOGICS: Information resources on Eucalyptus cultivation worldwide.
Irons, J., Oswood, M., Stout, R., Pringle, C., 1994. Latitudinal patterns in leaf litter breakdown: is temperature really important? Freshwat. Biol. 32, 401–411.
Jabiol, J., McKie, B.G., Bruder, A., Bernadet, C., Gessner, M.O., Chauvet, E., 2013. Trophic complexity enhances ecosystem functioning in an aquatic detritus-based model system. J. Anim. Ecol. 82, 1042–1051.
Jackson, M.C., Weyl, O.L.F., Altermatt, F., Durance, I., Friberg, N., Dumbrell, A.J., Piggott, J.J., Tiegs, S.D., Tockner, K., Krug, C.B., Leadley, P.W., Woodward, G., 2016. Recommendations for the next generation of global freshwater biological monitoring tools. Adv. Ecol. Res. 55, 615–636.
Jax, K., 2010. Ecosystem Functioning. Cambridge University Press, Cambridge, UK.
Jenkins, G.B., Woodward, G., Hildrew, A.G., 2013. Long-term amelioration of acidity accelerates decomposition in headwater streams. Glob. Chang. Biol. 19, 1100–1106.
Jinggut, T., Yule, C.M., Boyero, L., 2012. Stream ecosystem integrity is impaired by logging and shifting agriculture in a global megadiversity center (Sarawak, Borneo). Sci. Total Environ. 437, 83–90.
Johnson, M.J., Giller, P.S., O'Halloran, J., O'Gorman, K., Gallagher, M.B., 2005. A novel approach to assess the impact of landuse activity on chemical and biological parameters in river catchments. Freshwat. Biol. 50, 1273–1289.
Johnson, K.S., Thompson, P.C., Gromen, L., Bowman, J., 2014. Use of leaf litter breakdown and macroinvertebrates to evaluate gradient of recovery in an acid mine impacted stream remediated with an active alkaline doser. Environ. Monit. Assess. 186, 4111–4127.
Jonsson, M., 2006. Species richness effects on ecosystem functioning increase with time in an ephemeral resource system. Acta Oecol. 29, 72–77.

Jonsson, M., Malmqvist, B., 2000. Ecosystem process rate increases with animal species richness: evidence from leaf-eating, aquatic insects. Oikos 89, 519–523.
Jonsson, M., Malmqvist, B., 2003. Mechanisms behind positive diversity effects on ecosystem functioning: testing the facilitation and interference hypotheses. Oecologia 134, 554–559.
Kampfraath, A.A., Hunting, E.R., Mulder, C., Breure, A.M., Gessner, M.O., Kraak, M.H.S., Admiraal, W., 2012. DECOTAB: a multipurpose standard substrate to assess effects of litter quality on microbial decomposition and invertebrate consumption. Freshwat. Sci. 31, 1156–1162.
Karr, J.R., 1999. Defining and measuring river health. Freshwat. Biol. 41, 221–234.
Keuskamp, J.A., Dingemans, B.J.J., Lehtinen, T., Sarneel, J.M., Hefting, M.M., 2013. Tea Bag Index: a novel approach to collect uniform decomposition data across ecosystems. Methods Ecol. Evol. 4, 1070–1075.
Kohler, S.L., Wiley, M.J., 1997. Pathogen outbreaks reveal large-scale effects of competition in stream communities. Ecology 78, 2164–2176.
Kominoski, J.S., Marczak, L.B., Richardson, J.S., 2011. Riparian forest composition affects stream litter decomposition despite similar microbial and invertebrate communities. Ecology 92, 151–159.
Kominoski, J.S., Shah, J.J.F., Canhoto, C., Fischer, D.G., Giling, D.P., González, E., Griffiths, N.A., Larrañaga, A., Leroy, C.J., Mineau, M.M., Mcelarney, Y.R., Shirley, S.M., Swan, C.M., Tiegs, S.D., 2013. Forecasting functional implications of global changes in riparian plant communities. Front. Ecol. Environ. 11, 423–432.
Kominoski, J.S., Rosemond, A., Benstead, J.P., Gulis, V., Maerz, J.C., 2015. Low-to-moderate nitrogen and phosphorus concentrations accelerate microbially driven litter breakdown rates. Ecol. Appl. 25, 856–865.
Kreutzweiser, D.P., Good, K.P., Capell, S.S., Holmes, S.B., 2008. Leaf-litter decomposition and macroinvertebrate communities in boreal forest streams linked to upland logging disturbance. J. N. Am. Benthol. Soc. 27, 1–15.
Laćan, I., Resh, V.H., McBride, J.R., 2010. Similar breakdown rates and benthic macroinvertebrate assemblages on native and Eucalyptus globulus leaf litter in Californian streams. Freshwat. Biol. 55, 739–752.
Langhans, S.D., Tiegs, S.D., Gessner, M.O., Tockner, K., 2008. Leaf-decomposition heterogeneity across a riverine floodplain mosaic. Aquat. Sci. 70, 337–346.
Larrañaga, S., Larrañaga, A., Basaguren, A., Elosegi, A., Pozo, J., 2014. Effects of exotic eucalypt plantations on organic matter processing in Iberian streams. Int. Rev. Hydrobiol. 99, 363–372.
Lawton, J.K., 1994. What do species do in ecosystems? Oikos 71, 367–374.
Lecerf, A., Chauvet, E., 2008a. Diversity and functions of leaf-decaying fungi in human-altered streams. Freshwat. Biol. 53, 1658–1672.
Lecerf, A., Chauvet, E., 2008b. Intraspecific variability in leaf traits strongly affects alder leaf decomposition in a stream. Basic Appl. Ecol. 9, 598–605.
Lecerf, A., Richardson, J.S., 2010a. Biodiversity–ecosystem function research: insights gained from streams. River Res. Appl. 26, 45–54.
Lecerf, A., Richardson, J.S., 2010b. Litter decomposition can detect effects of high and moderate levels of forest disturbance on stream condition. For. Ecol. Manage. 259, 2433–2443.
Lecerf, A., Dobson, M., Dang, C.K., Chauvet, E., 2005. Riparian plant species loss alters trophic dynamics in detritus-based stream ecosystems. Oecologia 146, 432–442.
Lecerf, A., Usseglio-Polatera, P., Charcosset, J.-Y., Lambrigot, D., Bracht, B., Chauvet, E., 2006. Assessment of functional integrity of eutrophic streams using litter breakdown and benthic macroinvertebrates. Arch. Hydrobiol. 165, 105–126.
Lecerf, A., Patfield, D., Boiché, A., Riipinen, M.P., Chauvet, E., Dobson, M., 2007a. Stream ecosystems respond to riparian invasion by Japanese knotweed (Fallopia japonica). Can. J. Fish. Aquat. Sci. 64, 1273–1283.

Lecerf, A., Risnoveanu, G., Popescu, C., Gessner, M.O., Chauvet, E., 2007b. Decomposition of diverse litter mixtures in streams. Ecology 88, 219–227.
Lecerf, A., Marie, G., Kominoski, J.S., Leroy, C.J., Bernadet, C., Swan, C.M., 2011. Incubation time, functional litter diversity, and habitat characteristics predict litter-mixing effects on decomposition. Ecology 92, 160–169.
Lepori, F., Palm, D., Malmqvist, B., 2005. Effects of stream restoration on ecosystem functioning: detritus retentiveness and decomposition. J. Appl. Ecol. 42, 228–238.
LeRoy, C.J., Whitham, T.G., Wooley, S.C., Marks, J.C., 2007. Within-species variation in foliar chemistry influences leaf-litter decomposition in a Utah river. J. N. Am. Benthol. Soc. 26, 426–438.
Lillehammer, A., 1988. Stoneflies (Plecoptera) of Fennoscandia and Denmark. E.J. Brill & Scandinavian Science Press Ltd., Leiden
Lopes, M.P., Martins, R.T., Silveira, L.S., Alves, R.G., 2015. The leaf breakdown of Picramnia sellowii (Picramniales: Picramniaceae) as index of anthropic disturbances in tropical streams. Braz. J. Biol. 75, 846–853.
Loreau, M., Hector, A., 2001. Partitioning selection and complementarity in biodiversity experiments. Nature 412, 72–76.
Loreau, M., Downing, A., Emmerson, M., Gonzalez, A., Hughes, J., Inchausti, P., Joshi, J., Norberg, J., Sala, O., 2002. A new look at the relationship between diversity and stability. In: Loreau, M., Naeem, S., Inchausti, P. (Eds.), Biodiversity and Ecosystem Functioning: Synthesis and Perspectives. Oxford University Press, Oxford, UK, pp. 79–91.
Lorenzo, P., González, L., Reigosa, M.J., 2010. The genus Acacia as invader: the characteristic case of Acacia dealbata Link in Europe. Ann. For. Sci. 67, 1–11.
Majdi, N., Boiché, A., Traunspurger, W., Lecerf, A., 2014. Predator effects on a detritus-based food web are primarily mediated by non-trophic interactions. J. Anim. Ecol. 83, 953–962.
Malmqvist, B., Rundle, S., 2002. Threats to the running water ecosystems of the world. Environ. Conserv. 29, 134–153.
Mancinelli, G., Mulder, C., 2015. Detrital dynamics and cascading effects on supporting ecosystem services. Adv. Ecol. Res. 53, 97–160.
Mancinelli, G., Sangiorgio, F., Scalzo, A., 2013. The effects of decapod crustacean macroconsumers on leaf detritus processing and colonization by invertebrates in stream habitats: a meta-analysis. Int. Rev. Hydrobiol. 98, 206–216.
Manning, D.W.P., Rosemond, A.D., Kominoski, J.S., Gulis, V., Benstead, J.P., Maerz, J.C., 2015. Detrital stoichiometry as a critical nexus for the effects of streamwater nutrients on leaf litter breakdown rates. Ecology 96, 2214–2224.
Martínez, A., Larrañaga, A., Pérez, J., Descals, E., Basaguren, A., Pozo, J., 2013. Effects of pine plantations on structural and functional attributes of forested streams. For. Ecol. Manage. 310, 147–155.
McKie, B.G., Malmqvist, B., 2009. Assessing ecosystem functioning in streams affected by forest management: increased leaf decomposition occurs without changes to the composition of benthic assemblages. Freshwat. Biol. 54, 2086–2100.
McKie, B.G., Petrin, Z., Malmqvist, B., 2006. Mitigation or disturbance? Effects of liming on macroinvertebrate assemblage structure and leaf-litter decomposition in the humic streams of northern Sweden. J. Appl. Ecol. 43, 780–791.
McKie, B.G., Woodward, G., Hladyz, S., Nistorescu, M., Preda, E., Popescu, C., Giller, P.S., Malmqvist, B., 2008. Ecosystem functioning in stream assemblages from different regions: contrasting responses to variation in detritivore richness, evenness and density. J. Anim. Ecol. 77, 495–504.
McKie, B.G., Schindler, M., Gessner, M.O., Malmqvist, B., 2009. Placing biodiversity and ecosystem functioning in context: environmental perturbations and the effects of species richness in a stream field experiment. Oecologia 160, 757–770.

Medeiros, A.O., Rocha, P., Rosa, C.A., Graça, M.A.S., 2008. Litter breakdown in a stream affected by drainage from a gold mine. Fundam. Appl. Limnol./Arch. Hydrobiol. 172, 59–70.

Menéndez, M., Descals, E., Riera, T., Moya, O., 2013. Do non-native Platanus hybrida riparian plantations affect leaf litter decomposition in streams? Hydrobiologia 716, 5–20.

Molinero, J., Pozo, J., Gonzalez, E., 1996. Litter breakdown in streams of the Agüera catchment: influence of dissolved nutrients and land use. Freshwat. Biol. 36, 745–756.

Mora-Gómez, J., Elosegi, A., Mas-Martí, E., Romaní, A., 2015. Factors controlling seasonality in leaf-litter breakdown in a Mediterranean stream. Freshwat. Sci. 34, 1245–1258.

Murphy, J.F., Giller, P.S., 2000. Seasonal dynamics of macroinvertebrate assemblages in the benthos and associated with detritus packs in two low-order streams with different riparian vegetation. Freshwat. Biol. 43, 617–631.

Murphy, J.M., Giller, P.S., 2001. Detrital inputs to two low-order streams differing in riparian vegetation. Verh. Int. Ver. Theor. Angew. Limnol. 27, 1351–1356.

Murphy, J., Giller, P.S., Horan, M.A., 1998. Spatial scale and the aggregation of stream macroinvertebrates associated with leaf packs. Freshw. Biol. 39, 325–339.

Myers, R., 2003. The Basics of Chemistry. Greenwood Press, Westport, CT.

Niyogi, D., Simon, K., Townsend, C., 2003. Breakdown of tussock grass in streams along a gradient of agricultural development in New Zealand. Freshwat. Biol. 48, 1698–1708.

O'Connor, N.E., Donohue, I., 2013. Environmental context determines multi-trophic effects of consumer species loss. Glob. Chang. Biol. 19, 431–440.

Odum, H.T., 1956. Primary production in flowing waters. Limnol. Oceanogr. 1, 102–117.

Palmer, M.A., Febria, C.M., 2012. The heartbeat of ecosystems. Science 336, 1393–1394.

Palmer, M.A., McDonough, O., 2013. Ecological restoration to conserve and recover river ecosystem service. In: Sabater, S., Elosegi, A. (Eds.), River Conservation: Challenges and Opportunities. Foundation BBVA, Bilbao, Spain, pp. 279–300.

Pascoal, C., Cássio, F., 2004. Contribution of fungi and bacteria to leaf litter decomposition in a polluted river. Appl. Environ. Microbiol. 70, 5266–5273.

Pascoal, C., Pinho, M., Cássio, F., Gomes, P., 2003. Assessing structural and functional ecosystem condition using leaf breakdown: studies on a polluted river. Freshwat. Biol. 48, 2033–2044.

Pérez, J., Basaguren, A., Descals, E., Larrañaga, A., Pozo, J., 2013. Leaf-litter processing in headwater streams of northern Iberian Peninsula: moderate levels of eutrophication do not explain breakdown rates. Hydrobiologia 718, 41–57.

Perkins, D.M., Mckie, B.G., Malmqvist, B., Gilmour, S.G., Reiss, J., Woodward, G., 2010. Environmental warming and biodiversity–ecosystem functioning in freshwater microcosms: partitioning the effects of species identity, richness and metabolism. Adv. Ecol. Res. 43, 177–209.

Perkins, D.M., Bailey, R.A., Dossena, M., Gamfeldt, L., Reiss, J., Trimmer, M., Woodward, G., 2015. Higher biodiversity is required to sustain multiple ecosystem processes across temperature regimes. Glob. Chang. Biol 21, 396–406.

Petchey, O.L., Downing, A.L., Mittelbach, G.G., Persson, L., Steiner, C.F., Warren, P.H., Woodward, G., 2004. Species loss and the structure and functioning of multitrophic aquatic systems. Oikos 3, 467–478.

Piggott, J.J., Niyogi, D.K., Townsend, C.R., Matthaei, C.D., 2015. Multiple stressors and stream ecosystem functioning: climate warming and agricultural stressors interact to affect processing of organic matter. J. Appl. Ecol. 52, 1126–1134.

Pozo, J., 1993. Leaf litter processing of alder and eucalyptus in the Agüera system (North Spain) I. Chemical changes. Arch. Hydrobiol. 127, 299–317.

Pozo, J., Basaguren, A., Elósegui, A., Molinero, J., Fabre, E., Chauvet, E., 1998. Afforestation with Eucalyptus globulus and leaf litter decomposition in streams of northern Spain. Hydrobiologia 373 (374), 101–109.

Pozo, J., Casas, J., Menénedez, M., Mollá, S., Arostegui, I., Basaguren, A., Casado, C., Descals, E., Garcia-Avilés, J., González, J.M., Larranaga, A., Lopez, E., Lusi, M., Moya, O., Pérez, J., Riera, T., Roblas, N., Salinas, M.J., 2011. Leaf-litter decomposition in headwater streams: a comparison of the process among four climatic regions. J. N. Am. Benthol. Soc. 30, 935–950.

Raffaelli, D., Van Der Putten, W., Persson, L., Wardle, D., Petchey, O., Koricheva, J., Van Der Heijden, M., Mikola, J., Kennedy, T., 2002. Multi-trophic dynamics and ecosystem processes. In: Loreau, M., Naeem, S., Inchausti, P. (Eds.), Biodiversity and Ecosystem Functioning: Synthesis and Perspectives. Oxford University Press, Oxford, UK, pp. 147–154.

Ramírez, A., Pringle, C.M., Molina, L., 2003. Effects of stream phosphorus levels on microbial respiration. Freshwat. Biol. 48, 88–97.

Raymond, P.A., Hartmann, J., Lauerwald, R., Sobek, S., Mcdonald, C., Hoover, M., Ciais, P., Guth, P., 2013. Global carbon dioxide emissions from inland waters. Nature 503, 355–359.

Reisinger, A.J., Tank, J.L., Dee, M.M., 2016. Regional and seasonal variation in nutrient limitation of river biofilms. Freshwat. Sci. 35 (2), 474–489.

Reiss, J., Bridle, J.R., Montoya, J.M., Woodward, G., 2009. Emerging horizons in biodiversity and ecosystem functioning research. Trends Ecol. Evol. 24, 505–514.

Reiss, J., Bailey, R.A., Perkins, D.M., Pluchinotta, A., Woodward, G., 2011. Testing effects of consumer richness, evenness and body size on ecosystem functioning. J. Anim. Ecol. 80, 1–10.

Richardson, J.S., Béraud, S., 2014. Effects of riparian forest harvest on streams: a meta-analysis. J. Appl. Ecol. 51, 1712–1721.

Riipinen, M.P., Davy-Bowker, J., Dobson, M., 2009. Comparison of structural and functional stream assessment methods to detect changes in riparian vegetation and water pH. Freshwat. Biol. 54, 2127–2138.

Riipinen, M.P., Fleituch, T., Hladyz, S., Woodward, G., Giller, P., Dobson, M., 2010. Invertebrate community structure and ecosystem functioning in European conifer plantation streams. Freshwat. Biol. 55, 346–359.

Robinson, C.T., Gessner, M.O., 2000. Nutrient addition accelerates leaf breakdown in an alpine springbrook. Oecologia 122, 258–263.

Roon, D.A., Wipfli, M.S., Wurtz, T.L., 2014. Effects of invasive European bird cherry (Prunus padus) on leaf litter processing by aquatic invertebrate shredder communities in urban Alaskan streams. Hydrobiologia 736, 17–30.

Rosemond, A.D., Pringle, C.M., Ramírez, A., 1998. Macroconsumer effects on insect detritivores and detritus processing in a tropical stream. Freshwat. Biol. 39, 515–523.

Rosemond, A.D., Pringle, C.M., Ramírez, A., Paul, M.J., Meyer, J.L., 2002. Landscape variation in phosphorus concentration and effects on detritus-based tropical streams. Limnol. Oceanogr. 47, 278–289.

Rosemond, A.D., Benstead, J.P., Bumpers, P.M., Gulis, V., Kominoski, J.S., Manning, D.W.P., Suberkropp, K., Wallace, J.B., 2015. Experimental nutrient additions accelerate terrestrial carbon loss from stream ecosystems. Science 347, 318–321.

Rosenfeld, J.S., 2002. Functional redundancy in ecology and conservation. Oikos 98, 156–162.

Roussel, H., Chauvet, E., Bonzom, J.-M., 2008. Alteration of leaf decomposition in copper-contaminated freshwater mesocosms. Environ. Toxicol. Chem. 27, 637–644.

Royer, T.V., Minshall, G., 2001. Effects of nutrient enrichment and leaf quality on the breakdown of leaves in a hardwater stream. Freshwat. Biol. 46, 603–610.

Ruesink, J.L., Srivastava, D.S., 2001. Numerical and per capita responses to species loss: mechanisms maintaining ecosystem function in a community of stream insect detritivores. Oikos 93, 221–234.

Salinas, N., Malhi, Y., Meir, P., Silman, M., Roman Cuesta, R., Huaman, J., Salinas, D., Huaman, V., Gibaja, A., Mamani, M., 2011. The sensitivity of tropical leaf litter decomposition to temperature: results from a large-scale leaf translocation experiment along an elevation gradient in Peruvian forests. New Phytol. 189, 967–977.

Sanpera-Calbet, I., Lecerf, A., Chauvet, E., 2009. Leaf diversity influences in-stream litter decomposition through effects on shredders. Freshwat. Biol. 54, 1671–1682.

Schindler, M.H., Gessner, M.O., 2009. Functional leaf traits and biodiversity effects on litter decomposition in a stream. Ecology 90, 1641–1649.

Schulze, D.J., Walker, K.F., 1997. Riparian eucalypts and willows and their significance for aquatic invertebrates in the River Murray, South Australia. Regul. Rivers: Res. Manage. 577, 557–577.

Serra, M.N., Albariño, R., Villanueva, V.D., 2013. Invasive *Salix fragilis* alters benthic invertebrate commun ities and litter decomposition in northern Patagonian stream. Hydrobiologia 701, 173–188.

Silva-Junior, E.F., Moulton, T.P., Boëchat, I.G., Gücker, B., 2014. Leaf decomposition and ecosystem metabolism as functional indicators of land use impacts on tropical streams. Ecol. Indic. 36, 195–204.

Slocum, M.G., Roberts, J., Mendelssohn, I.A., 2009. Artist canvas as a new standard for the cotton-strip assay. J. Plant Nutr. Soil Sci. 172, 71–74.

Spänhoff, B., Meyer, E.I., 2004. Breakdown rates of wood in streams. J. N. Am. Benthol. Soc. 23, 189–197.

Spehn, A.E.M., Hector, A., Joshi, J., Schmid, B., Beierkuhnlein, C., Caldeira, M.C., Diemer, M., Dimitrakopoulos, P.G., Finn, J.A., Giller, P.S., Good, J., Harris, R., Högberg, P., Jumpponen, A., Leadley, P.W., Loreau, M., Minns, A., Mulder, C.P.H., Donovan, G.O., Otway, S.J., Palmborg, C., Pereira, J.S., Pfisterer, A.B., Prinz, A., Read, D.J., Schulze, E., 2005. Ecosystem effects of biodiversity manipulations in European grasslands. Ecol. Monogr. 75, 37–63.

Srivastava, D.S., Cardinale, B.J., Downing, A.L., Duffy, J.E., Jouseau, C., Sankaran, M., Wright, J.P., 2009. Diversity has stronger top-down than bottom-up effects on decomposition. Ecology 90, 1073–1083.

Stallcup, L.A., Ardón, M., 2006. Does nitrogen become limiting under high-P conditions in detritus-based tropical streams? Freshwat. Biol. 51, 1515–1526.

Stamm, C., Räsänen, K., Burdon, F.J., Altermatt, F., Jokela, J., Joss, A., Ackermann, M., Eggen, R.I.L., 2016. Unravelling the impacts of micropollutants in aquatic ecosystems: interdisciplinary studies at the interface of large-scale ecology. Adv. Ecol. Res. 55, 183–223.

Stevenson, R.J., Esselman, P.C., 2013. Nutrient pollution: a problem with solutions. In: Sabater, S., Elosegi, A. (Eds.), River Conservation: Challenges and Opportunities. Foundation BBVA, Bilbao, Spain, pp. 77–103.

Thompson M.S.A., Bankier C., Bell T., Dumbrell A.J., Gray C., Ledger M.E., Lehmann K., McKew B.A., Sayer C.D., Shelley F., Trimmer M., Warren S.L. and Woodward G., 2015. Gene-to-ecosystem impacts of a catastrophic pesticide spill: testing amultilevel bioassessment approach in a river ecosystem, Freshwat. Biol, http://dx.doi.org/10.1111/fwb.12676.

Tiegs, S.D., Langhans, S.D., Tockner, K., Gessner, M.O., 2007. Cotton strips as a leaf surrogate to measure decomposition in river floodplain habitats. J. N. Am. Benthol. Soc. 26, 70–77.

Tiegs, S.D., Peter, F.D., Robinson, C.T., Uehlinger, U., Gessner, M.O., 2008. Leaf decomposition and invertebrate colonization responses to manipulated litter quantity in streams. J. N. Am. Benthol. Soc. 27, 321–331.

Tiegs, S.D., Akinwole, P.O., Gessner, M.O., 2009. Litter decomposition across multiple spatial scales in stream networks. Oecologia 161, 343–351.

Tiegs, S.D., Clapcott, J.E., Griffiths, N.A., Boulton, A.J., 2013. A standardized cotton-strip assay for measuring organic-matter decomposition in streams. Ecol. Indic. 32, 131–139.

Tilman, D., Isbell, F., Cowles, J.M., 2014. Biodiversity and ecosystem functioning. Annu. Rev. Ecol. Syst. 45, 471–493.

Tolkkinen, M., Mykrä, H., Markkola, A., Aisala, H., Vuori, K., Tolkkinen, M., Mykr, H., Muotka, T., Lumme, J., Maria, A., 2013. Decomposer communities in human-impacted streams: species dominance rather than richness affects leaf decomposition. J. Appl. Ecol. 50, 1142–1151.

Tolkkinen, M., Mykrä, H., Annala, M., Markkola, A.M., Vuori, K.M., Muotka, T., 2015. Multi-stressor impacts on fungal diversity and ecosystem functions in streams: natural vs. anthropogenic stress. Ecology 96, 672–683.

Truchy, A., Angeler, D.G., Sponseller, R.A., Johnson, R.K., Mckie, B.G., 2015. Linking biodiversity, ecosystem functioning and services, and ecological resilience: towards an integrative framework for improved management. Adv. Ecol. Res. 53, 55–96.

Tylianakis, J.M., Rand, T.A., Kahmen, A., Klein, A.-M., Buchmann, N., Perner, J., Tscharntke, T., 2008. Resource heterogeneity moderates the biodiversity–function relationship in real world ecosystems. PLoS Biol. 6, e122.

Usseglio-Polatera, P., Bournaud, M., Richoux, P., Tachet, H., 2000. Biological and ecological traits of benthic freshwater macroinvertebrates: relationships and definition of groups with similar traits. Freshwat. Biol. 43, 175–205.

Vannote, R., Minshall, G., Cummins, K., Sedell, J., Cushing, C., 1980. The river continuum concept. Can. J. Fish. Aquat. Sci. 37, 130–137.

Violle, C., Navas, M., Vile, D., Kazakou, E., Fortunel, C., Garnier, E., Oikos, S., May, N., Violle, C., Navas, M., Vile, D., Kazakou, E., Fortunel, C., 2007. Let the concept of trait be functional!. Oikos 116, 882–892.

Vörösmarty, C.J., McIntyre, P.B., Gessner, M.O., Dudgeon, D., Prusevich, A., Green, P., Glidden, S., Bunn, S.E., Sullivan, C.A., Liermann, C.R., Davies, P.M., 2010. Global threats to human water security and river biodiversity. Nature 467, 555–561.

Wallace, J.B., Webster, J.R., 1996. The role of macroinvertebrates in stream ecosystem function. Annu. Rev. Entomol. 41, 115–139.

Wallace, J.B., Eggert, S.L., Meyer, J.L., Webster, J.R., 1997. Multiple trophic levels of a forest stream linked to terrestrial litter inputs. Science 277, 102–104.

Webster, J.R., Benfield, E.F., 1986. Vascular plant breakdown in fresh-water ecosystems. Annu. Rev. Ecol. Syst. 17, 567–594.

Wensink, S., Tiegs, S.D., 2016. Shoreline hardening alters freshwater shoreline ecosystems. Freshwat. Sci. 35 (3), 764–777.

WFD, 2000. Directive 2000/60/EC of the European Parliament and of the Council of 23 October 2000 establishing a framework for Community action in the field of water policy. Off. J. Eur. Communities 1–72.

Whiles, M.R., Wallace, J.B., 1997. Leaf litter decomposition and macroinvertebrate communities in headwater streams draining pine and hardwood catchments. Hydrobiologia 353, 107–119.

Wilsey, B.J., Polley, H.W., 2004. Realistically low species evenness does not alter grassland species-richness-productivity relationships. Ecology 85, 2693–2700.

Woodward, G., 2009. Biodiversity, ecosystem functioning and food webs in fresh waters: assembling the jigsaw puzzle. Freshw. Biol. 54, 2171–2187.

Woodward, G., Hildrew, A.G., 2002. Food web structure in riverine landscapes. Freshwat. Biol. 47, 777–798.

Woodward, G., Ebenman, B., Emmerson, M., Montoya, J.M., Olesen, J.M., Valido, A., Warren, P.H., 2005a. Body size in ecological networks. Trends Ecol. Evol. 20, 402–409.

Woodward, G., Speirs, D.C., Hildrew, A.G., 2005b. Quantification and resolution of a complex, size-structured food web. Adv. Ecol. Res. 36, 85–135.

Woodward, G., Papantoniou, G., Lauridsen, R.B., 2008. Trophic trickles and cascades in a complex food web: impacts of a keystone predator on stream community structure and ecosystem processes. Oikos 117, 683–692.

Woodward, G., Perkins, D.M., Brown, L.E., 2010. Climate change and freshwater ecosystems: impacts across multiple levels of organization. Philos. Trans. R. Soc. Lond. Ser. B Biol. Sci. 365, 2093–2106.

Woodward, G., Gessner, M.O., Giller, P.S., Gulis, V., Hladyz, S., Lecerf, A., Malmqvist, B., McKie, B.G., Tiegs, S.D., Cariss, H., Dobson, M., Elosegi, A., Ferreira, V., Graça, M.A.S., Fleituch, T., Lacoursière, J.O., Nistorescu, M., Pozo, J., Risnoveanu, G., Schindler, M., Vadineanu, A., Vought, L.B.-M., Chauvet, E., 2012. Continental-scale effects of nutrient pollution on stream ecosystem functioning. Science 336, 1438–1440.

Wright, J.G., Sutcliffe, D.W., Furse, M.T., 2000. Assessing the Biological Quality of Freshwaters. RIVPACS and Other Techniques. The Freshwater Biological Association, Ambleside, UK.

Young, R.G., Collier, K.J., 2009. Contrasting responses to catchment modification among a range of functional and structural indicators of river ecosystem health. Freshw. Biol. 54, 2155–2170.

Young, R.G., Townsend, C.R., Matthaei, C.D., 2004. Functional Indicators of River Ecosystem Health—An Interim Guide for Use in New Zealand. Cawthron Report 870. Ministry for the Environment.

Young, R.G., Matthaei, C.D., Townsend, C.R., 2008. Organic matter breakdown and ecosystem metabolism: functional indicators for assessing river ecosystem health. J. N. Am. Benthol. Soc. 27, 605–625.

CHAPTER FOUR

Unravelling the Impacts of Micropollutants in Aquatic Ecosystems: Interdisciplinary Studies at the Interface of Large-Scale Ecology

C. Stamm[*,1], K. Räsänen[*], F.J. Burdon[*], F. Altermatt[*,†], J. Jokela[*,‡],
A. Joss[*], M. Ackermann[*,§], R.I.L. Eggen[*,¶]

[*]Eawag, Swiss Federal Institute of Aquatic Science and Technology, Dübendorf, Switzerland
[†]University of Zurich, Zürich, Switzerland
[‡]ETH Zürich, Institute of Integrative Biology, Zürich, Switzerland
[§]ETH Zürich, Environmental Systems Science, Zürich, Switzerland
[¶]ETH Zürich, Institute of Biogeochemistry and Pollutant Dynamics, Zürich, Switzerland
[1]Corresponding author: e-mail address: christian.stamm@eawag.ch

Contents

1. Large-Scale Ecology and Human Impacts on Ecosystems	184
1.1 Micropollutant (MP) Impacts at Different Levels of Biological Organization	188
2. Water Management as a *Real-World Experiment*	193
2.1 Making Use of *Real-World Experiments* to Understand MP Impacts	193
2.2 The *EcoImpact* Project as a Case Study	196
3. Outlook: Potential of Combining *Real-World* and *Research-Led Experiments*	211
3.1 Planned Changes in Urban Infrastructure as *Real-World Experiments*	211
3.2 *Real-World* and *Research-Led Experiments* in Large-Scale Ecology	212
Acknowledgements	215
References	216

Abstract

Human-induced environmental changes are causing major shifts in ecosystems around the globe. To support environmental management, scientific research has to infer both general trends and context dependency in these shifts at global and local scales. Combining replicated *real-world experiments*, which take advantage of implemented mitigation measures or other forms of human impact, with *research-led* experimental manipulations can provide powerful scientific tools for inferring causal drivers of ecological change and the generality of their effects. Additionally, combining these two approaches can facilitate communication with stakeholders involved in implementing management strategies. We demonstrate such an integrative approach using the case study *EcoImpact*, which aims at empirically unravelling the impacts of wastewater-born micropollutants on aquatic ecosystems.

1. LARGE-SCALE ECOLOGY AND HUMAN IMPACTS ON ECOSYSTEMS

Rapid environmental change during the 'anthropocene' is a global phenomenon that has substantially altered terrestrial and aquatic ecosystems (Emmerson et al., 2016; Gray et al., 2016; Lewis and Maslin, 2015). These anthropogenic impacts range from global climate change and habitat fragmentation to invasive species and pollution (Millennium Ecosystem Assessment, 2005). To help support appropriate management decisions in face of such a multitude of large-scale changes, a good mechanistic understanding of how ecosystems respond to external drivers is important (Evans et al., 2013; Rockström et al., 2009). However, predicting the response of complex ecological systems to environmental change (natural or anthropogenic) is not trivial (Mouquet et al., 2015).

A key challenge in this context arises from several simultaneously acting and potentially interacting factors that can lead to multiple-stress situations (Tockner et al., 2010). This holds true at different spatial scales, from global (Vörösmarty et al., 2010) and regional (Hering et al., 2012) to local (Robinson et al., 2014). Describing patterns of ecosystem change is easier than understanding the causal drivers—not to mention being able to predict future ecological change (Burkhardt-Holm et al., 2005). Predicting the future state of ecosystems requires an ability to disentangle the effects of single factors from their possible multifactorial interactions, and a mechanistic understanding across all scales of observation (Evans et al., 2013). As a potential solution to this conundrum, the concept of the 'ecological forecast horizon' has been proposed as a quantitative link between the needed forecast quality and the actual predictability of ecological variables (Petchey et al., 2015). This then raises the question: which level of detail is necessary to make reliable predictions at different spatial and temporal scales?

From a management perspective, large-scale predictions are generally necessary to support policy changes that concern entire regions, countries or continents (e.g. how to preserve biodiversity and/or ecosystem services (Durance et al., 2016)). More detailed local scale models may be required to justify a particular management decisions with local effects, such as how to modify stream morphology to achieve an optimum for societal needs (Hostmann et al., 2005).

The issue of how much detail is needed for sound management decisions—and at which spatial scales—can be illustrated by the example of river restoration (Palmer, 2010). Restoration activities have often been triggered by ecological insight (Palmer et al., 2005). The ecological paradigm that 'loss of habitat diversity decreases biodiversity' was the foundation of many river restoration activities (Ward et al., 2002), whereby riverine biodiversity was predicted to improve if stream morphology was restored to a more natural state. As data from restoration projects accumulated, however, critical syntheses revealed that the expected success of restoration projects was achieved much less frequently than expected (Bernhardt and Palmer, 2011; Kail et al., 2015).

One of the main reasons for the low success rate of restoration activities was that the spatial scales of action were insufficient (Bernhardt and Palmer, 2011). Specifically, restoration has been typically undertaken at the reach or segment scale (Frissell et al., 1986), but it turned out that the paradigm 'diversity begets diversity' (MacArthur and MacArthur, 1961) was too simplistic. This is because habitat morphology is not the only important factor promoting local biodiversity (Sundermann et al., 2011). In particular, regional processes (including species distributions or water quality) are critically important for enabling establishment of organisms at restored habitat sites (Sundermann et al., 2011). These river restoration examples clearly illustrate that both regional and local drivers may be key predictors for ecological responses that should be considered in concert.

The problem of multiple stressors acting upon an ecosystem is also pertinent when dealing with ecological effects of water pollution, which is one of the major threats for biological diversity (Millennium Ecosystem Assessment, 2005). Water quality issues have a large-scale component, firstly because of the global distribution of the human population and hence the large number of streams that are affected by chemicals, and secondly due to the persistence of many chemicals in rivers and the long-range downstream effects of pollution (Ruff et al., 2015). On the other hand, as we will emphasize later, local variation in both chemical exposure and the ecosystem (e.g. source of the ecological community) may play a dominant role in how specific ecosystems respond.

In what follows, we show how the use of replicated, interdisciplinary *real-world experiments*, combined with targeted manipulative experiments to infer causality, can help to tackle questions on multiple stressor effects in the context of large-scale ecology. We illustrate the approach with a case study that aims at unravelling the impacts of wastewater-borne micropollutants (MPs) on stream ecosystems. We focus on these chemicals because they are prevalent in treated wastewater (WW; see Boxes 1 and 2) that influence water quality of many surface waters of different characteristics. For understanding the ecological effects of MPs, it is therefore critical to

BOX 1 Treatment of Wastewater and the Removal of Micropollutants (MPs) in Wastewater Treatment Plants (WWTPs)

Since the 1950s, wastewater (WW) treatment started being systematically implemented in urban areas of developed countries to cope with the waste of the growing human population. Today's conventional treatment primarily targets the removal of (i) particulate matter to reduce the microbial and pathogen load, (ii) degradable organic compounds to reduce the oxygen depletion in receiving waters, (iii) nutrients (i.e. nitrogen and phosphorus) to avoid eutrophication, and (iv) specific macropollutants, such as ammonia and nitrite (Neumann et al., 2015). This is generally achieved via a sewage network conveying the wastewater to a centralized wastewater treatment plant (WWTP). The main goal of implementing the required sewer and WW treatment is to protect drinking water resources, to achieve bathing water quality in recreational areas and to protect the aquatic environment from ecological degradation.

In Switzerland more than 95% of the municipal WW undergoes treatment in centralized WWTPs. Effluent from such WWTPs is still enriched with nutrients, other pollutants and microorganisms that are all discharged as complex mixture into the receiving waters (Fig. 7). Accordingly, WW effluent can still be a major input of pollutants to water bodies. Additionally, storm water overflows discharge (diluted) untreated WW during larger rainfall events. Such overflows are built to keep infrastructure costs of WWTPs at an affordable level, but come at the risk of releasing pollutants into the environment.

The trend in today's legislation in Europe (e.g. EU water framework directive; http://ec.europa.eu/environment/water/water-framework/index_en.html), as well as in developing WW infrastructure, is to avoid negative impacts in general rather than simply comply with thresholds set for specific quality parameters. The implementation of MP removal is part of this trend (Box 3): ozonation and activated carbon filtration are two technical means for significantly reducing the load of many MPs in WW (Hollender et al., 2009). For the sake of cost effectiveness, these technologies are mostly employed as an additional final step after state-of-the-art biological WW treatment.

BOX 2 Micropollutants

Micropollutants (MPs) can be defined as anthropogenic chemicals that occur in the (aquatic) environment well above a (potential) natural background level due to human activities but with concentrations remaining at trace levels (i.e. up to the microgram per litre range) (Fig. B.1). Thus, MPs are defined by their anthropogenic origin and their occurrence at low concentrations. Thousands of chemicals fall into this category (Schwarzenbach et al., 2006) and hundreds of them have been found at *EcoImpact* sites. MPs can consist of purely synthetic chemicals, such as strongly halogenated molecules (e.g. fluorinated surfactants), or of natural compounds such as antibiotics (e.g. penicillines) or oestrogens. MPs may originate from a wide range of sources (e.g. agriculture, households, traffic networks or industries) and enter water bodies through diverse entry paths. Depending on the source, MP transfer occurs as diffuse (e.g. agricultural land use) or as point source pollution, for which (municipal) WWTPs are important examples (see Box 1).

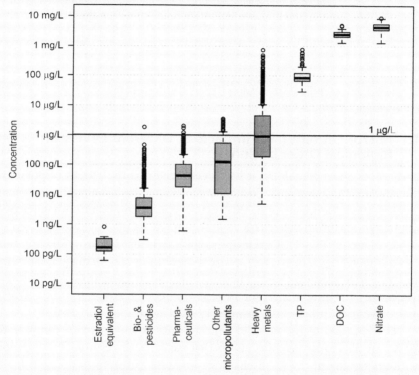

Fig. B.1 Boxplots of the ranges of measured pollutant concentrations (or effect concentrations for estradiol) illustrating the distinction between micropollutants (MPs, *blue* (*grey* in the print version)) and macropollutants (*green* (*light grey* in the print version)) across the 24 *EcoImpact* sites. Heavy metals (*red* (*grey* in the print version)) fall in-between the two classes. *DOC*, dissolved organic carbon; *TP*, total phosphorus. The *red* (*grey* in the print version) *line* indicates the concentration that is generally used to discriminate MPs from other pollutants. Box size corresponds to the first and third quartiles (25% and 75% quartiles); whiskers extend to the most extreme data point which is no more than 1.5 times the length of the box away from the box; *circles* depict outliers. The details of the methods are provided in Box 5.

Continued

BOX 2 Micropollutants—cont'd

The concentrations of MPs in aquatic environments generally make up only a tiny fraction of the dissolved organic matter content (see Fig. B.1). Why should we then be concerned about MPs? The answer to this is that generally these compounds are designed to exert very specific (biological) effects. These include curing a certain disease (e.g. bacterial infections), avoiding growth of unwanted weeds (e.g. herbicides), or providing a calorie free alternative to sugar (artificial sweeteners). In short, the biological activity of MPs is generally by orders of magnitude higher than that of average dissolved organic matter. How MPs act upon organisms—their mode of action—is often very specific and may predominantly affect specific groups of organisms (e.g. only vertebrates, only fungi, etc. (Fig. 2)). In this regard, MPs differ fundamentally from nutrients, which are essential to all organisms.

evaluate what general (large-scale) trends are, as opposed to environmentally contingent effects where local context dominates (Table 1).

1.1 Micropollutant (MP) Impacts at Different Levels of Biological Organization

Early on in aquatic ecology, it was recognized that the release of pollutants into streams can have acute and severe consequences for aquatic ecosystems (Hynes, 1963; Neumann et al., 2015). The effects of pollutants can range from singular catastrophic spills to the continuous release of contaminants at lower concentrations (Fig. 1). Catastrophic spills typically lead to concentrations causing acute toxic effects and capture the attention of both the general public and research communities. These events can trigger subsequent studies to quantify the extent and type of environmental damage and the recovery process (e.g. Güttinger and Stumm, 1992; Thompson et al., 2015).

The continuous release of chemicals, such as the discharge of untreated WW, can also cause negative effects that are easily discernible. For example, without treatment, WW discharges can cause eutrophication or fish kills (Hynes, 1963), but also outbreaks of human diseases caused by pathogens (Neumann et al., 2015). In developed countries, such effects have been successfully mitigated by the introduction of wastewater treatment plants (WWTPs) (Vaughan and Ormerod, 2012). WWTPs are designed to remove pathogens, nutrients, easily degradable organics and particulates (Box 1). However, toxic or highly biologically active chemicals are still discharged in substantial amounts (Petrie et al., 2015; Singer et al., 2016; Ternes,

Table 1 'The Big Question': *How Do MPs Affect Aquatic Ecosystems* from an Interdisciplinary Perspective

Research Fields	General Trends?	Context Dependency?	Key Approaches (as Exemplified from the Project *EcoImpact*)
Environmental chemistry	• How many MPs? • What kind of MPs? • At what concentrations do MPs occur?	• How do MPs vary locally depending on WWTP?	• Sampling of multiple sites with a different WWTP technology • Spatially and temporally replicated water 'spot' sampling • Passive samplers • Broad analytical window (high-resolution mass spectrometry)
Ecotoxicology	• Are MPs toxic? • What is the role of MP mixtures? • What are the cellular mechanisms of MP toxicity?	• How does MP toxicity depend on taxa? • How does MP toxicity differ between laboratory and field?	• Lab and field assays of MPs with different modes of action (affecting algae, invertebrates and fish): • PICT assays → community tolerance of periphyton (algae, bacteria) • *Gammarus* feeding assays → indicator of toxicity • Fish gene expression → upregulation of detoxification genes
Ecology	How do MPs affect: • Population demography • Community diversity • Ecosystem function	• How do species diversity responses to MPs depend on landscape use? • How do MPs interact with nutrients, temperature and/or pathogens? • How do MP effects differ across tropic levels?	• Spatially and temporally replicated field surveys • Manipulative flume experiments (WW dilution or MP spiking) • eDNA analyses for microbial diversity • *Gammarus* size distribution • Species diversity indices (EPT, SPEAR, Saprobic Index) • Organic matter processing assays (leaf litter and cotton strips) • *Gammarus*–leaf litter assays • Ecosystem respiration

The table illustrates how 'large-scale' (general trends) and 'local scale' (context dependent) questions may be approached in different research fields. To address 'the big question', expertise and synthesis across fields are needed. In the context of *EcoImpact*, these questions are addressed combining two kinds of *real-world experiments* (upstream–downstream of WWTP (Fig. 3) and before–after (BACI) WWTP upgrade with field and flume experiments (Box 4).

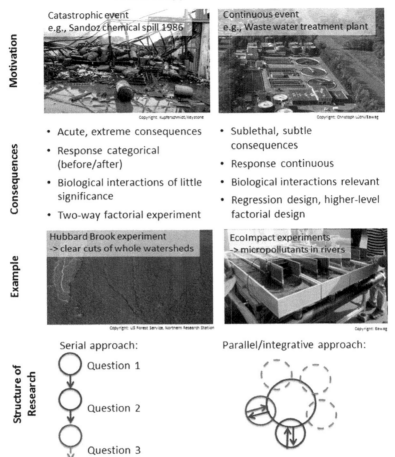

Fig. 1 Conceptual links between different types of environmental 'triggers' (catastrophic events vs continuous impacts) that reflect different types of human impacts on aquatic ecosystems, and how the type of impact may motivate research and project organization.

1998; Box 2). Moreover, the increasing use of chemicals in agriculture, households and industry is causing the number and amount of chemicals in WW effluent to increase across the globe (Keller et al., 2014).

Many of these compounds constitute MPs that occur frequently in high numbers (hundreds to thousands) but at low concentrations (Box 2; Schwarzenbach et al., 2006). This form of pollution is widespread, occurs continuously over time and is much more commonplace than catastrophic events. Due to the widespread presence of wastewater-born MPs in aquatic environments (Kolpin et al., 2002) and the societal desire for pristine water resources, the ecological consequences of MPs in streams at all levels of biological organization have received considerable research attention over the last three decades (Brodin et al., 2013; Schwarzenbach et al., 2006; Ternes, 1998). Ecotoxicological tests have demonstrated the toxicity of a wide range of single MPs, MP mixtures (Carvalho et al., 2014; Silva et al., 2002) and treated WW effluents (Bundschuh and Schulz, 2011) on aquatic organisms.

These pollutants not only harm individual organisms, but also effects propagate to higher levels of biological organization. For example, at the population level, treated WW affected sex ratios of gammarid amphipods (Peschke et al., 2014). At the community level, pollution-induced community tolerance (PICT) experiments suggest that periphyton communities may acquire tolerance to MPs through environmental filtering, physiological acclimation and, potentially, evolutionary processes (Rotter et al., 2011; Tlili et al., 2015). Field surveys of macroinvertebrate communities indicate that pesticides from WWTPs may decrease the fraction of sensitive species through environmental filtering (Bunzel et al., 2013; Burdon et al., 2016). At the ecosystem level, pharmaceuticals can induce behavioural changes in fish leading to more intense predation on prey communities in aquatic food webs (Brodin et al., 2013; Heynen et al., 2016) and key processes, such as in-stream metabolism, can be affected in complex ways due to subsidy effects of nutrients and stress effects of MPs (Aristi et al., 2015; Rosi-Marshall et al., 2013).

During the last two decades, progress has been made in merging concepts from ecotoxicology and ecology to predict how chemical pollution might affect ecosystems (Fischer et al., 2013). Concepts like 'stress ecology' (Van den Brink, 2008; van Straalen, 2003) have been developed, there have been advances in assessing indirect MPs effects in food webs through behavioural ecology (Brodin et al., 2014), and community ecology has been demonstrated to allow for predictions how aquatic mesocosm communities respond to mixtures of different chemicals (Halstead et al., 2014).

Although these studies demonstrate the ecological relevance of (wastewater-born) MPs, predicting changes in entire food webs and ecosystems is still a major challenge because:

(1) *MPs from WWTPs generally occur at sublethal concentrations.* Due to their low concentrations in the environment, effects of exposure to these compounds may strongly depend on the ability of individual organisms to physiologically acclimate and populations to adapt phenotypically (either through phenotypic plasticity or through genetic changes (Capy et al., 2000)). Realized effects may also be environmentally contingent, depending on resource availability, presence of other stressors (including other MPs), and food web interactions (see Fig. 2).

(2) *The occurrence of wastewater-born MPs is highly correlated with the input of other WW constituents*, particularly nutrients and microorganisms (Box 1). This highlights the problem of isolating MP effects when

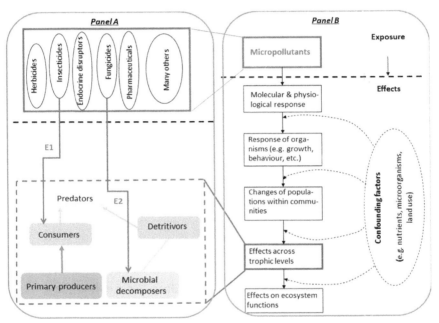

Fig. 2 Conceptual representation of how exposure to micropollutants (MPs) may cause effects in ecosystems at different levels of biological organization. MPs consist of diverse chemical groups (panel A, *red (grey* in the print version) *rectangle* at the top showing a selection of MPs) with specific modes of action that affect different compartments in a food web (exemplified with two *red (grey* in the print version) *arrows* (*E1, E2*) in panel A). The biological effects of MPs can reach different hierarchical levels (panel B) that are interlinked across a given food web and ecosystem. Simple predictions can be made for given types of MPs. For instance, if the key MPs consist of insecticides, then sensitive invertebrate species may disappear from the community (*arrow E1*). Likewise, if the key MPs consist of fungicides (*arrow E2*), they may have negative effects on microbial decomposers, thereby potentially reducing breakdown of organic matter. These asymmetric effects may then cascade differentially across the food web. However, the predictions become more complicated given that hundreds to thousands of MPs enter natural ecosystems and due to many trophic interactions (Box 2).

acting in concert with other co-occurring stressors, which may mask or exacerbate pollutant effects (Folt et al., 1999).

(3) *WW generally contains a very large number of different MPs*, which poses two main challenges for predicting the effects of MPs. First, quantifying all MPs present is a scientific challenge due to the immense number of chemicals, thus requiring cutting-edge instrumentation and data-processing tools (Schymanski et al., 2014). Second, the diverse modes of action of these chemicals (see Fig. 2) mean that not all organisms are affected equally and hence only certain parts of the food web may be directly impacted. These effects may subsequently propagate to other organisms via trophic interactions (indirect effects).

Hence, the question how to predict ecological effects due to MPs poses a general ecological problem: How much detail is required—and at which spatial and temporal scales—to enable accurate predictions of how ecosystems respond to (anthropogenic) stressors (see Table 1)? Making scientific progress that allows for accurate ecological predictions in real-world multiple stress scenarios requires not only advances of ecological theory but also sound empirical data (Mouquet et al., 2015). Below, we present the case study *EcoImpact* as one approach to make progress on this empirical aspect.

2. WATER MANAGEMENT AS A *REAL-WORLD EXPERIMENT*

2.1 Making Use of *Real-World Experiments* to Understand MP Impacts

Diverse empirical approaches can be used to tackle the question *How do MPs affect aquatic ecosystems?*, which all have their strengths and weaknesses. On one hand, correlative field surveys allow establishing patterns in nature—but are often accompanied by multiple confounding factors, necessitating a high level of replication to detect general trends, and not allowing direct tests of underlying causal mechanisms (Robinson et al., 2014). On the other hand, highly controlled experiments conducted under simple laboratory conditions may allow inference of the direct impact of specific factors (Benton et al., 2007), but suffer from a lack of ecological realism that makes it difficult to generalize the experimental findings to natural ecosystems (Carpenter, 1996). Experimental manipulations of entire ecosystems (e.g. entire lakes) can be seen as a kind of 'gold standard' and have been pivotal in demonstrating the ecological effects of different external drivers, such as phosphorus (Schindler et al., 2008), endocrine disruptors (Kidd et al., 2007), or general ecological regime shifts (Carpenter et al., 2011). Such manipulations are, however, rare and generally prevented by economic, logistical or ethical reasons.

Yet, ecosystem manipulations are common for 'nonscientific' reasons—such as management activities aiming to improve the ecological status of water bodies (Bernhardt et al., 2005). This is the basis for what we call *real-world experiments*, which make use of intentionally or unintentionally altered ecological conditions in 'natural' settings, and may allow powerful inferences by combining the benefits of ecologically realistic conditions, experimental manipulations and replication. This approach shares similarities to natural experiments as described by Diamond (1983) but extends this concept explicitly to anthropogenically influenced situations. Of particular relevance is that the monitoring of ecological consequences of specific mitigation measures (e.g. river restoration or WWTP upgrades) can be designed in a (quasi-)experimental manner (as if WWTPs were experimental treatments). Unfortunately, these opportunities have not been systematically capitalized on in the past, as noted for the example of river restoration (Bernhardt et al., 2005).

This is a missed scientific opportunity for applied large-scale ecology. Because ecological mitigation measures are generally implemented (distributed) across large spatial areas and in different ecological contexts, well-designed *real-world experiments* could aid in testing scientific hypotheses about causality and the relative importance of different influencing factors on the ecological status of ecosystems. Furthermore, with such experiments one can systematically evaluate how successfully money has been spent for achieving societal goals by implementing mitigation action (Bernhardt et al., 2005).

The Swiss water sector currently offers a prime opportunity for such real-world experiments. In Switzerland, WWTP infrastructure will be upgraded over the next 20 years to reduce the input of MPs from WWTPs by applying ozonation or powder-activated carbon as an additional treatment step (Box 3). Additionally, many smaller WWTPs will be decommissioned and their WW diverted to larger plants. Following these alterations, water quality in the rivers is expected to change in well-defined ways (e.g. MPs will either be reduced or removed entirely; Fig. 3).

These alterations should result in a measurable biological responses (Fig. 2). With this in mind, we present an interdisciplinary project (*EcoImpact*) that aims to unravel the role of MPs in stream ecosystems. More specifically, *EcoImpact* aims first at establishing the type and magnitude

BOX 3 Upgrading the Swiss WWTP Infrastructure for MP Removal

MPs include an enormous variety of chemical compounds and can enter the environment from various sources (Box 2). To reduce MPs entering the environment, multiple approaches are implemented, reaching from actions at the source (e.g. ban certain compounds) to actions at the end of the WW pipe. One of these measures is upgrading of WWTP infrastructure with an additional treatment step to remove MPs (Abbeglen and Siegrist, 2012). The Swiss government has recently decided that 100 of the 700 Swiss WWTPs will be upgraded during the next 20 years, with the first upgraded plants already being operational. This decision was based on (i) full-scale pilot studies, (ii) social and political acceptance analysis and (iii) technical and economic feasibility assessments (Eggen et al., 2014). Treatment at WWTPs allows the removal of unknown mixtures of MPs to a large extent. With the upgrading, about 50% of Swiss WW will undergo additional treatment.

The Swiss implementation strategy is currently based on two technically feasible and sufficiently cost-effective treatments: ozonation and powder-activated carbon. These two technologies have shown to reduce the total MPs by over 80% (Hollender et al., 2009). As added value of the upgrading, pathogen loads will be reduced by one to three orders of magnitude (Abbeglen and Siegrist, 2012).

At some sites (especially smaller WWTPs), the closure of the plant and transfer of the WW to another, typically larger and neighbouring WWTP is an alternative to upgrading. Locally this should result in an even better water quality at the cost of less discharge in the small streams.

The costs for these additional treatments depend on the condition and size of the WWTP, the technology chosen and the timing of the upgrading. The annual costs for urban drainage and WW treatment in Switzerland are expected to increase by about 6% and its energy consumption between 5% and 30% of its current demand.

(i.e. effect size) of individual to ecosystem-level responses to wastewater-born MPs in streams and environmental contingencies under real-world conditions. Second, it shall elucidate causal relationships between MPs and selected biological endpoints through artificial flume experiments.

In the following sections we will:

- Present a general research strategy where we treat existing WWTPs and their future upgrading as a *real-world experiment* to critically evaluate the ecological relevance of wastewater-born MPs.
- Describe how *EcoImpact* combines highly replicated, interdisciplinary field surveys with manipulative experiments to disentangle the effects of MPs in

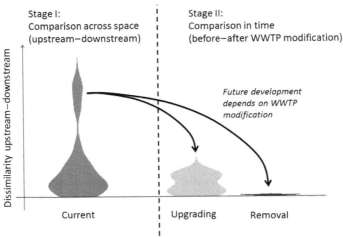

Fig. 3 Conceptual basis for the spatiotemporal design using existing WW treatment plant (WWTP) infrastructure and their planned modifications as *real-world experiments* for testing effects of MPs on aquatic ecosystems. Stage I (US–DS comparison) is based on the spatial comparison between the upstream and downstream locations of the WWTPs (see Fig. 5). The chemical and biological dissimilarities between these locations are quantified (represented by the *red* (*grey* in the print version) *violin plot* for the current conditions). The role of WW and MPs is statistically inferred from the spatially replicated design. Stage II (BACI comparison) compares the dissimilarities before and after the modification of the WWTP infrastructure. Upon upgrading and WW removal, it is hypothesized that the dissimilarities decrease (*green* (*light grey* in the print version) and *blue* (*dark grey* in the print version) *violins*). Because upgrading (only reducing MP load) and removal (reducing MP and nutrient loads) affect water quality differently, this stage allows to disentangle effects of these major WW constituents.

an ecologically relevant context. We show how the study is designed so as to be able to infer general trends (large scale) and context dependency (local scale and interactions among covarying factors, like nutrients) of MP effects (Table 1), and provide first results as proof of principle.

- Provide an outlook for how interdisciplinary, experimental research approaches can address human-induced environmental change in the context of large-scale ecology. We will also consider how inferential power can be increased by combining *real-world* and *research-led experiments*.

2.2 The *EcoImpact* Project as a Case Study

To make use of the opportunity provided by the upgrading of the Swiss WWTP infrastructure, *EcoImpact* was initiated by several departments of Eawag, the Swiss Federal Institute of Aquatic Science and Technology, in close collaboration with relevant stakeholders (e.g. federal and cantonal authorities, WWTP operators and personnel, etc.). This project combines

(i) replicated *spatial field comparisons* to isolate the effects of WW from other factors of influence (upstream–downstream design),
(ii) replicated *temporal field comparisons* following WWTP upgrading to isolate the effects of MPs from other influence factors (before–after or BACI design; Underwood, 1994). Explicitly, upon upgrading or complete removal of WW, the ecosystems at downstream (DS) sites should become more similar to their upstream (US) reference sites, the extent of convergence depending on the type of treatment (Fig. 3),
(iii) *manipulative experimental approaches* to increase mechanistic understanding and to infer causality, and
(iv) the *expertise of different scientific disciplines* (from engineering and environmental chemistry to ecotoxicology and ecology).

Such a combination of spatial and temporal comparisons offers powerful means for testing the effects of WWTP modifications on relevant ecological endpoints. However, the treatments of this *real-world experiment* are only coarsely defined (e.g. with regard to water quality) and cannot be modified according to scientific demands. Therefore, the field monitoring is complemented by targeted experiments in artificial flumes (described later) to establish causal links between exposure to MPs and the ecological responses.

Given that the environmental driver of interest here is chemical water quality and the expected responses are ecotoxicological and ecological, establishing effects along the effect chain (Fig. 2B) requires the combination of different disciplines and fields of expertise.

2.2.1 Design of the Field Survey

The US–DS comparison of the *EcoImpact* field survey is replicated across 24 sites (Fig. 4). This survey was designed so as to have two US locations as in-stream reference for the impacted DS location at each study site (Fig. 5). The two US reference locations provide a 'null model' for the differences in community composition that could be explained by distance alone (within a given stream) (Fig. 3; Burdon et al., 2016). In the BACI part of the study (Stage II), these US sites will provide the 'null model' for the temporal changes that may occur irrespective of the WWTP modifications. In combination, this design enables us to test both general responses to wastewater discharge and the environmental contingency of the effects.

In the field survey, well-established chemical, ecotoxicological and ecological methods are implemented (Table 3; Box 5). In doing so, these data

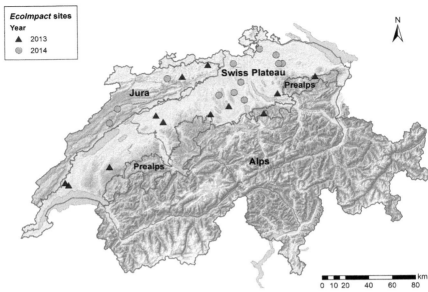

Fig. 4 Map of the 24 *EcoImpact* field sites distributed across three biogeographical regions Jura, Swiss Plateau, and Pre-Alps. *Swisstopo: 5704 000 000/DHM25@2003; Vector200@2015; swissTLM3D@2014, reproduced with permission of swisstopo/JA100119.*

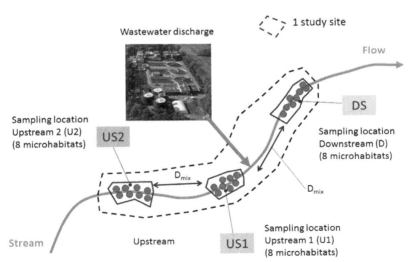

Fig. 5 Schematic of the layout of the sampling design for the US–DS comparisons. In each of the 24 streams, a study site consists of one DS location (*DS*), where the WW is fully mixed across the river cross-section, and two US locations (*US1, US2*). The US locations are placed at equidistant distances upstream of the WWTP and control for the effect of geographic distance and thereby allow for a comparison with the downstream location (DS) as well as for an in-stream control (US2–US1). Further details are given in the text and in Burdon et al. (2016). *Picture WWTP: Copyright/Author: Christoph Lüthi, Sandec and Eawag.*

Table 2 Quantitative Criteria Used for the Selection of Sites

Category	Site Property	Description	Threshold
Wastewater	Effluent upstream	Proportion of discharge (at Q_{347})	0%
	Effluent downstream	Proportion of discharge (at Q_{347})	>20%
Land use	Urban areas	Areal fraction of watershed	<21%
	Special crops (e.g. orchards, vegetables)	Areal fraction of watershed	<10%

Q_{347}: Discharge that is statistically exceeded at 347 days per year (95% low-flow conditions).

can also be put in context with landscape-level processes driving local and regional biodiversity (Altermatt et al., 2013; Burdon et al., 2016).

The 24 sites are located in the Swiss Plateau, Jura Mountains and Pre-Alps regions of Switzerland (Fig. 4). The high site replication is essential to account for effects across a wide range of environmental contexts (e.g. different catchment characteristics) upstream of the WWTPs, including different biogeographical regions. The sites were chosen so that the differences in the fraction of WW between US and DS locations of the selected WWTPs were as large as possible (Burdon et al., 2016). These requirements were realized by (i) applying specific criteria relating to land use, discharge and regional coverage during site selection (Table 2) and (ii) ensuring that conditions between US and DS sampling locations within a given location were as similar as possible with regard to surrounding land use, stream morphology, and riparian vegetation. Other confounding factors between sampling locations (e.g. confluence with a tributary) were similarly avoided. Because of these criteria, all sites consist of lower-order streams (because larger rivers invariably receive WW above potential WWTPs). For logistic reasons, the US-DS (Stage I) field study was conducted over 2 years (12 sites in 2013 and 12 sites in 2014).

2.2.1.1 Quantifying Environmental Drivers

To quantify pollutant exposure as well as putative confounding/correlated factors of water quality, water samples were regularly (mostly monthly/bi-monthly) collected for the determination of 20 basic water chemistry parameters (e.g. major ions, concentrations of MPs and heavy metals during base flow conditions, analytical method see Box 5). Benthic habitat characteristics were also recorded to account for context dependency (methods see

Box 5). These measurements enable testing of the association of physico-chemical environmental parameters with biological endpoints.

2.2.1.2 Biological Endpoints

A variety of biological responses relevant for assessing ecological effects of pollution were measured across four levels of biological organization from individuals, to populations, communities to ecosystem functions (Fig. 2). For instance, samples for in vitro ecotoxicological assays were taken once each year (May/June) simultaneously with the regular water samples. These assays covered endpoints that are relevant for different trophic levels: algae (inhibition of photosynthesis (Escher et al., 2008)), invertebrates (inhibition of acetylcholinesterase (Hamers et al., 2000)) and vertebrates (endocrine disruptors; YES-test (Routledge and Sumpter, 1996)). The ecological endpoints ranged from microbial and macroinvertebrate community descriptors (Burdon et al., 2016), and age structure of *Gammarus* amphipods, to leaf litter decomposition and microbial-mediated cotton strip breakdown rates (see Table 3).

2.2.1.3 First Insights

The design of the field study outlined before resulted in first outcomes for the Stage I of the project, which is the spatial US–DS comparison. The BACI design of Stage II can only be realized once a number of WWTPs have actually undergone upgrading.

The selection of the field sites indeed resulted in a broad range of ecological conditions upstream of the WWTPs. For example, catchment areas covered by forest and grassland ranged from 10% to 80% and arable land varied between 0% and 50% (see also Burdon et al., 2016). Correspondingly, also water quality varied strongly. We illustrate this aspect with variation in nitrate concentrations at the US locations (Fig. 6): the more the catchment was dominated by grassland and forests (i.e. the less arable land), the lower the nitrate concentrations. Importantly, these US landscape conditions correlated with the composition of the macroinvertebrate community at the US sampling locations: The larger the fraction of grassland or forest—land use types without substantial use of pesticides—the larger the observed fraction of species considered sensitive towards pesticides as indicated by the SPEAR Index (Fig. 6; see also Burdon et al., 2016).

This observed variation across the 24 sites offers the possibility to study how ecological effects caused by discharge of WW may depend on the ecological context (Burdon et al., 2016). To answer this question, one has to

Table 3 Key Ecological Responses Measured in the US–DS Field Surveys and Three Flume Experiments in Ecolmpact

Biological Level	Response	Method and/or Organism(s) Used	Description	References for Methods	Field	Flumes
Individuals/ population	Secondary production	*Ancylus*, *Gammarus*	Individual growth (% change in body length)	Hicklen et al. (2006), Watts et al. (2002)	No	Yes
	Fitness	*Ancylus*, *Gammarus*	Survival and condition of invertebrates (condition factor length:mass ratio)	Gerhardt (2011), Rosés et al. (1999)	No	Yes
	Demography	*Baetis*, *Gammarus*	Hess sampling, size distribution and population density	Ladewig et al. (2005)	Yes	No
Community	Detrital consumption	Leaf discs, *Gammarus*	Leaf disc consumption	Mancinelli (2012), O'Neal et al. (2002)	No	Yes
	Biofilm consumption	Plastic slides, *Ancylus*	AFDM of biofilm consumed per day	Rosemond et al. (2000)	No	Yes
	Invertebrate community composition	Semiquantitative sampling	Handnet used to collect invertebrates from standardized area and locations in channels, different diversity measures (e.g. EPT taxa, SPEAR Index, Saprobic Index)	Stucki (2010)	Yes	Dilution experiment only
	Algal (diatom) community composition	Semiquantitative sampling	Sample collected from stones, tiles and glass slides before preservation and identification	Biggs and Kilroy (2000)	Yes	Spiking experiment only

Continued

Table 3 Key Ecological Responses Measured in the US–DS Field Surveys and Three Flume Experiments in *EcoImpact*—cont'd

Biological Level	Response	Method and/or Organism(s) Used	Description	References for Methods	Field	Flumes
Ecosystem	Biofilm production	Tiles	AFDM of standing biomass, primary producers (chlorophyll *a*), resource quality (C:P, C:N ratios)	Biggs and Kilroy (2000), Lamberti and Resh (1985)	Yes	Yes
	Detrital processing	Leaf packs (coarse and fine)	AFDM loss per day measured for standard litter source (Black Alder (*Alnus glutinosa*) leaves)	Bärlocher (2005), Benfield (2007)	Yes	Yes
	Decomposition	Cotton strips (standardized carbon source cellulose)	Respiration (microbial activity), mass loss and tensile strength loss	Slocum et al. (2009), Tiegs et al. (2013)	Yes	Yes
	Ecosystem metabolism	Single-station oxygen logging	Change of oxygen concentrations as ecosystem response	Bott (1996), Stephens and Jennings (1976)	Yes	Yes

The chosen response variables are standard parameters used in ecological studies and reflect different parameters from the individual to ecosystem levels. Study organisms were chosen so as to represent taxa that are sensitive to water quality (e.g. diatom or macroinvertebrate communities) and/or play different key roles in stream ecosystems (e.g. periphyton and *Gammarus*–leaf litter interaction). *AFDM*, ash-free dry mass. See Box 4 for details on flume experiments.

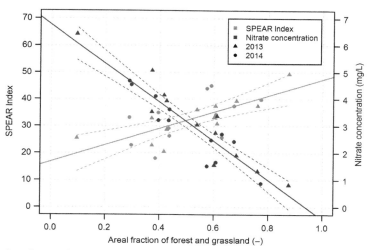

Fig. 6 Correlations between upstream land use (*x*-axis) and (A) the nitrate concentrations and (B) the SPEAR Index (a measure how many taxa sensitive against pesticides are present in a macroinvertebrate community (Liess and von der Ohe, 2005)) at the upstream locations of the 24 *EcoImpact* sites with 95% confidence intervals for the expected values (*dashed lines*). *Triangles* depict the 2013 sites, *circles* the 2014 sites. Details of the methods are provided in Burdon et al. (2016) and in Box 5.

consider to which extent water quality differs between the US and the respective DS location. The impact of WW discharge on DS water quality depends on the composition of the discharged WW, its fraction to total flow in the stream and on the US water quality. As expected, concentrations of chemicals (e.g. nutrients and MPs) as well as the microbial load increased in the DS section at all 24 *EcoImpact* sites (Fig. 7). The magnitude of the changes however differs among specific WW components and is particularly striking for some MPs (such as pharmaceuticals) which increased up to 30-fold (Fig. 7). Despite this general increase downstream, a closer look reveals that WW composition with regard to MPs and nutrients varied considerably between sites (Fig. 8).

The macroinvertebrate communities at DS locations clearly responded to WW input. They were characterized by reduced abundance of sensitive taxa such as the EPT (Ephemeroptera, Plecoptera, Trichoptera) fauna (Burdon et al., 2016). On the other hand, DS sites showed a pronounced increase in oligochaete worms (Burdon et al., 2016), a general pattern seen in classical WW studies (e.g. Hynes, 1963), which seems to persist despite the advances in (conventional) WW treatment over the last decades.

The context dependency, however, varied between the metrics for characterizing the macroinvertebrate communities. For example, the more

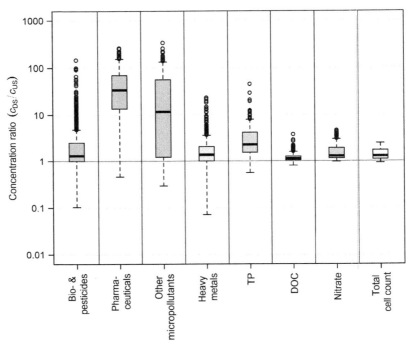

Fig. 7 *Boxplots* depicting the ratios between downstream (*DS*) concentrations and upstream concentrations (*US*) for different water quality parameters at the 24 EcoImpact sites across all sampling dates. The *red line* (*grey* in the print version) (ratio = 1) indicates that there is no difference between DS and US sites within a given stream. Values >1 indicate that the concentration of a given parameter is higher DS than US. *Bio- and pesticides*, concentration sum of all measured biocides and pesticides (31 compounds); *cell counts*, concentration of bacterial cells; *DOC*, dissolved organic carbon; *heavy metals*, concentration sums of the 10 most abundant heavy metals; *other MPs*, concentration sums of 5 compounds; *pharmaceuticals*, concentration sums of 21 compounds; *TP*, total phosphorus. Box size corresponds to the first and third quartiles (25% and 75% quartiles); whiskers extend to the most extreme data point which is no more than 1.5 times the length of the box away from the box. Details of the methods are provided in Box 5.

pristine the US conditions, the larger the change in the Saprobic Index at DS locations (Burdon et al., 2016). This variation in community-level responses is associated with background nutrient levels and, hence, the agricultural land use US of the WWTPs. In contrast, a larger fraction of WW leads to a more pronounced loss of sensitive macroinvertebrates at DS locations (i.e. SPEAR Index), irrespective of US conditions, indicating that some changes caused by discharged WW may not be environmentally contingent.

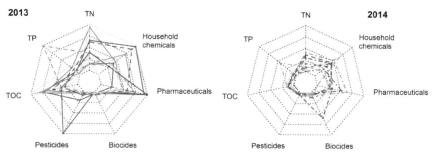

Fig. 8 *Spider diagrams* depicting the WW composition at the 24 *EcoImpact* sites. The *left* figure shows the results for the 2013 sites (one *colour* (*grey* in the print version) per site), the *right* for the 2014 sites. For each axis, the average concentrations of the respective January and June samples per site are normalized to the maximum value of all sites. *TOC*, total organic carbon; *TN*, total nitrogen; *TP*, total phosphorus; household chemicals (5 MPs); pharmaceuticals (21 MPs); biocides (3 MPs); pesticides (28 MPs).

Future analyses on the other biological endpoints (Table 3) will yield additional insights how effects of (specific) MPs may depend on the ecological context (ongoing work).

2.2.2 Inferring Causality (Flume Experiments)

Despite the merits of the *real-world experiment* approach described earlier, its potential for gaining insights into causal links between MPs and ecological effects is enhanced in combination with *research-led experiments*. To that end, we developed an artificial stream system (*Maiandros*) where contaminant concentrations can be manipulated through either dilution or spiking experiments (Box 4). The system is placed directly at a WWTP, which provides direct access to treated WW (allowing dilution experiments) and to WW treatment (allowing disposal of pollutants in spiking experiments).

The development of such experiments requires a substantial simplification compared to the natural system and is not a trivial task. This holds true both for water quality and for the biological endpoints under study. First, as there are thousands of MPs in streams receiving WW discharge (Box 2), how does one choose which ones to include? Second, as the impacts of MPs with different modes of action (Fig. 2) as well as their expected ecological relevance strongly depend on the study organism, which ecological endpoints are relevant to study? These quandaries touch upon the fundamental questions of how holistic experiments in ecology can actually be, and if ecology is inherently a reductionist science (Bergandi and Blandin, 1998). Here,

BOX 4 Experimental Flume System: *Maiandros*

Artificial flume systems allow experimental tests of causality of different water constituents on stream ecosystems and have been implemented by different institutions (Bruder et al., 2015; Grantham et al., 2012). Following such ideas, the *EcoImpact* team designed the flume system *Maiandros* that allows experimental manipulation of WW and replicated randomized experiments. *Maiandros* currently consists of 16 parallel channels (Fig. B.2) for running surface water experiments. The system allows for quantitatively mixing two different influents (e.g. local surface water with biologically treated WWTP effluent at different dilutions, or spiking experiments with known concentrations of MPs). Four mixing units are available; the blended influent of one water quality is distributed evenly to four flow channels. Each channel is doughnut shaped (length: 4 m, width: 0.15 m, water depth: 0.1 m) and is equipped with a paddle wheel. The typical hydraulic retention time is about 7 min.

Fig. B.2 The *Maiandros* flume system consisting of 16 flumes arranged in four blocks before launching the first experiment. Each of the four treatment combinations (e.g. four levels of dilutions or different combinations of MP and nutrients) is present in each block; within the blocks, the treatments are randomly assigned to the flumes. *Picture/copyright: A. Joss, Eawag.*

The design and implementation of such a flume system for MP studies in an ecologically relevant context is a technological challenge that requires expertise across all disciplines. Aspects to be addressed range from technical (e.g. how to keep mixtures of compounds in solution and which materials have the right mechanical and chemical properties) to ecological aspects (e.g. appropriate flow velocities, which organisms are suitable for such experiments and how to establish them in the flumes). Hence, as for the conceptual aspects and methodological of the project as a whole, interdisciplinarity is a key factor for a successful implementation.

EcoImpact takes a reductionist view in that we try to reproduce key findings from the field by testing causal hypotheses of potential drivers of change.

To that end, we followed a stepwise procedure. First, dilution experiments maintain the full chemical complexity of WW by exposing selected organisms to a range of mixtures between river water and treated WW (20%, 50%, 80%) and compare the biological effects to a control consisting of river water only (Exp. I). The selected biological endpoints included periphyton biomass and community tolerance (PICT assays; Tlili et al., 2015), performance (growth and survival) of freshwater crustaceans (gammarid amphipods) and molluscs (ancyclid snails) and ecosystem functions (see Table 3). Because several of these endpoints (e.g. cotton strip degradation, PICT assays; Table 3) were also measured at the field sites, this experiment allows a comparison of effects in situ in the field and in the flumes.

Second, spiking experiments substitute real WW by artificial mixture of chemicals and thereby allows disentangling the effects of MPs and nutrients (the main correlative factor in WW; Exp. II and Exp. III). In our case, we chose a mixture of 17 compounds (see Box 5) that are representative of MPs from our field sites and, hence, shared sufficient similarity with real-world mixtures to be of ecological relevance. In particular, frequently detected compounds with different modes of action (e.g. pesticides or pharmaceuticals) were selected. Exp. II, consisted of fully factorial combination of MP (present or not) and nutrient (present or not) treatments, compared to a control treatment (100% stream water). In Exp. III, the 'nutrients only' treatment was replaced by a technical control, to account for the possible effects of methanol used as solvent for the MPs. In all of these experiments, we applied a constant dilution or concentration treatment during 4 weeks.

In the field, such a disentangling of MP and nutrient effects will only be possible during Stage II (Fig. 3) once there is a sufficient number of WWTPs that have either been upgraded or shut down. In the flumes, this time lag can be avoided and we have obtained first results disentangling effects of MPs and nutrients (Exp. II and Exp. III). Results from Exp. II demonstrate that diatom communities that established on glass slides (method description see Box 5) in river water spiked with MPs have a reduced average biovolume as compared to the controls (Fig. 9) (Reyes, 2015). Spiking only nutrients to the river water, however, caused larger biovolumes. These findings suggest that MPs exert effects on ecologically important groups such as diatoms that

BOX 5 Methods Used in *EcoImpact*

The results briefly presented in our contribution, range from data on water chemistry and macroinvertebrate community responses in the field surveys to diatom community responses in Exp. II in the flumes (MP spiking experiment).

Water chemistry: Water chemistry was characterized both in terms of general composition (e.g. hardness, nutrients and major ions) and in terms of different types of pollutants (e.g. heavy metals, organic MPs, and oestrogens). General water chemistry, including nutrient levels, was quantified using standard methods of the Swiss National River Monitoring and Survey Programme (NADUF; http://www.bafu.admin.ch/wasser/13462/14737/15108/15109).

For the analysis of heavy metals, samples were acidified (0.6% HNO_3) and quantified with high-resolution inductively coupled plasma mass spectrometer (HR-ICP-MS; Element2, Thermo, Switzerland). Details can be found in Meylan et al. (2003). Here, we report the data of unfiltered samples (Figs 7 and B.1).

For the analysis of organic MPs (see Figs 7, 8 and B.1), the samples of the June 2013 campaign were enriched with offline solid-phase extraction (SPE) as described in Kern et al. (2009) using manually packed mixed-mode cartridges (Schollee et al., 2015), whereas all the other samples were enriched with an automated online SPE similar to Huntscha et al. (2012). Liquid chromatography high-resolution mass spectrometry (LC-HRMS) and the subsequent quantification were performed as described in Schollee et al. (2015) and Huntscha et al. (2012).

Oestrogen levels (Fig. B.1) were determined with the Yeast Oestrogen Screen test ('YES Test'; Routledge and Sumpter, 1996). This method relies on genetically modified yeast cells (*Saccharomyces cerevisiae*) containing the gene for the human oestrogen receptor coupled with a 'reporter gene' (LacZ). Oestrogenically active substances bind to the oestrogen receptor in the cell activating the reporter gene. This gene encodes for an enzyme (beta-galactosidase), which converts a dye from yellow to red. This process allows for photometric determination of the concentration of oestrogenically active substances.

Macroinvertebrate survey: The invertebrate sampling methods followed standard protocols for benthic macroinvertebrate biomonitoring in Switzerland (Stucki, 2010). Details for the methods can be found in Burdon et al. (2016) which shows responses for sites sampled in 2013. The presented results (Fig. 6) combine both 2013 and 2014 data. Briefly, at each sampling location (DS, US1, US2) in Spring 2013 and 2014 benthic invertebrates were sampled using pooled kick nets (25 × 25 cm opening, 500-μm mesh size) with a standardized sampling effort covering all major microhabitats found within each reach. Samples were stored in 80% ethanol prior to identification to the family level using relevant literature (Stucki, 2010). Macroinvertebrate communities were described using diversity-based indices, and two trait-based indices (Saprobic Index and SPEAR Pesticides Index) (Burdon et al., 2016).

BOX 5 Methods Used in *EcoImpact*—cont'd

The Saprobic Index ranges between 1 and 4 and increases with greater fractions of taxa that are tolerant to low oxygen conditions (Bunzel et al., 2013). Following the methods outlined in Burdon et al., 2016, the Saprobic Index was calculated using German saprobic trait values obtained from: http://www.freshwaterecology.info. This online resource is a taxa and autecology database for freshwater organisms (Version 5.0, Date accessed: 26.03.15; for more information, see Schmidt-Kloiber and Hering, 2015). The SPEAR Pesticides Index (SPEAR Index) describes the percentage of taxa susceptible to pesticides and was calculated using the SPEAR Calculator (Version 0.9.0, downloaded 17.2.16). Lower relative abundances of SPEAR taxa indicate pesticide stress and it is used extensively as an index of stream health in Europe (Beketov et al., 2009; von der Ohe and Liess, 2004).

Spiking experiments: Results from two spiking experiment in the *Maiandros* flumes (Box 4) are mentioned in the text (Exps II and III). In these two experiments, the MP mixture consisted of the following compounds (with nominal spiking concentration (ng L^{-1}) in the flumes given in parentheses): amisulpride (106), atenolol (217), benzotriazole (1098), candesartan (722, Exp. II only), carbamazepine (283), citalopram (55), clarithromycin (63), diazinon (4570 Exp. II, 457 Exp. III), diclofenac (603), diuron (75), β-estradiol (0.35), fexofenadine (322), tebuconazole (24), iopromide (1629), metformin (5014), sucralose (1175), triclosan (55), valsartan (677, Exp. III only) and Zn (6040). The increase of nutrient concentrations in the flumes due to spiking amounted to 0.05 P, 0.076 NH_4-N and 1.55 nitrate-N (values in mg L^{-1}).

Diatom data: Diatoms are sensitive bioindicators and readily respond to pollution (Visco et al., 2015). In *Maiandros* Exp. II, three glass slides (75 × 50 × 1.1 mm) were placed in each of the 16 channels, completely submerged and oriented in flow direction (16 October 2014). Prior to the experimental treatments, the slides were conditioned in the flumes with flowing river water for 1 week, so that all treatments would have a similar initial biofilm community. Subsequently, the glass slides (i.e. the biofilm community) was exposed to a combination of MPs and nutrients for 4 weeks (from 23 October to 11 November 2014). Subsequently, the biofilms were scratched off from both sides of the slides with a spatula into a centrifuge tube and the biofilm was immediately fixed with formaldehyde (4%). (Details can be found in Reyes, 2015.) One sample per channel (i.e. four replicates per treatment) was randomly selected and prepared for morphological identification of diatoms according to Hürlimann and Niederhauser (2007). Briefly, samples were first homogenized, subsampled and freed from coarse matter with hydrochloric acid. Subsequently, the samples were oxidized with sulphuric acid and potassium nitrate in order to remove all organic matter, leaving only the silica shells of the diatoms for later identification and measuring of cells (i.e. biovolume). Finally, the diatom valves were embedded with Naphrax into a glass

Continued

BOX 5 Methods Used in *EcoImpact*—cont'd

slide for microscopy. A total of 500 diatom valves per sample were identified to species level (when possible) or genus level after (Hofmann et al., 2013) with an inverted Leica Microscope and a x1000 magnification.

In order to characterize the biovolume of the community the dataset from Rimet and Bouchez (2012) was used. The volume for each species was calculated as *Length* × *Width* × *Height*. Size was corrected for all species accounting for at least 10% of abundance in each treatment with the mean value of 10 valves.

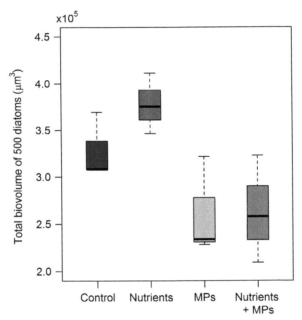

Fig. 9 Comparison of diatom biovolumes in a MP spiking experiment (Exp. II). In this experiment, glass slides were placed in the *Maiandros* flumes and diatoms allowed to establish under the four water treatments during 4 weeks (fall 2014). The treatments consisted of a control (C = River water only); MPs only (MP), MP+ with nutrients (nitrogen N, phosphorus, P) and nutrients only (N, P). The MP treatment consisted of spiking with a known concentration of mix of 17 MPs (see Box 5). Box sizes correspond to the first and third quartiles (25% and 75% quartiles); whiskers extend to the most extreme data point which is no more than 1.5 times the length of the box away from the box. *Data from Reyes, M., 2015. Auswirkungen von Mikroverunreinigungen und Nährstoffen von Kläranlagen auf Populationsstrukturen von Kieselalgen. Zertifikatsarbeit CAS Zürcher Hochschule für angewandte Wissenschaften, Wädenswil.*

override subsidy effects of nutrients. Upcoming work will analyse this aspect in more depth (including results from Exp. I to Exp. III).

3. OUTLOOK: POTENTIAL OF COMBINING *REAL-WORLD* AND *RESEARCH-LED EXPERIMENTS*

3.1 Planned Changes in Urban Infrastructure as *Real-World Experiments*

The current comparison of US to DS locations at 24 *EcoImpact* sites allows inferences on how WW may affect stream ecosystems—from physiological responses to ecosystem functions (Fig. 2; Table 3). In the future, upgrading of the WWTPs (Box 3) will provide an explicit opportunity to study the effects of altered water quality (i.e. removal of MPs or composite WW) on ecosystems across a large spatial range over time (Fig. 3). Due to the in-stream controls (Fig. 5), such temporal changes should be detectable against simultaneously occurring intrinsic changes that are independent of the changes related to WW (i.e. temporal changes that also occur in the US sections).

As promising as this approach looks, it comes with challenges that are of general relevance and can be exemplified with *EcoImpact*. First, to provide mechanistic insight, it is necessary to monitor key variables of interest in the long term both because not all responses may be immediate and to test how the initial effects change over time. Hence, one major issue to be solved is the long-term establishment of systematic observations. But also human interventions, such as implemented mitigation measures, used for the *real-world experiment* are not all realized simultaneously. For instance, it will take years for implementing all of foreseen WWTP modifications. Hence, Stage II of *EcoImpact* (BACI design) will last many years, a time span that goes beyond those of typical research projects. Due to financial and methodological constraints, monitoring programs of public authorities may not fully cover such needs. We emphasize that research institutions may benefit from contributing to such long-term observation given the opportunity this provides for gaining relevant long-term data (Lindenmayer et al., 2012). This in turn would allow, for example, studies on the interactions between ecological and evolutionary processes (Hairston et al., 2005; Moya-Laraño et al., 2014; Travis et al., 2014) of natural systems due to altered pollution patterns. Additional potential for long-term observations could be realized through the contributions of 'citizen scientists' (Bonney et al., 2009; Huddart et al., 2016). The usefulness of this approach has been recently demonstrated in following the ecological consequences of a pesticide spill in the United Kingdom (Thompson et al., 2015).

However, for really large-scale efforts, citizen science would likely not be sufficient. Here large-scale collaborations using highly standardized approaches, such as in the Nutrient Network (a globally distributed experimental network testing the effects of nutrient additions and food-web manipulations across terrestrial biomes) (Borer et al., 2014), may provide powerful means for answering big questions at a large scale. Given that WWTPs are common features of urban infrastructure in many countries around the globe and WW upgrades are conducted in several developed countries, such studies would allow detecting general trends and context dependency of MP impacts at large scales, as well as increase our understanding of ecosystem responses to environmental change in general.

3.2 *Real-World* and *Research-Led Experiments* in Large-Scale Ecology

We started off with the observation that environmental scientists often face a two-fold challenge in providing reliable answers at large scales to inform policy change on one hand, and local scale management advice on the other. Typically, such problems consist of multiple stressor situations. Important questions are then: how to identify the drivers of change at large and local scales, and what are the general (large-scale) trends in the ecological responses to the drivers like nutrient dependencies of litter decomposition at continental scale (Woodward et al., 2012), and when does local context dominate?

Above, we presented an interdisciplinary research approach to tackle such questions in the context of WW-borne MPs. We propose that the combination of *real-world experiments*, which explicitly take advantage of the implementation of mitigation measures in practice, and *research-led experiments* (as characterized in Table 4) provides a powerful tool for detecting general patterns (trends), elucidating the role of context dependency (e.g. local deviations from such trends) and establishing causality between external drivers and the biological responses. Such an approach broadens the scope for gathering relevant ecological data at large temporal and spatial scales and can be generalized beyond the problem of the ecological effects MPs as a tool for increasing inferential power of studies on ecosystems responses to human impact.

The major advantages of combining these approaches can be described as follows:

1. *Benefit from real-world experiments.* Although human interventions often result in multiple changes simultaneously, there are situations where a

Table 4 Comparison of *Real-World* and *Research-Led Experiments*

	Real-World Experiments	**Researcher-Led Experiments**
Advantages	• Response observed under full complexity • One to many replicates • Broad range of ecological context • Allows for detection of real-world trends • Facilitates communication with practitioners and society	• Designed according to scientific interests • Targeted at understanding mechanisms and establishing causality • Interpretation improved due to reduced complexity • Correlated influence factors can be separated • Randomized designs
Disadvantages	• Not designed to address scientific research questions • Possible mismatch between practical and scientific interests • Treatment categories may be coarse • Limited possibility for elucidating mechanisms and establishing causality • Several influencing factors may be correlated	• Artificial (boundary) conditions may limit transferability to and relevance for real-world situations
Layout	• Selection from existing situations (see Table 2)	• Predesigned experiment

Some exceptional experiments like whole-lake manipulations (Kidd et al., 2007; Schindler et al., 2008) may share features of both categories.

single factor (i.e. main driver) differs between points in space or time. By carefully identifying such situations, it is possible to layout a (quasi-) experimental design that allows both to isolate the influence of a single factor and to obtain generality at large scales due to the observation a broad range of ecological context

Key features of such a design are:
  (i) Avoiding confounding factors (for example, changes of stream morphology between US and DS sites that are not due to WW input) by careful choice of study sites.
  (ii) Quantifying the drivers of change—as well as putative correlated and confounding factors—to the best possible degree (e.g. composition of WW to test whether specific ecological effects are caused by specific MPs).

(iii) Including proper controls to account for confounding factors that may cause temporal changes irrespective of the treatments (e.g. altered ecological conditions due to climate change or the appearance of nonnative species).

(iv) Replicating in space and time. Only replication will allow to establish generality of findings, to investigate context-dependencies and to minimize the risk to overinterpret idiosyncratic, site-specific results (Nakagawa and Parker, 2015).

2. *Combine real-world experiments with research-led experiments.* While the appropriate experimental design in any *real-world experiment* is mainly achieved by a careful selection of existing situations, complementary classical *research-led experiments* are designed by scientists on purpose to test specific hypotheses. A key advantage of such integration is the possibility to separate different influence factors that are highly correlated in the real world, by replicated and randomized experiments. To make such experiments complementary to the *real-world experiments*, they should be designed as plausible simplifications of the real-world situation but still share common biological endpoints. Due to the inherent complexity of ecosystems, different types of manipulative experiments (lab, mesocosm, and field experiments), as well as measurements of different biological endpoints at different hierarchical levels across the ecosystem are needed (see Fig. 2).

3. *Benefit from a broad interdisciplinary team.* When the pertinent questions at hand—typical for ecologically relevant questions—reflect different interacting processes (such as chemical water quality, resulting toxicity and different biological responses), it is necessary to apply different methods and to collect and analyse vastly different types of data. Hence, collaborations across different research fields are mandatory. The common framing of the specific scientific question(s) and of an explicit conceptual framework for linking the results from the different disciplines is essential.

4. *Benefits of combining hypothesis-driven and exploratory approaches.* As ecosystems are inherently complex systems, any framing of the scientific question(s) and the underlying hypotheses is contingent on the current level of understanding. From a heuristic perspective, combining hypothesis-driven and exploratory approaches that widen the scientific view beyond the initial research question enhances the chance to identify blind spots in the current knowledge.

Real-world experiments have substantial additional advantages that make them interesting in the context of large-scale ecology. The 'implementation' of these experiments (not the scientific analysis, of course) comes for free to researchers and can be considered a 'legal' intervention in the environment. The significance of this argument becomes obvious when contrasted with the scenario of designing a purely scientific experiment to manipulate water quality at such large scales in real systems for studying ecological responses. Although there are well-known manipulative experiments on streams and lakes which were essential for advancing our ecological understanding (Kidd et al., 2007; Schindler et al., 2008; Woker and Wuhrmann, 1957), such experiments are generally too costly and contradict legal requirements regarding water protection and may raise serious ethical issues. Hence, if such experiments are going to happen for other reasons, science should take the opportunity to benefit as well. Given the large number of interventions into aquatic ecosystems, using them as *real-world experiments* can also be considered as one way that turns ecology into a branch of 'big data' science (Hampton et al., 2013).

Finally, such *real-world experiments* facilitate—according to our experiences—the interactions between researchers and stakeholders. When communicating with practitioners, it is essential to demonstrate the impact of scientific findings for practical situations that stakeholders are familiar. This transfer process is often not easy to establish. When working in the context of *real-world experiments*, this obstacle largely vanishes because the research results are obtained from situations with which both researchers and practitioners are familiar with. This aspect may get strengthened further if (long-term) observational studies are carried out in collaboration with citizen scientists and practitioners.

ACKNOWLEDGEMENTS

EcoImpact is an endeavour to which many people and institutions have contributed and still do. The field survey was only possible with the tireless effort by Marta Reyes, the careful work of the entire Aquabug-team (Pascal Stucki, Nathalie Menétrey, Sandra Knispel, Remo Wenger), Stefan Achermann, Alfredo Alder, Birgit Beck, Pravin Ganesanandamoorthy, Felix Neff, Christoph Ort, Anja Taddei, and many other helping in the field. The cotton strip assays profited a lot from the exchange with and support by Scott Tiegs (Oakland University). We obtained great analytical support by the AUA lab and Madelaine Langmeier for the classical water chemistry parameters, all heavy metal analyses were carried out carefully by David Kistler, and all the MPs were analysed by the team around Heinz Singer with major contributions by Barbara Spycher, Fabian Deurer, Philipp Longrée, and Nicole Munz. The *Maiandros* flume experiments were strongly supported by Anita Wittmer, Jack Eugster, Simon Mangold, Adrian Müller, Richard Fankhauser, and Tobias Wyler. The data analyses and visualizations were supported by Raphael Thierrin and Urs Schönenberger.

The entire project benefited continuously from intensive discussions and the exchange of ideas among a larger group of people from Eawag and the Ecotox Center: Renata Behra, Helmut Bürgmann, Nadine Czekalski, Kristy Deiner, Stephan Fischer, Juliane Hollender, Hanna Hartikainen, David Johnson, Cornelia Kienle, Alexandra Kroll, Miriam Langer, Cresten Mansfeld, Christa McArdell, Tiina Salo, Kristin Schirmer, Nele Schuwirth, Otto Seppälä, Linn Sgier, Rosi Siber, Manu Tamminen, Ahmed Tlili, Etienne Vermeirssen, and Inge Werner.

We also acknowledge the support by all operators of the WWTPs included in the field survey with their staff. A great thanks goes especially to the WWTP of Fällanden (ARA Bachwis) with the team around M. Moos for hosting the flume experiments.

Finally, such a project needs a financial basis. The core project has been predominantly funded by Eawag with additional support by the Federal Office for the Environment (FOEN). Additional funding is from the Swiss National Science Foundation Grant PP00P3_150698 (to F.A.).

Chris Robinson and two anonymous reviewers were very valuable for improving the quality of the manuscript, for identifying blind spots, and for clarifying critical aspects of the text.

REFERENCES

Abbeglen, C., Siegrist, H., 2012. Mikroverunreinigungen aus kommunalem Abwasser. Verfahren zur weitergehenden Elimination auf Kläranlagen. Umwelt-Wissen Nr. 1214, Bundesamt für Umwelt, Bern. 210 pp.

Altermatt, F., Seymour, M., Martinez, N., 2013. River network properties shape α-diversity and community similarity patterns of aquatic insect communities across major drainage basins. J. Biogeogr. 40 (12), 2249–2260.

Aristi, I., von Schiller, D., Arroita, M., Barceló, D., Ponsatí, L., García-Galán, M.J., Sabater, S., Elosegi, A., Acuña, V., 2015. Mixed effects of effluents from a wastewater treatment plant on river ecosystem metabolism: subsidy or stress? Freshw. Biol. 60, 1398–1410.

Bärlocher, F., 2005. Leaf mass loss estimated by litter bag technique. In: Graça, M.S., Bärlocher, F., Gessner, M. (Eds.), Methods to Study Litter Decomposition. Springer, pp. 37–42.

Beketov, M.A., Foit, K., Schäfer, R.B., Schriever, C.A., Sacchi, A., Capri, E., Biggs, J., Wells, C., Liess, M., 2009. SPEAR indicates pesticide effects in streams—comparative use of species- and family-level biomonitoring data. Environ. Pollut. 157, 1841–1848.

Benfield, E.F., 2007. Decomposition of leaf material. In: Hauer, F.R., Lamberti, G.A. (Eds.), Methods in Stream Ecology. Academic Press, Burlington, MA, pp. 711–720.

Benton, T.G., Solan, M., Travis, J.M.J., Sait, S.M., 2007. Microcosm experiments can inform global ecological problems. Trends Ecol. Evol. 22, 516–521.

Bergandi, D., Blandin, P., 1998. Holism vs. reductionism: do ecosystem ecology and landscape ecology clarify the debate? Acta Biotheor. 46, 185–206.

Bernhardt, E.S., Palmer, M.A., 2011. River restoration: the fuzzy logic of repairing reaches to reverse catchment scale degradation. Ecol. Appl. 21, 1926–1931.

Bernhardt, E.S., Palmer, M.A., Allan, J.D., Alexander, G., Barnas, K., Brooks, S., Carr, J., Clayton, S., Dahm, C., Follstad-Shah, J., Galat, D., Gloss, S., Goodwin, P., Hart, D., Hassett, B., Jenkinson, R., Katz, S., Kondolf, G.M., Lake, P.S., Lave, R., Meyer, J.L., O'Donnell, T.K., Pagano, L., Powell, B., Sudduth, E., 2005. Synthesizing U.S. river restoration efforts. Science 308, 636–637.

Biggs, B.J.F., Kilroy, C., 2000. Stream Periphyton Monitoring Manual. NIWA, Christchurch.

Bonney, R., Cooper, C.B., Dickinson, J., Kelling, S., Phillips, T., Rosenberg, K.V., Shirk, J., 2009. Citizen science: a developing tool for expanding science knowledge and scientific literacy. BioScience 59, 977–984.

Borer, E.T., Harpole, W.S., Adler, P.B., Lind, E.M., Orrock, J.L., Seabloom, E.W., Smith, M.D., 2014. Finding generality in ecology: a model for globally distributed experiments. Methods Ecol. Evol. 5, 65–73.

Bott, T.L., 1996. Primary productivity and community respiration. In: Hauer, F.R., Lamberti, G.A. (Eds.), Methods in Stream Ecology. Academic Press, Burlington, MA, pp. 663–690.

Brodin, T., Fick, J., Jonsson, M., Klaminder, J., 2013. Dilute concentrations of psychiatric drug alter behavior of fish from natural populations. Science 339, 814–815.

Brodin, T., Piovano, S., Fick, J., Klaminder, J., Heynen, M., Jonsson, M., 2014. Ecological effects of pharmaceuticals in aquatic systems—impacts through behavioural alterations. Phil. Trans. R. Soc. B Biol. Sci. 369, 20130580.

Bruder, A., Salis, R.K., McHugh, N.J., Matthaei, C.D., 2015. Multiple-stressor effects on leaf litter decomposition and fungal decomposers in agricultural streams contrast between litter species. Funct. Ecol. 30, 1057–1266. http://dx.doi.org/10.1111/1365-2435.12598.

Bundschuh, M., Schulz, R., 2011. Ozonation of secondary treated wastewater reduces ecotoxicity to *Gammarus fossarum* (Crustacea; Amphipoda): are loads of (micro)pollutants responsible? Water Res. 45, 3999–4007.

Bunzel, K., Kattwinkel, M., Liess, M., 2013. Effects of organic pollutants from wastewater treatment plants on aquatic invertebrate communities. Water Res. 47, 597–606.

Burdon, F.J., Reyes, M., Alder, A.C., Joss, A., Ort, C., Räsänen, K., Jokela, J., Eggen, R.I.L., Stamm, C., 2016. Environmental context and disturbance influence differing trait-mediated community responses to wastewater pollution in streams. Ecol. Evol. 6, 3923–3939. http://dx.doi.org/10.1002/ece1003.2165.

Burkhardt-Holm, P., Giger, W., Güttinger, H., Ochsenbein, U., Peter, A., Scheurer, K., Segner, H., Staub, E., Suter, M.J., 2005. Where have all the fish gone? The reasons why the fish catches in Swiss rivers are declining. Environ. Sci. Technol. 39, 441A–447A.

Capy, P., Gasperi, G., Biémont, C., Bazin, C., 2000. Stress and transposable elements: co-evolution or useful parasites? Heredity 85, 101–106.

Carpenter, S.R., 1996. Microcosm experiments have limited relevance for community and ecosystem ecology. Ecology 77, 677–680.

Carpenter, S.R., Cole, J.J., Pace, M.L., Batt, R., Brock, W.A., Cline, T., Coloso, J., Hodgson, J.R., Kitchell, J.F., Seekell, D.A., Smith, L., Weidel, B., 2011. Early warnings of regime shifts: a whole-ecosystem experiment. Science 332, 1079–1082.

Carvalho, R.N., Arukwe, A., Ait-Aissa, S., Bado-Nilles, A., Balzamo, S., Baun, A., Belkin, S., Blaha, L., Brion, F., Conti, D., Creusot, N., Essig, Y., Ferrero, V.E.V., Flander-Putrle, V., Fürhacker, M., Grillari-Voglauer, R., Hogstrand, C., Jonáš, A., Kharlyngdoh, J.B., Loos, R., Lundebye, A.-K., Modig, C., Olsson, P.-E., Pillai, S., Polak, N., Potalivo, M., Sanchez, W., Schifferli, A., Schirmer, K., Sforzini, S., Stürzenbaum, S.R., Søfteland, L., Turk, V., Viarengo, A., Werner, I., Yagur-Kroll, S., Zounková, R., Lettieri, T., 2014. Mixtures of chemical pollutants at European legislation safety concentrations: how safe are they? Toxicol. Sci. 141, 218–233. http://dx.doi.org/10.1093/toxsci/kfu118.

Diamond, J.M., 1983. Laboratory, field and natural experiments. Nature 304, 586–587.

Durance, I., Bruford, M.W., Chalmers, R., Chappell, N.A., Christie, M., Cosby, B.J., Noble, D., Ormerod, S.J., Prosser, H., Weightman, A., Woodward, G., 2016.

The challenges of linking ecosystem services to biodiversity: lessons from a large-scale freshwater study. Adv. Ecol. Res. 54, 87–134.

Eggen, R.I.L., Hollender, J., Joss, A., Schärer, M., Stamm, C., 2014. Reducing the discharge of micropollutants in the aquatic environment: the benefits of upgrading wastewater treatment plants. Environ. Sci. Technol. 48, 7683–7689. http://dx.doi.org/10.1021/es500907n.

Emmerson, M., Morales, M.B., Oñate, J.J., Batáry, P., Berendse, F., Liira, J., et al., 2016. How agricultural intensification affects biodiversity and ecosystem services. Adv. Ecol. Res. 55, 43–97.

Escher, B.I., Bramaz, N., Mueller, J.F., Quayle, P., Rutishauser, S., 2008. Toxic equivalent concentrations (TEQs) for baseline toxicity and specific modes of action as a tool to improve evaluation of ecotoxicity tests on environmental samples. J. Environ. Monit. 10, 612–621.

Evans, M.R., Bithell, M., Cornell, S.J., Dall, S.R.X., Dìaz, S., Emmott, S., Ernande, B., Grimm, V., Hodgson, D.J., Lewis, S.L., Mace, G.M., Morecroft, M., Moustakas, A., Murphy, E., Newbold, T., Norris, K.J., Petchey, O., Smith, M., Travis, J.M.J., Benton, T.G., 2013. Predictive systems ecology. Proc. R. Soc Lond. B Biol. Sci. 280, 20131452.

Fischer, B.B., Pomati, F., Eggen, R.I.L., 2013. The toxicity of chemical pollutants in dynamic natural systems: the challenge of integrating environmental factors and biological complexity. Sci. Total Environ. 449, 253–259.

Folt, C.L., Chen, C.Y., Moore, M.V., Burnaford, J., 1999. Synergism and antagonism among multiple stressors. Limnol. Oceanogr. 44, 864–877.

Frissell, C.A., Liss, W.J., Warren, C.E., Hurley, M.D., 1986. A hierarchical framework for stream habitat classification: viewing streams in a watershed context. Environ. Manag. 10, 199–214.

Gerhardt, A., 2011. GamTox: a low-cost multimetric ecotoxicity test with Gammarus spp. for in and ex situ application. Int. J. Zool. 2011. 574536.

Grantham, T.E., Cañedo-Argüelles, M., Perrée, I., Rieradevall, M., Prat, N., 2012. A mesocosm approach for detecting stream invertebrate community responses to treated wastewater effluent. Environ. Pollut. 160, 95–102.

Gray, C., Hildrew, A.G., Lu, X., Ma, A., McElroy, D., Monteith, D., et al., 2016. Recovery and nonrecovery of freshwater food webs from the effects of acidification. Adv. Ecol. Res. 55, 475–534.

Güttinger, H., Stumm, W., 1992. An analysis of the Rhine pollution caused by the Sandoz chemical accident, 1986. Interdiscip. Sci. Rev. 17, 127–136.

Hairston Jr., N.G., Ellner, S.P., Geber, M.A., Yoshida, T., Fox, J.A., 2005. Rapid evolution and the convergence of ecological and evolutionary time. Ecol. Lett. 8, 1114–1127.

Halstead, N.T., McMahon, T.A., Johnson, S.A., Raffel, T.R., Romansic, J.M., Crumrine, P.W., Rohr, J.R., 2014. Community ecology theory predicts the effects of agrochemical mixtures on aquatic biodiversity and ecosystem properties. Ecol. Lett. 17 (8), 932–941.

Hamers, T., Molin, K.R.J., Koeman, J.H., Murk, A.J., 2000. A small-volume bioassay for quantification of the esterase inhibiting potency of mixtures of organophosphate and carbamate insecticides in rainwater: development and optimization. Toxicol. Sci. 58, 60–67.

Hampton, S.E., Strasser, C.A., Tewksbury, J.J., Gram, W.K., Budden, A.E., Batcheller, A.L., Duke, C.S., Porter, J.H., 2013. Big data and the future of ecology. Front. Ecol. Environ. 11, 156–162.

Hering, J.G., Hoehn, E., Klinke, A., Maurer, M., Peter, A., Reichert, P., Robinson, C., Schirmer, K., Schirmer, M., Stamm, C., Wehrli, B., 2012. Moving targets, long-lived infrastructure, and increasing needs for integration and adaptation in water management: an illustration from Switzerland. Environ. Sci. Technol. 46, 112–118.

Heynen, M., Fick, J., Jonsson, M., Klaminder, J., Brodin, T., 2016. Effect of bioconcentration and trophic transfer on realized exposure to oxazepam in 2 predators, the dragonfly larvae (*Aeshna grandis*) and the Eurasian perch (*Perca fluviatilis*). Environ. Toxicol. Chem. 35, 930–937.

Hicklen, R.S., Chadwick, M.A., Dobberfuhl, D.R., 2006. Effects of detrital food sources on growth of a physid snail. J. Mollus. Stud. 72, 435–438.

Hofmann, G., Lange-Bertalot, H., Werum, M., 2013. Diatomeen im Süsswasser-Benthos von Mitteleuropa, 2. Korrigierte Auflage Koeltz Scientific Books, Königstein.

Hollender, J., Zimmermann, S.G., Koepke, S., Krauss, M., McArdell, C.S., Ort, C., Singer, H., von Gunten, U., Siegrist, H., 2009. Elimination of organic micropollutants in a municipal wastewater treatment plant upgraded with a full-scale post-ozonation followed by sand filtration. Environ. Sci. Technol. 43, 7862–7869. http://dx.doi.org/10.1021/es9014629.

Hostmann, M., Bernauer, T., Mosler, H.-J., Reichert, P., Truffer, B., 2005. Multi-attribute value theory as a framework for conflict resolution in river rehabilitation. J. Multi Criteria Decis. Anal. 13, 91–102.

Huddart, J.E.A., Thompson, M.S.A., Woodward, G., Brooks, S.J., 2016. Citizen science: from detecting pollution to evaluating ecological restoration. Wiley Interdiscip. Rev. Water 3, 287–300.

Huntscha, S., Singer, H.P., McArdell, C.S., Frank, C.E., Hollender, J., 2012. Multiresidue analysis of 88 polar organic micropollutants in ground, surface and wastewater using online mixed-bed multilayer solid-phase extraction coupled to high performance liquid chromatography-tandem mass spectrometry. J. Chromatogr. A 1268, 74–83. http://dx.doi.org/10.1016/j.chroma.2012.1010.1032.

Hürlimann, J., Niederhauser, P., 2007. Methoden zur Untersuchung und Beurteilung der Fliessgewässer. Kieselalgen Stufe F (flächendeckend). Bundesamt für Umwelt, Bern. Umwelt-Vollzug, 130 pp.

Hynes, H.B.N., 1963. The Biology of Polluted Waters, second ed. Liverpool University Press, Liverpool.

Kail, J., Brabec, K., Poppe, M., Januschke, K., 2015. The effect of river restoration on fish, macroinvertebrates and aquatic macrophytes: a meta-analysis. Ecol. Indic. 58, 311–321.

Keller, V.D.J., Williams, R.J., Lofthouse, C., Johnson, A.C., 2014. Worldwide estimation of river concentrations of any chemical originating from sewage-treatment plants using dilution factors. Environ. Toxicol. Chem. 33, 447–452.

Kern, S., Fenner, K., Singer, H., Schwarzenbach, R., Hollender, J., 2009. Identification of transformation products of organic contaminants in natural waters by computer-aided prediction and high resolution mass spectrometry. Environ. Sci. Technol. 443, 7039–7046.

Kidd, K.A., Blanchfield, P.J., Mills, K.H., Palace, V.P., Evans, R.E., Lazorchak, J.M., Flick, R.W., 2007. Collapse of a fish population after exposure to a synthetic estrogen. Proc. Natl. Acad. Sci. U.S.A. 104, 8897–8901.

Kolpin, D.W., Furlong, E.T., Meyer, M.T., Thurman, E.M., Zaugg, S.D., Barber, L.B., Buxton, H.T., 2002. Pharmaceuticals, hormones, and other organic wastewater contaminants in U.S. streams, 1999-2000: a national reconnaissance. Environ. Sci. Technol. 36, 1202–1211.

Ladewig, V., Jungmann, D., Köhler, H.R., Schirling, M., Triebskorn, R., Nagel, R., 2005. Population structure and dynamics of *Gammarus fossarum* (Amphipoda) upstream and downstream from effluents of sewage treatment plants. Arch. Environ. Contam. Toxicol. 50, 370–383.

Lamberti, G.A., Resh, V., 1985. Comparability of introduced tiles and natural substrates for sampling lotic bacteria, algae and macro invertebrates. Freshw. Biol. 15, 21–30.

Lewis, S.L., Maslin, M.A., 2015. Defining the anthropocene. Nature 519, 171–180.

Liess, M., von der Ohe, P.C., 2005. Analyzing effects of pesticides on invertebrate communities in streams. Environ. Toxicol. Chem. 24, 954–965.

Lindenmayer, D.B., Likens, G.E., Andersen, A., Bowman, D., Bull, C.M., Burns, E., Dickman, C.R., Hoffmann, A.A., Keith, D.A., Liddell, M.J., Lowe, A.J., Metcalfe, D.J., Phinn, S.R., Russell-Smith, J., Thurgate, N., Wardle, G.M., 2012. Value of long-term ecological studies. Austral Ecol. 37, 745–757.

MacArthur, R.H., MacArthur, J.W., 1961. On bird species diversity. Ecology 42, 594–598.

Mancinelli, G., 2012. To bite, or not to bite? A quantitative comparison of foraging strategies among three brackish crustaceans feeding on leaf litters. Estuar. Coast. Shelf Sci. 110, 125–133.

Meylan, S., Behra, R., Sigg, L., 2003. Accumulation of copper and zinc in periphyton in response to dynamic variations of metal speciation in freshwater. Environ. Sci. Technol. 37, 5204–5212.

Millennium Ecosystem Assessment, 2005. Ecosystems and Human Well-Being: Synthesis. Island Press, Washington, DC.

Mouquet, N., Lagadeuc, Y., Devictor, V., Doyen, L., Duputié, A., Eveillard, D., Faure, D., Garnier, E., Gimenez, O., Huneman, P., Jabot, F., Jarne, P., Joly, D., Julliard, R., Kéfi, S., Kergoat, G.J., Lavorel, S., Le Gall, L., Meslin, L., Morand, S., Morin, X., Morlon, H., Pinay, G., Pradel, R., Schurr, F.M., Thuiller, W., Loreau, M., 2015. Predictive ecology in a changing world. J. Appl. Ecol. 52, 1293–1310. http://dx.doi.org/10.1111/1365-2664.12482.

Moya-Laraño, J., Bilbao-Castro, J.R., Barrionuevo, G., Ruiz-Lupión, D., Casado, L.G., Montserrat, M., Melián, C.J., Magalhães, S., 2014. Eco-evolutionary spatial dynamics: rapid evolution and isolation explain food web persistence. Adv. Ecol. Res. 50, 76–143.

Nakagawa, S., Parker, T.H., 2015. Replicating research in ecology and evolution: feasibility, incentives, and the cost-benefit conundrum. BMC Biol. 13, 88. http://dx.doi.org/10.1186/s12915-12015-10196-12913.

Neumann, M.B., Rieckermann, J., Hug, T., Gujer, W., 2015. Adaptation in hindsight: dynamics and drivers shaping urban wastewater systems. J. Environ. Manage. 151, 404–415.

O'Neal, M.E., Landis, D.A., Isaacs, R., 2002. An inexpensive, accurate method for measuring leaf area and defoliation through digital image analysis. J. Econ. Entomol. 95, 1190–1194.

Palmer, M.A., 2010. Water resources: beyond infrastructure. Nature 467, 534–535.

Palmer, M.A., Bernhardt, E.S., Allan, J.D., Lake, P.S., Alexander, G., Brooks, S., Carr, J., Clayton, S., Dahm, C.N., Follstad Shah, J., Galat, D.L., Loss, S.G., Goodwin, P., Hart, D.D., Hassett, B., Jenkinson, R., Kondolf, G.M., Lave, R., Meyer, J.L., O'Donnell, T.K., Pagano, L., Sudduth, E., 2005. Standards for ecologically successful river restoration. J. Appl. Ecol. 42, 208–217.

Peschke, K., Geburzi, J., Köhler, H.-R., Wurm, K., Triebskorn, R., 2014. Invertebrates as indicators for chemical stress in sewage-influenced stream systems: toxic and endocrine effects in gammarids and reactions at the community level in two tributaries of Lake Constance, Schussen and Argen. Ecotoxicol. Environ. Saf. 106, 115–125.

Petchey, O.L., Pontarp, M., Massie, T.M., Kéfi, S., Ozgul, A., Weilenmann, M., Palamara, G.M., Altermatt, F., Matthews, B., Levine, J.M., Childs, D.Z., McGill, B.J., Schaepman, M.E., Schmid, B., Spaak, P., Beckerman, A.P., Pennekamp, F., and Pearse, I.S., 2015. The ecological forecast horizon, and examples of its uses and determinants. Ecol. Lett. 18, 597–611.

Petrie, B., Barden, R., Kasprzyk-Hordern, B., 2015. A review on emerging contaminants in wastewaters and the environment: current knowledge, understudied areas and recommendations for future monitoring. Water Res. 72, 3–27.

Reyes, M., 2015. Auswirkungen von Mikroverunreinigungen und Nährstoffen von Kläranlagen auf Populationsstrukturen von Kieselalgen. Zertifikatsarbeit CAS Zürcher Hochschule für angewandte Wissenschaften, Wädenswil.
Rimet, F., Bouchez, A., 2012. Life-forms, cell-sizes and ecological guilds of diatoms in European rivers. Knowl. Manag. Aquat. Ecosyst. 406. 01.
Robinson, C.T., Schuwirth, N., Baumgartner, S., Stamm, C., 2014. Spatial relationships between land-use, habitat, water quality and lotic macroinvertebrates in two Swiss catchments. Aquat. Sci. 76, 375–392. http://dx.doi.org/10.1007/s00027-014-0341-z.
Rockström, J., Steffen, W., Noone, K., Persson, A., Chapin, F.S., Lambin, E.F., Lenton, T.M., Scheffer, M., Folke, C., Schellnhuber, H.J., Nykvist, B., de Wit, C.A., Hughes, T., van der Leeuw, S., Rodhe, H., Sorlin, S., Snyder, P.K., Costanza, R., Svedin, U., Falkenmark, M., Karlberg, L., Corell, R.W., Fabry, V.J., Hansen, J., Walker, B., Liverman, D., Richardson, K., Crutzen, P., Foley, J.A., 2009. A safe operating space for humanity. Nature 461, 472–475.
Rosemond, A.D., Mulholland, P.J., Brawley, S.H., 2000. Seasonally shifting limitation of stream periphyton: response of algal populations and assemblage biomass and productivity to variation in light, nutrients, and herbivores. Can. J. Fish. Aquat. Sci. 57, 66–75.
Rosés, N., Poquet, M., Muñoz, I., 1999. Behavioural and histological effects of atrazine on freshwater molluscs (*Physa acuta* drap. and *Ancylus fluviatilis* Müll. gastropoda). J. Appl. Toxicol. 19, 351–356.
Rosi-Marshall, E.J., Kincaid, D.W., Bechtold, H.A., Royer, T.V., Rojas, M., Kelly, J.J., 2013. Pharmaceuticals suppress algal growth and microbial respiration and alter bacterial communities in stream biofilms. Ecol. Appl. 23, 583–593.
Rotter, S., Sans-Piché, F., Streck, G., Altenburger, R., Schmitt-Jansen, M., 2011. Active bio-monitoring of contamination in aquatic systems—an in situ translocation experiment applying the PICT concept. Aquat. Toxicol. 101, 228–236. http://dx.doi.org/10.1016/j.aquatox.2010.10.001.
Routledge, E.J., Sumpter, J.P., 1996. Estrogenic activity of surfactants and some of their degradation products assessed using a recombinant yeast screen. Environ. Toxicol. Chem. 15, 241–248.
Ruff, M., Müller, M.S., Loos, M., Singer, H.P., 2015. Quantitative target and systematic non-target analysis of polar organic micro-pollutants along the river Rhine using high-resolution mass-spectrometry—identification of unknown sources and compounds. Water Res. 87, 145–154. http://dx.doi.org/10.1016/j.watres.2015.09.017.
Schindler, D.W., Hecky, R.E., Findlay, D.L., Stainton, M.P., Parker, B.R., Paterson, M.J., Beaty, K.G., Lyng, M., Kasian, S.E.M., 2008. Eutrophication of lakes cannot be controlled by reducing nitrogen input: results of a 37-year whole-ecosystem experiment. Proc. Natl. Acad. Sci. U.S.A. 105, 11254–11258.
Schmidt-Kloiber, A., Hering, D., 2015. www.freshwaterecology.info—an online tool that unifies, standardises and codifies more than 20,000 European freshwater organisms and their ecological preferences. Ecol. Indic. 53, 271–282.
Schollee, J.E., Schymanski, E.L., Avak, S.E., Loos, M., Hollender, J., 2015. Prioritizing unknown transformation products from biologically-treated wastewater using high-resolution mass spectrometry, multivariate statistics, and metabolic logic. Anal. Chem. 87, 12121–12129.
Schwarzenbach, R.P., Escher, B.I., Fenner, K., Hofstetter, T.B., Johnson, C.A., von Gunten, U., Wehrli, B., 2006. The challenge of micropollutants in aquatic systems. Science 313, 1072–1077.
Schymanski, E.L., Singer, H.P., Longrée, P., Loos, M., Ruff, M., Stravs, M.A., Ripollés Vidal, C., Hollender, J., 2014. Strategies to characterize polar organic contamination in wastewater: exploring the capability of high resolution mass spectrometry. Environ. Sci. Technol. 48, 1811–1818.

Silva, E., Rajapakse, N., Kortenkamp, A., 2002. Something from 'Nothing'—eight weak estrogenic chemicals combined at concentrations below NOECs produce significant mixture effects. Environ. Sci. Technol. 36, 1751–1756.

Singer, H.P., Wössner, A.E., McArdell, C.S., Fenner, K., 2016. Rapid screening for exposure to 'non-target' pharmaceuticals from wastewater effluents by combining HRMS-based suspect screening and exposure modeling. Environ. Sci. Technol. 50, 6698–6707.

Slocum, M.G., Roberts, J., Mendelssohn, I.A., 2009. Artist canvas as a new standard for the cotton-strip assay. J. Plant Nutr. Soil Sci. 172, 71–74.

Stephens, D.W., Jennings, M.E., 1976. Determination of primary productivity and community metabolism in streams and lakes using diel oxygen measurements. U.S. Geological Survey Computer Contribution, 100 pp.

Stucki, P., 2010. Methoden zur Untersuchung und Beurteilung der Fliessgewässer. Makrozoobenthos Stufe F. Umwelt-Vollzug Nr. 1026, Bundesamt für Umwelt, Bern. 61 pp.

Sundermann, A., Stoll, S., Haase, P., 2011. River restoration success depends on the species pool of the immediate surroundings. Ecol. Appl. 21, 1962–1971.

Ternes, T.A., 1998. Occurrence of drugs in German sewage treatment plants and rivers. Water Res. 32, 3245–3260.

Thompson, M.S.A., Bankier, C., Bell, T., Dumbrell, A.J., Gray, C., Ledger, M.E., Lehmann, K., McKew, B.A., Sayer, C.D., Shelley, F., Trimmer, M., Warren, S.L., Woodward, G., 2015. Gene-to-ecosystem impacts of a catastrophic pesticide spill: testing a multilevel bioassessment approach in a river ecosystem. Freshw. Biol.. http://dx.doi.org/10.1111/fwb.12676.

Tiegs, S.D., Clapcott, J.E., Griffiths, N.A., Boulton, A.J., 2013. A standardized cotton-strip assay for measuring organic-matter decomposition in streams. Ecol. Indic. 32, 131–139.

Tlili, A., Berard, A., Blanck, H., Bouchez, A., Cássio, F., Eriksson, K.M., Morin, S., Montuelle, B., Navarro, E., Pascoal, C., Pesce, S., Schmitt-Jansen, M., Behra, R., 2015. Pollution-induced community tolerance (PICT): towards an ecologically relevant risk assessment of chemicals in aquatic systems. Freshw. Biol.. http://dx.doi.org/10.1111/fwb.12558.

Tockner, K., Pusch, M., Borchardt, D., Lorang, M.S., 2010. Multiple stressors in coupled river–floodplain ecosystems. Freshw. Biol. 55 (Suppl. 1), 135–151.

Travis, J., Reznick, D., Bassar, R.D., López-Sepulcre, A., Ferriere, R., Coulson, T., 2014. Understand nature? Answers from studies of the Trinidadian guppy. Adv. Ecol. Res. 50, 1–40.

Underwood, A.J., 1994. On beyond BACI: sampling designs that might reliably detect environmental disturbances. Ecol. Appl. 4, 3–15.

Van den Brink, P.J., 2008. Ecological risk assessment: from book-keeping to chemical stress ecology. Environ. Sci. Technol. 42, 8999–9004.

van Straalen, N.M., 2003. Ecotoxicology becomes stress ecology. Environ. Sci. Technol. 37, 324A–330A.

Vaughan, I.P., Ormerod, S.J., 2012. Large-scale, long-term trends in British river macroinvertebrates. Glob. Chang. Biol. 18, 2184–2194.

Visco, J.A., Apothéloz-Perret-Gentil, L., Cordonier, A., Esling, P., Pillet, L., Pawlowski, J., 2015. Environmental monitoring: inferring the diatom index from Next-Generation Sequencing data. Environ. Sci. Technol. 49, 7597–7605.

von der Ohe, P.C., Liess, M., 2004. Relative sensitivity distribution of aquatic invertebrates to organic and metal compounds. Environ. Toxicol. Chem. 23, 150–156.

Vörösmarty, C.J., McIntyre, P.B., Gessner, M.O., Dudgeon, D., Prusevich, A., Green, P., Glidden, S., Bunn, S.E., Sullivan, C.A., Liermann, C.R., Davies, P.M., 2010. Global threats to human water security and river biodiversity. Nature 467, 555–561.

Ward, J.V., Tockner, K., Arscott, D.B., Claret, C., 2002. Riverine landscape diversity. Freshw. Biol. 47, 517–539.
Watts, M.M., Pascoe, D., Carroll, K., 2002. Population responses of the freshwater amphipod Gammarus pulex (L.) to an environmental estrogen, 17α-ethinylestradiol. Environ. Toxicol. Chem. 21, 445–450.
Woker, H., Wuhrmann, K., 1957. Die Reaktion der Bachfauna auf Gewässervergiftungen. Rev. Suisse Zool. 64, 253–262.
Woodward, G., Gessner, M.O., Giller, P.S., Gulis, V., Hladyz, S., Lecerf, A., Malmqvist, B., McKie, B.G., Tiegs, S.D., Cariss, H., Dobson, M., Elosegi, A., Ferreira, V., Graça, M.A.S., Fleituch, T., Lacoursière, J.O., Nistorescu, M., Pozo, J., Risnoveanu, G., Schindler, M., Vadineanu, A., Vought, L.B.M., Chauvet, E., 2012. Continental-scale effects of nutrient pollution on stream ecosystem functioning. Science 336, 1438–1440.

PART II

Large/Long Temporal Scale Ecology and Model Systems

CHAPTER FIVE

The Colne Estuary: A Long-Term Microbial Ecology Observatory

D.B. Nedwell[1], G.J.C. Underwood, T.J. McGenity, C. Whitby, A.J. Dumbrell

School of Biological Sciences, University of Essex, Colchester, United Kingdom
[1]Corresponding author: e-mail address: nedwd@essex.ac.uk

Contents

1. Introduction	228
1.1 Ecological Importance of Estuaries	228
1.2 Anatomy of a 'Model' Estuary—Microbial Ecology	229
2. Study Site Description	231
2.1 Description of the Catchment and Estuary	231
2.2 Nitrogen Inputs	233
2.3 Phosphate Inputs and N:P Ratios	234
2.4 Physical Factors	236
3. Functional Ecology of Estuarine Microbes	238
3.1 Primary Production	238
3.2 The Importance of MPB in Muddy Estuarine Systems	240
3.3 Organic Matter Breakdown and Recycling	247
3.4 Nitrate Respiration	251
3.5 Nitrification in the Colne Estuary	260
3.6 Archaea	262
4. Estuarine Saltmarshes	263
4.1 The Role of Saltmarshes	263
5. Estuaries and Climatically Important Trace Gases	265
5.1 Trace Gas Production in the Estuary	265
5.2 Sulphur Gases	266
5.3 Isoprene Cycling	267
6. Stressors and Pollution	268
6.1 Crude Oil Degradation	268
6.2 Engineered Nanoparticles	269
7. Future Directions	270
Acknowledgements	273
References	273

Abstract

Research spanning over 40 years has examined many aspects of the microbial ecology of the Colne estuary (Essex, United Kingdom) and it is arguably the most comprehensively understood temperate estuary in the Northern hemisphere. The hypernutrified Colne estuary exhibits strong gradients of nutrient concentrations (nitrate, ammonium, phosphate, dissolved organic nitrogen) from river and treated sewage inputs at the top of the estuary, with concentrations decreasing towards the estuary mouth. These strong concentration gradients facilitate the study of the biogeochemical process rates and the microbial communities responsible. Planktonic primary production is at an oligotrophic level because of light limitation in the turbid water, but a mixed water column maintains planktonic photosynthesis despite low light. Dense microphytobenthic biofilms occur throughout the estuary, with high rates of annual primary production typical of NW European nutrient-rich estuaries, but benthic primary production accounts for only a small proportion of the N load to the estuary. Organic matter degradation is focussed in the estuarine sediments, with greatest organic content in the upper estuary, decreasing towards the mouth. Benthic biogeochemical processes, leading to organic mineralisation and element recycling, including O_2 uptake, sulphate reduction, methanogenesis, nitrate reduction, denitrification and anammox have all been quantified along the estuary, together with chemolithotrophic nitrification. Benthic denitrification removes a significant proportion of the N load to the estuary. Molecular techniques, including profiling, identification and quantification of 16S rRNA and key functional genes, have provided an understanding of the microbial communities in relation to position and biogeochemical activity. The concentrations and emissions of biological volatile compounds (nitrous oxide, methane, hydrogen sulphide and isoprene) have also been measured, and their ecological significance elucidated. Tidal exchange between the saltmarshes and the main estuary channel shows seasonal variation. When nitrate in tidal water is high in spring, the marsh sediments remove it, but tidally export ammonium and small particles of organic nitrogen to the estuary in summer when nitrogen in coastal water is low. Overall, the saltmarshes show a balanced nitrogen cycle suggestive of a stable climax community, but capable of responding to and removing increased nitrate concentrations in estuarine waters. These data, originating from the long-term study of a single system, are unique and this paper highlights how the Colne estuary *microbial ecology observatory* has contributed to our understanding of estuarine microbial ecology and biogeochemistry.

1. INTRODUCTION
1.1 Ecological Importance of Estuaries

Globally, estuaries are among the most productive of all ecosystems and provide an abundant food source for many commercially important fish species (Elliott et al., 2007; Platell et al., 2006). Estuaries also provide connectivity between the catchment and the sea and are an essential route through which anadromous fish species migrate from their main feeding areas in the sea to their spawning grounds in freshwaters (Gillanders, 2005; Pollard, 1981). The productivity of estuaries is also exploited by many bird species as they migrate to and from breeding grounds (Elliott and McLusky, 2002). The mudflats and saltmarshes associated with estuaries provide an important buffer zone between land and water, with the associated plants and microphytobenthos (MPB) stabilising sediments (Underwood and Paterson, 2003) and generally limiting the risk of flooding. Yet, estuarine zones are under increasing threat from anthropogenic activities following human population growth and industrialisation. In addition, large inputs of nutrients from agricultural land run off have led to some estuaries becoming highly eutrophic (Flindt et al., 1999; Howarth et al., 2011). Estuaries are influenced directly by their physical, chemical and geological characteristics, which, in turn, influence the ecology and function of the microorganisms driving in situ processes. Despite the great importance of estuaries to a range of biota and fisheries, more information is still needed to better understand their complex, dynamic and variable nature, and to better understand and manage the ecosystem services that estuaries provide.

1.2 Anatomy of a 'Model' Estuary—Microbial Ecology

The estuary of the River Colne, Essex, provides an excellent model system, typical of the muddy estuaries on the east coast of England, and has been the focus of long-term research, primarily at the University of Essex, spanning over 40 years. This long-term study focussed on what are probably the most poorly understood, yet arguably most ecologically important groups of living organisms—microbes. Early studies focussed on developing and applying suitable methods for measuring the rates of different biogeochemical activities within the estuary, and the emergence of molecular biological techniques then allowed the identification and quantification of the key groups of microorganisms responsible for these activities (Fig. 1). During this 40 years period, most of the major groups of microorganisms have been examined and together have contributed to one of the most comprehensively understood temperate estuaries in the Northern Hemisphere. Moreover, the knowledge gleaned from these unique long-term data sets from the Colne estuary can be translated to understand the ecology and function of estuaries elsewhere.

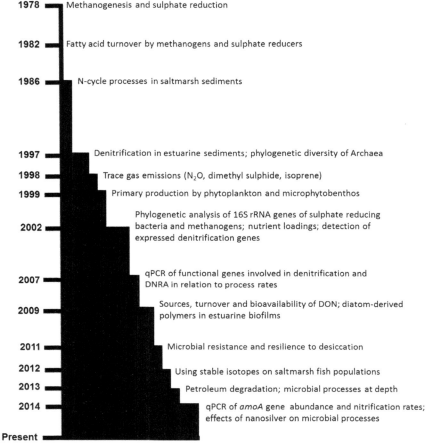

Fig. 1 Examples of some key research areas and focus over the 40-year-long study of the Colne estuary as a microbial observatory. Early studies focused on ecological processes and process rates, but in later years the focus shifted to identifying key microbial players using novel molecular techniques and linking these to ecological processes.

This chapter is designed to provide an ecological map describing the microbial anatomy of a model estuarine ecosystem and is based around long-term research on microbial communities, their constituent populations and the ecological processes and function that they regulate. To start, the physicochemical properties of the Colne estuary are described in depth (Section 2), in order to: contextualise the microbial diversity and biogeochemical process data presented, allow a deeper mechanistic understanding of the abiotic and biotic factors driving ecosystem processes, and provide comparative data for readers to assess the similarity of the Colne estuary (a model temperate Northern Hemisphere estuary) to estuarine systems globally. The majority of this chapter is then devoted to the functional ecology of estuarine microbes (Section 3), with particular attention on their role in primary production, the ecological importance of the MPB, and the central role microbes play in regulating major macronutrient cycles, especially the nitrogen cycle. Following from this, the estuary and its microbial ecology are placed within the context of the wider landscape, by exploring the linkages with saltmarsh habitats (Section 4), before moving onto its global significance as a regulator of climatically important trace-gas fluxes (Section 5). As common with almost all ecosystems, estuaries are subjected to multiple anthropogenic stressors, which may ultimately change their biodiversity and associated ecosystem functioning. This chapter explores the ecological impacts of two main contemporary stressors that have the potential to alter microbial communities and processes (Section 6). Finally, we finish with a speculative look into the future and explore the role of the Colne estuary as a model system for the next generation of estuarine research, describing new and future research projects and how this long-term study site will contribute to these (Section 7).

2. STUDY SITE DESCRIPTION

2.1 Description of the Catchment and Estuary

The Colne estuary is approximately 16 km long, from the upper limit of tidal influence defined by East Mill Weir (51.890168N, 0.916437E), to the mouth of the estuary where it enters the North Sea at Brightlingsea (51.804963N, 1.012553E) (Fig. 2). The estuary exhibits strong longitudinal gradients of salinity and solute concentrations, making biological responses to these gradients easy to detect, and its relatively small size makes sampling coverage relatively easy.

The river drains a catchment of approximately 350 km^2 in which agriculture is the predominant activity, particularly growth of cereal crops. The human population in the catchment was 216,000 at the last census in 2010 (Office for National Statistics: http://www.nationalarchives.gov.uk), the largest centre of population being Colchester, an ancient town founded in AD 46 by the Romans. The estuary covers an area of 14.5 km^2 and is a turbid hypernutrified mesotidal estuary, with a depth range from 1.5 m at the head to >15 m at the mouth and tidal amplitude >4 m. The estuary channel is surrounded by approximately 900 ha of saltmarsh, with creeks

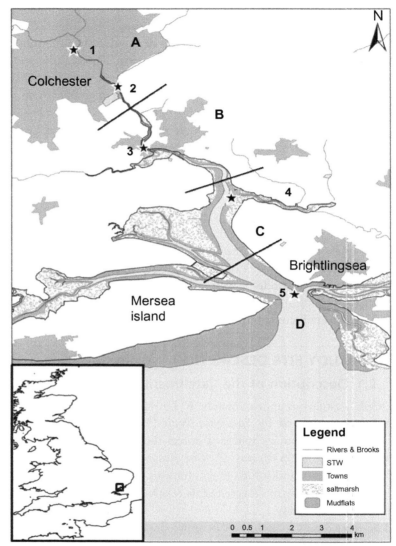

Fig. 2 Location of the Colne estuary and the sampling sites used. (1) The river enters the estuary at East Mills; (2) Hythe Bridge: location of Colchester sewage treatment works outflow; (3) Rowhedge ferry in mid-estuary; (4) Alresford Creek; (5) the mouth of the estuary at Bateman's Tower, Brightlingsea. Zones A–D represent major salinity regions of the estuary: A, 0.5–5‰; B, 5–18‰; C, 18–30‰; D, >30‰.

draining into the main estuary channel. The Colne–Blackwater Estuary Ecosystem has been assigned a moderate quality status under the E.U. Water Framework Directive. These geomorphological, hydrodynamic and catchment features mean that the Colne is a representative example of a large proportion of temperate estuarine systems worldwide (Elliott and McLusky, 2002), making it an ideal model system for ecological research. The Colne estuary is designated a Site of Special Scientific Interest, a National Nature Reserve, a Special Protected Area, a Special Area of Conservation, a Ramsar Site and a Marine Conservation Zone.

The largest inputs to the estuary are from the river flow, although the gauged inputs to the estuary are complicated by irregular unmetered abstractions from the river below the gauging station for water conservation purposes. River flow varies seasonally, increasing during winter due to high rainfall and decreasing during the summer months.

2.2 Nitrogen Inputs

The major fluvial N load to the estuary is from nitrate which varies seasonally and from year to year (Figs 3 and 4).

Nitrate loads are high during autumn and winter when rainfall washes nitrate from the soil, while ammonium is an order of magnitude smaller (Table 1). The years 1989–92 and 1996–98 were low river flow years when inputs from the STW were particularly significant.

The second major load of nutrients to the estuary is from Colchester sewage treatment work (STW; see Fig. 5), which discharges a relatively constant flow of tertiary-treated effluent that can be as much as 50% of total flow to estuary during summer months. A further significant N load to the estuary is dissolved organic nitrogen (DON; Table 1; Agedah et al., 2009). About 60% of the total N load to the estuary is from the river (particularly nitrate and DON) and about 40% from the STW (particularly ammonium; Table 1). The fluvial N load varies seasonally, with most occurring during winter when biological activity is reduced, whereas the ammonium load showed little seasonal signal (Fig. 5).

Nutrient concentrations at different sites along the estuary have changed over the last 30 years (Fig. 6). Highest concentrations of dissolved inorganic nitrogen (DIN) are at East Mills, decreasing down the estuary. In contrast, ammonium is maximal at Hythe Bridge, reflecting input of ammonium from the STW. A trend of reducing ammonium concentrations at the Hythe since 2000 reflects improved nitrification-based treatment processes in the STW.

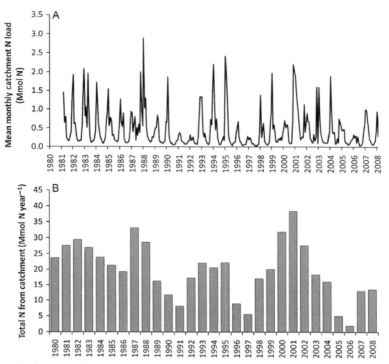

Fig. 3 Catchment N load entering the estuary (1981–2008): (A) monthly mean DIN load from catchment entering the estuary at East Mills; (B) total annual N load to estuary from catchment. *From McMellor, S., Underwood, G.J.C., 2014. Water policy effectiveness: 30 years of change in the hypernutrified Colne estuary, England. Mar. Pollut. Bull. 81, 200–209, with permission from Elsevier.*

In particular, phosphate concentrations at the Hythe have also decreased markedly since 2000 when phosphate stripping was introduced to the STW, but the low concentrations of phosphate at both Rowhedge and Brightlingsea indicate rapid removal of phosphate in the estuary.

2.3 Phosphate Inputs and N:P Ratios

The concentrations of inorganic N and P in the estuary can be extremely high, as much as 1 mM nitrate and 50 μM phosphate near the head of the estuary where the STW outfall is situated. Nutrient concentrations in the water column decline down the estuary as river water is mixed tidally with low nutrient coastal seawater, peaks of concentration at Hythe Bridge corresponding to increase near the STW outflow. Moreover, the N:P ratios tend to increase down the estuary towards the mouth indicating more rapid

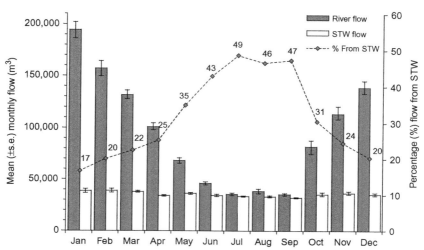

Fig. 4 Mean monthly flow 1981–2010 to estuary (m^3 ± S.E.) from river (*grey bars*), sewage treatment works (*white bars*) and percentage of estuarine flow from STW (*diamonds*). From McMellor, S., Underwood, G.J.C., 2014. Water policy effectiveness: 30 years of change in the hypernutrified Colne estuary, England. Mar. Pollut. Bull. 81, 200–209, with permission from Elsevier.

Table 1 Annual Loads (Mmol N year^{-1}) and Percentages of DON (Dissolved Organic Nitrogen), Nitrate and Ammonium Loads from River Colne and Colchester Sewage Treatment Works (STW) in 2003–04

	Nitrate (Mmol N year^{-1})	Ammonium (Mmol N year^{-1})	DON (Mmol N year^{-1})	Total N (Mmol N year^{-1})
River	38.0 (68%)	2.4 (19%)	12.0 (74%)	52.4 (62%)
STW	17.5 (32%)	10.2 (81%)	4.3 (32%)	32.0 (38%)

Percentages of each component from river or STW are shown in parentheses.
From Agedah, E.C., Binalaiyifa, H.E., Ball, A.S., Nedwell, D.B., 2009. Sources, turnover and bioavailability of dissolved organic nitrogen (DON) in the Colne estuary, UK. Mar. Ecol. Prog. Ser. 382, 23–33.

removal of phosphate than nitrate, presumably due to removal of phosphate by adsorption to particulates in the turbid estuary water (e.g. Prastka et al., 1998). Generally, it is assumed that a molar N:P ratio >10 indicates a likelihood for any algae becoming ultimately P limited as they use up nutrients during growth. This does not mean that algal growth rates are actually P limited at such high N:P ratios, as concentrations of both N and P may be high and saturating. It is only when concentrations of nutrients become depleted that the N:P ratio indicates which nutrient is likely to become

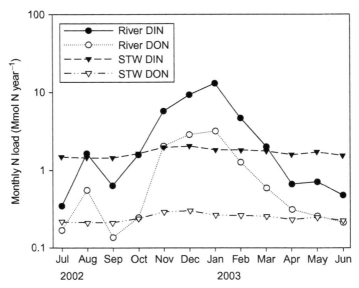

Fig. 5 Monthly loads of dissolved inorganic nitrogen (*DIN*) and dissolved organic nitrogen (*DON*) to the Colne estuary from the river and from the sewage treatment works 2002–03. From Agedah, E.C., Binalaiyifa, H.E., Ball, A.S., Nedwell, D.B., 2009. Sources, turnover and bioavailability of dissolved organic nitrogen (DON) in the Colne estuary, UK. Mar. Ecol. Prog. Ser. 382, 23–33, with permission from Inter-Research.

limiting to further algal biomass formation. The N:P ratios in the River Colne tend to be very high during winter when nitrate concentrations increase due to leaching of nitrate from catchment soils by winter rains, but decrease during summer to much lower values <10–15 at lower river flow rates with low nitrate. At the head of the estuary the effluent from the STW contains large amounts of phosphate, sufficient to change the water to a potentially N-limited condition with N:P ratios near 20 (Kocum et al., 2002a,b), while N:P values <10 near the estuary mouth indicate the potential for N limitation in the coastal seawater. For additional landscape-scale context, Nedwell et al. (2002) show data on N and P loads and N:P ratios for all UK estuaries.

2.4 Physical Factors

Elliot et al. (1994) reported a tidal flushing time of 0.7 days for the Colne estuary, and at mean low tide approximately 30% of the estuarine area is emerged, while 70% is permanently immersed (Robinson et al., 1998). A more useful parameter than tidal flushing to illustrate the residence time

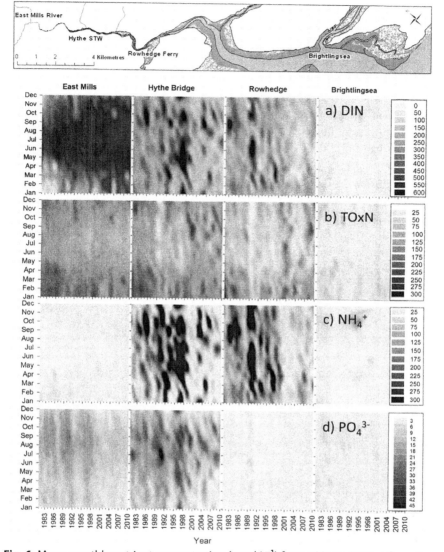

Fig. 6 Mean monthly nutrient concentration (μmol L^{-1}) from 1981 to 2010 at four sites down the Colne estuary from East Mills at the head to Brightlingsea at the mouth. (A) DIN concentration, (B) TOxN concentration, (C) NH_4^+ concentration and (D) PO_4^{3-} concentration. Note the varied scales for the different nutrient concentrations. *From McMellor, S., Underwood, G.J.C., 2014. Water policy effectiveness: 30 years of change in the hypernutrified Colne estuary, England. Mar. Pollut. Bull. 81, 200–209, with permission from Elsevier.*

of nutrients within the estuary is the freshwater flushing time (FWFT), which expresses the time required for the river flow to replace the freshwater lens present in the estuary (e.g. Trimmer et al., 2000). Sage (unpublished data) reported values of FWFT of 5–14 days for the Colne estuary, varying seasonally with rainfall and river flow. The longer the FWFT, the greater will be the interaction between the solutes in the water with the sediments beneath and in general the longer FWFT occurs in the low-flow period during the summer. The estuarine water, like the shallow southern North Sea, is always turbid due to suspended particulates, with average light attenuation coefficients (K, m^{-1}) of 3.2 at the top of the estuary to 1.4 at the mouth. This high turbidity has important impact on the productivity of the estuary (see later). There is usually no distinct turbidity maximum (a zone of particularly high settlement of particulates) in the water, which is in contrast to other turbid estuaries such as the Thames (Trimmer et al., 2000). The water column is not stratified, but well mixed throughout the estuary.

3. FUNCTIONAL ECOLOGY OF ESTUARINE MICROBES

3.1 Primary Production

The first study of primary production in the water column of the estuary was carried out in the 1990s (Kocum et al., 2002a,b). Despite the very high nutrient concentrations in the estuary water, the rates of primary production by phytoplankton in the Colne were at the lower end of the range of estuarine productivity values given by Underwood and Kromkamp (1999), and within the range suggested by Nixon (1995) as typical of oligotrophic estuaries. Phytoplankton production as measured by $^{14}CO_2$ uptake, and phytoplankton biomass as indicated by chlorophyll concentration, ranged between, respectively, 6.4×10^{-4}–17 µg C L^{-1} h^{-1} and 0.5–37.5 µg chl $a L^{-1}$. Contrary to what is seen in coastal waters, primary production rates were maximal where the water was most turbid (with very high light attenuation coefficients) at the top of the estuary. Estimates based on pseudo in situ measurements of annual primary production at four sites along the estuary (Brightlingsea, Aldboro Point, Wivenhoe and Hythe) gave average primary productivity values of 5.5, 4.2, 4.3 and 160.4 g C m^{-2} $year^{-1}$, respectively (Kocum et al., 2002a). When the area in each sector of the estuary was taken into account, this gave an average estimate for annual phytoplanktonic primary production in the estuary of 8.9 g C m^{-2} $year^{-1}$. The mid-tide surface area of water for the estuary is 3.75×10^6 m^2 (Robinson, 1996), indicating a total phytoplankton primary production of 3.34×10^7 g C $year^{-1}$ for the whole estuary. Using an average 106 C:16 N:1 P:1 Si Redfield atom ratio for algal biomass, this phytoplanktonic primary production in the estuary would account for 4.2×10^5 mol N, compared to an annual input of DIN (equivalent to nitrate, nitrite and ammonium) in 2003–04 of 68 Mmol N (Agedah et al., 2009): equivalent to only 0.6% of the DIN load to the estuary. Therefore, there is no doubt that despite the hypernutrified status of the estuary, it is not eutrophicated: i.e. the trophic status of the estuary biota does not reflect an overenriched community, but why should this be so when nutrient loads are so high?

The sources of inorganic nitrogen, nitrate or ammonium, assimilated by algae, can be measured using uptake of either ^{15}N-nitrate or ^{15}N-ammonium to determine the F-ratio. Values of $F > 0.5$ indicate preferential uptake of nitrate by the algae, while $F < 0.5$ indicate preferential uptake of ammonium. Kocum et al. (2002b) showed that the F-ratio of phytoplankton in the estuary was almost always <0.5, indicating preferential usage of ammonium rather than nitrate by the phytoplankton, despite ammonium concentration being an order of magnitude lower than nitrate. Even so, the primary production represented only $0.42/12.6 \times 100 = 3.3\%$ of the ammonium load to the estuary. Agedah et al. (2009) also showed that the large DON load was essentially inert within the time period that it was retained within the estuary, being neither broken down nor removed by assimilation within the water column, although it could be removed slowly by the bottom sediments. Only 2.2–5.2% of the DON was bioavailable at any one time. Similarly, using the Redfield C:P ratio of 106:1, the annual primary production was equivalent to only 3.93×10^5 g P year^{-1}. As so much of the P load to an estuary is by P adsorbed to particles (Prastka et al., 1998), which has not been measured for the Colne, we cannot estimate the percent of P input assimilated during primary production, but even compared solely to the soluble P input it is small. Generally, these data illustrated that the phytoplankton in the estuary utilised only a small part of the nutrient load to the estuary and was not limited by low N or P concentrations, but by other factors.

Phytoplankton biomass indicated by chlorophyll concentrations was lowest between January and March and then increased to maximum values in July, with highest concentrations in the upper estuary and decreasing towards the mouth. The phytoplankton communities were predominantly flagellated euglenophytes, with very small proportions of diatoms. Some resuspended benthic diatoms were present such as *Cylindrotheca closterium*, *Nitzschia* spp. and *Navicula* spp. particularly in spring, but decreasing into summer when wind speeds, turbulence and resuspension were likely to be lower. The low abundance of diatoms was probably associated with the Redfield N:Si ratios >1, indicating a potential for silicate limitation which would impact diatoms because of their requirement for silicate in their frustules, but not flagellates which do not have silicate in their cell walls (e.g. Justic et al., 1995). The low abundance of diatoms may have impact upon the nature of the consumers in higher trophic levels (e.g. Humborg et al., 1997) as many species such as fish will graze on diatoms but not on the much smaller flagellates.

The factor which was most significant to water column primary production in the turbid Colne estuary was the light regime. The vertical light attenuation coefficients were high (mean K 3.2 m^{-1}), indicating fast attenuation of light with depth but primary production is also influenced by the time that algal cells remain within the euphotic zone. This is indicated by the critical mixing ratio (CMR), which is the depth of the mixed zone of water (Z_{mix}) divided by the depth of the euphotic zone (Z_{eu}). Z_{mix} in the Colne is the water column depth as there is no stratification, while Z_{eu} is the measured depth of the euphotic zone (Kocum et al., 2002a). Grobbellar (1990) suggested that the CMR was the most important factor affecting phytoplankton production in turbid waters and values <5–20 were proposed to be necessary before net primary production (NPP) occurred. The ratio varied only 0.85–8.5 in the Colne estuary indicating only low potential for NPP. Furthermore, the higher CMR in the lower estuary indicated that photosynthesis was actually more light limited in the less turbid waters at the estuary mouth that in the more turbid upper estuary, because in the latter the shallow water column stopped phytoplankton dropping out of the euphotic zone. In situ primary production was significantly lower than the potential for production as indicated by measured P^B_{max} values, showing that primary production was light limited for most of the year. Therefore, it can be concluded that phytoplankton production in the estuary is limited by low light levels due to high light attenuation by turbidity, and this limits the phytoplankton primary production and degree of eutrophication. In contrast, the low values of CMR indicated that the shallow well-mixed water column helped maintain primary production despite the high light attenuation because phytoplankton remained within the euphotic zone.

3.2 The Importance of MPB in Muddy Estuarine Systems

MPB is a collective term for the assemblages of photosynthetic organisms (eukaryotic and prokaryotic), and their associated microbial flora, that inhabit the surface layers of illuminated sediments (Underwood and Kromkamp, 1999). The success of MPB living in muds is attributable to a number of adaptations, key of which is motility that allows photosynthetic cells to (re)position themselves in favourable light climates. Motile MPB are primarily pennate diatoms and euglenophytes, and some taxa of filamentous cyanobacteria (Underwood and Kromkamp, 1999). The more motile forms are dominant on finer sediments (termed epipelon). On sandier sediments, other taxa able to stick to sand grains (epipsammon) occur (Barnett et al., 2015). In estuaries such as the Colne, with extensive intertidal mud and sand flats, and a turbid water column that limits phytoplankton activity, the MPB are a major contributor to estuarine primary production (Underwood and Kromkamp, 1999). Thornton et al. (2002) estimated benthic primary production to range between 95 and 1199 g C m^{-2} $year^{-1}$ at the head of the estuary to 25–127 g C m^{-2} $year^{-1}$ at the mouth, with annual production of MPB within the estuary between 1.17 and 8.52×10^8 g C $year^{-1}$; an order of magnitude greater than that calculated for phytoplankton in the estuary. Using the Redfield C:N ratio the mean MPB production value for the estuary equated to 4.43 Mmol N $year^{-1}$, or about 5.5% of the annual DIN load to the estuary. The wide range of estimates reflect both the spatial and temporal heterogeneity of MPB abundance and productivity of these types of estuarine and coastal habitats, and different methods of scaling up to annual values. However, the rates and values for the Colne estuary are of the same magnitude as those recorded in other similar sheltered, fine-grained sediment estuarine systems, and the gradient down the estuary reflects the general pattern of higher values of MPB production within estuarine embayments compared to open coast or more nutrient-limited MPB-rich habitats.

MPB carbon drives the microbial loop within sediments and contributes to water-column carbon as cells and dissolved organic carbon that are resuspended on incoming tides, and is a significant source of organic carbon for grazing benthic invertebrates and higher trophic levels (Grangeré et al., 2012; Green et al., 2012).

The distribution, species composition and productivity of MPB in the Colne estuary have been studied during numerous research projects since 1992. There are few other estuaries in north-west Europe with a 23-year series of sediment Chl *a* concentration measurements, and because this data set covers a period of nutrient reduction and varying freshwater inputs (McMellor and Underwood, 2014), it allows consideration of how significant these factors are in determining broad-scale MPB biomass patterns.

Grouping data from multiple studies over the period 1992–2015 into four broad spatial regions (zones 1–4, Fig. 2) down the estuarine salinity gradient (Underwood et al., 1998) shows that average benthic sediment Chl *a* concentrations exhibit a high level of variability, yet within relatively consistent boundaries (Fig. 7A). Average sediment Chl *a* concentrations over the whole period varied between 72.7 (standard error (S.E.) ±12.8) and 138 mg m^{-2} (±17.2), with no strong patterns of increase or decline over this 23-year period (Fig. 7A) despite some significant changes in nutrient concentrations and loadings to the estuary during this time (McMellor and

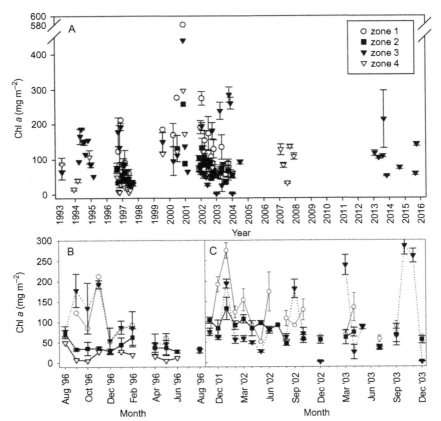

Fig. 7 (A) Sediment microphytobenthic Chl *a* (mg m^{-2}, *top* 2 mm of sediment) at various zones in the Colne estuary between 1993 and 2015. (B) Monthly variability at four stations between August 1996 and August 1997. (C) Monthly variability at three stations between November 2001 and December 2003. All values mean ± standard error, *n* varies with study zones indicated in Fig. 1.

Underwood, 2014). Blanchard et al. (2006) proposed a model where mudflat biofilms become temporally nutrient limited at Chl a concentrations in excess of 150–200 mg m^{-2}, regardless of overlying water concentrations, and the biomass distribution in the Colne (Fig. 7) conforms to this model in the broadest sense (with occasional outliers, particularly in the high nutrient concentration regions of the upper estuary).

There have been very few long-term studies of MPB running over more that 2–3 years, but those that exist appear to indicate some underlying broad drivers of biomass. De Jonge et al. (2012) synthesised sediment Chl a data in the Ems estuary (Netherlands) over the period from 1976 to 1999 (data mainly 1970s and 1990s). This showed a long-term positive relationship between biomass and annual air temperatures, with higher Chl a concentration during the 1990s. Similar interannual patterns of biomass at different stations within the estuary were found. Van der Wal et al. (2010) used remote sensing data to determine MPB biomass on mudflats using normalised differential vegetation index (NVDI) for seven estuarine and coastal mudflat regions in the southern North Sea region over the period 2001–09. Their study showed broad synchrony in the patterns of occurrence and biomass between estuaries, though stronger relationships were present within regional sets (e.g. within Dutch estuaries). Sediment type and exposure/tidal position were main drivers of sediment NVDI, but weather factors (windiness, and to a lesser extent air temperature) were also significant factors influencing biomass. A similar pattern of strong influences by weather, and also summer temperatures, was found by Benyoucef et al. (2014) using NVDI data from 1993 to 1998 and between 2006 and 2010 from two locations in the Loire estuary (France). McMellor and Underwood (2014) showed that there was a relationship between the North Atlantic Oscillation (NAO) climate signal and nutrient inputs into the Colne estuary over a 30-year period. The NAO influence was strongest at the freshwater end of the estuary and decreased towards the marine portion of the estuary. Despite the reduction in nutrient loading to the Colne over this 30-year period, and particularly in the period from 1995 to 2005 (McMellor and Underwood, 2014), there was not a significant decline in MPB biomass in the upper reaches of the Colne estuary. This probably reflects the superabundance of nutrients in the Colne system (hypernutrified status), effectively buffering the MPB from potential nutrient limitation effects. In addition, potential positive effects of climate-NAO drivers (such as warmer periods, *sensu* De Jonge et al., 2012; Van der Wal et al., 2010) could provide compensating drivers, resulting in limited overall changes in MPB biomass.

Long-term patterns (when present) are partially confounded by high levels of spatial and temporal patchiness, which is a characteristic of MPB in intertidal mudflat and sandflat environments, and also by influences on other elements of the estuarine system, such as grazers (De Jonge et al., 2012) that in turn influence MPB. Determining the potential effects of climate or environmental change on the MPB requires further good quality long-term data sets, sampled at an appropriate resolution.

Sediment Chl a is usually measured using minicores (diameter 1–2 cm) and patchiness at this scale contributes up to 36–60% of all measures of variability (Spilmont et al., 2011; Taylor et al., 2013; Weerman et al., 2011). Numerous daily, monthly and seasonal studies have been carried out on MPB distribution, noting seasonal changes and decreases in biomass across individual mudflats from high to low shore (Forster et al., 2006; Underwood and Paterson, 1993; van der Wal et al., 2010). MPB biomass in the Colne shows similar spatial and seasonal variability amongst sites along the estuarine gradient (Thornton et al., 2002; Fig. 7B and C), matching those broader patterns identified by van der Wal et al. (2010) and Benyoucef et al. (2014).

Correlations between nutrient concentrations and MPB biomass are often reported. While it is logical to assume that such correlations may reflect a dependency of nutrient loadings, the confidence level of such assumptions will depend on the trophic status of the particular system. There is no doubt that low inorganic nutrient concentrations (N, P, Si) limit MPB activity in estuaries in low-nutrient ecosystems, and that biofilm productivity can be temporally nutrient limited over diel exposure cycles (Miles and Sundbäck, 2000; Thornton et al., 1999). However, the Colne is generally hypernutrified (McMellor and Underwood, 2014), such that even though nutrient concentrations and loads have declined, there has not been a measurable decrease in MPB biomass over the last 23 years. The high organic content of sediments and active mineralisation maintains high porewater concentrations of nutrients in the sediments of the estuary, and MPB, because of their ability to vertically migrate, can access these sources of nutrients. High concentrations of particular compounds, e.g., ammonium and sulphide (due to STW output or organic enrichment), can be both inhibitory and selective for particularly resistant taxa. Peletier (1996) found significant changes in Chl a and species composition in the Ems Dollard estuary between 1979, 1980 and 1993, after the reduction of inputs from local potato starch industries that had been discharging organic waste onto adjacent mudflats. Small-scale (10–100 m scales) spatial patterns in biomass and species composition have been found within the Colne system

(Thornton et al., 2002; Underwood et al., 1998), which can be linked to species-specific preferences and tolerances (Underwood and Provot, 2000), and tolerance to sulphide and anoxia (McKew et al., 2013).

The main drivers of MPB photosynthesis and biomass accumulation are light, environmental stress, nutrients, physical disturbance and grazing (van Colen et al., 2014). The ability to move permits individual cells to position themselves within the sediment, maximising their exposure to light (during tidal emersion), but also migrating away from damaging light conditions (Underwood et al., 2005). Many MPB taxa are physiologically flexible, and able to use rapid photochemical and nonphotochemical quenching, longer-term acclimation of Chl a and other photopigments, to maintain high rates of primary production in a rapidly varying light climate over tidal emersion and during seasonal cycles (Barnett et al., 2015; Juneau et al., 2015; Underwood et al., 2005). These factors combine to produce general patterns of higher biomass on upper intertidal flats, and in sheltered regions of estuaries (Underwood and Kromkamp, 1999). More detailed studies (Thornton et al., 2002, 2007) demonstrated a trend for higher biomass on upper shores, and within the upper reaches of the estuary (Fig. 7B—Thornton et al., 2002; Fig. 7C—unpublished data). However, because many environmental gradients in estuaries covary, care needs to be taken in overinterpreting spatial patterns as being 'driven' by a key variable, and such conclusions should be tested experimentally or with robust models. It is worth noting that the lower estuary samples in Thornton et al. (2002) (zone 4 in Fig. 7B) were taken from a mixed shore, with higher proportions of sand (Point Clear). A more recent study between 2007 and 2008 of a cohesive mudflat, 1 km away on the opposite side of the estuary mouth (Pyefleet Channel, Taylor et al., 2013, zone 4 samples in Fig. 7A), measured sediment Chl a concentrations that are not significantly different from other mudflats further within the estuary (Fig. 7A). This is a good example from the Colne system of the interplay between sediment type, exposure, nutrients and other factors that influence MPB biomass and activity in estuarine systems worldwide.

These various drivers of productivity combine to generate longer-term seasonal patterns. In some estuaries, clear seasonal patterns, with summer peaks in MPB activity (Miles and Sundbäck, 2000; van der Wal et al., 2010), have been recorded, but other systems are much more stochastic, without any clear 'seasonal' pattern, or enhanced biomass during autumn and winter months (Tagus, Thames, the Wash). The Colne tends to fall into this latter category, with winter peaks of biomass during short-term periods of favourable conditions. Even if conditions are favourable, then 'top-down' factors, for example, sudden increases in grazers, e.g. *Hydrobia*, can cause significant biomass changes over a timescale of hours to days (Bellinger et al., 2009; Orvain et al., 2004). Our current state of knowledge means that we can make some qualitative predictions about the impact of changing climate on intertidal MPB activity. These would include a reduction in MPB activity, due to increasing temperatures and potential desiccation stresses during summer and increased storminess and wind events which can significantly disrupt biomass accumulation (De Jonge and van Beusekom, 1995; N. Redzuan and G.J.C. Underwood, unpublished observation). Further research on these issues is required.

MPB also produce extracellular polymeric substances (EPS), mainly polysaccharides, that stick sediment particles together and form polysaccharide sheaths and biofilms which increase sediment stability (Underwood and Paterson, 2003), a recognised 'supporting ecosystem service' but also provide a protective mechanism against salinity stress and desiccation (McKew et al., 2011; Steele et al., 2014). The presence of biofilms reduces resuspension of biomass during tidal inundation (Underwood and Paterson, 2003), with resuspension (wash away) during a period of normal tidal cover able to remove approximately between 25% and 65% of the biofilm material (Blanchard et al., 2006; Hanlon et al., 2006). However, sediment biostabilisation can also lead to mass failure, for example, during periods of wind-driven wave formation. In such circumstances, the MPB can then contribute significantly to the water column 'phytoplankton' (de Jonge and van Beusekom, 1995; Ubertini et al., 2012). The EPS and other carbohydrate fractions produced within the sediment by MPB are rapidly utilised by a wide range of heterotrophic bacteria (Haynes et al., 2007; Hofmann et al., 2009; McKew et al., 2013), and some taxa (a subset of Alphaproteobacteria and Gammaproteobacteria) are particularly adapted to utilise diatom EPS before the rest of the assemblage (Taylor et al., 2013). Microphytobenthic-derived DOC components contribute between 30% and 50% of the total organic matter in the sediments (Bellinger et al., 2009) and are therefore a key source of labile organic carbon, compared to the refractory nature of other sediment organic matter constituents. The relative importance of this carbon source in different intertidal habitats (sands to muds; temperate to tropical) appears to vary (Cook et al., 2007; Oakes et al., 2010, 2012), and its changing contribution to carbon turnover or carbon sequestration (Blue Carbon) in response to environmental drivers needs further research (Luisetti et al., 2014; van Colen et al., 2014).

3.3 Organic Matter Breakdown and Recycling

The organic matter present in the estuary derives from pelagic and benthic primary production, together with any organic matter carried in from the catchment or coastal sea and deposited within the estuary. There is a longitudinal gradient with muddy sediments at the top of the estuary having relatively high silt/clay fraction and organic matter content 2–3.7% by dry weight, declining down the estuary to Brightlingsea with sandy sediments with lower silt/clay and organic matter content 0.4–0.5% dry weight (Dong et al., 2000; Thornton et al., 2002). Only a very small proportion (<2%) of this organic matter will be labile and provide substrates immediately available for microorganisms (e.g. Nedwell, 1987). Organic matter degradation occurs by predominantly microbial activity which within the water column will be by aerobic metabolism. Oxygen penetrates only a few mm into the sediments, because it is removed rapidly by aerobic respiration by benthic biota. This surface oxic layer is always <1 mm in the highly organic silty sediments in the upper estuary, increasing to 4–5 mm depth in the sandier, more porous sediments near the estuary mouth, and also varies seasonally, becoming deeper in the winter when low temperature reduces the rate of benthic removal of oxygen, permitting its deeper penetration into the sediment (see Dong et al., 2000; Ogilvie et al., 1997). The sediment below this shallow surface oxic layer is anoxic, and organic matter degradation there is driven by fermentation and anaerobic respiratory processes that use a variety of electron acceptors other than O_2 such as nitrate, sulphate, CO_2, Fe^{3+}. Bioturbation by species such as the polychaete *Hediste* (*Nereis*) *diversicolor*, the amphipods *Carophium* spp. and the gastropods *Hydrobia* spp. is common, effectively increasing the surface area of the sediment, creating lateral redox zones from the wall of the burrows into sediments, thus expanding the depth from the surface in which aerobic processes occur and to which terminal electron acceptors such as nitrate can penetrate (Papaspyrou et al., 2014).

Thornton et al. (2002) measured O_2 uptake rates by the bottom sediments along the estuary and showed that when illuminated there was emission of O_2 to the water because of photosynthesis by benthic microalgae, while in the dark O_2 uptake from the water by the sediments was stimulated. The annual O_2 uptake in the dark varied from 37.5 mol O_2 m^{-2} year^{-1} at the Hythe to only 12.4 mol O_2 m^{-2} year^{-1} at the mouth at Brightlingsea. This is due to the sandy sediments having a relatively low content of organic matter. The benthic O_2 uptake is driven both by direct aerobic metabolism of organic matter in the sediment and by the reoxidation in the surface oxic layer of reduced end products of anaerobic metabolism such as ammonium from nitrate reduction, sulphide from sulphate reduction.

The concentration of oxygen dissolved in seawater is low (≈ 200 μM at 10°C and 30 salinity), whereas there is a comparatively large amount of sulphate (20 mM), so that organic matter breakdown in sediments is less likely to be limited by low sulphate concentration than by low O_2 concentration. Below the oxic layer of sediment sulphate respiring bacteria (SRB) break down organic matter anaerobically using sulphate as their electron acceptor, which is reduced to sulphide according to:

$$2CH_2O + SO_4^{2-} \rightarrow H_2S + 2HCO_3^-$$

One of the key conclusions of research in the Colne has been that a large part of organic degradation and recycling in the sediments is driven by sulphate reduction, and this is the case in most other estuaries. McKew et al. (2013) found that sulphate reduction and methanogenesis were key processes in the degradation of EPS produced by MPB biofilms, with the more structurally complex EPS components undergoing significant degradation under anaerobic conditions, but not in oxic slurries. Sulphate concentrations in these estuarine sediments tend to be high, suggesting that sulphate reduction rates are likely to be regulated by availability of electron donors (substrates) and temperature rather than sulphate concentrations (e.g. Nedwell and Abram, 1979). Sulphate reduction can account for 50% of organic matter degradation in high-sulphate environments such as marine or estuarine sediments (Jørgensen, 1982), while methanogenesis (methane formation by methanogenic Archaea (MA)) accounts for a similar proportion in freshwater sediments where sulphate concentrations are very low. These two groups of microorganisms, important in the terminal steps of organic matter breakdown, compete for common substrates such as acetate and H_2, which arise in anoxic sediments as products of anaerobic organic decay. Where sulphate is available, the SRB outcompete MA for acetate and H_2 because of their higher affinities for these substrates, but in the absence of sulphate MA win out. Using fluoroacetate, a specific inhibitor of acetate metabolism, Banat et al. (1981) showed that about 60% of sulphate reduction in the Colne sediments was driven by acetate oxidation, the remaining 40% being supported by other substrates such as H_2, propionate, butyrate and other fatty acids that SRB species, but not MA, are able to metabolise. Turnover of fatty acids in sediment from Colne Point saltmarsh was completely inhibited by 20 mM sodium molybdate, a specific inhibitor of SRB (Balba and Nedwell, 1982), confirming the major role of SRB in the terminal steps of organic matter degradation in these anoxic sediments.

Nedwell et al. (2004) demonstrated that sulphate reduction and methane formation were both measurable in bottom sediments right along the Colne estuary, sulphate reduction even in the upper estuary where sulphate was comparatively low, and methanogenesis even near the mouth where sulphate concentrations were high, but sulphate reduction was always almost two orders of magnitude greater than methanogenesis (see Table 2). Furthermore, the amount of methane formed in the sediments was generally greater than the amount emitted to the air, which suggested significant methane oxidation in the oxic layer by methanotrophic bacteria. Webster et al. (2015) reported methanotrophic ANME-2A genes from *Euryarchaeota* related to *Methanosarcinales* and *Methanomicrobiales*, which obtain energy from anaerobic oxidation of methane linked to sulphate reduction, but there is no data on aerobic methanotrophy in the Colne sediments.

Both sulphate reduction and methanogenesis showed significant seasonality, with maximum rates in summer (Nedwell et al., 2004). Probing of 16S rRNA extracts of the sediments (Purdy et al., 2003) showed the presence of acetate-using *Desulfobacter* spp., H_2-using *Desulfovibrio*, and *Desulfobacterium*, which apart from sulphate can also respire nitrate, which is present in high concentration in the upper estuary. The survival of *Desulfobacterium* in the lower-sulphate upper estuary sediments may be facilitated, therefore, by its ability to also respire nitrate if sulphate is low or absent. Slurry experiments showed that low-sulphate sediment from East Mills produced methane at a constant rate and when amended with acetate methanogenesis increased, accompanied by increase in the *Methanosarcinales*-targeted probe

Table 2 Annual Totals (1994–95) of Methane Formation, Methane Emission and Sulphate Reduction (All mmol m^{-2} $year^{-1}$, $\pm$ SE) at Three Sites Along the Colne Estuary From the Marine End (Colne Point Saltmarsh, Marsh Top and Open Mud in a Creek), Mid Estuary at Alresford and the Brackish Upper Estuary at the Hythe

Site	Methane Emission	Methane Formation	Sulphate Reduction
Colne Point, marsh top	25.3 ± 1.0	105 ± 8.8	9165 ± 495
Colne Point, creek	25.0 ± 1.2	105 ± 2.5	3720 ± 330
Alresford	17.7 ± 0.5	22.1 ± 1.1	3000 ± 620
The Hythe	22.3 ± 0.6	51.7 ± 5.1	4730 ± 260

Modified from Nedwell, D.B., Embley, T.M., Purdy, K.J., 2004. Sulphate reduction, methanogenesis and phylogenetics of the sulphate reducing bacterial communities along an estuarine gradient. Aquat. Microb. Ecol. 37, 209–217.

signal. Addition of sulphate immediately reduced methane formation in the slurries, confirming the presence of SRB even in these low sulphate upper estuary sediments and of competition between SRB and MA for these resources. If sulphate is present, methane formation is inhibited.

Using high-throughput sequencing Oakley et al. (2012) examined niche partitioning along the estuary of two examples of biogeochemically important genera; the SRB *Desulfobulbus* and the MA *Methanosaeta*, which are both terminal oxidisers of organic material in these anoxic sediments. *Methanosaeta* uses solely acetate, while *Desulfobulbus* is metabolically versatile, able to respire sulphate when oxidising small molecules such as propionate and also able to use nitrate as an alternative electron acceptor to sulphate. *Methanosaeta* genotypes were generally divided between those found solely in the marine habitat (about 30% of operational taxonomic units (OTUs)) and those which were distributed across all or most of the estuary (about 70% of OTUs). In contrast, for *Desulfobulbus* there were many more genotypes, reflecting a wide range of niche widths, from specialised genotypes found only at a single site to generalist genotypes found in all 10 sites sampled along the estuary. The majority of *Methanosaeta* sequences were cosmopolitan, found at many sites along the estuarine gradient, whereas most *Desulfobulbus* sequences were found only at particular sampling sites.

O'Sullivan et al. (2013) performed detailed depth profiles at fully marine and intermediate locations as well as at the head of the estuary. In the marine and intermediate locations, distinct changes of populations of methanogens and SRBs were found with depth, despite only modest changes in sulphide and methane concentrations over the same 30-cm depth profile. In contrast, at the brackish head of the estuary population structure remained relatively constant with depth, while there were distinct zones of sulphate reduction and methanogenesis (O'Sullivan et al., 2013). The lower sulphate concentration at the estuary head resulted in sulphate depletion just below the surface sediment and severely reduced rates of sulphate reduction. However, sulphate reduction increased below 25 cm in this sediment, possibly due to a cryptic sulphur cycle with sulphate being produced biologically by anaerobic sulphide oxidation then rapidly consumed by SRBs, or sulphate derived from subsurface seawater incursions. Overall, these findings demonstrate the dynamic nature of estuarine sediments where geochemical gradients are overturned then reformed more rapidly than specific functional clades of Bacteria and Archaea can reestablish (O'Sullivan et al., 2013).

Webster et al. (2015) examined the distribution of Archaea both along the estuary and with depth in the sediment down to about 30 cm. Cell counts, indicating the size of microbial communities, decreased with depth down the sediment profile and also decreased along the estuary towards the mouth. Bacteria were the dominant group at all sites, but Archaea were a substantial part of the sediment communities, constituting a greater proportion of the prokaryotic community at the estuary mouth despite being at lower absolute abundance. Phylogenetic analysis of the archaeal 16S rRNA gene libraries showed that the majority of the Archaea in the Colne estuary belonged to clades with no known isolates. 'Bathyarchaeota', Thaumarchaeota and methanogenic Euryarchaeota were the dominant groups of Archaea. These authors reported that MA from low-salinity Hythe sediments were dominated by acetotrophic *Methanosaeta* and putatively hydrogenotrophic Methanomicrobiales. In contrast, the marine-dominated Brightlingsea site was characterised by *mcrA* genes attributed to methylotrophic *Methanococcoides*, the metabolically versatile *Methanosarcina* and the methanotrophic ANME-2a group.

3.4 Nitrate Respiration

Nitrate in the estuarine water exchanges into the bottom sediments at rates dependent on the concentration gradient across the sediment–water interface (Ogilvie et al., 1997; Thornton et al., 2007) and also contributes to the oxidation of organic matter in the sediments. These nitrate exchange rates tend to be greatest in the upper estuary where nitrate concentration in the water is highest, and decline towards the estuary mouth. Indeed, during winter nitrate may be emitted to the water from the sandy sediments near the mouth (Ogilvie et al., 1997), suggesting a nitrate source within the sediment, probably from nitrification. The rate of vertical transport of nitrate into the sediment may be enhanced by the activity of the benthic fauna such as oligochaetes, polychaetes or bivalves, increasing the rates of nitrate transport into the anoxic zone, where it may be used as an electron acceptor by nitrate respiring microorganisms. Nogales et al. (2002) performed MPN counts of nitrate-reducing bacteria (as determined by reduction of nitrate to nitrite) in bottom sediments and reported high numbers; approximately 4×10^6 cells g^{-1} dry weight of sediment mid-estuary at Alresford and 3.5×10^7 cells g^{-1} dry weight of sediment in the upper estuary (Hythe), reflecting the higher nitrate concentrations in the upper estuary. Generally, therefore, benthic nitrate respiration rates and numbers of benthic nitrate respirers in estuarine sediments are likely to be stimulated both by higher nitrate concentrations in estuarine water and by sediment–water exchange rates enhanced by benthic biota. Unlike the phylogenetically cohesive SRB or MA, nitrate respirers are a very diverse group, with a corresponding diversity of metabolic end products of nitrate reduction (Fig. 8).

Complete reduction of nitrate to ammonium (nitrate ammonification, also called dissimilatory nitrate reduction to ammonium (DNRA)) is brought about predominantly by bacteria with a fermentative type of metabolism, whereas partial reductions of nitrate to gases (denitrification, DN), either N_2O or N_2, occurs in bacteria with respiratory metabolism. The balance of these two possible pathways is important as reduction of nitrate to N_2 or N_2O will result in loss of N from the water to the atmosphere, which may be of ecological benefit in a hypernutrified estuary (albeit that N_2O is a greenhouse gas with adverse impact on global warming), whereas reduction to ammonium conserves N within the aquatic system. Previous work (King and Nedwell, 1985) suggested that the balance between the availability of electron donors and electron acceptors influenced the balance between DN:DNRA. High nitrate concentrations (suggesting a low electron donor:electron acceptor ratio) will favour DN and nitrate removal, while

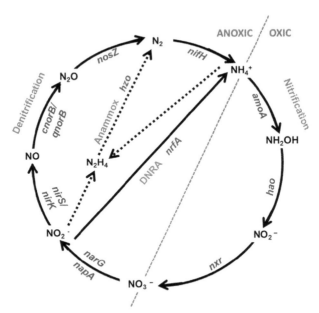

Fig. 8 Processes of nitrogen cycling in estuarine sediments, showing steps in denitrification, dissimilatory nitrate reduction to ammonia (*DNRA*) and anammox in anoxic sediments; and nitrification in oxic sediments, together with the genes coding for the key enzymes in these processes.

low nitrate concentrations (indicating a high electron donor:electron acceptor ratio) will tend to favour DNRA and conservation of N as ammonium. Ammonium within the sediment is, therefore, derived from both degradation of organic matter when organic N is mineralised to ammonium (ammonification), and ammonium formed as an end product of DNRA.

A number of studies were carried out examining the fate of nitrate inputs to the estuary. Early work (Ogilvie et al., 1997) established the exchange of nitrate between water and the sediments, and measured benthic denitrification with sediment cores, using acetylene to inhibit N_2O reduction and measure denitrification by N_2O accumulation. This technique may underestimate total denitrification as acetylene also inhibits nitrification that may be a significant source of nitrate within the sediment. However, they showed that denitrification was active with highest rates in the upper estuary and decreasing to low levels at the mouth. The turnover of the nitrate pool in the sediment was rapid and nitrate concentrations in the sediment porewater disappeared rapidly (within 40 min) after tidal exposure. Allowing for this decrease after exposure, they proposed an annual estimate of denitrification of nitrate derived from the overlying water of 6.2 Mmol N year^{-1} in the whole estuary, which is equivalent to between 34% and 49% of the total oxidised nitrogen (TOxN) load to the estuary. Linked nitrification–denitrification in sediments was also significant and, if also allowed for, benthic denitrification could account for 18–27% of the total N flux through the estuary. Thus, denitrification can lead to significant attenuation of the N load entering the estuary and the degree of attenuation would also be directly related to the retention time of nitrate in the estuary; the longer the FWFT the greater will be the attenuation of the estuarine N load. However, stratification in the water column within an estuary may restrict exchange of nitrate between water and sediment so that the impact of benthic denitrification in any estuary will be a function of both FWFT and stratification. Furthermore, the data suggested that N_2O production from denitrification was <2% of the total denitrification.

The development of the 'paired isotope technique' (Nielsen, 1992) with $^{15}NO_3^-$ added to sediment samples, incubated and the ^{15}N appearing in both N_2 and N_2O measured, permitted the determination of denitrification of both nitrate from the overlying water (D_w) and nitrate derived from nitrification within the sediment (D_n). Dong et al. (2000, 2006) measured DN under both dark and light conditions. Both D_w and D_n occurred throughout the estuary (Fig. 9), decreasing seaward, and denitrification in the sediments was always submaximal, nitrate concentration limiting the rates of

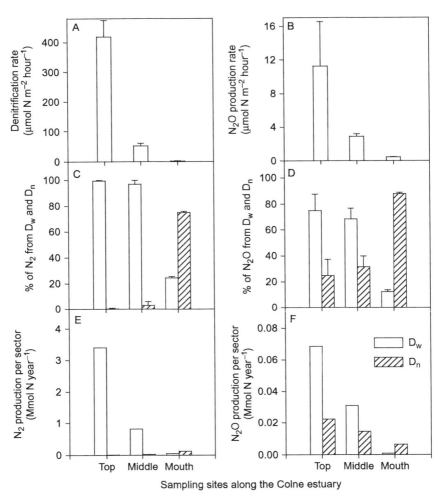

Fig. 9 (A) Rates of denitrification, (B) rates of N_2O production, (C) percentages of N_2 gas from D_w and D_n, (D) percentages of N_2O from D_w and D_n, (E) annual production of N_2 in each sector of estuary and (F) annual production of N_2O in each sector in the Colne estuary during June 2003. *From Dong, L.F., Nedwell, D.B., Stott, A., 2006. Sources of nitrogen used for denitrification and nitrous oxide formation in sediments of the of the hypernutrified Colne, the nitrified Humber and the oligotrophic Conway estuaries, United Kingdom. Limnol. Oceanogr. 51, 545–557.*

denitrification. In other words, the rates of denitrification were always less than the potential seen if there were higher nitrate concentrations. This indicated a large reserve capacity for denitrification in the estuary sediments. Fig. 9 illustrates for one occasion denitrification at three sites along the estuary (top at the Hythe, middle estuary at Alresford and the estuary mouth at Brightlingsea). Denitrification to both N_2 and N_2O decreased down the estuary commensurate with the decrease of nitrate in the water column, although N_2O was only a minor product of DN compared to N_2. The significance of D_n as a percentage of total DN increased towards the mouth (Fig. 9C and D) as did the percent of N_2O from Dn, emphasising the increasing significance of nitrification in the sandier, more oxic sediment near the estuary mouth. Dong et al. (2000) multiplied up these area rates to give total DN in each sector of estuary (Fig. 9E and F) to give total D_w for the Colne estuary as 4.73 Mmol N year^{-1}, which was 32% of the TOxN load during 1997–98 (a low-flow period). This estimate was very close to that of Ogilvie et al. (1997), indicating interannual comparability in denitrification in the estuary. While nitrate from the water column is removed only through D_w, D_n also removes significant amounts of nitrogen from the estuarine system by linked nitrification–denitrification, i.e., oxidation of ammonium to nitrate by nitrification in the surface oxic layer of sediment creating a subsurface nitrate peak, which nitrate then diffuses both back down into the anoxic sediment where it is denitrified and upwards towards the sediment–water interface and out into the water. Dong et al. (2000) demonstrated that the benthic microflora also stimulated D_n during periods of illumination by producing O_2 during photosynthesis and extending the depth of the oxic layer in which nitrification could occur. Total D_n for the estuary sediments was calculated as 2.2 Mmol N year^{-1}, which was one-third the amount of N removed by D_w, but nonetheless a further significant removal of N from the system. The conclusion from these studies was that the benthic denitrification provided a significant ecosystem service by removing within the estuary a considerable part of the nitrogen load and therefore reduced the impact of the discharged nutrient in the coastal sea.

Fig. 10 illustrates the inputs and transformations of nitrogen within the estuary on a yearly basis. The rates of transformations may be derived from different years, but give an overall indication of the relative magnitudes of the various processes. Importantly, it is apparent that the biological processing within the estuary takes up only a small proportion of the total N input to the estuary. Denitrification removes a significant but still minor part of the TOxN load, but primary production both by phytoplankton and MPB removes only a small part of the N inputs. In one way, this is beneficial in preventing overproduction of organic matter in the estuary and consequent eutrophication of the estuary, a fate which has impacted other nutrified estuaries.

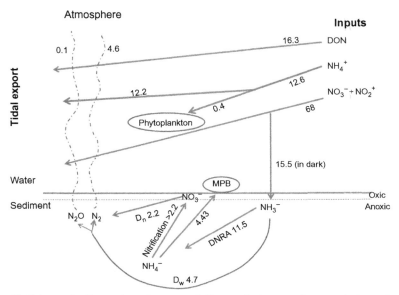

Fig. 10 Schematic illustrating the annual flows of nitrogen into and through the Colne estuary. Flows are in Mmol N year^{-1}. D_n, denitrification of nitrate from nitrification; *DNRA*, dissimilatory reduction of nitrate to ammonium; *DON*, dissolved organic nitrogen; D_w, denitrification in the sediment of nitrate from the water column; *MPB*, microphytobenthos.

Since microbes capable of nitrate reduction are taxonomically diverse, 16S rRNA-based methods are generally not suitable for their estimation. Thus, studies have generally focussed on using functional genes such as *narG* and *napA* as biomarkers of nitrate reduction (see Fig. 8). A subsequent step in the denitrification pathway is the reduction of nitrite. Two structurally different but functionally equivalent nitrite reductase enzymes catalyse nitrite reduction. One enzyme is a copper nitrite reductase (NirK) encoded by the *nirK* gene and the other is a cytochrome cd_1 nitrite reductase (NirS) encoded by the *nirS* gene (Fig. 8) (Zumft, 1997). Generally, the *nirS* gene is more widely distributed among denitrifiers (Braker et al., 1998). Nitric oxide reductase catalyses the reduction of NO to N_2O and is encoded by the *norB* gene (Fig. 8), which is widespread in denitrifying bacteria (Casciotti and Ward, 2005). The final step in denitrification is the reduction of N_2O to N_2, catalysed by N_2O reductase, which is encoded by the *nosZ* gene. Although the *nosZ* gene is largely unique to denitrifying bacteria, some denitrifiers have been found to lack this gene (Throbäck et al., 2004). The *nrfA* gene encodes for nitrite reductase, which catalyses the conversion of nitrite to ammonia (DNRA).

An early study by Nogales et al. (2002) used PCR and Southern blot hybridisation methods to detect *narG*, *napA*, *nirS*, *nirK* and *nosZ* genes in DNA extracted from sediments from the Hythe and Alresford. In the same study, *nirS* and *nosZ* mRNAs were also detected, suggesting that active expression of these genes was occurring at the time of sampling (Nogales et al., 2002). The advent of the quantitative polymerase chain reaction (qPCR)-based approaches permitted the quantification of the key functional genes of the different pathways, raising the possibility of measuring the amounts of these genes in natural samples and relating them to measured rates of the corresponding reactions in situ. For example, Smith et al. (2007, 2015) examined the rates of DN and DNRA along the estuary, in parallel with measurements of the key functional genes for these pathways. Smith et al. (2007) used sequence analysis of cloned PCR-amplified *narG*, *napA* and *nrfA* sequences and showed that the indigenous nitrate-reducing communities were phylogenetically diverse and different from previously described nitrate reduction sequences in soils and offshore sediments. The majority of clones were related to similar sequences in Deltaproteobacteria. A suite of qPCR primers was then developed to quantify these genes in sediments from three sites along the estuary (Fig. 11). Both nitrate and nitrite reductase gene copy numbers generally decreased significantly down the estuary, and reverse transcription PCR assays showed mRNA transcripts for three of five phylotypes quantified, demonstrating the in situ expression of these genes. However, there were large differences between the numbers of different gene homologues: for example, *narG-2* was much more abundant than *narG-1*; *napA-1* than *napA-2*. Furthermore, while *napA-3* was present in very high numbers, there was no significant decrease in copy numbers along the estuary. While we can now describe the distribution of these nitrate reduction genes in these estuarine sediments, we still cannot yet link them directly with process rates, nor distinguish why and how particular genes or gene homologues are adaptive—indeed, linking functional gene abundances to measured rates across most biogeochemical process remains a major challenge at the interface of molecular and functional ecology, which requires considerable further research. For example, Richardson (2000) has suggested that *nap* is adaptive for nitrate reduction where nitrate is at low concentration as NAP has a higher affinity for nitrate than NAR, but NAR coded by *nar* is more effective when nitrate is abundant. However, the distributions of these genes along the estuary fail to reflect such trends. How and why the different *napA* or *narG* homologues are adaptive remains to be elucidated.

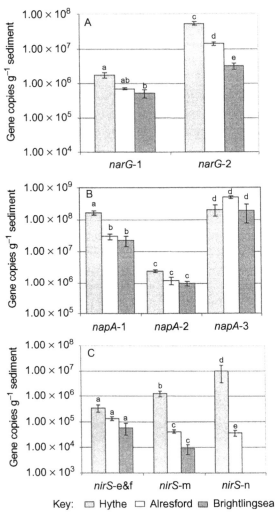

Fig. 11 Gene abundance (copy number g^{-1} sediment ($\pm$ SE) of nitrate reduction genes at three sites along the Colne estuary, October 2006). (A) narG, (B) napA and (C) nirS. Statistical differences between sites indicated by *different letters* for each gene phylotype. *Adapted from Smith, C.J., Nedwell, D.B., Dong L.F., Osborn A.M., 2007. Diversity and abundance of nitrate reductase genes (narG and napA), nitrite reductase genes (nirs and nrefA), and their transcripts in estuarine sediments. Appl. Environ. Microbiol. 73, 3612–3622.*

The recognition in the 1990s (van de Graaf et al., 1995) of a new pathway of anaerobic ammonium oxidation, the anammox pathway carried out by highly enriched but so far unisolated members of the Planctomycetes, provided a further pathway potentially leading to removal of N from sediments. In anammox (AN), ammonium is oxidised to N_2 using nitrite as a terminal electron acceptor which is itself reduced to N_2 (see Fig. 8). Dong et al. (2009) measured all three process rates (DN, DNRA, AN) and their appropriate functional genes along the Colne estuary showing again that in situ DN and DNRA rates both declined towards the estuary mouth as nitrate concentrations in the water declined, and slurry experiments suggested that as nitrate concentration declined the potential for DN decreased, but potential for DNRA increased. Anammox activity was only detected in sediments at the top of the estuary at the Hythe where both ammonium and nitrite were present in abundance, and there anammox contributed about 30% of N_2 formation. 16S rRNA sequences from Planctomycetes usually associated with anammox were also only detected at the Hythe. Anammox therefore seemed to be significant only in the upper estuary where high concentrations of both ammonium and nitrite in the water and sediment would tend to favour it.

In a further study, measuring the potentials for DN and DNRA both along the estuary and with depth in the sediments, Papaspyrou et al. (2014) confirmed that DN potential decreased both towards the estuary mouth and with increased depth in the sediments, but DNRA potential increased down the estuary, as suggested by Dong et al. (2009). More surprisingly, a potential for denitrification and presence of nitrate reductase genes continued to be measurable in the sediments well below the depth at which nitrate was detectable in the sediment porewater. Moreover, a large nitrate pool was detected in the deeper (>5 cm depth) sediment after either freezing of sediment samples or by KCl extraction. This seemed to be derived from intracellular nitrate pools that were released either by freeze fracturing of cells or by KCl extraction. Sequencing analysis of extracted DNA failed to detect known nitrate-accumulating microorganisms such as *Thioploca* or *Beggiatoa*, leaving other novel bacteria, microbenthic algae or foraminiferans as possible other candidates. In this case, the microorganisms with high intracellular nitrate loads were presumably using it as a transportable electron acceptor supply at depths in the sediment where there was no soluble nitrate in the porewater for them to use. The microorganisms responsible for this intracellular nitrate accumulation remain to be identified, further highlighting the need for increased resolution and application of novel molecular approaches.

3.5 Nitrification in the Colne Estuary

Nitrification is the oxidation of ammonium to nitrate by nitrifying bacteria, which is a key process in maintaining the nitrogen cycle. Ammonia oxidation is the first and rate-limiting step of nitrification and is carried out by two distinct groups of autotrophic microorganisms, ammonium-oxidising bacteria (AOB) and ammonium-oxidising archaea (AOA). Both AOB and AOA contain the membrane-bound enzyme ammonium monooxygenase, which is encoded by the *amoA* gene (Konneke et al., 2005; Rotthauwe et al., 1997). Early work (Ogilvie et al., 1997), based on export of nitrate from the sediment, demonstrated the occurrence of nitrification in the sandy sediments at the estuary mouth, but the general significance of nitrification in the estuary was poorly understood. The use of ^{15}N to measure denitrification (e.g. Dong et al., 2009) also scaled the magnitude of D_n, a measure of linked nitrification–denitrification, and D_n therefore gives a lower limit (2.2 Mmol N year^{-1}) for the amount of nitrification occurring in the estuary. This value will not include any ammonium oxidised to nitrite or nitrate that is not subsequently denitrified, but lost from the sediment to the water. Examination of the potential for nitrification (Li et al., 2015), measured using slurries of sediment, showed that there was no detectable nitrification in the water column and maximum nitrification potential occurred consistently in mid-estuary sediments at Alresford where there is abundant ammonium within the sediment porewater, derived from degradation of organic matter, combined with a reasonably deep surface oxic layer within which nitrification can occur (Fig. 12). There was also a seasonal effect with maximum nitrification potentials in January when water temperature was low and the depth of the surface oxic zone in which nitrification occurred was maximal. Addition of allyl thiourea (ATU), an inhibitor of autotrophic nitrification, showed that autotrophic nitrification was always the dominant process, but ATU-insensitive heterotrophic nitrification (Kuenen and Robertson, 1987) could be as much as 30% of the total during the summer in the organic sediments in the upper estuary at the Hythe.

It has been suggested that ammonium concentrations may define different ecological niches of AOA and AOB (Martens-Habbena et al., 2009), whereby AOA may dominate in some environments with low ammonium concentrations (Bernhard et al., 2005; Caffrey et al., 2007). However, until recently, little was known about estuarine benthic AOA/AOB communities and their potential links with biogeochemical function. Webster et al. (2015) reported the presence of *Thaumarchaeota*, a group of aerobic autotrophic ammonia oxidisers, which were found predominantly in surface sediments along the

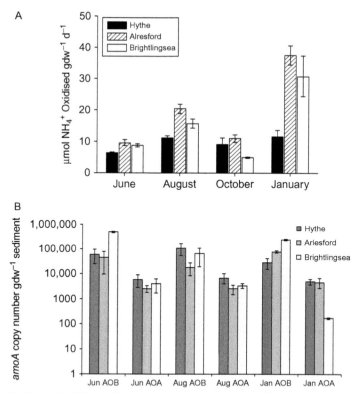

Fig. 12 (A) Mean (±SE) nitrification potential rates in sediments along the Colne estuary. (B) AOB and AOA *amoA* gene abundance (mean±SE) at Hythe, Alresford, Brightlingsea in June 2009 to January 2010. *Modified from Li, J., Nedwell, D.B., Beddow, J., Dumbrell, A.J., McKew, B.A., Thorpe, E.L., Whitby, C., 2015. amoA gene abundances and nitrification potential rates suggest that benthic ammonia-oxidizing bacteria and not archaea dominate N cycling in the Colne estuary, United Kingdom. Appl. Environ. Microbiol. 81, 159–165.*

estuary, commensurate with the oxic environment required for aerobic nitrification. A reduction in *Thaumarchaeota* 16S rRNA genes at the Alresford and Hythe sites compared to Brightlingsea may be associated with higher ammonia concentrations and dominant Betaproteobacteria nitrifiers at the two former sites.

Li et al. (2015) used qPCR analysis of the *amoA* genes for AOB and AOA in sediments along the estuary and showed that AOB *amoA* gene abundance was significantly greater (by 100-fold) than that of AOA. This showed that in the Colne estuary, with very high levels of inorganic N (Dong et al., 2000), nitrification was driven by AOB which predominated over AOA.

Furthermore, the ratio of AOB:AOA *amoA* gene copies in surface sediments increased from the head to the mouth of the Colne estuary. It might be argued that the significance of AOA could be expected to increase downstream as ammonium in the water declined, but the increased AOB:AOA ratios downstream contradicted such a trend. However, the nitrifiers are operating not in the water column but within the surface sediment where the porewater ammonium concentrations are usually much higher than in the overlying water, and thus AOB in the sediment may still be able to outcompete AOA for ammonium even in sediments at the estuary mouth.

Seasonal changes in AOB *amoA* gene abundance were also found, with higher *amoA* gene copies present in June (across all sites) compared to AOA *amoA* gene copy numbers which remained relatively unchanged and low throughout the year (Li et al., 2015). Using PCR-DGGE analysis of the *amoA* and 16S rRNA genes, AOB communities from the Hythe were found to be more distinct than those from Alresford and Brightlingsea. Specifically, DGGE profiles of the *amoA* gene showed a greater number of bands at the estuary head, which decreased downstream and with time. For both the *amoA* and 16S rRNA genes, there were a number of unique DGGE bands that had high sequence identity to *Nitrosomonas* spp. PCR-DGGE analysis of the AOB community identified several sequences that were related to *Nitrosomonas* spp., suggesting that in this nutrified estuary, AOA were likely to be inhibited by high ammonium concentrations, and AOB (possibly *Nitrosomonas* spp.) were of major significance in nitrogen cycling. High-throughput next-generation sequencing (NGS) of the 16S rRNA genes showed that both bacterial and archaeal communities changed in composition along the estuary, but not across seasons. In general, the bacterial communities were twice as rich as those of the Archaea. OTUs assigned to known AOB species represented only a small proportion (0.07%) of the total 16S rRNA bacterial sequence libraries, and nitrite-oxidising bacteria were more abundant than AOB. OTUs assigned to *Nitrospira* and *Nitrospina* accounted for 0.66% and 0.85% of 16S rRNA sequences (Li et al., 2015). The presence of nitrite-oxidising bacteria may be stimulated at the estuary head by nitrite present in the water column, exchanging down into the sediment.

3.6 Archaea

qPCR analysis of bacterial and archaeal 16S rRNA genes showed that bacterial 16S rRNA genes in Hythe and Wivenhoe sediments were approximately 50-fold more abundant than archaeal: $8.1–8.6 \times 10^8$ bacterial 16S rRNA gene copies g^{-1} dry sediment, compared to $1.4–2.0 \times 10^7$ archaeal 16S rRNA gene copies g^{-1} dry sediment (Poole, 2013). Munson et al. (1997) examined 16S rRNA gene libraries created with *Archaea*-specific primers to examine the sediments from saltmarsh creek and marsh top sites where methanogenesis occurred. A wide range of *Euryarchaeota* were recovered, but no sequences for any *Crenarchaeota*, nor *Methanococcus* or *Methanobacterium*. Genera that produce methane from reduction of CO_2 by H_2 such as *Methanoculleus* and *Methanogenium* were detected, and also *Methanosaeta* which is an acetoclastic methanogen. Thus, examples of the MA that use either acetate or H_2, both important substrates in anaerobic organic degradation, and for which MA compete with SRB, were detected. However, many of the groups of MA detected used substrates such as methylamines, methanol and other methyl-containing substrates which are not used by SRB, and therefore are substrates which MA may use without competition with SRB. This may permit the survival of these MA, albeit at low rates of activity, even in these sulphate-rich sediments in which SRB otherwise dominate.

A variety of sequences related to different halophilic Archaea (Haloarchaea) were also detected in the gene libraries from the saltmarsh sites, which was surprising as such halophiles had been reported to be obligate and requiring at least 1.5 M sodium chloride for growth. Haloarchaea were previously known only from extremely high-salinity environments such as the Dead Sea and solar salt pans in salt and soda lakes where they are often the dominant heterotrophs (e.g. McGenity and Oren, 2012). However, 16S rRNA gene sequences related to haloarchaea were recovered from tidal marine and saltmarsh sediments from the Colne estuary (Munson et al., 1997), suggesting that haloarchaea may exist at normal seawater salinity. Attempts to isolate them from creek or saltmarsh pan sediments on either glucose or glycerol, with antibiotics to inhibit Bacteria and eukaryotes, over a range of salinity from 2.5% to 30% sodium chloride, resulted in three groups of Haloarchaea being isolated (Purdy et al., 2004), capable of slow growth at seawater salinity (2.5%, w/v), even though their optimum salinity for growth was higher (10%). While their environmental activity and significance in the estuarine environment currently remain obscure, their continued existence in such 'nonextreme' environments is a paradox which requires further study.

4. ESTUARINE SALTMARSHES
4.1 The Role of Saltmarshes

The Colne estuary channel is surrounded by saltmarshes which are inundated tidally twice per day. Approximately 11% of the English east coast estuary area is saltmarsh (http://jncc.defra.gov.uk/protectedsites). In the Colne Point saltmarsh the total area divides up as creek area 38%, saltmarsh pans 2%, marsh top with a *Puccinellia/Halimione* plant association 38% and marsh top with *Puccinellia/Spartina* association 21% (Azni and Nedwell, 1986b; Hussey, 1980). Colne Point saltmarsh has an area of 3.5 km^2 and was the site of extensive early study on saltmarsh primary production (e.g. Hussey, 1980; Hussey and Long, 1982). Total assimilation of N during primary production in the dominant *Puccinellia/Halimione* association was 1.16 mol N m^{-2} year^{-1} of which 97% was by *Puccinellia maritima*. Simultaneous measurements of degradation of the detritus of these same primary producers (Azni and Nedwell, 1986a) showed that annual mineralisation of plant detritus varied seasonally, principally influenced by temperature, but on an annual basis broadly balanced the assimilation of N during primary production, indicating a stable cycle of production and degradation in these saltmarsh communities. Nitrogen fixation also input available N to the sediments, the greatest rates by an order of magnitude being in those saltmarsh pans with a benthic film of cyanobacteria (*Oscillatoria* sp.). Even in these pans the annual N$_2$ fixation was only equivalent to about 9% of the annually mineralised N in that sediment, and as pans represented only 2% of the marsh surface, was trivial overall. In the other sites, N$_2$ fixation was only 0–4% of annually mineralised detrital N, so generally N$_2$ fixation was not a major input of N to the system. Azni and Nedwell (1986b) also examined tidal exchange in Colne Point saltmarsh of both soluble N components such as nitrate, nitrite, ammonium and DON and particulate organic nitrogen (PON) over complete tidal cycles which overtopped the marsh in each season. Total oxidised nitrogen (TOxN equivalent to nitrate + nitrite) showed a seasonal change, with smaller amounts present during summer, but PON showed a little seasonal change. From these data the total amounts of tidal exchange of N components were estimated, taking into account the number of times in a season the marsh was overtopped by tides. On an annual basis, there was little net exchange of nitrogen between the marsh and estuary, although there was net annual export of ammonium, DON and organic N in large particulate material, but net imports into the marsh of TOxN, and organic N in small particulate material. That is, on an annual basis export of some N components balanced import of other N components. Rates of export to the local estuary of ammonium and dissolved organic N were greatest during summer when the rates of detrital degradation and ammonification within the sediments were at a maximum (Azni and Nedwell, 1986a) and available nutrients in the coastal seawater at a minimum. These results emphasised the overall stable nature of the N budgets of the saltmarsh on an annual basis, consistent with a climax ecosystem, but also emphasised the ecological role of the marshes and their interaction with the estuary channel communities. Tidal export of ammonium and DON occurred during summer when the rates of detrital degradation on the marsh (degradation which contributed to the ammonium and DON pools) were highest, but also when the coastal and estuary mouth phytoplankton are most likely to be N limited, as illustrated by low N:P ratios during this period (Kocum et al., 2002b). Export of N from the marsh areas to the estuary channel may therefore be a significant input to coastal production during this N-depleted period. Furthermore, the net removal of TOxN from tidal water by the marsh sediments also represents a mechanism by which the saltmarshes may respond to anthropogenically elevated nitrate concentrations in tidal estuary water and remove it by both stimulation of growth of saltmarsh plants and by benthic denitrification.

5. ESTUARIES AND CLIMATICALLY IMPORTANT TRACE GASES

5.1 Trace Gas Production in the Estuary

The work of Ogilvie et al. (1997) demonstrated that the trace gas N_2O was formed during DN in the estuary, and further work (Robinson et al., 1998) established the significance of hypernutrified estuaries to the global atmospheric budget as N_2O as the third most important greenhouse gas with a global warming potential 270 times that of CO_2 over a 20 year interval. Water samples were taken at 20 stations along the Colne estuary and dissolved N_2O concentrations and saturations measured, along with atmospheric N_2O and nitrate, and ammonium concentrations. From these in situ measurements water-air fluxes of N_2O were calculated (Scranton, 1983). There were strong concentrations of dissolved N_2O ranging from 5 to 79 nmol N_2O N L^{-1} at the seaward end to 158–1300 nmol N_2O-N L^{-1} in the mid and upper estuary. The high N_2O concentrations occurred in the low-salinity, high nitrate section of the estuary, and water column N_2O concentrations were positively correlated with nitrate and nitrite concentrations. Water column N_2O saturations (relative to air equilibration) were almost always greater than 100% even in the high-salinity, low nitrate region of the lower estuary and were up to 45-fold air equilibration in the mid and upper estuary confirming the strong source of N_2O in the estuary. The annual water–air flux of N_2O was calculated by using an average of the measured water–air fluxes and allowing for the area of water surface in each sector, and tidal exposure, giving an estimate of 1.2×10^5 mol N_2O–N for 1993–94 and 0.85×10^5 mol N_2O–N for 1994–95. These emissions accounted for only 0.5% of the TOxN load to the estuary, or 0.3% of the Total Nitrogen load, but if extrapolated globally would account for 0.13–0.45 Tg N_2O–N year^{-1} a globally significant amount. Robinson et al. (1998) demonstrated that when sediment was tidally exposed, the sediment nitrate pool was rapidly removed by nitrate reduction, and simultaneously N_2O emission also stopped. Addition of nitrate rapidly restored N_2O emission confirming the origin of the N_2O in denitrification and also emphasising the importance of the link between sediment and water column to N_2O formation. In essence, it was only when nitrate-rich estuarine water covered the sediment that rapid replenishment of the sedimentary nitrate pool permitted formation of N_2O. A question remains as to the relative significance of nitrification in the N_2O budget of the estuary as it also produces N_2O. For example, N_2O formation in soils is generally from nitrification, unless the soil becomes water-logged when N_2O from denitrification then dominates (e.g. Skiba et al., 1993). While the correlation between N_2O concentrations and concentrations of nitrate and nitrite in the estuarine water may support the contention that the dominant process giving N_2O is denitrification, there have been no direct simultaneous measurements of both N_2O sources in the Colne, although new techniques may provide this differentiation. More recent work (e.g. Sutka et al., 2006) has shown that the intramolecular distribution of N isotopes in the N_2O molecule might permit the direct evaluation of the sources of N_2O from nitrification or denitrification.

5.2 Sulphur Gases

A further trace gas of importance in global warming is dimethyl sulphide (DMS), which when released to the atmosphere may influence climate by acting as cloud condensation nuclei and thereby increase cloud cover (e.g. Charlson et al., 1987). DMS is produced in marine environments by decomposition of the presumed osmolyte dimethylsulphoniopropionate which occurs in phytoplankton, macroscopic algae and higher plants. DMS is present in seawater in concentrations sufficient to maintain a flux from seawater to the atmosphere. While DMS concentrations have been reported for oceanic and coastal water (e.g. Nedwell et al., 1994), there was little known about levels in estuarine systems. The fluxes of DMS, hydrogen sulphide (H$_2$S) and carbonyl sulphide (COS) from saltmarsh pans, the vegetated saltmarsh surface and from tidal creek sediment were made in Alresford Creek (mid-Colne estuary) in order to examine the possibility of S-gas emissions from estuarine systems. The emissions of all three gases were detected, the largest emission rates being of H$_2$S and DMS. Both DMS and H$_2$S, but not COS, emissions showed strong seasonal signals, significantly correlated with seasonal temperature but not correlated with illumination. The fastest rates of emissions were from saltmarsh pans rather than the vegetated marsh top, but as pans represented only about 2% of the surface area of the marsh their overall contribution was marginal. The annual emissions per unit area were smaller than reported from North American saltmarshes, although similar to those from Danish marshes (Jørgensen and Okholm-Hansen, 1985), and it was concluded that the magnitude of emissions of DMS from the Colne saltmarsh sediments to the atmosphere was minute compared, for example, to that from the North Sea waters. Sulphate reduction to H$_2$S within the Colne estuarine and saltmarsh sediments has been measured (Nedwell et al., 2004; Senior et al., 1982), and when the measured H$_2$S emissions were compared to sulphate reduction rates, it became apparent that $\ll 1\%$ of H$_2$S from sulphate reduction was emitted to the atmosphere. This may be the result of oxidation of H$_2$S prior to emission by chemolithotrophic sulphide-oxidising bacteria in the oxic surface layer of saltmarsh sediments, or in the oxic rhizosphere around the roots of higher plants, which can have major influence in attenuating emissions of gases (e.g. Watson et al., 1997). Certainly, 16S rRNA gene libraries from the mouth of the estuary show that putative sulphur-oxidising bacteria are highly abundant in surface mudflat sediments (McKew et al., 2011). Alternatively, H$_2$S can equilibrate with the precipitated sulphide minerals abundant within the sediments (e.g. Nedwell and Takii, 1988). In either case, although sulphate reduction to H$_2$S in intertidal sediments has the potential to impact the atmosphere, in fact the impact of H$_2$S on atmospheric processes is greatly attenuated by H$_2$S removal by either reoxidation or immobilisation as sulphide minerals.

5.3 Isoprene Cycling

Isoprene (2-methyl-1,3-butadiene; C$_5$H$_8$) is a major biogenic volatile hydrocarbon, with an annual flux similar in magnitude to that of methane (Guenther et al., 2012). It is highly reactive and its complex atmospheric reactions have diverse climatic consequences. For example, isoprene:

(1) reacts with oxidised molecules that would otherwise react with methane, thereby extending the residence time of methane and enhancing global warming; (2) reacts with nitrogen oxides to create tropospheric ozone by photolysis; (3) like DMS, can create cloud condensation nuclei, leading to a cooling effect (see Pacifico et al., 2009 for a summary). Although most isoprene is released into the atmosphere by terrestrial plants, there is increasing evidence of abundant production in the marine environment, particularly in coastal zones (Exton et al., 2015), including temperate estuaries (Exton et al., 2012). All microalgal species tested to date produce isoprene (Exton et al., 2013), whereas <9% of bacterial isolates from the Colne estuary and North Sea produced isoprene (T.J. McGenity and S. Cummins, unpublished data). To our knowledge, bacterial species from other coastal and marine environments have not been tested for isoprene production. The Colne served as a model system for understanding isoprene production processes in an estuary (Exton et al., 2012) and was the first example of a nonterrestrial environment in which microbially mediated isoprene consumption was demonstrated (Acuña Alvarez et al., 2009). Exton et al. (2012) found that water column isoprene concentrations were higher at low tide than high tide, and at the head compared with the mouth of the estuary. However, isoprene production rates from the sediment were similar along the estuary's length. Thus, it was surmised that the MPB was the primary source of isoprene in the estuary, and that low water column concentrations at high tide and at the estuary mouth were a consequence of dilution by coastal seawater. Isoprene production was highest in the warmest months, with production showing a strong positive correlation with light and temperature (Exton et al., 2012). Acuña Alvarez et al. (2009) showed that ex situ isoprene degradation rates were faster in samples from the head compared with the mouth of the estuary and from sediment compared to the water column. Several isoprene-degrading pure cultures were obtained, including *Rhodococcus* sp., which was abundant in microcosms and is the model isoprene-degrading genus (Crombie et al., 2015). Using a coculture of isoprene-producing algae and a mixture of estuarine bacterial isolates, it was also shown that isoprene degradation occurred at environmentally relevant concentrations (Acuña Alvarez et al., 2009).

6. STRESSORS AND POLLUTION
6.1 Crude Oil Degradation

Salt marshes and mudflats are very susceptible to crude oil pollution, owing to their low-tidal energy, soft fine-grained sediments and frequent proximity to shipping lanes, oil refineries and recreational boat traffic (McGenity, 2014). The estuary is susceptible to oil pollution by passing boat traffic and has been the subject of several investigations into the capacity to respond and recover from an oil spill. Studies have employed tidal mesocosms, using sediment cores taken from near the mouth of the estuary. Coulon et al. (2012) found that degradation of hydrocarbons in weathered crude oil was relatively rapid, with known aerobic obligate hydrocarbonoclastic bacteria, such as *Alcanivorax*, *Cycloclasticus* and *Oleibacter* spp. constituting 8% of the bacterial community in the upper layers of sediment after 12 days of incubation. Biofilms that floated from the surface were dominated by obligate hydrocarbonoclastic bacteria, with *Alcanivorax borkumensis* constituting almost half of the total bacterial community. In the same experiment, Chronopoulou et al. (2013) found that the MPB increased in abundance in oil-polluted mesocosm sediments, with a 10-fold higher abundance of cyanobacteria after 21 days. This was attributed to reduced grazing pressure and possibly diminished nutrient levels encouraging the growth of nitrogen-fixing cyanobacteria. Sanni et al. (2015) performed a similar experiment, in which hydrocarbon degradation was much more rapid. An oil-induced shift in the community composition of bacteria was also seen, but the archaeal community was not significantly affected by crude oil treatment. This suggests a fundamental difference in the susceptibility and resilience of the bacterial and archaeal communities to perturbation by hydrocarbons.

6.2 Engineered Nanoparticles

In addition to oil pollution, there is also the potential for the Colne and other similar estuaries to be exposed to engineered nanoparticles, via effluent discharge from STWs, where they pose a risk to the in situ microbial communities and their processes. Of particular concern are ionic silver and silver nanoparticles (AgNPs), which due to their antimicrobial properties could detrimentally impact on microbial processes such as hydrocarbon degradation. Any nanotoxicity effects are likely to be greater at the estuary head where nanoparticles may accumulate in sediments. This has significant implications to the in situ sediment microbial communities and their activities, given that microbial enzyme activity (as measured by fluorescein diacetate hydrolysis) is greater in sediments from the estuary head compared to further downstream (Poole, 2013). For example, Beddow et al. (2016) showed that AgNPs inhibited nitrification and reduced ammonia-oxidising bacterial but not archaeal *amoA* gene abundance in sediments from the estuary. In contrast, Beddow et al. (2014) examined the effects of AgNPs and their ions on the hydrocarbon-degrading communities in the estuarine sediments and found that the hydrocarbon-degrading microbial communities were surprisingly resilient to high concentrations of AgNPs. Significant shifts in bacterial community structure were also found, suggesting that AgNPs impacts on bacterial community diversity and activity which may have potential implications for other important microbial-mediated processes in estuaries in the future.

7. FUTURE DIRECTIONS

Ecosystems such as the Colne estuary, with comprehensively described microbiology, are rare but urgently needed model systems as ecological studies move towards a greater interrogation of microbial ecology across hierarchies of biological organisation and spatiotemporal scales (also see Maček et al., 2016). As the technology and analytical approaches required for this integration advance, initial applications within well-defined ecological systems allow assessments of the working limits of these approaches. To date, integrating microbial and macrobiota into a systems-level understanding of ecology has been relatively atypical, even some of the most comprehensively described food webs omit microbial taxa, and those that consider them will often take a black-box approach and describe the wealth (potentially tens of thousands) of microbial species as single nodes (Woodward et al., 2005). This is not unsurprising, as before the advent of NGS approaches, cataloguing the diversity of microbial species from ecosystems was a time consuming and expensive undertaking, with financial limitations often reducing sample sizes and prohibiting large-scale replication. This is certainly no longer the case, and NGS surveys of microbial diversity are now common place and routinely carried out in long-term field sites such as the Colne estuary (Coulon et al., 2012; Li et al., 2015; McKew et al., 2011, 2013; Papaspyrou et al., 2014). Arguably, it is this step-change in technology, along with ever decreasing costs, that now allows microbial data to be recorded alongside, and integrated with, other ecological variables, but challenges still remain.

NGS data describing microbial communities are fundamentally different to morphological identifications of other taxa, using different concepts to define species and provide quantifications of their relative abundances. This makes direct comparisons and/or quantification of interactions between morphologically and molecularly defined species challenging. A potential solution in aquatic systems, such as the Colne estuary, is to move all ecological surveys towards NGS methods focusing on total environmental DNA (eDNA—Jackson et al., 2016), and not just targeting microbes. eDNA is ubiquitous in ecosystems, originating from faeces, urine, saliva, loose cells (including microbes) and other sources. This ubiquity of eDNA, coupled with the assumption that all taxa produce it, has led to an emergent field that heralds the quantification of eDNA as the answer to providing rapid, noninvasive detection of species from any environment (Bohmann et al., 2014), and a method that will capture both microbial species and macroorganisms using a common recording currency. To date, most eDNA studies have focused on single species, although there are exceptions (e.g. Mächler et al., 2014; Thomsen et al., 2012), individually quantifying a range of phylogenetically diverse populations; including fish (Eichmiller et al., 2014), amphibians (Thomsen et al., 2012), mammals (Foote et al., 2012), microbes and invertebrates (Goldberg et al., 2013). The potential power of eDNA approaches lies in the presumed ability to rapidly and simultaneously quantify all populations within a community, enabling us to reconstruct entire food webs (Bohmann et al., 2014). However, this has yet to be fully tested, and there remains a large amount of uncertainty associated with eDNA predictions regarding the presence, absence and abundance of species within a multispecies context across all trophic levels. This is because there are probably species-specific biases in the presence and detectability of their eDNA. Once DNA is free within the environment, exposure to UV light and extracellular enzymatic activity from microbes cleave DNA into shorter fragments until it is impractical to investigate or is completely degraded (Rees et al., 2014). Spatial limits to detection are therefore determined by how far eDNA diffuses or actively travels (i.e. via currents) before it is fully degraded. However, eDNA persistence is still relatively poorly understood, and other physicochemical properties will interact with UV exposure and microbial activity, with such effects being both environment (Barnes et al., 2014) and species dependent (Deiner and Altermatt, 2014). This result is unsurprising; molecular biologists have long known that the structural difference between DNA sequences (e.g. mol% GC content) and fragment length affect degradation properties and these are a function of the nucleotide composition of individual species genomes. Moreover, degradation biases may not just depend on species-specific difference in DNA structure but also on the presence of other species' DNA (e.g. microbial enzymes may preferentially degrade some species' DNA over others). Thus although eDNA approaches hold massive potential for multispecies ecosystem-level research, there are still large knowledge gaps in our understandings of these methods which require further validation. It is in ecosystem such as the Colne estuary, where this validation is likely to be found, as a comprehensive understanding of biodiversity of the system is required to cross-check eDNA, and especially its microbial diversity given the role this plays in eDNA degradation.

While targeting eDNA via NGS may provide data describing which taxa (including microbes) are spatially or temporally co-occurring, it does not provide information on species interactions within the system; these either have to be directly recorded (e.g. gut content analysis) or inferred from the data or existing literature. This is particularly important for understanding ecosystem functioning and the associated role of microbes, as ecosystem processes and functions are driven not just by multiple species in isolation (i.e. diversity) but also by the interactions between these different species. Describing species interaction based on previous published work is a reasonably well-established approach, but from a microbiologist viewpoint suffers from a paucity of published data on microbial interactions. Thus, inference of species interactions, and by extension integration of microbes with macroorganisms, from NGS data provides an alternative (Vacher et al., 2016). Broadly speaking these approaches utilise two methods, (i) statistical inference of networks and (ii) logic-based machine-learning algorithms (both of which are reviewed in Vacher et al., 2016). These methods produce descriptive networks highlighting the strength of species interactions based on assessing properties of species' co-occurrence. Moreover, the Bayesian Networks, and Bayesian Dynamic Networks behind some statistical inference approaches, are commonly employed in exploring gene interaction networks within single-species genome and transcriptome studies. This opens the door to incorporating microbial functional gene networks from NGS data when full metagenome (as opposed to target amplicon sequencing) information is available, providing an ideal future method to integrating microbial data in assessments of biodiversity–ecosystem functioning relationships.

Although the earlier methods provide emerging opportunities for integrating microbial data into ecosystem-level research, they suffer from a common and unavoidable problem: microbes and microbial processes operate on different spatiotemporal scales from those of macroorganisms. To overcome these scale dependencies, spatially explicit studies operating across spatial scales are required to identify both the links and disconnect between microbial and other ecological data. Within the Colne estuary, the current Coastal Biodiversity and Ecosystem Service Sustainability (CBESS) project funded by the Natural Environment Research Council's (NERC; United Kingdom) Biodiversity and Ecosystem Service Sustainability (BESS) programme (Raffaelli et al., 2014) aims to provide this spatially explicit microbial link. CBESS uses a hierarchical sampling design across the Colne estuary's sediments and salt marshes, to scale relationships between biodiversity (from microbes to macroorganisms) and ecosystem functions and services with space (described in Raffaelli et al., 2014). By capitalising on fundamental ecological properties of how natural communities change through space (e.g. species area relationships) and associated taxon dependencies, CBESS will be able to identify appropriate scales for integration between microbial and nonmicrobial ecological data. In future studies, the coupling of information gained from CBESS with the novel NGS, eDNA and network approaches will allow the Colne estuary to remain at the forefront of microbial estuarine ecology for years to come.

ACKNOWLEDGEMENTS

We wish to acknowledge the support of the research in the Colne estuary over 40 years by the Natural Environment Research Council of the U.K. through a large number of research grants and studentships. We specifically acknowledge the current Grants: NE/K001914/1 (Data synthesis and management of marine and coastal carbon) to G.J.C.U. and T.J.M.; NE/J009555/1 (Microbial degradation of isoprene in the terrestrial environment) to T.J.M., and NE/J01561X/1 (CBESS: A hierarchical approach to the examination of the relationship between biodiversity and ecosystem service flows across coastal margins) to G.J.C.U., A.J.D. and T.J.M.

REFERENCES

Acuña Alvarez, L., Exton, D.A., Suggett, D.J., Timmis, K.N., McGenity, T.J., 2009. Characterization of marine isoprene-degrading communities. Environ. Microbiol. 11, 3280–3291.

Agedah, E.C., Binalaiyifa, H.E., Ball, A.S., Nedwell, D.B., 2009. Sources, turnover and bioavailability of dissolved organic nitrogen (DON) in the Colne estuary, UK. Mar. Ecol. Prog. Ser. 382, 23–33.

Azni, S. bin Abd. Aziz, Nedwell, D.B., 1986a. The nitrogen cycle of an East Coast, U.K., saltmarsh: I. Nitrogen assimilation during primary production; detrital mineralisation. Estuar. Coast. Shelf Sci. 22, 559–575.

Azni, S. bin Abd. Aziz, Nedwell, D.B., 1986b. The nitrogen cycle of an East Coast, U.K. saltmarsh: II Nitrogen fixation, nitrification, tidal exchange. Estuar. Coast. Shelf Sci. 22, 689–704.

Balba, M.T., Nedwell, D.B., 1982. Microbial metabolism of acetate, propionate and butyrate in anoxic sediments from the Colne Point saltmarsh, Essex, U.K. J. Gen. Microbiol. 128, 1415–1422.

Banat, I.M., Lindström, E.B., Nedwell, D.B., Balba, M.T., 1981. Evidence for coexistence of two distinct functional groups of sulfate-reducing bacteria in salt marsh sediment. Appl. Environ. Microbiol. 42, 985–992.

Barnes, M.A., Turner, C.R., Jerde, C.L., Renshaw, M.A., Chadderton, W.L., Lodge, D.M., 2014. Environmental conditions influence eDNA persistence in aquatic systems. Environ. Sci. Technol. 48 (2014), 1819–1827.

Barnett, A., Méléder, V., Blommaert, L., Lepetit, B., Gaudin, P., Vyverman, W., Sabbe, K., Dupuy, C., Lavaud, J., 2015. Growth form defines physiological photoprotective capacity in intertidal benthic diatoms. ISME J. 9, 32–45.

Beddow, J., Stolpe, B., Cole, P.A., Lead, J.R., Sapp, M.S., Lyons, B.P., McKew, B., Steinke, M., Benyahia, F., Colbeck, I., Whitby, C., 2014. Estuarine sediment hydrocarbon-degrading microbial communities demonstrate resilience to nanosilver. Int. J. Biodeter. Biodegr. 96, 206–215.

Beddow, J., Stolpe, B., Cole, P.A., Lead, J.R., Sapp, M., Lyons, B.P., Colbeck, I., Whitby, C., 2016. Nanosilver inhibits nitrification and reduces ammonia-oxidizing bacterial but not archaeal amoA gene abundance in estuarine sediments. Environ. Microbiol. http://dx.doi.org/10.1111/1462-2920.13441.

Bellinger, B.J., Underwood, G.J.C., Ziegler, S.E., Gretz, M.R., 2009. Significance of diatom-derived polymers in carbon flow dynamics within estuarine biofilms determined through isotopic enrichment. Aquat. Microb. Ecol. 55, 169–187.

Benyoucef, I., Blandin, E., Lerouxel, A., Jesus, B., Rosa, P., Méléder, V., Launeau, P., Barillé, L., 2014. Microphytobenthos interannual variations in a north-European estuary (Loire estuary, France) detected by visible infrared multispectral remote sensing. Estuar. Coast. Shelf Sci. 136 (2014), 43–52.

Bernhard, A.E., Donn, T., Giblin, A.E., Stahl, D.A., 2005. Loss of diversity of ammonia-oxidizing bacteria correlates with increasing salinity in an estuary system. Environ. Microbiol. 7, 1289–1297.

Blanchard, G.F., Agion, T., Guarini, J.-M., Herlory, O., Richard, P., 2006. Analysis of the short-term dynamics of microphytobenthic biomass on intertidal mudflats. In: Kromkamp, J. (Ed.), Functioning of Microphytobenthos in Estuaries: Proceedings of the Microphytobenthos Symposium, August 2003. Royal Netherlands Academy of Arts and Sciences, Amsterdam, The Netherlands, pp. 85–97.

Bohmann, K., Evans, A., Gilbert, M.T.P., Carvalho, G.R., Creer, S., Knapp, M., Yu, D.W., de Bruyn, M., 2014. Environmental DNA for wildlife biology and biodiversity monitoring. Trends Ecol. Evol. 29, 358–367.

Braker, G., Fesefeldt, A., Witzel, K.P., 1998. Development of PCR primer systems for amplification of nitrite reductase genes (*nirK* and *nirS*) to detect denitrifying bacteria in environmental samples. Appl. Environ. Microbiol. 64, 3769–3775.

Caffrey, J.M., Bano, N., Kalanetra, K., Hollibaugh, J.T., 2007. Ammonia oxidation and ammonia-oxidizing bacteria and archaea from estuaries with different histories of hypoxia. ISME J. 1, 660–662.

Casciotti, K.L., Ward, B.B., 2005. Phylogenetic analysis of nitric oxide reductase gene homologues from aerobic ammonia-oxidizing bacteria. FEMS Microbiol. Ecol. 52, 197–205.

Charlson, R.J., Lovelock, J.E., Andreae, M.O., Warren, S.G., 1987. Oceanic phytoplankton, atmospheric sulphur, cloud albedo and climate. Nature 326, 655–661.

Chronoupolou, P.-M., Fahy, A., Coulon, F., Paissé, S., Goñi-Urriza, M.S., Peperzak, L., Acuña-Alvarez, L., McKew, B.A., Lawson, T., Timmis, K.N., Duran, R., Underwood, G.J.C., McGenity, T.J., 2013. Impact of a simulated oil spill on benthic phototrophs and nitrogen-fixing bacteria. Environ. Microbiol. 15, 241–252.

Cook, P., Veuger, B., Böer, S., Middelburg, J.J., 2007. Effect of nutrient availability on carbon and nitrogen incorporation and flows through benthic algae and bacteria in nearshore sandy sediment. Aquat. Microb. Ecol. 49, 165–180.

Coulon, F., Chronoupolou, P.-M., Fahy, A., Païssé, S., Goñi-Urriza, M.S., Peperzak, L., Acuña-Alvarez, L., McKew, B.A., Brussard, C., Underwood, G.J.C., Timmis, K.N., Duran, R., McGenity, T.J., 2012. Central role of dynamic tidal biofilms dominated

by aerobic hydrocarbonoclastic bacteria and diatoms in the biodegradation of hydrocarbons in coastal mudflats. Appl. Environ. Microbiol. 78, 3638–3648.
Crombie, A.T., El Khawand, M., Rhodius, V.A., Fengler, K.A., Miller, M.C., Whited, G.M., McGenity, T.J., Murrell, J.C., 2015. Regulation of plasmid-encoded isoprene metabolism in *Rhodococcus*, a representative of an important link in the global isoprene cycle. Environ. Microbiol. 17, 3314–3329.
Deiner, K., Altermatt, F., 2014. Transport distance of invertebrate environmental DNA in a natural river. PLoS One 9 (2), e88786.
De Jonge, V.N., van Beusekom, J.E.E., 1995. Wind- and tide-induced resuspension of sediment microphytobenthos from the tidal flats in the Ems Estuary. Limnol. Oceanogr. 40 (4), 766–778.
De Jonge, V.N., de Boer, W.F., de Jong, D.J., Brauer, V.S., 2012. Long-term mean annual microphytobenthos chlorophyll a variation correlates with air temperature. Mar. Ecol. Prog. Ser. 468, 43–56.
Dong, L.F., Thornton, D.C.O., Nedwell, D.B., Underwood, G.J.C., 2000. Denitrification in sediments of the River Colne estuary, England. Mar. Ecol. Prog. Ser. 73, 109–112.
Dong, L.F., Nedwell, D.B., Stott, A., 2006. Sources of nitrogen used for denitrification and nitrous oxide formation in sediments of the of the hypernutrified Colne, the nitrified Humber and the oligotrophic Conway estuaries, United Kingdom. Limnol. Oceanogr. 51, 545–557.
Dong, L.F., Smith, C.J., Papaspyrou, S., Stott, A., Osborn, A.M., Nedwell, D.B., 2009. Changes in benthic denitrification, nitrate ammonification, and Annamox process rates and nitrate and nitrite reductase gene abundances along an estuarine gradient (the Colne estuary, United Kingdom). Appl. Environ. Microbiol. 75, 3171–3179.
Eichmiller, J.J., Bajer, P.G., Sorensen, P.W., 2014. The relationship between the distribution of common carp and their environmental DNA in a small lake. PLoS One 9, e112611.
Elliot, M., de Jonge, V.N., Burrell, K.L., Johnson, M.W., Phillips, G.L., Turner, T.M., 1994. Trophic status of the Ore/Ald, Deben, Stour and Colne estuaries: Report to the National Rivers Authority. University of Hull.
Elliott, M., McLusky, D.S., 2002. The need for definitions in understanding estuaries. Estuar. Coast. Shelf Sci. 55, 815–827.
Elliott, M., Whitfield, A.K., Potter, I.C., Blaber, S.J., Cyrus, D.P., Nordlie, N.G., Harrison, T.D., 2007. The guild approach to categorizing estuarine fish assemblages: a global review. Fish Fish. 8, 241–268.
Exton, D.A., Suggett, D.J., Steinke, M., McGenity, T.J., 2012. Spatial and temporal variability of biogenic isoprene emissions from a temperate estuary. Global Biogeochem. Cycles 26, GB2012.
Exton, D.A., Suggett, D.J., McGenity, T.J., Steinke, M., 2013. Chlorophyll-normalized isoprene production in laboratory cultures of marine microalgae and implications for global models. Limnol. Oceanogr. 58, 1301–1311.
Exton, D.A., McGenity, T.J., Steinke, M., Smith, D.J., Suggett, D.J., 2015. Uncovering the volatile nature of tropical coastal marine ecosystems in a changing world. Glob. Chang. Biol. 21, 1383–1394.
Flindt, M.R., Pardal, M.A., Lillebo, A.I., Martins, I., Marques, J.C., 1999. Nutrient cycling and plant dynamics in estuaries: a brief review. Acta Oecol. 20, 237–248.
Foote, A.D., Thomsen, P.F., Sveegaard, S., Wahlberg, M., Kielgast, J., et al., 2012. Investigating the potential use of environmental DNA (eDNA) for genetic monitoring of marine mammals. PLoS One 7 (8), e41781.
Forster, R.M., Creach, V., Sabbe, K., Vyverman, W., Stal, L.J., 2006. Biodiversity-ecosystem function relationship in microphytobenthic diatoms of the Westerschelde estuary. Mar. Ecol. Prog. Ser. 311, 192–201.

Gillanders, B.M., 2005. Using elemental chemistry of fish otoliths to determine connectivity between estuarine and coastal habitats. Estuar. Coast. Shelf Sci. 64, 45–67.

Goldberg, C.S., Sepulveda, A., Ray, A., Baumgardt, J., Waits, L.P., 2013. Environmental DNA as a new method for early detection of New Zealand mudsnails (Potamopyrgus antipodarum). Freshw. Sci., 792–800.

Grangeré, K., Lefebvre, S., Blin, J.-L., 2012. Spatial and temporal dynamics of biotic and abiotic features of temperate coastal ecosystems as revealed by a combination of ecological indicators. Estuar. Coast. Shelf Sci. 108, 109–118.

Green, B.C., Smith, D.J., Grey, J., Underwood, G.J.C., 2012. High site fidelity and low site connectivity in temperate salt marsh fish populations: a stable isotope approach. Oecologia 168, 245–255.

Grobbellar, J.U., 1990. Modelling algal productivity in large outdoor cultures and waste treatment systems. Biomass 21, 279–314.

Guenther, A.B., Jiang, X., Heald, C.L., Sakulyanontvittaya, T., Duhl, T., Emmons, L.K., Wang, X., 2012. The Model of Emissions of Gases and Aerosols from Nature version 2.1 (MEGAN2.1): an extended and updated framework for modeling biogenic emissions. Geosci. Model Dev. 5, 1471–1492.

Hanlon, A.R.M., Bellinger, B., Haynes, K., Xiao, G., Hofmann, T.A., Gretz, M.R., Ball, A.S., Osborn, A.M., Underwood, G.J.C., 2006. Dynamics of EPS production and loss in an estuarine, diatom-dominated, microalgal biofilm over a tidal emersion immersion period. Limnol. Oceanogr. 51, 79–93.

Haynes, K., Hofmann, T.A., Smith, C.J., Ball, A.S., Underwood, G.J.C., Osborn, A.M., 2007. Diatom-derived carbohydrates as factors affecting bacterial community composition in estuarine sediments. Appl. Environ. Microbiol. 73, 6112–6124.

Hofmann, T., Hanlon, A.R.M., Taylor, J.D., Ball, A.S., Osborn, A.M., Underwood, G.J.C., 2009. Dynamics and compositional changes in extracellular carbohydrates in estuarine sediments during degradation. Mar. Ecol. Prog. Ser. 379, 45–58.

Howarth, R., Chan, F., Conley, D.J., Garnier, J., Doney, S.C., Marino, R., Billen, G., 2011. Coupled biogeochemical cycles: eutrophication and hypoxia in temperate estuaries and coastal marine ecosystems. Front. Ecol. Environ. 9, 18–26.

Humborg, C., Ittekkot, V., Coclasu, A., Bodunges, B.V., 1997. Effect of Danube River dam on Black Sea biogeochemistry and ecosystem structure. Nature 386, 385–388.

Hussey, A., 1980. The net primary production of an Essex saltmarsh, with particular reference to *Puccinellia maritima*. PhD thesis, University of Essex.

Hussey, A., Long, S.P., 1982. Seasonal changes in weight of above- and below-ground vegetation and dead plant material in a salt marsh at Colne Point, Essex. J. Ecol. 70, 757–771.

Jackson, M.C., Weyl, O.L.F., Altermatt, F., Durance, I., Friberg, N., Dumbrell, A.J., Piggott, J.J., Tiegs, S.D., Tockner, K., Lehmann, A., Narwani, A.J., Krug, C.B., Leadley, P.W., Woodward, G., 2016. Recommendations for the next generation of global freshwater biological monitoring tools. Adv. Ecol. Res. 55, 607–630.

Jørgensen, B.B., 1982. Mineralisation of organic matter in the sea bed: the role of sulphate reduction. Nature 296, 643–645.

Jørgensen, B.B., Okholm-Hansen, B., 1985. Emissions of biogenic sulphur gases from a Danish estuary. Atmos. Environ. 19, 1737–1749.

Juneau, P., Barnett, A., Meleder, V., Dupuy, C., Lavaud, J., 2015. Combined effect of high light and high salinity on the regulation of photosynthesis in three diatom species belonging to the main growth forms of intertidal flat inhabiting microphytobenthos. J. Exp. Mar. Biol. Ecol. 463, 95–104.

Justic, D., Rabelais, N.N., Turner, R.E., Dortch, Q., 1995. Changes in nutrient structure of river-dominated coastal waters: stoichiometric nutrient balance and its consequences. Estuar. Coast. Shelf Sci. 40, 339–356.

King, D., Nedwell, D.B., 1985. The influence of nitrate concentration upon the end-products of nitrate dissimilation by bacteria in anaerobic saltmarsh sediment. FEMS Microbiol. Ecol. 31, 23–28.

Kocum, E., Underwood, G.J.C., Nedwell, D.B., 2002a. Simultaneous measurement of phytoplanktonic primary production, nutrient and light availability along a turbid, eutrophic UK east coast estuary (the Colne estuary). Mar. Ecol. Prog. Ser. 231, 1–12.

Kocum, E., Nedwell, D.B., Underwood, G.J.C., 2002b. Regulation of phytoplankton primary production along a hypernutrified estuary. Mar. Ecol. Prog. Ser. 231, 13–22.

Konneke, M., Bernhard, A.E., de la Torre, J.R., Walker, C.B., Waterbury, J.B., Stahl, D.A., 2005. Isolation of an autotrophic ammonia-oxidizing marine archaeon. Nature 437, 543–546.

Kuenen, J.G., Robertson, L.A., 1987. Ecology of nitrification and denitrification. In: Cole, J.-A., Ferguson, S.J. (Eds.), The Nitrogen and Sulphur Cycles. Cambridge University Press, Cambridge, pp. 162–218.

Li, J., Nedwell, D.B., Beddow, J., Dumbrell, A.J., McKew, B.A., Thorpe, E.L., Whitby, C., 2015. amoA gene abundances and nitrification potential rates suggest that benthic ammonia-oxidizing bacteria and not archaea dominate N cycling in the Colne estuary, United Kingdom. Appl. Environ. Microbiol. 81, 159–165.

Luisetti, T., Turner, R.K., Jickells, T., Andrews, J., Elliott, M., Schaafsma, M., et al., 2014. Coastal Zone Ecosystem Services: from science to values and decision making; a case study. Sci. Total Environ. 493, 682–693.

Maček, I., Vodnik, D., Pfanz, H., Low-Décarie, E., Dumbrell, A.J., 2016. Locally extreme environments as natural long-term experiments in ecology. Adv. Ecol. Res. 55, 283–324.

Mächler, E., Deiner, K., Steinmann, P., Altermatt, F., 2014. Utility of environmental DNA for monitoring rare and indicator macroinvertebrate species. Freshw. Sci. 33, 1174–1183.

Martens-Habbena, W., Berube, P.M., Urakawa, H., de la Torre, J.R., Stahl, D.A., 2009. Ammonia oxidation kinetics determine niche separation of nitrifying Archaea and Bacteria. Nature 46 (1), 976–979.

McGenity, T.J., 2014. Hydrocarbon biodegradation in intertidal wetland sediments. Curr. Opin. Biotechnol. 27, 46–54.

McGenity, T.J., Oren, A., 2012. Hypersaline environments. In: Bell, E.M. (Ed.), Life at Extremes: Environments, Organisms and Strategies for Survival. CAB International, Wallingford, UK, pp. 402–437.

McKew, B.A., Taylor, J.D., McGenity, T.J., Underwood, G.J.C., 2011. Resistance and resilience of benthic biofilm communities from a temperate saltmarsh to desiccation and rewetting. ISME J. 5, 30–41.

McKew, B.A., Dumbrell, A., Taylor, J.D., McGenity, T.J., Underwood, G.J.C., 2013. Differences between aerobic and anaerobic degradation of microphytobenthic biofilm-derived organic matter within intertidal sediments. FEMS Microbiol. Ecol. 84, 495–509.

McMellor, S., Underwood, G.J.C., 2014. Water policy effectiveness: 30 years of change in the hypernutrified Colne estuary, England. Mar. Pollut. Bull. 81, 200–209.

Miles, A., Sundbäck, K., 2000. Diel variation in microphytobenthic productivity in areas of different tidal amplitude. Mar. Ecol. Prog. Ser. 205, 11–22.

Munson, M.A., Nedwell, D.B., Embley, T.M., 1997. Phylogenetic diversity of Archaea in sediment samples from a coastal salt marsh. Appl. Environ. Microbiol. 63, 4729–4733.

Nedwell, D.B., 1987. Distribution and pool sizes of microbially available carbon in sediment measured by a microbiological assay. FEMS Microbiol. Ecol. 45, 47–52.

Nedwell, D.B., Abram, J.W., 1979. Relative influence of temperature and electron donor and electron acceptor concentrations on bacterial sulphate reduction in saltmarsh sediment. Microb. Ecol. 5, 67–72.

Nedwell, D.B., Takii, S., 1988. Bacterial sulphate reduction in sediments of a European salt marsh: acid-volatile and tin-reducible products. Estuar. Coast. Shelf Sci. 26, 599–606.

Nedwell, D.B., Shabbeer, M.T., Harrison, R.M., 1994. Dimethyl sulphide in North Sea waters and sediments. Estuar. Coast. Shelf Sci. 39, 209–217.

Nedwell, D.B., Dong, L.F., Sage, A.S., Underwood, G.J.C., 2002. Variations of the nutrient loads to the mainland U.K. estuaries: correlation with catchment areas, urbanisation and coastal eutrophication. Estuar. Coast. Shelf Sci. 54, 951–970.

Nedwell, D.B., Embley, T.M., Purdy, K.J., 2004. Sulphate reduction, methanogenesis and phylogenetics of the sulphate reducing bacterial communities along an estuarine gradient. Aquat. Microb. Ecol. 37, 209–217.

Nielsen, L.P., 1992. Denitrification in sediment determined from nitrogen isotope pairing. FEMS Microbiol. Ecol. 86, 357–362.

Nixon, S.W., 1995. Coastal marine eutrophication: a definition, social causes and future concerns. Ophelia 41, 199–219.

Nogales, B., Timmis, K.N., Nedwell, D.B., Osborn, A.M., 2002. Detection and diversity of expressed denitrification genes in estuarine sediments after reverse-transcription-PCR amplification from mRNA. Appl. Environ. Microbiol. 68, 5017–5025.

Oakes, J.M., Eyre, B.D., Middelburg, J.J., Boschker, H.T.S., 2010. Composition, production, and loss of carbohydrates in subtropical shallow subtidal sandy sediments: rapid processing and long-term retention revealed by 13C-labeling. Limnol. Oceanogr. 55, 2126–2138.

Oakes, J.M., Eyre, B.D., Middelburg, J.J., 2012. Transformation and fate of microphytobenthos carbon in subtropical shallow subtidal sands: a 13C-labeling study. Limnol. Oceanogr. 57, 1846–1856.

Oakley, B.B., Carbonero, F., Dowd, S.E., Hawkins, R.J., Purdy, K.J., 2012. Contrasting patterns of niche partitioning between two anaerobic terminal oxidizers of organic matter. ISME J 6, 905–914.

Ogilvie, B., Nedwell, D.B., Harrison, R.M., Robinson, A., Sage, A., 1997. High nitrate, muddy estuaries as nitrogen sinks: the nitrogen budget of the River Colne estuary (United Kingdom). Mar. Ecol. Prog. Ser. 150, 217–228.

Orvain, F., Sauriau, P.-G., Sygut, A., Joassard, L., Le Hir, P., 2004. Interacting effects of Hydrobia ulvae bioturbation and microphytobenthos on the erodibility of mudflat sediments. Mar. Ecol. Prog. Ser. 278, 205–223.

O'Sullivan, L.A., Sass, A.M., Webster, G., Fry, J.C., Parkes, R.J., Weightman, A.J., 2013. Contrasting relationships between biogeochemistry and prokaryotic diversity depth profiles along an estuarine sediment gradient. FEMS Microbiol. Ecol. 85, 143–157.

Pacifico, F., Harrison, S.P., Jones, C.D., Sitch, S., 2009. Isoprene emissions and climate. Atmos. Environ. 43, 6121–6135.

Papaspyrou, S., Smith, C.J., Dong, L.F., Whitby, C., Dumbrell, A.J., Nedwell, D.B., 2014. Nitrate reduction functional genes and nitrate reduction potentials persist in deeper estuarine sediments. Why? PLoS One 9 (4), 1–13.

Peletier, H., 1996. Long-term changes in intertidal estuarine diatom assemblages related to reduced input of organic waste. Mar. Ecol. Prog. Ser. 137, 265–271.

Platell, M.E., Orr, P.A., Potter, I.C., 2006. Inter- and intraspecific partitioning of food resources by six large and abundant fish species in a seasonally open estuary. J. Fish Biol. 69, 243–262.

Pollard, D.A., 1981. Estuaries are valuable contributors to fisheries production. Aust. Fish. 40, 7–9.

Poole, J., 2013. Impact of engineered nanoparticles on aquatic microbial processes. PhD thesis, University of Essex, United Kingdom.

Prastka, K.E., Sanders, R., Jickells, T., 1998. Has the role of estuaries as sources or sinks of dissolved inorganic phosphate changed over time? Results of a Ks study. Mar. Pollut. Bull. 36, 718–728.

Purdy, K.J., Cresswell-Maynard, T.D., Nedwell, D.B., McGenity, T.J., Grant, W.D., Timmis, K.N., Embley, T.M., 2004. Isolation of haloarchaea that grow at low salinities. Environ. Microbiol. 6, 591–595.

Purdy, K.J., Munson, M.A., Embley, T.M., Nedwell, D.B., 2003. Use of 16S rRNA-targeted oligonucleotide probes to investigate function and phylogeny of sulphate-reducing bacteria and methanogenic archaea in a UK estuary. FEMS Microbiol. Ecol. 44, 361–371.

Raffaelli, D.G., Bullock, J.M., Cinderby, S., Durance, I., Emmett, B., Harris, J., Hicks, K., Oliver, T.H., Paterson, D., White, P.C.L., 2014. Big data and ecosystem research programmes. Adv. Ecol. Res. 51, 41–77.

Rees, H.C., Maddison, B.C., Middleditch, D.J., Patmore, K.J., Gough, R.M.C., 2014. The detection of aquatic animal species using environmental DNA—a review of eDNA as a survey tool in ecology. J. Appl. Ecol. 51, 1450–1459.

Richardson, D.J., 2000. Bacterial respiration: a flexible process for a changing environment. Microbiology 146, 551–571.

Robinson, A.D., 1996. The Colne estuary as a source of N_2O and NOx gases to the atmosphere. PhD thesis, University of Essex.

Robinson, A.D., Nedwell, D.B., Harrison, R.M., Ogilvie, B.G., 1998. Hypernutrified estuaries as sources of N_2O emission to the atmosphere: the estuary of the River Colne, Essex, UK. Mar. Ecol. Prog. Ser. 164, 59–71.

Rotthauwe, J.-H., Witzel, K.P., Liesack, W., 1997. The ammonia-monoxygenase structural gene amoA as a functional marker: molecular fine-scale analysis of natural ammonia-oxidising populations. Appl. Environ. Microbiol. 63, 4704–4712.

Sanni, G.O., Coulon, F., McGenity, T.J., 2015. Dynamics and distribution of bacterial and archaeal communities in oil-contaminated temperate coastal mudflat mesocosms. Environ. Sci. Pollut. Res. Int. 22, 15230–15247.

Scranton, M.I., 1983. Gaseous nitrogen compounds in the marine environment. In: Carpenter, E.J., Capone, D.G. (Eds.), Nitrogen in the Marine Environment. Academic Press, New York, pp. 37–64.

Senior, E., Linström, E.B., Banat, I.M., Nedwell, D.B., 1982. Sulfate reduction and methanogenesis in the sediment of a saltmarsh on the East Coast of the United Kingdom. Appl. Environ. Microbiol. 43, 987–996.

Skiba, U., Smith, K.A., Fowler, D., 1993. Nitrification and denitrification as sources of nitric oxide and nitrous oxide in a sandy loam soil. Soil Biol. Biochem. 25, 1527–1536.

Smith, C.J., Nedwell, D.B., Dong, L.F., Osborn, A.M., 2007. Diversity and abundance of nitrate reductase genes (narG and napA), nitrite reductase genes (nirs and nrefA), and their transcripts in estuarine sediments. Appl. Environ. Microbiol. 73, 3612–3622.

Smith, C.J., Dong, L.F., Wilson, J., Stott, A., Osborn, A.M., Nedwell, D.B., 2015. Seasonal variation in denitrification and dissimilatory nitrate reduction to ammonia, process rates and corresponding key functional genes along an estuarine nitrate gradient. Front. Microbiol. 5, 1–11. http://dx.doi.org/10.3389/fmicb.2015.00542.

Spilmont, N., Seuront, L., Meziane, T., Welsh, D.T., 2011. There's more to the picture than meets the eye: sampling microphytobenthos in a heterogeneous environment. Estuar. Coast. Shelf Sci. 95, 470–476.

Steele, D., Franklin, D.J., Underwood, G.J.C., 2014. Protection of cells from salinity stress by extracellular polymeric substances (EPS) in diatom biofilms. Biofouling 30, 987–998.

Sutka, R.L., Ostrom, N.E., Ostrom, P.H., Breznak, J.A., Gandhi, H., Pitt, A.J., Li, F., 2006. Distinguishing nitrous oxide production from nitrification from denitrification on the basis of isotopomer abundances. Appl. Environ. Microbiol. 72, 638–644.

Taylor, J.D., McKew, B.A., Kuhl, A., McGenity, T.J., Underwood, G.J.C., 2013. Microphytobenthic extracellular polymeric substances (EPS) in intertidal sediments fuel both generalist and specialist EPS-degrading bacteria. Limnol. Oceanogr. 58, 1463–1480.

Thomsen, P.F., Kielgast, J., Iversen, L.L., et al., 2012. Monitoring endangered freshwater biodiversity using environmental DNA. Mol. Ecol. 21, 2565–2573.

Thornton, D.C.O., Underwood, G.J.C., Nedwell, D.B., 1999. Effect of illumination and emersion period on the exchange of ammonium across the sediment-water interface. Mar. Ecol. Prog. Ser. 184, 11–20.

Thornton, D.C.O., Dong, L.F., Underwood, G.J.C., Nedwell, D.B., 2002. Factors affecting microphytobenthic biomass, species composition and production in the Colne estuary (UK). Aquat. Microb. Ecol. 27, 285–300.

Thornton, D.C.O., Dong, L.F., Underwood, G.J.C., Nedwell, D.B., 2007. Sediment-water inorganic nutrient exchange and nitrogen budgets in the Colne estuary (UK). Mar. Ecol. Prog. Ser. 337, 63–77.

Throbäck, I.N., Enwall, K., Jarvis, A., Hallin, S., 2004. Reassessing PCR primers targeting nirS, nirK and nosZ genes for community surveys of denitrifying bacteria with DGGE. FEMS Microbiol. Ecol. 49, 401–417.

Trimmer, M., Nedwell, D.B., Sivyer, D.B., Malcolm, S.J., 2000. Seasonal benthic organic matter mineralisation measured by oxygen uptake and denitrification along a transect of the inner and outer River Thames estuary, UK. Mar. Ecol. Prog. Ser. 197, 103–119.

Ubertini, M., Lefebvre, S., Gangnery, A., Grangeré, K., Le Gendre, R., et al., 2012. Spatial variability of benthic-pelagic coupling in an estuary ecosystem: consequences for microphytobenthos resuspension phenomenon. PLoS One 7 (8), e44155. http://dx.doi.org/10.1371/journal.pone.0044155.

Underwood, G.J.C., Kromkamp, J., 1999. Primary production by phytoplankton and microphytobenthos in estuaries. Adv. Ecol. Res. 29, 93–153.

Underwood, G.J.C., Paterson, D.M., 1993. Seasonal changes in diatom biomass, sediment stability and biogenic stabilisation in the Severn Estuary, UK. J. Mar. Biol. Assoc. U. K. 73, 871–887.

Underwood, G.J.C., Paterson, D.M., 2003. The importance of extracellular carbohydrate production by marine epipelic diatoms. Adv. Botan. Res. 40, 184–240.

Underwood, G.J.C., Provot, L., 2000. Determining the environmental preferences of four estuarine epipelic diatom taxa—growth across a range of salinity, nitrate and ammonium conditions. Eur. J. Phycol. 35, 173–182.

Underwood, G.J.C., Phillips, M., Saunders, K., 1998. Distribution of estuarine benthic diatom species along salinity and nutrient gradients. Eur. J. Phycol. 33, 173–183.

Underwood, G.J.C., Perkins, R.G., Consalvey, M., Hanlon, A.R.M., Oxborough, K., Baker, N.R., Paterson, D.M., 2005. Patterns in microphytobenthic primary productivity: species-specific variation in migratory rhythms and photosynthetic efficiency in mixed-species biofilms. Limnol. Oceanogr. 50, 755–767.

Vacher, C., Tamaddoni-Nezhad, A., Kamenova, S., Peyrard, N., Moalic, Y., Sabbadin, R., Schwaller, L., Chiquet, J., Smith, M.A., Vallance, J., Fievet, V., Jakuschkin, B., Bohan, D.A., 2016. Learning ecological networks from next-generation sequencing data. Adv. Ecol. Res. 54, 1–39.

Van Colen, C., Underwood, G.J.C., Serôdio, J., Paterson, D.M., 2014. Ecology of intertidal microbial biofilms: mechanisms, patterns and future research needs. J. Sea Res. 92, 2–5.

Van de Graaf, A.A., Mulder, A., de Bruijn, P., Jetten, M.S., Robertson, L.A., Kuenen, J.G., 1995. Anaerobic oxidation of ammonium is a biologically mediated process. Appl. Environ. Microbiol. 61, 1246–1251.

van der Wal, D., Wielemaker-van den Dool, A., Peter, M.J., Herman, P.M.J., 2010. Spatial synchrony in intertidal benthic algal biomass in temperate coastal and estuarine ecosystems. Ecosystems 13, 338–351.

Watson, A., Stephen, K.D., Nedwell, D.B., Arah, J.R.M., 1997. Oxidation of methane in peat: kinetics of CH_4 and O_2 removal and the role of plant roots. Soil Biol. Biochem. 29, 1257–1267.

Webster, G., Sullivan, L.A., Meng, Y., Williams, A.S., Sass, A.M., Watkins, A.J., Parkes, R.J., Weightman, A.J., 2015. Archaeal community diversity and abundance changes along a natural salinity gradient in estuarine sediments. FEMS Microbiol. Ecol. 91, 1–18.

Weerman, E.J.P., Herman, P.M.J., van de Koppel, J., 2011. Top-down control inhibits spatial self-organization of a patterned landscape. Ecology 92, 487–495.

Woodward, G., Speirs, D.C., Hildrew, A.G., 2005. Quantification and resolution of a complex, size-structured food web. Adv. Ecol. Res. 46, 85–135.

Zumft, W.G., 1997. Cell biology and molecular basis of denitrification. Microbiol. Mol. Rev. 61, 533–616.

CHAPTER SIX

Locally Extreme Environments as Natural Long-Term Experiments in Ecology

I. Maček*,†,1, D. Vodnik*, H. Pfanz‡, E. Low-Décarie§, A.J. Dumbrell§
*Biotechnical Faculty, University of Ljubljana, Ljubljana, Slovenia
†Faculty of Mathematics, Natural Sciences and Information Technologies (FAMNIT), University of Primorska, Koper, Slovenia
‡Lehrstuhl für Angewandte Botanik, University Duisburg-Essen, Essen, Germany
§School of Biological Sciences, University of Essex, Colchester, United Kingdom
1Corresponding author: e-mail address: irena.macek@bf.uni-lj.si

Contents

1. Introduction	284
2. Locally Extreme Environments as Long-Term Experiments	285
2.1 Space-for-Time Substitutions	287
2.2 Space for Space	288
2.3 Tractable Natural Model Systems	289
3. Case Study: Mofettes	290
3.1 Specific Abiotic Factors	295
3.2 Mofettes as a Natural Long-Term Analogue to Free Air Carbon-Dioxide Enrichment (FACE) Experiments?	298
3.3 Plant Ecophysiology in Mofette Areas	299
3.4 Community Ecology	303
3.5 CCS: What Can We Learn from Mofettes?	312
3.6 Exploring Mofette Food Webs and Biological Networks	313
4. Conclusions	314
Acknowledgements	316
References	316

Abstract

Many natural phenomena and ecological processes take place extremely slowly, requiring both long-term observations and experiments to investigate them. An alternative is to investigate natural systems that have long-term and stable environmental conditions that are opposed to those of the surrounding ecosystem. Locally extreme environments provide an example of this, and are a powerful tool for the study of slower ecological and evolutionary processes, allowing the investigation of longer term mechanisms at logistically tractable spatial and temporal scales. These systems can be used to gain insight into adaptation of natural communities and their ecological networks. We present a case study and review the literature investigating biological communities at terrestrial mofettes—natural sites with constant geogenic CO_2 exhalations and consequent soil hypoxia. Mofettes are often used as natural analogues to future conditions predicted by current climate change scenarios, as model ecosystems for environmental impact assessments of carbon capture and storage systems and for the investigation of physiological, ecological and evolutionary studies of a range of phylogenetically distinct organisms across spatial scales. The scientific power of locally extreme environments is just starting to be harnessed and these systems are bound to provide growing insight into long-term ecological processes, which will be essential for our capacity to adequately manage ecosystems and predict ecological and evolutionary responses to global change.

1. INTRODUCTION

Many important ecological processes occur over decades or even centuries, and long-term observations and experiments in representative sites around the globe are required to gain insight into these slower processes, which include succession, nutrient cycling, response to gradual environmental change and evolutionary adaptation. Conducting long-term observations of natural systems are challenging, as they are highly dynamic and noisy, with short-term variation obscuring longer term trends. Yet in long-term experiments, a number of variables are directly controlled, which allows the causes of slower ecological responses to be established. However, long-term studies lasting more than a few decades are extremely rare (Owens, 2013; but see Storkey et al., 2016), particularly in ecology. In part, this is because such set-ups demand stable funding and the persistence of several generations of highly motivated researchers. Moreover, even the longest running experiment is too short to study many ecological processes (e.g. succession or adaptive change of organisms with longer generation times) in different ecosystems.

Locally extreme environments can serve as long-term natural experiments in ecology and evolution. These systems could potentially provide broad insights into a range of questions, from the mechanisms of adaptation of large long-lived organisms, to the biogeography of microorganisms. In extreme environments, selective pressures are often permanent and relatively constant, providing ideal conditions for examining genetic adaptation of biological communities to specific conditions. Research into one such model system, natural CO_2 springs or vents—also called mofettes—is presented here as a case study. This covers a wide range of topics, from long-term models of climate change and plant physiological responses to elevated CO_2

(Sections 3.2 and 3.3), the use of mofettes as models for the environmental impact assessments (EIA) of CO_2 leakage from carbon capture and storage (CCS) systems (Section 3.5), to more ecological, evolutionary and functional aspects of plant, faunal and soil microbial communities (Section 3.4). Importantly, a section is reserved that covers new insights into microbial community ecology, which has historically lagged behind research on higher taxa, but where significant progress and advances using long-term mofette ecosystems have been made (Section 3.4.6). Finally, future directions and advances in the use of locally extreme environments with well defined and constant selective pressures for long-term ecological studies are discussed in depth; the research potential for these systems is broad, from acclimation to adaptation, from communities of macroorganisms to microorganisms, and from individual taxa to ecological networks.

2. LOCALLY EXTREME ENVIRONMENTS AS LONG-TERM EXPERIMENTS

Extreme environments have been defined as having one or more environmental parameters showing values permanently close to the lower or upper limits known for life (ESF Report, 2007; also see Torossian et al., 2016). However, defining extreme environments remain difficult and contentious. Clearly, an extreme for humans or higher (multicellular) organisms may not be an extreme for the microbes able to use a greater range of different metabolic pathways to survive in that environment. In an analogous context, our current high oxygen atmosphere may be considered extreme for anaerobic organisms that dominated for a large period of the existence of life. Extreme environments may also be defined in comparison to the local range of conditions in the region or experienced by a specific species, with conditions far from the regional mean or mean of a species' range considered extreme.

Extreme environments are hotspots of biological discovery that have served as a rich source of biotechnological compounds. New discoveries are promised in the future by the rapidly growing number of sequenced extremophile genomes (e.g. Ferrer et al., 2007). These systems allow us to investigate the very limits of biological processes under extreme conditions, in environments that are themselves highly diverse. They can be characterised by extremes of temperature, radiation, pressure, vacuums, desiccation, salinity, pH, oxygen tension and/or other physicochemical extremes. The diversity of extreme ecosystems is also large, from the deep sea to deserts (Rothschild and Mancinelli, 2001). Ecological investigation, including the loss of biodiversity and response of extreme ecosystems to global change, is growing in importance. While extreme ecosystems may seem marginal, in terms of their spatial extent, some (e.g. permafrost, polar oceans, deep sea) represent large areas or volumes on Earth, and can have potentially large contributions to biogeochemical cycling across the planet, and by extension, they could play key roles in shaping global responses to climate change.

Environments that are locally extreme relative to the average regional conditions may hold less promise in terms of biological discovery, yet they provide some of the most powerful natural experiments for the study of ecology and evolution (Fig. 1). These systems allow the conceptual compression of both space and time, allowing the investigation of important questions at manageable spatiotemporal scales that would otherwise be experimentally unfeasible (e.g. O'Gorman et al., 2012, 2014). Moreover,

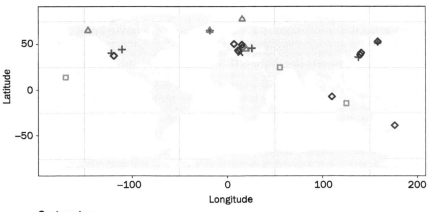

Fig. 1 Geographic distribution of example locally extreme environments used as natural long-term experiments: adaptation to high temperature in corals (Barshis et al., 2010; Bauman et al., 2014; Schoepf et al., 2015), archaea and cyanobacteria dispersal and local evolution in hot springs (Papke et al., 2003; Whitaker et al., 2003), shift in composition of communities and adaptation to ocean acidification in marine mofettes (Calosi et al., 2013; Hall-Spencer et al., 2008), algae, fungi, fauna and flora adaptation to high CO_2 in terrestrial mofettes (Collins and Bell, 2006; Fordham et al., 1997; Hohberg et al., 2015; Maček et al., 2011; Onoda et al., 2009; Pfanz et al., 2004), shift in composition of communities with warming in geothermal streams (Gudmundsdottir et al., 2011; O'Gorman et al., 2014; Živić et al., 2013).

locally extreme environments also tend to have lower biodiversity, providing a more tractable system to investigate.

2.1 Space-for-Time Substitutions

Long-term studies are required for the observation and understanding of slow ecological processes, long-lived organisms, rare events or capturing episodic phenomena (Franklin, 1989). However, for many of the slower processes, such as the study of ecological succession, equilibrium dynamics or speciation, the length of the required study is logistically prohibitive. For the investigation of these processes, a space-for-time approach can be used. This assumes that processes will follow a similar course across systems and that study systems with different ages are adequate substitutes to the study of a single system through time. Alternatively, they are also often used under the assumption that conditions in an extreme contemporary system can adequately simulate the conditions of a normal system in the past or in the future. Furthermore, ecosystems that have been perturbed at different time points have been instrumental in the study of recovery and species succession (Banet and Trexler, 2013; Pickett, 1989).

Extreme environments provide a unique context in which to apply the space-for-time approach, as they allow the study of the responses of communities that have been exposed to alternative environments at different time points through geological time. Deep sea hydrothermal vents of the fast-spreading East Pacific ridge get annihilated and created by regular volcanic eruptions, representing temporary habitats with a range of ages that undergo colonisation by organisms that can adapt locally (Van Dover, 2002). Natural CO_2 vents that represent local extremes can be used as contemporary proxies of high CO_2, acidified, low O_2 and sometimes high temperature environments that were both common in the geological past before the rise of photosynthesis or the evolution and expansion of land plants (Lomax, 2012), and that will become more common in the future following climate change (Hall-Spencer et al., 2008).

The space-for-time approach is particularly important in the context of understanding and predicting the evolutionary effect of pollution and anthropogenic global change, as the 'missing' chronosequences can be recaptured as spatial snapshots across the landscape. Terrestrial and marine CO_2 vents are used to study the evolutionary response to long-term CO_2 enrichment (Collins and Bell, 2006; Fordham et al., 1997) and ocean acidification (Calosi et al., 2013). Natural hydrocarbon seeps help determine the biodiversity and ecology of oil degraders and the long-term effects of spills (Head et al., 2006). Geothermal freshwater systems are potential tools in the study of changes in the composition of freshwater communities and evolutionary ecology of species with global warming (Gudmundsdottir et al., 2011; O'Gorman et al., 2012, 2014; Živić et al., 2013).

2.2 Space for Space

Steep environmental gradients reduce the spatial scale required for observation and experimentation, and working at smaller spatial scales is beneficial for many investigations because it can remove confounding biogeographical effects. An important qualifier, however, is that it is often essential to employ reciprocal transplants, to account for potential local adaptive responses that can then be disentangled from those operating at more proximate ecological scales (Kawecki and Ebert, 2004). Altitudinal gradients are widely seen as among the most powerful natural experiments for the investigation of ecological responses to physical conditions (notably temperature), though with obvious limitations that must also be considered (Körner, 2007). Altitude has often been used as a proxy for latitude, allowing the study of biogeographical processes that occur at the scale of the planet over a short vertical gradient (Körner, 2000). Observations of altitudinal range shifts, particularly range contractions, with global warming are far easier to characterise than the corollary of large-scale latitudinal shifts (Jump et al., 2009).

Locally extreme environments also generate steep ecological gradients that are ideal for experimentation, including reciprocal transplants. The huge global diversity of these environments provides the potential to study responses to just about any physiochemical variable, including temperature, hypoxia, pH and salinity. The study of these gradients over the range from 'normal' to 'extreme' also allows the investigation of tipping points and potential nonlinear relationships between ecological responses and environmental conditions varying along the gradient. For example, bacterial diversity appears to be maximal at intermediate levels of salinity that are above the regional average, but below the maximum local value (Schapira et al., 2009).

Locally extreme environments, such as hydrothermal vents and cold seeps, can be treated as relatively isolated patches within the context of a generally benign ecosystem, allowing the investigation of island biogeographical processes at reduced spatial scales (Dawson, 2015). The close proximity of similar patches from which dispersal can be tracked, provides an ideal context to investigate related metapopulation and metacommunity dynamics.

Locally extreme environments are often aggregated within a small geographic area, with relatively small between-site variance in geophysical parameters. This can allow for spatial replication while still maintaining a feasible spatial scale of study. This replication is essential for the study of the repeatability of ecological succession or evolutionary processes. The capacity to thrive (growth and reproduction) in extreme environments may have arisen independently multiple times, as dispersal between extreme environments may be very limited. For example, cyanobacterial communities of hot springs (Papke et al., 2003) and hyperthermophilic archaeal communities (Whitaker et al., 2003) show clear patterns of geographic endemism even at small spatial scales. However, extreme environments, such as mofettes, can also host communities containing both endemic and cosmopolitan organisms, suggesting that endemism may in itself not be a diagnostic signal of limited dispersal (Herbold et al., 2014). High dispersal into extreme environments, and with it gene flow, could increase the probability of an existing adapted organism establishing itself in the extreme environment, increasing the effective population size on which local selection pressures can act, although this gene flow may also limit the scope of local adaptation.

Organisms adapted to locally extreme environments provide a unique system to study geographic dispersal of species and gene flow, particularly in the context of microorganisms. If local adaptation to the extreme conditions requires trade-offs in the ability to grow in other environments, the abundance of organisms arriving in regionally average habitats from the local extremes could easily be examined through enrichment studies, selecting for the capacity to thrive in the extreme environment, but not in the locally benign/average environment. The prevalence of extremophiles (e.g. thermophilic bacteria) has been used in the measurement of microbial dispersal to the arctic seabed and is suggested to support the widespread dispersal of thermophiles from extreme environments (Bartholomew and Paik, 1966; de Rezende et al., 2013; Hubert et al., 2009).

2.3 Tractable Natural Model Systems

In addition to the smaller geographic scale, the lower species diversity associated with extreme environments makes them tractable natural model systems (Wall and Virginia, 1999). Thus making it feasible to identify and track all (or most) species and their interactions, particularly in microbial communities which tend to be too diverse in most other environments to characterise fully. These systems share many of the conceptual features of natural microcosms that have been instrumental in testing contemporary ecological theory, including metacommunity theory and biodiversity–function relationships (Reiss et al., 2010; Srivastava et al., 2004).

3. CASE STUDY: MOFETTES

Extreme environments characterised by specific gas composition are exemplars that are increasingly used as natural experiments (e.g. Maček et al., 2005, 2011; Rothschild and Mancinelli, 2001). Hypoxic (low O_2 concentration compared to atmospheric concentrations) or anoxic (devoid of O_2) environments are mostly found at high altitude, in marine or freshwater environments and in soils (underground borrows, waterlogged soils, compacted soils and microenvironments in soil aggregates), where low O_2 levels can be permanent or temporary, the latter being more common (Hourdez, 2012). Mofette fields (Figs 3 and 7) are areas with CO_2 gas vents occurring in tectonically or volcanically active sites, and environments characterised by both high CO_2 and low O_2. The study of these systems has greatly contributed to our understanding of the impact of long-term high CO_2 concentrations and associated decreased O_2 levels on biological communities. For example, mofette research has provided insight into a range of biological processes from acclimation to adaptation, and by revealing the main environmental drivers of community assembly in these systems (e.g. Frerichs et al., 2013; Hohberg et al., 2015; Maček, 2013; Maček et al., 2011; Šibanc et al., 2014—see Section 3.4). Mofettes have allowed a diversity of questions across the life sciences to be addressed (Fig. 2).

Terrestrial mofettes are sites characterised by diffuse degassing at ambient temperature of CO_2 of deep mantle origin (>99%), and traces of other gases, including methane (CH_4), nitrogen (N_2), hydrogen sulphide (H_2S) or noble gases. Geogenic CO_2 displaces O_2 from soil pore spaces, therefore soil hypoxia, or even anoxia is a common soil characteristic at mofettes. Ascending CO_2 dominates the airspace of the soil and leads to aboveground CO_2 enrichment of the atmosphere and atmospheric accumulation of extremely high, even sometimes deadly, CO_2 concentrations can occur under certain topographic (Kies et al., 2015) or weather conditions (e.g. lack of air mixing). As it is heavier than air, mofette CO_2 may accumulate within depressions in the landscape, which if large enough may even form a gas lake with concentrations ranging from 5% to nearly 100%. Depending on

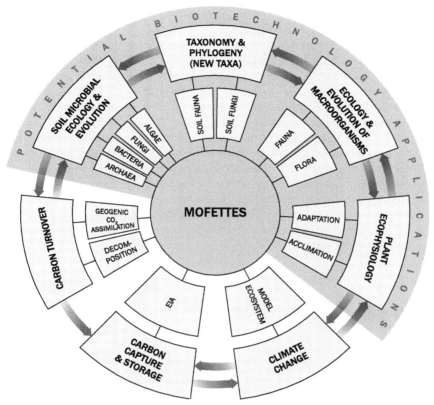

Fig. 2 A schematic framework of the existent research in mofette fields across the life sciences, and the potential directions for future research in currently unexplored disciplines (e.g. biotechnology). *EIA*, Environmental Impact Assessments.

depression morphology, weather conditions and time of the day, the gas lake may be transient or stable, stratified or homogeneous, and with or without a distinct border to the adjacent atmosphere (see Kies et al., 2015). Such gas lakes can range from several dm^2 to >1000 m^2 (Bossoleto, Italy, Fig. 8) (see also Barry et al., 2013). The environment of mofettes can therefore capture both extremely high CO_2 levels and extremely low O_2 concentrations relative to the wider environs (Box 1), dramatically influencing and limiting the local ecology of these sites (e.g. Beulig et al., 2016; Maček et al., 2005, 2011; Šibanc et al., 2014).

Natural degassing of CO_2 is known to occur at many volcanic–hydrothermal systems, but also from rift areas worldwide. Mofettes cooccur

BOX 1 Mofettes in Antiquity

In antiquity, mofettes were culturally important sites around the world, often seen as the entrance to the underworld (Pfanz et al., 2014). Mofette locations had probably been found by shepherds or herdsmen who witnessed strange behaviour of their cattle (Ustinova, 2009). Dead animals can be found frequently at these locations and vegetation growth is suppressed, both being indicative of mofette microlocation (Figs 3 and 9). In Greek mythology, Pluton (a.k.a. Hades) was the king of the shadows of the underworld, and caves and fractures mostly with foul-smelling exhalations (H_2S) were often thought to be entrances to Hades (Pfanz et al., 2014; Zwingmann, 2012). If priests came to know about such places, they declared those sites as sacred and often temples and sanctuaries were built in the surrounding area (Pfanz et al., 2014; Zwingmann, 2012)—and this can help us gauge the longevity, and hence the potential evolutionary timescale of adaptation, of contemporary mofettes found in these areas.

Fig. 3 Stavešinci grassland mofette area in northeast Slovenia. On the *left*, inhibition in vegetation growth can be seen centrally around the mofette area exposed to high levels of geogenic CO_2 (*white line*). On the *right*, the *hole* in the ground (mofette) is a result of high groundwater levels and soil erosion due to continuous bubbling of geological gas at the same spot.

with seismic structures and in pre- and postvolcanic areas (Fig. 1). In Europe, they can be found in Slovenia (Radenci area), Italy (Tuscany, regions with volcanoes), Germany (Eifel, Rhön, Teutoburger Forest, NW Franconia), in the Czech Republic (Cheb basin), Iceland, Romania (Hargitha Mountains), Hungary and France (Massif Central). Worldwide they are found for instance within the caldera of the Yellowstone volcano or in the Inyo crater range, in the Cascades range (United States), in geothermal fields of New Zealand (Bloomberg et al., 2012), Japan, Kamchatka and Indonesia (Djeng Plateau).

Mofettes have been used to study a diversity of organisms, representing all major taxonomic groups, ranging from macroorganisms—plants, animals (e.g. Maček et al., 2005; Pfanz et al., 2004, 2007) to mesofauna—Collembola, Nematoda (Frerichs et al., 2013), and microorganisms—fungi (Maček et al., 2011), microalgae (Beulig et al., 2016; Collins and Bell, 2006) and archaea and bacteria (e.g. Beulig et al., 2016; Krüger et al., 2011; Šibanc et al., 2014). These groups have distinct responses to a range of abiotic factors found in mofettes, and can be categorised into two groups: facultative and obligatory extremophiles. The former can tolerate the extreme conditions, but have optima closer to that of benign conditions and become abundant in extreme environments due to competitive release. The latter have optimal growth in the extreme environment, but may not be able to grow in benign environments (Bell and Callaghan, 2012). The gas regime of soil at mofettes, and soil hypoxia in particular, can have a strong impact on the communities of obligatory aerobic eukaryotic organisms like plants, soil fauna (e.g. Hohberg et al., 2015) and arbuscular mycorrhizal (AM) fungi (Maček et al., 2011). Soil O_2 concentration is the strongest abiotic predictor of the composition of soil archaeal and bacterial communities in the Stavešinci mofette area in Slovenia (Šibanc et al., 2014) with secondary effects of other soil factors like CO_2 concentrations, soil pH and nutrient availability. However, further investigation is urgently needed to understand complex biological processes and ecological interactions in mofettes (Fig. 4), including the study of:

(1) Mechanisms regulating the relative abundance of different taxa within communities that could be distinct in an environment with strong long-term, and permanent (press) disturbance compared with a benign (control) environment (see Section 3.4.6).
(2) Mofette-specific food webs and biological networks (see Section 3.6), and carbon turnover.
(3) Other nontrophic interactions among organisms (e.g. facilitation), mediated through changes in the abiotic environment—for example, plant-axial transfer of O_2 via aerenchyma into the root system and rhizosphere could drive aerobic metabolism in specific groups of obligatory aerobic eukaryotes (e.g. mycorrhizal fungi, fauna).
(4) Bioprospecting and biotechnological potential of the endemic biota that has evolved to the specific conditions of mofette areas.

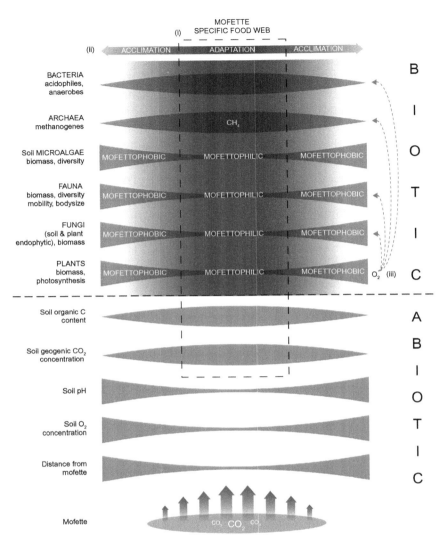

Fig. 4 Schematic of the effect of CO_2 venting at mofettes on abiotic and biotic factors. The gradients in different factors are indicated by the thickness of the *lens-shaped* forms. A *biconvex lens* shape is indicative of a higher value of the specific parameter closer to the mofette centre (e.g. CO_2 concentration), a *biconcave shape* indicates the opposite (e.g. O_2 concentration). Some complex biological processes and ecological interactions are specifically indicated. (i) Components of the mofette-specific food web, with the expansion into part of the abiotic environment. Geogenic CO_2 is being largely integrated into the food web by light dependant and dark CO_2 fixation, while decomposition is slowed down, possible by the permanent exclusion of meso- to macroscopic eucaryotes (Beulig et al., 2016). (ii) A gradient between potential for adaptation and acclimation in a range of phylogenetically distant groups of organisms is indicated.

3.1 Specific Abiotic Factors
3.1.1 Gaseous Regime
The gaseous regimes at different mofettes have been characterised by measuring soil CO_2 efflux rates (e.g. Kämpf et al., 2013; Thomalla, 2015; Vodnik et al., 2009), soil CO_2 concentrations (Pfanz et al., 2004; Saßmannshausen, 2010; Thomalla, 2015; Vodnik et al., 2006), dissolved CO_2 concentration in aquatic systems (Hall-Spencer et al., 2008) and atmospheric CO_2 concentrations (Kies et al., 2015; van Gardingen et al., 1995). At large geographic scales, patterns of degassing from mofettes depend on the movement of geological faults, which influence gaseous resistance. Spatial patterns in degassing at the local scale depend on soil permeability, which can be tightly related to soil water content, and to weather conditions (e.g. air pressure). In areas of high seismic activity, these local patterns were found to be related to the seismic activity (Thomalla, 2015; Weinlich et al., 2013). At mofettes, CO_2 can be the dominant gas in the upper 0.5 m of soil. The Slovenian, German and Czech mofettes are known for emitting very pure CO_2, consisting of 99.9 vol.% CO_2 (Beulig et al., 2015; Pfanz et al., 2004; Vodnik et al., 2006), which is enriched in ^{13}C (Beulig et al., 2015; Vodnik et al., 2006), a minor fraction of N_2 (<0.15 vol.%), and traces of Ar, O_2, CH_4 and He (ppm range) (Beulig et al., 2015). Italian mofettes are often contaminated with reduced sulphur gas (H_2S) or in swamp areas with methane (CH_4). Noble gases, predominantly the isotopes are always found within the mofette emissions. Geophysicists are using He isotopes to determine the origin of these geogenic gases (Bräuer et al., 2013).

Aboveground CO_2 concentrations are lower and more variable. At 1–2 m the atmosphere is most frequently only slightly enriched, but can reach several thousand ppm (Kies et al., 2015; van Gardingen et al., 1995). Higher atmospheric CO_2 concentrations are reported for mofettes with unique topography, for example, in the Italian Bossoleto mofette

---

The highest potential for adaptation is in the locally extreme mofette environment, with constant long-term abiotic selective pressure central to the mofette in contrast to a more variable transition zone. (iii) *Arrows* between plants and other groups of soil organisms indicate a beneficial nontrophic interaction (expanded concept of cross-trophic and cross-system facilitation) by reducing physical stress (hypoxia), and by creating new, oxygenated habitats via aerenchyma formation and axial transfer of O_2 from aboveground into the root system and rhizosphere. In the context of an extreme environment, this means that some plant species may modify conditions sufficiently to make life more hospitable for organisms that otherwise would not survive.

Fig. 5 Geological CO_2 emissions as seen in wet mofettes. Bosoletto mofette in Tuscany, Italy (*left*), Stavešinci mofette in Slovenia (*right*).

(Kies et al., 2015), which prevent mixing of the air mass. Though atmospheric CO_2 concentrations can be variable at mofette sites (daily fluctuations), the soil enrichment by geogenic CO_2 can be fairly stable over longer time periods (measurements conducted over decades) (Saßmannshausen, 2010; Thomalla, 2015; Vodnik et al., 2006).

Geogenic CO_2 influences several soil characteristics. It displaces O_2 from pores, which leads to hypoxic or even anoxic conditions. Redox potential of mofette soil can be severely decreased, in particular at mofettes where O_2 availability is further reduced by a high water table (Mehlhorn et al., 2014). Thus mofette soils can be classified as natural reductosols (Blume and Felix-Henningsen, 2009) (Fig. 5).

3.1.2 Soil pH

Another characteristic of the mofette soil is low pH, which results from acidic dissolution of CO_2. At temperatures above 25°C, the solubility of CO_2 in water is low (<1.5 ppt) and will have a smaller effect on the pH of the solution. pH will affect the carbonic acid equilibrium: the relative concentrations of CO_2, carbonic acid (H_2CO_3), bicarbonate (HCO_3^-) and carbonate (CO_3^{2-}) in solution. In acidic water (pH < 6), $CO_{2(aq)}$ dominates (>50%) while bicarbonate dominates in basic water (pH > 8) (Mook, 2000). For example, in the Slovenian Stavešinci mofette area, soil pH is between 4.6 (high CO_2 exposed areas) and 6.0 (control areas not affected by CO_2 vents) (Fig. 6). Therefore, the majority of the geogenic CO_2 is released as undissolved gas and the dissolved fraction is dominated by aqueous CO_2. However, some of the CO_2 does dissolve leading to a decrease in pH with increasing CO_2 concentrations in soils (Fig. 6, see also Saßmannshausen, 2010; Thomalla, 2015), and indeed in

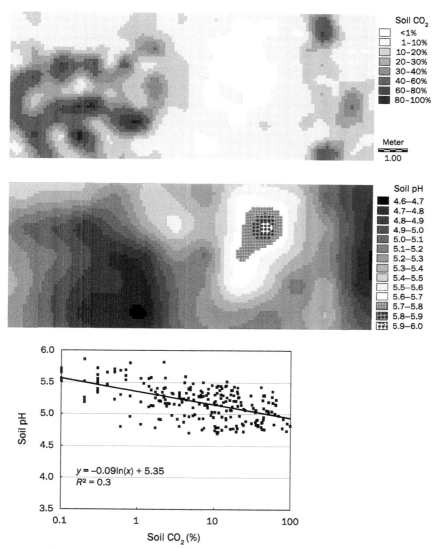

Fig. 6 Soil CO_2 concentration (depth of 20 cm) and pH of the soil (H_2O) within the Stavešinci mofette area (Slovenia). Correlation between both parameters is shown on the chart. For the site description and methodology, see Vodnik et al. (2006).

the water column at marine mofettes (Hall-Spencer et al., 2008). Moreover, stable spatial patterns of pH in different parts of the season support the conclusion that soil enrichment is relatively constant (Thomalla, 2015; Vodnik et al., 2006).

3.1.3 Mineral Nutrients

Many mechanisms control the availability of mineral nutrients in soils at mofettes. The exclusion of meso- and macroscopic eukaryotes (Beulig et al., 2016) and decreased microbial activity (e.g. Maček et al., 2009; Ross et al., 2001) is expected to reduce decomposition and mineralisation of organic matter. In addition, interactions of soil organic matter (SOM) with the mineral fraction of the soil are reported to change, which influences stability of different SOM fractions (Rennert and Pfanz, 2015; Rennert et al., 2011). Furthermore, chemical speciation of mineral nutrients is affected by hypoxic (or anoxic) conditions, low redox potential and the pH of the soil. In mofettes, these conditions promote reduction of the oxidised forms of essential nutrients, lowering their availability (e.g. N) or exposing their toxicity (e.g. Mn_2^+, Fe_2^+ and H_2S) to living organisms (Elzenga and van Veen, 2010). Thus, the response of organisms in mofette sites can be influenced directly by the high CO_2 concentrations and also indirectly by being affected by associated changes in nutrient bioavailability.

3.2 Mofettes as a Natural Long-Term Analogue to Free Air Carbon-Dioxide Enrichment (FACE) Experiments?

With growing concerns about climate change driven by rising atmospheric CO_2 concentrations, Italian CO_2 springs where some of the first mofettes to be used in environmental and biological studies (e.g. Miglietta et al., 1993; Raschi et al., 1997). Experimental systems aimed at studying the effects of rising atmospheric CO_2, which include the global network of Free Air Carbon-dioxide Enrichment (FACE) experiments (e.g. Ainsworth and Long, 2005; Cotton et al., 2015; Lewin et al., 1994), are relatively recent and cannot (yet) be used to investigate the long-term impacts of rising CO_2 across multiple plant and animal generations (but see Andresen et al., 2016). Mofettes, however, offer an attractive natural analogue where acclimation and adaptation of organisms to elevated CO_2 can be studied (Fig. 4). Climate change-related mofette studies have mainly investigated photosynthesis and transpiration responses to elevated atmospheric CO_2 concentration (see Section 3.3.1), aiming to improve predictions of physiological and growth performance of plants in the future. The main benefits of this approach include: long-term CO_2 enrichment, CO_2 gradients over space, natural conditions, free CO_2 (low-cost experiments), the possibility to introduce selected vegetation or to use existing vegetation as a source of plant material for experiments under controlled conditions (Paoletti et al., 2005).

Though environmental conditions at mofettes are useful proxies to examine ecological responses to global climate change, studies have often failed to show clear patterns in these sites. Atmospheric conditions in mofettes can be highly variable in space and time, with short-term CO_2 fluctuations and vertical CO_2 gradients, and in addition, hypoxia being common in mofette soils and effecting plant root function. Conditions other than CO_2, including soil mineral composition, can also deviate substantially between areas affected by the mofettes and control areas (Paoletti et al., 2005—see also Section 3.3). Thus, to harness the ability of mofette sites to simulate the future conditions predicted in a globally changing climate, sites must be carefully selected and paired with adequate reference/control sites. In addition, the response of the whole ecosystem, including belowground conditions and processes, must be accounted for.

3.3 Plant Ecophysiology in Mofette Areas

Elevated atmospheric CO_2 mainly affects photosynthesis, and indirectly evapotranspiration by controlling stomatal opening. Photosynthesis is the main mechanism for the production of organic carbon in most ecosystems and is crucial for productivity, carbon balance and sequestration potential. For this reason, the effects of elevated CO_2 on photosynthesis have been intensively studied, mainly in controlled experiments (growth chambers, open top chambers, FACE systems), but also at sites with natural CO_2 enrichment (e.g. Raschi et al., 1997, see Section 3.2). However, at mofette sites, the CO_2 levels in soil can be high enough to cause hypoxia in the rhizosphere, thus directly affecting respiration for subsurface organisms and for organs such as roots. In rare cases, plants can also change their growth form. For example, plants that are usually lying along the ground grow semierect to avoid direct contact with high CO_2 concentrations.

3.3.1 Photosynthesis and Transpiration

Elevated CO_2 affects carbon assimilation across different organisational levels and temporal scales. Under short-term conditions, plants respond to elevated CO_2 by decreasing stomatal conductivity and enhancing net photosynthesis. Plants exposed to step changes of ambient CO_2 concentration (from 400 to 700 µmol CO_2 mol^{-1} or vice versa) can achieve a new, steady-state level of stomatal conductivity within minutes and the response time is shorter in C_4 than in C_3 plants (Hladnik et al., 2009; Vodnik et al., 2013). On the other hand, populations of mofette plants, which experience fluctuating CO_2 concentrations, are not characterised by faster stomatal reaction relative to those grown under normal gaseous environments. Once leaf cells are exposed to high CO_2, their metabolism can be affected by changes in pH (see Section 3.3.2), but measurements of atmospheric CO_2 concentrations and chlorophyll fluorescence (via a PAM fluorometer) indicate that it does not acutely reduce photochemical efficiency of photosynthetic apparatus.

Physiological and morphological changes that occur over days to weeks reflect acclimation. Meta-analyses of FACE experiments have revealed that growth under elevated CO_2 decreased specific leaf area, stomatal conductivity and it increased photosynthesis. However, the long-term increase of photosynthetic rate is lower than that of plants grown at ambient CO_2 when instantaneously exposed to elevated CO_2 (Ainsworth and Long, 2005). Acclimation of net photosynthesis to high CO_2 is caused by a decrease in maximum carboxylation velocity and decreased investment in RuBisCO (Rogers and Humphries, 2000) and is more pronounced when plants are nitrogen limited. Inadequate nitrogen supply could restrict the development of new sinks and therefore exacerbate the source–sink imbalance in plants grown under elevated CO_2 (Stitt and Krapp, 1999). Another explanation for accentuated acclimation under low nitrogen conditions is that the decrease in RuBisCO may reflect a general decrease in leaf protein caused by reallocation of nitrogen to younger leaves or earlier leaf senescence in nitrogen-limited plants (Stitt and Krapp, 1999).

To a large extent, the process of acclimation, including reduced CO_2 stimulation of photosynthesis due to insufficient mineral nutrition, can be observed in plants growing near mofettes. Most mofette studies report on acclimation via the downregulation of photosynthesis (see review by Pfanz et al., 2004; Vodnik et al., 2002a,b), yet contrasting studies show no difference in maximum net photosynthesis between CO_2 enriched and control sites (Barták et al., 1999; Stylinski et al., 2000). Variation in the response of plants to CO_2 at mofette sites could be caused by differences in other environmental factors influencing photosynthetic performance, such as irradiance, which often has the largest effect on leaf morphological and photosynthetic traits and should therefore be carefully considered in 'natural experiments' (Osada et al., 2010). Soil nitrogen availability can also vary between mofette and reference sites. For example, leaf nitrogen content in *Erica arborea* growing at a mofette was interactively affected by atmospheric CO_2 and soil nitrogen concentration (Bettarini et al., 1995). Plant nitrogen status strongly affects photosynthetic response. Pfanz et al. (2007) reported on photosynthetic downregulation, exhibited as reduced photosynthetic capacity, decreased carboxylation efficiency and decreased CO_2 compensation point (derived from a photosynthetic CO_2 response curves), in plants growing in CO_2 enriched but nitrogen poor soil. Similarly Osada et al. (2010) showed that maximum carboxylation velocity of *Polygonum sachalinense* was affected by the interaction of CO_2 and soil N, suggesting that downregulation of photosynthesis at elevated CO_2 was more evident at lower soil N availability. Reduction of net photosynthesis in plants growing at the highly CO_2 enriched locations exceeds the rate expected for acclimation and would therefore better to be termed inhibition. In this case, reduction of net photosynthesis is more likely resulting from primary effects of soil CO_2 concentrations acting on chemical processes in the soil and on roots.

There is no general evidence that long-term exposure to natural CO_2 enrichment results in photosynthetic adaptations of plants to mofette conditions. Fordham et al. (1997) suggest that long-term adaptation of *Agrostis canina* ssp. *monteluccii* plants may be associated with a potential for increased growth, but this does not appear to be linked with differences in the intrinsic capacity for photosynthesis. In an in situ reciprocal transplant experiment with plants originating from mofette sites (Onoda et al., 2009) adaptation to high CO_2 was observed on the level of higher water use efficiency, an associated reduction in stomatal conductance and a lower starch concentration. However, there was no genotypic improvement in photosynthetic nitrogen use efficiency.

3.3.2 Mineral Nutrition, Root Respiration and Aerenchyma Formation

Roots are the primary plant organ affected by increasing geogenic CO_2 concentrations, via soil hypoxia, changes in mineral nutrients and/or pH (Maček et al., 2005). The high root density near the surface is hypothetically caused by a withdrawal of roots from deeper (hypoxic) horizons and new root growth close to the soil surface. Species with shallow roots, and those with a higher capacity for new (secondary) root formation, are thus favoured at the expense of deep rooted species. In the case of severe hypoxia, root respiration will also be affected. Root respiration significantly decreases only when plant roots are exposed to concentrations of CO_2 sufficiently high to limit O_2 availability (over 50% CO_2 causing a decrease in O_2 saturation in liquid of below 10%, Maček et al., 2005). Active nutrient uptake is an energetically costly process that requires metabolic activity. Soil hypoxia can inhibit ion uptake by roots because of limited aerobic metabolism and a higher dependency on less-efficient anaerobic metabolism. In nonadapted plants, due to the lack of O_2, fermentation in root cells would occur, leading to the production of lactic acid and ethanol. Nutrient uptake can also be limited by other aspects of high CO_2 environments, such as reduced stomatal conductivity, transpiration and sap flow rate, which hinders ion transport to the shoot (McGrath and Lobell, 2013). The content of essential mineral nutrients is reduced in plants exposed to geogenic CO_2 enrichment—as observed for N, S, P, K and Zn in *Phleum pratense* leaves at the Stavešinci mofette (Pfanz et al., 2007).

High CO_2 can also directly affect root metabolism by lowering intracellular pH (Bown, 1985). In contrast to aboveground portions of plants, which are covered by a cuticule with a low permeability for gases and in which gas is mainly regulated by the openings of stomata, young roots have no cuticule, and thus no clear barrier to gas exchange. Gaseous and dissolved CO_2 move freely through rhizodermal and apoplastic spaces (free diffusion space outside the plasma membrane, mostly consisting of cell walls and intercellular air spaces) into the root tissue and cells. Therefore, root cells are directly impacted by the high levels of soil CO_2 in mofette areas resulting in potential cytoplasm acidification. If the concentration of CO_2 is high enough that the decrease in pH cannot be compensated by cellular mechanisms that stabilise intracellular pH, metabolism is impaired (Pfanz et al., 1987). Similar mechanisms of direct CO_2 effects can be expected also for soil meso- and microfauna, and soil microbes (see Section 3.4).

In plants, the formation of aerenchyma can allow relatively high levels of O_2 in the rhizosphere and sustain aerobic respiration of the roots even in the context of very high soil CO_2 concentrations. Aerenchyma can substantially reduce internal impedance for the transport of O_2, N_2 and various metabolically generated gases such as CO_2 and ethylene between roots and shoots. Aerenchyma are typical adaptive traits of wetland plant species, but can arise as response to flooding or mineral deficiency (Marschner, 2012), and there has been extensive development of aerenchyma reported in maize grown in the Slovenian Stavešinci mofettes (Videmšek et al., 2006). In addition, plant-mediated changes in the rhizosphere (e.g. O_2 axial transport and leakage from roots) can affect other organisms living in hypoxic systems (e.g. soil and endophytic microbes and fauna). Facilitation is often used to describe beneficial (nontrophic) interactions between physiologically independent plants and is mediated through changes in the abiotic environment (Brooker and Callaway, 2009). The concept is still a matter of some debate (Brooker and Callaway, 2009), yet mofettes offer a useful model ecosystem to study plant-mediated O_2 transport into rhizosphere among plants as well as other groups of organisms (Fig. 4).

To conclude, high CO_2 concentrations may not act directly as a selective agent and cause an evolutionary response in plants, yet mofette plants need to adapt to hypoxia with a suite of traits that enable their growth, development and existence in the challenging gaseous environment. These traits include structural and morphological adaptations which: (i) support aerobic respiration in roots, (ii) avoid anoxia and (iii) mitigate problems with mineral nutrition.

3.4 Community Ecology

Many plants and animals are sensitive to soil hypoxia and are effectively excluded from mofette habitats. Larger and more mobile animals have the option to avoid the hypoxic conditions of mofettes, which are otherwise lethal for many animals, including those as big as birds and mammals (Fig. 9). Among the hypoxia-tolerant plants, some taxa can exploit an 'escape strategy' based on a suite of inducible morphological and anatomical traits that allow growth in such environments (Armstrong and Armstrong, 1988, 1990; Perata et al., 2011; Vartapetian and Jackson, 1997). Plants in mofette areas form three categories (Pfanz, 2008): (1) mofettophobic plants that strictly avoid geogenic CO_2 at concentrations above 2–3%; (2) mofettophilic plants that grow above high CO_2 concentrations (Figs 7 and 8) and (3) mofettovague plants that grow in both degassing and reference/control areas (see also Saßmannshausen, 2010; Thomalla, 2015). Similar categorisation could be applied to other groups of organisms (animals and microbes) and has recently been used for some collembolans (Hohberg et al., 2015; Russell et al., 2011) (Fig. 4).

Most multicellular (macro)organisms are aerobes. However, microorganisms have far greater breadth in O_2 requirements and tolerance, with species ranging from obligate aerobes to obligate anaerobes, where O_2 is toxic. Microaerophiles require O_2 for energy production, but are harmed by atmospheric concentrations of O_2 and aerotolerant anaerobes do not use O_2 but tolerate its presence. The majority of initial mofette research has been done on plant physiological responses to geogenic CO_2 (see Section 3.3), and only in the recent years has focus shifted to researching the communities of flora, fauna and microorganisms—which arguably provide greater insights into ecosystem-level responses to both elevated CO_2 and localised hypoxia and anoxia.

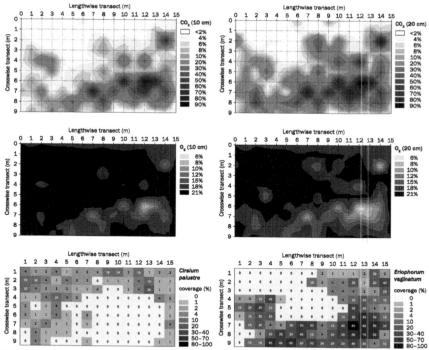

Fig. 7 Hill mofette in Plesna Valley (Czech Republic). Soil CO_2 (*above*) and O_2 (*middle*) concentrations in 10 and 20 cm soil depth (measured with portable gas analyser GA2000+, Ansyco, Germany) and plant coverage of the area with mofettophobic *Cirsium palustre* (*bottom left*) and mofettophilic *Eriophorum vaginatum* (*bottom right*).

Fig. 8 Mofettophilic plants in hypoxic habitats. Cotton grass (*Eriophorum vaginatum*) growing around a mofette at Plesna Valley, Czech Republic (*left*). Reeds (*Phragmites australis*) growing in a nocturnal CO_2 gas lake in the Bossoleto mofette in Tuscany, Italy (*right*).

3.4.1 Flora

Specific assemblages of plant species are found in mofette areas. Cotton grass (*Eriophorum vaginatum*) and heather (*Calluna vulgaris*) are examples of typical mofettophilic species, whereas marsh thistle (*Cirsium palustre*) is mofettophobic (Figs 7 and 8). These distributions of plants in response to CO_2/soil hypoxia, suggest species composition can be used as a bioindicator at mofette sites. Despite some contingent and site-specific responses among certain taxa, in general, mofettophilic plants also tend to be helophytic and adapted to stagnant waters (Pfanz, 2008). Across mofette sites in Europe, Northern America and New Zealand, the abundance of bog and swamp helophytic species of the genera *Juncus* and *Carex* is tightly correlated with CO_2 concentration. The capacity of these plants to transport O_2 from aboveground organs to roots (Armstrong and Armstrong, 1988, 1990; Vartapetian and Jackson, 1997) allows them to grow in hypoxic soils (Pfanz, 2008). Helophytes can even survive within a diurnal 80% CO_2 gas lake in which the atmosphere becomes saturated with CO_2 at night. In the Bossoleto mofette in Italy, reeds (*Phragmites australis*) grow in a transient CO_2 lake (Fig. 8); during the day, CO_2 concentrations are closer to normal atmospheric concentrations, but during the night, the vegetation is submerged in nearly pure CO_2. Plants within this area have yellowed leaves and reach only 25–35% of the height of those in reference sites. During the day, these plants use Venturi ventilation and facilitated diffusion to supply atmospheric O_2 to rhizomes and roots (Armstrong et al., 1992). Thus, mofette environments could be considered model systems for transient hypoxia (e.g. waterlogging), future climate conditions and the interaction between the two, which may provide insights into how ecosystems respond to episodic flooding in a high CO_2 world.

3.4.2 Fauna

Gas lakes act as CO_2 traps that can serve as sampling sites for the local mofette faunal community. Some animals may be attracted to high-CO_2 sites either by the local conditions, the warmer temperature within the gas lake, or by the abundance of carrion or other food sources. Within a Czech meadow mofette, a CO_2 trap of ~600 cm^2 contained more than 500 individual dead animals representing 140 different species from 17 orders, 64 families and 105 different genera. As decomposition rates are reduced most specimens are well preserved (Beulig et al., 2016) and it may even be possible to reconstruct realist food chains and trophic modules (see Section 3.6). The high concentrations of CO_2 in gas lakes can affect animals both via displacement of atmospheric O_2 and via dissolved CO_2 leading to reduced internal pH. In the first case, the physiological effects are elevated blood CO_2 concentrations (hypercapnia) or O_2 deficiencies like hypoxia or even anoxia (Lehwess-Litzmann, 1943). In the latter case, cellular compartments become acidified (Pfanz and Heber, 1989) and enzymatic activity is reduced or blocked (Pfanz, 1994). Consequently, extreme CO_2 concentrations are not tolerated by aerobes; in mammals a hypoxic response starts at an O_2 level of about 6% (Simon and Keith, 2008), from CO_2 concentrations of about 8–10%, they asphyxiate, and death ensues after prolonged exposure to 15–20% (Fig. 9).

Some animals can apparently perceive and actively avoid extremely high CO_2 environments, including some birds and subterranean mammals, such as cliff swallows and European moles (*Talpa europaea*), whereas others are able to persist. Collembolan and nematodes seem to adapt to enhanced CO_2 and/or decreased O_2 concentrations. Yet, although 18 different collembolan species ($\approx$31,000 individuals) can be found in a 1 m^2 in the upper 5 cm soil horizon in 2% CO_2, this falls to just five mofettophilic species in 20% CO_2 (Hohberg et al., 2015; Russell et al., 2011). Some species are apparently endemic to mofettes, including the collembola *Folsomia mofettophila* in some Czech meadows (Schulz and Potapov, 2010). At the Italian Bossoleto mofette site, living collembola, nematodes and arachnids are found inside the gas lake at 80% CO_2, where they may be able to access photosynthetically derived O_2 diffusing from plant roots to form oxic microhabitats (Fig. 4). Similar to mofette plant species, animals adapted to living submerged in stagnant water are best adapted to the local

Fig. 9 CO_2 traps serving as sampling sites for the local mofette faunal community. Bossoleto mofette, Italy (*left*) and Stavešinci mofette, Slovenia (*right*).

hypoxic conditions, especially via adaptation of the O_2 carrier pigments (haemoglobins and haemocyanins) with very high intrinsic O_2 affinities (Hourdez and Lallier, 2007; Hourdez, 2012).

3.4.3 Soil Microalgae

The importance of primary production by microalgae in soils is not well established, but microalgae are the dominant eukaryotes in mofettes (Beulig et al., 2016). Microbial autotrophs including microalgae should benefit from the high CO_2 concentrations of mofettes, with increased photosynthetic rates and accelerated growth. As observed for plants, microalgae may be expected to acclimate and adapt to the high CO_2 concentrations. With their large population size and rapid generation time, they are ideal models for studying adaptation and evolution in photosynthetic organisms, yet those in mofettes do not appear to be specifically adapted to high CO_2, though some strains have poor growth at ambient CO_2 (Collins and Bell, 2006). This apparent lack of adaptation might reflect swamping of gene flow from the surrounding environments, although adaptation to high CO_2 concentrations in microalgae appears to be unlikely due to factors including spatial and temporal heterogeneity, a potential lack of evolutionary cost for the genetic capacity to grow at low CO_2 when growing at high CO_2 (Low-Décarie et al., 2013). More generally, adaptation to conditions that increase growth rate, and thus do not represent a selective pressure (Low-Décarie et al., 2014), may be unlikely even beyond the context of mofettes and CO_2, such as with the addition of nutrients.

3.4.4 Fungi

Most fungi are considered to be aerobes, although their habitats can often be hypoxic or even anoxic (e.g. due to soil infiltration with water or metabolic activity during infection of other organisms). As with multicellular eukaryotes, fungi have a hypoxic response below 6% O_2 (Simon and Keith, 2008). Some saprophytic fungi are facultative anaerobes specialised in growth in hypoxic environments, including several moulds, typically found in soil and decaying organic matter like *Aspergillus fumigatus* and *Fusarium oxysporum* (Grahl et al., 2012). The diversity of fungi at mofette sites remains mostly unexplored (Table 1), although a new species of hypoxia-tolerant yeast has recently been described from Slovenia (Šibanc et al., unpublished). As many biotechnological applications, such as industrial fermentation, require the capacity to grow in high-CO_2 low-O_2 environments, mofettes are likely

Table 1 A Summary of Current Studies on Different Aspects of Ecology and Diversity of Soil Archaea, Bacteria and Fungi in Mofette Soils

Microbial Group	Gene Region and/or Methodology Used	Mofette Location	References
AM fungi	18S rRNA pyrosequencing	Stavešinci, Slovenia, Bossoleto, Italy, Cheb basin, Czech Republic	Šibanc et al. (unpublished)
Soil/water yeasts	18S rRNA, ITS sequencing, isolation, growth rate response, metabolic tests	Stavešinci, Slovenia	Šibanc et al. (unpublished)
Soil microbiome, carbon turnover	Isotope geochemistry, soil activity analyses, metatranscriptomics, metagenomics	Cheb Basin, Czech Republic	Beulig et al. (2016)
Soil archaea and bacteria	Microcosm study with anoxic incubation, 16S rRNA pyrosequencing, qPCR, DNA-SIP analysis	Cheb Basin, Czech Republic	Beulig et al. (2015)
Soil archaea and bacteria	16S rRNA, T-RLFP, clone libraries, sequencing	Stavešinci, Slovenia	Šibanc et al. (2014)
Soil archaea and bacteria	16S rRNA, DGGE, activity measurements	Larcher See, Germany	Frerichs et al. (2013)
Soil microbes	16S-23S spacer region, ITS, ARISA, qPCR, PLFA, enzymatic analyses	Mammoth Mountain, USA	McFarland et al. (2013)
AM fungi	18S rRNA RFLP, clone libraries, sequencing	Stavešinci, Slovenia	Maček et al. (2011)
Soil microbes	qPCR and activity measurements, $nirK$ genes DGGE	Larcher See, Germany	Krüger et al. (2009, 2011)
Soil CO_2-fixing bacteria	$cBBl$ genes, RFLP	Stavešinci, Slovenia	Videmšek et al. (2009)
Soil microbes	Lipid biomarkers and ^{13}C analyses, qPCR, biomass and activity analyses	Latera Caldera, Italy	Beaubien et al. (2008) and Oppermann et al. (2010)

ideal locations for bioprospecting for industrially important fungi. To date, the biotechnological and medical potential of mofette sites and their adapted biota remains largely unknown and untapped (Fig. 2).

Mofette fungal ecology has largely focused on functionally important and ubiquitous soil fungi from the phylum Glomeromycota, which form arbuscular mycorrhizae with over 70% of all vascular plants (Brundrett, 2009). AM fungi increase plant nutrient uptake (Smith and Read, 2008) and the fungal partner relies completely on plant assimilated carbon. Roots of plant species in Slovenian mofettes are still colonised by AM fungi, despite the prolonged stress and carbon cost to the plant of the symbiosis (Maček et al., 2011, 2012). It is not yet clear how the AM fungi cope, but the extraradical mycelium is likely to be severely restricted in highly hypoxic soils and it is unknown whether the plants benefit from the symbiosis in this environment. Fungal hyphae are finer than roots by at least an order of magnitude and the costs to a plant of acquiring nutrients symbiotically will be lower than via new root growth, but only if sufficient nutrients are provided by the fungal symbiont. As with all mycelial fungi, AM fungi acquire resources in numerous spatially dispersed locations, and move them within the mycelium to support growth at favoured locations (Bago et al., 2002). However, the mofette fungal types are probably not being subsidised by mycelium in surrounding soil, which might explain how these aerobic fungi survive as they must be adapted to, or at least competitive in, hypoxic conditions, presumably either by tolerating low O_2 or acquiring sufficient O_2 from the roots (Maček et al., 2011). Acquiring O_2 from roots is a relatively new concept in AM fungal symbiosis, not only relevant for mofette fungi, but also for AM fungi colonising submerged plants in aquatic environments. Submerged environments are known to contain specialised species of AM fungi (e.g. Sudová et al., 2015). For example, the AM fungi *Rhizoglomus melanum* has been discovered in lake sediments and was isolated from the rhizosphere of two aquatic macrophytes (Sudová et al., 2015) that form small-submerged rosettes with a well-developed root system (Beck-Nielsen and Madsen, 2001; Farmer, 1985). These provide rapid O_2 diffusion from the shoots to the roots and subsequent high O_2 losses into the sediment (Smolders et al., 2002). The AM fungi–plant interaction may thus extend beyond the exchange of nutrients to the provision of O_2, which allows the AM fungi to survive in hypoxic environments. This type of facilitation through local changes in the abiotic environment by plants or other organisms (Brooker and Callaway, 2009) may be common in mofette environments (Fig. 10).

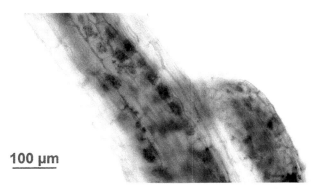

Fig. 10 C4 grass *Setaria pumila* roots that are highly colonised with arbuscular mycorrhizal (AM) fungi at extremely high-geogenic CO_2 concentrations at Stavešinci mofette.

In the Slovenian mofettes, the AM fungal community has high turnover (beta diversity) between high-geogenic CO_2 exposed and control locations and specific taxa are consistently dominant in hypoxic soils (Maček et al., 2011). Some AM fungi species are more strongly associated with soil environmental conditions than with their host plants' distribution (Dumbrell et al., 2010; Maček et al., 2011). Thus the strong environmental gradients of mofette sites, which include extreme and lethal conditions, make them ideal models for further study of community assembly rules, temporal dynamics (e.g. inter annual variability), facilitation and symbiosis in AM fungi, and provide an insight into different pathways of plant mineral assimilation and the role of the fungal partner in this process in hypoxic environments.

3.4.5 Archaea and Bacteria

The diversity of soil archaea and bacteria is vast, yet still little is known about the importance of abiotic factors in regulating these communities (but see Nedwell et al., 2016). Bacterial and archaeal communities from mofette areas have been studied over the last decade using the full range of molecular microbial ecology approaches, ranging from T-RFLP profiling, clone libraries and Sanger sequencing to current Next-Generation Sequencing (NGS) and multiomics analyses (see Table 1). Microbial diversity generally does not differ across sites with different geological CO_2 exposure: diverse archaeal and bacterial communities exist in mofette soils and in the most extreme conditions (Šibanc et al., 2014). Among all the factors, soil O_2 concentration was consistently the strongest predictor of the compositional changes across both the archaeal and bacterial communities in Slovenian mofettes (Šibanc et al., 2014). Shifts towards anaerobic, and in some cases, acidophilic groups of microbes in high CO_2 exposed soils (Šibanc et al., 2014) have also been observed in other European mofettes (e.g. Frerichs et al., 2013; Krüger et al., 2009, 2011; Oppermann et al., 2010). Importantly, a significant increase in the relative abundance of strictly anaerobic methanogenic archaea from the group Methanomicrobia in high CO_2 exposed subsurface soil is reported (Šibanc et al., 2014). Energy metabolism within the Methanomicrobia is restricted to the formation of methane (CH_4) from CO_2 and H_2 (Thauer et al., 2008). This has implications associated with increased biogenic CH_4 production and thus important consequences for ecosystem functioning, as well as wider significance in climate change research (see Section 3.5).

3.4.6 Advances in Community Ecology Research: Lessons from Mofette Environments

Most of the existing mofette studies examining the community composition of different organisms represent single snapshots or, at best, a few time points. Yet, recent independent studies on soil microbial communities, performed at different times, and in at least three different mofette sites distributed across Europe draw similar conclusions. Microbial communities from all mofettes exhibit increased abundance of methanogenic and anaerobic taxa, and in some cases also acidophiles (Section 3.4.5). This suggests a general and relatively stable pattern in the development of soil microbial communities. Moreover, this also appears to extend to other groups of organisms (e.g. fungi and soil invertebrates with recently described novel yeast and collembolan species; see Sections 3.4.2 and 3.4.4). This high consistency and replication of responses makes it possible to predict the temporal assembly of mofette microbial communities with far greater precision and accuracy than is typically the case for more benign ecosystems.

This higher temporal predictability (stability) is also evident from the preliminary results on AM fungal communities, which suggest that under permanent (long-term) selective pressure, community composition is more constant relative to control sites. In the latter, stochastic processes and other environmental factors (e.g. vegetation) play a much bigger role in structuring communities over time. The major shifts in AM fungal community composition within and between consecutive years happen each spring, when the winter community supported by low photosynthetic carbon flux into roots is shifted to the summer community (with high photosynthetic carbon flux into roots) and the pattern of new community assembly during each year is largely stochastic (Dumbrell et al., 2011). Yet this pattern may be less prominent in extreme environments—such as mofettes.

Drawing on these multiple lines of growing evidence, we suggest that extreme, persistent and directed abiotic selective pressure results in a more stable system with highly specific microbial communities, dominated by locally adapted and tolerant taxa that are persistent and abundant in soils exposed to a long-term press disturbance. The case of AM fungal community composition shows the potential of the mofettes to serve as model ecosystems to study some of the major unresolved questions in community ecology. Importantly, the observed stability may be more universal than currently acknowledged in many other environments with long-term disturbances or specific selective pressures. Soil microbial community composition is still difficult to predict in nature: in 'normal' soils this is especially difficult because there are so many different environmental factors simultaneously shaping the community, yet none of the factors dominates in isolation. Since mofettes are a permanent but relatively small spatial scale perturbation to the normal environment, they are ideal for tracking adaptive traits of relatively small organisms (microbes, micro- and mesofauna and flora). Despite their small size, these organisms dominate soil diversity and represent most of the living 'dark matter' that we still know very little about, but that is critical for the functioning of the whole planet.

Recently developed molecular tools (e.g. NGS) provide sufficient resolution to look into microbial communities with experimental designs that can disentangle spatial and temporal patterns of their composition with enough replication to deliver firm and robust conclusions. Several important fundamental questions in microbial ecology can be addressed here: for example, 'How do soil microbes respond to long-term (press) disturbance or environmental change?', 'What stability and assembly rules govern observed community compositions?' and the still largely unanswered question of 'What are the main environmental predictors of microbial community composition?'. Importantly, questions about long-term-related changes in soil microbial communities are relevant to many other anthropogenic drivers in a more general sense, including climate change, nutrient input, land-use change, pollution and many more. However, in particular, questions about the stability of natural communities hit by press perturbations have received relatively little attention so far.

3.5 CCS: What Can We Learn from Mofettes?

In the last few years, the focus of research has largely shifted to support the technology of CCS (Frerichs et al., 2013; Lal, 2008; Noble et al., 2012), as mofette sites provide natural model ecosystems for EIA of gas leakage from underground carbon storage systems.

CCS is the process of capturing waste CO_2 from large point sources, and depositing it underground. It is proposed as one of the possible measures for limiting the global rise of atmospheric CO_2 and consequently mitigating global warming. Mofettes and other naturally occurring CO_2 accumulations can be used to study how CO_2 will behave during and after storage and to develop and calibrate methods for monitoring its leakage (Holloway et al., 2007). A key benefit of these natural analogues of CCS is that at these sites the processes related to CO_2 accumulation and degassing occur over long, even geological timescales. Parameters used in monitoring can include soil CO_2 concentrations or soil CO_2 efflux, pH, redox potential and other physical and chemical properties of the soil affected by elevated CO_2. Biological methods involve plant surveys, bacterial counts and DNA/RNA-based molecular methods, which have recently been applied to characterise microbial communities and functional traits that may change due to CO_2 leakage (Noble et al., 2012).

New studies on microbial community composition involving different mofette sites have revealed that a severe CO_2 leakage from CCS systems leading to localised soil hypoxia could cause a shift in the soil microbial community composition towards methanogenic taxa dominating the community (see Section 3.4.5). This results in higher production of biogenic methane (CH_4), a far more potent greenhouse gas in terms of warming potential than CO_2 (Šibanc et al., 2014), highlighting how CO_2 leakage from CCS systems could create positive feedback loops with climate warming.

3.6 Exploring Mofette Food Webs and Biological Networks

Mofettes are characterised by a permanent exclusion of higher trophic levels and their associated physical and ecological traits from the local food webs for two reasons: first, because higher-level multicellular organisms die in mofette environments, and second, because their dead bodies persist for longer than in control environments, which implies impaired decomposition rates. This could be a direct consequence of exclusion of larger consumers from the mofette food webs (e.g. Beulig et al., 2016). Among the potentially excluded aerobic eukaryotes are fungi that are known to have an important role in SOM decomposition and especially fresh litter degradation (Schneider et al., 2012). In mofettes, hypoxia can appear even in shallow soil (up to 10 cm; Fig. 7) and recent metatranscriptomic data suggest that, in deeper layers, litter decomposition is driven by enzymes with bacterial origins, especially among the acidophiles and anaerobes (Beulig et al., 2016, see Section 3.4.5).

Beulig et al. (2016) report that high quantities of mofette soil carbon originate from the assimilation of geogenic CO_2 (up to 67%) via plant primary production and subsurface microbial CO_2 fixation, thus this is an important carbon input at the base of the mofette food web (Fig. 4). High concentrations of geogenic CO_2 can promote dark CO_2 fixation by CO_2-utilising prokaryotes (Beulig et al., 2015, 2016; Šibanc et al., 2014), which also contribute to mofette-specific breakdown of SOM, generating an unusual biogeochemical profile in these environments (Beulig et al., 2015). The reduction of the soil community to the microbial component makes the mofette a valuable model environment for studying diversity effects on specific soil functions. However, high similarity in biogeochemical potentials of SOM degradation in mofette and reference sites, many imply a degree of redundancy in soil biomes for specific enzymatic traits (Beulig et al., 2016).

This all highlights a need for studies examining mofette systems within a broader food–web context. As a large portion of the specific ecology of these systems is microbial, mofettes provides an ideal opportunity to explore network-based approaches for incorporating NGS-based data into food–web ecology (e.g. Vacher et al., 2016). In addition, by taking a network or food–web-based approach classic concepts of community stability, something that is postulated to be regulated within mofette systems can be explored further.

4. CONCLUSIONS

Many ecological and evolutionary processes are slow and long-term studies are required for their investigation. Locally extreme environments are arguably among the most powerful natural experiments for the study of ecology and evolution (Maček et al., 2011), which would otherwise be experimentally unfeasible due to a huge demand for both temporal and financial commitments. Mofettes are one such extreme ecosystem with a long-term directed selective pressure due to a continuous exposition of the system to high CO_2 concentrations (Vodnik et al., 2006). Ancient civilisations believed mofettes were gateways to a hidden underworld (Box 1; Pfanz et al., 2014), and even in our modern age this still holds: these still relatively unknown extreme ecosystems may hold the key to many fundamental ecological and evolutionary questions that have yet to be answered (Maček et al., 2011). The highly dynamic and noisy character of many 'normal' soil systems makes it hard to draw general conclusions as to how they function, and the ecological networks in such systems are very complex. Natural ecosystems that are under long-term and stable impacts of a specific, well-characterised abiotic factor make such observations far more straightforward.

In general, such extreme environments could provide broad insights into a range of questions, from the mechanisms of adaptation of large long-lived organisms, evolution and biogeography of microorganisms to the classic concepts of succession and community stability. Mofettes still have to be carefully considered in terms of providing an ecological long-term model that goes beyond the frame of an investigation and description of yet another extreme environment, with all its specificities and contingencies. Therefore, for each such potential model system, researchers must first fully understand its abiotic parameters, their temporal dynamics and variability, before focus is shifted towards understanding their ecological complexity. For example, early mofette research in the 1990s proposed these systems as ideal natural models for climate change research (e.g. Miglietta et al., 1993; Raschi et al., 1997), but due to their complex soil geochemistry (e.g. Beulig et al., 2016; Vodnik et al., 2006), and high variability of the atmospheric conditions this is not always the case (Paoletti et al., 2005). In fact, the mofette research focus has in the last decade largely shifted to ecological and evolutionary studies in adaptive and functional traits of relatively small-sized organisms (microbes, micro- and mesofauna and flora) living in mofette specific and hypoxic soils (e.g. Beaubien et al., 2008; Beulig et al., 2016; Krüger et al., 2011; Maček et al., 2011; Šibanc et al., 2014).

In mofettes, as in other extreme environments, further investigation is needed and here we highlight some promising research avenues (see Figs 2 and 4). Notwithstanding the complexity and specificity of each particular system, general insights are clearly emerging, with more on the horizon that could help understand other forms of anthropogenically induced change via long-term (press) disturbances. Similar patterns in community composition should, in theory, be manifested, and indeed, at least for the ubiquitous soil AM fungi, this has already been confirmed (Maček et al., 2011). The most abundant and apparently competitive taxa in AM fungal communities in mofette soils are also widespread in other soil types exposed to long-term disturbance (e.g. heavy metal polluted soils, soils under intensive agriculture and submerged soils).

The expansion of extreme environments research can extend further still, and even beyond the confines of our own planet. Harnessing the natural experiment potential of locally extreme environments will undoubtedly lead to growing insights at the frontiers of ecology and evolution, including exobiology aimed at understanding the potential for life beyond Earth. Extreme environments can serve as useful analogues for extraterrestrial habitats (DiRuggiero et al., 2002; Friedmann, 1993) and can provide insight into novel strategies at the limits of life. First, though, we need to address pressing fundamental and applied questions in ecology important for the maintenance of diversity on our planet, and these model systems provide an ideal means of contributing to that goal.

ACKNOWLEDGEMENTS

This work was supported by the Slovenian Research Agency (ARRS) projects J4-5526 and J4-7052, research programme P4-0085 and a Swiss National Science Foundation project (SCOPES). We gratefully acknowledge all of the given support. We would like to thank Žiga Stanonik for the help with figure preparation and the anonymous referees and Prof. Guy Woodward for very constructive comments and feedback on this work.

REFERENCES

Ainsworth, E.A., Long, S.P., 2005. What have we learned from 15 years of free-air CO_2 enrichment (FACE)? A meta-analytic review of the responses of photosynthesis, canopy properties and plant production to rising CO_2. New Phytol. 165, 351–372.

Andresen, L.C., Müller, C., de Dato, G., Dukes, J.S., Emmett, B.A., Estiarte, M., Jentsch, A., Kröel-Dulay, G., Lüscher, A., Niu, S., Peñuelas, J., Reich, P.B., Reinsch, S., Ogaya, R., Schmidt, I.K., Schneider, M.K., Sternberg, M., Tietema, A., Zhu, K., Bilton, M.C., 2016. Shifting impacts of climate change: long-term patterns of plant response to elevated CO_2, drought, and warming across ecosystems. Adv. Ecol. Res. 55, 435–473.

Armstrong, J., Armstrong, W., 1988. Phragmites australis: a preliminary study of soil-oxidizing sites and internal gas transport pathways. New Phytol. 108, 373–382.

Armstrong, J., Armstrong, W., 1990. Light-enhanced convective throughflow increases oxygenation in rhizomes and rhizosphere of Phragmites australis (Cav.) Trin. ex Steud. New Phytol. 114, 121–128.

Armstrong, J., Armstrong, W., Beckett, P.M., 1992. Phragmites australis: venturi- and humidity-induced convections enhance rhizome aeration and rhizosphere oxidation. New Phytol. 120, 197–207.

Bago, B., Pfeffer, P.E., Zipfel, W., Lammers, P., Shachar-hill, Y., 2002. Tracking metabolism and imaging transport in arbuscular mycorrhizal fungi. Metabolism and transport in AM fungi. Plant Soil 244, 189–197.

Banet, A.I., Trexler, J.C., 2013. Space-for-time substitution works in everglades ecological forecasting models. PLoS One 8, e81025. http://dx.doi.org/10.1371/journal.pone.0081025.

Barry, P.H., Hilton, D.R., Fischer, T.P., de Moor, J.M., Mangasini, F., Ramirez, C., 2013. Helium and carbon isotope systematics of cold "mazuku" CO_2 vents and hydrothermal gases and fluids from Rungwe Volcanic Province, southern Tanzania. Chem. Geol. 339, 141–156.

Barshis, D.J., Stillman, J.H., Gates, R.D., Toonen, R.J., Smith, L.W., Birkeland, C., 2010. Protein expression and genetic structure of the coral *Porites lobata* in an environmentally extreme Samoan back reef: does host genotype limit phenotypic plasticity? Mol. Ecol. 19, 1705–1720.

Barták, M., Raschi, A., Tognetti, R., 1999. Photosynthetic characteristics of sun and shade leaves in the canopy of *Arbutus unedo* L. trees exposed to in situ long-term elevated CO_2. Photosynthetica 37, 1–16.

Bartholomew, J.W., Paik, G., 1966. Obligate thermophilic ocean basin cores isolation and identification of obligate thermophilic spore forming bacilli from ocean basin cores. J. Bacteriol. 92, 635–638.

Bauman, A.G., Baird, A.H., Burt, J.A., Pratchett, M.S., Feary, D.A., 2014. Patterns of coral settlement in an extreme environment: the southern Persian Gulf (Dubai, United Arab Emirates). Mar. Ecol. Prog. Ser. 499, 115–126.

Beaubien, S.E., Ciotoli, G., Coombs, P., Dictor, M., Krüger, M., Lombardi, S., Pearce, J., West, J., 2008. The impact of a naturally occurring CO_2 gas vent on the shallow ecosystem and soil chemistry of a Mediterranean pasture (Latera, Italy). Int. J. Greenh. Gas Con. 2, 373–387.

Beck-Nielsen, D., Madsen, T.V., 2001. Occurrence of vesicular-arbuscular mycorrhiza in aquatic macrophytes from lakes and streams. Aquat. Bot. 71, 141–148.

Bell, E.M., Callaghan, T.V., 2012. What are extreme environments and what lives in them? In: Bell, E.M. (Ed.), Life at Extremes: Environments, Organisms, and Strategies for Survival. CAB International, Wallingford, pp. 1–12.

Bettarini, I., Calderoni, G., Miglietta, F., Raschi, A., Ehleringer, J., 1995. Isotopic carbon discrimination and leaf nitrogen content of *Erica arborea* L. along a CO_2 concentration gradient in a CO_2 spring in Italy. Tree Physiol. 15, 327–332.

Beulig, F., Heuer, V.B., Akob, D.M., Viehweger, B., Elvert, M., Herrmann, M., Hinrichs, K.-U., Küsel, K., 2015. Carbon flow from volcanic CO_2 into soil microbial communities of a wetland mofette. ISME J. 9, 746–759.

Beulig, F., Urich, T., Nowak, M., Trumbore, S.E., Gleixner, G., Gilfillan, G.D., Fjelland, K.E., Küsel, K., 2016. Altered carbon turnover processes and microbiomes in soils under long-term extremely high CO_2 exposure. Nat. Microbiol. 1, 1–9.

Bloomberg, S., Rissmann, C., Mazot, A., Oze, C., Horton, T., Kennedy, B., Werner, C., Christenson, B., Pawson, J., 2012. Soil gas flux exploration at the Rotokawa geothermal field on White Island, New Zealand. In: Thirty-Sixth Workshop on Geothermal Reservoir Engineering, January 30, pp. 1–11.

Blume, H.-P., Felix-Henningsen, P., 2009. Reductosols: natural soils and technosols under reducing conditions without an aquic moisture regime. J. Plant Nutr. Soil Sci. 172, 808–820.

Bown, A.W., 1985. CO_2 and intracellular pH. Plant Cell Environ. 8, 459–465.

Bräuer, K., Kämpf, H., Niedermann, S., Strauch, G., 2013. Indications for the existence of different magmatic reservoirs beneath the Eifel area (Germany): a multi-isotope (C, N, He, Ne, Ar) approach. Chem. Geol. 356, 193–208.

Brooker, R.W., Callaway, R.M., 2009. Facilitation in the conceptual melting pot. J. Ecol. 97, 1117–1120.

Brundrett, M.C., 2009. Mycorrhizal associations and other means of nutrition of vascular plants: understanding the global diversity of host plants by resolving conflicting information and developing reliable means of diagnosis. Plant Soil 320, 37–77.

Calosi, P., Rastrick, S.P.S., Lombardi, C., de Guzman, H.J., Davidson, L., Giangrande, A., Hardege, J.D., Schulze, A., Spicer, J.I., Jahnke, M., Gambi, M., Giangr, A.E., 2013. Adaptation and acclimatization to ocean acidification in marine ectotherms: an in situ transplant experiment with polychaetes at a shallow CO_2 vent system. Philos. Trans. R. Soc. Lond. B Biol. Sci. 368, 1–15.

Collins, S., Bell, G., 2006. Evolution of natural algal populations at elevated CO_2. Ecol. Lett. 9, 129–135.
Cotton, T.E.A., Fitter, A.H., Miller, R.M., Dumbrell, A.J., Helgason, T., 2015. Fungi in the future: inter-annual variation and effects of atmospheric change on arbuscular mycorrhizal fungal communities. New Phytol. 205, 1598–1607.
Dawson, M.N., 2015. Island and island-like marine environments. Glob. Ecol. Biogeogr. 25, 831–846. http://dx.doi.org/10.1111/geb.12314.
de Rezende, J.R., Kjeldsen, K.U., Hubert, C.R.J., Finster, K., Loy, A., Jørgensen, B.B., 2013. Dispersal of thermophilic *Desulfotomaculum* endospores into Baltic Sea sediments over thousands of years. ISME J. 7, 72–84.
DiRuggiero, J., Nandakumar, R., Eisen, J., Schwartz, M., Thomas, R., Davila, J., Ofman, L., Robb, F., 2002. Genomic and physiological studies on extremophiles: model systems for exobiology. In: Workshop "Astrobiology in Russia".
Dumbrell, A.J., Nelson, M., Helgason, T., Dytham, C., Fitter, A.H., Nelson, H., Dytham, C., Fitter, A.H., 2010. Relative roles of niche and neutral processes in structuring a soil microbial community. ISME J. 4, 337–345.
Dumbrell, A.J., Ashton, P.D., Aziz, N., Feng, G., Nelson, M., Dytham, C., Fitter, A.H., Helgason, T., 2011. Distinct seasonal assemblages of arbuscular mycorrhizal fungi revealed by massively parallel pyrosequencing. New Phytol. 190, 794–804.
Elzenga, J.T., van Veen, H., 2010. Waterlogging and plant nutrient uptake. In: Mancuso, S., Shabala, S. (Eds.), Waterlogging Signalling and Tolerance in Plants. Springer-Verlag, Berlin, Heidelberg, pp. 23–35.
European Science Foundation (ESF), 2007. Annual report. http://www.esf.org/fileadmin/Public_documents/Publications/AnnualReport2007.pdf.
Farmer, A.M., 1985. The occurrence of vesicular-arbuscular mycorrhiza in isoetid-type submerged aquatic macrophytes under naturally varying conditions. Aquat. Bot. 21, 245–249.
Ferrer, M., Golyshina, O., Beloqui, A., Golyshin, P.N., 2007. Mining enzymes from extreme environments. Curr. Opin. Microbiol. 10, 207–214.
Fordham, M., Barnes, J.D., Bettarini, I., Polle, A., Slee, N., Raines, C., Miglietta, F., Raschi, A., 1997. The impact of elevated CO_2 on growth and photosynthesis in *Agrostis canina* L. ssp. *monteluccii* adapted to contrasting atmospheric CO_2 concentrations. Oecologia 110, 169–178.
Franklin, J.F., 1989. Importance and justification of long-term studies in ecology. In: Likens, G.E. (Ed.), Long-Term Studies in Ecology: Approaches and Alternatives. Springer-Verlag, New York, pp. 3–19.
Frerichs, J., Oppermann, B.I., Gwosdz, S., Möller, I., Herrmann, M., Krüger, M., 2013. Microbial community changes at a terrestrial volcanic CO_2 vent induced by soil acidification and anaerobic microhabitats within the soil column. FEMS Microbiol. Ecol. 84, 60–74.
Friedmann, E.I., 1993. Extreme environments and exobiology. Plant Biosyst. 127, 369–376.
Grahl, N., Shepardson, K.M., Chung, D., Cramer, R.A., 2012. Hypoxia and fungal pathogenesis: to air or not to air? Eukaryot. Cell 11, 560–570.
Gudmundsdottir, R., Gislason, G.M., Palsson, S., Olafsson, J.S., Schomacker, A., Friberg, N., Woodward, G., Hannesdottir, E.R., Moss, B., 2011. Effects of temperature regime on primary producers in Icelandic geothermal streams. Aquat. Bot. 95, 278–286.
Hall-Spencer, J.M., Rodolfo-Metalpa, R., Martin, S., Ransome, E., Fine, M., Turner, S.M., Rowley, S.J., Tedesco, D., Buia, M.C., 2008. Volcanic carbon dioxide vents show ecosystem effects of ocean acidification. Nature 454, 96–99.
Head, I.M., Jones, D.M., Röling, W.F.M., 2006. Marine microorganisms make a meal of oil. Nat. Rev. Microbiol. 4, 173–182.

Herbold, C.W., Lee, C.K., McDonald, I.R., Cary, S.C., 2014. Evidence of global-scale aeolian dispersal and endemism in isolated geothermal microbial communities of Antarctica. Nat. Commun. 5, 3875.

Hladnik, J., Eler, K., Kržan, K., Pintar, M., Vodnik, D., 2009. Short-term dynamics of stomatal response to sudden increase in CO_2 concentration in maize supplied with different amounts of water. Photosynthetica 47, 422–428.

Hohberg, K., Schulz, H.-J., Balkenhol, B., Pilz, M., Thomalla, A., Russell, D.J., Pfanz, H., 2015. Soil faunal communities from mofette fields: effects of high geogenic carbon dioxide concentration. Soil Biol. Biochem. 88, 420–429.

Holloway, S., Pearce, J.M., Hards, V.L., Ohsumi, T., Gale, J., 2007. Natural emissions of CO_2 from the geosphere and their bearing on the geological storage of carbon dioxide. Energy 32, 1194–1201.

Hourdez, S., 2012. Hypoxic environments. In: Bell, E.M. (Ed.), Life at Extremes: Environments, Organisms and Strategies for Survival. Gutenberg Press, Malta, pp. 438–453.

Hourdez, S., Lallier, F.H., 2007. Adaptation to hypoxia in hydrothermal vent and cold-sep invertebrates. Rev. Environ. Sci. Biotechnol. 6, 143–159.

Hubert, C., Loy, A., Nickel, M., Arnosti, C., Baranyi, C., Brüchert, V., Ferdelman, T., Finster, K., Christensen, F.M., Rosa de Rezende, J., Vandieken, V., Jørgensen, B.B., 2009. A constant flux of diverse thermophilic bacteria into the cold Arctic seabed. Science 325, 1541–1544.

Jump, A.S., Mátyás, C., Peñuelas, J., 2009. The altitude-for-latitude disparity in the range retractions of woody species. Trends Ecol. Evol. 24, 694–701.

Kämpf, H., Bräuer, K., Schumann, J., Hahne, K., Strauch, G., 2013. CO_2 discharge in an active, non-volcanic continental rift area (Czech Republic): characterisation (δ13C, 3He/4He) and quantification of diffuse and vent CO_2 emissions. Chem. Geol. 339, 71–83.

Kawecki, T.J., Ebert, D., 2004. Conceptual issues in local adaptation. Ecol. Lett. 7, 1225–1241.

Kies, A., Hengesch, O., Tosheva, Z., Raschi, A., Pfanz, H., 2015. Diurnal CO_2-cycles and temperature regimes in a natural CO_2 gas lake. Int. J. Greenh. Gas Con. 37, 142–145.

Körner, C., 2000. Why are there global gradients in species richness? Mountains might hold the answer. Trends Ecol. Evol. 15, 513–514.

Körner, C., 2007. The use of "altitude" in ecological research. Trends Ecol. Evol. 22, 569–574.

Krüger, M., West, J., Frerichs, J., Oppermann, B., Dictor, M.-C., Jouliand, C., Jones, D., Coombs, P., Green, K., Pearce, J., May, F., Möller, I., 2009. Ecosystem effects of elevated CO_2 concentrations on microbial populations at a terrestrial CO_2 vent at Laacher See, Germany. Energy Procedia 1, 1933–1939.

Krüger, M., Jones, D., Frerichs, J., Oppermann, B.I., West, J., Coombs, P., Green, K., Barlow, T., Lister, R., Shaw, R., Strutt, M., Möller, I., 2011. Effects of elevated CO_2 concentrations on the vegetation and microbial populations at a terrestrial CO_2 vent at Laacher See, Germany. Int. J. Greenh. Gas Con. 5, 1093–1098.

Lal, R., 2008. Carbon sequestration. Philos. Trans. R. Soc. Lond. B Biol. Sci. 363, 815–830.

Lehwess-Litzmann, I., 1943. Kohlensäure-Vergiftungen. Arch. Toxicol. 12, 22–57.

Lewin, K.F., Hendrey, G.R., Nagy, J., Lamorte, R.L., 1994. Design and application of a free-air carbon dioxide enrichment facility. Agric. For. Meteorol. 70, 15–29.

Lomax, B.H., 2012. Past extremes. In: Bell, E.M. (Ed.), Life at Extremes: Environments, Organisms and Strategies for Survival. CABI, Wallingford, pp. 13–35.

Low-Décarie, E., Jewell, M.D., Fussmann, G.F., Bell, G., 2013. Long-term culture at elevated atmospheric CO_2 fails to evoke specific adaptation in seven freshwater phytoplankton species. Proc. R. Soc. B Biol. Sci. 280, 20122598.

Low-Décarie, E., Fussmann, G.F., Bell, G., 2014. Aquatic primary production in a high-CO_2 world. Trends Ecol. Evol. 29, 1–10.

Maček, I., 2013. A decade of research in mofette areas has given us new insights into adaptation of soil microorganisms to abiotic stress. Acta Agric. Slov. 101, 209–217.

Maček, I., Pfanz, H., Francetič, V., Batič, F., Vodnik, D., 2005. Root respiration response to high CO_2 concentrations in plants from natural CO_2 springs. Environ. Exp. Bot. 54, 90–99.

Maček, I., Videmšek, U., Kastelec, D., Stopar, D., Vodnik, D., 2009. Geological CO_2 affects microbial respiration rates in Stavešinci mofette soil. Acta Biol. Slov. 52, 41–48.

Maček, I., Dumbrell, A.J., Nelson, M., Fitter, A.H., Vodnik, D., Helgason, T., 2011. Local adaptation to soil hypoxia determines the structure of an arbuscular mycorrhizal fungal community in roots from natural CO_2 springs. Appl. Environ. Microbiol. 77, 4770–4777.

Maček, I., Kastelec, D., Vodnik, D., 2012. Root colonization with arbuscular mycorrhizal fungi and glomalin-related soil protein (GRSP) concentration in hypoxic soils from natural CO_2 springs. Agric. Food Sci. 21, 62–71.

Marschner, P., 2012. Marschner's Mineral Nutrition of Higher Plants, third ed. Academic Press, San Diego. 672 pp.

McFarland, J.W., Waldrop, M.P., Haw, M., 2013. Extreme CO_2 disturbance and the resilience of soil microbial communities. Soil Biol. Biochem. 65, 274–286.

McGrath, J.M., Lobell, D.B., 2013. Reduction of transpiration and altered nutrient allocation contribute to nutrient decline of crops grown in elevated CO_2 concentrations. Plant Cell Environ. 36, 697–705.

Mehlhorn, J., Beulig, F., Küsel, K., Planer-Friedrich, B., 2014. Carbon dioxide triggered metal(loid) mobilisation in a mofette. Chem. Geol. 382, 54–66.

Miglietta, F., Raschi, A., Bettarini, I., Resti, R., Selvi, F., 1993. Natural CO_2 springs in Italy: a resource for examining long-term response to rising atmospheric CO_2 concentrations. Plant Cell Environ. 16, 873–878.

Mook, W., 2000. Chemistry of carbonic acid in water. In: Environmental Isotopes in the Hydrological Cycle: Principles and Applications. INEA/UNESCO, Paris, pp. 143–165.

Nedwell, D.B., Underwood, G.J.C., McGenity, T.J., Whitby, C., Dumbrell, A.J., 2016. The Colne estuary: a long-term microbial ecology observatory. Adv. Ecol. Res. 55, 227–281.

Noble, R.R.P., Stalker, L., Wakelin, S.A., Pejcic, B., Leybourne, M.I., Hortle, A.L., Michael, K., 2012. Biological monitoring for carbon capture and storage: a review and potential future developments. Int. J. Greenh. Gas Con. 10, 520–535.

O'Gorman, E.J., Pichler, D.E., Adams, G., Benstead, J.P., Cohen, H., et al., 2012. Impacts of warming on the structure and functioning of aquatic communities: individual-to ecosystem-level responses. Adv. Ecol. Res. 47, 81–176. http://dx.doi.org/10.1016/b978-0-12-398315-2.00002-8.

O'Gorman, E.J., Benstead, J.P., Cross, W.F., Friberg, N., Hood, J.M., Johnson, P.W., Sigurdsson, B.D., Woodward, G., 2014. Climate change and geothermal ecosystems: natural laboratories, sentinel systems, and future refugia. Glob. Chang. Biol. 20, 3291–3299.

Onoda, Y., Hirose, T., Hikosaka, K., 2009. Does leaf photosynthesis adapt to CO_2-enriched environments? An experiment on plants originating from three natural CO_2 springs. New Phytol. 182, 698–709.

Oppermann, B.I., Michaelis, W., Blumenberg, M., Frerichs, J., Schulz, H.M., Schippers, A., Beaubien, S.E., Krüger, M., 2010. Soil microbial community changes as a result of long-term exposure to a natural CO_2 vent. Geochim. Cosmochim. Acta 74, 2697–2716.

Osada, N., Onoda, Y., Hikosaka, K., 2010. Effects of atmospheric CO_2 concentration, irradiance, and soil nitrogen availability on leaf photosynthetic traits of *Polygonum sachalinense* around natural CO_2 springs in northern Japan. Oecologia 164, 41–52.

Owens, B., 2013. Long-term research: slow science. Nature 495, 300–303.
Paoletti, E., Pfanz, H., Raschi, A., 2005. Pros and cons of CO_2 springs as experimental sites. In: Omasa, K., Nauchi, I., De Kok, L.J. (Eds.), Plant Responses to Air Pollution and Global Change. Springer-Verlag, Tokyo, pp. 195–202.
Papke, R.T., Ramsing, N.B., Bateson, M.M., Ward, D.M., 2003. Geographical isolation in hot spring cyanobacteria. Environ. Microbiol. 5, 650–659.
Perata, P., Armstrong, W., Voesenek, L.A.C.J., 2011. Plants and flooding stress. New Phytol. 190, 269–273.
Pfanz, H., 1994. Apoplastic and symplastic proton concentrations and their significance for metabolism. In: Schulze, E.-D., Caldwell, M.M. (Eds.), Ecophysiology of Photosynthesis. Springer Verlag, Berlin, Heidelberg, pp. 103–122.
Pfanz, H., 2008. Mofetten—Kalter Atem Schlafender Vulkane. RVDL-Verlag, Köln. 85 pp.
Pfanz, H., Heber, U., 1989. Determination of extra- and intracellular pH values in relation to the action of acidic gases in cells. In: Linskens, H.F., Jackson, J.F. (Eds.), Modern Methods of Plant Analysis, New Series. Gases in Plant and Microbial Cells, vol. 9. Springer, Berlin, pp. 322–343.
Pfanz, H., Martinoia, E., Lange, O.L., Heber, U., 1987. Flux of sulfur dioxide into leaf cells and cellular acidification by sulfur dioxide. Plant Physiol. 85, 928–933.
Pfanz, H., Vodnik, D., Wittmann, C., Aschan, G., Raschi, A., 2004. Plants and geothermal CO_2 exhalations—survival in and adaption to high CO_2 environment. Prog. Bot. 65, 499–538.
Pfanz, H., Vodnik, D., Wittmann, C., Aschan, G., Batič, F., Turk, B., Maček, I., 2007. Photosynthetic performance (CO_2-compensation point, carboxylation and net photosynthesis) of timothy grass (*Phleum pratense* L.) is affected by elevated carbon dioxide in post-volcanic mofette areas. Environ. Exp. Bot. 61, 41–48.
Pfanz, H., Yüce, G., D'Andria, F., D'Alessandro, W., Pfanz, B., Manetas, Y., Papatheodorou, G., 2014. The ancient gates to hell and their relevance to geogenic CO_2. In: Wexler, P. (Ed.), History of Toxicology and Environmental Health. Toxicology in Antiquity, vol. I. Academic Press, Amsterdam, pp. 92–113.
Pickett, S.T.A., 1989. Space-for-time substitution as an alternative to long-term studies. In: Likens, G.E. (Ed.), Long-Term Studies in Ecology. Springer, New York, NY, pp. 110–135.
Raschi, A., Miglietta, F., Tognetti, R., van Gardingen, P., 1997. Plant Responses to Elevated CO_2: Evidence from Natural Springs. Cambridge University Press, Cambridge. 272 pp.
Reiss, J., Fernanda Cássio, R.A.B., Woodward, G., Pascoal, C., 2010. Assessing the contribution of micro-organisms and macrofauna to biodiversity-ecosystem functioning relationships in freshwater microcosms. Adv. Ecol. Res. 43, 151–176.
Rennert, T., Pfanz, H., 2015. Geogenic CO_2 affects stabilization of soil organic matter. Eur. J. Soil Sci. 66, 838–846.
Rennert, T., Eusterhues, K., Pfanz, H., Totsche, K.U., 2011. Influence of geogenic CO_2 on mineral and organic soil constituents on a mofette site in the NW Czech Republic. Eur. J. Soil Sci. 62, 572–580.
Rogers, A., Humphries, S.W., 2000. A mechanistic evaluation of photosynthetic acclimation at elevated CO_2. Glob. Chang. Biol. 6, 1005–1011.
Ross, D.J., Tate, K.R., Newton, P.C.D., Wilde, R.H., Clark, H., 2001. Carbon and nitrogen pools and mineralization in a grassland organic soil at a New Zealand carbon dioxide spring. Soil Biol. Biochem. 33, 849–852.
Rothschild, L.J., Mancinelli, R.L., 2001. Life in extreme environments. Nature 409, 1092–1101.
Russell, D.J., Schulz, H.-J., Hohberg, K., Pfanz, H., 2011. Occurrence of collembolan fauna in mofette fields (natural carbon-dioxide springs) of the Czech Republic. Soil Org. 83, 489–505.

Saßmannshausen, F., 2010. Vegetationsökologische Charakterisierung terrestrischer Mofettenstandorte am Beispiel des west-tschechischen Plesná-Tals. Doctoral dissertation, Universität Duisburg-Essen, Essen, Germany.

Schapira, M., Buscot, M.J., Leterme, S.C., Pollet, T., Chapperon, C., Seuront, L., 2009. Distribution of heterotrophic bacteria and virus-like particles along a salinity gradient in a hypersaline coastal lagoon. Aquat. Microb. Ecol. 54, 171–183.

Schneider, T., Keiblinger, K.M., Schmid, E., Sterflinger-Gleixner, K., Ellersdorfer, G., Roschitzki, B., Richter, A., Eberl, L., Zechmeister-Boltenstern, S., Riedel, K., 2012. Who is who in litter decomposition? Metaproteomics reveals major microbial players and their biogeochemical functions. ISME J. 6, 1749–1762.

Schoepf, V., Stat, M., Falter, J.L., McCulloch, M.T., 2015. Limits to the thermal tolerance of corals adapted to a highly fluctuating, naturally extreme temperature environment. Sci. Rep. 5, 1–14.

Schulz, H.-J., Potapov, M.B., 2010. A new species of *Folsomia* from mofette fields of the Northwest Czechia (Collembola, Isotomidae). Zootaxa 2553, 60–64.

Šibanc, N., Dumbrell, A.J., Mandić-Mulec, I., Maček, I., 2014. Impacts of naturally elevated soil CO_2 concentrations on communities of soil archaea and bacteria. Soil Biol. Biochem. 68, 348–356.

Simon, M.C., Keith, B., 2008. The role of oxygen availability in embryonic development and stem cell function. Nat. Rev. Mol. Cell Biol. 4, 285–296.

Smith, S.E., Read, D.J., 2008. Mycorrhizal Symbiosis, third ed. Academic Press, London, pp. 11–145.

Smolders, A.J.P., Lucassen, E., Roelofs, J.G.M., 2002. The isoetid environment: biogeochemistry and threats. Aquat. Bot. 73, 325–350.

Srivastava, D.S., Kolasa, J., Bengtsson, J., Gonzalez, A., Lawler, S.P., Miller, T.E., Munguia, P., Romanuk, T., Schneider, D.C., Trzcinski, M.K., 2004. Are natural microcosms useful model systems for ecology? Trends Ecol. Evol. 19, 379–384.

Stitt, M., Krapp, A., 1999. The interaction between elevated carbon dioxide and nitrogen nutrition: the physiological and molecular background. Plant Cell Environ. 22, 553–621.

Storkey, J., Macdonald, A.J., Bell, J.R., Clark, I.M., Gregory, A.S., Hawkins, N.J., Hirsch, P.R., Todman, L.C., Whitmore, A.P., 2016. The unique contribution of Rothamsted to ecological research at large temporal scales. Adv. Ecol. Res. 55, 1–42.

Stylinski, C.D., Oechel, W.C., Gamon, J.A., Tissue, D.T., Miglietta, F., Raschi, A., 2000. Effects of lifelong [CO_2] enrichment on carboxylation and light utilization of *Quercus pubescens* Willd. examined with gas exchange, biochemistry and optical techniques. Plant Cell Environ. 23, 1353–1362.

Sudová, R., Sýkorová, Z., Rydlová, J., Čtvrtlíková, M., Oehl, F., 2015. *Rhizoglomus melanum*, a new arbuscular mycorrhizal fungal species associated with submerged plants in freshwater lake Avsjøen in Norway. Mycol. Prog. 14, 1–8.

Thauer, R.K., Kaster, A.-K., Seedorf, H., Buckel, W., Hedderich, R., 2008. Methanogenic archaea: ecologically relevant differences in energy conservation. Nat. Rev. Microbiol. 6, 579–591.

Thomalla, A., 2015. Boden- und vegetationskundliche Untersuchungen zur Charakterisierung der Ausgasung- und Vegetationsdynamik zweier trockener Mofetten im west-tschechischen Plesnatal. Doctoral thesis, University Duisburg-Essen.

Torossian, J.L., Kordas, R.L., Helmuth, B., 2016. Cross-scale approaches to forecasting biogeographic responses to climate change. Adv. Ecol. Res. 55, 371–433.

Ustinova, Y., 2009. Cave experiences and ancient Greek oracles. Time Mind 2, 265–286.

Vacher, C., Tamaddoni-Nezhad, A., Kamenova, S., Peyrard, N., Moalic, Y., Sabbadin, R., Schwaller, L., Chiquet, J., Smith, M.A., Vallance, J., Fievet, V., Jakuschkin, B.,

Bohan, D.A., 2016. Learning ecological networks from next-generation sequencing data. Adv. Ecol. Res. 54, 1–39.
van Dover, C.L., 2002. Evolution and biogeography of deep-sea vent and seep invertebrates. Science 295, 1253–1257.
van Gardingen, P.R., Grace, J., Harkness, D.D., Miglietta, F., Raschi, A., 1995. Carbon dioxide emissions at an Italian mineral spring: measurements of average CO_2 concentration and air temperature. Agric. For. Meteorol. 73, 17–27.
Vartapetian, B.B., Jackson, M.B., 1997. Plant adaptations to abiotic stress. Ann. Bot. 79 (Suppl. A), 3–20.
Videmšek, U., Turk, B., Vodnik, D., 2006. Root aerenchyma—formation and function. Acta Agric. Slov. 87, 445–453.
Videmšek, U., Hagn, A., Schloter, M., Vodnik, D., 2009. Abundance and diversity of CO_2-fixing bacteria in grassland soils close to natural carbon dioxide springs. Microb. Ecol. 58, 1–9.
Vodnik, D., Pfanz, H., Maček, I., Kastelec, D., Lojen, S., Batič, F., 2002a. Photosynthesis of cockspur (*Echinochloa crus-galli* (L.) Beauv.) at sites of naturally elevated CO_2 concentration. Photosynthetica 40, 575–579.
Vodnik, D., Pfanz, H., Wittmann, C., Maček, I., Kastelec, D., Turk, B., Batič, F., 2002b. Photosynthetic acclimation in plants growing near a carbon dioxide spring. Phys. Chem. Chem. Phys. 42, 239–244.
Vodnik, D., Kastelec, D., Pfanz, H., Maček, I., Turk, B., 2006. Small-scale spatial variation in soil CO_2 concentration in a natural carbon dioxide spring and some related plant responses. Geoderma 133, 309–319.
Vodnik, D., Videmšek, U., Pintar, M., Maček, I., Pfanz, H., 2009. The characteristics of soil CO_2 fluxes at a site with natural CO_2 enrichment. Geoderma 150, 32–37.
Vodnik, D., Hladnik, J., Vrešak, M., Eler, K., 2013. Interspecific variability of plant stomatal response to step changes of $[CO_2]$. Environ. Exp. Bot. 88, 107–112.
Wall, D.H., Virginia, R.A., 1999. Controls on soil biodiversity: insights from extreme environments. Appl. Soil Ecol. 13, 137–150.
Weinlich, F.H., Stejskal, V., Teschner, M., Poggenburg, J., 2013. Geodynamic processes in the NW Bohemian swarm earthquake region, Czech Republic, identified by continuous gas monitoring. Geofluids 13, 305–330.
Whitaker, R.J., Grogan, D.W., Taylor, J.W., 2003. Geographic barriers isolate endemic populations of hyperthermophilic archaea. Science 301, 976–978.
Živić, I., Živić, M., Milošević, D., Bjelanović, K., Stanojlović, S., Daljević, R., Marković, Z., 2013. The effects of geothermal water inflow on longitudinal changes in benthic macroinvertebrate community composition of a temperate stream. J. Therm. Biol. 38, 255–263.
Zwingmann, N., 2012. Antiker Tourismus in Kleinasien und auf den vorgelagerten Inseln. Habelt Verlag, Bonn.

CHAPTER SEVEN

Climate-Driven Range Shifts Within Benthic Habitats Across a Marine Biogeographic Transition Zone

N. Mieszkowska[*,†,1], H.E. Sugden[‡]
[*]School of Environmental Sciences, University of Liverpool, Liverpool, United Kingdom
[†]The Marine Biological Association of the UK, Plymouth, United Kingdom
[‡]School of Marine Science and Technology, The Dove Marine Laboratory, Newcastle University, Cullercoats, United Kingdom
[1]Corresponding author: e-mail address: nova@mba.ac.uk

Contents

1. Introduction	326
2. The Rise of Natural History and Species Recording	328
3. History and Development of Biogeographic Research in the Northeast Atlantic	330
4. Patterns of Change Across the Boreal–Lusitanian Biogeographic Breakpoint in the Northeast Atlantic	332
5. Factors Setting Biogeographic Range Limits	336
5.1 Environmental Conditions	336
5.2 Biological Processes	337
5.3 Ecological Factors	338
5.4 Defining the Habitat of a Species	338
6. Long-Term Time-Series for Benthic Ecosystems in the Northeast Atlantic and Regional Seas	341
6.1 Marine Biodiversity and Climate Change: MarClim	341
6.2 European Kelp Forests	343
6.3 North Sea Soft Sediment Benthos	343
7. Observed Changes in the Physical Environment	344
8. Impacts of Climate Change on Intertidal Benthic Species	347
8.1 Biogeographic Range Shifts	348
8.2 Changes in Population Dynamics	349
8.3 Biological Mechanisms	350
9. Future Advances in Quantifying and Modelling Distributional Responses to Climate Change	352
9.1 Standardizing the Recording and Availability of Data on Species	352

9.2 Developing Scientific Methodologies for Quantifying Previous, and Modelling Future Changes in Species Distributions in Response to Climate Change   354
References   356

Abstract

Anthropogenic climate change is causing unprecedented rapid responses in marine communities, with species across many different taxonomic groups showing faster shifts in biogeographic ranges than in any other ecosystem. Spatial and temporal trends for many marine species are difficult to quantify, however, due to the lack of long-term datasets across complete geographical distributions and the occurrence of small-scale variability from both natural and anthropogenic drivers. Understanding these changes requires a multidisciplinary approach to bring together patterns identified within long-term datasets and the processes driving those patterns using biologically relevant mechanistic information to accurately attribute cause and effect. This must include likely future biological responses, and detection of the underlying mechanisms in order to scale up from the organismal level to determine how communities and ecosystems are likely to respond across a range of future climate change scenarios. Using this multidisciplinary approach will improve the use of robust science to inform the development of fit-for-purpose policy to effectively manage marine environments in this rapidly changing world.

1. INTRODUCTION

The world's oceans encompass 71% of the surface of the planet from the poles to the tropics, and have a natural, year-round, global sea temperature gradient that is evident between the warm tropical areas and the cold, polar regions. At the regional oceanographic scale, warm and cold currents also influence seawater movement and temperature trends within the oceans, including long-term, large-scale oceanographic water circulation patterns such as the Gulf Stream, transferring seawater from the east coast of the United States in a northeasterly direction across the Atlantic Ocean to the western and northwestern coastlines of the United Kingdom and Europe. Additionally, large-scale, pervasive climate patterns and modes such as the North Atlantic Oscillation (NAO), Atlantic Multidecadal Oscillation, Pacific Decadal Oscillation and the El Niño Southern Oscillation occur over multiannual time periods, affecting not only oceanic and atmospheric processes and patterns in temperature, but also precipitation and oceanic storm tracks across large areas of the globe (Hurrell, 1995). These long-term atmospheric pressure differences have implications for the oceanography and long-term impacts on the diverse range of marine biota and ecosystems. In addition, latitudinal gradients in oceanic temperature and large-scale oceanic circulation also influence marine species and ecosystems. In the Northeast Atlantic the NAO consistently explains variation in the growth of marine fish, their abundances and assemblage compositions (Attrill and Power, 2002), while in the North Sea the NAO has been linked to changes in the biodiversity and carrying capacity of the pelagic ecosystem resulting in an abrupt regime shift (Beaugrand et al., 2008).

While there are naturally occurring short-scale weather, and longer-term environmental patterns in the marine oceanic environment, the anthropogenically driven, globally important effect of climate change is superimposed on these natural sources of variability, with warming of around 1°C in the northeast Atlantic between the 1980s and 2010s exceeding the mean global ocean warming of 0.06°C (Levitus et al., 2000; Southward et al., 2004a). In addition to the global oceans being warmed as a direct result of pervasive, human-induced greenhouse gas emissions, natural environmental events and cycles outlined above are also being altered in duration, frequency and intensity during recent decades, with anthropogenic climate change being the primary driver of these changes (IPCC, 2014).

The monitoring and understanding of oceanic provinces (biogeographic areas categorized by the differences in biogeochemical processes and biodiversity between ocean regions), cycles and the species that inhabit the global seas have long been of interest to people in both the scientific and amateur naturalist communities around the world. The wealth of available data extending back across the last two centuries stems from the long tradition of recording the natural history across natural habitats. The United Kingdom was one of the pioneering nations that instigated such studies, including many of the early investigations of the marine ecosystems that were focused on the Northeast Atlantic ocean from the late 1800s onwards (including Bergmann, 1847; Burrows et al., 2002; Crisp, 1964; Crisp and Fisher-Piette, 1959; Fischer-Piette, 1953, 1955; Fischer-Piétte, 1963; Forbes, 1853, 1858; Hutchins, 1947; Lewis, 1956, 1976, 1986; Mieszkowska et al., 2006a, 2009, 2014b; Orton, 1920; Southward and Crisp, 1954a,b; Southward et al., 2004a).

Global climate change is the largest and most pervasive stressor now acting on marine systems globally, severely and rapidly impacting marine species and ecosystems. Rapid alterations in the marine climate are forcing species to adapt to climate-driven changes, move to track the rates of climatic changes or become locally extinct, with a high degree of spatial and temporal heterogeneity in the resultant impacts on marine communities. Such a wide spatiotemporal range of effects can only be detected through the analysis of data collected at many locations spanning large sectors of the biogeographic range of marine species, across multiple years and decades. It is fortunate that several long term, sustained observation programmes of marine life were initiated before the onset of the current period of climate change. Today's researchers can utilize these long-term time-series as resources to segregate natural and multiple sources of anthropogenically induced change in marine ecosystems (e.g. Burrows et al., 1992; Holme, 1961; Mieszkowska et al., 2009, 2014a; Simpson et al., 2011; Southward, 1967; Southward et al., 1995), and to track trends and pervasive shifts around the global oceans across decadal and century-scale timeframes.

Here we explore the value of rare, sustained observations of the marine environment, ecosystems and component species with respect to the detection of long-term trends, and shifts in biogeographic distributions of coastal benthic species across a major marine biogeographic boundary between Boreal and Lusitanian provinces in the Northeast Atlantic. Such time-series are invaluable due to their applicability for the determination of spatial and temporal heterogeneity in climate change impacts, which are of greater spatiotemporal magnitude than the majority of research and monitoring for the marine environment that only covers specific areas for a short time period. The importance of studying and recording the distribution and abundance of species across large sectors of their biogeographic distributions, and the use of sustained time-series datasets in the development of predictive modelling of future changes to marine species are discussed.

2. THE RISE OF NATURAL HISTORY AND SPECIES RECORDING

Our coastal habitats contain a wealth of biodiversity, and there has been a long history of recreational exploration of the coastal habitats across the coastal ecosystems within the Northeast Atlantic region. In the 19th century, natural history became popular and fashionable among the gentry and a new generation of amateur naturalists were born. Several books on the natural history of nearshore and intertidal habitats around the UK were published and the collection of specimens and pressing of seaweeds were pursued by many, reflecting a rich heritage of descriptive studies on British rocky shores (Burrows et al., 2014; Hawkins et al., 2010; Mieszkowska et al., 2006a; Sugden et al., 2009).

During this time, major exploratory voyages were funded by the British government to gather scientific data on a wide range of oceanic features, most notably the Challenger Expedition which set out to gather data on ocean temperatures, seawater chemistry, currents, marine life and the geology of the seafloor (Thomson and Murray, 1873–1876). This voyage resulted in the first systematic plots of currents and temperatures in the oceans and built on previous work by Maury (1855) that brought together thousands of data points on winds and currents and identified the major trade routes at the time. These UK expeditions received overwhelming support from the general public and further swelled enthusiasm for marine habitats.

The rise in interest in the natural world and success of such voyages led many to believe that closer study of the marine environment was needed and that dedicated laboratories should be established in order to facilitate this. The Marine Biological Association of the United Kingdom was established in 1884 to meet this growing need. Initially focusing on fisheries and overexploitation, several additional laboratories quickly followed suit and were established in the following years. Many amateur naturalists, who had undertaken the description of species within the marine environment, quickly became associated with these laboratories and bestowed their pioneering collections upon their institutions (e.g. Alder and Hancock collected specimens between 1845 and 1855 which are now held at the Dove Marine Laboratory and the Great North Museum: Hancock).

Museum collections such as those donated by pioneering naturalists not only serve to classify and describe the abundance of species found in coastal and marine habitats at the time but also to act as a repository of information providing data on the morphological characteristics, that can be related where historical records of physical conditions exist to environmental conditions at the time of collection. In areas where studies on benthic species are sparse and evidence for changes in the climate and its impact on species assemblages are limited, these collections can provide information regarding the functioning of ecosystems and changes in communities and geographical distributions as a response to global climate change (Wernberg et al., 2011), although such data cannot provide information on ecosystem functioning or community change as most samples were collected systematically, with only a limited part of the community being sampled during a survey.

The groundswell of enthusiasm for marine biology has not waned and with the rise of recent citizen science initiatives around the world, there has never been a greater opportunity for members of the general public to contribute to the field of ecology. Citizen science has an important role to play in the collection of data and the building of evidence bases in ecological sciences (Dickinson et al., 2010; Roy et al., 2012). Identifying and understanding the patterns which occur at broad geographic scales are the first step in looking at natural vs anthropogenic variability and change. This requires large datasets across large spatial scales, collected for species from many taxonomic groups, with surveys across all of the seasons, for multiple years. Citizen science can contribute directly to this goal by identifying biogeographic patterns for marine species, even without aiming to understand the mechanistic drivers. Citizen science surveys can fill datagaps that exist between the few marine scientific surveys that span large geographical areas over long time periods, and help to address the urgent need to detect, track and predict the rapid effects of global climate change on marine species, communities and ecosystems (see also chapter "Recommendations for the next generation of global freshwater biological monitoring tools" by Jackson et al., in this volume). Citizen science can be complementary to additional experimental science to support emerging questions about the distribution of organisms across space and time (Conrad and Hilchey, 2011; Dickinson et al., 2010), but does not replace scientific research with respect to taxonomic expertise, quality assurance and surveyor accuracy or methodological consistency.

3. HISTORY AND DEVELOPMENT OF BIOGEOGRAPHIC RESEARCH IN THE NORTHEAST ATLANTIC

Here we outline some of the major scientific developments and advances in knowledge that facilitated the field of marine biogeography within the Northeast Atlantic since the start of systematic scientific and recording in the 1800s, and served as novel examples for the rest of the world's oceans.

The development of the field of marine biogeography in the late 19th century is evidenced in the British scientist Forbes' pioneering biogeographical map published in 1858, where he mapped the boundary between the cold-temperate Boreal and warm-temperate Lusitanian regions bisecting the UK, and related these to geographical changes (Forbes, 1858). Forbes' observations that species were grouped according to these areas, with distributional limits occurring in the vicinity of the transition zone are still broadly applicable today.

Forbes made three observations of lasting value to marine research: (1) each zoogeographic province is an area where there was a 'special manifestation of creative power' (a geographic location where many species evolved and dispersed from), and the animals originally formed there can become mixed with emigrants from other provinces; (2) each species was created only once, and individuals tend to migrate outwards from their centre of origin to new locations (the abundant centre hypothesis) and (3) provinces must, to be understood, be traced back, like species, to their origins in past time (track their evolutionary history). These points are of value to phylogeographic studies and are useful tools for understanding how climate change will impact a specific species with respect to their thermal evolutionary origins.

The American phycologist Setchell (1915) proposed a series of nine biogeographic zones bounded by summer temperatures, based on the historical background of regional terrestrial faunistics and floristics. These zones he then demarcated into subzones on the basis of the winter temperatures within them to explain the distribution of marine macroalgae. This viewpoint was to be debated in later decades, when this expectation of a single system of zonation to summarize the empirical diversity of biological distributions was disputed, and considered inappropriate as an interpretive summary for marine species distributions (Hutchins, 1947). This heuristic tool can aid our understanding of large-scale trends or shifts in the biogeographic distribution of species; however, it has limited applications with respect to the prediction of climate change impacts due to the coarse spatiotemporal scale of the zonation scheme to classify biogeographic provinces.

Building on the ideas presented by Forbes, the Swedish scientist Ekman undertook the huge task of analysing all of the pertinent literature available on marine animal distribution at that time, leading to the publication of his book *Tiergeographie des Meeres* (Ekman, 1935). This was the first systematic treatise on historical marine biology to be published (Kafanov, 2006) and was followed by a revised English edition, *Zoogeography of the Sea* (Ekman and Palmer, 1953). Ekman and Palmer (1953) considered the marine world to be composed of a series of large regions or subregions. For the continental shelf, they described regions located in warm, temperate and polar waters; their separation by zoogeographic barriers and their endemism. He proposed a new, qualitative 'Eckman Index' modified from the Jackard Similarity Index, Pesenko (1982) that has served as the basis for a continually developing topic, comparing the similarity and diversity of species within a sample, that is still relevant in modern scientific research today. The American researcher Briggs (1974) divided the continental shelf further, into a series of large biogeographic regions that, in turn, contained smaller provinces, defined on the basis of endemism (Briggs and Bowen, 2013).

Today, several versions of global marine classifications exist, dividing the world's oceans into between 54 and 62 marine biogeographic provinces, providing large-scale delineations of major faunistic regions (e.g. Berlanger et al., 2012; Briggs and Bowen, 2013; Spalding et al., 2007; VLIZ, 2009). These are based on the wealth of knowledge and data collected throughout the modern history of marine science, combining long-term time-series, field-based records and observations, experimental research, remote sensed oceanographic and biological data, and understanding of the biogeographic distributions and abundances of marine algae, invertebrates and vertebrates. Differences in the exact number of provinces can be traced back to the specific faunal list and species groups with which each classification was determined, however, they are still in major agreement with each other at the global scale. With these data and biogeographic classifications, the natural and anthropogenic factors that set and change the distributions of marine species across a range of spatial and temporal scales can be studied at the ocean-basin scale. Such coarse classifications are a useful starting point from which to track the movement of the range edges of species that are located at the edges of these biogeographic provinces by the comparison of current distributions to historical locations. Such a large-scale approach, however, does not capture the patterns of species distributions, including the latitudinal variation in range boundaries across marine taxa, the smaller scale variation between populations within the biogeographic range, the biological processes shaping the range edges and the overlaps in distributions of species originating in neighbouring but differing biogeographic provinces that are known to occur. Such data are essential in order to accurately forecast future changes to the biogeographic distributions of marine species arising from pervasive anthropogenic climate change against the background of long-term natural evolution and changes to marine ecosystems.

4. PATTERNS OF CHANGE ACROSS THE BOREAL–LUSITANIAN BIOGEOGRAPHIC BREAKPOINT IN THE NORTHEAST ATLANTIC

A wealth of small-scale, species or taxa-specific studies date back through the 1900s and late 1800s for intertidal and shallow subtidal benthic systems and species around the English and French coastlines, with occasional studies along the Atlantic coastlines of Spain and Portugal. Many surveys in England and Wales were completed by a small group of researchers at the Marine Biological Association in Plymouth, Robin Hood's Bay Marine Laboratory and collaborative universities (Boalch, 1987; Crisp and Southward, 1958; Lewis, 1986; Russell, 1973; Southward, 1967) in the early to mid-1900s (Southward et al., 2004a, for review).

French surveys were predominantly carried out by researchers based at, or with links to the Station Biologique de Roscoff, which was founded in 1877. These field campaigns were predominantly focused on the English Channel and Brittany region (Fig. 1; e.g. Ancellin et al., 1969; Audouin and Milne-Edwards, 1832; Beauchamp, 1914; Crisp and Fisher-Piette, 1959; Davy de Virville, 1940; Fischer-Piette, 1932, 1934, 1936; Fischer-Piette and Gaillard, 1956; Gaillard, 1965; Plessis, 1961; Prenant, 1927). Most were isolated studies with restricted geographical or taxonomic focus, different recording methodologies and lack of cross-calibration between surveyors working on the same species or families. The lack of integration

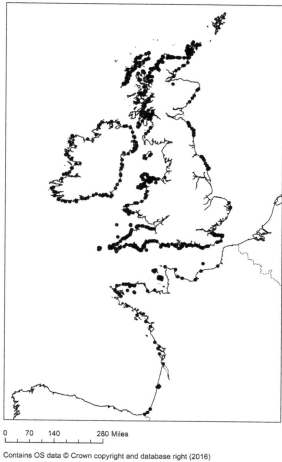

Fig. 1 Map of the UK and France showing the MarClim long-term survey locations.

or use of common methodologies between these studies prevents the quantitative assessments of range shifts, and metaanalyses of community-level changes difficult to undertake due to the lack of standard data collection techniques.

Some continuity between surveyors exists over this time period, with Fischer-Piétte completing broadscale surveys of rocky intertidal species along the coastlines of France, Spain and Portugal in the 1950s (Fischer-Piette, 1953, 1955, 1958; Fischer-Piétte, 1963), and a survey of the French Atlantic coastline with Crisp (Crisp and Fisher-Piette, 1959), who also carried out the UK surveys with Southward (Crisp and Southward, 1958; Orton, 1920; Southward and Crisp, 1954b). The extensive time-series observations for coastal systems in the northeast Atlantic dating back several decades has facilitated research into the detection and quantification of species range shifts both at the Boreal/Lusitanian transition zone bisecting the UK and the Atlantic coastline of Europe (Lima et al., 2007a,b; Mieszkowska et al., 2006b, 2014b).

Multidecadal-scale shifts in species distributions, and the resultant changes to community composition at benthic coastal survey sites within the region of the Boreal–Lusitanean biogeographic marine boundary in the Northeast Atlantic have been studied using these comprehensive datasets with extensive geographic coverage spanning over four degrees of latitude for key structural and functional species of rocky intertidal invertebrates and macroalgae.

For the Galician coast, species distributions for kelps were compiled from a range of studies spanning the 1920s to 2000s and along the Portuguese coast, observational data (presence/absence) were available for kelps and trends assessed by comparing literature records reporting the presence of the species in the 1960s (Ardré, 1970) and observational data in 2008/2010 (Assis et al., 2009, 2013).

Differences between distributions of animals on the north and south coasts of the channel are less conspicuous than those between the east and west basins. There is a clearly marked east-to-west trend in the distributions of many of the animals and plants which has been described. In nearly all such species, not only is the general trend from east to west the same, but also the range and even the detailed features show that a remarkable similarity exists between the British and French coasts.

The differences that exist are due to the presence of more southern forms on the French coast (Crisp and Southward, 1958). In the 1950s, *Gibbula umbilicalis* and *Littorina neritoides* had a greater range eastward on the French side: and *Bifurcaria tuberculata*, *Laminaria ochroleuca* and *Gibbula magus* were generally more abundant in France. Several species were found exclusively or almost exclusively on the French side, such as *Gibbula pennanti*, *Haliotis tuberculata*, *Paracentrotus lividus*, and also species not included in the present survey such as *Pollicipes cornucopia* (now *Pollicipes pollicipes*) and *Pachygrapsus marmoratus*, which was at one time present in the Roscoff area. The only species which appeared to be more abundant on the British coast was *Chthamalus stellatus* (Crisp and Southward, 1958).

The distribution and density/abundance of individual native kelp species are declining in southern European areas (Northwestern Iberian Peninsula, Gulf of Biscay and Mediterranean Sea), with the exception of the Southern Iberian Peninsula where no trends were observed (Araújo, pers. comm.). French experts reported a decreasing trend or stability of *Laminaria hyperborea* beds in Brittany and a general decrease in *Laminaria digitata* and *Saccharina latissima* in Northern France (eastern English Channel and Dover Strait), even if some areas were characterized by a relative stable kelp distribution, such as Iroise/Ushant Sea and North Brittany (Billot et al., 2003). German experts reported an increase in *L. hyperborea* in the southern North Sea and concomitant slight decline of *S. latissima*.

In central Europe, trends of the most abundant kelp species vary according to species identity and geographical area. Kelp abundances around the UK have shown a shift towards increasing abundance of Lusitanian species including *Sacchoriza polyschides* and *Laminaria ochroleucha* in recent decades in response to warming of the marine climate (Brodie et al., 2014; Mieszkowska et al., 2006b; Smale et al., 2015). Boreal kelps have not shown a significant decline in abundance to date, however, they are predicted to decline in abundance and undergo range retractions with continued climate change (Brodie et al., 2014). A metaanalysis of intertidal brown macroalgae around the UK coastline found regional differences in abundance trends, with declines in the southern region, but no change or increases in central and northern regions of the UK (Yesson et al., 2015).

Around the coast of France, kelp abundances vary independently of the latitude. Brittany constitutes a mosaic of contrasting conditions, with the western and northwestern regions being colder and less affected by climate change than the other three regions (Derrien-Courtel et al., 2013; Gallon et al., 2014). Signs of maladaptive response (alteration of meiosis) of *L. digitata* at its southern edge of its distribution (southern Brittany) became apparent where genetic diversity has declined (Oppliger et al., 2014). Such a response means that this European kelp species is at risk of local extinction as predicted by ecological niche models under global change scenarios (Assis et al., 2015; Raybaud et al., 2013).

In the Iberian Peninsula, quantitative data are scarce for most of the species but in the southwest of Portugal and along the Bay of Biscay, a trend of decreasing abundance was verified for *S. polyschides*, *L. hyperborea* and *L. ochroleuca*. These results are in accordance with recent publications reporting on range contractions and/or changes in abundance in recent years, at the southern and eastern distributional ranges of these species (Assis et al., 2013, 2015; Diez et al., 2012; Fernández, 2011; Martinez et al., 2015; Voerman et al., 2013). Global warming was the main driver of changes in kelp abundance identified for this region which is in agreement with recent studies relating the recent retreat in kelp distribution with the global trend of increasing sea surface temperature (Bartsch et al., 2012; Diez et al., 2012; Voerman et al., 2013) and with modelling approaches (Bartsch et al., 2012).

Comparative species distribution modelling found that subtidal red species in the English Channel spanning the biogeographic breakpoint showed differing responses to increased sea temperature between the survey periods 1992–98 and 2010–12. A decrease in the distribution of most species occurred in the three easterly sectors, whereas the western and northwestern Brittany sectors showed more stability and were proposed as potential future refuges for this phylum under future climate change (Gallon et al., 2014). Subtidal algal communities around the Iberian Peninsula and the Canary Islands have also been documented to be changing in composition to favour more Mediterranean species (Martinez et al., 2015).

5. FACTORS SETTING BIOGEOGRAPHIC RANGE LIMITS
5.1 Environmental Conditions

The biogeographical investigations that were initiated in the late 1800s and early 1900s on the species present in studies spanning geographical regions were additionally able to generate a new awareness of the role that the environment played in setting both (in intertidal habitats) vertical and latitudinal boundaries (Appellof, 1912; Couthouy, 1844; Dana, 1890; Orton, 1920; Parr, 1933; Runnström, 1929, 1936; Schmidt, 1909; Setchell, 1893, 1915, 1917, 1920a,b, 1922). Hutchins' seminal paper (Hutchins, 1947) expanded further on this concept, proposing biogeographic provinces across the entire northern hemisphere, based on the distribution of coastal species. He proposed two general theories on the factors setting range limits that are still widely acknowledged today: first, that the distributional range edges for a species will occur where seasonal temperatures become too extreme for the survival of individuals, such as mortality due to cold winter temperature, or an inability to survive hot summer temperatures. Second, critical temperatures for the completion of life cycle stages—such as a trigger to induce spawning, and growing seasons of adequate duration as well as intensity—may set range boundaries via nonlethal, biological, mechanistic responses of individuals experiencing the environmental thermal regime. The effects of temperature acting on physiological mechanisms can be seen on a local scale at the population level, and such changes have been shown contribute to regional-scale variation in species distributions and range limits. Temporary populations can exist beyond range edges, however, they do not become established, either due to rare extremes or insufficient larval supply. Thus, distributions may be limited both towards the poles and towards the equator by either lethal or sublethal thermal tolerances or requirements, which can be related to summer and/or winter seasonal conditions (Hutchins, 1947). The broad pattern observed in species distributions for coastal marine systems along the latitudinal extent of European rocky shores was observed to be related to environmental temperature, but also to environmental gradients in physical factors such as water coverage, wave exposure, air and sea temperature and irradiance (Connell, 1975).

Temperature is a primary driver setting the distribution of marine species globally (Atkinson et al., 1987; Cain, 1944; Parmesan and Yohe, 2003; Stenseth et al., 2002). Climate is a pervasive influence at all levels of organization in biotic systems because of temperature-dependent processes from enzyme reactions through to ecosystems, and the locations of range limits for marine species have been linked to the boundaries between adjacent marine biogeographic provinces via seasonal mean sea surface temperature isotherms (van den Hoek, 1982).

5.2 Biological Processes

In addition to the geographical factors that set biogeographic ranges, all species are also subject to controls on their distribution by the biological processes that govern an organism's ability to tolerate environmental and physical conditions. Survival, reproduction and dispersal are fundamental biological processes determining where an organism and thus populations will become established and maintained. In addition, the tolerance of a species to environmental conditions will have profound effects on the extent and location of the distributional range. Additional biological processes including metabolism and ontogeny processes, and the plasticity of the underlying mechanisms all contribute to the physiological performance of an organism. Geographic distributions are thus a function of a range of morphological and physiological adaptations of individual organisms, characteristics acquired through the process of natural selection.

5.3 Ecological Factors

Further complicating the determination of the driving factors setting species distributions are ecological factors, such as interactions between species. A species invading a new location will encounter other species with which it has never had contact, leading to the potential for competitive, predatory or compensatory interactions. When these adaptations confer a positive effect for an organism, it will survive and have the opportunity to proliferate in this new location, whereas in locations where such adaptations have a negative impact, the distribution will be constrained, ultimately causing a distributional limit or gap in the distributional range.

5.4 Defining the Habitat of a Species

The role of biotic vs abiotic factors in determining the distribution and abundance of marine animals, and whether the importance of the different factors and their significance for population density varied across a species' range, were two important issues that were further studied over the second half of the 20th century by researchers on both sides of the North Atlantic (Andrewartha and Birch, 1954; Crisp and Southward, 1958; Cushing and Dickson, 1976; Edwards et al., 2001; Nicholson, 1954; Southward, 1995; Southward et al., 1988). The habitat of any species defines both the abiotic (environmental and physical) and biotic (biological and ecological) parameters that promote survival of an organism, and their variation across space and time at the biogeographic scale determines the distributional range of a species.

Early into the development of biogeographic studies on the marine environment, researchers focusing on North Atlantic coastal marine ecosystems identified temporal and geographical patterns in the abundances of benthic and pelagic species at sites where repeated surveys were carried out. They also began to discuss environmental 'features' of habitats that were potentially setting geographical distributions for coastal marine species, with sea and air temperature, salinity, nutrients and hydrodynamics that comprised a habitat being proposed as the likely drivers (Crisp and Southward, 1958).

The abundances of cooccurring species of cold-temperate, high-latitude and warm-temperate, mid-latitude species reflected fluctuations between a warm period in the 1950s and a cooler phase across the 1960s and 1970s (Southward et al., 2004a, for review). These changes correlated with natural cycles in oceanic conditions (Cushing and Dickson, 1976; Russell et al., 1971; Southward, 1991) and the understanding of the role of climate, and how changes in environmental temperature could alter species distributions.

Much of this early work was carried out in the UK, providing a wealth of data on temporal fluctuations in population abundances for coastal species in the vicinity of the major marine boundary between Boreal and Lusitanian provinces. Investigations on the distribution of barnacles around the UK coastline in the 1950s prompted the initiation of the first repeated, broad-scale surveys of the abundance and distribution of invertebrates and macroalgae around the coastlines of Ireland and the English and French sides of the English Channel by Crisp and Southward (1958) and Southward and Crisp (1954a). Complementary studies were carried out around the UK by J.R. Lewis and colleagues at the Robin Hood's Bay Laboratory in the northeast of England during the 1980s (Lewis, 1976, 1986; Lewis et al., 1982). The data from these studies were collated, and resurveys started at The Marine Biological Association of the UK from 2001, and form one of the most spatiotemporally extensive intertidal time-series in the world today (Mieszkowska et al., 2006a, 2014b; Southward et al., 2004a).

Van den Hoek (1982) and Breeman (1988) mapped broadscale distributions of macroalgae across the North Atlantic. They noted that high-latitude range edges occurred close to winter sea surface temperature isotherms corresponding to species' thermal minima, and low-latitude limits were located where the highest summer temperatures correlated with upper thermotolerance limits. The distributional limits of most algal species in this oceanographic region seemed to be set by sublethal effects of temperature via the prevention of growth and/or reproduction, rather than by lethal effects. Moreover, these sublethal effects would act on species of colder, Boreal and cold-temperate species at different times to warm-temperate 'warm water' species. The poleward expansion of most warm water species would be limited by the absence of suitable high temperatures during the summer, while the equatorial range extension of coldwater species would be hindered by the effects of warm winters. More recent studies of the status, distribution and abundance of kelps along the seaboards of the Northeastern Atlantic have shown a general trend that supports these theories at the continental scale, with decreasing abundance of some native kelp species at their southern distributional range limits and increasing abundance in northern, higher latitude sectors of their distribution (Araújo et al., 2016).

The temporal frequency of observations for such time-series were mostly seasonal or annual, providing long-term insights into trends or oscillations of marine benthic species at the decadal scale, but being of insufficient temporal resolution to show shorter term and/or small spatial scale events, and natural stochasticity within marine systems. The general theory of the distributions of coastal marine species shifting to higher latitudes in response to chronic warming of the global oceans did not account for extreme, shorter term events such as in 1962/3 in the UK, however, where an unusually cold winter caused the rocky intertidal zone across most of Wales to freeze over, resulting in the contraction of the poleward range limits of many intertidal ectothermic species of Lusitanian biogeographic origins by hundreds of kilometres to lower, warmer latitudes (Crisp, 1964). The range limits did not reextend to similar latitudes until after several decades of warmer climatic conditions and no severe winter seasonal thermal events (Mieszkowska et al., 2006a).

The accumulated knowledge and scientific understanding from these biogeographic scale, field-based studies of the distribution of marine species spanning the 20th century has provided much of the evidence for the movement of species ranges during periods of both cooler, and the current warming of the global climate (Mieszkowska et al., 2014a,b; Southward, 1995). Most of the datasets from sustained observations time-series have been correlated with sea temperature to establish links between ongoing climate change and the geographic patterns of impacts on marine ecosystems or species (Crisp and Fisher-Piette, 1959; Crisp and Southward, 1958; Cushing and Dickson, 1976; Fischer-Piette, 1955, 1958; Forbes, 1858; Southward, 1963, 1967; Southward et al., 2004b). This approach has advanced the knowledge of how benthic marine algae, invertebrates and vertebrates have, and are likely to respond to climate change; however, correlative analyses and models are unable to determine causal drivers or mechanisms underpinning biogeographic shifts. In addition, correlative approaches including species distribution models (SDMs) assume that species are at equilibrium with their environments, and that environmental gradients have been adequately sampled, are thus unable to accurately forecast future changes in species distributions in novel climates that have not been experienced in the history of any extant marine species (Elith and Leathwick, 2009). Correlative SDMs and metaanalyses have been useful for demonstrating broadscale trends in how ecological systems are responding to climate change and are useful frameworks with which to estimate the magnitude of future impacts (Helmuth et al., 2015a).

The environmental factors controlling the abundance and distribution of species have long been of interest to ecologists. Nevertheless, better understanding is required of the biological mechanisms enabling species to survive and the environmental drivers of these processes in light of the current period of global warming (Holbrook et al., 1997; Lodge, 1993; Lubchenco et al., 1993; Mooney, 1991; Mooney et al., 1993). Biogeographical time-series, combined with more recent, targeted, experimental research to determine the biological mechanisms by which species are responding to environmental temperatures have advanced our understanding of which environmental factors, and biological responses set biogeographical range limits for marine species inhabiting the Northeast Atlantic coastal seas. Climate impacts research is undergoing a paradigm shift facilitated by emergent physiological measurement technologies and newly available environmental data at microclimatic scales relevant to individual organisms and populations. With the development of modern experimental, molecular and mathematical techniques in recent decades, the biological processes and mechanisms underpinning species-specific responses to both range limits set by natural environmental conditions and alterations to biogeographic distributions as a direct or indirect result of anthropogenic climate change are now being quantitatively documented, and forecasts of future change statistically modelled (see Section 8 for further information).

6. LONG-TERM TIME-SERIES FOR BENTHIC ECOSYSTEMS IN THE NORTHEAST ATLANTIC AND REGIONAL SEAS

6.1 Marine Biodiversity and Climate Change: MarClim

The value of repeated observations of biological and environmental parameters was first recognized in the mid-1950s with the establishment of broadscale surveys around the coastal habitats of the UK to investigate the movement of invertebrates at their range edges (Lewis, 1976). Southward (1980) stated that '*The need for sustained biological monitoring and development of methods of prediction is obvious*' (Southward, 1980; Southward et al., 1995), really highlighting this need for sustained observations.

Surveys of the rocky intertidal zone conducted around the coastlines of France and England during the 1950s and sporadically throughout the 1960s and 1970s (Crisp and Fisher-Piette, 1959; Crisp and Southward, 1958; Southward and Crisp, 1954a), the 1980s (Hawkins, unpublished data) and 2000–15 (Mieszkowska et al., 2006a, 2014b) have been combined by the MarClim Project, run by the Marine Biological Association of the UK. This combined dataset represents the most spatiotemporally extensive time-series for intertidal systems globally and continues the annual recording of species abundances and distributions at 120 time-series sites to date (Mieszkowska et al., 2014b). The data were collected by trained surveyors, experienced in rocky intertidal taxonomic identification, using the same methodology for all surveys across the decades. Each surveyor trained and cross-calibrated the next generation of surveyors, ensuring comparability between data from Crisp and Southward to Hawkins and Mieszkowska, then Sugden, all of whom were cross-calibrated with at least two of these surveyors in the field.

Species abundances at each site were collected using the categorical ACFOR scale developed by Crisp and Southward (1958), whereby species were assigned a category of either abundant, common, frequent, occasional, rare or not seen (absent) based on the percentage cover for sessile species and the number of individuals per m^2 for mobile species. This methodology has been shown to be an accurate means of assessing both abundance and changes in abundance (Mieszkowska et al., 2006a) and is amenable to analysis using modern statistical methodologies including density-structured dynamic species distribution modelling (Mieszkowska et al., 2014b) and hierarchical ANOVA at nested spatial scales (Burrows et al., 2009).

The data collected to form this time-series are now being utilized to inform both UK and EU policy, most notably as an indicator of Good Environmental Status for the Marine Strategy Framework Directive. It has been analysed to develop indicators for Descriptor 1 (Biological diversity is maintained. Quality and occurrence of habitats and distribution and abundance of species are in line with prevailing physiographic, geographic and climatic conditions) and Descriptor 2 (nonindigenous species introduced by human activities are at levels that do not adversely alter the ecosystems) (Burrows et al., 2014). This reinforces the need for large spatial and temporally extensive datasets not only to ascertain the biological mechanisms driving the patterns of change but also to determine the impacts of anthropogenic change in order that it can be appropriately managed through the creation of fit-for-purpose legislation based on robust scientific techniques.

6.2 European Kelp Forests

Kelp forests along the North Atlantic coastlines of Europe and the UK support complex coastal food webs via the provision of habitat, primary production, detrital resources and carbon export to subtidal soft sediment systems (Araújo et al., 2016; Chung et al., 2013; Krumhansl and Scheibling, 2012). The most extensive sustained observations have been carried out in the intertidal or shallow subtidal coastal habitats of Norway, France and the UK, although the data are limited in their vertical and geographical coverage of kelp forests (Araújo et al., 2016). The available data show that there has been very little shift in dominance between kelp species of cold Boreal, warm Lusitanian and invasive biogeographic origins within communities located in the low eulittoral/infralittoral along the coastal regions from Norway, south to Portugal. In subtidal areas, there has been an almost complete loss of native Boreal, cold water species along the mainland coast of Europe. This has resulted in an epibiotic community 10 times less rich in biodiversity than seen supported by native colder water kelps that provide large complex structures (Brattegard and Holthe, 2001; Breuer and Schramm, 1988; Brodie et al., 2014; Smale et al., 2013). Boreal kelps have not declined in abundance or distribution around the UK coastline with the exception of *Alaria esculenta* (Greville) that has undergone a retraction of the southern range limit in southwest England (Mieszkowska et al., 2006a; Smale et al., 2013). The Lusitanian kelp *S. polyschides* has shown extensions of the northern distributional limit around the UK during the 2000s, however, between 2010 and 2014 the abundance of all Lusitanian kelps declined at MarClim long-term survey sites around the coastlines of England and Wales (Mieszkowska, 2015a,b; Mieszkowska et al., 2014a,b). These time-series data suggest that geographical limits and survivorship are based on a species thermotolerance to extreme winter temperatures.

6.3 North Sea Soft Sediment Benthos

Benthic sediment habitats have been investigated in the North Sea since the 1960s. The data collected at these stations have been used to investigate multidecadal dynamics in the North Sea ecosystem. Patterns have been established between the abundance of zooplankton and the abundance of benthos and linked to these decadal periods of warming and cooling (Frid et al., 2009). Furthermore, these climatic oscillations are thought to drive changes in the benthic assemblages and in some instances extirpation of species in the central North Sea (Robinson and Frid, 2008).

It is proposed that despite this climate forcing within the ecosystem, normal functioning can be maintained over time, with aperiodic disruption under some conditions. Alterations to affect trait composition occurred when taxonomic composition shifted at the same time at both time-series stations in the central North Sea. This event coincided with the North Sea climatic regime shift (Beaugrand, 2004), and therefore implied alteration to benthic functioning may have been driven by broadscale hydroclimatic forcing or the associated changes to the North Sea ecosystem. The community composition of plankton changed around this time in 1987 (Beaugrand, 2004) and detrital flux to the seabed surrounding the stations was estimated to be particularly low during this period (Buchanan, 1993). It is therefore suggested that climatic variability can cause temporary changes to benthic functioning, even with systems exhibiting long-term functional stability (Frid and Caswell, 2015; Neumann and Kroncke, 2011).

Similar patterns of functional change and recovery have been reported for epifauna responding to a cold winter in the German Bight (Neumann and Kroncke, 2011) and macrofauna responding to hypoxia events over a 19-year period at a site in the southwestern Baltic Sea (Gogina et al., 2014). Benthic invertebrate communities have also been shown to exhibit a level of seasonal stability in trait expression, despite taxonomic composition being highly variable on this seasonal timescale (Beche et al., 2006; Munari, 2013); while recent analysis suggests similar patterns emerge over a period of millennia (Caswell and Frid, 2013; Frid and Caswell, 2015). Substitutions of characteristically similar benthic species may therefore occur widely and across multiple temporal scales acting to conserve ecological functioning (Frid and Caswell, 2015).

7. OBSERVED CHANGES IN THE PHYSICAL ENVIRONMENT

Anthropogenic climate change has been altering the temperature of the world's atmosphere and oceans for several decades, with persistent alterations in the global climate recorded since the early 1980s (IPCC, 2014). Records dating back over a century show how the global annual temperature has responded to the release of greenhouse gases into the atmosphere as a result of human activities, with surface temperatures increasing at an average rate of 0.07°C per decade since 1880, and at an average rate of 0.17°C (0.31°F) per decade since 1970 (IPCC, 2014). Detailed analyses of long-term increases in global sea surface temperatures have revealed that this warming has been seasonally consistent across the second half of the 20th and start of the 21st centuries, with increases in temperature most pronounced in the winter months with temperatures increasing on average by 1.5°C over the past 50 years (IPCC, 2014). This warming has caused many sessile and mobile marine species to shift their range limits northwards towards cooler marine environments matching the observed rate of climate change (Mieszkowska et al., 2006a, 2014a). The record warmth in the global oceans has been a major contribution to this increase in global surface temperatures, with the average annual temperature for global ocean surfaces being 0.74°C higher than the 20th century average (NOAA, 2016).

The recognition that this unprecedented rate of warming would affect the survival of life on earth initiated a global research effort at the end of the 20th century to document and track climate change within the world's oceans and associated climatic impacts on the distributions of both marine and terrestrial species. Ecological theory indicated that climate impacts would likely be first manifest at the distributional range limits for many species, where organisms were already living close to thermal tolerance limits latitudes (Helmuth et al., 2006; Hughes, 2000; Parmesan and Yohe, 2003). This would result in retractions of low-latitude range limits in warmer environmental regimes, and extensions of the high-latitude range edges where colder thermal environments were increasing to more thermally tolerable values, causing a polewards shift of biogeographic ranges to higher, cooler latitudes. As these range limits are often located close to major biogeographic boundaries, these regions were proposed in the early days of marine climate impacts research as likely 'hotspots' for changes to marine biodiversity.

New analyses and advanced modelling of marine thermal data show that shifts in the thermal environment have differed across the global oceans since the onset of global warming, with resultant geographically varying rates in the velocity of climate change (the geographic shift of sea surface isotherms over time) (Burrows et al., 2011; http://www.noaa.gov/climate). In addition to these temporal inconsistencies in warming there is high spatial and temporal heterogeneity in the velocity of this change (Lima and Wethey, 2012) due to the shallow mixed layer depth at tropical latitudes vs the deep mixed layer in polar regions (Schneider and Thompson, 1981). Significant changes in regional oceanic temperature can be detected within 10 years in the tropics, but can take an order of magnitude longer to be observed at the poles (Schneider and Thompson, 1981; Washington and Meehl, 1989). The temperate seas of the Northeast Atlantic region have experienced phases of warming (e.g. 1920s to 1950s) and cooling (early 1960s to mid-1980s) throughout the 20th century (Southward, 1980; Southward and Butler, 1972), with a 'hiatus' in warming being evident for the first half of the 2010s (IPCC, 2014). The North Atlantic has shown some of the largest increases in sea surface temperatures globally (Hawkins et al., 2003; Scharf et al., 2004; Southward et al., 2004a; Woehrlings et al., 2005) which is twice the rate of any previous warming event on record (Mann et al., 1998, 1999). In the western English Channel off Plymouth, a 1°C increase in mean sea surface temperature has occurred since 1990. This increase exceeds any other change recorded over the past 100 years (Hawkins et al., 2003) and is most apparent in winter months. The rate of warming cannot be predicted with certainty but models based on medium–high CO_2 emissions scenarios indicate that sea surface temperatures around Britain will increase by between 0.5 and 5°C by 2080 (Hulme et al., 2002).

In addition to increases in global surface temperatures, increased emissions of CO_2 into the global atmosphere as a result of anthropogenic activities are resulting in increased ocean acidification (Gattuso et al., 2015; IPCC, 2014; Schiermeir, 2004; Turley et al., 2004). Rising atmospheric CO_2 is tempered by oceanic uptake as are rising global temperatures. This uptake of CO_2 leads to reductions in the pH of ocean waters and alterations in fundamental chemical balance. The acidification of surface waters is well documented (Feely et al., 2008) and leads to lower $CaCO_3$ saturation (Caldeira and Wickett, 2003). Changes in ocean pH are seen as a threat to marine biota as many utilize $CaCO_3$ is the calcification of hard structures (Orr et al., 2005; Smith and Buddermeier, 1992). Emergent research is demonstrating that the biological effects are far more complex than the limitation of calcification levels alone and can include alterations to immune, growth and reproductive systems (Bibby et al., 2008; Gazeau et al., 2010; Kroeker et al., 2013; Kurihara, 2008). There is high variability in the vulnerability of different species to ocean acidification, with some species showing positive responses to increased pCO_2 (Kroeker et al., 2010). This leads to a complex set of interactions acting on marine organisms whose responses may vary considerably from species to species, however, the biological mechanisms involved, and the extent of impacts at the suborganismal level is still largely unknown, and much further research is required to accurately determine the biological responses of marine species to ocean acidification.

Temperature is a pervasive influence at all levels of organization in biotic systems from the temperature-dependent enzyme reactions through to the structure and functioning of ecosystems (Atkinson et al., 1987; Cain, 1944; Parmesan and Yohe, 2003; Stenseth, 2008). Concerns regarding the responses of natural systems changes in environmental temperature due to climate change in recent decades have prompted a vast research effort to assess and understand the type and extent of responses that have already occurred, and to develop the ability to predict future changes as global warming continues. Marine species have been shown to track the local climate velocities, demonstrating the need for a thorough understanding of both the spatiotemporal variation in climate change, and the individual, population and species-level responses to such change (Pinsky et al., 2013).

Climate change may affect species via such long-term, low-amplitude directional shifts in temperature (Russell, 1973; Southward, 1963, 1991), but also by acute, short-term events including changes in temperature, droughts and flooding which have spatially extensive impacts that last for decades (Bailey, 1955; Crisp, 1964; Hawkins and Holyoak, 1998; Huntsman, 1946; Mattson and Haack, 1987; Smale and Wernberg, 2013).

Extensive disruptions of most terrestrial (Parmesan and Yohe, 2003; Peters and Lovejoy, 1992; Walther et al., 2002) and marine (Fields et al., 1993; Graus and MacIntyre, 1998; Hoegh-Guldberg, 1999; Holbrook et al., 1997; Peters and Lovejoy, 1992; Precht and Aronson, 2004; Ray et al., 1992; Schneider, 1993; Vitousek, 1994; Vitousek et al., 1996) ecological assemblages are expected to continue during the 21st century as species are continually forced to move, adapt or suffer extirpation (Holt, 1990) in response to the unprecedented levels of global climate change predicted by global climate models (Houghton et al., 1995, 2001).

8. IMPACTS OF CLIMATE CHANGE ON INTERTIDAL BENTHIC SPECIES

Intertidal ecosystems are one of the most extensively studied marine habitats due to their easy access, and diurnal exposure to surveyors due to the tidal cycle. The ecology and biology of the species and communities within these habitats are also well understood, and underlie much of the existing knowledge and theories that are used across ecological research and teaching today. They are familiar systems to scientific, governmental and lay audiences and provide important insights into the effects of climate change that can be disseminated to nonspecialist audiences and decision makers around the globe.

Intertidal ecosystems differ from other marine environments, however, because they exhibit enormous thermal variability in time and space, with most of the variation occurring within and because of diurnal and tidal cycles

(Helmuth et al., 2002). Consequently, average temperature estimates, even at the very small spatial scale of an individual habitat, do not accurately encapsulate the habitat thermal heterogeneity and total range of body temperatures that organisms experience on a regular basis, which can be much higher than the temperature of the surrounding air, even under moderate levels of solar radiation (Helmuth et al., 2002; Kearney et al., 2012, 2013; Marshall et al., 2010). In these spatially and temporally heterogeneous intertidal thermal environments, with multiple daily changes in tidal cover, wind, precipitation and solar radiation, selection associated with temperature extremes is likely to play a major role, and behavioural thermoregulation may constitute a crucial strategy to ameliorating the impact of stressful temperatures (Bonebrake and Deutsch, 2012; Kearney et al., 2009, 2012). The detection of responses of benthic species to climate change, and a comprehensive understanding of the biological mechanisms underpinning shifts in distributional ranges must therefore combine large scale, long-term recording of the abundance and biogeographic ranges with experimental physiological research to populate biophysical models capable of predicting biologically realistic predictions of future responses of species to climate change.

8.1 Biogeographic Range Shifts

Species-specific responses can be highly variable, raising concerns regarding broadscale generalizations and metaanalyses of polewards range shifts. Range shifts are occurring faster in the marine environment than in terrestrial ecosystems (Poloczanska et al., 2013). This likely to be due to a combination of factors including: greater connectivity within the marine system (Menge, 2000; Thorson, 1950), less habitat fragmentation and fewer land-use changes (Pearson and Dawson, 2003), coupled with the shorter lifespan of many marine species, the predominance in many benthic species of sessile adults that are unable to move away from unsuitable conditions (Newell, 1979) and pelagic larval stages with high dispersal potential (Burrows et al., 2011; Gaines and Bertness, 1992; Roughgarden and Feldman, 1975).

A generalized prediction is that biogeographic ranges of species will shift polewards in response to the polewards movement of seasonal isotherms as the global climate warms (Lodge, 1993; Lubchenco et al., 1993). Suitable habitat exists beyond the distributional limits of many species of plants (Grace, 1987; Woodward, 1987), mammals (Andrewartha and Birch, 1954; Graham and Grimm, 1990), birds (Root, 1988) and marine invertebrates

(Kendall and Lewis, 1986; Lewis, 1964) but the unsuitability of environmental conditions currently prevents their colonization and therefore the ranges are assumed to be limited by climate. Alternatively, the range edge may lie some distance inside the 'envelope' of suitable climate space due to local factors such as a lack of suitable habitat, poor dispersal and connectivity of suitable habitat space (Kendall, 1987) or if biological interactions are important in setting distributional limits.

Global 'fingerprints' of climate change do indeed show coherent patterns of range shifts (Root et al., 2003) at an average rate for 857 species of 72.0 ± 13.5 km per decade of the leading range edge and 15.4 ± 8.7 km of the trailing range edge towards the poles (Poloczanska et al., 2013), illustrating that the impacts of global warming are already apparent. The rate of change to intertidal species and systems is greater than for terrestrial systems (Cheung et al., 2009; Mieszkowska et al., 2014a,b; Poloczanska et al., 2013), however, the response to a global rise in temperature can be spatiotemporally variable and species specific, highlighting the problems associated with generalizations for predicting the pattern of movement within a given species range limit.

8.2 Changes in Population Dynamics

Temporal changes in species abundances within populations have also been recorded for coastal birds (Lusk et al., 2001; Thompson and Ollason, 2001; Veit, 1997; Veit et al., 1996), mammals (Stenseth et al., 1999), zooplankton (Roemmich and MacGowan, 1995), intertidal invertebrates (Barry et al., 1995; Murray et al., 2001; Sagarin et al., 1999; Southward et al., 1995), fish (Beaugrand et al., 2003; Genner et al., 2004; McFarlane et al., 2000) and corals (Sheppard, 2003). Increasing numbers of species from warm climatic regions are expected to replace those with colder climate affinities, leading to alterations in the composition of local assemblages (Barry et al., 1995; Holbrook et al., 1997; McGowan et al., 1996; Sagarin et al., 1999; Southward et al., 1995). These local scale changes will also facilitate the poleward spread of species by altering the ratio of extinction to colonization events within range edge populations (Parmesan et al., 1999). There is a general pattern evident across a wide range of marine and terrestrial taxa of the highest population densities occurring at the centre of distribution of a species, with abundances decreasing towards the range edges (Brown, 1984). A positive relationship between abundance and geographic distribution has been identified for many coastal marine species; however, the spatial and temporal coverage of past data have often made it difficult to resolve whether increases or decreases in species abundance represent actual changes in biogeography, or merely fluctuating population dynamics within a species range, raising questions as to the general validity of this hypothesis for marine benthic systems (Barry et al., 1995; Murray et al., 2001; Parmesan et al., 2005; Veit, 1997). There are also several exceptions to this pattern in marine species, including kelps, with populations located in the centre of the biogeographic distribution being as vulnerable to climate change as populations at the range edges (Bennett et al., 2015; Wernberg et al., 2016).

The goals of current research into biological processes by which species respond to climate change are to unravel the mechanisms by which each species is responding to multiple stressors, and to apply this knowledge to modelling past, current and future responses of benthic species to climate change and ocean acidification within the marine environment.

8.3 Biological Mechanisms

The possibility of differential rates and extents of climate warming between local and regional scales may also result in variations in the biological responses (Schneider and Thompson, 1981; Washington and Meehl, 1989). In order to accurately predict the rate and extent of future biogeographic shifts in species distributions, the biological mechanisms driving these changes need to be better understood. Physical, ecological, evolutionary and physiological factors acting on the processes of reproduction, birth, dispersal, recruitment and mortality are all involved in shaping species' ranges (Brown et al., 1996; Carter and Prince, 1981; Lennon et al., 1997; Lodge, 1993) and must also be considered when studying the effects of a changing environment. These processes operate predominantly at the local scale, and the effects of environmental change will be most apparent at the population and metapopulation level (McCarty, 2001). Close to the poleward distributional limits of a species, populations will be exposed to environmental temperatures (i.e. cold) which approach their thermal tolerance limits more often than at locations in the centre of the range (Bertness et al., 1999; Fields et al., 1993; Wethey, 1984). Physiological tolerances of individuals to adverse conditions will determine individual survival and maintenance of the local population, and may shape the range limits as a direct result (Bauer, 1992; Lewis, 1986; Lewis et al., 1982; Newell, 1979). In addition, range limits may also be affected by the timing of low tides within intertidal environments, as low water springs occurring during the centre of the day will expose species to higher temperatures, whereas night-time low water spring tides will result in exposure to cold stress at the high latitude, leading range edges for intertidal species (Helmuth et al., 2002; Mieszkowska et al., 2007).

The role of climate signals in the timing of phenological events in terrestrial plant and animal species is already well understood (Walther et al., 2002). Similar changes are being documented in marine systems, with peak abundance of phytoplankton and zooplankton in the North Sea (Edwards and Richardson, 2004), prespawning migration events in squid (Sims et al., 2001), spawning in fish (Sims et al., 2004) and intertidal limpets (Moore et al., 2011) occurring earlier each year in the 2000s than in the 1980s. Nearly all phenophases show strong correlations with spring temperature with a 1-month time lag, indicating that these shifts are being driven by climate warming (Walther et al., 2002). The variation in the timing of reproductive events suggests that many responses may be species specific, and may lead to changes in timing of life stages in relation to the availability of food sources and alterations in competitive, predatory and mutualistic interactions between species (Bertness et al., 1999; Davies et al., 1998).

Critical temperatures are directly connected with the process of reproduction at the larval stages and in those species that metamorphose to sessile benthic stages (Hutchins, 1947; Orton, 1920; Thorson, 1946). Adaptations to heat tolerance have been evidenced across geographic locations or associated with the transitions from one habitat to another, through the role of biological mechanisms and behavioural thermoregulation ameliorating the impacts of extreme temperatures. Despite this, the vulnerability to continued warming will always differ across species, habitats and localities (Marshall et al., 2013) where some species will suffer negatively (Bartsch et al., 2013).

Cold water species are thought to be more resilient than warm water species and are better able to cope with longer periods of adverse conditions (Helmuth et al., 2006; Mieszkowska et al., 2006a; Southward, 1967; Southward et al., 1995). In these situations, they may grow slower, ceasing their reproductive growth and offspring production, but may still persist in an ecosystem with their reproductive output sufficient to maintain local populations if conditions become amenable for them to do so. This phenomenon was described by Southward (1967) and Southward et al. (1975, 1995), who observed southward range expansions of cold water intertidal barnacles associated with cooling SST pulses, followed by warmer periods during which these populations were able to maintain their extended ranges. It is known that warm water species generally grow faster and reproduce earlier during their shorter life extent compared with cold water congeners (Lewis, 1986). Thus, it is reasonable to expect that geographical responses to climate warming will be noticed earlier for warm water species, because their life cycle characteristics confer them a more opportunistic character. Also, in a warming scenario, the new environmental conditions will allow southern species to expand their range northwards, thus releasing them from competitor and predator pressures, although potentially exposing them to new competitive or predatory species interactions. Conversely, native cold water species still have to interact with coevolved competitors, predators and diseases in addition to the invaders, which might decrease their fitness (Sax and Brown, 2000).

It has been shown that the distributions of both terrestrial (e.g. Brooker, 2006) and marine organisms (e.g. Connell, 1961; Wethey, 2002) are conditioned not only by the physical environment but also by biological interactions, and that the responses to the environment can themselves be affected by biotic factors like competition and predation (Richardson and Schoeman, 2004; Sanford, 2002). An increase in competition between cold- and warm water species is expected if warm water species are shifting polewards while cold water species are not moving or only moving by chance both southwards and northwards (Lima et al., 2007a).

9. FUTURE ADVANCES IN QUANTIFYING AND MODELLING DISTRIBUTIONAL RESPONSES TO CLIMATE CHANGE

9.1 Standardizing the Recording and Availability of Data on Species

There are still few coordinated, standardized monitoring programmes for intertidal and subtidal rocky and soft sediment habitats and ecosystems that span multiple countries across the seas of the Northeast Atlantic region, resulting in limited monitoring and data for species distributions at local, regional or national scales (Mieszkowska et al., 2006b, 2014b; Raybaud et al., 2013). Data and knowledge on the current distribution, temporal trends and important drivers, is fragmented and often outdated, with differing methodologies across countries, habitats and faunal groups, creating difficulties when attempting temporal assessments of biodiversity at the national and continental scales (Hummel et al., 2015; Steneck et al., 2002; Wells et al., 2007).

Several multinational organizations have established networks, online information resources and databases to combine the available data and promote collective work at the international level to address the problem of conserving biodiversity in the face of global challenges such as climate change. At the global scale, the International Union for Conservation of Nature (IUCN), the world's oldest and largest global environmental organization, is a neutral forum for governments, NGOs, scientists, business and local communities to find practical solutions to conservation and development challenges. At the regional level, organizations include the OSPAR Commission, comprised of member states on the western coasts and catchments of Europe, together with the European Union (OSPAR, 2016). OSPAR has been allocated the role of the competent regional organization guiding international cooperation on the protection of the marine environment of the North-East Atlantic. The role of OSPAR is to 'harmonize policies and strategies, including the drawing up of programmes and measures, for the protection of the marine environment' and to undertakes and publishes at regular intervals joint assessments of the quality status of the marine environment and of the effectiveness of the measures taken and planned. The OSPAR North-East Atlantic Environment Strategy is taking forward work related to the implementation of the Ecosystem Approach to partially address the data issues at the species-specific level. In addition, there are now organizations and resources such as the extant data repository EurOBIS (EurOBIS, 2016), that are providing a central point to search for data and the UK-run MEDIN data partnership for marine observing systems (MEDIN, 2016) and previous datasets such as the Marbef funded LargeNet project (LargeNet, 2016).

Assessments from these data repositories and wider searches of the academic and governmental publications and databases show that large-scale spatial trends for many marine species in Europe are difficult to identify as a result of one or more reasons: (i) the lack of available long-term quantitative datasets in large parts of the geographical distributions across part, or all of Europe; and (ii) the occurrence of small-scale spatial variability due to either natural or anthropogenic forcing factors, with some species increasing in parts of their geographical distribution but decreasing in other areas, in some cases occurring only a few kilometres apart from each other. Additionally, contrasting trends for the same species have documented at different depths or due to local small-scale variations, e.g. wave exposure, even within the same habitat or coastline area.

9.2 Developing Scientific Methodologies for Quantifying Previous, and Modelling Future Changes in Species Distributions in Response to Climate Change

Understanding and predicting current and future responses to global climate change is one of the greatest challenges facing science today. This chapter has already dealt with the developments and issues facing the collection of broadscale and long-term field data on species distributions, and touched on the need for new, biologically mechanistic information that is also required to accurately attribute cause and effect within the complex problem of determining species responses to global climate change.

Forecasts of future biological responses to climate change are often required to incorporate the interaction of multiple climatic and nonclimatic stressors at far smaller spatiotemporal scales than provided by international climate models and scenarios, in order to be of use to today's scientists, policymakers and managers. The desire for generalizations in climatic change and biological responses has meant that often, scientific predictions of ecological responses to climate change, and the design of experiments to understand underlying mechanisms, are too often based on broadscale trends and averages that at a proximate level may have little relevance to the vulnerability of organisms and their ecosystems (Helmuth et al., 2015b).

Most existing models of climate change, field-based recording or monitoring studies and many laboratory experimental systems, are, however, too coarse in resolution, and too simplistic to incorporate data at the microclimate scale. Such small spatiotemporal resolution information on climatic data, however, is required to provide realistic environmental changes at the scale at which marine organisms are reacting to changes in the thermal environment, and which will ultimately drive geographic-scale changes in species distribution patterns (chapter "Cross-scale approaches to forecasting biogeographic responses to climate change" by Torassian et al. in this volume). In environments where temperature is a limiting factor for biological processes and organismal survival and reproduction, small differences in microtopography within the habitat mosaic can create strong microclimatic differentials over short distances and allow persistent microclimatic refuges to develop (Coulson et al., 1995).

There is thus an urgent need for the research community to create biologically relevant metrics and models of climate change and climate change impacts, incorporating both the processes by which large-scale climate change trains the smaller scale, but organism relevant weather systems, and information on how organisms, and ultimately ecosystems respond to these climatically driven changes (Helmuth et al., 2015b). Such an approach is likely to offer relevant information on those physical aspects of climate change will be most relevant to monitor and predict, and also increase the ability to communicate the impacts of climate change to governmental, conservation and public sectors engaged in the development and acceptance of climate change policies (Spencer et al., 2012; chapter "Cross-scale approaches to forecasting biogeographic responses to climate change" by Torassian et al. in this volume).

The coastline and shallow marine ecosystems around the British coastline are some of the most extensively studied systems globally, providing some of the longest biological time-series globally on changes in distribution and abundance for the coastal marine species (Mieszkowska et al., 2005, 2014b; Southward et al., 2004a). Less spatiotemporally extensive or continuous survey data exist for specific taxa or habitats along the wider spatial extent of the European coastline, as described in Section 7. The central, online repository of these datasets is advancing the understanding of species distributions, ecosystem structuring and impacts of climate change from the small scale, individual site-based level up to the larger, continental scale.

In addition to these developments in survey and time-series data availability are the development of both climatic data and models at spatiotemporal scales relevant to individual organisms and changes in the thermal habitat over small areas of metres to kilometres (e.g. microclimatic data) (Helmuth et al., 2006; Kearney et al., 2012), and physiological models capable of translating the organismal response to changes in climate up to the biogeographic scale relevant to entire species. These biophysical-ecologically coupled dynamic energy budget models are now being developed for marine species in the Northeast Atlantic region, and appear promising in their predictions of current and future distributions of both commercial and ecologically important marine species (Kearney et al., 2010, 2012; Sarà et al., 2011, 2014). This trait-based bioenergetic mechanistic approach for predictions of life history traits in marine organisms show great promise for addressing the problems currently faced by scientists, conservationists, reserve managers and policymakers when attempting to determine the spatiotemporal effects of global climate change, and consolidates the various scientific disciplines required to record, test and forecast climate-driven changes to marine species.

Long-term observations enable the prediction and exploration of future scenarios via experimental testing in the field and laboratory as well as the creation of correlative predictive models of future locations of biogeographic distributions for species. Sustained observations are essential to disentangle natural long-term variability from anthropogenic drivers of change. In addition, targeted physiological studies to determine and quantify biological mechanisms underpinning the responses of marine benthic species to climate change and ocean acidification will provide a new understanding of how individual organisms respond to changes in their thermal environment, and quantify performance across a gradient of sea and air temperatures within coastal ecosystems. These mechanistic data can be input into biomechanistic models, capable of producing biologically relevant measures of organismal performance when exposed to specific thermal regimes at any geographic location across the biogeographic range of a species. This integrated research concept provides new and innovative multidisciplinary approach to comprehensively track and understand species responses to current and future global climate change.

REFERENCES

Ancellin, J., Gall, P.L., Texier, C., Vilquin, A., Vilquin, C., 1969. Observations sur la distribution de la fauna et la flora dans la zone de balancement des marees le long du littoral du nordouest du Cotentin. Memoires de la Societe Nationale des Sciences Naturelles et Mathematiques de Cherbourg, 52, pp. 139–199.

Andrewartha, H.G., Birch, L.C., 1954. The Distribution and Abundance of Animals. University of Chicago Press, Chicago, IL.

Appellof, A., 1912. Invertebrate bottom fauna of the Norwegian Sea and North Atlantic. In: Murray, J., Hjort, J. (Eds.), The Depths of the Ocean. Macmillan, London, pp. 457–560.

Araújo, R.M., Assis, J., Aguillar, R., Airoldi, L., Bárbara, I., Bartsch, I., Bekkby, T., Christie, H., Davoult, D., Derrien-Courtel, S., Fernandez, C., 2016. Status, trends and drivers of kelp forests in Europe: an expert assessment. Biodivers. Conserv. 25, 1–30.

Ardré, F., 1970. Contribution à l'étude des algues marines du Portugal. I. La flore. Portugaliae acta biologica. Serie B, Sistematica, ecologia, biogeografia e paleontologia, p. 423.

Assis, J., Tavares, D., Tavares, J., Cunha, A., Alberto, F., Serrao, E.A., 2009. Findkelp, a GIS-based community participation project to assess Portuguese kelp conservation status. J. Coast. Res. 3, 1469–1473.

Assis, J., Coelho, N.C., Alberto, F., Valero, M., Raimondi, P., Reed, D., Serrao, E.A., 2013. High and distinct rangeedge genetic diversity despite local bottlenecks. PLoS One 8, e68646.

Assis, J., Zupan, M., Nicastro, K.R., Zardi, G.I., McQuaid, C.D., Serrao, E.A., 2015. Oceanographic conditions limit the spread of a marine invader along Southern African shores. PLoS One 10, e0128124.

Atkinson, T.C., Briffa, K.R., Coope, G.R., 1987. Seasonal temperatures in Britain during the past 22,000 years, reconstructed using beetle remains. Nature 325, 587–592.

Attrill, M.J., Power, M., 2002. Climate influence on a marine fish assemblage. Nature 417, 275–278.

Audouin, J.V., Milne-Edwards, H., 1832. Recherches pour servir l'histoire naturelle du littoral de la France. Crochard, libraire, Paris.

Bailey, R.M., 1955. Differential mortality from high temperature in a mixed population of fishes in southern Michigan. Ecology 36, 526–528.

Barry, J.P., Baxter, C.H., Sagarin, R.D., Gilman, S.E., 1995. Climate-related, long-term faunal changes in a California rocky intertidal community. Science 267, 672–674.

Bartsch, I., Wiencke, C., Laepple, T., 2012. Global seaweed biogeography under a changing climate: the prospected effects of temperature. In: Wiencke, C., Bischof, K. (Eds.), Seaweed Biology: Novel Insights into Ecophysiology, Ecology and Utilization. Springer, Heidelberg.

Bartsch, I., Vogt, J., Pehlke, C., Hanelt, D., 2013. Prevailing sea surface temperatures inhibit summer reproduction of the kelp *Laminaria digitata* at Helgoland (North Sea). J. Phycol. 49, 1061–1071.

Bauer, R.T., 1992. Testing generalisations about latitudinal variation in reproduction and recruitment patterns with sicyoniid and caridean shrimp species. Invertebr. Reprod. Dev. 22, 193–202.

Beauchamp, P., 1914. Les greves de Roscoff. Etude sur la repartition des etres dans la zone de balancement des marees Paris.

Beaugrand, G., 2004. The North Sea regime shift: evidence, causes, mechanisms and consequences. Prog. Oceanogr. 60, 245–262.

Beaugrand, G., Brander, K.M., Lindley, J.A., Souissi, S., Reid, P.C., 2003. Plankton effect on cod recruitment in the North Sea. Nature 426, 661–664.

Beaugrand, G., Edwards, M., Brander, K., Luczak, C., Ibanez, F., 2008. Causes and projection of abrupt climate-driven ecosystem shifts in the North Atlantic. Ecol. Lett. 11, 1157–1168.

Beche, L.A., McElravy, E.P., Resh, V.H., 2006. Long-term seasonal variation in the biological traits of benthic-macroinvertebrates in two Mediterranean-climate streams in California, U.S.A. Freshw. Biol. 51, 56–75.

Bennett, S., Wernberg, T., Bijo, A.J., de Bettignies, T., Campbell, A.H., 2015. Central and rear edge populations can be equally vulnerable to warming. Nat. Commun. 6, 10280.

Bergmann, C., 1847. Uber die verhaltnisse der warmeokonomie der thiere zu ihrer grosse. Gottinger Studien 1, 595–708.

Berlanger, C.L., Jablonski, D., Roy, K., Berke, S.K., Krug, A.Z., Valentine, J.W., 2012. Global environmental predictors of benthic marine biogeographic structure. Proc. Natl. Acad. Sci. 109, 14046–14051.

Bertness, M.D., Leonard, G.H., Levine, J.M., Bruno, J.F., 1999. Climate-driven interactions among rocky intertidal organisms caught between a rock and a hot place. Oecologia 120, 446–450.

Bibby, R., Widdicombe, S., Parry, H., Spicer, J., Pipe, R., 2008. Effects of ocean acidification on the immune response of the blue mussel *Mytilus edulis*. Aquat. Biol. 2 (1), 67–74.

Billot, C., Engel, C.R., Rousvoal, S., Kloareg, B., Valero, M., 2003. Current patterns, habitat discontinuities and population genetic structure: the case of the kelp *Laminaria digitata* in the English Channel. Mar. Ecol. Prog. Ser. 253, 111–121.

Boalch, G.T., 1987. Changes in the phytoplankton of the western English Channel in recent years. Br. Phycol. J. 22, 225–235.

Bonebrake, T.S., Deutsch, C.A., 2012. Climate heterogeneity modulates impacts of warming on tropical insects. Ecology 93, 449–455.

Brattegard, T., Holthe, T., 2001. Distribution of marine, benthic macro-organisms in Norway. A tabulted catelogue. Research Report No. 1997-1.

Breeman, A.M., 1988. Relative importance of temperature and other factors in determining geographical boundaries of seaweeds: experimental and phenological evidence. Helgoländer Meeresun. 42, 199–241.

Breuer, G., Schramm, W., 1988. Changes in macroalgal vegetation in the Kiel bight (Western Baltic Sea) during the past 20 years. Kiel. Meeresforsch. Sonderh. 6, 241–255.
Briggs, J.C., 1974. Marine Zoogeography. McGraw-Hill, New York.
Briggs, J.C., Bowen, B.W., 2013. Marine shelf habitat: biogeography and evolution. J. Biogeogr. 40, 1023–1035.
Brodie, J., Williamson, C.J., Smale, D.A., Kamenos, N.A., Mieszkowska, N., Santos, R., Cunliffe, M., Steinke, M., Yesson, C., Anderson, K.M., Asnaghi, V., Brownlee, C., Burdett, H.L., Burrows, M.T., Collins, S., Donohue, P.J.C., Harvey, B., Foggo, A., Noisette, F., Nunes, J., Ragazzola, F., Raven, J.A., Schmidt, D.N., Suggett, D., Teichberg, M., Hall-Spencer, J.M., 2014. The future of the northeast Atlantic benthic flora in a high CO_2 world. Ecol. Evol. 4, 2787–2798.
Brooker, R.W., 2006. Plant–plant interctions and environmental change. New Phytol. 171, 271–284.
Brown, J.H., 1984. On the relationship between abundance and distribution of species. Am. Nat. 124, 255–279.
Brown, J.H., Stevens, G.C., Kaufman, D.M., 1996. The geographic range: size, shape, boundaries, and internal structure. Annu. Rev. Ecol. Syst. 27, 597–623.
Buchanan, J.B., 1993. Evidence of Benthic pelagic coupling at a station off the Northumberland Coast. J. Exp. Mar. Biol. Ecol. 172, 1–10.
Burrows, M.T., Hawkins, S.J., Southward, A.J., 1992. A comparison of reproduction in co-occurring chthamalid barnacles, Chthamalus stellatus (Poli) and Chthamalus montagui Southward. J. Exp. Mar. Biol. Ecol. 160, 229–249.
Burrows, M.T., Moore, J.J., James, B., 2002. Spatial synchrony of population changes in rocky shore communities in Shetland. Mar. Ecol. Prog. Ser. 240, 39–48.
Burrows, M.T., Harvey, R., Robb, L., Poloczanska, E.S., Mieszkowska, N., Moore, P., Leaper, R., Hawkins, S.J., Benedetti-Cecchi, L., 2009. Spatial scales of variance in abundance of intertidal species: effects of region, dispersal mode, and trophic level. Ecology 90, 1242–1254.
Burrows, M.T., Schoeman, D.S., Buckley, L.B., Moore, P., Poloczanska, E.S., Brander, K.M., Brown, C., Bruno, J.F., Duarte, C.M., Halpern, B.S., Holding, J., 2011. The pace of shifting climate in marine and terrestrial ecosystems. Science 334, 652–655.
Burrows, M.T., Mieszkowska, N., Hawkins, S.J., 2014. Marine strategy framework directive indicators for UK rocky shores. Report 522.
Cain, S.A., 1944. Foundations of Plant Geography. Harper Brothers, New York, London.
Caldeira, K., Wickett, M.E., 2003. Anthropogenic carbon and ocean pH. Nature 425, 365.
Carter, R.N., Prince, S.D., 1981. Epidemic models used to explain biogeographical distribution limits. Nature 293, 644–645.
Caswell, B.A., Frid, C.J.L., 2013. Learning from the past: functional ecology of marine benthos during eight million years of aperiodic hypoxia, lessons from the Late Jurassic. Oikos 122, 1687–1699.
Cheung, W.W.L., Lam, V.W.Y., Sarmiento, J.L., Kearney, K., Watson, R., Pauly, D., 2009. Projecting global marine biodiversity impacts under climate change scenarios. Fish Fish. 10 (3), 235–251.
Chung, I.K., Oak, J.H., Lee, J.A., Shin, J.A., Kim, J.G., Park, K.S., 2013. Installing kelp forests/seaweed beds for mitigation and adaptation against global warming: Korean Project Overview. ICES J. Mar. Sci. 70, 1038–1044.
Connell, J.H., 1961. The influence of interspecific competition and other factors on the distribution of the barnacle Chthamalus stellatus. Ecology 42, 710–723.
Connell, J.H., 1975. Some mechanisms producing structure in natural communities: a model and evidence from field experiments. Ecology and Evolution of Communities, University Press, Cambridge, pp. 460–490.
Conrad, C.C., Hilchey, K.G., 2011. A review of citizen science and community-based environmental monitoring: issues and opportunities. Environ. Monit. Assess. 176, 273–291.

Coulson, S.J., Hodkinson, I.D., Strathdee, A.T., Block, W., Webb, N.R., Bale, J.S., Worland, M.R., 1995. Thermal environments of arctic soil organisms during winter. Arct. Alp. Res. 27, 364–370.

Couthouy, J.P., 1844. Remarks upon the coral formations in the Pacific; with suggestions as to the causes of their absence in the same parallels of latitude on the coast of South America. Boston J. Nat. Hist. 4, 66–105.

Crisp, D.J., 1964. The effects of the severe winter of 1962–63 on marine life in Britain. J. Anim. Ecol. 33, 165–210.

Crisp, D.J., Fisher-Piette, E., 1959. Repartition des principales especes interastidales de la Cote Atlantique Francaise en 1954–1955. Ann. Inst. Oceanogr. Monaco 36, 276–381.

Crisp, D.J., Southward, A.J., 1958. The distribution of intertidal organisms along the coasts of the English Channel. J. Mar. Biol. Assoc. UK 37, 157–208.

Cushing, D.H., Dickson, R.R., 1976. The biological response in the sea to climatic changes. Adv. Mar. Biol. 14, 1–122.

Dana, J.D., 1890. Corals and Coral Islands. Dodd, Mead & Co., New York.

Davies, A.J., Jenkinson, L.S., Lawton, J.H., Shorrocks, B., Wood, S., 1998. Making mistakes when predicting shifts in species range in response to global warming. Nature 391, 783–786.

Davy de Virville, A., 1940. Les zones de vegetation sur le littoral atlantique. Soc. Biogeogr. 7, 257–295.

Derrien-Courtel, S., Gall, A.L., Grall, J., 2013. Regional-scale analysis of subtidal rocky shore community. Helgol. Mar. Res. 67, 697–712.

Dickinson, J.L., Zuckerberg, B., Bonder, D.N., 2010. Citizen Science as an ecological and research tool: challenges and benefits. Annu. Rev. Ecol. Evol. Syst. 41, 149–172.

Diez, J.M., Ibabez, I., Rushing, A.J.M., Mazer, S.J., Crimmins, T.M., Crimmins, M.A., Bertelsen, C.D., Inouye, D.W., 2012. Forecasting phenology: from species variability to community patterns. Ecol. Lett. 15, 545–553.

Edwards, M., Richardson, A.J., 2004. Impact of climate change on marine pelagic phenology and trophic mismatch. Nature 430, 881–884.

Edwards, M., Reid, P., Planque, B., 2001. Long-term and regional variability of pytoplankton biomass in the Northeast Atlantic 1960–1995. ICES J. Mar. Sci. 58, 3–49.

Ekman, S., 1935. Tiergeographie des Meeres. Academic Verlagsges, Leipzig.

Ekman, S., Palmer, E., 1953. Zoogeography of the Sea. Sidgwick & Jackson, London.

Elith, J., Leathwick, J.R., 2009. Species distribution models: ecological explanation and prediction across space and time. Annu. Rev. Ecol. Evol. Syst. 40 (1), 677–697.

EurOBIS, 2016. European Ocean Biogeographic Information System. http://www.eurobis.org/.

Feely, R.A., Sabine, C.L., Lee, K., Berelson, W., Kleypas, J., Fabry, V.J., Millero, F.J., 2008. Impact of anthropogenic CO_2 on $CaCO_3$ system in the ocean. Science 305, 362–366.

Fernández, C., 2011. The retreat of large brown seaweeds on the north coast of Spain: the case of *Saccorhiza polyschides*. Eur. J. Phycol. 46, 352–360.

Fields, P.A., Graham, J.N., Rosenblatt, R.H., Somero, G.N., 1993. Effects of expected global climate change on marine faunas. Trends Ecol. Evol. 8, 361–367.

Fischer-Piette, E., 1932. Repartition des principales especes fixees sur les rochers battus des cotes et des iles de La Manche, de Lannion Fecamp. Ann. Inst. Oceanogr. 12, 105–213.

Fischer-Piette, E., 1934. Sur la distribution verticale des organismes fixes dans la zone de fluctuation des marees. CR Acad. Sci. Paris 198, 1721–1723.

Fischer-Piette, E., 1936. Études sur la biogéographie intercotidale des deux rives de la manche. J. Linn. Soc. London, Zool. 40, 181–272.

Fischer-Piette, E., 1953. Repartition de quelques mollusques intercotidaux communs le long des cotes septentrionales de l'Espagne. J. Conchyliologie 93, 1–39.

Fischer-Piette, E., 1955. Repartition. Le long des cotes septentrionale de l'Espagne. Des principles especes peuplant les roches intercotidaux. Ann. Inst. Oceanogr. Monaco NS 31, 37–124.

Fischer-Piette, E., 1958. Sur l'ecologie intercotidale Ouest-iberique. CR Acad. Sci. Paris 246, 1301–1303.

Fischer-Piétte, E., 1963. La distribution des principaux organisms nord–ibériques en 1954–55. Ann. Inst. Oceanogr. (Paris) 40, 165–311.

Fischer-piette, P., Gaillard, J.M., 1956. Sur l'ecologie de *Gibbula umblicalis* da Costa et *Gibbula pennanti*. Phil. J. Conchy. Paris 96, 115–118.

Forbes, E., 1853. The Natural History of Europe's Seas. John van Vorst, London.

Forbes, E., 1858. The Distribution of Marine Life, Illustrated Chiefly by Fishes and Molluscs and Radiata. William Blackwood and Sons, Edinburgh.

Frid, C.J.L., Caswell, B.A., 2015. Does ecological redundancy maintain functioning of marine benthos on centennial to millennial time scales? Mar. Ecol. 37 (2), 392–410.

Frid, C.L.J., Garwood, P.R., Robinson, L.A., 2009. Observing change in a North Sea benthic system: a 33 year time series. J. Mar. Syst. 77, 227–236.

Gaillard, J.M., 1965. Aspects qualitatifs et quantitatifs de la croissance de la coquille de quelques Mollusques Prosobranches en fonction de la latitude et des conditions ecologiques. Muséum National d'Histoire Naturelle, Paris.

Gaines, S.D., Bertness, M.D., 1992. Dispersal of juveniles and variable recruitment in sessile marine species. Nature 360, 579–580.

Gallon, R.K., Robuchon, M., Leroy, B., Gall, L.L., Valero, M., Feunteun, E., 2014. Twenty years of observed and predicted changes in subtidal red seaweed assemblages along a biogeographical transition zone: inferring potential causes from environmental data. J. Biogeogr. 41, 2293–2306.

Gattuso, J.P., Magnan, A., Bille, R., Cheung, W.W.L., Howes, E.L., Joos, F., Allemand, D., Bopp, L., Cooley, S.R., Eakin, C.M., Hoegh-Guldberg, O., 2015. Contrasting futures for ocean and society from different anthropogenic CO_2 emissions scenarios. Science 349. acc4772.

Gazeau, F., Gattuso, J.-P., Dawber, C., Pronker, A.E., Peene, F., Peene, J., Heip, C.H.R., Middelburg, J.J., 2010. Effect of ocean acidification on the early life stages of the blue mussel *Mytilus edulis*. Biogeosciences 7 (7), 2051–2060.

Genner, M.J., Sims, D.W., Wearmouth, V.J., Southall, E.J., Southward, A.J., Henderson, P.A., Hawkins, S.J., 2004. Regional climatic warming drives long-term community changes of British marine fish. Proc. Biol. Sci. 271, 655–661.

Gogina, M., Darr, A., Zettler, M.L., 2014. Approach to assess consequences of hypoxia disturbance events for benthic ecosystem functioning. J. Mar. Syst. 129, 203–213.

Grace, J., 1987. Climatic tolerance and the distribution of plants. New Phytol. 106, 113–130.

Graham, R.W., Grimm, E., 1990. Effects of global climate change on the patterns of terrestrial biological communities. Trends Ecol. Evol. 5, 289–292.

Graus, R.R., MacIntyre, I.G., 1998. Global warming and the future of Caribbean coral reefs. Carbonates Evaporites 13, 43–47.

Hawkins, B.A., Holyoak, M., 1998. Transcontinental crashes of insect populations? Am. Nat. 152, 480–484.

Hawkins, S.J., Southward, A.J., Genner, M.J., 2003. Detection of environmental change in a marine ecosystem—evidence from the western English Channel. Sci. Total Environ. 310, 245–256.

Hawkins, S.J., Sugden, H.E., Moschella, P.S., Mieszkowska, N., Thompson, R.C., Burrows, M.T., 2010. The seashore. In: Maclean, N. (Ed.), Silent Summer: The State of the Wildlife in Britain and Ireland. Cambridge University Press, Cambridge.

Helmuth, B., Harley, C.D.G., Halpin, P., O'Donnell, M., Hofmann, G.E., Blanchette, C., 2002. Climate change and latitudinal patterns of intertidal thermal stress. Science 298, 1015–1017.

Helmuth, B., Broitman, B.R., Blanchette, C.A., Gilman, S., Halpin, P., Harley, C.D.G., O'Donnell, M.J., Hofmann, G.E., Menge, B., Strickland, D., 2006. Mosaic patterns of thermal stress in the rocky intertidal zone: implications for climate change. Ecol. Monogr. 76, 461–479.

Helmuth, B., Russell, B.D., Connell, S.D., Dong, Y., Harley, C.D., Lima, F.P., Sará, G., Williams, G.A., Mieszkowska, N., 2015a. Beyond long-term averages: making biological sense of a rapidly changing world. Clim. Chang. Res. 1, 1.

Helmuth, B.T., Russell, B.D., Connnell, S., Dong, Y., Harley, C.D.G., Lima, F.P., Sarà, G., Williams, G.A., Mieszkowska, N., 2015b. Climate profiling: making biological sense of long term averages in a changing world. Clim. Change Rev. 1 (1), 6.

Hoegh-Guldberg, O., 1999. Climate change, coral bleaching and the future of the world's coral reefs. Mar. Freshw. Res. 50, 839–866.

Holbrook, S.J., Schmitt, R.J., Stevens, J.S.J., 1997. Changes in an assemblage of reef fishes associated with a climate shift. Ecol. Appl. 7, 1299–1310.

Holme, N.A., 1961. The bottom fauna of the English Channel. J. Mar. Biol. Assoc. UK 41 (2), 397–461.

Holt, R.D., 1990. The microevolutionary consequences of climate change. Trends Ecol. Evol. 5, 311–315.

Houghton, J.T., Filho, L.G.M., Callender, B.A., Harris, N., Kattenberg, A., Maskell, K., 1995. Climate Change 1995: The Science of Climate Change. Cambridge University Press, Cambridge.

Houghton, J.T., Ding, Y., Griggs, D.J., Noguer, M., van der Linden, P.J., Dai, X., Maskell, K., Johnson, C.A., 2001. Climate Change 2001: The Scientific Basis. Cambridge University Press, New York.

Hughes, L., 2000. Biological consequences of global warming: is the signal already apparent? Trends Ecol. Evol. 15, 56.

Hulme, M., Jenkins, G.J., Lu, X., Turnpenny, J.R., Mitchell, T.D., Jones, R.G., Lowe, J., Murphy, J.M., Hassell, D., Boorman, P., McDonald, R., Hill, S., 2002. Climate Change Scenarios for the United Kingdom: The UKCIP02 Scientific Report. Tyndall Centre for Climate Change Research, School of Environmental Science, University of East Anglia, Norwich, UK.

Hummel, H., Frost, M., Juanes, J.A., Kochmann, J., Bolde, C.F.C.P., Aneiros, F., Vandenbosch, F., Franco, J.N., Echavarri, B., Guinda, X., Puente, A., 2015. A comparison of the degree of implementation of marine biodiversity indicators by European countries in relation to the Marine Strategy Framework Directive (MSFD). J. Mar. Biol. Assoc. UK 95, 1519–1531.

Huntsman, A.G., 1946. Heat stroke in Canadian maritime stream fishes. J. Fish. Res. Board Can. 6, 476–482.

Hurrell, J.W., 1995. Decadal trends in the North Atlantic Oscillation: regional temperatures and precipitation. Science 269, 676–679.

Hutchins, L.W., 1947. The bases for temperature zonation in geographical distribution. Ecol. Monogr. 17, 325–335.

IPCC, 2014. Climate Change 2014: Synthesis Report. Contribution of Working Groups I, II and III to the Fifth Assessment Report of the Intergovernmental Panel on Climate Change. IPCC, Geneva, Switzerland.

Kafanov, A.I., 2006. Sven Ekman: on the 130th anniversary of his birth. Russ. J. Mar. Biol. 32, 137–139.

Kearney, M., Shine, R., Porter, W.P., 2009. The potential for behavioural thermoregulation to buffer "cold-blooded" animals against climate warming. Proc. Natl. Acad. Sci. U. S. A. 106, 3835–3840.

Kearney, M., Simpson, S.J., Raubenheimer, D., Helmuth, B., 2010. Modelling the ecological niche from functional traits. Philos. Trans. R. Soc. Lond., B, Biol. Sci. 365, 3469–3483.

Kearney, M.R., Matzelle, A., Helmuth, B., 2012. Biomechanics meet the ecological niche: the importance of temporal data resolution. J. Exp. Biol. 215, 922–933.

Kearney, M.R., Simpson, S.J., Raubenheimer, D., Kooijman, S.A., 2013. Balancing heat, water and nutrients under environmental change: a thermodynamic niche framework. Funct. Ecol. 27 (4), 950–966.

Kendall, M.A., 1987. The age and size structure of some Northern populations of the trochid gastropod *Monodonta lineata*. J. Molluscan Stud. 53, 213–222.

Kendall, M.A., Lewis, J.R., 1986. Temporal and spatial patterns in the recruitment of *Gibbula umbilicalis*. Hydrobiologia 142, 15–22.

Kroeker, K.J., Kordas, R.L., Crim, R.N., Singh, G.G., 2010. Meta-analysis reveals negative yet variable effects of ocean acidification on marine organisms. Ecol. Lett. 13 (11), 1419–1434.

Kroeker, K.J., Kordas, R.L., Crim, R., Hendriks, I., Ramajo, L., Singh, G.S., Duarte, C.M., Gattuso, J.P., 2013. Impacts of ocean acidification on marine organisms: quantifying sensitivities and interaction with warming. Glob. Chang. Biol. 19, 1884–1896.

Krumhansl, K., Scheibling, R., 2012. Production and fate of kelp detritus. Mar. Ecol. Prog. Ser. 467, 281–302.

Kurihara, H., 2008. Effects of CO_2-driven ocean acidification on the early developmental stages of invertebrates. Mar. Ecol. Prog. Ser. 373, 275–284.

LargeNet, 2016. Large-scale and long-term networking of observations of global change and its impact on Marine Biodiversity. http://www.marbef.org/projects/largenet/index.php

Lennon, J.J., Turner, J.R.G., Connell, D., 1997. A metapopulation model of species boundaries. Oikos 78, 486–502.

Levitus, S., Antonov, J.I., Boyer, T.P., Stephens, C., 2000. Warming of the world ocean. Science 287 (5461), 2225–2229.

Lewis, J.R., 1956. X.—Intertidal communities of the Northern and Western coasts of Scotland. Trans. R. Soc. Edinb. 63 (1), 185–220.

Lewis, J.R., 1964. The Ecology of Rocky Shores. English Universities Press, London.

Lewis, J.R., 1976. The role of physical and biological factors in the distribution and stability of rocky shore communities. Biology of Benthic Organisms. 11th European Symposium of Marine Biology, Galway, pp. 417–423.

Lewis, J.R., 1986. Latitudinal trends in reproduction, recruitment and population characteristics of some rocky littoral molluscs and cirripedes. Hydrobiologia 142, 1–13.

Lewis, J.R., Bowman, R.S., Kendall, M.A., Williamson, P., 1982. Some geographical components in population dynamics: possibilities and realities in some littoral spp. Neth. J. Sea Res. 16, 18–28.

Lima, F.P., Wethey, D.S., 2012. Three decades of high-resolution coastal sea surface temperatures reveal more than warming. Nat. Commun. 3, 1–13.

Lima, F.P., Ribeiro, P.A., Queiroz, N., Hawkins, S.J., Santos, A.M., 2007a. Do distributional shifts of northern and southern species of algae match the warming pattern? Glob. Chang. Biol. 13, 2592–2604.

Lima, F.P., Ribeiro, P.A., Queiroz, N., Xavier, R., Tarroso, P., Hawkins, S.J., Santos, A.M., 2007b. Modelling past and present geographical distribution of the marine gastropod *Patella rustica* as a tool for exploring responses to environmental change. Glob. Chang. Biol. 13, 2065–2077.

Lodge, D.M., 1993. Species invasions and deletions: community effects and responses to climate and habitat change. In: Karieva, P.M., Kingsolver, J.G., Huey, R.B. (Eds.), Biotic Interactions and Global Change. Sinauer, Sunderland, MA, pp. 367–387.

Lubchenco, J., Navarette, S.A., Tissot, J., Castilla, C., 1993. Possible ecological responses to global climate change: nearshore benthic biota of northeastern Pacific coastal ecosystems. In: Mooney, H.A., Fuentes, E.R., Kronberg, B.I. (Eds.), Earth System Responses to Global Change. Academic Press, New York, pp. 147–166.

Lusk, J.J., Guthery, F.S., Maso, S.J.D., 2001. Northern bobwhite (Colinus virginianus) abundance in relation to yearly weather and long-term climate patterns. Ecol. Model. 146, 3–15.

Mann, M.E., Bradley, R.S., Hughes, M.K., 1998. Global-scale temperature patterns and climate forcing over the past six centuries. Nature 392, 779–787.

Mann, M.E., Bradley, R.S., Hughes, M.K., 1999. Northern hemisphere temperatures during the past millennium: inferences, uncertainties, and limitations. Geophys. Res. Lett. 26, 759–762.

Marshall, D.J., McQuaid, C.D., Williams, G.A., 2010. Non-climatic thermal adaptation: implications for species' responses to climate warming. Biol. Lett. 6, 669–673.

Marshall, D.J., Baharuddin, N., McQuaid, C.D., 2013. Behaviour moderates climate warming vulnerability in high-rocky-shore snails: interactions of habitat use, energy consumption and environmental temperature. Mar. Biol. 160, 2525–2530.

Martinez, B., Afonso-Carrillo, J., Anadón, R., Araújo, R., Arenas, F., Arrontes, J., Bárbara, I., Borja, A., Díez, I., Duarte, L., Fernández, C., Tasende, M.G., Gorostiaga, J.M., Juanes, J.A., Peteiro, C., Puente, A., Rico, J.M., Sangil, C., Sansón, M., Tuya, F., Viejo, R.M., 2015. Regresión de las algas marinas en la costa atlántica de la Península Ibérica y en las Islas Canarias por efecto del cambio climático. ALGAS, Boletín Informativo Sociedad Española Ficología 49, 5–12.

Mattson, W.J., Haack, R.A., 1987. The role of drought in outbreaks of plant-eating insects. Bioscience 37, 110–118.

Maury, M., 1855. The Physical Geography of the Sea. Harper & Brothers, New York.

McCarty, J.P., 2001. Ecological consequences of recent climate change. J. Soc. Conserv. Biol. 15, 320–331.

McFarlane, G.A., King, J.R., Beamish, R.J., 2000. Have there been recent changes in climate? Ask the fish. Prog. Oceanogr. 47, 147–169.

McGowan, J.A., Chelton, D.B., Conversi, A., 1996. Plankton patterns, climate, and change in the California Current. CalCOFI 37.

MEDIN, 2016. Marine Environmental Data & Information Network. http://www.oceannet.org/.

Menge, B.A., 2000. Top-down and bottom-up community regulation in marine rocky intertidal habitats. J. Exp. Mar. Biol. Ecol. 250, 257–289.

Mieszkowska, N., 2015a. MarClim Annual Welsh Intertidal Climate Monitoring Survey 2014. Natural Resources Wales 050-MFG-08.

Mieszkowska, N., 2015b. Marine Biodiversity and Climate Change Monitoring in the UK. Final Report to Natural England on the MarClim Annual Survey 2014. Natural England.

Mieszkowska, N., Leaper, R., Moore, P., Kendall, M.A., Burrows, M.T., Lear, D., Poloczanska, E., Hiscock, K., Moschella, P.S., Thompson, R.C., Herbert, R.J., Laffoley, D., Baxter, J., Southward, A.J., Hawkins, S.J., 2005. Assessing and predicting the influence of climatic change using rocky shore biota. J. Mar. Biol. Assoc. UK 20, 701–752.

Mieszkowska, N., Kendall, M.A., Hawkins, S.J., Leaper, R., Williamson, P., Hardman-Mountford, N.J., Southward, A.J., 2006a. Changes in the range of some common rocky shore species in Britain—a response to climate change? Hydrobiologia 555, 241–251.

Mieszkowska, N., Leaper, R., Moore, P., Kendall, M.A., Burrows, M.T., Lear, D., Poloczanska, E., Hiscock, K., Moschella, P.S., Thompson, R.C., Herbert, R.J., Laffoley, D., Baxter, J., Southward, A.J., Hawkins, S.J., 2006b. Marine Biodiversity and Climate Change: Assessing and Predicting the Influence of Climate Change Using Intertidal Rocky Shore Biota. Marine Biological Association of the United Kingdom. ROAME No. F01AA402.

Mieszkowska, N., Hawkins, S.J., Burrows, M.T., Kendall, M.A., 2007. Long-term changes in the geographic distribution and population structures of *Osilinus lineatus* (Gastropoda: Trochidae) in Britain and Ireland. J. Mar. Biol. Assoc. UK 89, 537–545.

Mieszkowska, N., Genner, M.G., Sims, D.W., 2009. Climate change and fishing impacts on Atlantic cod Gadhus morhua (Linneus) in the North Sea. Adv. Mar. Biol. 56, 214–249.

Mieszkowska, N., Burrows, M.T., Pannacciulli, F., Hawkins, S.J., 2014a. Multidecadal signals within co-occuring intertidal barnacles *Semibalanus balanoides* and *Chthamalus* spp. linked to the Atlantic Multidecadal Oscillation. J. Mar. Syst. 133, 70–76.

Mieszkowska, N., Sugden, H., Firth, L., Hawkins, S.J., 2014b. The role of sustained observations in tracking impacts of environmental change on marine biodiversity and ecosystems. Phil. Trans. R. Soc. A 372, 1–13.

Mooney, H.A., 1991. Biological response to climate change: an agenda for research. Ecol. Appl. 1, 112–117.

Mooney, H.A., Fuentes, E., Kronberg, B.I., 1993. Earth System Responses to Global Change: Contrasts Between North and South America. Academic Press, San Diego.

Moore, P.J., Thompson, R.C., Hawkins, S.J., 2011. Phenological changes in intertidal conspecific gastropods in response to climate warming. Glob. Chang. Biol. 17 (2), 709–719.

Munari, C., 2013. Benthic community and biological trait composition in respect to artificial coastal defence structures: a study case in the northern Adriatic Sea. Mar. Environ. Res. 90, 47–54.

Murray, S.N., Goodson, J., Gerrard, A., Luas, T., 2001. Long-term changes in rocky intertidal seaweed populations in urban Southern California. J. Phycol. 37, 37–38.

Neumann, H., Kroncke, I., 2011. The effect of temperature variability on ecological functioning of epifauna in the German Bight. Mar. Ecol. 32, 49–57.

Newell, R.C., 1979. Biology of Intertidal Animals. Marine Ecological Surveys Ltd, Faversham, Kent.

Nicholson, A.J., 1954. An outline of the dynamics of animal populations. Aust. J. Zool. 2, 9–65.

NOAA, 2016. http://www.ncdc.noaa.gov/sotc/global/201513.

Oppliger, L.V., Pv, Dassow, Bouchemousse, S., Robuchon, M., Valero, M., Correa, J.A., Mauger, S., Destombe, C., 2014. Alteration of sexual reproduction and genetic diversity in the kelp species *Laminaria digitata* at the Southern limit of its range. PLoS One 9, e102518.

Orr, J.C., Fabry, V.J., Aumont, O., Bopp, L., Doney, S.C., Feely, R.A., Gnanadesikan, A., Gruber, N., Ishida, A., Joos, F., Key, R.M., Lindsay, K., Maier-Reimer, E., Matear, R., Monfray, P., Mouchet, A., Najjar, R.G., Plattner, G.-K., Rodgers, K.B., Sabine, C.L., Sarmiento, J.L., Schlitzer, R., Slater, R.D., Totterdell, I.J., Weirig, M.-F., Yamanaka, Y., Yool, A., 2005. Anthropogenic ocean acidification over the twenty-first century and its impact on calcifying organisms. Nature 437, 681–686.

Orton, J.H., 1920. Sea-temperature, breeding and distribution in marine animals. J. Mar. Biol. Assoc. UK 12, 339.

OSPAR, 2016. OSPAR Commission Protecting and Conserving the North-East Atlantic and Its Resources. http://www.ospar.org/.

Parmesan, C., Yohe, G., 2003. A globally coherent fingerprint of climate change impacts across natural systems. Nature 421, 37–42.

Parmesan, C., Ryrholm, N., Stefanescu, C., Hill, J.K., Thomas, C.D., Descimon, H., Huntley, B., Kaila, L., Kullberg, J., Tammaru, T., Tennent, W.J., Thomas, J.A., Warren, M., 1999. Poleward shifts in geographical ranges of butterfly species associated with regional warming. Nature 399, 579–583.

Parmesan, C., Gaines, S., Gonzales, L., Kaufman, D.M., Kingsolver, J., Peterson, A.T., Sagarin, R., 2005. Empirical perspective on species borders: from traditional biogeography to global change. Oikos 108, 58–75.

Parr, A.E., 1933. A Geographical–Ecological Analysis of the Seasonal Changes in Temperature Conditions in Shallow Water Along the Atlantic Coast of the United States. Bull. Bingham Oceanogr. Coll., Literary Licensing, LLC.

Pearson, R.G., Dawson, T.P., 2003. Predicting the impacts of climate change on the distribution of species: are bioclimate envelope models useful? Glob. Ecol. Biogeogr. 12, 361–371.

Pesenko, Y.A., 1982. Printsipy i metody kolichestvennogo analiza v faunistichkikh issledovaniyakh (Principles and Methods of Quantitative Analysis in Faunistic Investigations). Nauka, Moskow.

Peters, R.L., Lovejoy, T.E., 1992. Global Warming and Biological Diversity. Yale University Press, New Haven.

Pinsky, M.L., Worm, B., Fogarty, M.J., Sarmiento, J.L., Levin, S.A., 2013. Marine taxa track local climate velocities. Science 341, 1239–1242.

Plessis, Y., 1961. Ecologie de l'estran rocheux: etude des biocenose et recherches experimentales. Ann. Inst. Oceanogr. 5, 410–511.

Poloczanska, E.S., Brown, C.J., Sydeman, W.J., Kiessling, W., Schoeman, D.S., Moore, P.J., Bruno, J.F., Buckley, L.B., Burrows, M.T., Duarte, C.M., 2013. Global imprint of climate change on marine life. Nat. Clim. Chang. 3 (10), 919–925.

Precht, W.F., Aronson, R.B., 2004. Climate flickers and range shifts of reef corals. Front. Ecol. Environ. 2, 307–314.

Prenant, M., 1927. Notes éthologiques sur la faune marine sessile des environs de Roscoff: Spongiaires, Tuniciers, Anthozoaires, associations de la faune fixée. II. les Presses universitaires de France, Paris.

Ray, G.C., Hayden, B.P., Bulger, A.J., McCormick-Ray, M.G., 1992. Effects of global warming on the biodiversity of coastal marine zones. In: Peters, R.L., Lovejoy, T.E. (Eds.), Global Warming and Biological Diversity. Yale University Press, New Haven, CT, pp. 91–104.

Raybaud, V., Beaugrand, G., Goberville, E., Delebecq, G., Destombe, C., Valero, M., Davoult, D., Morin, P., Gevaert, F., 2013. Decline in kelp in west Europe and climate. PLoS One 8, e66044.

Richardson, A., Schoeman, D., 2004. Climate impact on plankton ecosystems in the northeast Atlantic. Science 305, 1609–1612.

Robinson, L.A., Frid, C.L.J., 2008. Historical marine ecology: examining the role of fisheries in changes in North Sea benthos. Ambio 37, 368–371.

Roemmich, D., MacGowan, J., 1995. Climate warming and the decline of zooplankton in the California current. Science 267, 324–326.

Root, T., 1988. Environmental factors associated with avian distributional boundaries. J. Biogeogr. 15, 489–505.

Root, T.L., Price, J.T., Hall, K.R., Schneider, S.H., Rosenzweig, C., Pounds, A., 2003. Fingerprints of global warming on wild animals and plants. Nature 421, 57–60.

Roughgarden, J., Feldman, M., 1975. Species packing and predation pressure. Ecology 56, 489–492.

Roy, H.E., Pocock, M.J.O., Preston, C.D., Roy, D.B., Savage, J., Tweddle, J.C., Robinson, L.D., 2012. Understanding citizen science and environmental monitoring: final report on behalf of UK Environmental Observation Framework. NERC/Centre for Ecology & Hydrology, Wallingford.

Runnström, S., 1929. Weitere Studien über die Temperaturanpassung der Fortpflanzung und Entwicklung mariner Tiere. Grieg.

Runnström, S., 1936. Die Anpassung der Fortpflanzumg und Entwicklung mariner Tiere an die Temperaturverhaltnisse verschiedener verbreitungsgebiete. Bergen Aarbog 3, 1–46.

Russell, F.S., 1973. A summary of the observations on the occurrence of planktonic stages of fish off Plymouth 1924–1952. J. Mar. Biol. Assoc. UK 53, 347–356.

Russell, F.S., Southward, A.J., Boalch, G.T., Butler, E.I., 1971. Changes in biological conditions in the English Channel off Plymouth during the last half century. Nature 234, 468–470.

Sagarin, R.D., Barry, J.P., Gilman, S.E., Baxter, C.H., 1999. Climate-related change in an intertidal community over short and long time scales. Ecol. Monogr. 69, 465–490.

Sanford, E., 2002. Water temperature, predation, and the neglected role of physiological rate effects in Rocky Intertidal communities. Integr. Comp. Biol. 42, 881–891.

Sarà, G., Kearney, M., Helmuth, B., 2011. Combining heat-transfer and energy budget models to predict thermal stress in Mediterranean intertidal mussels. Chem. Ecol. 27, 135–145.

Sarà, G., Rinaldi, A., Montalto, V., 2014. Thinking beyond organism energy use: a trait-based bioenergetic mechanistic approach for predictions of life history traits in marine organisms. Mar. Ecol. 35, 506–515.

Sax, D.F., Brown, J.H., 2000. The paradox of invasion. Glob. Ecol. Biogeogr. 9, 363–371.

Scharf, F.S., Manderson, J.P., Fabrizio, M.C., Pessutti, J.P., Rosendale, J.E., Chant, R.J., Bejda, A.J., 2004. Seasonal and interannual patterns of distribution and diet of bluefish within a middle Atlantic bight estuary in relation to abiotic and biotic factors. Estuar. Coast. Shelf Sci. 27, 426–436.

Schiermeir, Q., 2004. Researchers seek to turn the tide on acid seas. Nature 430, 802.

Schmidt, J., 1909. The distribution of pelagic fry and spawning regions of the gadoids in the North Atlantic from Iceland to Spain.

Schneider, S.H., 1993. Scenarios of global qarming. In: Kareiva, P.M., Kingsolver, J., Huey, R.B. (Eds.), Biotic Interactions and Global Change. Sinauer Associates, Sunderland, MA, pp. 9–23.

Schneider, S.H., Thompson, S.L., 1981. Atmospheric CO_2 and climate—importance of the transient-response. J. Geophys. Res. Oceans Atmos. 86, 3135–3147.

Setchell, W.A., 1893. On the classification and geographical distribution of the Laminariaceae. Trans. Connecticut Acad. Arts Sci. 9, 333–375.

Setchell, W.A., 1915. The law of temperature connected with the distribution of the marine algae. Ann. Mo. Bot. Gard. 2, 287–305.

Setchell, W.A., 1917. Geographical distribution of the marine algae. Science 45, 197–204.

Setchell, W.A., 1920a. Stenothermy and zone-invasion. Am. Nat. 54, 385–397.

Setchell, W.A., 1920b. The temperature interval in the geographical distribution of marine algae. Science 52, 187–190.

Setchell, W.A., 1922. Cape Cod in its relation to the marine flora of New England. Rhodora 24, 1–11.

Sheppard, C.R.C., 2003. Predicted recurrences of mass coral mortality in the Indian Ocean. Nature 425, 294–297.

Simpson, S.D., Jennings, S., Johnson, M.P., Blanchard, J.L., Schön, P.J., Sims, D.W., Genner, M.J., 2011. Continental shelf-wide response of a fish assemblage to rapid warming of the sea. Curr. Biol. 21, 1565–1570.

Sims, D.W., Genner, M.J., Southward, A.J., Hawkins, S.J., 2001. Timing of squid migration reflects North Atlantic climate variability. Proc. R. Soc. Lond. 268, 2607–2611.

Sims, D.W., Wearmouth, V.J., Genner, M.J., Southward, A.J., Hawkins, S.J., 2004. Low-temperature-driven early spawning migration of a temperate marine fish. J. Anim. Ecol. 73, 333–341. http://dx.doi.org/10.1111/j.0021-8790.2004.00810.x.

Smale, D.A., Wernberg, T., 2013. Extreme climatic event drives range contraction of a habitat-forming species. Proc. R. Soc. B 280, 20122829. http://dx.doi.org/10.1098/rspb.2012.2829.
Smale, D.A., Burrows, M.T., Moore, P., O'Connor, N., Hawkins, S.J., 2013. Threats and knowledge gaps for ecosystem services provided by kelp forests: a northeast Atlantic perspective. Ecol. Evol. 3 (11), 4016–4038.
Smale, D.A., Wernberg, T., Yunnie, A.L., Vance, T., 2015. The rise of *Laminaria ochroleuca* in the Western English Channel (UK) and comparisons with its competitor and assemblage dominant *Laminaria hyperborea*. Mar. Ecol. 36 (4), 1033–1044.
Smith, S.V., Buddermeier, R.W., 1992. Global change and coral reef ecosystems. Annu. Rev. Ecol. Syst. 23, 89–118.
Southward, A.J., 1963. The distribution of some plankton animals in the English Channel and approaches. III. Theories about long term biological changes, including fish. J. Mar. Biol. Assoc. UK 43, 1–29.
Southward, A.J., 1967. Recent changes in abundance of intertidal barnacles in south-west England: a possible effect of climatic deterioration. J. Mar. Biol. Assoc. UK 47, 81–95.
Southward, A.J., 1980. The western English Channel—an inconstant ecosystem? Nature 285, 361–366.
Southward, A.J., 1991. Forty years of changes in species composition and population density of barnacles on a rocky shore near Plymouth. J. Mar. Biol. Assoc. UK 71, 495–513.
Southward, A.J., 1995. The importance of long time-series in understanding the variability of natural systems. Helgoländer Meeresun. 45, 329–333.
Southward, A.J., Butler, E.I., 1972. Further changes of sea temperature in Plymouth Area. J. Mar. Biol. Assoc. UK 52, 931–937.
Southward, A.J., Crisp, D.J., 1954a. The distribution of certain intertidal animals around the Irish coast. Proc. R. Ir. Acad. 57, 1–29.
Southward, A.J., Crisp, D.J., 1954b. Recent changes in the distribution of the intertidal barnacles *Chthamalus stellatus* Poli and *Balanus balanoides* L. in the British Isles. J. Anim. Ecol. 23, 163–177.
Southward, A.J., Butler, E.I., Pennycuick, L., 1975. Recent cyclic changes in climate and in abundance of marine life. Nature 253, 714–717.
Southward, A.J., Boalch, G.T., Maddock, L., 1988. Fluctuations in the herring and pilchard fisheries of Devon and Cornwall linked to change in climate since the 16th century. J. Mar. Biol. Assoc. UK 68, 423–445.
Southward, A.J., Hawkins, S.J., Burrows, M.T., 1995. Seventy years' observations of changes in distribution and abundance of zooplankton and intertidal organisms in the western English Channel in relation to rising sea temperature. J. Therm. Biol. 20, 127–155.
Southward, A.J., Langmead, O., Hardman-Mountford, N.J., Aiken, J., Boalch, G.T., Dando, P.R., Genner, M.J., Joint, I., Kendall, M.A., Halliday, N.C., Harris, R.P., Leaper, R., Mieszkowska, N., Pingree, R.D., Richardson, A.J., Sims, D.W., Smith, T., Walne, A.W., Hawkins, S.J., 2004a. Long-Term Oceanographic and Ecological Research in the Western English Channel Advances in Marine Biology. Academic Press, Amsterdam, The Netherlands, pp. 1–105.
Southward, A.J., Langmead, O., Hardman-Mountford, N.J., Aiken, J., Boalch, G.T., Dando, P.R., Genner, M.J., Joint, I., Kendall, M., Halliday, N.C., Harris, R.P., Leaper, R., Mieszkowska, N., Pingree, R.D., Richardson, A.J., Sims, D.W., Smith, T., Walne, A.W., Hawkins, S.J., 2004b. Long-term biological and environmental researches in the western English Channel. Adv. Mar. Biol. 47, 1–105.
Spalding, M.D., Fox, H.E., Allen, G.R., Davidson, N., Ferdaña, Z.A., Finlayson, M.A.X., Halpern, B.S., Jorge, M.A., Lombana, A., Lourie, S.A., Martin, K.D., McManus, E., Molnar, J., Recchia, C.A., Robertson, J., 2007. Marine ecoregions of the world: a bioregionalization of coastal and shelf areas. BioScience 57, 573–583.

Spencer, M., Mieszkowska, N., Robinson, L.A., Simpson, S.D., Burrows, M.T., Birchenough, S.N.R., Capasso, E., Cleall-Harding, P., Crummy, J., Duck, C., Eloire, D., Frost, M., Hall, A.J., Hawkins, S.J., Johns, D.G., Sims, D.W., Smyth, T.J., Frid, C.J., 2012. Region-wide changes in marine ecosystem dynamics: state-space models to distinguish trends from step changes. Glob. Chang. Biol. 18, 1270–1281.

Steneck, R.S., Graham, M.H., Bourque, B.J., Corbett, D., Erlandson, J.M., Estes, J.A., Tegner, M.J., 2002. Kelp forest ecosystems: biodiversity, stability, resilience and future. Environ. Conserv. 29, 436–459.

Stenseth, N.C., 2008. Effects of climate change on marine ecosystems. Climate Res. 37, 121–122.

Stenseth, N.C., Chan, K.S., Tong, H., Boonstra, R., Boutin, S., Krebs, C.J., Post, E., O'Donaghue, M., Yokkoz, N.G., Forchammer, M.C., Hurrell, J.W., 1999. Common dynamic structure of Canada lynx populations within three climatic regions. Science 285, 1071–1073.

Stenseth, N.C., Mysterud, A., Ottersen, G., Hurrell, J.W., Chan, K.S., Lima, M., 2002. Ecological effects of climate fluctuations. Science 297, 1292–1296.

Sugden, H.E., Underwood, A.J., Hawkins, S.J., 2009. The aesthetic value of littoral hard substrata and an ethical framework for appreciation and conservation. In: Wahl, M. (Ed.), Hard-Bottom Communities: Patterns, Scales, Dynamics, Functions, Shifts Ecological Studies. Springer Verlag, Dordrecht, NY.

Thompson, P.M., Ollason, J.C., 2001. Lagged effects of ocean climate change on fulmar population dynamics. Nature 413, 417–420.

Thomson, C.W., Murray, J., 1873–1876. Report on the Scientific Results of the Exploring Voyage of the HMS Challenger 1873–1876. Order of Her Majesty's Government, London.

Thorson, G., 1946. Reproduction and larval development of Danish marine bottom invertebrates. Medd Komm Danmarks Fiskeri-Og Havunders, Serie Plankton 4, 1–529.

Thorson, G., 1950. Reproductive and larval ecology of marine bottom invertebrates. Biol. Rev. 25, 1–45.

Turley, C., Blackford, J., Widdicombe, S., Lowe, D., Gilbert, F., Nightingale, P., 2004. Reviewing the Impact of Increased CO_2 on Oceanic pH and the Marine Ecosystem. Plymouth Marine Laboratory, UK.

van den Hoek, C., 1982. Phytogeographic distribution groups of benthic marine algae in the North Atlantic Ocean. A review of experimental evidence from life history studies. Helgoländer Meeresun. 35, 153–214.

Veit, R.R., 1997. Apex marine predator declines ninety percent in association with changing oceanic climate. Glob. Chang. Biol. 3, 23–38.

Veit, R.R., Pyle, P., McGowan, J.A., 1996. Ocean warming and long-term change in pelagic bird abundance within the California current system. Mar. Ecol. Prog. Ser. 139, 11–18.

Vitousek, P.M., 1994. Beyond global warming: ecology and global change. Ecology 75, 1861–1876.

Vitousek, P.M., D'Antonio, C.M., Loope, L.L., Westbrooks, R., 1996. Biological invasions as global environmental change. Am. Sci. 84, 218–228.

VLIZ, 2009. Longhurst Biogeographical Provinces. http://www.marineregions.org.

Voerman, S.E., Llera, E., Rico, J.M., 2013. Climate driven changes in subtidal kelp forest communities in NW Spain. Mar. Environ. Res. 90, 119–127.

Walther, G., Post, E., Convey, P., Menzel, A., Parmesan, C., Beebee, T.J.C., Fromentin, J., Hoegh-Guldberg, O., Bairlein, F., 2002. Ecological responses to climate change. Nature 416, 389–395.

Washington, W.M., Meehl, G.A., 1989. Climate sensitivity due to increased CO_2: experiments with a coupled atmosphere and ocean general circulation model. Climate Dynam. 4, 1–38.

Wells, E., Wilkinson, M., Wood, P., Scanlan, C., 2007. The use of macroalgal species richness and composition on intertidal rocky seashores in the assessment of ecological quality under the European Water Framework Directive. Mar. Pollut. Bull. 55, 151–161.

Wernberg, T., Russell, B.D., Thomsen, M.S., Frederico, C., Gurgel, D., Bradshaw, C.J.A., Poloczanska, E.S., Connell, S.D., 2011. Seaweed communities in retreat from ocean warming. Curr. Biol. 21, 1828–1832.

Wernberg, T., Bennett, S., Babcock, R.C., de Bettignies, T., Cure, K., Depczynski, M., Dufois, F., Fromont, J., Fulton, C.J., Hovey, R.K., Harvey, E.S., 2016. Climate-driven regime shift of a temperate marine ecosystem. Science 353 (6295), 169–172.

Wethey, D.S., 1984. Sun and shade mediate competition in the barnacles *Chthamalus* and *Semibalanus*: a field experiment. Biol. Bull. 167, 176–185.

Wethey, D.S., 2002. Biogeography, competition, and microclimate: the Barnacle *Chthamalus fragilis* in New England. Integr. Comp. Biol. 42, 872–880.

Woehrlings, D., Lefebvre, A., Fevre-Lehoerff, G.L., Delesmont, R., 2005. Seasonal and longer term trends in sea temperature along the French North Sea coast, 1975 to 2002. J. Mar. Biol. Assoc. UK 85, 39–48.

Woodward, F.I., 1987. Climate and Plant Distribution. Cambridge University Press, Cambridge.

Yesson, C., Bush, L.E., Davies, A.J., Maggs, C.A., Brodie, J., 2015. Large brown seaweeds of the British Isles: evidence of changes in abundance over four decades. Estuar. Coast. Shelf Sci. 155, 167–175.

CHAPTER EIGHT

Cross-Scale Approaches to Forecasting Biogeographic Responses to Climate Change

J.L. Torossian[*,1], R.L. Kordas[†], B. Helmuth[*]
[*]Northeastern University, Marine Science Center, Nahant, MA, United States
[†]Imperial College London, Ascot, Berkshire, United Kingdom
[1]Corresponding author: e-mail address: torossian.j@husky.neu.edu

Contents

1. Introduction	372
1.1 Gazing into the Crystal Ball: How Do We Forecast the Future by Looking at the Present and Past?	372
1.2 Ecological Forecasting and Inconsistent Conceptions of the Ecological Niche	376
1.3 Mechanism or Correlation?	379
2. Common Pitfalls and Their Unintended Consequences	383
2.1 Scale and Data Resolution	384
2.2 Inclusion of Relevant Biological Details	388
2.3 All the World's a Stage: From Autecology to Synecology and Beyond	401
3. Moving Forward: How Do We Make Useful Forecasts While Recognizing Limitations?	407
3.1 Hybrid Models	408
3.2 Physiologically and Trait-Based Indicators	410
3.3 Transparency in Assumptions, Ensemble Forecasting and Tests of Model Skill: Accuracy Is in the Eye of the Stakeholder?	411
4. Conclusions	413
Acknowledgements	414
References	414

Abstract

The emphasis in recent scientific studies has gradually shifted from merely documenting the numerous biological impacts of global climate change to developing predictive tools that help forecast which organisms, ecosystems and locations are most (and least) likely to be affected. This work often focuses on two very different scales of approach: the impacts of environmental change on individual organisms and their physiological vulnerability; and large-scale, biogeographic shifts in patterns of species distributions. While both scales of research are important and potentially informative of one another, with a few key exceptions, biogeographic (large scale) and physiological (small scale) approaches often operate in isolation from one another. There is a general consensus that a better understanding of mechanistic drivers will likely improve our ability to develop a more predictive framework. To this end, experimental and theoretical research has begun to tease apart the complexities of how changes in climate (as reflected in weather) ultimately translate into changes in growth, survival, productivity, species distribution and abundances, and the provision of ecosystem services. Yet, considerable debate still remains over how much detail is required to make effective predictions. Many authors have raised concerns about the implicit assumptions inherent in biological forecasts and while we cannot wait for perfect models before acting, oversimplifications can lead to unintended consequences. Additionally, what makes one forecasting method "good enough" or one approach "better" than another may depend largely on the application to which the predictions are being made. This review explores and summarizes the concerns raised by researchers working on problems at diverse scales and offers suggestions of how cross-scale research can best avoid potential pitfalls while preparing society for the ongoing impacts of climate change.

1. INTRODUCTION

1.1 Gazing into the Crystal Ball: How Do We Forecast the Future by Looking at the Present and Past?

The impacts of global climate change on organisms and ecosystems are now incontrovertible (Pacifici et al., 2015) as are the resulting consequences for human society (Fenichel et al., 2016). Shifts in patterns of distribution, abundance and biodiversity of many taxa are well documented, and monitoring studies have shown that many species are, on average, shifting to higher latitudes and elevations in response to increasing thermal stress (Root et al., 2003; Somero, 2005, 2010; U.S. Environmental Protection Agency, 2008; Wilson et al., 2005). More and more studies, however, are recognizing the inherent spatial and temporal heterogeneity in such responses. Many have emphasized that effective adaptation strategies—which typically occur on local or regional scales—cannot be based solely on simple generalizations and trends such as poleward distributional shifts or a reliance purely on observed correlations between large-scale environmental drivers and ecological response (Evans et al., 2015; Heller et al., 2015; Rapacciuolo et al., 2014; Turner et al., 2012). At the same time, however, there is broad recognition that by waiting to include extensive levels of detail, we may be inadvertently delaying action due to imperfect knowledge of how complex physical and ecological systems operate (García-Callejas and Araújo, 2016;

Yokomizo et al., 2014). Finding the required balance between these two competing demands is of vital importance.

Advances are being made on multiple fronts. Physiologists are teasing apart the complex interactions of multiple stressors acting on organisms (Crain et al., 2008; Firth and Williams, 2009; Gunderson et al., 2016; Jackson et al., 2016; Molinos and Donohue, 2010; Piggott et al., 2015), and exploring the physiological and genetic underpinnings of vulnerability that make some organisms "winners" and others "losers" in the face of rapid environmental change (Loya et al., 2001; Somero, 2010). Ecologists are gaining a better understanding of drivers and consequences of biodiversity loss (Srivastava and Vellend, 2005; Wernberg et al., 2011; Widdicombe and Spicer, 2008), the provision of ecosystem services (Turner et al., 2012) and ecosystem phase shifts (Dudgeon et al., 2010; Möllmann et al., 2015; Selkoe et al., 2015). Biogeographic modellers are becoming increasingly more sophisticated in including biological details and spatial and temporal complexity in drivers (Ackerly et al., 2010; Evans et al., 2013). Palaeontologists are attempting to discern the underlying drivers that have led to large-scale changes in the past, with the goal of placing ongoing changes in context (Lieberman, 2003; McGuire and Davis, 2014). And, climate modellers are developing forecasts that have spatial and temporal resolution more relevant to ecological responses (Kodra and Ganguly, 2014).

In recent years the focus of climate change ecology has thus gradually shifted from one of asking *if* climate change is impacting natural and managed ecosystems to *how, when* and *where* such impacts are most likely to occur (Drake and Griffen, 2010; Helmuth et al., 2014; Selkoe et al., 2015; Tribbia and Moser, 2008; Wernberg et al., 2011). These shifts in scientific focus have gone hand-in-hand with an ever-increasing need for science-based, often spatially and temporally explicit products that can be used in decision making (Halpin, 1997; Howard et al., 2013; Petes et al., 2014; U.S. Environmental Protection Agency, 2008; Yokomizo et al., 2014). Specifically, as policy discussions shift to include not just mitigation of greenhouse gas emissions but also societal adaptation to inevitable global change (Glick et al., 2009, 2011; Janetos et al., 2012; Preston et al., 2011), more emphasis is being placed on how we might use our understanding of biology to forecast future ecological changes (Gunderson et al., 2016; National Academy of Sciences, 2016; Pörtner and Peck, 2010; Somero, 2010; Subbey et al., 2014). By estimating which species, ecosystems and locations are most vulnerable to environmental change (Ackerly et al., 2015; Rapacciuolo et al., 2014; Somero, 2010), and by understanding the likely drivers of ecological tipping points (Lubchenco and Petes, 2010; Monaco and Helmuth, 2011; Selkoe et al., 2015; Veraart et al., 2012), we may be able to prevent some of the more damaging ecological impacts from occurring, by ameliorating nonclimatic stressors, or at a minimum to put in place alternatives to the ecosystem services that will likely be lost (Petchy et al., 2015).

Successful adaptation strategy will require data and knowledge from multiple, often quite disparate fields of inquiry (Root and Schneider, 1995). Perhaps not surprisingly, however, advances in techniques and theoretical frameworks do not always permeate disciplinary boundaries and in some cases appear to be evolving in parallel among scientists working in different habitats or at different levels of biological organization. Making such connections is daunting given the immense body of knowledge being generated, but not doing so comes at the risk of unintended consequences when failing to recognize common pitfalls that may have already been raised by researchers working in other circles. Our intention in this review is to facilitate crosstalk among climate modellers, biogeographers, ecologists, physiologists and resource managers to address these global-scale issues. Drawing lessons from a wide range of habitats and taxa, we explore how forecasts of population and species abundances and distributions are most commonly used, and provide an overview of recent work focusing on the hidden snares and potential sources of error that researchers working broadly in the arena of climate change ecology have uncovered.

Forecasting changes in species distribution and abundance at diverse scales is needed to inform policy and management decisions related to changes in multiple climate drivers. Altered responses of populations, species and communities to climate change can have significant economic and environmental consequences that impact a wide variety of stakeholder needs ranging in scale from the well-being of local (human) communities that rely on natural resources, to the economies and food security status of nations, to the global economy (Defries et al., 2012). The effective prediction of these impacts can aid in their possible prevention, or at least in the deployment of mitigation strategies to reduce their overall societal impact. Prioritization of which species and habitats require the most scrutiny, the temporal and spatial scale of the prediction and the tolerance for error all vary according to the stakeholder. Several researchers have explored the concept of how stakeholder needs affect the acceptability of false positives (predictions of events that do not happen) and false negatives (failure to predict an event; Anderegg et al., 2014; Araújo et al., 2005b; Giuliani and Castelletti, 2016). For example, fisheries managers constantly wrestle with the competing consequences of failing to predict the impending decline of stocks while not placing undue burden on fishers by being overly conservative (e.g. Quataert et al., 2007).

Recent years have thus witnessed a healthy debate over the applications and shortcomings of various modelling approaches (Araújo and Guisan, 2006; Araújo and New, 2007; Araújo et al., 2005a; Hijmans and Graham, 2006; Sieck et al., 2011; Smith, 2002; Woodin et al., 2013). Discussions have focused on both how we might minimize uncertainty regarding the biological and socioeconomic impacts of global change (Estes et al., 2013; Evans et al., 2013; McGregor, 2015; McMahon et al., 2009; Subbey et al., 2014), and, conversely, how we can make effective decisions even in the face of uncertainty (Anderegg et al., 2014; Giuliani and Castelletti, 2016; Tribbia and Moser, 2008; Yokomizo et al., 2014). To a large extent, therefore, an end-users' tolerance and requirements for different forms of error can inform the types and levels of "detail", that need to be included in initial model development, and thus why "off-the-shelf" model forecasts need to be used with extreme caution or at least awareness of assumptions made during their creation. Ability to deal with model inaccuracies may depend on the spatial scale and question to which the model is being applied. Researchers studying changes in global biodiversity may have a much higher tolerance for error in predictions at the level of the individual species, as their concern is for the net effect of the entire portfolio, as compared to conservation biologists attempting to prevent the extinction of a threatened species at one or more regional locations (Dawson et al., 2011). Throughout the review, we point to the most likely practical consequences of including or ignoring different biological and physical details. Notably, our goal is not to identify which approaches are "right" or "wrong" or even to compare levels of accuracy among modelling frameworks, per se. Instead, our central thesis is that because no modelling framework is likely to be ideal for every application (Hein et al., 2006), an understanding of the trade-offs inherent in various modelling approaches is needed to acknowledge the strengths and weaknesses inherent in a given model during both model development and application by diverse end-users (Berkhout et al., 2014; Root and Schneider, 1995). In particular, we discuss the potential dangers inherent in assuming that the spatial and temporal scales of model variables can be automatically defined by the scale of output (Bennie et al., 2014; Evans et al., 2013; Harwood et al., 2014; Potter et al., 2013). In other words, we explore why small-scale processes traditionally studied by physiological ecologists may have significant consequences for large-scale patterns studied by biogeographers (Gillingham et al., 2012; Harwood et al., 2014; Zwart et al., 2015), and conversely why biogeographic context may be critical for defining what occurs at the level of organisms (Gouhier et al., 2010; Wethey et al., 2011a).

1.2 Ecological Forecasting and Inconsistent Conceptions of the Ecological Niche

At their core, attempts to forecast biological responses to climate change rest on the assumption that the survival, distribution and abundance of organisms are driven, either directly or indirectly, by their tolerance to environmental conditions (Andrewartha and Birch, 1954). Ultimately, species distributions under global change will be determined by the differential survival and reproduction of interacting species undergoing physiological stress (Somero, 2005), alterations in the strength (Kordas et al., 2011; Wethey, 2002) and in some cases direction (Crain, 2008) of species interactions, and the ability of organisms to disperse to new habitats and/or successfully repopulate an area following a disturbance (Travis et al., 2013). Understanding the limitations of different modelling approaches thus first requires a recognition that not all models are designed to encapsulate the same aspect of a species' ecological niche, and that in some cases this distinction is not even recognized (Soberón, 2007). The niche is a concept that while central to much of ecological theory over the last century, nevertheless, continues to be an active topic of discussion, especially among climate biologists (Anderson, 2013; Austin, 2007; Beaugrand and Helaouët, 2008; Boogert et al., 2006; Kearney et al., 2010a; Schoener, 2009). Assumptions of what defines a species niche (or niche space) remain an often unrecognized source of confusion between those making predictions at the levels of individuals or populations and those working at large scales, typically using correlational methods based on GIS data (Kearney, 2006; Soberón, 2007). How the definition and usage of the niche concept has evolved from a primarily physiological construct to one where environmental parameters are mapped on to space (Box 1) has direct relevance for modern forecasts of species range shifts in response to global environmental change (Guisan and Zimmerman, 2000; Jeschke and Strayer, 2008; Kearney and Porter, 2009; Stoffels et al., 2015; Wiens and Graham, 2005).

Models and experimental approaches conducted at the level of organisms and populations typically focus on determinants of a species *fundamental niche space*, by exposing organisms to suites of environmental stressors under controlled conditions and then extrapolating to both current and projected environmental conditions in the field (Pörtner and Peck, 2010; Somero, 2005). Studies such as these have informed our understanding of which

BOX 1 Historic Constructions of the Niche

Since the concept of the niche was first promulgated in the scientific literature in the early 20th century by Grinnell (1917), the application and even definition of the word continue to serve as fodder for an enormous body of research and a source of discussion and debate (Chase and Leibold, 2003; Kearney et al., 2010a,b; Pocheville, 2015; Schoener, 2009). Grinnell first used the term to describe everything that fostered or limited the existence of a species at a particular location, including not only abiotic factors but also food, suitable habitat, predators and competitors (Pocheville, 2015). More specifically, as the term implies, it was used to define a species' "nook" or "recess" in an ecological community (Schoener, 2009). Grinnell (1928) incorporated the concept of the niche into an ecological hierarchy where, at small scales, the term described the biotic factors that governed species associations separate from large-scale geography (Pocheville, 2015). He primarily used the term as a vehicle for exploring "ecological equivalents" or species that occupy similar niches in different locations (Schoener, 2009). Patterns of species distributions at larger spatial scales (higher levels in the hierarchy) did have a direct biogeographic connotation (i.e. could be mapped explicitly on to space) and were associated primarily with abiotic drivers. Elton (1927) described niche as the functional role of the species in its community, including in his definition its relationship to resource availability and predation. In line with Grinnell's hierarchical definition of how the physical environment defines the biogeographic distribution of species, Hutchins (1947) explored how temperatures could be used to define the distributions of species along a latitudinal gradient, as determined by their physiological tolerances.

Perhaps the most well-known definition of niche by today's standards is the population-persistence niche (Schoener, 2009) formulated by Hutchinson (1957) as a quantitative description of the tolerable range of environmental conditions that permit the persistence of a population at a given location (Schoener, 2009). Often described as an "*n*-dimensional hypervolume" the Hutchinsonian niche refers to the multidimensional set of ecological conditions which define where a population can or cannot exist. Hutchinson's definition (which aligns more with Grinnell's definition than with the Eltonian meaning) further allows one to distinguish between a species *fundamental niche* (the portion of the niche defined by limits of tolerance to abiotic environmental variables) and the *realized niche* of a species, the portion of the fundamental niche space to which a species is limited by interactions with other species such as predators and competitors. The fundamental niche of a species should be broader than its realized niche because the former represents the entire range of environmental conditions where survival is theoretically possible, and the realized niche describes the conditions where a species is actually found in nature (Chase and Leibold, 2003).

species and populations are likely to be most (and least) vulnerable to environmental change (Somero, 2010), including native and nonnative species (Schneider and Helmuth, 2007), commercially important organisms (Pörtner and Peck, 2010; Ruckelshaus et al., 2013), critically threatened and endangered species and habitats (Sieck et al., 2011) and ecological keystones (Sanford, 1999). They have also offered guidance into what factors influence resilience or vulnerability. For example, recent studies have emphasized that local adaptation may confer greater resilience to populations than previously expected (Kuo and Sanford, 2009; Sanford and Kelly, 2011) but that conversely, low levels of genetic variability within populations can make them highly vulnerable to change (Kellermann et al., 2009; Pearson et al., 2009). Populations of a variety of different species may have a reduced capacity to handle change given that many display lower genetic diversity than previously predicted by neutral theory, likely due to a combination of factors such as population size, linked selection and life history traits (Ellegren and Galtier, 2016).

"Macrophysiological" studies which extend these controlled experiments to patterns of distribution in the field have provided critical links between physiology and biogeography (Chown and Gaston, 2008; Gaston et al., 2009; Pörtner and Peck, 2010). In some cases, however, these exquisitely detailed physiological measures are mapped on to physical space with comparatively coarse measurements of environmental parameters, which may or may not serve as effective proxies for what ultimately drives the growth, survival and reproduction of organisms (discussed in Hannah et al., 2014; Helmuth et al., 2014; Maclean et al., 2015; Potter et al., 2013). For example, daily mean air temperature is known to be a poor predictor of thermal stress in many ectothermic plants and animals (Scherrer and Koerner, 2010), especially in environments with high daily variability (Seabra et al., 2015). Even in environments typically considered to be relatively thermally stable such as coral reefs, sea surface temperatures (SSTs) can differ from subsurface temperatures by several degrees over the course of a day (Castillo and Lima, 2010; Leichter et al., 2006). The use of annual mean values is an even more egregious but yet commonly used method for estimating physiological thermal stress. Moreover, while an increasing number of studies are beginning to show how altered environmental conditions affect rates of predation (Pincebourde et al., 2008; Sanford, 1999) and competition (Wethey, 2002), these too are often not considered when making predictions of species distributions based only on measurements of physiological tolerance (Soberón, 2007). We explore these concepts in more detail below.

At the opposite end of the spectrum, statistical ("climate envelope") models typically are based on correlations made between species presence and absence and measurements of large-scale environmental conditions (Hampe, 2004; Peterson, 2001). This approach therefore implicitly measures a species *realized (or ecological) niche space* because limiting factors are determined through correlation, even though they are often assumed to detect only the direct influence of climate on species limits (Araújo and Rozenfeld, 2014). Correlative approaches are inherently based in space: the environmental data at the specific location(s) where species disappear define the limits to the ecological niche. In some cases, underlying mechanisms can be drawn from robust, repeatable patterns of coincidence (Peterson, 2001). For example, satellite measurements of SST anomalies (degree heating weeks above a locally defined threshold) have served as a highly effective means of assessing and predicting the severity of coral bleaching over large scales (Barton and Casey, 2005; Gleeson and Strong, 1995; Strong et al., 1997). More commonly, however, underlying mechanisms are more difficult to untangle. For example, over a range of scales a species' range limit may be set not by its own physiological niche limits, but those of strong competitors (e.g. Araújo and Rozenfeld, 2014; Connell, 1961; Wethey, 1983). And more importantly, existing environmental data may not encapsulate the likely novel conditions presented by future climate change, restricting the range of correlation that can be made (Ackerly et al., 2010).

Later in this review we explore hybrid approaches that have attempted to bridge this apparent divide, but the fundamental question remains: how much mechanistic detail is required to make useful forecasts without impeding our ability to make timely and useful predictions (Adler et al., 2012; Martínez et al., 2015; Seebacher and Franklin, 2012; Zwart et al., 2015)? In the next section, we explore the long-standing debate over the utility of mechanistic vs correlative approaches.

1.3 Mechanism or Correlation?

The relative merits of correlative and process-based approaches have been discussed in detail by many authors (Estes et al., 2013; Gutt et al., 2012; Hijmans and Graham, 2006), but most arguments focus on concerns regarding stationarity (i.e. that a model developed for one location can be applied at any other location) and the related concept of space-for-time substitution (i.e. that models developed under contemporary environmental conditions are informative of responses in the future under what may be nonanalogue climatic conditions; Austin, 2007; Brown et al., 2011; Casati et al., 2008; Craig, 2010; Helmuth et al., 2014; Woodin et al., 2013).

The advantage of correlative methods is that they can make predictions quickly, even when little is known about the underlying processes driving species distributions. An analogy to the application of climate indices, a topic that we explore in detail later in the paper, is instructive in this regard. In brief, indices such as the El Niño Southern Oscillation (ENSO), the Pacific Decadal Oscillation (PDO) and other phenomena describe "packages of weather" (Stenseth et al., 2003), i.e., the regular coincidence of high or low rainfall, high or low temperature, etc., in a region (Forchhammer and Post, 2004). The advantage of these indices is that they are holistic, i.e., they encapsulate the interaction of multiple variables which can then be correlated with biological response. The downside is that they are proxies for the actual drivers of biological response and work only as long as correlations among variables hold (Stenseth et al., 2003). Similarly, correlation-based approaches are adept at capturing suites of conditions that interact in the spatial domain to drive survival and hence distribution, including both direct impacts on physiology and survival and indirect effects through impacts on competitors, predators and prey (Araújo and Rozenfeld, 2014). A major advantage of both climate envelope modelling and correlations with climate indices is that they do not require a priori knowledge of the underlying mechanisms that determine historical limits and events, as long as the suite of variables used includes those which actually elicit a biological response (Araújo and Guisan, 2006; Hampe, 2004). In other words, they are typically a "black box" approach which may limit extrapolations into the future. While a post hoc analysis of the strength of each variable (e.g. through principal components analysis; Coppens d'Eeckenbrugge and Lacape, 2014) provide some insight into the relative importance of different variables, connections to process are generally elusive (but see Kroeker et al., 2016).

Of greater concern is the risk that environmental factors that limit species distributions historically may not fully capture the suite of stressor combinations in the future (i.e. space-for-time substitution), under what will likely be "nonanalogue" climatic conditions (Fitzpatrick and Hargrove, 2009; Seabra et al., 2015; Williams and Jackson, 2007). Not only are the magnitude, duration and frequency of future extremes such as heat waves likely to exceed those seen today (Kodra and Ganguly, 2014; Woodward et al., 2016), but because most environmental drivers do not change simultaneously in space and time, organisms are likely to be exposed to novel combinations of stressors such as heat, drought, hypoxia, precipitation and (in aquatic systems) pH (Jentsch et al., 2007; Rahmstorf and Coumou, 2011; Thompson et al., 2013). These interactions among multiple environmental variables can be complex and have been shown to act additively but also synergistically and even antagonistically (Christensen et al., 2006; Crain et al., 2008; Piggott et al., 2015; Sokolova, 2013; Woodward et al., 2016; described in further detail later).

Correlative models also often assume that current distributions are relatively static, i.e., that current range boundaries reflect the climatic limits to a species' distribution. The assumption that distribution limits are in equilibrium may not always hold. For example, correlative models may fail when contemporary range limits are set not by climate but by physical barriers to dispersal (Gaylord and Gaines, 2000; Kershaw et al., 2013; Peterson et al., 1999). Further, when rare extreme events cause large-scale mortality, it may take many decades before the species returns to its original geographic limits (Wethey et al., 2011b; Woodward, 1990). All of these issues call into question assumptions of model stationarity when a model developed in one location is extended to make predictions over broader scales.

Attempts to reduce errors due to a failure of stationarity when using correlative approaches are made by training models with a portion of the data and testing the model outputs with the remainder (Araújo et al., 2005a; Watling et al., 2013). In some cases, predictions made using data from one location (e.g. continent) are validated against observations made in a disparate location (Thuiller et al., 2005). For example, models are constructed for ancestral populations, and the model is then applied to a derivative population, on another continent. When these models fail to predict the observed distribution in the new location, this has been interpreted as evidence of a "niche-shift": the assumption is that the model did not fail, but rather the population is assumed to have evolved in to a new niche space (Broennimann et al., 2007; Gallagher et al., 2010; Medley, 2010; Tingley et al., 2014). Whether such comparisons truly are indicative of evolutionary shifts in biological traits, or whether they are the result of dissimilarity in drivers of distribution in the two populations, remains a point of contention. For example, Duncan and colleagues (2009) used data on the invasion of five species of South African dung beetle to Australia using climate envelope models. They found that for two of the species, the model developed in the native range worked as well as models using data from the introduced range. For the other three species, however, the model developed for South African animals performed very poorly when applied in Australia.

In contrast, mechanistic models include data collected on the physiology and behaviour of populations and species (Botkin et al., 2007; Briscoe et al., 2016; Brown et al., 2011; Buckley and Kingsolver, 2012; Chown and Gaston, 2008). The main advantage is that data can be determined experimentally and observationally using multiple combinations of variables, including those outside the range observed today (Botkin et al., 2007; Brown et al., 2011). They also can include variability in physiological vulnerability that may occur among populations, for example, as a result of local adaptation (Kuo and Sanford, 2009), or acclimatization (Williams et al., 2016), although notably these processes are still seldom considered in most modelling efforts (but see Dong et al., 2015). Mechanistic approaches thus potentially avoid problems with space-for-time substitutions (Fitzpatrick and Hargrove, 2009). When combined with approaches such as energetics modelling (Woodin et al., 2013), mechanistic approaches are particularly effective at accounting for the interactive roles of multiple stressors, and for the potential mitigating influence of resource availability (Kearney et al., 2010a; Sokolova, 2013).

A disadvantage of mechanistic models, however, is that they can be extremely data intensive and therefore, time consuming. As a consequence, results can be slow to be disseminated and incorporated into policy, especially when there are data gaps. This could potentially have devastating consequences for threatened species (Banks-Leite et al., 2014; Longcore et al., 2010). Although mechanistic models incorporate more biology than correlative models, they do not often incorporate much ecology (e.g. indirect effects of the changing abiotic environment on a species' predators and food supply; Kordas et al., 2011; Louthan et al., 2015; but see Soberón and Peterson, 2005; Tingley et al., 2014)—factors that are implicitly included in correlative approaches, although generally assumed to be unimportant (Araújo and Luoto, 2007). In essence, mechanistic models are based on a species fundamental niche space, which can be much broader than its actual realized niche (Kearney and Porter, 2009). In order to be effective, such models must include mechanisms that actually constrain species range boundaries, and these are not always identical across a species' global distribution. For example, Sarà et al. (2011) found that both mortality and nonlethal reproductive failure defined the geographic range limits of Mediterranean mussels. In cases such as these, mechanistic models based only on lethal limits will not exhibit stationarity when applied to other locations (Woodin et al., 2013).

Direct comparisons of results generated from correlative and mechanistic approaches have produced divergent results. In some instances, little difference has been observed in their predictive power (Robertson et al., 2003), especially when species functional traits are included in correlative models (Kearney et al., 2010b). Other modelling efforts have shown large differences in predictions made by the two approaches (Phillips et al., 2008), whereas different studies have suggested that the inclusion of ecological traits such as migratory ability and trophic level can increase the accuracy of species distribution models (SDMs) (McPherson and Jetz, 2007). In both situations, however, several authors have stressed that the relative accuracy of the approaches is likely affected by the spatial and temporal scales of resolution being used, suggesting that congruence between approaches is most likely to appear at coarse scales. As a result, researchers have suggested a hierarchical approach that applies correlative modelling at larger spatial and temporal scales, and methods that can account for factors such as microclimate at smaller scales (Pearson and Dawson, 2003). An increasing number of modelling approaches that are essentially hybrids of correlative and mechanistic models have been developed (Brown et al., 2011; Buckley and Kingsolver, 2012; Buckley et al., 2010; Kearney and Porter, 2009; Rougier et al., 2015). For example, physiological data can be combined with statistical approaches (Tingley et al., 2014).

Predicting the future based on the present is challenging, and virtually impossible to completely validate without waiting for years or decades to see if predictions materialize. How "accurate" must these predictions be before we act, and what are the consequences if they are wrong? Are some model outputs "right" and others "wrong", or does the choice of model depend largely on the application to which it is applied (Allouche et al., 2006; Petchy et al., 2015)? Considering tests of model accuracy within this context can help inform which level of detail can be safely ignored for the question being asked.

2. COMMON PITFALLS AND THEIR UNINTENDED CONSEQUENCES

Here we provide an overview of recent studies, from diverse organisms and habitats, that have identified common pitfalls that occur due to a failure to recognize the unintended consequences of common assumptions, most often as a result of not considering key ideas outside of one's own discipline. We then explore the potential theoretical and practical consequences of failing to consider each pitfall. We hope to (a) facilitate crosstalk among researchers using, in some cases, vastly different approaches to study similar questions; (b) advocate for an integration of these approaches when possible; and (c) argue for a need to explicitly recognize and acknowledge sources of uncertainty inherent in different modelling methods. Specifically, by highlighting many of the most common (but often unrecognized) assumptions and pitfalls, our intention is to promote an open conversation about where the field is heading. We advocate for a shift from discussions of which approach is "best" to one where multiple models are kept in the forecasting toolbox, each with their own limitations and strengths. Importantly, we argue that it is often dangerous to assume that the scale of one's application defines the scale of the processes being considered: biogeographers need to consider the roles of genetics, physiology, ecological interactions and microclimates just as physiologists, ecologists and geneticists need to understand large-scale geographic context. In many cases, the central challenge will thus be to overcome scale mismatches between the temporal and spatial resolution of the data being used (Kearney et al., 2012), the temporal and spatial resolutions of the predictions being made (Benedetti-Cecchi et al., 2005; Burrows et al., 2009; Gouhier et al., 2010; Menge et al., 2011) and the applications for which they are used (Benedetti-Cecchi et al., 2003; Cumming et al., 2006).

2.1 Scale and Data Resolution

What exactly are we modelling when we use climate models and weather data to estimate the effects of environmental change on organisms (Kearney, 2006)? Drawing on previous definitions of these terms (Mitchell, 2005), Kearney (2006) defined *habitat* as a qualitative descriptor of the physical and biotic features of an organism's surroundings (either actual or potential), suggesting that this term should be used only for correlative approaches that do not seek a mechanistic understanding. Importantly, a habitat can be described without reference to an organism, i.e., the characteristics of a habitat are independent of the organisms. In contrast to this description of the "stage" on which organisms play, *environment* was used to describe the biotic and abiotic phenomena that actually interact with an organism. As a result, environment must be defined with reference to the organism of interest. For example, nocturnal and diurnal animals experience very different environmental (e.g. solar radiation) and biotic factors (e.g. predators), even when living in the same habitat. Environment also includes the ability of an organism to modify its physical habitat, for example, nesting and burrowing animals. Finally, *niche* was used to describe the subset of environmental conditions that drive an organism's fitness, for example, its body temperature, which can differ substantially from environmental temperature (Helmuth, 2002). These are far from mere semantic issues and have far-reaching consequences for our understanding of how changes in climate ultimately—through weather—drive the survival and physiological performance of organisms, which ultimately results in changes in distribution, abundance and biodiversity (Williams et al., 2016; Woodward et al., 2016).

2.1.1 Site vs Body Temperature

Temperature is among the most universally important drivers related to climate change as virtually all biochemical and physiological processes are driven to at least some degree by temperature (Somero, 2005). Notably, it is the temperature of an organism's body, tissues and cells that ultimately drive physiological responses—not the temperature of its surroundings, per se. This distinction is not trivial, as the body temperature of ectothermic plants and animals can be radically different from local air temperature, especially in the sun (Denny et al., 2006; Huey et al., 1989; Mitchell, 1976; Pincebourde and Casas, 2006). With the exception of birds and mammals, the vast majority of organisms are ectotherms and produce minimal metabolic heat. In air, body temperatures are driven by multiple environmental factors including solar radiation, air temperature, wind speed and cloud cover (Denny et al., 2006; Kearney, 2006; Kingsolver, 2002) and are affected by the characteristics of the organism such as shape, colour, mass and material properties (Helmuth, 2002). As a result, the temperature of ectothermic animals can be substantially different from that of the surrounding air or substrate (Kearney, 2006), in some cases by 15°C or more (Fitzhenry et al., 2004). Species-specific heat-flux models can generate estimates of body temperatures given a number of pertinent weather variables and individual traits described earlier (Helmuth, 1998). In practical terms, the mechanics of heat exchange mean that organisms in environments with any appreciable variation in environmental conditions have body temperatures that can fluctuate fairly rapidly. As shown in Fig. 1, temperatures can be highly variable

Fig. 1 Visible (*left side*) and infrared (*right side*) imagery reveals thermal heterogeneity both between and within adjacent terrestrial (A) and marine (B) habitats. In these examples, temperatures exhibit a 40°C range over the scale of several meters.

over even small spatial scales, especially in the sun (Scherrer and Koerner, 2010). Failure to explicitly account for differences between air temperature and body temperature can lead to both overestimation and underestimation of species range limits, population resilience, critical habitat and therefore overall threat level under climate change. It also can lead to potential misinterpretations when laboratory studies that explore the effects of *body* temperature on fitness and rate processes are extrapolated to the field using *environmental* temperatures.

2.1.2 Microhabitats

The potential importance of microhabitat-scale variability in environmental conditions has emerged as a recent subject of discussion and debate (Maclean et al., 2015; Potter et al., 2013). It has now been shown for several ecosystems that variability at small spatial scales—often less than a square metre—can rival or exceed differences observed over much larger biogeographic scales (Denny et al., 2011; Rapacciuolo et al., 2014; Seabra et al., 2011) and that differences in environments at these scales have significant physiological consequences (Helmuth and Hofmann, 2001; Miller et al., 2015; Williams and Somero, 1996). An organism's ability to move among microhabitats can thus significantly affect its vulnerability to environmental change (Bennett et al., 2014; Chapperon and Seuront, 2011; Hayford et al., 2015; Huey et al., 1989; Kearney et al., 2009; Mathot et al., 2015; Miller et al., 2015; Monaco et al., 2015; Otero et al., 2015; Sunday et al., 2014). This has several potential consequences for our predictions of how climate change may drive biogeographic pattern (Bennett et al., 2014; Storlie et al., 2014). Microhabitats may serve as "rescue sites" that allow organisms to repopulate an area following an extreme event that otherwise kills off other organisms (Hannah et al., 2014; Low-Décarie et al., 2015). This buffering capacity may also be important in determining true range edges. A recent study by Lima and colleagues (Lima et al., 2016) showed that the equatorial range limit of limpets was defined not by climate per se, but by the last location where the presence of microhabitats could no longer provide sufficient buffering for the species to survive. Similar arguments (and debates) have emerged recently among plant ecologists (De Frenne et al., 2014; Harwood et al., 2014; Maclean et al., 2015). In the long term, rescue sites that contain organisms that are preadapted to different climate conditions could serve as a source for repopulation, such as with geothermally heated streams or lakes (O'Gorman et al., 2014).

At larger scales, several studies have now suggested that what otherwise appear as latitudinal clines (when considering only coarse environmental data) may in fact present organisms with geographic mosaics (Ackerly et al., 2015) in factors such as temperature (Helmuth et al., 2006; Mislan et al., 2009), pH (Kroeker et al., 2016), precipitation (Ackerly et al., 2015) and snow cover (Holtmeier and Broll, 1992). This can be the result of both the influence of microhabitat complexity (Potter et al., 2013) and of local factors such as coastal tides (Helmuth et al., 2002), and cloud cover (Wilson and Jetz, 2015) that can override otherwise latitudinal gradients. When mosaics of multiple environmental stressors are then overlain with one another, the end result is a geographically complex pattern of environmental stress that is arguably only measurable using knowledge of an organism's physiological tolerances (Kroeker et al., 2016; Williams et al., 2016).

The necessity of including high-spatial resolution data in models thus remains a point of contention between scientists working at different temporal and spatial scales (Buckley et al., 2010; De Frenne et al., 2014;

Harwood et al., 2014; Kearney et al., 2012, 2014). Incorporating this small-scale information requires knowledge of the frequency and distribution of microclimates that are usually far smaller in size than the resolution of pixels from remote sensing observations (Kearney et al., 2014; Potter et al., 2013) and requires additional environmental data that go beyond more common variables such as air temperature (Shi et al., 2015; Wilson and Jetz, 2015). To this end, several researchers have begun to develop tools for downscaling to subpixel levels (Pappas et al., 2015; Vance-Borland et al., 2009) by accounting for factors such as exposure to solar radiation (Kearney et al., 2014; Mislan and Wethey, 2015; Wethey et al., 2011a), and through direct measurements of organism temperature in the field (Helmuth et al., 2006; Lathlean et al., 2014; Lima et al., 2011; Rey et al., 2015).

2.2 Inclusion of Relevant Biological Details
2.2.1 Weather, Climate and Climate Indices

Explicitly recognizing the difference between weather and climate (generally defined as a 30+-year trend in weather conditions) is crucial, yet the two phenomena are often conflated when designing experiments and making model predictions. At a proximal level, the physiology and survival of an organism are driven by weather patterns, which are the local manifestation of long-term changes in climate. For example, most models predict an increase in average surface temperature between 2°C and 6°C by the end of the century (IPCC, 2013), but increases of this magnitude are already occurring on shorter timescales in some locations. Experimental studies which purport to foreshadow "conditions by 2100" by making simple comparisons of physiological responses under constant conditions based on contemporary average temperatures (or any other environmental factors with high spatial and temporal variability) against conditions with a 2°C mean increase thus ignore the true nature of what climate change entails (Reid and Ogden, 2006).

During a heatwave in the summer of 2012, water temperatures in the Gulf of Maine were up to 3°C hotter than the norm (Pershing et al., 2015), which had significant socioeconomic impacts such as the movement of lobsters to coastal waters which dramatically increased catch rates (Mills et al., 2013). These extreme events can have lasting consequences for species distributions (Jentsch and Beierkuhnlein, 2008; Woodward et al., 2016). The contemporary geographic distribution of several species of marine invertebrates in Europe, for example, reflects the result of a severe winter that occurred in 1962–63 (Wethey et al., 2011b). Similarly, climate envelope models that include climatic extremes can improve model performance when predicting spatial distributions of trees (Zimmermann et al., 2009). For this reason, biogeographers are increasingly calling for the use of variability and extremes in modelling efforts (Thompson et al., 2013; van de Pol et al., 2010; Vasseur et al., 2014; Woodward et al., 2016). Species limits can also be set by the cumulative effects of more chronic exposure to stress, which can lead to reproductive failure (Kearney et al., 2012; Kingsolver and Woods, 2016; Sarà et al., 2011). The magnitude and timing of extreme events can also interact with chronic exposures to physiological stressors to drive organism fitness (Helmuth et al., 2010; Kearney et al., 2012; Montalto et al., 2014; Woodward et al., 2016).

Findings such as these potentially have profound implications for how we predict future responses to environmental change, especially when using temporally and spatially coarse environmental data and climate indices. For example, an anomalously "warm year" based only on average annual temperature may show strong correlations with biological responses, but there are many ways to arrive at a high annual mean. For example, if the increase in mean is the result of an increased number of heat waves coupled with otherwise "normal" temperatures, observed biotic responses could be the result of short-term extreme events. In contrast, a year where many days are slightly warmer than average, but the incidence of extreme heat waves is low, could lead to the same observed change in annual mean, in which case biological effects are more likely the results of cumulative stress occurring over longer-time periods. Whether a high annual mean is truly diagnostic of the observed biological response therefore depends entirely on how well correlated it is with the actual underlying driver (Stenseth et al., 2003). This argues for a need for extreme caution when extrapolating from trends in space or time to an actual mechanism (Box 2). Seabra et al. (2015), for example, reported a latitudinal gradient in average temperature along the coast of Europe. They also quantified spatial patterns in other biologically relevant environmental variables such as maxima, minima and daily range over the same gradient. Their results showed that spatial patterns in these biologically important parameters did not follow the same pattern as average temperature. In this case, estimates of geographic patterns based only on averages miss crucial biological detail, and efforts to smooth data may in fact mask biologically significant patterns (Box 2).

Whether and when biological detail is really needed for distribution modelling remains a point of discussion and debate. The performance of correlational SDMs does not necessarily improve with the addition of more environmental inputs, but it can be difficult to make a priori assessments about which parameters to include in a model. Several analyses of SDMs

> **BOX 2 Trends vs Predictive Relationships**
> Even the most carefully calibrated models have the potential to produce enormous errors when outputs are used for applications for which they were not originally designed. As described throughout this review, perhaps the most common example of this problem is when climate models are used as literal interpretations of future weather. But there are also numerous examples where statistical correlations and trends are mistakenly equated with predictive models, especially when they are "smoothed" using averaging. Consider the correlation shown in panel A below, which shows a statistically significant, negative relationship between response Y and driver X. Values of Y, for example, might be temperature and X latitude. Certainly it would surprise no one that there is a trend towards lower temperatures at higher latitudes. But do the data shown here really represent what might be called an "environmental gradient" and how much predictive power does this relationship imply? It is clear from the coefficient of determination (R^2) of the relationship (0.21) that latitude has only limited explanatory power, and that other more local factors play a much larger role. So what can data such as these tell us, and what cannot they?
>
> In the example shown here, higher latitude is more *permissive* of colder temperatures, but there are also sites, especially at more mid-latitudes, with temperatures that are as warm as those at equatorial sites. It would therefore be inappropriate to estimate temperature as a function of latitude. It is not at all unusual, however, to take trends similar to those shown in panel A and to smooth them, either with a running average or else a mean value over a series of bins, here shown in panel B. This greatly increases the R^2 of the relationship (0.58), has the effect of removing extremes and also makes the trend appear far more "predictive". Technically, the results are just as valid as the raw data: *average* temperature tends to decrease with increasing latitude. But do these capture biologically relevant variables, and is the variability that is removed really due to measurement error, or does it represent real variability in underlying drivers? Seabra et al. (2015) studied this question directly, asking how do a number of biologically significant temperature variables (e.g. maximum and minimum temperature) correspond to latitudinal gradients? They found that while average temperature followed a latitudinal gradient, very few of the other ecologically

BOX 2 Trends vs Predictive Relationships—cont'd

relevant parameters did. This observation is of particular concern for climate change studies, because it suggests that simple metrics and averaged values are *only* diagnostic to the extent that they reflect underlying drivers, relationships that may be breaking down as we enter novel climatic conditions.

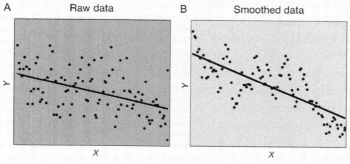

show that "bioclimatic" indicators, such as seasonal maximum and minimum values that are presumably more directly related to a species' biology, do not necessarily improve model performance over monthly means. Watling et al. (2012), for instance, looked at the use of bioclimatic indicators in climate envelope models on several common Floridian taxa of terrestrial vertebrates. They found that on average the use of bioclimatic indicators did not improve predictions of models over monthly means of temperature and precipitation, but cautioned that the model performance was not consistent for each of the 12 species examined. In the case of the wood stork, a species with a broad geographic range where vastly different conditions can be experienced in the same month among different locations in the species range, bioclimatic indications produced a much better model for the two algorithms (Generalized Linear Model and Random Forest) used (Watling et al., 2012). In an assessment of 243 avian species, Barbet-Massin and Jetz (2014) found that the parameter selection influenced model predictions with several correlated temperature variables identified as the most important predictors in SDMs. Of these correlated variables, potential evapotranspiration, mean annual temperature and growing degree-days were the most important. The uncorrelated mean diurnal temperature and the moisture index were both shown to be important predictors as well (Barbet-Massin and Jetz, 2014).

Some newer approaches have taken one step closer to incorporating mechanisms by using derivatives of climate that are likely to have more direct relevance to actual drivers of population-level responses. Climate change velocity, for instance, has recently been used as a metric for the impact of the rate of temperature change on the ability of populations and species to cope (Lima and Wethey, 2012). The instantaneous velocity of temperature change, or climate-change velocity, is calculated as the ratio of the temporal gradient of mean annual temperature to the spatial gradient of mean temperature between neighbouring grid cells (Loarie et al., 2009). It can be calculated as a taxon-specific metric (Pinsky et al., 2013) to account for irregular (i.e. nongridded) species movements. This metric describes the rate and direction of changing temperatures and can help predict not only the magnitude and direction of species range shifts but also the capacity for species to both adapt to and track changing climates. Velocities can also be calculated for thermal extremes and nonthermal variables such as evapotranspiration and climatic water deficit (Dobrowski et al., 2013) although more work still needs to be done to understand how well these metrics can track species shifts in terrestrial systems (Burrows et al., 2014; Pinsky et al., 2013). In both terrestrial and marine environments, climate velocities reveal geographic pathways and barriers that facilitate and prevent migration, respectively (Burrows et al., 2014; Pinsky et al., 2013), and in freshwater streams, isotherm shift rates are used to estimate distributional upstream as conditions warm (Isaak and Rieman, 2013). These studies that compare movement of taxa relative to climate velocity can identify species that are matching a shifting climate or those that are lagging behind. Pinsky et al. (2013) examined range shifts in 360 different marine taxa and demonstrated that climate velocities were better able to predict both the direction and magnitude of a species range shift than species-specific trait data. However, this assumption may not apply generally across taxa and locations as seen in a study by Sunday et al. (2015) on marine species in the Tasman Sea, where climate velocity explained some variation in species movement but species traits, such as trophic position and mobility, were more important predictors (also see Zwart et al., 2015 for a phytoplankton example). Adding in species traits provides insights into the mechanisms that cause the observed patterns, for example, traits related to range size and metapopulation dynamics (Sunday et al., 2015). Climate velocity therefore is, like other indirect metrics, predictive of biological response only to the extent that it correlates with underlying mechanistic drivers. However, unlike static metrics such as annual means or extremes that may capture the magnitude of environmental stress, climate velocity (or isotherm shift rate) additionally is an indirect measure of the likelihood that populations and species can acclimatize or adapt.

Importantly, some studies have shown that the absence of mechanistic detail can lead to complete failure of modelling efforts. Jones et al. (2009, 2010) studied the geographic distribution of mussels on the east coast of the United States, which showed a strong correlation with lethal temperatures. When the same model was applied to the coast of Europe, the assumption of stationarity did not hold and the model overpredicted the distribution of the species by a factor of 2. However, when the cumulative effects of chronic stress on energetics were included, the model performed well by matching predicted regions with actual regions of survival (Woodin et al., 2013). In instances such as these the importance of including multiple metrics of environmental stressors (e.g. means, extremes, time history) is thus only evident given knowledge of the biology of the organism in question, in this case the importance of chronic exposure to sublethal temperatures. Woodin et al. (2013) argued that in the Western Atlantic, the spatial velocity of change was sufficiently high (i.e. temperatures changed over very short distances at the geographic limit to the species equatorial distribution) that inclusion of mechanism was probably irrelevant. Mussels would likely die at locations very similar to those where they would experience energetic failure. In contrast, in the Eastern Atlantic, where the spatial gradient in temperatures was broad (i.e. the spatial velocity was low), the inclusion of mechanism (energetics) was critical for interpreting the observed pattern.

An informative example of the potential importance of understanding underlying mechanisms is provided by Hallett et al. (2004), who compared model predictions of Soay sheep (*Ovis aries*) population dynamics made using climate indices vs local weather data. Previous studies of the population (Catchpole et al., 2000) had shown that dynamics could be explained using estimates of rainfall based on the strength of the North Atlantic Oscillation (NAO), and that, paradoxically, this climate index appeared to serve as a better predictor than local weather data. Taken at face value, this result could be interpreted as an indication that understanding mechanism provided little gain in predictive power. Noting that the effects of the NAO were in part dependent on population density of the sheep, however, Hallett and colleagues ran simulations that accounted for the ways in which sheep density (and intraspecific competition) influenced the susceptibility of the population to winter storms. Because winter weather affected both animal energetic condition and food availability, they developed a model that accounted for both the magnitude and relative timing of these events. They found that models based on local weather data which included the magnitude and temporal sequence of several drivers of physiological condition—strong winds, cold temperatures and heavy rainfall—had better explanatory power than coarser predictive variables such as NAO. Thus, while climate indices such as NAO may be a good "catch all" descriptor of generalized stress (Forchhammer and Post, 2004; Stenseth et al., 2002), the inclusion of biological realism increased predictive ability when using temporally finer-scale weather data. When the inclusion of biological detail fails to improve model performance, it is therefore not always clear whether such details are truly irrelevant, or whether instead the model did not include *enough* or the *right kinds* of biological detail. Below we explore in more depth what some of those details may be, and examples of how some researchers have been able to incorporate them in statistical models.

2.2.2 Performance Curves

Thermal performance curves (TPCs) are a standard method of estimating physiological performance as a function of body temperature. A typical TPC (Fig. 2) is unimodal and asymmetric over a physiological temperature range (PTR). At very cold temperatures, performance is near zero (CT_{min}); as temperature increases, performance rises until it reaches a maximum level at some optimal temperature (T_{opt}), followed by a sharp decline towards the critical maximum (CT_{max}) after which performance becomes negative (Angilletta, 2006; Huey et al., 2012; Schulte et al., 2011). If the organism spends sufficient time in negative energy balance (below CT_{min} or above CT_{max}) or exceeds lethal limits (LT_{min} and LT_{max}), it will die (Kingsolver and Woods, 2016). The performance breadth is an arbitrarily defined performance threshold around the optimum with some studies opting for 80% (Angilletta et al., 2010), while others define performance breath at 69% of the optimum related to activation and inactivation states of enzymes (Freitas et al., 2007). Often metrics such as growth rate, escape velocity, respiration, heart rate, and feeding rate are used to determine how well an organism performs at a particular temperature (Angilletta, 2009). Thermal performance curves (or metrics derived from them) can be used to make ecologically relevant forecasts under current and changing environmental conditions related to population growth and fecundity (Estay et al.,

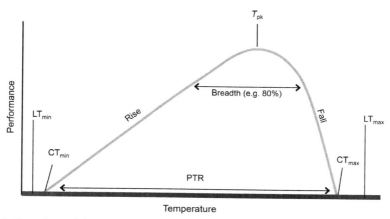

Fig. 2 Hypothetical thermal performance curve (TPC) over the physiological temperature range (PTR). The curve is typically unimodal and left-skewed (Dell et al., 2011). CT_{min} is the minimum critical temperature, when performance is less than or equal to zero at cold temperatures. As temperature increases, performance rises up to T_{pk}, the temperature when performance is at a maximum. Performance then declines sharply to CT_{max}, the maximum critical temperature, when performance is less than or equal to zero at hot temperatures. Mortality occurs rapidly at temperatures above LT_{max} (lethal maximum temperature) or below LT_{min} (lethal minimum temperature). Performance breadth describes the width of the curve at an arbitrary point, usually a value between 69% and 80% of maximum performance (Angilletta et al., 2010).

2014), species range shifts (Angert et al., 2011; Buckley, 2008), global patterns of thermal tolerance (Vasseur et al., 2014) and large-scale predictions about impacts of climate change on species performance (Clusella-Trullas et al., 2011).

TPCs are therefore potentially very useful for investigating the likely determinants of species range boundaries and patterns of mortality. Pörtner and Peck (2010), for example, have put forth the oxygen and capacity limitation of thermal tolerance (OCLTT) hypothesis that posits that death occurs when an organism's metabolic demand (which is temperature dependent) is unable to be met by supply, which for marine organisms is also temperature dependent. However, the duration of exposure to otherwise sublethal temperatures can affect performance (Kingsolver and Woods, 2016), and recent studies have focused on the role of nutritional state in altering thermal sensitivity (Connell et al., 2013; Kroeker et al., 2016). This suggests that factors other than oxygen such as energetics and the cumulative effects of cellular damage may also be key. When applied to ecological problems TPCs are typically meant to define one axis of a species' fundamental niche space, but (as we explore below) organisms are often faced with multiple interacting stressors. The application of energetics models can potentially serve as a way of integrating all of these factors (Brown et al., 2004; Kooijman, 2009; Sokolova, 2013) and recent studies have applied dynamic energy budgets to understanding species distributions (Freitas et al., 2009; Kearney and White, 2012; Montalto et al., 2014; Sarà et al., 2011, 2012; Thomas et al., 2011).

Critically, though, we still lack an understanding of how most species perform over their entire thermal ranges, with most studies focused only on mortality. A meta-analysis of thermal responses by Dell and colleagues (2011) reveals that out of 684 independently sampled thermal responses for microbes, plants, and animals, only 35% of them captured the optimum and almost none of them measured the full curve. More studies are beginning to explicitly address this gap, but our lack of understanding of how curves are shaped at the extremes remains a challenge to truly understanding how changes in temperature impact performance (Dowd et al., 2015).

Even with complete performance curves in hand, scaling up from individual performance to population-level performance is not always straightforward: when measured on the same individual, different metrics can diverge in their relationship to temperature. Stoffels et al. (2015), for example, measured aerobic scope (maximum-basal metabolic rate) and burst locomotor performance (driven primarily by anaerobic pathways) in crayfish and showed that the thermal optimum for locomotor performance was lower than that of aerobic scope. Widdows (1973) observed no correlation between heart beat rate and oxygen consumption by marine mussels. Furthermore, the shape of a curve based on physiological rates such as respiration is unlikely to match that based on traits that integrate multiple rates (e.g. growth; Dell et al., 2011). TPCs can also vary for the same individual depending on life history stage (Telemeco, 2014), the amount of thermal variation they are exposed to (Gilchrist, 1995), and acclimatization effects due to seasonal changes (Deutsch et al., 2008). Dispersal limitation coupled with strong selection can lead to local adaptation where physiological vulnerability varies among populations (Dong et al., 2015; Gaitan-Espitia et al., 2014; Sanford and Kelly, 2011). This is especially true of species with fast generation times and can occur, for example, in tropical fish (Donelson et al., 2011). Even individuals in the same population can display different curves due to underlying genetic polymorphisms or plastic responses to microhabitat effects (Schmidt et al., 2000). Performance curves are also thus unlikely to be temporally or spatially static for a given species, but rather may change across their geographic range.

2.2.3 Multiple Stressors

Making predictions based on temperature alone comes with a suite of predictive challenges but thermal stress does not operate in isolation. The abiotic environment is multidimensional, with shifts almost never occurring in one dimension alone. In addition to changes in air, water and soil temperature, climate change is predicted to affect both aquatic (Crain et al., 2008; Howard et al., 2013) and terrestrial habitats (Aber et al., 2001; Dietl et al., 2015; Kerns et al., 2016) through shifts in multiple environmental parameters including solar radiation, humidity, precipitation, wind (Moore and Watt, 2015), smog effects, nutrient regimes/eutrophication (Jonard et al., 2015; Smith, 2003), anoxia/hypoxia (Diaz and Rosenberg, 2008), salinity (Garza and Robles, 2010; Nielsen et al., 2003), ocean acidification (Pörtner and Farrell, 2008) and food availability (Place et al., 2012; Sarà et al., 2013).

Experimentalists increasingly realize that single stressor studies do not fully represent how organisms are affected by multiple stressors in nature. A spate of recent physiological studies has begun to explore the diverse and often counter-intuitive ways in which multiple stressors affect physiological performance (Molinos and Donohue, 2010; Firth and Williams, 2009; Gunderson et al., 2016; Jackson et al., 2016). Since an organism physiologically responds to various environmental conditions in different ways, predictions are difficult to make. For example, an organism may respond in a linear or exponentially declining fashion to some environmental conditions (e.g. toxins), where its physiological response worsens as the stressor increases. However, the same organism may respond in a nonlinear, unimodal manner to a different environmental parameter (e.g. temperature, pH, precipitation), where some intermediate value is preferred and extreme values produce stressful responses. Additionally, multiple stressors can work together to increase the either positive or negative biological effect (i.e. synergistic or additive), work against each other to lessen the effect (i.e. antagonistic), or a stressor can completely reverse the effect of another, although the most appropriate definitions for these effects are still under consideration (Piggott et al., 2015). For example, warming can reverse the effects of excess nutrient enrichment on freshwater organisms by increasing metabolic rates or enzyme conversion rates, which can eliminate toxins from the body more quickly (Jackson et al., 2016). Several meta-analyses have summarized the results of multiple stressor studies both at the level of individual organisms and on intact assemblages (Crain et al., 2008; Jackson et al., 2016; Piggott et al., 2015) and have found that antagonistic effects are the most common type, representing about 40% of 2 stressor studies.

Generalities remain elusive in this field; for example, combinations of the same stressors (e.g. pH and warming) do not consistently result in the same outcome on different species. In fact, many multistressor effects are highly species specific (Connell et al., 2013; Denman et al., 2011; Doney et al., 2009; Fabry, 2008; Kroeker et al., 2010) and influenced by spatial and temporal coincidence of stressors (Pincebourde et al., 2012), making overall effects at both the population (Brennan and Collins, 2015) and community levels difficult to predict. Patterns can become even more complex due to differential vulnerability among interacting species. For example, the outcome of a temperature plus precipitation study using interacting species A and B is probably not predictable from single stressor studies on each of the species, nor single stressor studies on the pair. These changes in both physiological rates and species interactions mediated by multistressors can ultimately lead to changes in food web dynamics, local abundance, community structure, diversity (Crain et al., 2008) and biogeographic patterns (Gunderson et al., 2016; Kroeker et al., 2016). A promising approach for studying the cumulative effects of multiple stressors lies in the application of energetics modelling (Matzelle et al., 2015; Sokolova, 2013), an idea we explore next.

2.2.4 Finding Common Currency: Energetics and Cumulative Stress

All of the above point to the complexities of basing physiological and ultimately ecological responses on patterns of only a few environmental variables, especially given the highly nonlinear and often interactive effects that they can have on organisms. While exposure to lethal conditions is important, a host of studies now point to the key influence of energetic failure, and how factors such as resource availability and assimilation rate may in some cases ameliorate stress. To this end, a number of studies have begun to explore the application of energetics modelling towards understanding the cumulative impacts of temporal and spatial variation in multiple environmental stressors (Kearney et al., 2010a; Matzelle et al., 2015; Sokolova, 2013). Energetics models (see Box 3) view an organism as an energy integrator where the balance of energy within it mediates processes such as

BOX 3 Energetics Models

Scope for growth (SFG): SFG is an accounting method that keeps track of energy inputs and outputs. When the amount of energy consumed through feeding is greater than the amount of energy lost to respiration and excretion, the animal has a positive energy balance and the organism is able to grow (increase in weight) and reproduce. A negative energy balance will result in a loss of reserves (decrease in weight). Allometric scaling can be used to account for animals of different sizes. A potential downside of SFG is that the approach uses several oversimplifying assumptions about metabolic processes. And, in contrast to DEB, estimates are static snapshots in time, and thus the incorporation of time history effects is very difficult. Although SFG models are good for predicting growth, they are not as adept at predicting reproduction and fecundity (Filgueira et al., 2011). SFG models have historically been used in aquatic environments for important fishery species but have also been applied to terrestrial insect species. Generally, SFG is applied as broadly across ecosystems as other energetics models.

Dynamic energy budget (DEB): A main difference between SFG and DEB is how they treat available energy (Filgueira et al., 2011). In DEB energy is first stored as reserves, which can then be mobilized. The flow of energy through an individual organism is modelled in terms of energy obtained through feeding and the energy available for maintenance of regular metabolic activity, growth and reproduction (van der Meer, 2006). As a result, DEB models are useful for describing organisms that experience variability in both food availability and abiotic stressors (Filgueira et al., 2011; Thomas et al., 2011) and can predict intraspecific and interspecific differences in metabolic rates. State variables are structural body size and reserves (van der Meer, 2006) which are not directly measureable (Nisbet et al., 2012). DEB models have been generated for a variety of different taxa and can be adapted to fit a multitude of life history strategies due to its vast theoretical framework.

Metabolic theory of ecology (MTE): Developed by Brown et al. (2004) to establish a universal energy budget that can scale from individuals to ecosystems, the primary state variable in MTE is body size, which is directly measureable. Limits to growth are based only on organismal assimilation rates, and there are no limits to food availability as feeding rate is assumed to increase at the same rate as metabolic demand. As a result, MTE is not adept at examining potential interactions between food and stress. However, in contrast to DEB, MTE calculates oxygen demand and thus may provide a means of estimating lethality when supply no longer meets demand (Kearney and White, 2012). While MTE also does not allow for intraspecific variations of metabolic scaling based on size, it does allow for interspecific metabolic scaling, but only at temperatures lower than a species' optimum.

Niche Mapper: Although primarily a biophysical model that relates environmental parameters with the material properties of the organism in question, Niche Mapper can also incorporate energetic/metabolic parameters. This framework is able to make predictions at the individual, population and potentially community levels (Briscoe et al., 2016; Kearney and Porter, 2009; Porter and Mitchell, 2006). Niche Mapper can make spatially and temporally explicit predictions when used in conjunction with a fine-scale microclimate model (Fuentes and Porter, 2013; Kearney et al., 2014).

growth and reproduction. Feeding and photosynthesis assimilate energy that can then be allocated to somatic growth and maintenance, development of reproductive tissue or parental care of offspring (Kooijman, 2009). Models including scope for growth (SFG), metabolic theory of ecology (MTE) and dynamic energy budgets (DEB) have all been used to explore how organisms are affected by their environment and in recent years have been used as powerful tools to examine the cumulative effects of environmental change (Kearney et al., 2010a,b; Sokolova, 2013). There have been many reviews of these modelling techniques including the relative limitations of each approach (Allen and Gillooly, 2007; Cyr and Walker, 2004; Filgueira et al., 2011; O'Connor et al., 2007a; Pawar et al., 2015; van der Meer, 2006). While such discussions remain highly contentious, some methods appear to be, at least in some cases, better suited to some applications than others. Kearney et al. (2012) showed the importance of considering time history effects, i.e., the return time of extreme events, in driving patterns of reproduction, and argued that DEB models were particularly well suited for this application given their ability to integrate such effects. In contrast, Filgueira et al. (2011) reported similar predictions of growth using both DEB and SFG models. Barillé et al. (2011) showed that the relative performance of DEB and SFG depended on the type of food used as a model input. In some cases, combinations of modelling approaches can be highly effective, for example, when parameters from a more data-rich modelling approach are used as inputs in a more generalizable modelling framework such as DEB (e.g. Nisbet et al., 2012). While surprisingly few direct comparisons of model outputs have been made (Kearney and White, 2012), current evidence suggests that a recognition of the assumptions underlying a model can help to define where it can be best put to good use (Kearney and White, 2012). For example, MTE models are based on metabolic demand as a function of body mass, which in turn corresponds to oxygen demand in line with the OCLTT hypothesis (Pörtner and Peck, 2010). As a result, they assume that metabolic demand increases as a power function with temperature, and ignore any decreases in demand that can occur at temperatures above optimal temperatures, as stereotypically shown in TPCs (Matzelle et al., 2014). While potentially useful in capturing lethality, they likely will not capture cumulative effects of exposure to sublethal conditions (Kingsolver and Woods, 2016) and their utility is therefore limited to temperatures below an organism's optimum. In contrast, DEB models are more focused on energetics and the potential impacts of energy supply and demand and thus are more in line with studies showing the effects of nutritional state on physiological tolerance to other stressors (Thomas et al., 2011). However, they do not implicitly calculate death due to exposure to lethal extremes, and the incidence of extremes must be calculated outside of the model. Again, the importance of including both lethal and sublethal responses to changing environmental conditions is not always straightforward (Woodin et al., 2013).

2.3 All the World's a Stage: From Autecology to Synecology and Beyond

Physiological studies have made major advances in understanding how environmental change will likely shape whole organism responses. There still remain many open questions, and among the largest is how these autecological responses will translate into synecological pattern and process at the level of assemblages, communities and ecosystems. In the next section, we explore how ecological processes are likely to significantly constrain our ability to understand and forecast biotic responses to environmental change.

2.3.1 The Importance of Dispersal

The accuracy and precision of forecasting models can be influenced by the inclusion of dispersal parameters. For example, the ability of many plant species to migrate is expected to be outpaced by the rate of climate change in many regions, but whether this happens depends on whether seeds can become established at sites distant from the moving front (Neilson et al., 2005). Similar arguments have been made for marine organisms that disperse via larvae (Hutchins, 1947). A recent focus on connectivity (Castorani et al., 2015) and metapopulation/metacommunity dynamics (Guichard et al., 2004) has begun to explore the critical role of dispersal, stepping stones (Storlie et al., 2014) and refugia (Ackerly et al., 2015) in driving ecological responses to environmental change. Such approaches may be especially critical in mosaic landscapes where reproductive failure and mortality can occur well within species range boundaries (Ackerly et al., 2015; Helmuth et al., 2002, 2006). However, despite many studies showing the importance of dispersal to biogeographic pattern (Clark et al., 2003; Pearson et al., 2006), most correlative (and many mechanistic) models predicting patterns of species distribution leave out dispersal altogether (Botkin et al., 2007) and when it is included, simplifying assumptions are often made. The ability of a species to successfully colonize a suitable location is important for predicting extinction vulnerability, range expansion potential, disentangling the drivers of species' presence/absence and understanding local population abundance patterns. Dispersal potential can be influenced by abiotic factors such as geological (Lieberman, 2005), hydrodynamical (Gaylord and Gaines, 2000; Herbert et al., 2009) or other biogeographical barriers (Rahel, 2007). In some cases, geographical breaks appear to be independent of physical barriers and instead are an emergent property of the process of speciation (Irwin, 2002). When the genetics of populations are poorly understood, what are often assumed to be contiguous populations of a single species can in fact be multiple populations of cryptic species that are unable to mate with one another, further confounding understanding of connectivity (Bickford et al., 2007). For example, Rissler and Apodaca (2007) combined ecological niche modelling with phylogenetic analyses of black salamanders (*Aneides flavipunctatus*) in California. Their analyses revealed the presence of multiple morphologically cryptic mitochondrial lineages—i.e., lineages that would otherwise have been classified as a single taxonomic unit—and showed that patterns were consistent with divergences in their ecological niche space.

Physiological responses of propagules to environmental conditions can also affect dispersal and predicted patterns of biogeographic distribution. The physiological tolerance of invertebrate larvae or resting stages, fungal spores and plant seeds to factors such as temperature (Abdelghany et al., 2010) and desiccation (Alpert, 2005) can be very different from that of adults. O'Connor et al. (2007b) suggested that dispersal of marine larvae can be affected by temperature-dependent development, where at higher temperatures larvae develop faster and become competent to settle more quickly, reducing dispersal distance. Boulangeat et al. (2012) incorporated dispersal into a multi-tiered two-step model to look at the factors driving Alpine plant distribution and abundance in Europe. The modelling framework suggested that dispersal was an important predictor of species presence/absence patterns which they attributed to dispersal limitation. Dispersal was also important for identifying local species abundance and may be particularly useful for understanding sink populations that exist in locations where abiotic conditions are thought to be unsuitable. This study also highlights the fact that it can be easy to conflate dispersal boundaries with limiting environmental conditions if dispersal breaks occur in coincidence with one or more limiting environmental stressors. Conversely, if dispersal is not considered at all, it can be easy to mistake dispersal limitation for environmental limitation (Herbert et al., 2009).

In terrestrial, lentic and marine systems, dispersal is often modelled as a (more or less) symmetrical probability density function where dispersal declines exponentially with distance from the source (e.g. D'Aloia et al., 2015). Symmetrical models imagine dispersal as a kernel where dispersal probability is highest closest to a point source—often assumed to be close to the centre of the species range—and then tapers off to zero at some distance away (e.g. at or near the species' range limit). Regional weather patterns on land, currents and nearshore flow in aquatic ecosystems, species interactions, life history traits and development time can alter species trajectory and the probability of arrival at a particular location. Moreover, environmental conditions do not always create smooth geographical gradients but rather mosaics of stress and suitability (Helmuth et al., 2006). The result is that traditional models based on symmetric dispersal from a single kernel likely tend to overestimate probability of settlement at some sites, but may also miss the presence of importance stepping stones and refugia (Hannah et al., 2014). A more comprehensive approach therefore would involve the application of multiple, overlapping dispersal kernels along a species' range that account for the discontinuous nature of larval/seed production, coupled with asymmetrical dispersal kernels. A similar approach has been taken, for example, when studying fish dispersal in fragmented stream landscapes (Pepino et al., 2012).

Although we still lack a complete understanding of how patterns of dispersal affect both biogeographical and community/population-level processes, it seems virtually certain that many current patterns of dispersal will shift under climate change. A review of marine dispersal studies by Gerber et al. (2014) identified two major ways that changes in oceanic conditions are likely to alter dispersal. The first is that larval dispersal distance is expected to decrease, due in large part to shorter development times in the water column (O'Connor et al., 2007b), which will reduce what they term *functional connectivity*. The second is that disturbances in the environment due to habitat degradation, restructuring or shifts in currents are likely to alter *structural connectivity*. The barriers to dispersal by plant species on land are different from those in the sea, but many of the same considerations apply. For long-lived species, such as trees, range shifts can be slow to occur and may not be able to track the rate of climate change (Christmas et al., 2015). Increased dispersal potential is thus often considered to correlate with lower extinction risk. Counter-intuitively, a conceptual model by (Norberg et al., 2012) showed that an overall increase in dispersal ability among multiple species can lead to more localized extinctions due to increased dispersal capabilities of competitors. Although in some cases dispersal distance is expected to decrease, this model highlights the fact that predictions of range shifts and dispersal should account for species interactions to more accurately predict the effects of a shifting climate (Pillai et al., 2012). From a management perspective, conservation sites that are closer together may in some cases enhance resilience, especially among organisms with short dispersal distances (Christmas et al., 2015; Gerber et al., 2014). However, the trade-off is an increased risk of mortality due to regional-scale stressors or facilitation of the spread of disease (Southwell et al., 2016). Network planning (Nathan, 2005) which takes into account the positive sides of connectivity, including the ability of some locations to serve as "rescue sites" following mortality events (Hannah et al., 2014; Harwood et al., 2014), and the potential downsides to connectivity, are now an active area of research in the conservation biology community (Becker and Hall, 2016; Southwell et al., 2016) but have yet to be fully integrated into many species range models (but see, e.g., Araújo, 1999; Guisan and Zimmerman, 2000).

2.3.2 Species Interactions

Environmental change affects community-level processes at multiple levels of biological organization. First, as described earlier, local environmental change has direct physiological effects on organisms, such as altered metabolic rates (Brown et al., 2004; Kooijman, 2009; Mathot et al., 2015; Miller et al., 2015; van der Meer, 2006), the induction of energetically costly defence mechanisms, such as heat-shock proteins (Krebs and Loeschcke, 1994; Petes et al., 2007, 2008; Willett, 2010), or under extreme conditions, mortality (Allen and Gillooly, 2007; Benmarhnia et al., 2014). Interspecific variability in vulnerability among interacting species can thus have effects at the level of assemblages and communities (Dell et al., 2014; Monaco and Helmuth, 2011). For example, Wethey (1983, 1984) experimentally demonstrated that a competitively dominant species of barnacle had a lower physiological tolerance for heat and desiccation stress, allowing a competitive subordinate to survive at higher intertidal elevations where conditions were more physically stressful. These local drivers also played out on biogeographical scales (Wethey, 2002). In coastal zones where wind speeds are predicted to increase due to great differentials between land and sea temperature, models suggest that higher winds may lower body temperatures of intertidal organisms due to increased convection (Helmuth et al., 2011), but may negatively affect some taxa such as algae as a result of increased desiccation (Bell, 1993, 1995). Interspecific variability in drag acting on corals can select for species with lower drag coefficients following large storms (Madin et al., 2006). Rollinson et al. (2016) showed that five species of trees differed in their response to temperature and precipitation, which in turn affected competitive ability and size. Differential susceptibility to environmental change can also dictate the probability that ecosystems will be invaded by nonnative species (Kearney et al., 2008; Kelley, 2014; Phillips et al., 2008; Schneider, 2008). Species also show differential phenological responses to climate change (Primack et al., 2009), which can create temporal mismatches between, e.g., plants and their pollinators (Memmott et al., 2007), and between consumers and their prey (Winder and Schindler, 2004). Environmental change can thus decouple trophic interactions, leading to significant impacts on food webs (Gilbert et al., 2014). For example, a phenological advance in the growth season of plants in West Greenland was shown to cause a trophic mismatch with migratory Caribou, which led to a fourfold reduction in offspring production by the herbivore (Post and Forchhammer, 2008).

The differential effects of environmental stress have been conceptualized as Environmental Stress Models (ESMs; Menge and Sutherland, 1987; Menge et al., 2002; Petes et al., 2008). While not necessarily developed with global climate change in mind, ESMs nevertheless have direct relevance for exploring the ecological impacts of a changing climate. Consumer Stress Models (Cheng and Grosholz, 2016; Menge and Sutherland, 1987) suggest that sensitivity to environmental stress is greater in mobile consumers (which may be too large or too slow to find refugia) compared to their sessile prey. In contrast, Prey Stress Models (Menge and Olson, 1990) suggest that prey experience higher levels of physiological stress than their consumers (which can presumably avoid stressful conditions through behaviour). The conditions under which each of these two models operates have been the subject of field tests using biochemical indicators of stress (Menge et al., 2002; Petes et al., 2008) and have pointed to the complexity of defining "stress" given the cascading mechanisms by which weather drives local environmental conditions, behaviour alters exposure to those conditions and physiology determines vulnerability (Monaco and Helmuth, 2011; Pawar et al., 2015).

In addition to direct (physiological) effects, environmental change also has indirect effects on species interactions, for instance by altering foraging rate (Bannerman et al., 2011; Cerdá et al., 1998; Hayford et al., 2015). The temperature dependence of metabolism has formed the basis of the Metabolic Theory of Ecology (MTE) which argues that metabolic demand, which varies with temperature and body size, ultimately controls ecological processes at all levels of biological organization (Box 3; Brown et al., 2004; O'Connor et al., 2007a; van der Meer, 2006). While MTE is backed by empirical support (Savage et al., 2004), it still remains a point of contention (O'Connor et al., 2007a), in part because of assumptions of how body size is related to metabolism (Hirst et al., 2014), but also because of the assumption that feeding rate follows an Arrhenius relationship with temperature (Vucic-Pestic et al., 2010), i.e., the somewhat adaptationist argument that animals feed more because they require more energy to offset higher metabolic rates (Savage et al., 2004). This latter assumption is especially problematic because, as an increasing number of experimental studies have shown, the relationship between foraging rate and temperature is (like a TPC) typically hump-shaped, increasing to some optimum before declining rapidly (Englund et al., 2011; Rall et al., 2012). For example, elevated water temperatures have been shown to increase foraging rates by seastars (Sanford, 1999, 2002). However, under more thermally stressful conditions experienced during aerial exposure at low tide, the same species showed up to a 40% decrease in foraging rates at elevated temperatures (Pincebourde et al., 2008). Simply being in the presence of a predator can significantly increase the metabolic rate of prey (Rovero et al., 1999; Trussell et al., 2006), while at the same time decreasing their foraging time (Stoks, 2001; Trussell and Schmitz, 2012). These nonconsumptive effects can rival the direct physiological influence of fairly substantial changes in temperature (Miller et al., 2014).

The ways in which temperature affects the behaviour of interacting species within intact food webs can be complex (Mustin et al., 2007), especially when thermal optima differ among interacting species (Dell et al., 2013, 2014; Freitas et al., 2007; Pawar et al., 2015). Allan and colleagues (2015) studied the effects of increased temperature on interaction between a piscivorous dottyback fish and its prey, a planktivorous damselfish. They showed that maximal attack speeds by the predator increased, but that predator avoidance behaviours by the prey species decreased. As a result, at least over the range of temperatures tested, predation rate increased exponentially. Barton (2014) examined the effects of wind speed (which has, on average, decreased worldwide due at least in part to global climate change; McVicar et al., 2012) on soybean aphids and their predators, multicoloured Asian ladybeetles. Barton (2014) experimentally showed that wind speed had no direct physiological effect on aphids, but when comparing field conditions where wind was either blocked or not, he showed that by moving plants, wind doubled the amount of time that it took predators to begin consuming prey. In plots where wind was experimentally blocked, aphid (prey) abundance was reduced by 40% relative to the control (Barton, 2014).

Barton and Ives (2014) further explored the direct and indirect effects of warming on corn leaf aphids, their predators and two species of aphid-tending ant mutualists which defend aphids from predators. They showed that increased temperatures had a positive direct effect on aphid population growth rate. However, winter ants (which as the name implies prefer cooler temperatures) were less aggressive in their defence of aphids under warmer conditions, leading to a decrease in aphid abundance when in the presence of their predators. In contrast, warming increased the number of cornfield ants, although they were less aggressive in their defence of aphids as winter ants (Barton and Ives, 2014).

When considered in the context of the other issues raised earlier, these results strongly caution against the application of "community-level" thermal sensitivity metrics based on mapping species distributions against local environmental conditions (Watson et al., 2013), especially when using broad-scale climatic averages (Stuart-Smith et al., 2015). For example, the optimal temperature of some species can lie very close to their lethal limits. As a result, there is some evidence to suggest that species prefer environments with temperatures that may be below their optima (Martin and Huey, 2008), in order to minimize the risk of exposure to lethal extremes (Vasseur et al., 2014; Woodin et al., 2013). Thermal habitat partitioning has also been shown to occur in many organisms (Brandt et al., 1980; Escoriza and Ben Hassine, 2015), especially in organisms living in highly thermally heterogeneous habitats (Broitman et al., 2009; Wethey, 2002). There is no physiological basis to the increasingly common assumption that a species' thermal preferences can be discerned simply from broad-scale measurements of environmental temperature.

3. MOVING FORWARD: HOW DO WE MAKE USEFUL FORECASTS WHILE RECOGNIZING LIMITATIONS?

All of the potential caveats, pitfalls and complexities explored above point to the challenges inherent in scaling from suborganismal physiology to ecosystems (Pawar et al., 2015) and provide ample evidence of the highly nonlinear ways that processes at suborganismal scales can lead to higher order ecological responses. Conversely, they also suggest that the application of "environmental data" (from remote sensing platforms, weather stations and climate models) is also far more complex than is generally appreciated by scientists working at organismal and suborganismal levels (Helmuth, 2002; Kearney, 2006). Underlying every modelling technique is a series of equations and algorithms that make a myriad of assumptions about the data being used and the way that a particular system functions. Many of these come in the form of prewritten packages that can be implemented in a variety of coding platforms and programs for use by different end-users with different goals. However, the selection of different model algorithms can create larger differences in predictions than differences in the resolution of the underlying climate variables for some model types (Watling et al., 2015). Furthermore, one model or suite of equations is not likely the best predictor of the future; there may be multiple models that would predict overlapping and dissimilar results when fed the same data set (Wenger et al., 2013). Nevertheless, as emphasized throughout this manuscript, as well as by others (e.g. Yokomizo et al., 2014) action on climate change needs to occur even in the face of uncertainty. In engineering and meteorological sciences, it is common practice to run multiple models and gauge model congruence before making predictions and to report model results with an explicit description of uncertainty. Understanding the range of possible predictions is imperative for understanding how likely our policies are to fail and under what circumstances (Wenger et al., 2013). Below we review promising new methods for quantitative forecasts and argue for a greater need to be fully transparent about the limitations and advantages of various modelling approaches, and to test model skill (the ability of a model to predict a defined set of events) based on metrics relevant to the ultimate application (Araújo and New, 2007; Pacifici et al., 2015).

3.1 Hybrid Models

One method of ecological forecasting to gain traction in recent years is the use of hybrid modelling approaches that overlay the predictions of different model types or incorporate both mechanistic and correlational information into a multistep model (Bocedi et al., 2014; Briscoe et al., 2016; Ehrlen and Morris, 2015; Martinez et al., 2015; Pacifici et al., 2015; Peterson et al., 2015). These methods allow for quantification of model uncertainty, both in terms of spatial and temporal predictions as well as in identification of both the geographic regions and mechanisms/processes that are most likely to be important under current and future scenarios. Added power is gained because the results from these models can be treated as independent due to the fact that the model types operate at different scales, rely on different assumptions and capture different processes limiting a species distribution. The regions of models overlap of predicted species suitability (or decline) are areas where we have higher confidence in those predictions than others derived from one model alone (Kearney et al., 2010b; Wenger et al., 2013).

A 2010 study by Kearney et al. used this approach to examine both the fecundity and the distribution of greater gliders, *Petauroides volans*, under current and future conditions. Niche Mapper (see Box 3) was used as the mechanistic model and both Maxent and Bioclim algorithms for the SDMs. Niche Mapper was fit with fine-scaled data downscaled using a microclimate model, while the SDMs were fit using mean annual temperature, precipitation of the warmest quarter and dominant vegetation type as predictor variables which were sufficient to get a high level of model fit. This study showed that both models performed similarly to predict the range of this small marsupial under current and future climate conditions. As expected, Niche Mapper (a mechanistic model) overpredicted the range because it did not account for interactions beyond the fundamental niche. Although these models generally agreed on range predictions, they had certain regions where congruence was greater. From a management scenario, the use of multiple model types can help confirm previous predictions that have been made to generate a list of species that are most at risk and geographical regions that must be maintained as they are crucial for the continued persistence of these species.

Predicting regions that *will become* critical habitat is important for land management and conservation because, as climates begin to change, there is no guarantee that habitat that is currently suitable for a given species will continue to be suitable in the future. Briscoe et al. (2016) used both SDMs and mechanistic models to look at the locations of predicted refugia for the koala, *Phascolarctos cinereus*, under climate change scenarios. For an endemic species, like the koala, model accuracy may mean the difference between persistence and extinction. As with the greater glider example (Kearney et al., 2010a,b), this paper used the Niche Mapper framework for the bioenergetics model. A correlational model developed using MaxEnt, a model adept at handling presence-only data, was fitted with both mean and extreme conditions of selected environmental parameters. All the models (Niche Mapper, MaxEnt means and MaxEnt extremes) had relatively high predictive power when compared to independently collected data on current distributions. Niche Mapper correlated better with the MaxEnt extremes models (i.e. weather-based models) over the model based on averages (i.e. climate-based models). Again, Niche Mapper overpredicted areas of suitable habitat due in part to its inability to incorporate the presence of Eucalyptus trees, the koala's primary food source and a parameter included in MaxEnt. The models' predictive abilities diverged further when predicting the species' response to future climate scenarios. In contrast to the Watling et al. (2012) study, which found that there was little effect of incorporating ecologically relevant variables into correlational models over variables categorizing mean conditions, the variables entered into the Maxent models caused predictions to differ significantly from one another—especially under future conditions. The authors cautioned that removing correlated variables from climate models could skew predictions under future conditions where correlations could be decoupled.

Although these approaches do provide a step forward for understanding how distributions are likely to change, they do not explicitly consider or predict changes in local abundance and also ignore several potential factors that affect the future viability of a given species, including possibility of evolutionary change as well as dispersal potential. RangeShifter (Bocedi et al., 2014) is an individual-based model that incorporates aspects of population dynamics, dispersal and interindividual variability. It can also be used to make a variety of predictions including: range shifts, species dynamics, ecoevolutionary processing and tests of potential management strategies. Ehrlen and Morris (2015) also presented a framework for incorporating local demography into several existing model types by accounting for density-dependent population growth rates to predict equilibrium abundance. Methods were based on recent demographic work that has developed independently from those working on predictive models.

3.2 Physiologically and Trait-Based Indicators

One potentially fruitful avenue that facilitates the inclusion of organism-level information is the use of physiologically or trait-based indicators (Evans et al., 2015; Jackson et al., 2016; Kearney et al., 2010a; Woodin et al., 2013; Zwart et al., 2015; see chapter "Recommendations for the next generation of global freshwater biological monitoring tools" by Jackson et al. in this volume). The application of "indicators" for forecasting vulnerability to climate change has now been applied across diverse levels of biological organization (Diamond et al., 2012; Matzarakis and Amelung, 2008; McLachlan et al., 2005; Philippart et al., 2011). Rather than making spatially and temporally explicit forecasts of where ecological impacts are most likely to occur, per se, these approaches seek characteristics that make species and ecosystems most vulnerable to environmental change and/or facilitate the tracking of changes in the environment (Janetos et al., 2012). In the ecological literature, indicators range widely in the amount of detail and amount of mechanistic understanding that they entail. For example, local adaptation to climate change is unlikely in populations with very low rates of intrinsic increase, or in populations with low abundance with low genetic variance (Pearson et al., 2009). Unfortunately, risk is likely to be compounded as several indicators are also used to describe already threatened and endangered species (Mace et al., 2008).

The use of trait-based indicators is useful in several regards. Foden et al. (2013) conducted a systematic survey of nearly 17,000 species of birds, amphibians and corals. They additionally divided traits into those that made species sensitive (e.g. reliance on specialized microhabitats, environmental tolerances that are likely to be exceeded, etc.), and/or likely to have low adaptive capacity (e.g. poor dispersal ability, low genetic diversity, etc.). Methods such as these have also been used at finer level of detail to explore vulnerability among species within clades (Diamond et al., 2012). Again, the level of mechanistic detail varies, ranging from the results of controlled physiological studies to expert opinion (e.g. Case and Lawler, 2016; Hare et al., 2016).

In addition to serving as a valuable tool for identifying threats to endangered species, these methods can also be applied to understanding vulnerability of commercially important species (Hare et al., 2016). While less common, the use of trait-based indicators can also be used to model and monitor the vulnerability of keystone species (Hoey and Bellwood, 2009; Paine, 1966; Power et al., 1996), and thus may provide a mechanism of forecasting the likelihood of ecological tipping points (Dakos et al., 2012; Monaco and Helmuth, 2011). Studies exploring the vulnerability of key organisms to suites of stressors (Kroeker et al., 2016) can not only provide targets for monitoring but also may also point to options for adaptive management through the amelioration of nonclimatic stressors (Selkoe et al., 2015).

3.3 Transparency in Assumptions, Ensemble Forecasting and Tests of Model Skill: Accuracy Is in the Eye of the Stakeholder?

The use of ensemble approaches, where multiple models are used to make projections of the same response variables, has gained traction in ecological modelling as a means of capturing various sources of uncertainty both in model input parameters and in model structure (Araújo and New, 2007; Brook et al., 2009; Casati et al., 2008; Wenger et al., 2013). Araújo and New (2007) provide a detailed overview of options for the use of model ensembles. In brief, forecast models used can vary in (a) initial conditions,

(b) model parameters (including estimates of uncertainty in each parameter), (c) boundary conditions (e.g. those based on projections of greenhouse gas emissions) and (d) the type (class) of model used. All of these options can relate in one form or another to the issues raised in this review and argue that a careful consideration of the inherent limitations and sources of error inherent in each model approach can be used to our advantage.

Traditionally, the central tendency of model ensembles is selected as the most likely outcome, the assumption being that error in either direction is centred on the mean or median model prediction. A probabilistic approach can also be applied, assigning likelihood to each model outcome depending on how many models yield results within a given range (Wenger et al., 2013). However, this form of consensus forecasting (Araújo and New, 2007) does not consider that models likely vary in skill according to the application to which they are employed (Giuliani and Castelletti, 2016; Pacifici et al., 2015), and the tolerance of the end-user for false positives and false negatives (Anderegg et al., 2014). A manager concerned over the possible extinction of a rare species may have grave concerns over the potential for a false negative (Type II error). The inclusion of a large amount of biological detail in such a case may be warranted, as would be the close detailed monitoring of weather conditions rather than reliance on annual averages or climate indices. At the opposite end of the spectrum, a palaeontologist interested in the broad-scale drivers of extinction in the past may care little if an estimate varies by many thousands of years, or if the details of the mechanism that caused the species' demise are known except in the broadest terms. Between these two extremes lies a range of applications each with varying requirements and tolerances for error (Fig. 3). In many cases, nested approaches may be useful. For example, if the primary modelling goal is to assess whether a particular region is likely to serve as suitable habitat for a nonnative species, or if it is likely to become unsuitable for a native species, then correlative approaches may be sufficient initially (Kearney et al., 2010b). These initial results can then be supplemented with finer-scale monitoring and modelling, with the goal of developing early warning systems (e.g. Kefi et al., 2014; Mills et al., 2013).

The primary point being made here is that models to be used in ensembles need not be selected randomly, or even that a broad array of approaches need to be used so as to reflect all possible contingencies. Rather, based on the primary focus of the user and her tolerance for (and the cost of) false positives or false negatives, models can be selected with these goals in mind

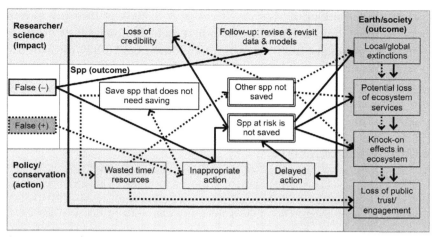

Fig. 3 Potential consequences of false positives and false negatives from predictive models when using forecasting models to inform management and policy decisions. *Solid*: pathways affected by a false negative; *dashed*: pathways affected by a false positive. *Double outline* denotes highest risk outcomes.

(Araújo and New, 2007). The "skill" of a forecasting model (or the robustness of the decisions that it informs), therefore, need not to be a static metric (Giuliani and Castelletti, 2016).

4. CONCLUSIONS

Throughout this review, we have explored the many reasons why forecasting the ecological impacts of climate change can be such a difficult venture. The field of decision support has provided much-needed guidance as to how action on climate change adaptation can be taken even in the face of large uncertainty (Selkoe et al., 2015; Smith, 2002; Yokomizo et al., 2014). Further, it is now widely accepted that solutions to climate change will need to involve diverse sectors, even within the scientific community. Our goal here is not to systematically denigrate the various approaches being used to explore large-scale changes in responses to environmental change, but rather to assemble concerns raised by researchers from diverse organisms, ecosystems and approaches with the hope of fostering further transdisciplinary research. While a "perfect" modelling solution does not exist, it is nevertheless vital to be transparent about what assumptions go into each, especially when these go unrecognized by the researchers using them: what may seem like a trivial detail by someone working at one scale may have catastrophic consequences for someone working at another. It is highly likely that in many cases by understanding the concerns raised by others trained in otherwise foreign disciplines, new insights can be gained in one's own field. Only through truly transdisciplinary research, and an explicit and transparent recognition of the strengths of weaknesses inherent in various forecasting approaches, can we move forward thoughtfully, yet with sufficient momentum to meet the ever-growing challenges presented by global climate change.

ACKNOWLEDGEMENTS

The authors thank Guy Woodward for his thoughtful comments on the manuscript. J.L.T. and B.H. were supported by US National Science Foundation Grant IOS-1557868, and R.L.K. was supported by US National Science Foundation fellowship DBI-1401656. This is contribution number 342 of the Northeastern University Marine Science Center.

REFERENCES

Abdelghany, A.Y., Awadalla, S.S., Abdel-Baky, N.F., El-Syrafi, H.A., Fields, P.G., 2010. Effect of high and low temperatures on the drugstore beetle (Coleoptera: Anobiidae). J. Econ. Entomol. 103 (5), 1909–1914.

Aber, J., Neilson, R.P., McNulty, S., Lenihan, J.M., Bachelet, D., Drapek, R.J., 2001. Forest processes and global environmental change: predicting the effects of individual and multiple stressors. Bioscience 51 (9), 735–751.

Ackerly, D.D., Loarie, S.R., Cornwell, W.K., Weiss, S.B., Hamilton, H., Branciforte, R., et al., 2010. The geography of climate change: implications for conservation biogeography. Divers. Distrib. 16, 476–487.

Ackerly, D.D., Cornwell, W.K., Weiss, S.B., Flint, L.E., Flint, A.L., 2015. A geographic mosaic of climate change impacts on terrestrial vegetation: which areas are most at risk? PLoS One 10 (6), e0130629.

Adler, P.B., Dalgleish, H.J., Ellner, S.P., 2012. Forecasting plant community impacts of climate variability and change: when do competitive interactions matter? J. Ecol. 100, 478–487.

Allan, B.J.M., Domenici, P., Munday, P.L., McCormick, M.I., 2015. Feeling the heat: the effect of acute temperature changes on predator–prey interactions in coral reef fish. Conserv. Physiol. 3, cov011.

Allen, A.P., Gillooly, J.F., 2007. The mechanistic basis of the metabolic theory of ecology. Oikos 116, 1073–1077.

Allouche, O., Tsoar, A., Kadmon, R., 2006. Assessing the accuracy of species distribution models: prevalence, kappa and the true skill statistic (TSS). J. Appl. Ecol. 43, 1223–1232.

Alpert, P., 2005. The limits and frontiers of desiccation-tolerant life. Integr. Comp. Biol. 45, 685–695.

Anderegg, W.R.L., Callaway, E.S., Boykoff, M.T., Yohe, G., Root, T.L., 2014. Awareness of both type 1 and 2 errors in climate science and assessment. Bull. Am. Meteorol. Soc. 95, 1445–1451.

Anderson, R.P., 2013. A framework for using niche models to estimate impacts of climate change on species distributions. Ann. N. Y. Acad. Sci. 1297, 8–28.

Andrewartha, H.G., Birch, L.C., 1954. The Distribution and Abundance of Animals. University of Chicago Press, Chicago, IL.

Angert, A.L., Sheth, S.N., Paul, J.R., 2011. Incorporating population-level variation in thermal performance into predictions of geographic range shifts. Integr. Comp. Biol. 51 (5), 733–750.

Angilletta, M.J., 2006. Estimating and comparing thermal performance curves. J. Therm. Biol. 31 (7), 541–545.
Angilletta, M.J., 2009. Thermal Adaptation: A Theoretical and Empirical Synthesis. Oxford University Press, New York, NY.
Angilletta, M.J., Huey, R.B., Frazier, M.R., 2010. Thermodynamic effects on organismal performance: is hotter better? Physiol. Biochem. Zool. 83 (2), 197–206.
Araújo, M.B., 1999. Distribution patterns of biodiversity and the design of a representative reserve network in Portugal. Divers. Distrib. 5, 151–163.
Araújo, M.B., Guisan, A., 2006. Five (or so) challenges for species distribution modelling. J. Biogeogr. 33, 1677–1688.
Araújo, M.B., Luoto, M., 2007. The importance of biotic interactions for modelling species distributions under climate change. Glob. Ecol. Biogeogr. 16, 743–753.
Araújo, M.B., New, M., 2007. Ensemble forecasting of species distributions. Trends Ecol. Evol. 22 (1), 42–47.
Araújo, M.B., Rozenfeld, A., 2014. The geographic scaling of biotic interactions. Ecography 37, 406–415.
Araújo, M.B., Pearson, R.G., Thuiller, W., Erhard, M., 2005a. Validation of species–climate impact models under climate change. Glob. Chang. Biol. 11, 1504–1513.
Araújo, M.B., Thuiller, W., Williams, P.H., Reginster, I., 2005b. Downscaling European species atlas distributions to a finer resolution: implications for conservation planning. Glob. Ecol. Biogeogr. 14, 17–30.
Austin, M., 2007. Species distribution models and ecological theory: a critical assessment and some possible new approaches. Ecol. Model. 200, 1–19.
Banks-Leite, C., Pardini, R., Boscolo, D., Cassano, C.R., Püttker, T., Barros, C.S., et al., 2014. Assessing the utility of statistical adjustments for imperfect detection in tropical conservation science. J. Appl. Ecol. 51, 849–859.
Bannerman, J.A., Gillespie, D.R., Roitberg, B.D., 2011. The impacts of extreme and fluctuating temperatures on trait-mediated indirect aphid-parasitoid interactions. Ecol. Entomol. 36 (4), 490–498.
Barbet-Massin, M., Jetz, W., 2014. A 40-year, continent-wide, multispecies assessment of relevant climate predictors for species distribution modelling. Divers. Distrib. 20, 1285–1295.
Barillé, L., Lerouxel, A., Dutertre, M., Haure, J., Barillé, A.-L., Pouvreau, S., et al., 2011. Growth of the Pacific oyster (Crassostrea gigas) in a high-turbidity environment: comparison of model simulations based on scope for growth and dynamic energy budgets. J. Sea Res. 66, 392–402.
Barton, B.T., 2014. Reduced wind strengthens top-down control of an insect herbivore. Ecology 95 (9), 2375–2381.
Barton, A.D., Casey, K.S., 2005. Climatological context for large-scale coral bleaching. Coral Reefs 24 (4), 536–554.
Barton, B.T., Ives, A.R., 2014. Direct and indirect effects of warming on aphids, their predators, and ant mutualists. Ecology 95 (6), 1479–1484.
Beaugrand, G., Helaouët, P., 2008. Simple procedures to assess and compare the ecological niche of species. Mar. Ecol. Prog. Ser. 363, 29–37.
Becker, D.J., Hall, R.J., 2016. Heterogeneity in patch quality buffers metapopulations from pathogen impacts. Theor. Ecol. 9 (2), 197–205.
Bell, E.C., 1993. Photosynthetic response to temperature and desiccation of the intertidal alga *Mastocarpus papillatus*. Mar. Biol. 117, 337–346.
Bell, E.C., 1995. Environmental and morphological influences on thallus temperature and desiccation of the intertidal alga *Mastocarpus papillatus* Kützing. J. Exp. Mar. Biol. Ecol. 191, 29–55.
Benedetti-Cecchi, L., Bertocci, I., Micheli, F., Maggi, E., Fosella, T., Vaselli, S., 2003. Implications of spatial heterogeneity for management of marine protected areas

(MPAs): examples from assemblages of rocky coasts in the northwest Mediterranean. Mar. Environ. Res. 55, 429–458.

Benedetti-Cecchi, L., Bertocci, I., Vaselli, S., Maggi, E., 2005. Determinants of spatial pattern at different scales in two populations of the marine alga *Rissoella verruculosa*. Mar. Ecol. Prog. Ser. 293, 37–47.

Benmarhnia, T., Sottile, M.-F., Plante, C., Brand, A., Casati, B., Fournier, M., et al., 2014. Variability in temperature-related mortality projections under climate change. Environ. Health Perspect. 122 (12), 1293–1298.

Bennett, N.L., Severns, P.M., Parmesan, C., Singer, M.C., 2014. Geographic mosaics of phenology, host preference, adult size and microhabitat choice predict butterfly resilience to climate warming. Oikos 124, 41–53.

Bennie, J., Wilson, R.J., Maclean, I.M.D., Suggitt, A.J., 2014. Seeing the woods for the trees—when is microclimate important in species distribution models? Glob. Chang. Biol. 20, 2699–2700.

Berkhout, F., van den Hurk, B., Bessembinder, J., de Boer, J., Bregman, B., van Drunen, M., 2014. Framing climate uncertainty: socio-economic and climate scenarios in vulnerability and adaptation assessments. Reg. Environ. Chang. 14 (3), 879–893.

Bickford, D., Lohman, D.J., Sohdi, N.S., Ng, P.K.L., Meier, R., Winker, K., et al., 2007. Cryptic species as a window on diversity and conservation. Trends Ecol. Evol. 22 (3), 148–155.

Bocedi, G., Palmer, S.C.F., Pe'er, G., Heikkinen, R.K., Matsinos, Y.G., Watts, K., et al., 2014. RangeShifter: a platform for modelling spatial eco-evolutionary dynamics and species' responses to environmental changes. Methods Ecol. Evol. 5 (4), 388–396.

Boogert, N.J., Paterson, D.M., Laland, K.N., 2006. The implications of niche construction and ecosystem engineering for conservation biology. Bioscience 56 (7), 570–578.

Botkin, D.B., Saxe, H., Araújo, M.B., Betts, R., Bradshaw, R.H.W., Cedhagen, T., et al., 2007. Forecasting the effects of global warming on biodiversity. Bioscience 57 (3), 227–236.

Boulangeat, I., Gravel, D., Thuiller, W., 2012. Accounting for dispersal and biotic interactions to disentangle the drivers of species distributions and their abundances. Ecol. Lett. 15 (6), 584–593.

Brandt, S.B., Magnuson, J.J., Crowder, L.B., 1980. Thermal habitat partitioning by fishes in Lake Michigan. Can. J. Fish. Aquat. Sci. 37 (10), 1557–1564.

Brennan, G., Collins, S., 2015. Growth responses of a green alga to multiple environmental drivers. Nat. Clim. Chang. 5 (9), 892–897.

Briscoe, N.J., Kearney, M.R., Taylor, C., Brendan, W.A., 2016. Unpacking the mechanisms captured by a correlative SDM to improve predictions of climate refugia. Glob. Chang. Biol. 22 (7), 2425–2439.

Broennimann, O., Treier, U., Muller-Scharer, H., Thuiller, W., Peterson, A., Guisan, A., 2007. Evidence of climatic niche shift during biological invasion. Ecol. Lett. 10, 701–709.

Broitman, B.R., Szathmary, P.L., Mislan, K.A.S., Blanchette, C.A., Helmuth, B., 2009. Predator–prey interactions under climate change: the importance of habitat vs body temperature. Oikos 118 (2), 219–224.

Brook, B.W., Akcakaya, H.R., Keith, D.A., Mace, G.M., Pearson, R.G., Araújo, M.B., 2009. Integrating bioclimate with population models to improve forecasts of species extinctions under climate change. Biol. Lett. 5 (6), 723–725.

Brown, J.H., Gillooly, J.F., Allen, A.P., Savage, V.M., West, G.B., 2004. Toward a metabolic theory of ecology. Ecology 85 (7), 1771–1789.

Brown, C.J., Schoeman, D.S., Sydeman, W.J., Brander, K., Buckley, L.B., Burrows, M., et al., 2011. Quantitative approaches in climate change ecology. Glob. Chang. Biol. 17 (12), 3697–3713.

Buckley, L.B., 2008. Linking traits to energetics and population dynamics to predict lizard ranges in changing environments. Am. Nat. 171 (1), E1–E19.

Buckley, L.B., Kingsolver, J.G., 2012. Functional and phylogenetic approaches to forecasting species' responses to climate change. Ann. Rev. Ecol. Evol. Syst. 43, 205–226.

Buckley, L.B., Urban, M.C., Angilletta, M.J., Crozier, L.G., Rissler, L.J., Sears, M.W., 2010. Can mechanism inform species distribution models? Ecol. Lett. 13, 1041–1054.

Burrows, M.T., Harvey, R., Robb, L., Poloczanska, E.S., Mieszkowska, N., Moore, P., et al., 2009. Spatial scales of variance in abundance of intertidal species: effects of region, dispersal mode, and trophic level. Ecology 90 (5), 1242–1254.

Burrows, M.T., Schoeman, D.S., Richardson, A.J., Molinos, J.G., Hoffmann, A., Buckley, L.B., et al., 2014. Geographical limits to species-range shifts are suggested by climate velocity. Nature 507 (7493), 492–495.

Casati, B., Wilson, L.J., Stephenson, D.B., Nurmi, P., Ghelli, A., Pocernich, M., et al., 2008. Forecast verification: current status and future directions. Meteorol. Appl. 15, 3–18.

Case, M.J., Lawler, J.J., 2016. Relative vulnerability to climate change of trees in western North America, Clim. Change 136 (2), 367–379.

Castillo, K.D., Lima, F.P., 2010. Comparison of in situ and satellite-derived (MODIS-Aqua/Terra) methods for assessing temperatures on coral reefs. Limnol. Oceanogr. Methods 8, 107–117.

Castorani, M.C.N., Reed, D.C., Alberto, F., Bell, T.W., Simons, R.D., Cavanaugh, K.C., et al., 2015. Connectivity structures local population dynamics: a long-term empirical test in a large metapopulation system. Ecology 96 (12), 3141–3152.

Catchpole, E.A., Morgan, B.J.T., Coulson, T.N., Freeman, S.N., Albon, S.D., 2000. Factors influencing Soay sheep survival. J. R. Stat. Soc. C-Appl. Stat. 49, 453–472.

Cerdá, X., Retana, J., Cros, S., 1998. Critical thermal limits in Mediterranean ant species: trade-off between mortality risk and foraging performance. Funct. Ecol. 12, 45–55.

Chapperon, C., Seuront, L., 2011. Space-time variability in environmental thermal properties and snail thermoregulatory behaviour. Funct. Ecol. 25 (5), 1040–1050.

Chase, J.M., Leibold, M.A., 2003. Ecological Niches: Linking Classical and Contemporary Approaches. University of Chicago Press, Chicago, IL.

Cheng, B.S., Grosholz, E.D., 2016. Environmental stress mediates trophic cascade strength and resistance to invasion. Ecosphere 7 (4), e10247.

Chown, S.L., Gaston, K.J., 2008. Macrophysiology for a changing world. Proc. R. Soc. B 275, 1469–1478.

Christensen, M.R., Graham, M.D., Vinebrooke, R.D., Findlay, D.L., Paterson, M.J., Turner, M.A., 2006. Multiple anthropogenic stressors cause ecological surprises in boreal lakes. Glob. Chang. Biol. 12, 2316–2322.

Christmas, M.J., Breed, M.F., Lowe, A.J., 2015. Constraints to and conservation implications for climate change adaptation in plants. Conserv. Genet. 17 (2), 305–320.

Clark, J.S., Lewis, M., McLachlan, J.S., HilleRisLambers, J., 2003. Estimating population spread: what can we forecast and how well? Ecology 84 (8), 1979–1988.

Clusella-Trullas, S., Blackburn, T.M., Chown, S.L., 2011. Climatic predictors of temperature performance curve parameters in ectotherms imply complex responses to climate change. Am. Nat. 177 (6), 738–751.

Connell, J.H., 1961. The influence of interspecific competition and other factors on the distribution of the barnacle *Chthamalus stellatus*. Ecology 42 (4), 710–723.

Connell, S.D., Kroeker, K.J., Fabricius, K.E., Kline, D.I., Russell, B.D., 2013. The other ocean acidification problem: CO_2 as a resource among competitors for ecosystem dominance. Philos. Trans. R. Soc. B 368, 20120442.

Coppens d'Eeckenbrugge, G., Lacape, J.-M., 2014. Distribution and differentiation of wild, feral, and cultivated populations of perennial upland cotton (*Gossypium hirsutum* L.) in Mesoamerica and the Caribbean. PLoS One 9 (9), e107458.

Craig, R.K., 2010. Stationarity is dead—long live transformation: five principles for climate change adaptation law. Harv. Env. Law Rev. 34, 9–75.

Crain, C.M., 2008. Interactions between marsh plant species vary in direction and strength depending on environmental and consumer context. J. Ecol. 96, 166–173.

Crain, C.M., Kroeker, K., Halpern, B.S., 2008. Interactive and cumulative effects of multiple human stressors in marine systems. Ecol. Lett. 11, 1304–1315.

Cumming, G.S., Cumming, D.H.M., Redman, C.L., 2006. Scale mismatches in social-ecological systems: causes, consequences, and solutions. Ecol. Soc. 11 (1), 14.

Cyr, H., Walker, S.C., 2004. An illusion of mechanistic understanding. Ecology 85 (7), 1802–1804.

Dakos, V., Carpenter, S.R., Brock, W.A., Ellison, A.M., Guttal, V., Ives, A.R., et al., 2012. Methods for detecting early warnings of critical transitions in time series illustrated using simulated ecological data. PLoS One 7 (7), e41010.

D'Aloia, C.C., Bogdanowicz, S.M., Francis, R.K., Majoris, J.E., Harrison, R.G., Buston, P.M., 2015. Patterns, causes, and consequences of marine larval dispersal. Proc. Natl. Acad. Sci. U.S.A. 112 (45), 13940–13945.

Dawson, T.P., Jackson, S.T., House, J.I., Prentice, I.C., Mace, G.M., 2011. Beyond predictions: biodiversity conservation in a changing climate. Science 332, 53–58.

De Frenne, P., Coomes, D.A., De Schrijver, A., Staelens, J., Alexander, J.M., Bernhardt-Römermann, M., Brunet, J., Chabrerie, O., Chiarucci, A., Ouden, J., Eckstein, R.L., 2014. Plant movements and climate warming: intraspecific variation in growth responses to nonlocal soils. New Phytol. 202 (2), 431–441.

DeFries, R.S., Ellis, E.C., Chapin, F.S., Matson, P.A., Turner, B.L., Agrawal, A., Crutzen, P.J., Field, C., Gleick, P., Kareiva, P.M., Lambin, E., 2012. Planetary opportunities: a social contract for global change science to contribute to a sustainable future. Bioscience 62 (6), 603–606.

Dell, A.I., Pawar, S., Savage, V.M., 2011. Systematic variation in the temperature dependence of physiological and ecological traits. Proc. Natl. Acad. Sci. U.S.A. 108 (26), 10591–10596.

Dell, A.I., Pawar, S., Savage, V.M., 2013. The thermal dependence of biological traits. Ecology 94, 1205.

Dell, A.I., Pawar, S., Savage, V., 2014. Temperature dependence of trophic interactions are driven by asymmetry of species responses and foraging strategy. J. Anim. Ecol. 83 (1), 70–84.

Denman, K., Christian, J.R., Steiner, N., Portner, H.-O., Nojiri, Y., 2011. Potential impacts of future ocean acidification on marine ecosystems and fisheries: current knowledge and recommendations for future research. ICES J. Mar. Sci. 68, 1019–1029.

Denny, M.W., Miller, L.P., Harley, C.D.G., 2006. Thermal stress on intertidal limpets: long-term hindcasts and lethal limits. J. Exp. Biol. 209, 2420–2431.

Denny, M.W., Dowd, W.W., Bilir, L., Mach, K.J., 2011. Spreading the risk: small-scale body temperature variation among intertidal organisms and its implications for species persistence. J. Exp. Mar. Biol. Ecol. 400, 175–190.

Deutsch, C.A., Tewksbury, J.J., Huey, R.B., Sheldon, K.S., Ghalambor, C.K., Haak, D.C., et al., 2008. Impacts of climate warming on terrestrial ectotherms across latitude. Proc. Natl. Acad. Sci. U.S.A. 105, 6668–6672.

Diamond, S.E., Sorger, D.M., Hulcr, J., Pelini, S.L., Del Toro, I., Hirsch, C., et al., 2012. Who likes it hot? A global analysis of the climatic, ecological, and evolutionary determinants of warming tolerance in ants. Glob. Chang. Biol. 18 (2), 448–456.

Diaz, R.J., Rosenberg, R., 2008. Spreading dead zones and consequences for marine ecosystems. Science 321 (5891), 926–929.

Dietl, G.P., Kidwell, S.M., Brenner, M., Burney, D.A., Flessa, K.W., Jackson, S.T., Koch, P.L., 2015. Conservation paleobiology: leveraging knowledge of the past to inform conservation and restoration. Annu. Rev. Earth Planet. Sci. 43 (1), 79–103.

Dobrowski, S.Z., Abatzoglou, J., Swanson, A.K., Greenberg, J.A., Mynsberge, A.R., Holden, Z.A., et al., 2013. The climate velocity of the contiguous United States during the 20th century. Glob. Chang. Biol. 19 (1), 241–251.

Donelson, J.M., Munday, P.L., McCormick, M.I., Pitcher, C.R., 2011. Rapid transgenerational acclimation of a tropical reef fish to climate change. Nat. Clim. Chang. 2 (1), 30–32.
Doney, S.C., Fabry, V.J., Feely, R.A., Kleypas, J.A., 2009. Ocean acidification: the other CO_2 problem. Ann. Rev. Mar. Sci. 1, 169–192.
Dong, Y., Han, G., Ganmanee, M., Wang, J., 2015. Latitudinal variability of physiological responses to heat stress of the intertidal limpet Cellana toreuma along the Asian coast. Mar. Ecol. Prog. Ser. 529, 107–119.
Dowd, W.W., King, F.A., Denny, M.W., 2015. Thermal variation, thermal extremes and the physiological performance of individuals. J. Exp. Biol. 218, 1956–1967.
Drake, J.M., Griffen, B.D., 2010. Early warning signals of extinction in deteriorating environments. Nature 467, 456–459.
Dudgeon, S.R., Aronson, R.B., Bruno, J.F., Precht, W.F., 2010. Phase shifts and stable states on coral reefs. Mar. Ecol. Prog. Ser. 413, 201–216.
Duncan, R.P., Cassey, P., Blackburn, T.M., 2009. Do climate envelope models transfer? A manipulative test using dung beetle introductions. Proc. R. Soc. B 267, 1449–1457.
Ehrlen, J., Morris, W.F., 2015. Predicting changes in the distribution and abundance of species under environmental change. Ecol. Lett. 18 (3), 303–314.
Ellegren, H., Galtier, N., 2016. Determinants of genetic diversity. Nat. Rev. Genet. 17 (7), 422–433.
Elton, C.S., 1927. Animal Ecology. The Macmillan Company, New York, NY.
Englund, G., Öhlund, G., Hein, C.L., Diehl, S., 2011. Temperature dependence of the functional response. Ecol. Lett. 14, 914–921.
Escoriza, D., Ben Hassine, J., 2015. Niche partitioning at local and regional scale in the North African Salamandridae. J. Herpetol. 49 (2), 276–283.
Estay, S.A., Lima, M., Bozinovic, F., 2014. The role of temperature variability on insect performance and population dynamics in a warming world. Oikos 123 (2), 131–140.
Estes, L.D., Bradley, B.A., Beukes, H., Hole, D.G., Lau, M., Oppenheimer, M.G., et al., 2013. Comparing mechanistic and empirical model projections of crop suitability and productivity: implications for ecological forecasting. Glob. Ecol. Biogeogr. 22, 1007–1018.
Evans, M.R., Grimm, V., Johst, K., Knuuttila, T., de Langhe, R., Lessells, C.M., et al., 2013. Do simple models lead to generality in ecology? Trends Ecol. Evol. 28 (10), 578–583.
Evans, T.G., Diamond, S.E., Kelly, M.W., 2015. Mechanistic species distribution modelling as a link between physiology and conservation. Conserv. Physiol. 3, 1–16.
Fabry, V.J., 2008. Marine calcifiers in a high-CO_2 ocean. Science 320, 1020–1022.
Fenichel, E.P., Levin, S.A., McCay, B., St Martin, K., Abbott, J.K., Pinsky, M.L., 2016. Wealth reallocation and sustainability under climate change. Nat. Clim. Chang. 6, 237–244.
Filgueira, R., Rosland, R., Grant, J., 2011. A comparison of scope for growth (SFG) and dynamic energy budget (DEB) models applied to the blue mussel (*Mytilus edulis*). J. Sea Res. 66 (4), 403–410.
Firth, L.B., Williams, G.A., 2009. The influence of multiple environmental stressors on the limpet Cellana toreuma during the summer monsoon season in Hong Kong. J. Exp. Mar. Biol. Ecol. 375, 70–75.
Fitzhenry, T., Halpin, P.M., Helmuth, B., 2004. Testing the effects of wave exposure, site, and behavior on intertidal mussel body temperatures: applications and limits of temperature logger design. Mar. Biol. 145 (2), 339–349.
Fitzpatrick, M.C., Hargrove, W.W., 2009. The projection of species distribution models and the problem of non-analog climate. Biodivers. Conserv. 18, 2255–2261.
Foden, W.B., Butchart, S.H.M., Stuart, S.N., Vie, J.C., Akcakaya, H.R., Angulo, A., et al., 2013. Identifying the world's most climate change vulnerable species: a systematic trait-based assessment of all birds, amphibians and corals. PLoS One 8 (6), 13.

Forchhammer, M.C., Post, E., 2004. Using large-scale climate indices in climate change ecology studies. Popul. Ecol. 46, 1–12.

Freitas, V., Campos, J., Fonds, M., Van der Veer, H.W., 2007. Potential impact of temperature change on epibenthic predator–bivalve prey interactions in temperate estuaries. J. Therm. Biol. 32 (6), 328–340.

Freitas, V., Cardoso, J., Santos, S., Campos, J., Drent, J., Saraiva, S., et al., 2009. Reconstruction of food conditions for Northeast Atlantic bivalve species based on Dynamic Energy Budgets. J. Sea Res. 62 (2-3), 75–82.

Fuentes, M.M.P.B., Porter, W.P., 2013. Using a microclimate model to evaluate impacts of climate change on sea turtles. Ecol. Model. 251, 150–157.

Gaitan-Espitia, J.D., Bacigalupe, L.D., Opitz, T., Lagos, N.A., Timmermann, T., Lardies, M.A., 2014. Geographic variation in thermal physiological performance of the intertidal crab *Petrolisthes violaceus* along a latitudinal gradient. J. Exp. Biol. 217 (Pt 24), 4379–4386.

Gallagher, R.V., Beaumont, L.J., Hughes, L., Leishman, M.R., 2010. Evidence for climatic niche and biome shifts between native and novel ranges in plant species introduced to Australia. J. Ecol. 98, 790–799.

García-Callejas, D., Araújo, M.B., 2016. The effects of model and data complexity on predictions from species distributions models. Ecol. Model. 326, 4–12.

Garza, C., Robles, C., 2010. Effects of brackish water incursions and diel phasing of tides on vertical excursions of the keystone predator *Pisaster ochraceus*. Mar. Biol. 157 (3), 673–682.

Gaston, K.J., Chown, S.L., Calosi, P., Bernardo, J., Bilton, D.T., Clarke, A., et al., 2009. Macrophysiology: a conceptual reunification. Am. Nat. 174 (5), 595–612.

Gaylord, B., Gaines, S.D., 2000. Temperature or transport? Range limits in marine species mediated solely by flow. Am. Nat. 155, 769–789.

Gerber, L.R., Mancha-Cisneros, M.D.M., O'Connor, M.I., Selig, E.R., 2014. Climate change impacts on connectivity in the ocean: implications for conservation. Ecosphere 5 (3), art33.

Gilbert, B., Tunney, T.D., McCann, K.S., DeLong, J.P., Vasseur, D.A., Savage, V., et al., 2014. A bioenergetic framework for the temperature dependence of trophic interactions. Ecol. Lett. 17 (8), 902–914.

Gilchrist, G.W., 1995. Specialists and generalists in changing environments. 1. Fitness landscapes of thermal sensitivity. Am. Nat. 146 (2), 252–270.

Gillingham, P.K., Huntley, B., Kunin, W.E., Thomas, C.D., 2012. The effect of spatial resolution on projected responses to climate warming. Divers. Distrib. 18 (10), 990–1000.

Giuliani, M., Castelletti, A., 2016. Is robustness really robust? How different definitions of robustness impact decision-making under climate change. Clim. Chang. 135 (3–4), 409–424.

Gleeson, M.W., Strong, A.E., 1995. Applying MCSST to coral reef bleaching. Adv. Space Res. 16 (10), 151–154.

Glick, P., Staudt, A., Stein, B., 2009. A New Era for Conservation: Review of Climate Change Adaptation Literature. National Wildlife Foundation, Washington, DC.

Glick, P., Stein, B., Edelson, N.A., 2011. Scanning the Conservation Horizon: A Guide to Climate Change Vulnerability Assessment. National Wildlife Foundation, Washington, DC.

Gouhier, T.C., Guichard, F., Menge, B.A., 2010. Ecological processes can synchronize marine population dynamics over continental scales. Proc. Natl. Acad. Sci. U.S.A. 107 (18), 8281–8286.

Grinnell, J., 1917. The niche-relationships of the California Thrasher. Auk 34, 427–433.

Grinnell, J., 1928. Presence and absence of animals. Univ. Calif. Chronicle 30, 429–450.

Guichard, F., Levin, S.A., Hastings, A., Siegel, D., 2004. Toward a dynamic metacommunity approach to marine reserve theory. Bioscience 54 (11), 1003–1011.

Guisan, A., Zimmerman, N.E., 2000. Predictive habitat distribution models in ecology. Ecol. Model. 135, 147–186.
Gunderson, A.R., Armstrong, E.J., Stillman, J.H., 2016. Multiple stressors in a changing world: the need for an improved perspective on physiological responses to the dynamic marine environment. Ann. Rev. Mar. Sci. 8, 12.1–12.22.
Gutt, J., Zurell, D., Bracegridle, T.J., Cheung, W., Clark, M.S., Convey, P., et al., 2012. Correlative and dynamic species distribution modelling for ecological predictions in the Antarctic: a cross-disciplinary concept. Polar Res. 31, 11091.
Hallett, T.B., Coulson, T., Pilkington, J.G., Clutton-Brock, T.H., Pemberton, J.M., Grenfell, B.T., 2004. Why large-scale climate indices seem to predict ecological processes better than local weather. Nature 430, 71–75.
Halpin, P., 1997. Global climate change and natural-area protection: management responses and research directions. Ecol. Appl. 7, 828–843.
Hampe, A., 2004. Bioclimate envelope models: what they detect and what they hide. Glob. Ecol. Biogeogr. 13, 469–476.
Hannah, L., Flint, L., Syphard, A.D., Moritz, M.A., Buckley, L.B., McCullough, I.M., 2014. Fine-grain modeling of species' response to climate change: holdouts, stepping-stones, and microrefugia. Trends Ecol. Evol. 29 (7), 390–397.
Hare, J.A., Morrison, W.E., Nelson, M.W., Stachura, M.M., Teeters, E.J., Griffis, R.B., et al., 2016. A vulnerability assessment of fish and invertebrates to climate change on the Northeast US continental shelf. PLoS One 11 (2), 30.
Harwood, T.D., Mokany, K., Paini, D.R., 2014. Microclimate is integral to the modeling of plant responses to macroclimate. Proc. Natl. Acad. Sci. U.S.A. 111 (13), E1164–E1165.
Hayford, H.A., Gilman, S.E., Carrington, E., 2015. Foraging behavior minimizes heat exposure in a complex thermal landscape. Mar. Ecol. Prog. Ser. 518, 165–175.
Hein, L., van Koppen, K., de Groot, R.S., van Ierland, E.C., 2006. Spatial scales, stakeholders and the valuation of ecosystem services. Ecol. Econ. 57, 209–228.
Heller, N.E., Kreitler, J., Ackerly, D.D., Weiss, S.B., Recinos, A., Branciforte, R., et al., 2015. Targeting climate diversity in conservation planning to build resilience to climate change. Ecosphere 6 (4), 65.
Helmuth, B.S.T., 1998. Intertidal mussel microclimates: predicting the body temperature of a sessile invertebrate. Ecol. Monogr. 68 (1), 51–74.
Helmuth, B., 2002. How do we measure the environment? Linking intertidal thermal physiology and ecology through biophysics. Integr. Comp. Biol. 42 (4), 837–845.
Helmuth, B.S., Hofmann, G.E., 2001. Microhabitats, thermal heterogeneity, and patterns of physiological stress in the rocky intertidal zone. Biol. Bull. 201 (3), 374–384.
Helmuth, B.S., Harley, C.D.G., Halpin, P., O'Donnell, M., Hofmann, G.E., Blanchette, C., 2002. Climate change and latitudinal patterns of intertidal thermal stress. Science 298, 1015–1017.
Helmuth, B., Broitman, B.R., Blanchette, C.A., Gilman, S., Halpin, P., Harley, C.D.G., et al., 2006. Mosaic patterns of thermal stress in the rocky intertidal zone: implications for climate change. Ecol. Monogr. 76 (4), 461–479.
Helmuth, B., Broitman, B.R., Yamane, L., Gilman, S.E., Mach, K., Mislan, K.A.S., et al., 2010. Organismal climatology: analyzing environmental variability at scales relevant to physiological stress. J. Exp. Biol. 213, 995–1003.
Helmuth, B., Yamane, L., Lalwani, S., Matzelle, A., Tockstein, A., Gao, N., 2011. Hidden signals of climate change in intertidal ecosystems: what (not) to expect when you are expecting. J. Exp. Mar. Biol. Ecol. 400, 191–199.
Helmuth, B., Russell, B.D., Connell, S.D., Dong, Y., Harley, C.D.G., Lima, F.P., et al., 2014. Beyond long-term averages: making biological sense of a rapidly changing world. Clim. Chang. Res. 1, 10–20.

Herbert, R.J.H., Southward, A.J., Clarke, R.T., Sheader, M., Hawkins, S.J., 2009. Persistent border: an analysis of the geographic boundary of an intertidal species. Mar. Ecol. Prog. Ser. 379, 135–150.

Hijmans, R.J., Graham, C.H., 2006. The ability of climate envelope models to predict the effect of climate change on species distributions. Glob. Chang. Biol. 12, 2272–2281.

Hirst, A.G., Glazier, D.S., Atkinson, D., 2014. Body shape shifting during growth permits tests that distinguish between competing geometric theories of metabolic scaling. Ecol. Lett. 17 (10), 1274–1281.

Hoey, A.S., Bellwood, D.R., 2009. Limited functional redundancy in a high diversity system: single species dominates key ecological process on coral reefs. Ecosystems 12, 1316–1328.

Holtmeier, F.-K., Broll, G., 1992. The influence of tree islands and microtopography on pedoecological conditions in the forest-alpine tundra ecotone on Niwot Ridge. Arct. Alp. Res. 24 (3), 216–228.

Howard, J., Babij, E., Griffis, R., Helmuth, B., Himes-Cornell, A., Niemier, P., et al., 2013. Oceans and marine resources in a changing climate. Oceanogr. Mar. Biol. Ann. Rev. 51, 71–192. London.

Huey, R., Peterson, C.R., Arnold, S.J., Porter, W.P., 1989. Hot rocks and not-so-hot rocks: retreat-site selection by garter snakes and its thermal consequences. Ecology 70 (4), 931–944.

Huey, R.B., Kearney, M.R., Krockenberger, A., Holtum, J.A., Jess, M., Williams, S.E., 2012. Predicting organismal vulnerability to climate warming: roles of behaviour, physiology and adaptation. Philos. Trans. R. Soc. Lond. B Biol. Sci. 367 (1596), 1665–1679.

Hutchins, L.W., 1947. The bases for temperature zonation in geographical distribution. Ecol. Monogr. 17, 325–335.

Hutchinson, G.E., 1957. Concluding remarks. Cold Spring Harbor Symp. Quant. Biol. 22, 415–427.

IPCC, 2013. Climate Change 2013: The Physical Science Basis. Contribution of working group 1 to the fifth assessment report of the intergovernmental panel on climate change, Cambridge University Press, United Kingdom and New York.

Irwin, D.E., 2002. Phylogeographic breaks without geographic barriers to gene flow. Evolution 56 (12), 2383–2394.

Isaak, D.J., Rieman, B.E., 2013. Stream isotherm shifts from climate change and implications for distributions of ectothermic organisms. Glob. Chang. Biol. 19 (3), 742–751.

Jackson, M.C., Loewen, C.J.G., Vinebrooke, R.D., Chimimea, C.T., 2016. Net effects of multiple stressors in freshwater ecosystems: a meta-analysis. Glob. Chang. Biol. 22, 180–189.

Janetos, A.C., Chen, R.S., Arndt, D., Kenney, M.A., 2012. National Climate Assessment Indicators: Background, Development, and Examples. Columbia University Academic Commons, Washington, DC.

Jentsch, A., Beierkuhnlein, C., 2008. Research frontiers in climate change: effects of extreme meteorological events on ecosystems. C. R. Geosci. 340 (9–10), 621–628.

Jentsch, A., Kreyling, J., Beirkuhnlein, C., 2007. A new generation of climate-change experiments: events, not trends. Front. Ecol. Environ. 5 (7), 365–374.

Jeschke, J., Strayer, D., 2008. Usefulness of bioclimatic models for studying climate change and invasive species. Ann. N. Y. Acad. Sci. 1134, 1–24.

Jonard, M., Fürst, A., Verstraeten, A., Thimonier, A., Timmermann, V., Potočić, N., et al., 2015. Tree mineral nutrition is deteriorating in Europe. Glob. Chang. Biol. 21 (1), 418–430.

Jones, S.J., Mieszkowska, N., Wethey, D.S., 2009. Linking thermal tolerances and biogeography: *Mytilus edulis* (L.) at its southern limit on the East Coast of the United States. Biol. Bull. 217, 73–85.

Jones, S.J., Lima, F.P., Wethey, D.S., 2010. Rising environmental temperatures and biogeography: poleward range contraction of the blue mussel, *Mytilus edulis* L., in the western Atlantic. J. Biogeogr. 37, 2243–2259.

Kearney, M., 2006. Habitat, environment and niche: what are we modelling? Oikos 115 (1), 186–191.

Kearney, M., Porter, W., 2009. Mechanistic niche modelling: combining physiological and spatial data to predict species ranges. Ecol. Lett. 12 (4), 334–350.

Kearney, M.R., White, C.R., 2012. Testing metabolic theories. Am. Nat. 180 (5), 546–565.

Kearney, M., Phillips, B.L., Tracy, C.R., Christian, K.A., Betts, G., Porter, W.P., 2008. Modelling species distributions without using species distributions: the cane toad in Australia under current and future climates. Ecography 31 (4), 423–434.

Kearney, M., Shine, R., Porter, W.P., 2009. The potential for behavioral thermoregulation to buffer "cold-blooded" animals against climate warming. Proc. Natl. Acad. Sci. U.S.A. 106 (10), 3835–3840.

Kearney, M., Simpson, S.J., Raubenheimer, D., Helmuth, B., 2010a. Modelling the ecological niche from functional traits. Philos. Trans. R. Soc. B 365, 3469–3483.

Kearney, M.R., Wintle, B.A., Porter, W.P., 2010b. Correlative and mechanistic models of species distribution provide congruent forecasts under climate change. Conserv. Lett. 3 (3), 203–213.

Kearney, M.R., Matzelle, A., Helmuth, B., 2012. Biomechanics meets the ecological niche: the importance of temporal data resolution. J. Exp. Biol. 215, 922–933.

Kearney, M.R., Isaac, A.P., Porter, W.P., 2014. microclim: global estimates of hourly microclimate based on long-term monthly climate averages. Sci. Data 1, 140006.

Kefi, S., Guttal, V., Brock, W.A., Carpenter, S.R., Ellison, A.M., Livina, V.N., et al., 2014. Early warning signals of ecological transitions: methods for spatial patterns. PLoS One 9 (3), 13.

Kellermann, V., van Heerwaarden, B., Sgrò, C.M., Hoffmann, A.A., 2009. Fundamental evolutionary limits in ecological traits drive Drosophila species distributions. Science 325 (5945), 1244–1246.

Kelley, A.L., 2014. The role thermal physiology plays in species invasion. Conserv. Physiol. 2 (1), cou045.

Kerns, B.K., Kim, J.B., Kline, J.D., Day, M.A., 2016. US exposure to multiple landscape stressors and climate change. Reg. Env. Chang. 16 (3), 1–12.

Kershaw, F., Waller, T., Micucci, P., Draque, J., Barros, M., Buongermini, E., Pearson, R.G., Mendez, M., 2013. Informing conservation units: barriers to dispersal for the yellow anaconda. Divers. Distrib. 19 (9), 1164–1174.

Kingsolver, J.G., 2002. Impacts of global environmental change on animals. In: Mooney, H.A., Canadell, J.G. (Eds.), The Earth System: Biological and Ecological Dimensions of Global Environmental Change. John Wiley & Sons, Chichester, pp. 56–66.

Kingsolver, J.G., Woods, H.A., 2016. Beyond thermal performance curves: modeling time-dependent effects of thermal stress on ectotherm growth rates. Am. Nat. 187 (3), 283–294.

Kodra, E., Ganguly, A.R., 2014. Asymmetry of projected increases in extreme temperature distributions. Nat. Sci. Rep. 4, 5884.

Kooijman, S.A.L.M., 2009. Dynamic Energy Budget Theory for Metabolic Organisation. Cambridge University Press, Great Britain.

Kordas, R.L., Harley, C.D.G., O'Connor, M.I., 2011. Community ecology in a warming world: the influence of temperature on interspecific interactions in marine systems. J. Exp. Mar. Biol. Ecol. 400, 218–226.

Krebs, R.A., Loeschcke, V., 1994. Costs and benefits of activation of the heat-shock response in *Drosophila melanogaster*. Funct. Ecol. 8 (6), 730–737.

Kroeker, K., Kordas, R.L., Crim, R.N., Singh, G.G., 2010. Meta-analysis reveals negative yet variable effects of ocean acidification on marine organisms. Ecol. Lett. 13, 1419–1434.

Kroeker, K.J., Sanford, E., Rose, J.M., Blanchette, C.A., Chan, F., Chavez, F.P., et al., 2016. Interacting environmental mosaics drive geographic variation in mussel performance and species interactions. Ecol. Lett. 19, 771–779.

Kuo, E.S.L., Sanford, E., 2009. Geographic variation in the upper thermal limits of an intertidal snail: implications for climate envelope models. Mar. Ecol. Prog. Ser. 388, 137–146.

Lathlean, J.A., Ayre, D.J., Coleman, R.A., Minchinton, T.E., 2014. Using biomimetic loggers to measure interspecific and microhabitat variation in body temperatures of rocky intertidal invertebrates. Mar. Freshw. Res. 66 (1), 86–94.

Leichter, J.J., Helmuth, B., Fischer, A.M., 2006. Variation beneath the surface: quantifying complex thermal environments on coral reefs in the Caribbean, Bahamas and Florida. J. Mar. Res. 64, 563–588.

Lieberman, B.S., 2003. Paleobiogeography: the relevance of fossils to biogeography. Annu. Rev. Ecol. Syst. 34, 51–69.

Lieberman, B.S., 2005. Geobiology and paleobiogeography: tracking the coevolution of the Earth and its biota. Palaeogeogr. Palaeoclimatol. Palaeoecol. 219 (1-2), 23–33.

Lima, F.P., Wethey, D.S., 2012. Three decades of high-resolution coastal sea surface temperatures reveal more than warming. Nat. Commun. 3, 704.

Lima, F.P., Burnett, N.P., Helmuth, B., Aveni-Deforge, K., Kish, N., Wethey, D.S., 2011. Monitoring the intertidal environment with bio-mimetic devices. In: George, A. (Ed.), In: Advances in Biomimetics, INTECH Publishing, Rijeka, Croatia.

Lima, F.P., Gomes, F., Seabra, R., Wethey, D.S., Seabra, M.I., Cruz, T., 2016. Loss of thermal refugia near equatorial range limits. Glob. Chang. Biol. 22, 254–263.

Loarie, S.R., Duffy, P.B., Hamilton, H., Asner, G.P., Field, C.B., Ackerly, D.D., 2009. The velocity of climate change. Nature 462 (7276), 1052–1055.

Longcore, T., Lam, C.S., Kobernus, P., Polk, E., Wilson, J.P., 2010. Extracting useful data from imperfect monitoring schemes: endangered butterflies at San Bruno Mountain, San Mateo County, California (1982–2000) and implications for habitat management. J. Insect Conserv. 14, 335–346.

Louthan, A.M., Doak, D.F., Angert, A.L., 2015. Where and when do species interactions set range limits? Trends Ecol. Evol. 30 (12), 780–792.

Low-Décarie, E., Kolber, M., Homme, P., Lofano, A., Dumbrell, A., Gonzalez, A., et al., 2015. Community rescue in experimental metacommunities. Proc. Natl. Acad. Sci. U.S.A. 112 (46), 14307–14312.

Loya, Y., Sakai, K., Yamazato, K., Nakano, Y., Sambali, H., van Woesik, R., 2001. Coral bleaching: the winners and the losers. Ecol. Lett. 4, 122–131.

Lubchenco, J., Petes, L.E., 2010. The interconnected biosphere: science at the ocean's tipping points. Oceanography 23, 115–129.

Mace, G.M., Collar, N.J., Gaston, K.J., Hilton-Taylor, C., Akcakaya, H.R., Leader-Williams, N., et al., 2008. Quantification of extinction risk: IUCN's system for classifying threatened species. Conserv. Biol. 22 (6), 1424–1442.

Maclean, I.M.D., Hopkins, J.J., Bennie, J., Lawson, C.R., Wilson, R.J., 2015. Microclimates buffer the responses of plant communities to climate change. Glob. Ecol. Biogeogr. 24, 1340–1350.

Madin, J., Black, K., Connolly, S., 2006. Scaling water motion on coral reefs: from regional to organismal scales. Coral Reefs 25, 635–644.

Martin, T.L., Huey, R.B., 2008. Why "suboptimal" is optimal: Jensen's inequality and ectotherm thermal preferences. Am. Nat. 171 (3), E102–E118.

Martínez, B., Arenas, F., Trilla, A., Viejo, R.M., Carreno, F., 2015. Combining physiological threshold knowledge to species distribution models is key to improving forecasts of the future niche for macroalgae. Glob. Chang. Biol. 21 (4), 1422–1433.

Mathot, K.J., Nicolaus, M., Araya-Ajoy, Y., Dingemanse, N.J., Kempenaers, B., 2015. Does metabolic rate predict risk-taking behaviour: a field experiment in a wild passerine bird. Funct. Ecol. 29, 239–249.

Matzarakis, A., Amelung, B., 2008. Physiological equivalent temperature as indicator for impacts of climate change on thermal comfort of humans. In: Thomson, M.C.,

Garcia-Herrera, R., Beniston, M. (Eds.), Seasonal Forecasts, Climatic Change and Human Health. Springer, Dordrecht, Netherlands, pp. 161–172.

Matzelle, A., Montalto, V., Sará, G., Zippay, M.L., Helmuth, B., 2014. Dynamic energy budget model parameter estimation for the bivalve *Mytilus californianus*: application of the covariation method. J. Sea Res. S1, 105–110.

Matzelle, A.J., Sarà, G., Montalto, V., Zippay, M., Trussell, G.C., Helmuth, B., 2015. A bioenergetics framework for integrating the effects of multiple stressors: opening a 'black box' in climate change research. Am. Malacol. Bull. 33 (1), 150–160.

McGregor, G., 2015. Climatology in support of climate risk management: a progress report. Prog. Phys. Geogr. 39 (4), 536–553.

McGuire, J.L., Davis, E.B., 2014. Conservation paleobiogeography: the past, present and future of species distributions. Ecography 37, 1092–1094.

McLachlan, J.S., Clark, J.S., Manos, P.S., 2005. Molecular indicators of tree migration capacity under rapid climate change. Ecology 86 (8), 2088–2098.

McMahon, S.M., Dietze, M.C., Hersh, M.H., Moran, E.V., Clark, J.S., 2009. A predictive framework to understand forest responses to global change. Ann. N. Y. Acad. Sci. 1162, 221–236.

McPherson, J.M., Jetz, W., 2007. Effects of species' ecology on the accuracy of distribution models. Ecography 30, 135–151.

McVicar, T.R., Roderick, M.L., Donohue, R.J., Li, L.T., Van Niel, T.G., Thomas, A., et al., 2012. Global review and synthesis of trends in observed terrestrial near-surface wind speeds: implications for evaporation. J. Hydrol. 416–417, 182–205.

Medley, K.A., 2010. Niche shifts during the global invasion of the Asian tiger mosquito, *Aedes albopictus* Skuse (Culicidae), revealed by reciprocal distribution models. Glob. Ecol. Biogeogr. 19, 122–133.

Memmott, J., Craze, P.G., Waser, N.M., Price, M.V., 2007. Global warming and the disruption of plant–pollinator interactions. Ecol. Lett. 10 (8), 710–717.

Menge, B.A., Olson, A.M., 1990. Role of scale and environmental factors in regulation of community structure. Trends Ecol. Evol. 5 (2), 52–57.

Menge, B.A., Sutherland, J.P., 1987. Community regulation: variation in disturbance, competition, and predation in relation to environmental stress and recruitment. Am. Nat. 130 (5), 730–757.

Menge, B.A., Olson, A.M., Dahlhoff, E.P., 2002. Environmental stress, bottom-up effects, and community dynamics: integrating molecular-physiological with ecological approaches. Integr. Comp. Biol. 42 (4), 892–908.

Menge, B.A., Gouhier, T.C., Freidenburg, T., Lubchenco, J., 2011. Linking long-term, large-scale climatic and environmental variability to patterns of marine invertebrate recruitment: toward explaining "unexplained" variation. J. Exp. Mar. Biol. Ecol. 400 (1-2), 236–249.

Miller, L.P., Matassa, C.M., Trussell, G.C., 2014. Climate change enhances the negative effects of predation risk on an intermediate consumer. Glob. Chang. Biol. 20, 3834–3844.

Miller, L.P., Allen, B.J., King, F.A., Chilin, D.R., Reynoso, V.M., Denny, M.W., 2015. Warm microhabitats drive both increased respiration and growth rates of intertidal consumers. Mar. Ecol. Prog. Ser. 522, 127–143.

Mills, K.E., Pershing, A.J., Brown, C.J., Chen, Y., Chiang, F.-S., Holland, D.S., et al., 2013. Fisheries management in a changing climate: lessons from the 2012 ocean heat wave in the Northwest Atlantic. Oceanography 26 (2), 191–195.

Mislan, K.A.S., Wethey, D.S., 2015. A biophysical basis for patchy mortality during heat waves. Ecology 96 (4), 902–907.

Mislan, K.A.S., Wethey, D.S., Helmuth, B., 2009. When to worry about the weather: role of tidal cycle in determining patterns of risk in intertidal ecosystems. Glob. Chang. Biol. 15 (12), 3056–3065.

Mitchell, J.W., 1976. Heat transfer from spheres and other animal forms. Biophys. J. 16, 561–569.
Mitchell, S.C., 2005. How useful is the concept of habitat? A critique. Oikos 110, 634–638.
Molinos, J.G., Donohue, I., 2010. Interactions among temporal patterns determine the effects of multiple stressors. Ecol. Appl. 20 (7), 1794–1800.
Möllmann, C., Folke, C., Edwards, M., Conversi, A., 2015. Marine regime shifts around the globe: theory, drivers and impacts. Philos. Trans. R. Soc. B 370, 20130260.
Monaco, C.J., Helmuth, B., 2011. Tipping points, thresholds, and the keystone role of physiology in marine climate change research. Adv. Mar. Biol. 60, 123–160.
Monaco, C.J., Wethey, D.S., Gulledge, S., Helmuth, B., 2015. Shore-level size gradients and thermal refuge use in the predatory sea star *Pisaster ochraceus*: the role of environmental stressors. Mar. Ecol. Prog. Ser. 539, 191–205.
Montalto, V., Sará, G., Ruti, P., Dell'Aquila, A., Helmuth, B., 2014. Testing the effects of temporal data resolution on predictions of bivalve growth and reproduction in the context of global warming. Ecol. Model. 278, 1–8.
Moore, J.R., Watt, M.S., 2015. Modelling the influence of predicted future climate change on the risk of wind damage within New Zealand's planted forests. Glob. Chang. Biol. 21 (8), 3021–3035.
Mustin, K., Sutherland, W.J., Gill, J.A., 2007. The complexity of predicting climate-induced ecological impacts. Clim. Res. 35 (1–2), 165–175.
Nathan, R., 2005. Long-distance dispersal research: building a network of yellow brick roads. Div. Dist. 11, 125–130.
National Academies of Sciences, Engineering and Medicine, 2016. Next Generation Earth System Prediction: Strategies for Subseasonal to Seasonal Forecasts. The National Academies Press, Washington, DC.
Neilson, R.P., Pitelka, L.F., Solomon, A.M., Nathan, R., Midgely, G.F., Fragoso, J.M.V., et al., 2005. Forecasting regional to global plant migration in response to climate change. Bioscience 55 (9), 749–759.
Nielsen, D.L., Brock, M.A., Rees, G.N., Baldwin, D.S., 2003. Effects of increasing salinity on freshwater ecosystems in Australia. Aust. J. Bot. 51 (6), 655–665.
Nisbet, R.M., Jusup, M., Klanjscek, T., Pecquerie, L., 2012. Integrating dynamic energy budget (DEB) theory with traditional bioenergetic models. J. Exp. Biol. 215, 892–902.
Norberg, J., Urban, M.C., Vellend, M., Klausmeier, C.A., Loeuille, N., 2012. Eco-evolutionary responses of biodiversity to climate change. Nat. Clim.Chang. 2 (10), 747–751.
O'Connor, M.P., Kemp, S.J., Agosta, S., Hansen, F., Sieg, A.E., Wallace, B.P., et al., 2007a. Reconsidering the mechanistic basis of the metabolic theory of ecology. Oikos 116, 1059–1073.
O'Connor, M.I., Bruno, J.F., Gaines, S.D., Halpern, B.S., Lester, S.E., Kinlan, B.P., et al., 2007b. Temperature control of larval dispersal and the implications for marine ecology, evolution, and conservation. Proc. Natl. Acad. Sci. U.S.A. 104 (4), 1266–1271.
O'Gorman, E.J., Benstead, J.P., Cross, W.F., Friberg, N., Hood, J.M., Johnson, P.W., Sigurdsson, B.D., Woodward, G., 2014. Climate change and geothermal ecosystems: natural laboratories, sentinel systems, and future refugia. Glob. Chang. Biol. 20 (11), 3291–3299.
Otero, L.M., Huey, R.B., Gorman, G.C., 2015. A few meters matter: local habitats drive reproductive cycles in a tropical lizard. Am. Nat. 186 (3), E72–E80.
Pacifici, M., Foden, W.B., Visconti, P., Watson, J.E.M., Butchart, S.H.M., Kovacs, K.M., et al., 2015. Assessing species vulnerability to climate change. Nat. Clim. Chang. 5 (3), 215–224.
Paine, R.T., 1966. Food web complexity and species diversity. Am. Nat. 100 (910), 368–378.
Pappas, C., Fatichi, S., Rimkus, S., Burlando, P., Huber, M.O., 2015. The role of local-scale heterogeneities in terrestrial ecosystem modeling. J. Geophys. Res. Biogeosci. 120, 341–360.

Pawar, S., Dell, A.I., Savage, V.M., 2015. From metabolic constraints on individuals to the dynamics of ecosystems. In: Belgrano, A., Woodward, G., Jacob, U. (Eds.), Aquatic Functional Biodiversity: An Ecological and Evolutionary Perspective. Academic Press, London, pp. 3–36.

Pearson, R.G., Dawson, T.P., 2003. Predicting the impacts of climate change on the distribution of species: are bioclimate envelope models useful? Glob. Ecol. Biogeogr. 12, 361–371.

Pearson, R.G., Thuiller, W., Araújo, M.B., Martinez-Meyer, E., Brotons, L., McClean, C., et al., 2006. Model-based uncertainty in species range prediction. J. Biogeogr. 33 (10), 1704–1711.

Pearson, G.A., Lago-Leston, A., Mota, C., 2009. Frayed at the edges: selective pressure and adaptive response to abiotic stressors are mismatched in low diversity edge populations. J. Ecol. 97, 450–462.

Pepino, M., Rodrigues, M.A., Magnan, P., 2012. Fish dispersal in fragmented landscapes: a modeling framework for quantifying the permeability of structural barriers. Ecol. Appl. 22 (5), 1435–1445.

Pershing, A.J., Alexander, M.A., Hernandez, C.M., Kerr, L.A., Le Bris, A., Mills, K.E., et al., 2015. Slow adaptation in the face of rapid warming leads to collapse of the Gulf of Maine cod fishery. Science 350 (6262), 809–812.

Petchy, O.L., Pontarp, M., Massie, T.M., Kéfi, S., Ozgul, A., Weilenmann, M., et al., 2015. The ecological forecast horizon, and examples of its uses and determinants. Ecol. Lett. 18, 597–611.

Peterson, A.T., 2001. Predicting species' geographic distributions based on ecological niche modeling. Condor 103, 599–605.

Peterson, A.T., Soberón, J., Sanchez-Cordero, V., 1999. Conservatism of ecological niches in evolutionary time. Science 285, 1265–1267.

Peterson, A.T., Papeş, M., Soberón, J., 2015. Mechanistic and correlative models of ecological niches. Eur. J. Ecol. 1 (2). http://dx.doi.org/10.1515/eje-2015-0014.

Petes, L.E., Menge, B.A., Murphy, G.D., 2007. Environmental stress decreases survival, growth, and reproduction in New Zealand mussels. J. Exp. Mar. Biol. Ecol. 351, 83–91.

Petes, L.E., Menge, B.A., Harris, A.L., 2008. Intertidal mussels exhibit energetic trade-offs between reproduction and stress resistance. Ecol. Monogr. 78 (3), 387–402.

Petes, L.E., Howard, J.F., Helmuth, B.S., Fly, E.K., 2014. Science integration into U.S. climate and ocean policy. Nat. Clim. Chang. 4, 671–677.

Philippart, C.J.M., Anadon, R., Danovaro, R., Dippner, J.W., Drinkwater, K.F., Hawkins, S.J., et al., 2011. Impacts of climate change on European marine ecosystems: observations, expectations and indicators. J. Exp. Mar. Biol. Ecol. 400 (1-2), 52–69.

Phillips, B.L., Chipperfield, J.D., Kearney, M., 2008. The toad ahead: challenges of modelling the range and spread of an invasive species. Wildl. Res. 35, 222–234.

Piggott, J.J., Townsend, C.R., Matthaei, C.D., 2015. Climate warming and agricultural stressors interact to determine stream macroinvertebrate community dynamics. Glob. Chang. Biol. 21, 1887–1906.

Pillai, P., Gonzalez, G., Loreau, M., 2012. Evolution of dispersal in a predator-prey metacommunity. Am. Nat. 179 (2), 204–216.

Pincebourde, S., Casas, J., 2006. Multitrophic biophysical budgets: thermal ecology of an intimate herbivore insect-plant interaction. Ecol. Monogr. 76 (2), 175–194.

Pincebourde, S., Sanford, E., Helmuth, B., 2008. Body temperature during low tide alters the feeding performance of a top intertidal predator. Limnol. Oceanogr. 53 (4), 1562–1573.

Pincebourde, S., Sanford, E., Casas, J., Helmuth, B., 2012. Temporal coincidence of environmental stress events modulates predation rates. Ecol. Lett. 15 (7), 680–688.

Pinsky, M.L., Worm, B., Fogarty, M.J., Sarmiento, J.L., Levin, S.A., 2013. Marine taxa track local climate velocities. Science 341, 1239–1242.

Place, S.P., Menge, B.A., Hofmann, G.E., 2012. Transcriptome profiles link environmental variation and physiological response of *Mytilus californianus* between Pacific tides. Funct. Ecol. 26 (1), 144–155.

Pocheville, A., 2015. The ecological niche: history and recent controversies. In: Heams, T., Huneman, P., Lecointre, G., Silberstein, M. (Eds.), Handbook of Evolutionary Thinking in the Sciences. Springer, Dorderecht, pp. 547–586.

Porter, W.P., Mitchell, J.W., 2006. Wisconsin Alumni Research Foundation, Method and system for calculating the spatial-temporal effects of climate and other environmental conditions on animals. U.S. Patent 7, 155, 377.

Pörtner, H.-O., Farrell, A.P., 2008. Physiology and climate change. Science 322 (5902), 690–692.

Pörtner, H.O., Peck, M.A., 2010. Climate change effects on fishes and fisheries: towards a cause-and-effect understanding. J. Fish Biol. 77 (8), 1745–1779. http://dx.doi.org/10.1111/j.1095-8649.2010.02783.x.

Post, E., Forchhammer, M.C., 2008. Climate change reduces reproductive success of an Arctic herbivore through trophic mismatch. Philos. Trans. R. Soc. B 363, 2367–2373.

Potter, K.A., Woods, H.A., Pincebourde, S., 2013. Microclimatic challenges in global change biology. Glob. Chang. Biol. 19, 2932–2939.

Power, M.E., Tilman, D., Estes, J.A., Menge, B.A., Bond, W.J., Mills, L.S., et al., 1996. Challenges in the quest for keystones. Bioscience 46 (8), 609–620.

Preston, B.L., Westaway, R.M., Yuen, E.J., 2011. Climate adaptation planning in practice: an evaluation of adaptation plans from three developed nations. Mitig. Adapt. Strateg. Glob. Chang. 16 (4), 407–438.

Primack, R.B., Ibáñez, I., Higuchi, H., Lee, S.D., Miller-Rushing, A.J., Wilson, A.M., et al., 2009. Spatial and interspecific variability in phenological responses to warming temperatures. Biol. Conserv. 142 (11), 2569–2577.

Quataert, P., Briene, J., Simoens, I., 2007. Evaluation of the European Fish Index: false-positive and false-negative error rate to detect disturbance and consistency with alternative fish indices. Fish. Manag. Ecol. 14 (6), 465–472.

Rahel, F.J., 2007. Biogeographic barriers, connectivity and homogenization of freshwater faunas: it's a small world after all. Freshw. Biol. 52, 696–710.

Rahmstorf, S., Coumou, D., 2011. Increase of extreme events in a warming world. Proc. Natl. Acad. Sci. U.S.A. 108 (44), 17905–17909.

Rall, B.C., Brose, U., Hartvig, M., Kalinkat, G., Schwarzmüller, F., Vucic-Pestic, O., et al., 2012. Universal temperature and body-mass scaling of feeding rates. Philos. Trans. R. Soc. B 367, 2923–2934.

Rapacciuolo, G., Maher, S.P., Schneider, A.C., Hammond, T.T., Jabis, M.D., Walsh, R.E., et al., 2014. Beyond a warming fingerprint: individualistic biogeographic responses to heterogeneous climate change in California. Glob. Chang. Biol. 20, 2841–2855.

Reid, M.A., Ogden, R.W., 2006. Trend, variability or extreme event? The importance of long-term perspectives in river ecology. River Res. Appl. 22 (2), 167–177.

Rey, B., Halsey, L.G., Hetem, R.S., Fuller, A., Mitchell, D., Rouanet, J.-L., 2015. Estimating resting metabolic rate by biologging core and subcutaneous temperature in a mammal. Comp. Biochem. Physiol. A 183, 72–77.

Rissler, L.J., Apodaca, J.J., 2007. Adding more ecology into species delimitation: ecological niche models and phylogeography help define cryptic species in the black salamander (Aneides flavipunctatus). Syst. Biol. 56 (6), 924–942.

Robertson, M.P., Peter, C.I., Villet, M.H., Ripley, B.S., 2003. Comparing models for predicting species' potential distributions: a case study using correlative and mechanistic predictive modelling techniques. Ecol. Model. 164 (2–3), 153–167.

Rollinson, C.R., Kaye, M.W., Canham, C.D., 2016. Interspecific variation in growth responses to climate and competition of five eastern tree species. Ecology 97 (4), 1003–1011.

Root, T.L., Schneider, S.H., 1995. Ecology and climate: research strategies and implications. Science 269 (5222), 334–341.
Root, T.L., Price, J.T., Hall, K.R., Schneider, S.H., Rosenzweigk, C., Pounds, J.A., 2003. Fingerprints of global warming on wild animals and plants. Nature 421, 57–60.
Rougier, T., Lassalle, G., Drouineau, H., Dumoulin, N., Faure, T., Deffuant, G., et al., 2015. The combined use of correlative and mechanistic species distribution models benefits low conservation status species. PLoS One 10 (10), e0139194.
Rovero, F., Hughes, R.N., Chelazzi, G., 1999. Cardiac and behavioural responses of mussels to risk of predation by dogwhelks. Anim. Behav. 58 (4), 707–714.
Ruckelshaus, M., Doney, S.C., Galindo, H.M., Barry, J.P., Chan, F., Duffy, J.E., et al., 2013. Securing ocean benefits for society in the face of climate change. Mar. Policy 40, 154–159.
Sanford, E., 1999. Regulation of keystone predation by small changes in ocean temperature. Science 283, 2095–2097.
Sanford, E., 2002. Water temperature, predation, and the neglected role of physiological rate effects in rocky intertidal communities. Integr. Comp. Biol. 42 (4), 881–891.
Sanford, E., Kelly, M.W., 2011. Local adaptation in marine invertebrates. Ann. Rev. Mar. Sci. 3, 509–535.
Sarà, G., Kearney, M., Helmuth, B., 2011. Combining heat-transfer and energy budget models to predict thermal stress in Mediterranean intertidal mussels. Chem. Ecol. 27 (2), 135–145.
Sarà, G., Reid, G.K., Rinaldi, A., Palmeri, V., Troell, M., Kooijman, S., 2012. Growth and reproductive simulation of candidate shellfish species at fish cages in the Southern Mediterranean: Dynamic Energy Budget (DEB) modelling for integrated multi-trophic aquaculture. Aquaculture 324, 259–266.
Sarà, G., Palmeri, V., Rinaldi, A., Montalto, V., Helmuth, B., 2013. Predicting biological invasions in marine habitats through eco-physiological mechanistic models: a case study with the bivalve *Brachidontes pharaonis*. Divers. Distrib. 19 (10), 1235–1247.
Savage, V.M., Gillooly, J.F., Brown, J.H., West, G.B., Charnov, E.L., 2004. Effects of body size and temperature on population growth. Am. Nat. 163 (3), 429–441.
Scherrer, D., Koerner, C., 2010. Infra-red thermometry of alpine landscapes challenges climatic warming projections. Glob. Chang. Biol. 16 (9), 2602–2613.
Schmidt, P.S., Bertness, M.D., Rand, D.M., 2000. Environmental heterogeneity and balancing selection in the acorn barnacle *Semibalanus balanoides*. Proc. R. Soc. Lond. B 267, 379–384.
Schneider, K.R., 2008. Heat stress in the intertidal: comparing survival and growth of an invasive and native mussel under a variety of thermal conditions. Biol. Bull. 215, 253–264.
Schneider, K.R., Helmuth, B., 2007. Spatial variability in habitat temperature may drive patterns of selection between an invasive and native mussel species. Mar. Ecol. Prog. Ser. 339, 157–167.
Schoener, T.W., 2009. The Niche. In: Levin, S. (Ed.), Princeton Guide to Ecology. Princeton University Press, Princeton, NJ, pp. 3–13.
Schulte, P.M., Healy, T.M., Fangue, N.A., 2011. Thermal performance curves, phenotypic plasticity, and the time scales of temperature exposure. Integr. Comp. Biol. 51 (5), 691–702.
Seabra, R., Wethey, D.S., Santos, A.M., Lima, F.P., 2011. Side matters: microhabitat influence on intertidal heat stress over a large geographical scale. J. Exp. Mar. Biol. Ecol. 400 (1–2), 200–208.
Seabra, R., Wethey, D.S., Santos, A.M., Lima, F.P., 2015. Understanding complex biogeographic responses to climate change. Sci. Rep. 5, 12930.
Seebacher, F., Franklin, C.E., 2012. Determining environmental causes of biological effects: the need for a mechanistic physiological dimension in conservation biology. Philos. Trans. R. Soc. B 367, 1607–1614.

Selkoe, K.A., Blenckner, T., Caldwell, M.R., Crowder, L.B., Erickson, A.L., Essington, T.E., et al., 2015. Principles for managing marine ecosystems prone to tipping points. Ecosyst Health Sustain. 1 (5), 17.

Shi, N.N., Tsai, C.-C., Camino, F., Bernard, G.D., Yu, N., Wehner, R.d., 2015. Keeping cool: enhanced optical reflection and radiative heat dissipation in Saharan silver ants. Science 349 (6245), 298–301.

Sieck, M., Ibisch, P.L., Moloney, K.A., Jeltsch, F., 2011. Current models broadly neglect specific needs of biodiversity conservation in protected areas under climate change. BMC Ecol. 11, 12.

Smith, L.A., 2002. What might we learn from climate forecasts? Proc. Natl. Acad. Sci. U.S.A. 99 (Suppl. 1), 2487–2492.

Smith, V.H., 2003. Eutrophication of freshwater and coastal marine ecosystems a global problem. Environ. Sci. Pollut. Res. 10 (2), 126–139.

Soberón, J., 2007. Grinnellian and Eltonian niches and geographic distributions of species. Ecol. Lett. 10, 1115–1123.

Soberón, J., Peterson, A.T., 2005. Interpretation of models of fundamental ecological niches and species distributional areas. Biodivers. Inform. 2, 1–10.

Sokolova, I.M., 2013. Energy-limited tolerance to stress as a conceptual framework to integrate the effects of multiple stressors. Integr. Comp. Biol. 53 (4), 597–608.

Somero, G.N., 2005. Linking biogeography to physiology: evolutionary and acclimatory adjustments of thermal limits. Front. Zool. 2 (1), 1–9.

Somero, G.N., 2010. The physiology of climate change: how potentials for acclimatization and genetic adaptation will determine 'winners' and 'losers'. J. Exp. Biol. 213, 912–920.

Southwell, D.M., Rhodes, J.R., McDonald-Madden, E., Nicol, S., Helmstedt, K.J., McCarthy, M.A., 2016. Abiotic and biotic interactions determine whether increased colonization is beneficial or detrimental to metapopulation management. Theor. Popul. Biol. 109, 44–53.

Srivastava, D.S., Vellend, M., 2005. Biodiversity-ecosystem function research: is it relevant to conservation? Ann. Rev. Ecol. Evol. Syst. 36, 267–294.

Stenseth, N.C., Mysterud, A., Ottersen, G., Hurrell, J.W., Chan, K.-S., Lima, M., 2002. Ecological effects of climate fluctuations. Science 297, 1292–1296.

Stenseth, N.C., Ottersen, G., Hurrell, J.W., Mysterud, A., Lima, M., Chan, K.-S., et al., 2003. Studying climate effects on ecology through the use of climate indices: the North Atlantic Oscillation, El Niño Southern Oscillation and beyond. Proc. R. Soc. Lond. B 270, 2087–2096.

Stoffels, R.J., Richardson, A.J., Vogel, M.T., Coates, S.P., Müller, W.J., 2015. What do metabolic rates tell us about thermal niches? Mechanisms driving crayfish distributions along an altitudinal gradient. Oecologia 180 (1), 45–54. http://dx.doi.org/10.1007/s00442-015-3463-7.

Stoks, R., 2001. Food stress and predator-induced stress shape developmental performance in a damselfly. Oecologia 127, 222–229.

Storlie, C., Merino-Viteri, A., Phillips, B., VanDerWal, J., Welbergen, J., Williams, S., 2014. Stepping inside the niche: microclimate data are critical for accurate assessment of species' vulnerability to climate change. Biol. Lett. 10 (9), 20140576.

Strong, A.E., Barrientos, C.S., Duda, C., Sapper, J. (Eds.), 1997. Proceedings of the 8th International Coral Reef Symposium. Tropical Research Institute, Panama.

Stuart-Smith, R.D., Edgar, G.J., Barrett, N.S., Kininmonth, S.J., Bates, A.E., 2015. Thermal biases and vulnerability to warming in the world's marine fauna. Nature 528, 88–92.

Subbey, S., Devine, J.A., Schaarschmidt, U., Nash, R.D.M., 2014. Modelling and forecasting stock–recruitment: current and future perspectives. ICES J. Mar. Sci. 71 (8), 2307–2322.

Sunday, J.M., Bates, A.E., Kearney, M.R., Colwell, R.K., Dulvy, N.K., Longino, J.T., et al., 2014. Thermal-safety margins and the necessity of thermoregulatory behavior across latitude and elevation. Proc. Natl. Acad. Sci. U.S.A. 111 (15), 5610–5615.

Sunday, J.M., Pecl, G.T., Frusher, S., Hobday, A.J., Hill, N., Holbrook, N.J., et al., 2015. Species traits and climate velocity explain geographic range shifts in an ocean-warming hotspot. Ecol. Lett. 18 (9), 944–953.

Telemeco, R.S., 2014. Immobile and mobile life-history stages have different thermal physiologies in a lizard. Physiol. Biochem. Zool. 87 (2), 203–215.

Thomas, Y., Mazurie, J., Alunno-Bruscia, M., Bacher, C., Bouget, J.F., Gohin, F., et al., 2011. Modelling spatio-temporal variability of *Mytilus edulis* (L.) growth by forcing a dynamic energy budget model with satellite-derived environmental data. J. Sea Res. 66 (4), 308–317.

Thompson, R.M., Beardall, J., Beringer, J., Grace, M., Sardina, P., 2013. Means and extremes: building variability into community-level climate change experiments. Ecol. Lett. 16 (6), 799–806.

Thuiller, W., Richardson, D., Pysek, P., Midgely, G., Hughes, G., Rouget, M., 2005. Niche-based modelling as a tool for predicting the risk of alien plant invasions at a global scale. Glob. Chang. Biol. 11, 2234–2250.

Tingley, R., Vallinoto, M., Sequeira, F., Kearney, M.R., 2014. Realized niche shift during a global biological invasion. Proc. Natl. Acad. Sci. U.S.A. 111 (28), 10233–10238.

Travis, J.M.J., Delgado, M., Bocedi, G., Baguette, M., Barton, K., Bonte, D., et al., 2013. Dispersal and species' responses to climate change. Oikos 122 (11), 1532–1540.

Tribbia, J., Moser, S.C., 2008. More than information: what coastal managers need to plan for climate change. Environ. Sci. Policy 11, 315–328.

Trussell, G.C., Schmitz, O.J. (Eds.), 2012. Species Functional Traits, Trophic Control, and the Ecosystem Consequences of Adaptive Foraging in the Middle of Food Chains. Cambridge University Press, Cambridge.

Trussell, G.C., Ewanchuk, P.J., Matassa, C.M., 2006. The fear of being eaten reduces energy transfer in a simple food chain. Ecology 87, 2979–2984.

Turner, M.G., Donato, D.C., Romme, W.H., 2012. Consequences of spatial heterogeneity for ecosystem services in changing forest landscapes: priorities for future research. Landsc. Ecol. 28 (6), 1081–1097.

U.S. Environmental Protection Agency (EPA), 2008. Effects of climate change for aquatic invasive species and implications for management and research. National Center for Environmental Assessment, Washington, DC. EPA/600/R-08/014. Available from the National Technical Information Service, Springfield, VA at http://www.epa.gov/ncea.

van de Pol, M., Ens, B.J., Heg, D., Brouwer, L., Krol, J., Maier, M., et al., 2010. Do changes in the frequency, magnitude and timing of extreme climatic events threaten the population viability of coastal birds? J. Appl. Ecol. 47 (4), 720–730.

van der Meer, J., 2006. Metabolic theories in ecology. Trends Ecol. Evol. 21 (3), 136–140.

Vance-Borland, K., Burnett, K., Clarke, S., 2009. Influence of mapping resolution on assessments of stream and streamside conditions: lessons from coastal Oregon, USA. Aquatic Conserv.: Mar. Freshw. Ecosyst. 19, 252–263.

Vasseur, D.A., DeLong, J.P., Gilbert, B., Grieg, H.S., Harley, C.D.G., McCann, K.S., et al., 2014. Increased temperature variation poses a greater risk to species than climate warming. Proc. R. Soc. Lond. B 281, 20132612.

Veraart, A.J., Faassen, E.J., Dakos, V., van Nes, E.H., Lürling, M., Scheffer, M., 2012. Recovery rates reflect distance to a tipping point in a living system. Nature 481, 357–404.

Vucic-Pestic, O., Ehnes, R.B., Rall, B.C., Brose, U., 2010. Warming up the system: higher predator feeding rates but lower energetic efficiencies. Glob. Chang. Biol. 17 (3), 1301–1310.

Watling, J.I., Romañach, S.S., Bucklin, D.N., Speroterra, C., Brandt, L.A., Pearlstine, L.G., et al., 2012. Do bioclimate variables improve performance of climate envelope models? Ecol. Model. 246, 79–85.

Watling, J.I., Bucklin, D.N., Speroterra, C., Brandt, L.A., Mazzotti, F.J., Romañach, S.S., 2013. Validating predictions from climate envelope models. PLoS One 8 (5), e63600.
Watling, J.I., Brandt, L.A., Bucklin, D.N., Fujisaki, I., Mazzotti, F.J., Romanach, S.S., et al., 2015. Performance metrics and variance partitioning reveal sources of uncertainty in species distribution models. Ecol. Model. 309, 48–59.
Watson, J.E.M., Iwamura, T., Butt, N., 2013. Mapping vulnerability and conservation adaptation strategies under climate change. Nat. Clim. Chang. 3, 989–994.
Wenger, S.J., Som, N.A., Dauwalter, D.C., Isaak, D.J., Neville, H.M., Luce, C.H., et al., 2013. Probabilistic accounting of uncertainty in forecasts of species distributions under climate change. Glob. Chang. Biol. 19 (11), 3343–3354.
Wernberg, T., Russell, B.D., Moore, P.J., Ling, S.D., Smale, D.A., Campbell, A., et al., 2011. Impacts of climate change in a global hotspot for temperate marine biodiversity and ocean warming. J. Exp. Mar. Biol. Ecol. 400, 7–16.
Wethey, D.S., 1983. Geographic limits and local zonation: the barnacles *Semibalanus* (*Balanus*) and *Chthamalus* in New England. Biol. Bull. 165, 330–341.
Wethey, D.S., 1984. Sun and shade mediate competition in the barnacles Chthamalus and Semibalanus: a field experiment. Biol. Bull. 167, 176–185.
Wethey, D.S., 2002. Biogeography, competition, and microclimate: the barnacle *Chthamalus fragilis* in New England. Integr. Comp. Biol. 42, 872–880.
Wethey, D.S., Brin, L.D., Helmuth, B., Mislan, K.A.S., 2011a. Predicting intertidal organism temperatures with modified land surface models. Ecol. Model. 222, 3568–3576.
Wethey, D.S., Woodin, S.A., Hilbish, T.J., Jones, S.J., Lima, F.P., Brannock, P.M., 2011b. Response of intertidal populations to climate: effects of extreme events versus long term change. J. Exp. Mar. Biol. Ecol. 400 (1–2), 132–144.
Widdicombe, S., Spicer, J.I., 2008. Predicting the impact of ocean acidification on benthic biodiversity: what can animal physiology tell us? J. Exp. Mar. Biol. Ecol. 366, 187–197.
Widdows, J., 1973. Effect of temperature and food on the heart beat, ventilation rate and oxygen uptake of *Mytilus edulis*. Mar. Biol. 20, 269–276.
Wiens, J.J., Graham, C.H., 2005. Niche conservatism: integrating evolution, ecology, and conservation biology. Ann. Rev. Ecol. Evol. Syst. 36, 519–539.
Willett, C.S., 2010. Potential fitness trade-offs for thermal tolerance in the intertidal copepod *Tigriopus californicus*. Evolution 69 (9), 2521–2534.
Williams, J.W., Jackson, S., 2007. Novel climates, no-analog communities, and ecological surprises. Front. Ecol. Environ. 5 (9), 475–482.
Williams, E.E., Somero, G.N., 1996. Seasonal-, tidal-cycle- and microhabitat-related variation in membrane order of phospholipid vesicles from gills of the intertidal mussel *Mytilus californianus*. J. Exp. Biol. 199, 1587–1596.
Williams, C.M., Buckley, L.B., Sheldon, K.S., Vickers, M., Pörtner, H.-O., Dowd, W.W., et al., 2016. Biological impacts of thermal extremes: mechanisms and costs of functional responses matter. Integr. Comp. Biol. 56, 73–84.
Wilson, A.M., Jetz, W., 2015. Remotely sensed high-resolution global cloud dynamics for predicting ecosystem and biodiversity distributions. PLoS Biol. 14 (3), e1002415.
Wilson, R.J., Gutiérrez, D., Gutiérrez, J., Martínez, D., Agudo, R., Monserrat, V.J., 2005. Changes to the elevational limits and extent of species ranges associated with climate change. Ecol. Lett. 8, 1138–1146.
Winder, M., Schindler, D.E., 2004. Climate change uncouples trophic interactions in an aquatic ecosystem. Ecology 85 (8), 2100–2106.
Woodin, S.A., Hilbish, T.J., Helmuth, B., Jones, S.J., Wethey, D.S., 2013. Climate change, species distribution models, and physiological performance metrics: predicting when biogeographic models are likely to fail. Ecol. Evol. 3 (10), 3334–3346.
Woodward, F.I., 1990. The impact of low temperatures in controlling the geographical distribution of plants. Philos. Trans. R. Soc. B 326, 585–593.

Woodward, G., Bonada, N., Brown, L.E., Death, R.G., Durance, I., Gray, C., et al., 2016. The effects of climatic fluctuations and extreme events on running water ecosystems. Philos. Trans. R. Soc. B 371 (1694), 20150274. http://dx.doi.org/10.1098/rstb.2015.0274.

Yokomizo, H., Coutts, S.R., Possingham, H.P., 2014. Decision science for effective management of populations subject to stochasticity and imperfect knowledge. Popul. Ecol. 56 (1), 41–53.

Zimmermann, N.E., Yoccoz, N.G., Edwards, J., Thomas, C., Meier, E.S., Thuiller, W., Guisan, A., et al., 2009. Climatic extremes improve predictions of spatial patterns of tree species. Proc. Natl. Acad. Sci. U.S.A. 106 (2), 19723–19728.

Zwart, J.A., Solomon, C.T., Jones, S.E., 2015. Phytoplankton traits predict ecosystem function in a global set of lakes. Ecology 96 (8), 2257–2264.

PART III

Large SpatioTemporal Scale Ecology

CHAPTER NINE

Shifting Impacts of Climate Change: Long-Term Patterns of Plant Response to Elevated CO_2, Drought, and Warming Across Ecosystems

L.C. Andresen[*,†,1], C. Müller[†,‡], G. de Dato[§], J.S. Dukes[¶], B.A. Emmett[‖],
M. Estiarte[#,**], A. Jentsch[††], G. Kröel-Dulay[‡‡], A. Lüscher[§§,¶¶], S. Niu[‖‖],
J. Peñuelas[#,**], P.B. Reich[##,***], S. Reinsch[‖], R. Ogaya[#,**],
I.K. Schmidt[†††], M.K. Schneider[¶¶], M. Sternberg[‡‡‡], A. Tietema[§§§],
K. Zhu[¶¶¶], M.C. Bilton[‖‖‖]

[*]University of Gothenburg, Gothenburg, Sweden
[†]Justus-Liebig-University Giessen, Gießen, Germany
[‡]School of Biology and Environmental Science, University College Dublin, Dublin, Ireland
[§]Council for Agricultural Research and Economics–Forestry Research Centre (CREA-SEL), Arezzo, Italy
[¶]Purdue University, West Lafayette, IN, United States
[‖]Center for Ecology and Hydrology (CEH), Bangor, United Kingdom
[#]CSIC, Global Ecology Unit CREAF-CSIC-UAB, Cerdanyola del Vallès, Barcelona, Catalonia, Spain
[**]CREAF, Cerdanyola del Vallès, Barcelona, Catalonia, Spain
[††]Bayreuth Center of Ecology and Environmental Research (BayCEER), Bayreuth, Germany
[‡‡]MTA Centre for Ecological Research, Institute of Ecology and Botany, Budapest, Hungary
[§§]ETH Zürich, Institute of Agricultural Sciences, Zürich, Switzerland
[¶¶]Institute for Sustainability Sciences, Agroscope, Zürich, Switzerland
[‖‖]Key Laboratory of Ecosystem Network Observation and Modelling, Institute of Geographic Sciences and Natural Resources Research, Chinese Academy of Sciences, Beijing, China
[##]University of Minnesota, Minneapolis, MN, United States
[***]Hawkesbury Institute for the Environment, Western Sydney University, Richmond, Australia
[†††]University of Copenhagen, København, Denmark
[‡‡‡]Tel Aviv University, Tel Aviv, Israel
[§§§]University of Amsterdam, ESS, Amsterdam, The Netherlands
[¶¶¶]Rice University, Houston, TX, United States
[‖‖‖]University of Tübingen, Tübingen, Germany
[1]Corresponding author: e-mail address: louise.andresen@gu.se

Contents

1. Introduction	439
2. Methods for Data Analysis	444
2.1 Field Site Experiments	444
2.2 Treatment Effect Size and Certainty	447
2.3 Data Analysis	448

2.4 Accumulated Approach Across Sites   449
2.5 Piecewise Approach Within Sites   449
3. Results   450
3.1 Accumulated Patterns Across Sites   450
3.2 Sites Grouped by Climate Parameters   452
3.3 Piecewise Regression Within Sites   454
4. Discussion   456
4.1 Response Pattern Types   456
4.2 Responses to Drought   457
4.3 Responses to Warming   458
4.4 Responses to Elevated CO_2 459
4.5 Biomass as a Response Parameter   461
5. Conclusions   462
Acknowledgements   462
Appendix A. Details of the Database I   463
Appendix B. Details of the Database II   464
Appendix C. Site Details   465
Appendix D. Site Groupings   467
References   469

Abstract

Field experiments that expose terrestrial ecosystems to climate change factors by manipulations are expensive to maintain, and typically only last a few years. Plant biomass is commonly used to assess responses to climate treatments and to predict climate change impacts. However, response to the treatments might be considerably different between the early years and a decade later. The aim of this data analysis was to develop and apply a method for evaluating changes in plant biomass responses through time, in order to provide a firm basis for discussing how the 'short-term' response might differ from the 'long-term' response. Across 22 sites situated in the northern hemisphere, which covered three continents, and multiple ecosystems (grasslands, shrublands, moorlands, forests, and deserts), we evaluated biomass datasets from long-term experiments with exposure to elevated CO_2 (eCO_2), warming, or drought. We developed methods for assessing biomass response patterns to the manipulations using polynomial and linear (piecewise) model analysis and linked the responses to site-specific variables such as temperature and rainfall. Polynomial patterns across sites indicated changes in response direction over time under eCO_2, warming, and drought. In addition, five distinct pattern types were confirmed within sites: 'no response', 'delayed response', 'directional response', 'dampening response', and 'altered response' patterns. We found that biomass response direction was as likely to change over time as it was to be consistent, and therefore suggest that climate manipulation experiments should be carried out over timescales covering both short- and long-term responses, in order to realistically assess future impacts of climate change.

1. INTRODUCTION

Predicted and observed increases in temperature and CO_2 concentration and changes in precipitation patterns (IPCC, 2013) have motivated experimental scientists to manipulate climate factors in situ at the ecosystem scale over the last three decades. Warming, increased atmospheric CO_2 concentration, and reduced rainfall became the main factors of global climate change research. Driven by research questions concerning ecosystem vulnerability and carbon sequestration, but limited by available funding and technical challenges, scientists attempt to use such experiments to determine climate change impact under future scenarios. Climate manipulation experiments can impose continuous and empirically comparable climatic impacts on both managed and natural ecosystems (Beier et al., 2004; Kröel-Dulay et al., 2015; Mikkelsen et al., 2008). Based on findings from these experiments it is becoming increasingly evident, that the temporal patterns in responses of plant communities to climatic factors are not straightforward. In climate manipulation studies, the vegetation parameters such as individual density, standing biomass, or annual net primary production (ANPP) can show no response, or a delayed, dampened, or intensified response to climate treatments (Estiarte et al., 2016; Körner, 2006; Smith et al., 2015). Indeed, it has been reported that there were contrasting effects of the climate treatment over the duration of the experiment, i.e. early vs late in the experiment (Mueller et al., 2016; Niu et al., 2010; Smith et al., 2015). The divergence within an experiment through time or between experiments could result in a criticism of the manipulation treatment technology or design. However, the development in the plant response can be attributable to evolutionary and ecological controls (Bilton et al., 2016; Kröel-Dulay et al., 2015; Niu and Wan, 2008). To investigate this further, there is a need to assess how common the changing response patterns are.

Highest attention in climate change research has been directed to ecosystem carbon balance, where aboveground plant biomass is often the major response variable studied, due to its relative ease of measurement, role in carbon sequestration (Dieleman et al., 2012), and potential insertion into climate model predictions (IPCC, 2013; Luo et al., 2015). Aboveground biomass also has multiple roles to play in forage quantity and quality in grazed ecosystems (Ruppert et al., 2015), and for many ecosystem services (Isbell et al., 2011), and therefore as an overall community parameter has often lead to its use as an indicator of community health, resilience (Ruppert et al., 2015), and general response of the community to climate change. This parameter does not give the full picture of community response to climate change, as root biomass (Arndal et al., 2013; Körner, 2006), species plastic responses (Liancourt et al., 2015), and species composition (Bilton et al., 2016) are important for interpretation. However, in general, aboveground biomass is a relatively well studied and consistent measure, which can be transferred across ecosystems and plant types (e.g. annuals, herbaceous perennials, and woody species) and will therefore be the focus of the current analysis.

In this regard, a general prediction for elevated CO_2 (eCO_2) was to increase biomass production in ecosystems, which has recently been observed across many long-term (7–11 years) FACE (free-air carbon dioxide enrichment) experiments (Feng et al., 2015). Similarly, a general positive biomass response across warming experiments has been shown in the short (Dieleman et al., 2012) and long term (Kaarlejärvi et al., 2012). In contrast, the vegetative response to reduced precipitation (drought) has been predicted to result in a decrease in biomass (Sala et al., 2012) which has been shown in some manipulation experiments (Kröel-Dulay et al., 2015). However, it has been commonly acknowledged that changes of treatment-related response patterns for long-term extended experiments are possible and must be investigated with caution, as effect size might change over time (Keuper et al., 2011; Körner, 2006; Leuzinger et al., 2011; Mueller et al., 2016; Smith et al., 2015). eCO_2, drought, and warming impacts on the plant community are likely to be highly context dependent, and are therefore likely to affect any response patterns. For example, warming may have a positive impact on biomass in cooler regions, but cause reductions in biomass—due to increased aridity—in hot and dry regions. Additionally, site climatic factors can define plant community composition and the plasticity of plant species inhabiting those regions through their adaptations to the specific climate, e.g. the specific leaf type and anatomical adaptations, which leads to different responses of communities to the climate change drivers. Under this notion, plants can be adapted to warm, dry, or variable climates, which can result in greater resistance of the community to climate change in these regions (Tielbörger et al., 2014), and therefore impact upon response patterns. Altogether, climate variables have a large impact on the prominence of biomass response patterns that are observed across ecosystems. Therefore, in the current study, sites were categorized by climatic variables and tested for their response to the climatic change factors to differentiate response direction and pattern type for ecosystems where differences between short- and long-term effects were most likely.

The first set of defined response types was presented by Körner (2006) for FACE experiments, however until now, response types for drought and warming systems have been less systematically addressed. Several types of plant response pattern to eCO_2 were suggested, and it was acknowledged that responses might change through time and across ecosystems. The first response pattern type suggested was a fixed positive or intensifying response. Körner (2006) predicted this would occur in FACE experiments situated in systems relatively unlimited by other resources (nutrients, space, and water). The second response pattern type was a transient response. Here, an effect (positive) would be observed in biomass under eCO_2 in the early years, but would peak and return to the original level, which would be observed as no treatment effect at a later point in time. This transient response type was predicted to occur in systems where resource limitations constrain the positive CO_2 response (Körner, 2006). Nitrogen (N) availability is one of the potentially limiting resources, however, a recent meta-analysis of FACE studies showed that at sites classified as nutrient poor, the gross N mineralization increased, which directly increases plant availability of inorganic N (Rütting and Andresen, 2015). Hence, sites with increased N mineralization, with N_2 fixation (Liang et al., 2016) or with ectomycorrhiza status (Terrer et al., 2016) would not return to a 'no response' due to N limitation. The progressive N limitation (Luo et al., 2004) would rather be avoided and a positive plant biomass response would be sustained. However, other patterns, such as transient responses, may emerge by regarding ecosystem-specific data in connection with resource limitations which may develop over time and thus preventing positive response to eCO_2 in the long term. Directly connected to resource limitations, climatic differences between sites, especially related to water availability (precipitation amount and timing) and temperature might cause variability in biomass response patterns. eCO_2, and also warming and drought, exert controls on access to the nutrients, and control mineralization (Sardans et al., 2008; Williams et al., 2012), the consequences of which will be explored in the current analysis.

Smith et al. (2015) suggested three types of response patterns (besides the 'no response' pattern) as general concepts for climate manipulation factors (specifically for N addition and water manipulations across the United States), where the first two types overlap with Körner's (2006) definitions: (1) continuous directional response (being positive or negative), (2) a transient response, and (3) a stepped response type. They proposed a series of mechanisms which cascade in a hierarchical fashion leading to ecosystem responses, ranging from initial and relatively rapid responses of individuals (physiological plasticity), to changes in community composition, and finally species immigration and loss (Smith et al., 2015; see also Grime et al., 2007). The authors suggested that such stepwise acclimations and adaptations, typically expressed by individuals in the next growing season or the next generation, could show up as a step in the response curve (and a stabilization at the new level). Indeed, a number of drought and warming manipulation studies have reported species composition changes over the course of an experiment, showing shifts in abundances and dominance of particular species and functional groups (Bilton et al., 2016; Harte and Shaw, 1995; Kröel-Dulay et al., 2015; Niu and Wan, 2008; Prieto et al., 2009). In our data analysis, we have no means to distinguish individual, community composition, or species level changes, however we still expect to see fingerprints of these compositional changes observed while analysing the commonly used parameter 'aboveground biomass'.

Although manipulation experiments may show clear impacts on the plant community by changing species composition, this can lead to a no-net effect at the total biomass level, as seen in a drought manipulation experiment in Israel (Bilton et al., 2016). Therefore, it was shown that reduced precipitation gradually selected for dry associated species and reduced the presence of wet associated species, but had no overall impact on total biomass. Similar conclusions were drawn using the classical passive warming chambers within the international tundra experiment (ITEX) network (Elmendorf et al., 2016). There the authors found a change in the plant community which they termed 'thermophilization', which revealed that the warming treatments selected for species from warmer niche distributions. This was despite the fact that there was no observed change in total biomass or productivity between treatments. Furthermore, species composition change may also account for transient patterns, where an initial impact of the climate manipulation either decreases or increases the presence of some species. This change triggers the emergence (or disappearance) of other species over time, forming a polynomial response shape for total aboveground biomass (Liu et al., personal communication). Furthermore, the biomass patterns may not return to the previous state, but continue to be polar (opposite treatment effect late compared to early) in the experiment. Next to no response, the stepwise response was the most frequent among sites within the Smith et al. (2015) meta-analysis, and therefore we also expected the stepwise pattern to be frequent response type among climate manipulation sites in this study. However, also the polynomial (transient) type is likely to be an important and frequent response pattern in long-term experiments (Leuzinger et al., 2011).

The terminology of transient, directional, and stepwise response types are applicable for our research questions, however here we apply our own specific

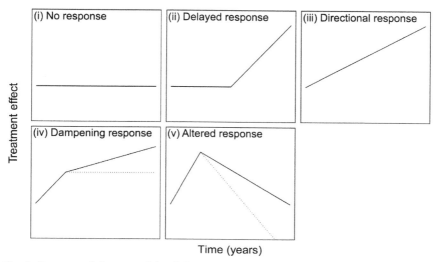

Fig. 1 Conceptual diagram of the defined types of response categories: (i) No response; (ii) Delayed response (early years no response then increase or decrease); (iii) Directional response (linear increase or -decrease); (iv) Dampening response (an early response, followed by a less steep or flat response); and (v) Altered response (initial shift to positive or negative followed by a response in opposing direction, possibly stabilizing).

definitions to allow for generalizations across FACE, drought, and warming biomass response patterns (Fig. 1). Following a hierarchy in responses, we expect to find (i) 'no response' will be common (Smith et al., 2015); (ii) a 'delayed response', i.e. absent early response will be followed by an increase or a decrease in biomass; (iii) the 'directional' (linear) response is similar to those of Körner (2006) and Smith et al. (2015); (iv) a 'dampening response', i.e. an initial response followed by a lesser slope (sometimes flat) than the early response direction (Fig. 1) may be similar to the stepwise type of Smith et al. (2015); and finally, (v) an 'altered response', i.e. a response which goes in one direction in early years and then changes direction for later years, is similar in principle to the transient response of Körner (2006) and Smith et al. (2015). However, the 'altered response' may continue beyond the original level, or may not reach the original level again (Fig. 1).

The overall goal of our data analysis method was to determine if response patterns of plant biomass to climate manipulations show any distinct trends through time, providing a mechanism for advancing our understanding of climate change response patterns. Furthermore, by acknowledging that a response trend can change from short- to long-term, we aimed to define whether responses are similar or contrasting under different climatic conditions. Ultimately, we aimed to answer the question: Can we identify 'short-term' vs 'long-term' responses of aboveground biomass to climate change manipulations?

2. METHODS FOR DATA ANALYSIS
2.1 Field Site Experiments

The data for this analysis was collected from 22 field sites across the northern hemisphere, (Fig. 2), from a wide variety of terrestrial ecosystems including grassland, shrubland, moorland, forests, and deserts. Most sites had continuous long-term single treatment and control designs, but a number of sites also had combined treatments with up to four factors. We therefore studied 13 sites with drought, 9 sites with elevated atmospheric CO_2 concentration, and 11 sites with warming experiments (Table 1, Appendices A and B).

Warming manipulations were implemented using: (i) warming by passive night-time warming done with retractable curtains (Beier et al., 2004; Mikkelsen et al., 2008), (ii) infrared heater lamp installations (Dukes et al., 2005; Niu et al., 2010; Zelikova et al., 2014), and (iii) thermocouples installed on the soil surface (Fridley et al., 2011).

Rainfall removal during the growing season was implemented using different types of coverage such as: (i) drought by retractable curtains (Beier et al., 2004; Mikkelsen et al., 2008), (ii) partial rain exclusion by PVC strips (Barbeta et al., 2015; Tielbörger et al., 2014) or by removable transparent shelters (Fridley et al., 2011; Reich et al., 2014), or (iii) drought by stable

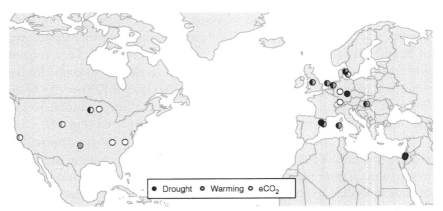

Fig. 2 Presentation of the field sites by *circle*, the *colour* (*different shades* in the print version) indicating the main treatment factor. National boundaries were generated using rworldmap (South, 2011).

Table 1 The Field Sites in the Meta-Analysis, With Locations Indicated in the Map (Fig. 2)

Site Name	Country	Treatments	Years Measured	Vegetation	References
Rhinelander	Wisconsin, USA	CO_2 in Aspen, +comb. Maple, Birch	1998–2008 (11)	Forest trees	Talhelm et al. (2014)
BioCON	Minnesota, USA	Drought (45%)	2006–2015 (10)	Grassland	Reich et al. (2014)
BioCON	Minnesota, USA	CO_2, NCO_2	1998–2015 (18)	Grassland	Reich et al. (2001, 2014)
Clocaenog	Wales, UK	Drought (25%), warming (curtains)	1998–2012 (12)	Shrubland	Kröel-Dulay et al. (2015)
Mols	Denmark	Drought (18%), warming (curtains)	1998–2012 (10)	Shrubland	Kröel-Dulay et al. (2015)
PHACE	Wyoming, USA	CO_2, warming (infrared lamp) +comb.	2005–2013 (9)	Grassland	Mueller et al. (2016)
Buxton	England, UK	Drought (20%), warming (thermoc.) irrigation (20%) +comb.	1994–2009 (13)	Grassland	Fridley et al. (2011)
Brandbjerg	Denmark	CO_2, drought (8%), warming (curtains)	2004–2012 (8)	Shrubland + comb. (3 factor)	Kröel-Dulay et al. (2015)
GiFACE	Germany	CO_2	1997–2014 (18)	Grassland	Kammann et al. (2005)
EVENT I	Germany	Drought	2004–2010 (7)	Grassland	Jentsch et al. (2007)
Oldebroeck	Netherlands	Drought (19%), warming (curtains)	1998–2011 (11)	Shrubland	Kopittke et al. (2013)
SwissFACE	Switzerland	CO_2, N, clipping	1993–2002 (10)	Grassland	Schneider et al. (2004)

Continued

Table 1 The Field Sites in the Meta-Analysis, With Locations Indicated in the Map (Fig. 2)—cont'd

Site Name	Country	Treatments	Years Measured	Vegetation	References
Kiskunsag	Hungary	Drought (22%), warming (curtains)	2001–2012 (12)	Grassland	Kröel-Dulay et al. (2015)
Porto Conte	Italy	Drought (16%), warming (curtains)	2002–2012 (6)	Shrubland	Kröel-Dulay et al. (2015)
Jasper Ridge	California, USA	CO_2, N, warming (infrared lamp), irrigation (50%), +comb. (4 factor)	1998–2014 (17)	Grassland	Dukes et al. (2005)
Oak Ridge	Tennessee, USA	CO_2	1998–2008 (11)	Forest trees	Norby et al. (2010)
Duke FACE	N. Carolina, USA	CO_2	1996–2002 (7)	Forest trees	Norby et al. (2005)
Prades	Catalonia, Spain	Drought (30%)	1999–2014 (17)	Forest	Barbeta et al. (2015)
Great plains	Oklahoma, USA	Warming (infrared lamp), clipping	2000–2008 (9)	Grassland	Niu et al. (2010)
Garraf	Catalonia, Spain	Drought (49%), warming (curtains)	1998–2015 ($d=18$)	Shrubland	Kröel-Dulay et al. (2015) ($w=9$)
Matta	Israel	Drought (30%)	2002–2014 (12)	Shrubland/an.	Tielbörger et al. (2014)
Lahav	Israel	Drought (30%)	2002–2014 (12)	Shrubland/an.	Sternberg et al. (2011)

The treatments with drought indicate percentage of rainfall removal, with irrigation the percentage addition of water; warming treatment type is indicated as being: thermocouples installed at the soil, retractable curtains giving passive night time warming, or infrared heating lamps installed over the vegetation; N indicates a combination with nitrogen fertilization. Combined treatments of the treatment factors (+comb.), clipping of vegetation or variation in plant species. For years measured, number in brackets indicates number of sampling years.

transparent roofs (Jentsch et al., 2007). Finally, experiments elevating atmospheric CO_2 concentration used the FACE technique (Hendrey and Miglietta, 2006).

An aboveground biomass parameter was estimated for each year using various methods across the sites. Typically tree stands were assessed using dimensional measures of the trunk and litterfall (Norby et al., 2010). Another nondestructive measurement was the 'point-intercept' method, where a pin is lowered into the vegetation and plant hits on the pin in a fixed grid are recorded. Plant pin hits were correlated to pin hits of plots where biomass was harvested as reference (Kröel-Dulay et al., 2015). At other sites, typically grasslands, biomass was harvested above defoliation height as part of the management type and hereby directly determined the agriculturally relevant forage production (Schneider et al., 2004). We have aligned these measures as equally valid estimates of annual biomass production in this analysis.

Grouping of sites into types of climatic categories (Appendices C and D) were done using the field site geographic coordinates to access site climate data from the 'WorldClim' database (Hijmans et al., 2005). For each site we extracted the temperature and rainfall parameters: mean annual temperature (MAT), mean diurnal temperature range (mean of monthly (max. temp. − min. temp.)), temperature seasonality, maximum temperature of warmest and minimum temperature of the coldest month, annual precipitation, precipitation of wettest (PWM) and driest month, and precipitation seasonality. For each manipulation factor, sites were put into one of three groups, based on their site climatic values (e.g. coldest MAT; intermediate MAT; warmest MAT sites).

2.2 Treatment Effect Size and Certainty

The data analysis mainly compared single treatment biomass to biomass in the control plots. In addition, for combined manipulations, the treatment combinations of interest were compared to the single factor treatment (e.g. to test the warming factor, we compared; warming + nitrogen combination vs nitrogen as 'control'), to get a single factor response (Appendices A and B). To compare different manipulation effects on biomass, the effect sizes were presented as a log ratio response (LRR, Eq. 1) (Hedges et al., 1999).

$$LRR = \ln(\text{treated}) - \ln(\text{control}) \qquad (1)$$

Log ratios have a number of strengths for the presented comparisons. Firstly, a log ratio is a relative response, in theory allowing for relative impacts of the manipulation to be observed. Secondly, an LRR is also symmetric, therefore making no prior assumptions as to the direction of the response to a manipulation, e.g. a doubling of biomass is expressed by the same value as a halving. Lastly, since biomass often has a log-normal distribution, we calculated a difference in treatment effect based on the statistically correct (logarithmic) scale.

For all sites where a pretreatment year was provided, we estimated the starting effect size at year 0 by taking the intercept value of a simple regression from year 0 to years 2 or 3. Using this method, in experiments where a clear treatment effect in early years was occurring, the intercept was very similar to the taking the year 0 value (therefore not hiding any patterns we were interested in). However, it was deemed a more suitable method than simply taking the year 0 data to provide an unbiased starting point that is independent of natural year-to-year variation (i.e. if effect sizes showed large fluctuations naturally, a year 0 value would be an inappropriate estimate of initial treatment plot differences). The starting effect size was subtracted from the other yearly values to normalize the dataset. Plotwise data was not always available, therefore, for each year, treatment means and standard deviations of the means were calculated on the normal (untransformed) scale, and then converted to log value. Certainty of the mean LRR values were then estimated on the normal scale by coefficient of variance of the standard error (CVse), adjusted for sample size (Eq. 2; Sokal and Rohlf, 1995). The reciprocal of the CVse's was then used to weight yearly mean LRR values throughout the analysis as follows (Eq. 3):

$$\text{CVse} = (\text{sd}/\text{mean}) \times (1 + (1/4 \times n)) \times (\text{sqrt}(n))^{-1} \quad (2)$$

$$\text{Weight} = (\text{CVse}_c + \text{CVse}_t)^{-1} \quad (3)$$

2.3 Data Analysis

We used two approaches to identify response patterns: (i) an accumulated approach to analyse multiple studies simultaneously within a group of treatment type or category of climatic zone and vegetation type and (ii) a segmented piecewise regression (break point analysis) to analyse individual experimental responses.

All statistical analyses were carried out in R version 3.1.1 (R Development Core Team, 2008), using statistical packages 'lme4' (Bates et al., 2015) for the accumulated approach, and package 'segmented' (Muggeo, 2008) for the piecewise regression approach.

2.4 Accumulated Approach Across Sites

For the accumulated approach, within the main treatment factors drought, warming, and eCO_2, a mixed model was applied to the experimental mean LRR values (Eq. 1) weighted using the weight values (Eq. 3), with site as a random factor nested within a categorical factor year. To estimate and display yearly mean values across all sites included in our analysis, the categorical fixed factor year was tested, and the coefficients used as mean estimates. To test response over time, the continuous fixed factor year was fitted as both a linear model (LRR $= a + b \times$ Year) and a polynomial model (LRR $= a + b \times$ Year $+ c \times$ Year^2). Model comparison for goodness of fit was made between the linear and polynomial models using log-likelihood estimation tested against the χ^2 distribution to obtain p-values (indicated in the text as significant ($P < 0.05$) differences or differences by tendency ($0.05 < P < 0.1$)).

A linear model as best fit, may indicate no response or a directional change. Whereas a significantly improved model using the polynomial equation, provides evidence that the relationship is unlikely to be linear. While more mechanistic approaches may provide better estimates for the exact relationship, our method suggests that when a polynomial curve is a better explanation, the response directions are different at the start and end of the experiments (e.g. show a 'delayed', 'dampening', or 'altered' response).

2.5 Piecewise Approach Within Sites

Average response across many sites accumulated has the potential to lead to inaccurate or misleading conclusions, so to confirm any trends we therefore analysed all individual experiments using segmented piecewise regression. Here, models of different degrees of complexity were tested: Model 1, simple linear regression, using weighted LRR as response (Eqs. 1 and 3), and the continuous variable year as an explanatory factor. Models 2–4 were then fitted to each dataset, as either 2-line, 3-line, or 4-line models. The best model was identified using model comparison with log-likelihood estimation on the F-distribution. If a more complex model provided a better fit

(*P*-value < 0.1), we assumed that there was a strong tendency for different response patterns within the experiments. Segmented regression uses maximum likelihood to minimize the differences between lines to form an *almost continuous* line (albeit with different slopes), returning estimates for individual line slopes (and associated *t*-values), and breakpoints of the connecting lines (determining the point of change within the experiment). Multiple simulations were run to avoid false convergence of the breakpoint estimate, and visual estimates were used to confirm the accuracy of the automated findings.

3. RESULTS
3.1 Accumulated Patterns Across Sites

When grouping information from all sites using the accumulated approach, the average response patterns of aboveground biomass to the three main climate manipulation factors (drought, warming, and eCO_2) indicated differing responses over time (Fig. 3). By the mid-way point in recordings (c.8–10 years) the drought manipulations decreased aboveground biomass, with a lowest yearly mean value in year 10 with ~1.14 times less biomass under drought than control. Contrastingly, warming and eCO_2 generally increased biomass, with the highest yearly values in year 8 for warming (~1.19 times more biomass under warming than control) and in year 10 for eCO_2 (~1.26 times more biomass under eCO_2 than control) (Fig. 3).

Importantly, taking all sites into consideration, these response directions changed during the course of the experiments. Under drought, the polynomial fit showed a tendency to provide a better fit than a linear model ($P=0.075$, $\chi^2=3.1767$); whereas under warming ($P=0.039$, $\chi^2=4.24$) and eCO_2 ($P=0.014$, $\chi^2=6.06$) polynomial fits were significantly the better descriptor of response over time. Under all manipulation factors, at the midpoint in recordings, the treatment response slope dampened and reversed in direction, sometimes resulting in mean values that indicated opposing effects in the later years compared to in the early time points (see Fig. 3 drought and eCO_2). Some care must be taken in interpreting these accumulated responses, because at later time-points variations in yearly effect size differed, depending on which experiments were included in the analysis, e.g. negative effect in year 13 in eCO_2, but highly positive in years 17 and 18 (Fig. 3). This suggests that not all experiments show these altering response patterns over time, but certainly that the majority of sites have opposing or no effects at the later time points.

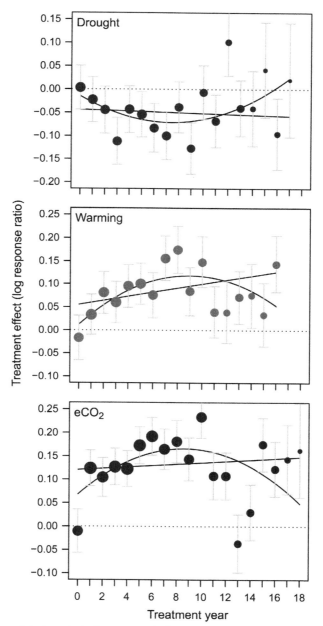

Fig. 3 Accumulated analysis from a total of 22 sites (68 experimental comparisons) showing response of total aboveground biomass to long-term climate manipulations. Three main climate manipulation factors were analysed with log ratio response (LRR; Eq. 1) control treatments to: drought (reduced precipitation), warming (increased temperature), and elevated CO_2. *Points* indicate mean estimates of yearly LRR given by categorical mixed model analyses, corrected for starting year. *Error bars* are standard errors (SE) given by the mixed models, and *point size* is the reciprocal of SE (indicating variance and number of sites compared). Superimposed are both linear and polynomial continuous mixed model fits.

3.2 Sites Grouped by Climate Parameters

To disentangle the response patterns by their climatic characteristics, sites were grouped according to temperature or rainfall categories (see Section 2.1 and Appendices C and D for details).

Looking at the biomass responses to drought, using MAT (Fig. 4) to group sites, the 'coldest MAT' and 'intermediate MAT' sites showed decreasing negative responses to the drought treatment, indicating a

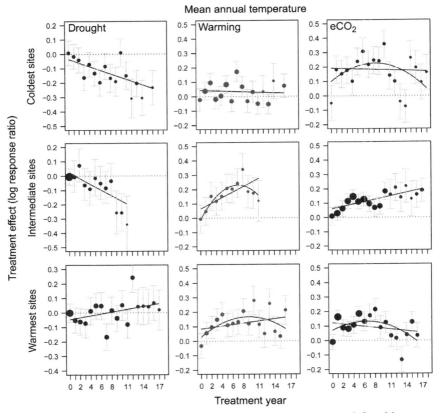

Fig. 4 Long-term experimental sites divided into three site groups defined by mean annual temperature (MAT); showing aboveground biomass responses (expressed as log ratio response, LRR; Eq. 1) to the climate manipulations drought, warming, and elevated CO_2. *Points* indicate mean estimates of yearly LRR given by categorical mixed model analyses, corrected for starting year. *Error bars* are standard errors (SE) given by the mixed models, and *point size* is the reciprocal of SE (indicating variance and number of sites compared). Superimposed are both linear continuous mixed model fits and polynomial fits, when model comparison ($P < 0.1$).

directional response type. Contrastingly, the 'warmest MAT' sites showed no response to the drought treatment over time ($P > 0.05$). In contrast, a division of the sites into three groups by 'precipitation of wettest month' (PWM) (Fig. 5) and annual precipitation (not shown), showed little in terms of consistent hierarchical responses across the site groupings. For PWM, it was the driest sites which showed a directional negative response ($t = -2.04$, $P = 0.04$), with neither wetter groups showing a response over time.

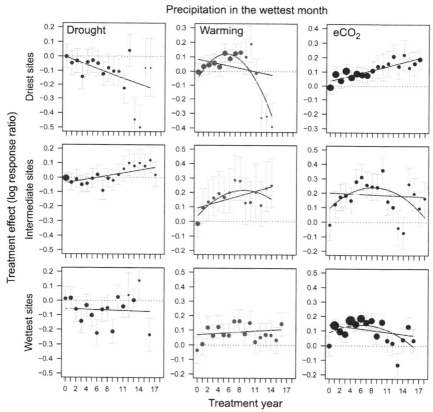

Fig. 5 Long-term experimental sites divided into three site groups defined by precipitation in the wettest month (PWM); showing aboveground biomass responses (expressed as log ratio response, LRR; Eq. 1) to the climate manipulations drought, warming, and elevated CO_2. *Points* indicate mean estimates of yearly LRR given by categorical mixed model analyses, corrected for starting year. *Error bars* are standard errors (SE) given by the mixed models, and *point size* is the reciprocal of SE (indicating variance and number of sites compared). Superimposed are both linear continuous mixed model fits and polynomial fits, when model comparison ($P < 0.1$).

However, for annual precipitation, the only directional response was found in the intermediate group (strongly decreasing), with a no response pattern in the driest and wettest site groups.

Different long-term biomass response patterns for the warming treatments were visible by dividing sites into three MAT groups (Fig. 4). The 'warmest' and 'intermediate MAT' sites had the most positive biomass responses in early years and had polynomial fits different from the linear fit ($P=0.042$ and 0.088, respectively), suggesting that the response pattern altered from increasing to decreasing. Contrastingly, the coldest group showed persistently no response. Furthermore, the driest group of the PWM (Fig. 5), and of the AP (not shown), both had a polynomial model as best fit (in both $P<0.001$), suggesting a change in response direction, reaching into negative effects, while the 'wetter' sites had simple constant positive response patterns.

The eCO_2 experiments showed the largest average biomass effect sizes across the three addressed climate factors. The climatic site division by PWM (Fig. 5) revealed two different pattern types, with a directional positive response type revealed at the driest sites. Contrastingly, a polynomial pattern expressing a maximum effect size in intermediate years, proved to be a better description of the response pattern for sites with intermediate precipitation ($P=0.039$) and at tendency in those sites with high precipitation ($P=0.051$). These two response pattern types start out with the same direction (positive), but at the 'wetter' sites, the direction changes to be dampening and negative over time. The same pattern for wetter and drier groups was found classifying sites using annual precipitation (not shown).

3.3 Piecewise Regression Within Sites

In order to gain an understanding of within-experiment response patterns for the main treatment factors CO_2, drought, and warming, the biomass response patterns were analysed by break point analysis using segmented piecewise linear regression. This technique revealed five distinct common response patterns over time (Fig. 1), and the number of experiments within each category (Table 2). These results confirmed some of the findings from the accumulated approach, and in addition add extra detail and accuracy to our interpretations.

In total, out of 68 experimental comparisons, we found that a great proportion of experiments (roughly half for all three main factors) showed a

Table 2 Response Pattern Trends in the Log Ratio Response (LRR) Time Series, the Main Factor Results Counted for Elevated CO_2, Drought, and Warming

Category	Short to Long Term	CO_2	Drought	Warming
No response	Same	12	5	10
Delayed response	Differ	0	3	2
Directional response	Same	1	3	3
Dampening response	Differ	3	3	1
Altered response	Differ	10	3	9

The response categories defined were: (i) No response; (ii) Delayed response (early years no response then increase or decrease); (iii) Directional response (linear increase or decrease), (iv) Dampening response (an early response, followed by a less steep or flat response); and (v) Altered responses (initial shift to positive or negative followed by a response in opposing direction, possibly with stabilising afterwards). The types are illustrated in Fig. 1.

change in response over time (Table 2). Similar to the findings of Smith et al. (2015), the most dominant response pattern observed across all experiments were that of 'no response'. 'Delayed responses' were rare, and only noted under drought and warming, and not under eCO_2 (Table 2). Perhaps surprisingly rare also was the consistent 'directional response', which occurred at similar frequency within experiments as the 'dampening' type (a change in the strength of slope over time).

Finally, and noteworthy, was the number and degree by which some of the biomass response patterns shifted over time becoming 'altered' in their direction. In most examples of altered response types, the reversal of the biomass response direction was in correspondence with the patterns found in the accumulated analysis: first increasing then decreasing under eCO_2 (7 out of 10) and warming (9 out of 9); first decreasing then increasing under drought (2 out of 3, and by far the strongest two in terms of effect magnitude and direction switch). This confirmed that these patterns are far from rare, and what we saw in the accumulated analysis was a fair reflection of what may be observed within a climate manipulation experiment. Interestingly, in some cases, the later directional responses continued so that the response switched from positive (or negative) in early stages to negative (or positive) in later years, establishing an entirely different, and opposing interpretation of biomass response depending on which treatment year is taken as being the conclusive observation.

4. DISCUSSION

With the two implemented forms of analysis (accumulative across sites, piecewise within sites), our study highlights that response direction of plant aboveground biomass often changes over the course of long-term climate manipulation experiments. Under the three manipulation treatments, a polynomial curve described vegetation change over time better than a simple linear relationship. This was highly supported by our piecewise regressions, whereby five distinct response pattern types were identified, and confirmed that (i) a differing response direction is as likely to occur in long-term experiments as a consistent response direction, (ii) altered directional patterns more commonly switched in direction similar to that noted in the accumulative analysis, and that (iii) sometimes effect sizes showed an opposite response early compared to later in the experiment. These findings emphasize that great care must be taken when interpreting climate change impacts on aboveground biomass from short-term experiments. The potential mechanisms involved in these response pattern types are discussed with relevance to climatic regions, adaptations, and the specific inference under the different treatment factors. Furthermore, we suggest cautions of the current analyses and areas of further investigation applicable for ultimately determining long-term consequences of climate change on plant communities around the world.

4.1 Response Pattern Types

Our piecewise regression analysis identified five distinct response pattern types occurring within the long-term climate manipulation experiments, which suggests that different mechanisms are associated with vegetation response under climate change. 'No response' was our most common finding and is similar to the pattern analysis results of Smith et al. (2015). A no response pattern is indicative of some form of resistance of the community to climate change (Grime et al., 2008; Tielbörger et al., 2014). In terms of maintaining a community of similar aboveground biomass—generally considered important for carbon sequestration (Dieleman et al., 2012), forage production in grazed systems (Golodets et al., 2013; Ruppert et al., 2015), and a large array of ecosystem services (Isbell et al., 2011)—for many of the long-term experiments included in our analysis, we show how little these communities may react with regard to climate change. This could be due to a nonsignificant response to the treatment which was not strong enough to impose a change in community. However, even though the total community biomass seems to be highly resistant to climate change or the manipulation imposed, a number of studies have shown that even with no-net response, the individuals and species represented within these communities may be changing considerably (Bilton et al., 2016; Elmendorf et al., 2016; Soussana and Lüscher, 2007). Indeed, species composition change is a factor which cannot be accounted for in our analysis, and it seems likely that under all these climate manipulation factors, ongoing selection will be occurring over time, gradually selecting for adaptations initially with respect to plastic responses, genotypes, and ultimately species (Grime et al., 2007; Ňuelas et al., 2013; Smith et al., 2015). This hierarchical order of community response to climate change is likely to manifest itself in all the different response pattern types identified in the current study, and the rates at which these occur are likely to be associated with the treatment factor, and the adaptations inherent in the community. Delayed responses, for example, are potentially mediated by the resistance of the plant community to climate change, but over time, the accumulated impacts start to affect the plant community, whether overcoming the inherent plastic adaptations of the individuals, or as the community changes considerably to affect overall biomass.

The 'dampening response' type suggests a mechanism which reduces the impact of the climate treatment on the plant biomass, which could be due to a limitation of nutrient resources or environmental constraints (Körner, 2006), or the community reaching a new stable state at a different level of resources, through a new community or adaptations (acclimation) (Herbert et al., 1999).

And finally, the altered response could have multiple interpretations and mechanisms. These could be accounted for by species composition changes as some species initially respond and others then respond later on once the community interactions have changed (Niklaus et al., 2001). It also seems likely that any initial responses may be due to plastic responses of the individuals, and that later as selection processes continue, they respond by acclimation or from those responses of a new community.

4.2 Responses to Drought

Ecosystems have been shown to differ in the sensitivity of their productivity (ANPP) to interannual precipitation variability: where the driest ecosystems tended to be more sensitive than the mesic (Huxman et al., 2004). However, such relationships may not be strong in the long term, as no supporting evidence was found in a review of 11 drought experiments (Estiarte et al., 2016). This is possibly due to high variation of precipitation conditions from year-to-year to which the communities are often exposed. This variation is often considerably greater than the variation in temperature and CO_2 levels, for which the plants have been selected (Jentsch et al., 2007). However, this is controversially discussed, with some studies showing or hypothesizing that there should be stronger responses to reduced precipitation in wetter regions (Bilton et al., 2016), whereas others suggest dry regions should be most affected (Golodets et al., 2015; Huxman et al., 2004). In our study, it seems that precipitation quantity at a site (or at least for the parameters tested) is not a consistent measure for determining community resistance to long-term reductions in precipitation. Although we identified changing pattern types using piecewise regressions—e.g. in one example an altered response pattern occurred, with an initial decrease in biomass to a midpoint but then greater biomass by the end of the recordings—there was no consistency in which precipitation regions these site patterns belonged.

Indeed, classifying the sites by temperature seemed to show the most consistent patterns. These findings suggest that MAT is grouping our particular sites well in terms of aridity, with the warmest sites being most resistant to climate change manipulations, as previously suggested (Bilton et al., 2016; Tielbörger et al., 2014). It is possible that plants growing at warmer sites are better adapted to aridity due to more regular exposure in their evolutionary history.

In natural systems, the evolutionary restrictions of the occurrence of plants in defined climatic zones are also likely to define the magnitude and direction of biomass responses to drought. But whether the investigated plant communities are resistant to drought or vulnerable and responding with composition changes cannot be resolved with this data analysis. However, it is likely that drought tolerant species gain biomass at the cost of drought sensitive species (Bilton et al., 2016; Kröel-Dulay et al., 2015; Lui et al., personal communication), which is potentially hidden from our analysis but could partly explain the 'no response to small positive response' seen at sites in warm and wet climates in our study in the long term.

4.3 Responses to Warming

In general, increased temperatures (warming) have a dual effect on ecosystems by increasing nutrient mobilization and extending the growing season in temperate regions (Ehrenfeld et al., 2005), the benefits of which may be negated by increases in water loss through evapotranspiration from a system (Luo, 2007). Altogether, in the accumulative analysis of our study, sites with indications of dry and warm climates were prone to changing the biomass response direction, from an increasing positive response over to a decreasing and negative biomass response midway through. Also, all altered response patterns identified using piecewise regression within sites pointed in this direction. The altered directional patterns experienced at the drier and warmer of the analysed sites, may be due to the increased mobilization of nutrients in the early phases of an experiment, but nutrient limitation later in the experiment, as predicted under eCO_2 (Körner, 2006; Luo et al., 2004). Moreover, effects on community composition may also be occurring, whereby those species which are generally more adapted to increased temperature also grow larger during the initial phases, and then outcompete other less adapted species.

Contrastingly, the coldest MAT sites showed persistently no warming response, refuting our hypothesis that colder regions would have the largest positive responses to warming. These observations were markedly different than results gleaned from provenance trial transfer functions, where trees from coldest origins benefited most with enhanced growth by transfer to slightly warmer sites (Reich and Oleksyn, 2008). Indeed, we expected that at the warmest sites, due to increased aridity, the warming treatment would decrease biomass. The groupings of 'maximum temperature of the warmest months' (not shown), supported the MAT findings for the warming induced biomass response, and both suggest that site selection may account for no biomass decrease in warmer sites. This highlights that more climate warming manipulation experiments should be performed in the driest regions to identify clear trends.

4.4 Responses to Elevated CO_2

The factor eCO_2 showed the largest magnitudes of biomass response of all three treatment factors tested. All climate manipulation experiments have an instantly imposed treatment effect that changes the climate parameter relative to ambient. Thereafter, the treatments are constant throughout the long-term experiment, compared to the 'natural' slower background increase in CO_2 concentration and temperature over decades. The initial step increase may impose an unrealistic response, particularly in the short term. However, the intense experimental treatments are needed in order to predict the community response to future climate change, which cannot be obtained by a simple monitoring of natural changes. Thus, biomass response during the first years of a CO_2 experiment have to be treated with caution as compared to biomass increase under a slowly developing change in climatic conditions (Rastetter et al., 1997). However, by deploying the climate treatments for many years in long-term experiments (beyond this initial phase), realistic acclimation responses to future climatic conditions are likely (Mueller et al., 2016; Schneider et al., 2004). Still, it is surprising that manipulations of eCO_2 commonly alter the direction of the biomass response over time. Sometimes these even manifest themselves as decreases in biomass at the end of the experimental duration, which contradicts many predictions of the effects of rising CO_2 levels on plant communities, and those values are often used to predict the effect of climate change in ecosystem models (IPCC, 2013).

For eCO_2 manipulations, there is little literature to our knowledge that links community composition change to other gradients of climatic factors, an area which would be of high relevance in future studies. Undoubtedly, there are a number of ecophysiological adaptations possessed by plants (e.g. stomatal closure) which are useful under drought conditions, but could also affect response under eCO_2. A connection between regional climatic adaptations and CO_2 elevation should therefore be possible. Our results, when categorizing sites by precipitation (PWM), at least suggested some connection, showing that consistent increasing responses were generally found in the drier sites, whereas an altered direction was more likely to occur in wetter sites.

The hypothesis about soil nutrient availability as a controlling factor for plant biomass response was not addressed by our investigation, but could be used as a means to categorize sites and searching for long- or short-term patterns in parallel to what has been described here. While applicable to all climate manipulation factors, it is often hypothesized that nutrient availability will change over time particularly within eCO_2 experiments (Körner, 2006; Luo et al., 2004). However, there is growing evidence that warming and drought may also play a role in nutrient mineralization (Luo, 2007). Linking the original state of nutrient availability to the changes in availability over time under treatments, may lead to improved interpretations of altered response pattern types seen in experimental studies, making results more applicable to future climate change predictions.

4.5 Biomass as a Response Parameter

Plant biomass is a popular measure for biotic carbon sequestration and is often a parameter in Earth system models (IPCC, 2013; Luo et al., 2015), in the form of a carbon pool that can increase in response to elevated atmospheric carbon dioxide. With our results, we suggest that modellers that use plant biomass as a factor for carbon sequestration, seek to use responses of long term, and not only short-term data series, because the response directions and patterns may change through time. One could argue that we only see half of the reality (or the 'tip of the iceberg') by looking at only aboveground biomass but not belowground biomass. Indeed, research has shown that belowground responses are important (Arndal et al., 2013; Körner, 2006), and especially in cases where aboveground responses were sparse, belowground responses might show different response patterns; we can only urge for more research of plant root biomass response in future work.

It is also worth noting that for ecosystems with annuals, biomass is reset annually and is equivalent to ANPP whereas for perennial ecosystems differences in biomass arise because of accumulation of differences in ANPP. Therefore, ecosystems showing positive ANPP–precipitation relationships under control treatment are more likely to accumulate differences in biomass under drought (Estiarte et al., 2016). This expectation would also be applicable for other climate change/manipulation factors and highlights that care must be taken when interpreting responses seen in standing aboveground biomass compared to an ANPP response, particularly for plants of different life history. To account for this, an early attempt was made in the current analysis to group some of these sites by their plant type characteristics (e.g. annuals, grasslands, shrublands, moorlands, and forests), whereby the rate and magnitude of response, especially with regard to overall biomass, was hypothesized to occur at different times. However, this broad scale categorization did not provide a clear indication, arguably either due to lack of replication or multiple interacting effects with climate adaptations. With a greater number of studies included in a similar analysis, and greater accuracy of lifespan estimates, we believe that interpretation by expected lifetime of an individual may prove to add significant understanding as to how plants may respond in the long term to climate change manipulations and climate change itself.

5. CONCLUSIONS

A large number of field long-term manipulation studies exist, and our survey is far from complete. The number of experiments are also growing as more are being conducted and being funded for longer time periods. However, it is noticeable that all of the sites for which we obtained data are from the northern hemisphere. Future research should include more long-term data series and rainfall addition experiments—a common prediction of climate change. Perhaps even more importantly, experiments should be set-up in climatic regions currently not covered, to build-up the global picture. In particular, sites in the southern hemisphere, warming at sites in arid zones, and manipulations in tropical regions are particularly sparse.

Despite these limitation, we can draw overall conclusions for a given climatic driver and show that a response pattern is as likely to vary through time as it is to be persistent. These findings highlight that care must be taken when interpreting climate change impacts on overall biomass from short-term experiments. We suggest that plant physiological changes and individual plastic responses, enforced by climate manipulations in the early years, is only one possible mechanism behind the response of the current community. However, longer-term experiments with multiple climate change factors are more likely to reveal how a new community may develop and how it will respond as climate change persists. Furthermore, long-term experiments are essential to provide input parameters for ecosystems models which are used to predict ecosystem dynamics under future climate change scenarios.

ACKNOWLEDGEMENTS

We would like to thank the following for making biomass data available for the analysis and setting up and running experimental field sites: Alan Talhelm, Rick Norby, J. Philip Grime, Andrew Askew, Jason Fridley, Teis Mikkelsen, Dana Blumenthal, Jack Morgan, Elise Pendall, and many more. Furthermore, Martin Erbs, Tobias Rütting, Cesar Terrer, and Katja Tielbörger helped with valuable details. We thank several sites for making unpublished data available and acknowledge that those findings will be made public elsewhere. The main authors would like to thank their funding agencies: L.C.A., C.M. (LOEWE excellence cluster FACE$_2$FACE), and M.C.B. (German Research Foundation, DFG; TI338_12-1); and jointly the ClimMani (ESSEM COST Action ES1308, Climate Change Manipulation Experiments in Terrestrial Ecosystems) for supporting a short-term scientific mission. Many funding bodies have made these experiments possible, for which we are very grateful, please refer to site-specific articles listed in Table 1 for details.

APPENDIX A. DETAILS OF THE DATABASE I

At each treatment year (0 is pretreatment year) the count of experiments within all sites, and in brackets the count of individual plots.

Year	Total	Drought	eCO$_2$	Warming
0	49 (686)	14 (252)	16 (208)	19 (226)
1	65 (1017)	17 (278)	23 (453)	25 (286)
2	66 (1023)	16 (272)	26 (471)	24 (280)
3	66 (973)	15 (216)	26 (471)	25 (286)
4	64 (1011)	15 (266)	26 (471)	23 (274)
5	68 (1059)	17 (278)	26 (495)	25 (286)
6	62 (1023)	14 (260)	26 (495)	22 (268)
7	62 (1019)	14 (256)	25 (489)	23 (274)
8	50 (875)	10 (208)	21 (441)	19 (226)
9	47 (841)	11 (214)	19 (421)	17 (206)
10	42 (709)	9 (112)	19 (421)	14 (176)
11	40 (673)	10 (118)	15 (373)	15 (182)
12	30 (594)	6 (86)	14 (368)	10 (140)
13	30 (566)	6 (46)	11 (350)	13 (170)
14	25 (516)	4 (26)	11 (350)	10 (140)
15	21 (492)	2 (14)	11 (350)	8 (128)
16	26 (542)	4 (34)	11 (350)	11 (158)
17	4 (228)	1 (6)	3 (222)	0
18	2 (216)	0	2 (216)	0

APPENDIX B. DETAILS OF THE DATABASE II

'Years' is duration of experiments and the count is number of experiments, and in brackets the number of sites.

Years	Total	Drought	eCO$_2$	Warming
6	2 (2)	1 (1)	1 (1)	0
7	12 (1)	4 (1)	4 (1)	4 (1)
8	5 (2)	0	2 (1)	3 (2)
9	3 (2)	1 (1)	0	2 (1)
10	6 (2)	1 (1)	4 (1)	1 (1)
11	5 (3)	2 (2)	1 (1)	2 (2)
12	5 (3)	2 (2)	3 (1)	0
14	4 (2)	2 (2)	0	2 (2)
16	22 (3)	3 (2)	8 (1)	11 (2)
17	2 (2)	1 (1)	1 (1)	0
18	2 (1)	0	2 (1)	0

APPENDIX C. SITE DETAILS

Long-Term manipulation experiments, showing climatic parameters as given by WORLDCLIM online database (average 50-year estimates between 1950 and 2000)

Site	Country	Latitude	Longitude	Mean Annual Temp. (°C)	Max. Temp. of Warmest Month (°C)	Temp. Seasonality (Standard Dev. × 100)	Annual Precip. (mm)	Precip. of Wettest Month (mm)	Precip. Seasonality (Coeff. of Variation)	Drought	Warming	eCO$_2$
Rhinelander	Wisconsin, USA	45.4	−89.37	4.4	26.0	1103.6	816	112	43	—	—	#
BioCON	Minnesota, USA	45.24	−93.12	6.8	28.4	1156.4	757	113	50	#	—	—
BioCON-FACE	Minnesota, USA	45.24	−93.12	6.8	28.4	1156.4	757	113	50	—	—	#
Clocaenog	Wales, UK	53.03	−3.28	7.4	17.7	452.1	1103	126	20	#	#	—
Mols	Denmark	56.23	10.57	7.5	19.9	586.6	592	62	21	#	#	—
PHACE	Wyoming, USA	41.11	−104.54	7.8	29.8	845.5	381	65	62	—	#	#
Buxton	England, UK	53.2	−1.92	8.0	19.1	469.1	1156	130	18	#	#	—
Brandbjerg	Denmark	55.53	11.58	8.2	19.3	605.5	600	63	20	#	#	#
GiFACE	Germany	50.32	8.41	8.3	22.1	625.6	745	72	12	—	—	#
EVENT I	Germany	49.92	11.59	8.3	22.9	652.7	643	75	21	#	—	—
Oldebroeck	Netherlands	52.24	5.55	9.2	21.2	535.9	786	76	16	#	#	—

Continued

Site	Country	Latitude	Longitude	Mean Annual Temp. (°C)	Max. Temp. of Warmest Month (°C)	Temp. Seasonality (Standard Dev. × 100)	Annual Precip. (mm)	Precip. of Wettest Month (mm)	Precip. Seasonality (Coeff. of Variation)	Drought	Warming	eCO$_2$
SwissFACE	Switzerland	47.27	8.41	9.4	23.6	635.2	1091	137	29	—	—	#
Kiskunsag	Hungary	46.53	19.23	11.1	27.3	777.3	554	70	23	#	#	—
Porto Conte	Italy	40.36	8.9	11.5	25.6	586.0	958	142	53	#	#	—
Jasper Ridge	California, USA	37.24	−122.14	13.2	25.7	372.3	907	185	88	—	#	#
Oak Ridge	Tennessee, USA	35.54	−84.2	14.2	30.9	764.6	1396	150	15	—	—	#
Duke forest FACE	N. Carolina, USA	35.58	−79.05	15.5	31.8	752.5	1148	125	15	—	—	#
Prades	Spain	41.21	1.2	15.7	27.5	522.4	555	74	34	#	—	—
Great plains	Oklahoma, USA	34.59	−97.31	16.3	34.8	838.6	908	139	38	—	#	—
Garraf	Spain	41.18	1.49	16.4	27.6	507.1	561	80	34	#	#	—
Matta	Israel	31.42	35.03	17.8	31.1	520.9	326	74	105	#	—	—
Lahav	Israel	31.23	34.54	19.7	31.5	475	204	46	102	#	—	—

Indicated are whether site contained (#) drought, warming, and/or elevated CO$_2$ manipulations.

APPENDIX D. SITE GROUPINGS

Classified by site-specific climatic parameters (See Appendix C). Group ranks from lowest to highest values for all parameters, e.g. for mean annual temperature, group 1 has the coldest sites, whereas group 3 has the warmest sites; for annual precipitation, group 1 has the driest sites, whereas group 3 has the wettest.

Table A.1 Drought (Reduced Precipitation) Manipulation Sites

Site	Country	Mean Annual Temp. (°C)	Max. Temp. of Warmest Month (°C)	Temp. Seasonality (Standard Dev. × 100)	Annual Precip. (mm)	Precip. of Wettest Month (mm)	Precip. Seasonality (Coeff. of Variation)
BioCON	Minnesota, USA	1	3	3	3	3	3
Clocaenog	Wales, UK	1	1	1	3	3	1
Mols	Denmark	1	1	3	2	1	1
Buxton	England, UK	1	1	1	3	3	1
Brandbjerg	Denmark	1	1	3	2	1	1
EVENT I	Germany	2	2	3	2	2	1
Oldebroeck	Netherlands	2	1	2	3	2	1
Kiskunsag	Hungary	2	2	3	1	1	2
Porto Conte	Italy	3	2	2	3	3	3
Prades	Spain	3	3	2	1	1	2
Garraf	Spain	3	3	1	1	2	2
Matta	Israel	3	3	1	1	1	3
Lahav	Palestine	3	3	1	1	1	3

Table A.2 Warming (Increased Temperature) Manipulation Sites

Site	Country	Mean Annual Temp. (°C)	Max. Temp. of Warmest Month (°C)	Temp. Seasonality (Standard Dev. × 100)	Annual Precip. (mm)	Precip. of Wettest Month (mm)	Precip. Seasonality (Coeff. of Variation)
Clocaenog	Wales, UK	1	1	1	3	2	1
Mols	Denmark	1	1	2	1	1	2
PHACE	Wyoming, USA	1	3	3	1	1	3
Buxton	England, UK	1	1	1	3	3	1
Brandbjerg	Denmark	2	1	3	2	1	1
Oldebroeck	Netherlands	2	2	2	2	2	1
Kiskunsag	Hungary	2	3	3	1	1	2
Porto Conte	Italy	3	2	2	3	3	3
Jasper Ridge	California, USA	3	2	1	2	3	3
Great plains	Oklahoma, USA	3	3	3	3	3	3
Garraf	Spain	3	3	1	1	2	2

Table A.3 Elevated CO_2 Manipulation Sites

Site	Country	Mean Annual Temp. (°C)	Max. Temp. of Warmest Month (°C)	Temp. Seasonality (Standard Dev. × 100)	Annual Precip. (mm)	Precip. of Wettest Month (mm)	Precip. Seasonality (Coeff. of Variation)
Rhinelander	Wisconsin, USA	1	2	3	2	2	2
BioCON-FACE	Minnesota, USA	1	2	3	2	2	3
PHACE	Wyoming, USA	1	3	3	1	1	3
Brandbjerg	Denmark	2	1	1	1	1	2
GiFACE	Germany	2	1	1	1	1	1
SwissFACE	Switzerland	2	1	2	3	3	2
Jasper Ridge	California, USA	3	2	1	2	3	3
Oak Ridge	Tennessee, USA	3	3	2	3	3	1
Duke forest FACE	N. Carolina, USA	3	3	2	3	2	1

REFERENCES

Arndal, M.F., Merrild, M.P., Michelsen, A., Schmidt, I.K., Mikkelsen, T.N., Beier, C., 2013. Net root growth and nutrient acquisition in response to predicted climate change in two contrasting heathland species. Plant Soil 369, 615–629.

Barbeta, A., Mejia-Chang, M., Ogay, A.R., Voltas, J., Dawson, T.E., Peñuelas, J., 2015. The combined effects of a long-term experimental drought and an extreme drought on the use of plant-water sources in a Mediterranean forest. Glob. Change Biol. 21, 1213–1225.

Bates, D., Maechler, M., Bolker, B., Walker, S., 2015. Fitting linear mixed-effects models using lme4. J. Stat. Softw. 67 (1), 1–48.

Beier, C., Emmett, B., Gundersen, P., Tietema, A., Peñuelas, J., Estiarte, M., Gordon, C., Gorissen, A., Llorens, L., Roda, F., Williams, D., 2004. Novel approaches to study climate change effects on terrestrial ecosystems in the field: drought and passive nighttime warming. Ecosystems 7, 583–597.

Bilton, M.C., Metz, J., Tielbörger, K., 2016. Climatic niche groups: a novel application of a common assumption predicting plant community response to climate change. Perspect. Plant Ecol. Evol. Syst. 19, 61–69.

Dieleman, W.I.J., Vicca, S., Dijkstra, F.A., Hagedorn, F., Hovenden, M.J., Larsen, K.S., Morgan, J.A., Volder, A., Beier, C., Dukes, J.S., King, J., Leuzinger, S., Linder, S., Luo, Y., Oren, R., de Angelis, P., Tinge, D., Hoosbeek, M., Janssens, I.A., 2012. Simple additive effects are rare: a quantitative review of plant biomass and soil process responses to combined manipulations of CO_2 and temperature. Glob. Change Biol. 18, 2681–2693.

Dukes, J.S., Chiariello, N.R., Cleland, E., Moore, L.A., Shaw, M.R., Thayer, S., Tobeck, T., Mooney, H.A., Field, C.B., 2005. Responses of grassland production to single and multiple global environmental changes. PLoS Biol. 3, e319.

Ehrenfeld, J.G., Ravit, B., Elgersma, K., 2005. Feedback in the plant-soil system. Annu. Rev. Environ. Resour. 30, 75–115.

Elmendorf, S.C., Henry, G.H.R., Hollister, R.D., Fosaa, A.M., Gould, W.A., Hermanutz, L., Hofgaard, A., Jónsdóttir, I.S., Jorgenson, J.C., Lévesque, E., Magnusson, B., Molau, U., Myers-Smith, I.H., Oberbauer, S.F., Rixen, C., Tweedie, C.E., Walker, M.D., 2016. Experiment, monitoring, and gradient methods used to infer climate change effects on plant communities yield consistent patterns. PNAS 112, 448–452.

Estiarte, M., Vicca, S., Peñuelas, J., Bahn, M., Beier, C., Emmett, B.A., Fay, P.A., Hanson, P.J., Hasibeder, R., Kigel, J., Kröel-Dulay, G., Larsen, K.S., Lellei-Kovacs, E., Limousin, J.M., Schmidt, I.K., Sternberg, M., Tielbörger, K., Tietema, A., Janssense, I.A., 2016. Few multiyear precipitation-reduction experiments find a shift in the productivity-precipitation relationship. Glob. Change Biol. 22, 2570–2581.

Feng, Z., Rütting, T., Pleijel, H., Wallin, G., Reich, P.B., Kammann, C.I., Newton, P.C.D, Kobayashi, K., Luo, Y., Uddling, J., 2015. Constraints to nitrogen acquisition of terrestrial plants under elevated CO_2. Glob. Change Biol. 21, 3152–3168.

Fridley, J.D., Grime, P., Askew, A.P., Moser, B., Stevens, C.J., 2011. Soil heterogeneity buffers community response to climate change in species-rich grassland. Glob. Change Biol. 17, 2002–2011.

Golodets, C., Sternberg, M., Kigel, J., Boeken, B., Henkin, Z., Seligman, N.G., Ungar, E.U., 2013. From desert to Mediterranean rangelands: will increasing drought and inter-annual rainfall variability affect herbaceous annual primary productivity? Clim. Chang. 119, 785–798.

Golodets, C., Sternberg, M., Kigel, J., Boeken, B., Henkin, Z., Seligman, N.G., Ungar, E.U., 2015. Climate change scenarios of biomass production along an aridity gradient: vulnerability increases with aridity. Oecologia 177, 971–979.

Grime, J.P., Hodgson, J.G., Hunt, R., 2007. Comparative Plant Ecology: A Functional Approach to Common British Species, second ed. Castlepoint Press, Colvend, UK.

Grime, J.P., Fridley, J.D., Askew, A.P., Thompson, K., Hodgson, J.D., Bennett, C.R., 2008. Long-term resistance to simulated climate change in an infertile grassland. PNAS 105, 10028–10032.

Harte, J., Shaw, R., 1995. Shifting dominance within a montane vegetation community: results of a climate-warming experiment. Science 267, 876–880.

Hedges, L.V., Gurevitch, J., Curtis, P.S., 1999. The meta-analysis of response ratios in experimental ecology. Ecology 80, 1150–1156.

Hendrey, G.R., Miglietta, F., 2006. FACE technology: past, present, and future. In: Nösberger, J., Long, S.P., Norby, R.J., Stitt, M., Hendrey, G.R., Blum, H. (Eds.), Managed Ecosystems and CO_2. Springer, Berlin, Heidelberg, New York, pp. 15–45.

Herbert, D.A., Rastetter, E.B., Shaver, G.R., Ågren, G.I., 1999. Effects of plant growth characteristics on biogeochemistry and community composition in a changing climate. Ecosystems 2, 367–382.

Hijmans, R.J., Cameron, S.E., Parra, J.L., Jones, P.G., Jarvis, A., 2005. Very high resolution interpolated climate surfaces for global land areas. Int. J. Climatol. 25, 1965–1978.

Huxman, T.E., Smith, M.D., Fay, P.A., Knapp, A.K., Shaw, M.R., Loik, M.E., Stanley, D., Smith, S.D., Tissue, D.T., Zak, J.C., Weltzin, J.F., Pockman, W.T., Sala, O.E., Haddad, B.M., Harte, J., Koch, G.W., Schwinning, S., Small, E.E., Williams, D.G., 2004. Convergence across biomes to a common rain-use efficiency. Nature 429, 651–654.

IPCC Intergovernmental Panel on Climate Change, 2013. AR5, Climate Change 2013: The Physical Science Basis.

Isbell, F., Calcagno, V., Hector, A., Connolly, J., Harpole, W.S., Reich, P.B., Scherer-Lorenzen, M., Schmid, B., Tilman, D., van Ruijven, D., Weigelt, A., Wilsey, B.J., Zavaleta, E.S., Loreau, M., 2011. High plant diversity is needed to maintain ecosystem services. Nature 477, 199–202.

Jentsch, A., Kreyling, J., Beierkuhnlein, C., 2007. A new generation of climate change experiments: events, not trends. Front. Ecol. Environ. 5, 365–374.

Kaarlejärvi, E., Baxter, R., Hofgaard, A., Hytteborn, H., Khitun, O., Molau, U., Sjögersten, S., Wookey, P., Olofsson, J., 2012. Effects of warming on shrub abundance and chemistry drive ecosystem level changes in a forest–tundra ecotone. Ecosystems 15, 1219–1233.

Kammann, C., Muller, C., Grunhage, L., Jäger, H.-J., 2005. Response of aboveground grassland biomass and soil moisture to moderate long-term CO_2 enrichment. Basic Appl. Ecol. 6, 351–365.

Keuper, F., Dorrepaal, E., van Bodegom, P.M., Aerts, R., Van Logtestijn, R.S.P., Callaghan, T.V., Cornelissen, J.H.C., 2011. A race for space? How Sphagnum fuscum stabilizes vegetaton compoition during long-term climate manipulations. Glob. Change Biol. 17 (6), 2162–2171.

Kopittke, G.R., van Loon, E.E., Kalbitz, K., Tietema, A., 2013. The age of managed heathland communities: implications for carbon storage? Plant Soil 369, 219–230.

Körner, C., 2006. Plant CO_2 responses: an issue of definition, time and resource supply. New Phytol. 172, 393–411.

Kröel-Dulay, G., Ransijn, J., Schmidt, I.K., Beier, C., De Angelis, P., de Dato, G., Dukes, J.S., Emmett, B., Estiarte, M., Garadnai, J., Kongstad, J., Kovacs-Lang, E., Larsen, K.S., Liberati, D., Ogaya, R., Riis-Nielsen, T., Smith, A.R., Sowerby, A., Tietema, A., Peñuelas, J., 2015. Increased sensitivity to climate change in disturbed ecosystems. Nat. Commun. 6, 6682. http://dx.doi.org/10.1038/ncomms7682.

Leuzinger, S., Luo, Y., Beier, C., Dieleman, W., Vicca, S., Körner, C., 2011. Do global change experiments overestimate impacts on terrestrial ecosystems? Trends Ecol. Evol. 26, 236–241.

Liancourt, P., Boldgiv, B., Song, D.S., Spence, L.A., Helliker, B.R., Petraitis, P.S., Casper, B.B., 2015. Leaf-trait plasticity and species vulnerability to climate change in a Mongolian steppe. Glob. Chang. Biol. 21, 3489–3498.

Liang, J., Qi, X., Souza, L., Luo, Y., 2016. Processes regulating progressive nitrogen limitation under elevated carbon dioxide: a meta-analysis. Biogeosciences 13, 2689–2699.

Liu, D., Peñuelas, J., Ogaya, R., Estiarte, M., Tielbörger, K., Slowik, F., Yang, X., Bilton, M.C. Contrasting species compositional shifts under long-term experimental warming and drought in a Mediterranean shrubland (personal communication).

Luo, Y., 2007. Terrestrial carbon-cycle feedback to climate warming. Annu. Rev. Ecol. Evol. Syst. 38, 683–712.

Luo, Y., Currie, W.S., Dukes, J.S., Finzi, A.C., Hartwig, U., Hungate, B.A., McMurtrie, R.E., Oren, R., Parton, W.J., Pataki, D.E., Shaw, M.R., Zak, D.R., Field, C.B., 2004. Progressive nitrogen limitation of ecosystem responses to rising atmospheric carbon dioxide. Biogeosciences 54, 731–739.

Luo, Y., Keenan, T.F., Smith, M., 2015. Predictability of the terrestrial carbon cycle. Glob. Chang. Biol. 21, 1737–1751.

Mikkelsen, T.N., Beier, C., Jonasson, S., Holmstrup, M., Schmidt, I.K., Ambus, P., Pilegaard, K., Michelsen, A., Albert, K., Andresen, L.C., Arndal, M.F., Bruun, N., Christensen, S., Danbæk, S., Gundersen, P., Jørgensen, P., Linden, L.G., Kongstad, J., Maraldo, K., Priemé, A., Riis-Nielsen, T., Ro-Poulsen, H., Stevnbak, K., Selsted, M.B., Sørensen, P., Larsen, K.S., Carter, M.S., Ibrom, A., Martinussen, T., Miglietta, F., Sverdrup, H., 2008. Experimental design of multifactor climate change experiments with elevated CO_2, warming and drought: the CLIMAITE project. Funct. Ecol. 22, 185–195.

Mueller, K.E., Blumenthal, D.M., Pendall, E., Carrillo, Y., Dijkstra, F.A., Williams, D.G., Follett, R.F., Morgan, J.A., 2016. Impacts of warming and elevated CO_2 on a semi-arid grassland are non-additive, shift with precipitation, and reverse over time. Ecol. Lett. 19, 956–966. http://dx.doi.org/10.1111/ele.12634.

Muggeo, V.M.R., 2008. Segmented: an R package to fit regression models with broken-line relationships. R News 8 (1), 20–25.

Niklaus, P.A., Leadley, P.W., Schmid, B., Körner, C., 2001. A long-term field study on biodiversity X elevated CO_2 interactions in grassland. Ecol. Monogr. 71, 341–356.

Niu, S., Wan, S., 2008. Warming changes species competitive hierarchy in a temperate steppe of northern China. J. Plant Ecol. 1, 103–110.

Niu, S., Sherry, R.A., Zhou, X., Wan, S., Luo, Y., 2010. Nitrogen regulation of the climate carbon feedback: evidence from a long-term global change experiment. Ecology 91, 3261–3273.

Norby, R.J., DeLucia, E.H., Gielen, B., Calfapietra, C., Giardina, C.P., King, J.S., Ledford, J., McCarthy, H.R., Moore, D.J.P., Ceulemans, R., De Angelis, P., Finzi, A.C., Karnosky, D.F., Kubiske, M.E., Lukac, M., Pregitzer, K.S., Scarascia-Mugnozza, G.E., Schlesinger, W.H., Oren, R., 2005. Forest response to elevated CO_2 is conserved across a broad range of productivity. PNAS 102, 18052–18056.

Norby, R.J., Iversen, C.M., Childs, J., Tharp, M.L., 2010. ORNL Net Primary Productivity Data. Carbon Dioxide Information Analysis Center (http://cdiac.ornl.gov), U.S. Department of Energy, Oak Ridge National Laboratory, Oak Ridge, TN.

Ñuelas, J.P., Sardans, J., Estiarte, M., Àogaya, R., Carnicer, J., Coll, M., Barbeta, A., Rivas-Ubach, A., Llusià, J., Garbulsky, M., Filella, I., 2013. Evidence of current impact of climate change on life: a walk from genes to the biosphere. Glob. Chang. Biol. 19, 2303–2338.

Prieto, P., Peñuelas, J., Lloret, F., Llorens, L., Estiarte, M., 2009. Experimental drought and warming decrease diversity and slow down post-fire succession in a Mediterranean shrubland. Ecography 32, 623–636.
R Development Core Team, 2008. R: A Language and Environment for Statistical Computing. R Foundation for Statistical Computing, Vienna, Austria. ISBN: 3-900051-07-0.
Rastetter, E.B., Ågren, G.I., Shaver, G.R., 1997. Responses of N-limited ecosystems to increased CO_2: a balanced-nutrition, coupled-element-cycles model. Ecol. Appl. 7, 444–460.
Reich, P.B., Oleksyn, J., 2008. Climate warming will reduce growth and survival of Scots pine except in the far north. Ecol. Lett. 11, 588–597.
Reich, P.B., Knops, J., Tilman, D., Craine, J., Ellsworth, D., Tjoelker, M., Lee, T., Wedink, D., Naeem, S., Bahauddin, D., Hendrey, G., Jose, S., Wrage, K., Goth, J., Bengston, W., 2001. Plant diversity enhances ecosystem responses to elevated CO_2 and nitrogen deposition. Nature 410, 809–812.
Reich, P.B., Hobbie, S.E., Lee, T.D., 2014. Plant growth enhancement by elevated CO_2 eliminated by joint water and nitrogen limitation. Nat. Geosci. 7, 920–924.
Ruppert, J.C., Harmoney, K., Henkin, Z., Snyman, H.A., Sternberg, M., Willms, W., Linstädter, A., 2015. Quantifying drylands' drought resistance and recovery: the importance of drought intensity, dominant life history and grazing regime. Glob. Chang. Biol. 21, 1258–1270.
Rütting, T., Andresen, L.C., 2015. Nitrogen cycle responses to elevated CO_2 depend on ecosystem nutrient status. Nutr. Cycl. Agroecosyst. 101, 285–294.
Sala, O.E., Gherardi, L.A., Reichmann, L., Jobbagy, E., Peters, D., 2012. Legacies of precipitation fluctuations on primary production: theory and data synthesis. Philos. Trans. R. Soc. B. Biol. Sci. 367, 3135–3144.
Sardans, J., Peñuelas, J., Estiarte, M., Prieto, P., 2008. Warming and drought alter C and N concentration, allocation and accumulation in a Mediterranean shrubland. Glob. Chang. Biol. 14, 2304–2316.
Schneider, M.K., Lüscher, A., Richter, M., Aeschlimann, U., Hartwig, U., Blum, H., Frossard, E., Nösberger, J., 2004. Ten years of free-air CO_2 enrichment altered the mobilization of N from soil in Lolium perenne L. swards. Glob. Change Biol. 10, 1377–1388.
Smith, M.D., La Pierre, K.J., Collins, S.L., Knapp, A.K., Gross, K.L., Barrett, J.E., Frey, S.D., Gough, L., Miller, R.J., Morris, J.T., Rustad, L.E., Yarie, J., 2015. Global environmental change and the nature of aboveground net primary productivity responses: insights from long-term experiments. Oecologia 177, 935–947.
Sokal, R.R., Rohlf, F.J., 1995. Biometry, third ed. Freeman, New York, ISBN: 0-7167-2411-1.
Soussana, J.F., Lüscher, A., 2007. Temperate grasslands and global atmospheric change: a review. Grass Forage Sci. 62, 127–134.
South, A., 2011. rworldmap: a new R package for mapping global data. R J. 3 (1), 35–43.
Sternberg, M., Holzapfel, C., Tielbörger, K., Sarah, P., Kigel, J., Lavee, H., Fleischer, A., Jeltsch, F., Köchy, M., 2011. The use and misuse of climatic gradients for evaluating climate impact on dryland ecosystems—an example for the solution of conceptual problems. In: Blanco, J., Kheradmand, H. (Eds.), Climate Change—Geophysical Foundations and Ecological Effects, pp. 361–374. ISBN 978-953-307-419-1.
Talhelm, A.F., Pregitzer, K.S., Kubiske, M.E., Zak, D., Campany, C.E., Burton, A.J., Dickson, R.E., Hendrey, G.R., Isebrands, J.G., Lewin, K.F., Nagy, J., Karnosky, D.F., 2014. Elevated carbon dioxide and ozone alter productivity and ecosystem carbon content in northern temperate forests. Glob. Change Biol. 20, 2492–2504.

Terrer, C., Vicca, S., Hungate, B.A., Phillips, R.P., Prentice, I.C., 2016. Mycorrhizal association as a primary control of the CO_2 fertilization effect. Science 353 (6294), 72–74.
Tielbörger, K., Bilton, M.C., Metz, J., Kigel, J., Holzapfel, C., Lebrija-Trejos, E., Konsens, I., Parag, H.A., Sternberg, M., 2014. Middle-Eastern plant communities tolerate 9 years of drought in a multi-site climate manipulation experiment. Nat. Commun. 5, 5102. http://dx.doi.org/10.1038/ncomms6102.
Williams, A.P., Allen, C.D., Macalady, A.K., Griffin, D., Woodhouse, C.A., Meko, D.M., Swetnam, T.W., Rauscher, S.A., Seager, R., Grissino-Mayer, H.D., Dean, J.S., Cook, E.R., Gangodagamage, C., Cai, M., McDowel, N.G., 2012. Temperature as a potent driver of regional forest drought stress and tree mortality. Nat. Clim. Chang. 3, 292–297.
Zelikova, T.J., Blumenthal, D.M., Williams, D.G., Souza, L., LeCain, D.R., Morgan, J., Pendall, E., 2014. Long-term exposure to elevated CO_2 enhances plant community stability by suppressing dominant plant species in a mixed-grass prairie. PNAS 111, 15456–15461.

CHAPTER TEN

Recovery and Nonrecovery of Freshwater Food Webs from the Effects of Acidification

C. Gray*,[†], A.G. Hildrew[†,‡], X. Lu[§], A. Ma[§], D. McElroy[¶], D. Monteith[∥], E. O'Gorman*, E. Shilland[#], G. Woodward*,[1]

*Imperial College London, Ascot, Berkshire, United Kingdom
[†]School of Biological and Chemical Sciences, Queen Mary University of London, London, United Kingdom
[‡]Freshwater Biological Association, The Ferry Landing, Ambleside, Cumbria, United Kingdom
[§]School of Electronic Engineering and Computer Science, Queen Mary University of London, London, United Kingdom
[¶]Centre for Research on Ecological Impacts of Coastal Cities, University of Sydney, Sydney, NSW, Australia
[∥]Centre for Ecology & Hydrology, Lancaster Environment Centre, Lancaster, United Kingdom
[#]Environmental Change Research Centre, University College London, London, United Kingdom
[1]Corresponding author: e-mail address: guy.woodward@imperial.ac.uk

Contents

1. Introduction 476
    1.1 Food Web Recovery Research 477
    1.2 Freshwater Acidification 477
    1.3 The Recovery of Acidified Food Webs 481
    1.4 The Acid Waters Monitoring Network 485
2. Methods 487
    2.1 Sites 487
    2.2 Chemistry 487
    2.3 Biota 487
    2.4 Food Web Construction 488
    2.5 Network Metrics 489
    2.6 Statistical Analyses 490
3. Results 491
    3.1 Effects of Acidity on Food Web Structure 491
    3.2 Directional Change in Food Web Structure 494
    3.3 Food Web Recovery from Acidification 496
4. Discussion 502
    4.1 Food Web Recovery Across the AWMN Sites and Acidity Gradient 503
    4.2 The Recovery of Freshwater Food Webs from Acidification 505
    4.3 Caveats and Future Directions 506
5. Conclusion 507
Appendix 508
Acknowledgment 528
References 528

Advances in Ecological Research, Volume 55
ISSN 0065-2504
http://dx.doi.org/10.1016/bs.aecr.2016.08.009

© 2016 Elsevier Ltd
All rights reserved.

475

Abstract

Many previous attempts to understand how ecological networks respond to and recover from environmental stressors have been hindered by poorly resolved and unreplicated food web data. Few studies have assessed how the topological structure of large, replicated collections of food webs recovers from perturbations. We analysed food web data taken from 23 UK freshwaters, sampled repeatedly over 24 years, yielding a collection of 442 stream and lake food webs. Our main goal was to determine the effect of acidity on food web structure and to analyse the way food web structure recovered from the effects of acidity over time.

Long-term monotonic reversals of acidification were evident at many of the sites, but the ecological responses were generally far less evident than chemical changes, or absent. Across the acidity gradient, food web linkage density and network efficiency declined with increasing acidity, while node redundancy (i.e. trophic similarity among species within a web) increased. Within individual sites, connectance, linkage density, trophic height, resource vulnerability and network efficiency tended to increase over time as sites recovered from acidification, while consumer generality and node redundancy tended to decrease. There was evidence for a lag in biological recovery, as those sites showing a recovery in both their biology and their chemistry were a nested subset of those which only showed a chemistry trend.

These findings support the notion that food web structure is fundamentally altered by acidity, and that inertia within the food web may be hindering biological recovery. This suggestion of lagged recovery highlights the importance of long-term monitoring when assessing the impacts of anthropogenic stressors on the natural world. This temporal dimension, and recognition that species interactions can shape community dynamics, is missing from most national biomonitoring schemes, which often rely on space-for-time proxies.

1. INTRODUCTION

Natural ecosystems are increasingly exposed to anthropogenic stressors, such as habitat modification, pollution and global climate change (Smith and Zeder, 2013; Steffen et al., 2011; Sutherland et al., 2016). A deeper understanding of how they respond to and recover from such perturbations is important if we are to manage our natural resources effectively in the coming decades (Pimm et al., 1995; Woodward et al., 2010a).

Biological recovery from the effects of stressors does not necessarily follow from the removal of that stressor, as there may be time lags or ecological hysteresis, even to the extent that alternative equilibria are possible for otherwise identical environmental conditions (Battarbee et al., 2014; Feld et al., 2011; Murphy et al., 2014; O'Neill, 1998; Scheffer and Carpenter, 2003). These may arise via species interactions, which can alter the rate and/or trajectory of recovery (Scheffer and Carpenter, 2003) and confound attempts to scale up predictions made from individuals or species populations to the whole community or ecosystem, because of the increasing scope for 'ecological surprises' to be manifested via complex indirect pathways in the food web (Ings et al., 2009; Thompson et al., 2012). For instance, artificially high nutrient concentrations can trigger regime shifts in shallow lakes, which may persist even long after nutrient loads have been reduced (Scheffer and Carpenter, 2003).

1.1 Food Web Recovery Research

Due to the difficulties in constructing highly resolved food webs, however, very few studies have examined how trophic network structure responds to, and recovers from, perturbations, and fewer still have a replicated design. Field experiments have revealed how replicated freshwater food webs respond to drought, through the loss of rare and rare-for-size consumers, as well as the larger taxa high in the food web (e.g. eight stream food webs; Ledger et al., 2012; Woodward et al., 2012). Other studies have economised on effort and increased their sample sizes, and hence ability to detect responses statistically, by making assumptions about the diet of consumers. Thus, O'Gorman and Emmerson (2010) used 144 marine food webs in a mesocosm study across a range of experimental treatments to investigate how their structure responded to the removal of keystone species. Very few examples exist where the recovery of replicated, natural food webs following a perturbation has been studied, although McLaughlin et al. (2013) constructed a collection of 96 terrestrial food webs which tracked the recovery of 16 riparian food webs after a flood over the course of a year. Often, a space-for-time substitution approach is used: for instance, Layer et al. (2010b) studied the structure of 20 freshwater food webs sampled once from 20 sites distributed across a wide pH gradient, and this was subsequently set in the context of long-term change in the single model system of Broadstone Stream over four decades of rising pH (Layer et al., 2011). However, no studies of which we are aware have analysed the long-term recovery from perturbation of replicated food webs distributed across wide ecological gradients.

1.2 Freshwater Acidification

Freshwater acidification is usually caused by atmospheric pollution (though there are naturally rather acidic systems), in which strong mineral acids emitted from industrial sources are deposited on the landscape (e.g. Driscoll et al., 2001). Where soils and geology have an insufficient supply of base cations to buffer acidity, run-off to streams and lakes becomes strongly acidic. At a pH of 5.5, alkalinity falls to zero and inorganic aluminium concentration rises to become toxic to many forms of life, including almost all fish (Sutcliffe and Hildrew, 1989). Such anthropogenic acidification has profound ecological impacts, including the loss of many acid-sensitive species from all trophic levels (e.g. Dillon et al., 1984; Schindler, 1988). Evidence for the causes and consequences of acidification and its effect on species assemblages is long-standing and overwhelming (e.g. Hildrew and Ormerod, 1995; Likens and Bormann, 1974; Schindler, 1988) and comes primarily from north-western Europe and much of north and eastern USA and Canada. This ranges from palaeolimnological reconstructions of lake pH (e.g. Battarbee et al., 1988), widespread surveys and descriptions (e.g. Henriksen et al., 1990; Townsend et al., 1983), experimental acidification of whole systems (e.g. Findlay et al., 1999; Hall et al., 1980; Webster et al., 1992) and biogeochemical modelling (e.g. Jenkins et al., 1990). In the UK, intensive research on the long-term and large-scale ecological consequences of acidification on running waters was concentrated in three main study systems: Llyn Brianne in south-west Wales (Durance and Ormerod, 2007), the Ashdown Forest of south-eastern England (Hildrew, 2009) and via the UK's Acid Waters Monitoring Network (UKAWMN), which interdigitated to a limited extent by sharing three sites with the other two (Layer et al., 2010a,b).

In the face of such evidence, reductions of polluting emissions were agreed upon in both Europe and North America, which have resulted in dramatically reduced depositions since the 1970s (RoTAP, 2012; Stoddard et al., 1999). In 1988, the UK government set up UKAWMN (incorporated within the recently expanded Upland Waters Monitoring Network—or UWMN, in 2014) to assess the effectiveness of these measures, which came at considerable economic and social costs. The network was designed explicitly to detect any recovery in the quality of surface waters at its 23 stream and lake sites, as well as any shifts in their biology and ecology that might be expected to accompany chemical recovery. The network's sites are distributed across acid-sensitive (base poor) regions of the UK (Fig. 1), mainly in the uplands areas of the north and west where precipitation and wet deposition of acidity tend to be high, although some are in small, acid-sensitive areas in the south and east. A few sites are located in the extreme north and west of Scotland and Northern Ireland which were thought sufficiently remote from industrial pollutant sources to have been significantly affected by acid deposition (Patrick et al., 1991).

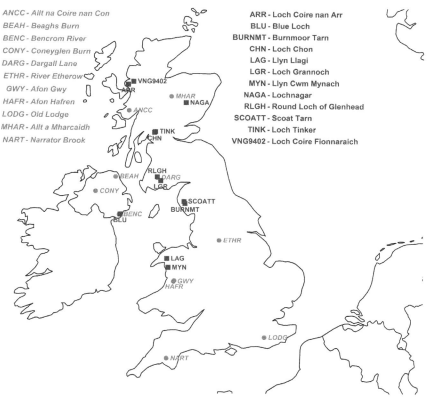

Fig. 1 The Upland Waters Monitoring sites, consisting of 11 lakes (*dark blue* (*dark grey* in the print version) *squares*) and 12 streams (*light blue* (*light grey* in the print version) *circles*).

Substantial (though not complete) chemical recovery from acidification has now occurred at most sites that had been acidified at the outset (Monteith et al., 2014). This has included large reductions in inorganic aluminium concentrations in the most acidified waters and more widespread but gradual increases in pH, while Acid Neutralising Capacity has increased in proportion to the rate of reduction in acid anions. Ubiquitous increases in dissolved organic carbon (DOC) concentration also appear to be part of the biogeochemical response, leading to a partial replacement of mineral acidity by organic acidity that has tempered the pH response (Evans et al., 2008). Evidence for biological recovery, in terms of the establishment of acid-sensitive assemblages of species, has been much less obvious (Battarbee et al., 2014), but most evident in the diatoms of the epilithon and from the colonisation of recovering sites by some species of macrophytes. In the last UWMN data interpretation report (Kernan et al., 2010), invertebrate assemblages showed signs of partial recovery at around half of the chemically recovering sites. Only two of those sites, both of which were particularly severely acidified at the onset of monitoring, showed any evidence of recovery of salmonid populations, another indicator of decreasing acidity (Malcolm et al., 2014; Murphy et al., 2014). Similar 'sluggish' biological recovery has also been reported in other acidified systems elsewhere globally (e.g. Keller et al., 2007; Nedbalová et al., 2006).

Several hypotheses have been put forward to explain these delays in the anticipated simple reversal of acidification, including dispersal limitations, pollutant legacies and attendant recurring acid episodes (e.g. Kowalik et al., 2007), interactions with other stressors (e.g. climate change; Johnson and Angeler, 2010) and indirect food web effects (Ledger and Hildrew, 2005; Monteith et al., 2005; Yan et al., 2003). These mechanisms are not necessarily mutually exclusive explanations and there are differing levels of support for each. For instance, the role of dispersal is still uncertain for differing biological components, with evidence for and against its role: Gray and Arnott (2011) suggested it may constrain the recovery of lake zooplankton in Canadian lakes, contrary to previous findings (Keller et al., 2002; Yan et al., 2004). Evidence for more mobile taxa, such as benthic insects with flying adults, however, suggests that dispersal is usually sufficient to allow rapid recolonisation and thus unlikely to explain delayed biological recovery (Hildrew, 2009; Masters et al., 2007). The type of waterbody also affects both its rate of chemical recovery and its recolonisation potential: lakes are larger and thus better able to absorb spikes of run-off that can create acid episodes than streams (Evans et al., 2001) and yet they show similarly limited biological recovery and so acidic episodes alone cannot explain the lag in biological recovery. In addition, these hypotheses may operate additively or even synergistically.

To date, no evidence has been found for differing rates of recovery between streams and lakes. Given the evidence that persistent acidic episodes in stream systems can limit biological recovery (Kowalik et al., 2007; Lepori and Ormerod, 2005) one might expect stream systems to recover more slowly than lakes, although Monteith et al., 2014 found that the magnitude of acid pulses at UWMN sites had declined at a similar rate to mean acidity. Conversely, the dynamic nature of streams, with a natural regime of frequent flow disturbances, and downstream connectivity to pools of less acid-sensitive species in the lower reaches might render them naturally more resilient (e.g. Hildrew and Giller, 1994). Thus, we had no clear a priori hypothesis or expectations of the relative rates of recovery in the communities of lakes and streams.

1.3 The Recovery of Acidified Food Webs

The possibility that species interactions within the food web might inhibit ecological recovery requires further testing (Frost et al., 1998; Webster et al., 1992). Circumstantial evidence from streams suggests that generalist, nonpredatory invertebrates (e.g. stoneflies of the families Leuctridae and Nemouridae), which are acid tolerant and often dominate the benthos of acid streams, may fill the feeding niche of specialist, acid-sensitive grazers and inhibit their return (Layer et al., 2013; Ledger and Hildrew, 2005). Further, dynamic modelling found that the reticulate acidified food webs are more robust, suggesting that they might be more inherently stable and thus less prone to (re)invasion (Layer et al., 2010b). Finally, the common reliance on space-for-time proxies may miss the transient dynamics and the possible existence of alternate stable states of a system responding to stress. For instance, if the strong effects of pH on food web properties reported in the space-for-time survey of 20 sites by Layer et al. (2010b) are not evident in systems undergoing actual changes in pH over time, then the expected mapping of the biota onto the environmental template might not be evident, and its absence would indeed indicate ecological inertia. This has wider implications for biomonitoring science in general, which is underpinned by space-for-time approaches and rarely has access to truly long-term (i.e. multidecadal) high-quality biological time series (Friberg et al., 2011). Mismatches between temporal and space-for-time data may provide evidence for time lags or hystereses resulting from the system's own internal dynamics, although the potential importance of biogeographic constraints should also not be discounted. On the other hand, if the two data types match perfectly, then indirect food web effects—which would otherwise reshape the simple biota–environment relationship—can effectively be discounted.

Repeated assessment of the topology of a large collection of food webs, as they recover from the effects of acidification, is needed to complement previous work, which dealt with single 'model' systems (e.g. Layer et al., 2010a, 2011) or relied on space-for-time substitutions. As an example of the latter, Layer et al. (2010b) suggested that the smaller but more interconnected acidified webs had more stable configurations of trophic linkages across a pH gradient of <5 to >8. Linkage density and connectance are both common measures of web complexity, and an abundance of 'redundant' interactions can help stabilise the network's structure in the face of perturbations, by preventing the secondary extinctions that arise when consumers are left without resources (Dunne et al., 2002; Thébault and Fontaine, 2010). Indeed, linkage density increased with stream pH across the spatial gradient of the 20 sites, which included 10 of the UWMN streams used in this study.

Mean food chain length (the number of steps between a basal resource and a particular consumer, see Box 1) gives a measure of the trophic height of the web as a whole (Williams and Martinez, 2004), which tends to shorten as environmental stress increases and productivity declines (Woodward et al., 2005a,b). Food chains are generally assumed to be shorter than six links, and omnivory, which is commonplace in aquatic systems, tends to truncate them further (Hildrew et al., 1985; Lawton, 1989; Pimm, 1980; Williams and Martinez, 2000; Yodzis, 1989). Although long-term data are still scarce, there is some evidence that suggests that increasing pH leads to higher productivity and an overall lengthening of food chains (Gerson et al., 2016; Grahn et al., 1974; Hildrew, 2009; Woodward et al., 2005a,b). For instance, progressively larger and more acid-tolerant predators have (re)invaded Broadstone Stream as pH has risen since the 1970s, culminating in the recent return of the apex predator, the brown trout (*Salmo trutta*) (Layer et al., 2011).

In addition to simple food chain metrics, the range of both resources and consumers to which each species is connected in the web has important implications for the overall network's dynamical and structural stability, and its ability to respond to or resist environmental change. *Generality* (see Box 1) is a substructural measure of the dietary breadth of a consumer, derived statistically from its number of resources. If a consumer is a specialist (i.e. narrow diet), then it might be more vulnerable to extinction as the loss of only a few species will leave it with insufficient resources. *Vulnerability* (see Box 1) is the converse measure of *Generality*; it is derived from the number of consumers feeding on a particular resource species, and indicates how important that resource is in terms of the consumers it supports. Freshwater predators are commonly gape-limited generalists, so the size and diversity of prey increase with consumer size (Woodward and Hildrew, 2002; Woodward et al., 2010b). Similarly, the herbivorous consumers in acidified streams are also generalists, feeding on a wide range of detritus

BOX 1 Definitions of Food Web Metrics Used in This Study

Food Web Metric	Definition
Connectance (C)	Number of links (L)/number of species $(S)^2$. The proportion of potential trophic links that do occur (Warren, 1994)
Linkage density	L/S. Number of links per taxon. A measure of average diet specialisation across the food web (Tylianakis et al., 2007)
Generality (G)	The mean number of prey per consumer (Schoener and Schoenerz, 1989)
Vulnerability (V)	The mean number of consumers per prey (Schoener and Schoenerz, 1989)
Mean food chain length	Average number of links found in a food chain across a food web (Levine, 1980; Williams and Martinez, 2000)
Maximum food chain length	The maximum number of links found in any food chain in a food web (Levine, 1980; Williams and Martinez, 2000)
Efficiency	How well connected a network is, as well as the distribution of those connections across a network. High efficiency indicates that the species of a food web are all closely connected to one another (Latora and Marchiori, 2001)
Redundancy	The trophic similarity among species within a web: high redundancy indicates that many of the species in a food webs are the same resources and consumers; many of the feeding pathways are the same (Briand and Cohen, 1984; Cohen and Briand, 1984)

and algae (e.g. Layer et al., 2013; Ledger and Hildrew, 2005). As acidity decreased and acid-sensitive, but more specialist, species reinvade, the average generality of consumers (i.e. normalised to the size of the food web) should decrease. The average vulnerability of resources (again normalised to the size of the food web) should increase with decreasing acidity as the consumer guild becomes more speciose (Layer et al., 2010b). However, Layer et al. (2010b) found that normalised consumer generality and resource vulnerability did not change systematically, and there was no relationship between either the variation (standard deviation) in consumer generality or resource vulnerability and pH, although the sample size was rather small.

Under acidified conditions, generalist primary consumers can partially occupy the niche left by the loss of specialist herbivorous species, potentially creating 'ecological inertia' within the food web by slowing the return of the latter as pH rises (Ledger and Hildrew, 2005). The effect of acidity on this redundancy of feeding pathways within the whole food web has not been investigated previously using network-based approaches. As acidified systems tend to be species poor and dominated by generalist consumer species and few specialists (e.g. Hämäläinen and Huttunen, 1996; Ledger and Hildrew, 2005), there should be greater trophic redundancy (i.e. species of acid streams should have more similar diets and share more predators than more speciose webs of relative specialists), which could make them resistant to perturbations and hence more robust to the loss of food resources than those at higher pH (Naeem, 1998; Peralta et al., 2014; Solé et al., 2003).

As with other features of the whole network (such as connectance) or parts of the network (e.g. generality) the so-called small-world properties of food webs have also been linked to stability (Montoya and Solé, 2002; Watts and Strogatz, 1998) and to the rate at which perturbations propagate (Montoya et al., 2006). Essentially, if species are highly connected to the rest of the food web in a 'small-world' network, then perturbations may spread (and dissipate) rapidly, but if there are less well connected (more degrees of separation) this may lead to longer-lived oscillatory dynamics and feedbacks that require a long time to reach equilibrium. Thus, it is not simply the linkage density or strength of connections that are important for determining stability or food web inertia, but their particular configuration. Even large food webs from circumneutral or higher pH systems can exhibit these properties, with most species being only 1–2 degrees of separation from the rest of the web (Thompson et al., in press). Network efficiency (see Box 1) is a measure of how well connected a network is, as well as the distribution of those connections across a network (Latora and Marchiori, 2001), and can enable inferences to be made about the small-world properties of food webs. Although rarely applied to date in food web studies, this metric derived from the wider field of network science could provide new insights into how these small-world aspects of food web topology itself might shape the trajectory of biological recovery (Layer et al., 2010b; Monteith et al., 2005). The lengthening of food chains associated with the reinvasion of consumers might be expected to increase the overall efficiency of the network, as the wider breadth of diet of new, large, top predators effectively reduces the distance (in terms of number of links) between resources.

1.4 The Acid Waters Monitoring Network

The data analysed here (from the AWMN, now renamed the Upland Waters Monitoring Network) consist of repeated observations on the same 23 sites (Fig. 1) over 24 years (1988–2012). Spatially, the sites encompass a wide range of pH (from 3.71 to 7.49), and include some that were strongly acidified at the onset of monitoring (e.g. Old Lodge) and others that were circumneutral and have changed little (e.g. Allt na Coire nan Con). The gradual long-term chemical recovery of many of these sites, particularly among those formerly the most acidic, provides a unique and large-scale picture of the chemical drivers and biological responses in the network over both space and time. Characterising the 'baseline' variation in food web structure in the near absence of changes in environmental stress is crucial for our understanding of how a community recovers from that stressor and, more broadly, for gauging its potential responses to future environmental changes. The food webs compiled from these data provide the replication and statistical power that has previously been insufficient for rigorous analysis of food web responses to acidification and chemical recovery.

We aimed to describe changes in food web structure as water chemistry recovered over three decades of chemical and biological monitoring. Our emphasis was on structural changes in binary networks of the presence/absence of nodes and links, rather than the effects of acidity on the dynamical stability of species populations per se. To determine the effect of environmental variables on food web structure, we tested the following hypotheses, using data gathered across all sites and years:

1. When analysed across the entire acidity gradient, food web structure will be directly affected by acidity. Food webs of less-acidified systems should exhibit higher linkage density, food chain length and network efficiency but lower connectance consumer generality and redundancy (Table 1).
2. There will be directional and predictable changes in food web structure at each site through time as acidity decreases (Table 1).
3. If indirect food web effects arising from biotic interaction are unimportant, changes in network structure through time will match recovery from acidification, i.e., those sites which experience change in their chemistry will also change at a comparable pace in their food web attributes (Fig. 2).

Additionally, in order to investigate if the rates of recovery were different between stream and lake sites, the two ecosystem types were analysed both together and separately.

Table 1 Summary of Predicted Relationships Between Acidity and Food Web Structure

Food Web Metric	Predicted Relationship with Decreasing Acidity
Connectance	↓
Linkage density	↑
Mean trophic height	↑
Maximum trophic height	↑
Resource vulnerability	↑
Consumer generality	↓
Standard deviation in vulnerability	No change
Standard deviation in generality	No change
Network efficiency	↑
Redundancy	↓

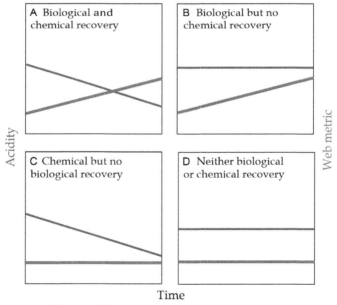

Fig. 2 A conceptual figure of the possible outcomes for biological ('web metric', *thick, blue line* (*grey* in the print version)) and chemical ('chemical', *thin, red line* (*dark grey* in the print version)) recovery at the UWMN sites over time.

2. METHODS
2.1 Sites

The UWMN consists of 11 streams and 12 lakes from across the UK (Fig. 1): full site descriptions and sampling methodologies are provided in Patrick et al. (1995) and Kernan et al. (2010). Water chemistry, epilithic diatom, macroinvertebrate and fish sampling began in spring 1988 and has continued annually at most sites up to 2012, except for a few exceptions when access was occasionally restricted (see Kernan et al., 2010). The sites are distributed along a latitudinal gradient across the UK, which can be interpreted as a proxy for the degree of acid deposition at the outset: sites at high latitude were exposed to relatively little acid deposition, while those sites at lower latitudes were generally more acidified (Patrick et al., 1991). One lake site, Loch Coire nan Arr, was affected by damming that increased the water level, so it was replaced in 2001 by the nearby Loch Coire Fionnaraich, which has comparable characteristics (Fig. 1).

2.2 Chemistry

Water samples for chemical analysis were taken in acid-rinsed bottles monthly from streams and quarterly from lakes. A large number of chemical determinands were recorded at each site: for more details see Kernan et al. (2010) and Monteith et al. (2014). In total 13 variables that are considered key drivers or indicators of acidification (Monteith et al., 2014) were used here: pH, alkalinity, H^+, conductivity, nitrate (NO_3), nonlabile aluminium, soluble aluminium, labile aluminium, DOC, sodium (Na^+), sulphate (SO_4^{2-}), calcium (Ca^{2+}) and chloride (Cl^-). With the exception of pH, we used the annual arithmetic mean of all chemical data as summary statistics for each site. Annual average pH was calculated by first converting pH to H^+ concentration, calculating the annual arithmetic mean and then converting back to pH.

2.3 Biota

Benthic diatoms, macroinvertebrates and fish were sampled annually from 1988 to 2012. Benthic diatoms were sampled according to standard UWMN protocols (Patrick et al., 1991), by selecting five cobble-sized stones from the streambed, or from the permanently submerged littoral zone in lakes. Three samples were taken from each site, in streams at the top, middle and bottom of fixed 50 m reaches, and in the lakes from three discrete locations on the shoreline (away from inflow or outflow streams). The biofilm on the surface of the stone was scrubbed into a funnel, washed into a plastic vial with stream or lake water, and immediately preserved in Lugol's iodine solution. In the laboratory, samples were prepared by digestion in hydrogen peroxide (H_2O_2) and diluted with distilled water. To enable subsequent examination by light microscopy at $1000 \times$ magnification, a subsample was pipetted onto a coverslip, dried and mounted onto a slide using Naphrax (Battarbee et al., 2001). Three hundred diatom valves from each sample were identified to species to give the diatom species assemblage per site per year.

Macroinvertebrates were sampled according to standard UWMN protocols (Patrick et al., 1991), by taking five separate 1-min kick samples with a standard hand net (330 µm mesh) from riffle sections of streams and from the dominant littoral habitat of lakes. Macroinvertebrates were subsequently sorted and preserved with 70% Industrial Methylated Spirit. Oligochaeta, Diptera and Bivalvia were identified to class, family and genus, respectively, while all others were identified to species. All taxa were counted and the counts from the five samples summed to represent the macroinvertebrate assemblage per site per year.

Annual electrofishing surveys were conducted according to standard UWMN protocols (Patrick et al., 1991). Surveys were undertaken between mid-September and mid-October at each stream site or, for lakes, in the outflow stream immediately downstream. It was assumed that composition of the fish assemblage in lake outflows could serve as a proxy for that of fish in the lake itself. Three 50-m reaches, distributed across 500 m of the stream or lake outflow, were isolated using stop nets and electrofished. Depletion electrofishing was employed and all salmonids were counted, while the presence of any other fishes was also recorded.

2.4 Food Web Construction

Species lists were compiled for each site in each year for which there was complete biological and chemical data, yielding 442 food webs. Binary food webs were constructed using the WebBuilder function in R (Gray et al., 2015) and the database of freshwater aquatic trophic interactions contained therein, based on the presence/absence of species (nodes) at each site in each year and the occurrence of a trophic linkage in the feeding link database. This method is based on the assumption that all feeding links between specific pairs of species that have been reported previously would be realised, wherever and whenever both species coexist at a study site (Hall and Raffaelli, 1991; Layer et al., 2010b; Martinez, 1991; Pocock et al., 2012), although many of the feeding links within that database were, in fact, derived from direct observation from previous UWMN surveys (Layer et al., 2010a,b, 2011; Ledger and Hildrew, 2005). When species-specific trophic interactions had not previously been described for some rare or understudied taxa (nodes), feeding links were assigned on the basis of taxonomic similarity, for instance, by assuming that different species within the same genus consumed and were consumed by the same species. This method is often used when constructing freshwater food webs as in these systems consumer diets tend to be highly generalist and determined primarily by the size of their prey (e.g. Layer et al., 2010a,b, 2013). Food webs built in this manner are structurally comparable to those built solely through analysis of consumers gut contents; for instance, the method has predicted the links of four well-documented freshwater food webs with 40–60% accuracy (on a scale from −100% to 100%, Gray et al., 2015).

2.5 Network Metrics

Several commonly used metrics were calculated for each food web, including: connectance ($C=L/S^2$; where $L=$ the number of trophic links and $S=$ the number of species), linkage density ($LD=L/S$), mean trophic height (after Levine, 1980: defined as 1 plus the mean trophic level of a consumer's resources, averaged across all consumers) and maximum trophic height (defined in the same way, except that the maximum value across all consumers was taken). Mean generality (G; number of resources per consumer, see Box 1) and vulnerability (V; number of consumers per resource, see Box 1) were calculated, as were normalised G and V for each taxon k, as:

$$G_k = \frac{1}{L/S} \sum_{i=1}^{S} a_{ik} \tag{1}$$

$$V_k = \frac{1}{L/S} \sum_{j=1}^{S} a_{jk} \tag{2}$$

where S is the number of nodes and L the number of links in a food web. $a_{ik}=1$ if taxon k consumes taxon i (otherwise $a_{ik}=0$), and $a_{jk}=1$ if taxon k is being consumed by taxon j (otherwise $a_{jk}=0$). Mean G_k and V_k in any given food web equal 1, making their standard deviations, which give an indication of the variability in G and V, respectively, across a network, comparable across networks of different size. These metrics were all calculated using the R package cheddar (Hudson et al., 2013).

Network efficiency (Latora and Marchiori, 2001, see Box 1) describes the 'reachability' of each node by any other node, and is a measure of overall connectivity, and was calculated using the sna R package (Butts, 2013) as:

$$E = \frac{1}{S(S-1)} \sum_{i \neq j \in G} \frac{1}{d_{ij}} \qquad (3)$$

where d_{ij} is the shortest path length between node i and j.

The proportional node redundancy (see Box 1) of each network was calculated by grouping nodes into trophic species (i.e. nodes with common resources and consumers) and then calculated as:

$$\text{redundancy} = 1 - \frac{T}{S} \qquad (4)$$

where T is the number of trophic species within the network. Redundancy was calculated using functions from the cheddar package (Hudson et al., 2013) in R.

2.6 Statistical Analyses

All statistical analyses were done in R version 3.1.1 (R Core Team, 2013), PRIMER-E with PERMANOVA + (2006). To simplify the chemical data, Principal Component Analysis (PCA) was performed on all 13 water chemistry variables across all sites and years, using a resemblance matrix constructed from Euclidean distances. As some of the variables were measured on different scales (i.e. NO_3 vs pH), and to reduce the influence of extremely large or small values, each variable were centred to zero and scaled by their standard deviations (van den Berg et al., 2006). Sample scores on the first PC axis (PC1) were extracted for use as a proxy for a general gradient in overall acidity in further analysis.

2.6.1 Effects of Acidity on Food Web Structure

For data visualisation purposes only, as we were unable to fully account for both temporal and spatial pseudoreplication of our data simultaneously using multivariate analysis, principal coordinates analysis was used on all data across all sites and time points. The resemblance matrix of food web metrics was constructed from square root transformed variables, using Bray–Curtis distances. More rigorous statistical inferences were drawn from univariate approaches, in which pseudoreplication was addressed within the variance structure of the relevant model(s), as explained later.

To assess the effect of acidity on food web structure (our first hypothesis), each network metric was regressed against the derived acidity gradient (PC1), and any trend assessed with Generalised Linear Mixed Effects models. Alongside acidity (PC1), site type (lake or stream) was fitted as a fixed effect, and any potential interactions with acidity (PC1) were assessed on the basis of stepwise model simplification and model AIC (Crawley, 2013). For each model, site was fitted as having a random effect on the intercept of the model, and year was fitted to have a random effect on the slope and intercept of the model.

2.6.2 Directional Change in Food Web Structure

Mann–Kendall trend tests were used to determine if there were significant monotonic trends in the acidity and food web structure over time at each site (our second hypothesis). The acidity gradient (PC1) extracted from the PCA above, and all the network metrics described earlier were calculated for each site in each year. These variables were then assessed for monotonic trends over time at each site.

2.6.3 Food Web Recovery from Acidification

To test our third hypothesis we used χ^2 contingency tests to assess the extent to which sites that exhibited clear decreases in acidity also showed evidence of directional change in their food web structure (as in Murphy et al., 2014). For acidity (PC1), and for each network metric, we counted the number of sites (out of 23) that exhibited (a) a biological and a chemical trend, (b) a biological but not a chemical trend, (c) a chemical but not a biological trend (i.e. evidence for ecological inertia) and (d) neither a biological nor a chemical trend (Fig. 2). The χ^2 test assessed whether the distribution of sites across these four categories was due to chance.

3. RESULTS

The PCA of the chemical variables revealed that the first axis was strongly correlated with pH, H^+ ions and SO_4. It was therefore used a proxy for the acidity-related stress to which each food web was exposed (Fig. 3), as it encompassed the variation in these pH-related chemistry variables. From here on PC1 is called 'acidity' and refers not only to pH but to the chemical stress associated with low pH.

3.1 Effects of Acidity on Food Web Structure

When the food web data were analysed at the regional (UK) scale, and modelled against the acidity gradient extracted from Fig. 3, some clear trends in food web structure emerged, although several did not match or even ran counter to our initial hypotheses (Fig. 4). Contrary to our expectations (Table 1) connectance and trophic height were unrelated to the acidity gradient (Table 2; more acid sites are to the right in Fig. 5). As predicted, linkage density increased with decreasing acidity (Table 2; Fig. 5) and lake

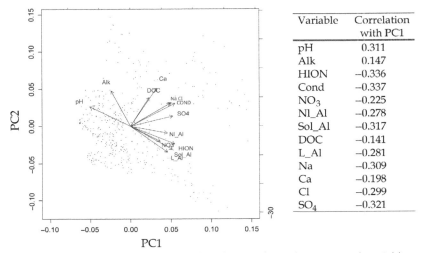

Fig. 3 PCA ordination of chemical variables. The correlation between each variable and acidity (PC1) is given in the accompanying table. The first axis, PC1, is strongly related to pH, SO_4 and aluminium such that PC1 can be interpreted as an 'acidity gradient' with acid stress increasing to the right.

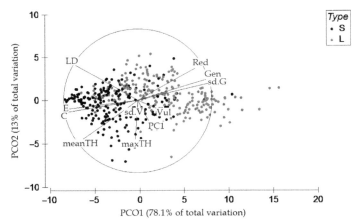

Fig. 4 Ordination of food webs (*points*) based on the resemblance matrix created from food web metrics using Bray–Curtis distances, which was analysed with PERMANOVA. Site type is shown as either *black points* (streams) or *grey points* (lakes). This plot allows comparison of food web metrics (*blue* (*grey* in the print version) vectors and names) with the predictor variable PC1 extracted from Fig. 3 in multivariate space. Longer vectors indicate a stronger correlation. The *blue* (*grey* in the print version) *circle* indicates the boundary for a correlation of 1. *C*, connectance; *LD*, linkage density; *meanTH*, mean trophic height; *maxTH*, maximum trophic height; *Vul*, resource vulnerability; *Gen*, consumer generality; *E*, network efficiency; *Red*, node redundancy.

Table 2 Statistics of Fit for the Multiple Mixed Effects Models

Response Variable	Predictor Variable	d.f.	F Value	p Value
Connectance	PC1	99	−1.537	0.1270
	Type	1	7.991	**<0.0001**
Linkage density	PC1	72	−3.902	**<0.0001**
	Type	1	2.686	**0.0130**
Mean trophic height	PC1	76	−0.017	0.9864
	Type	1	4.294	**0.0003**
Maximum trophic height	PC1	74	1.407	0.1640
	Type	1	−0.068	0.9460
Network efficiency	PC1	101	−2.306	**0.0231**
	Type	1	8.288	**<0.0001**
Normalised vulnerability	PC1	54	−0.264	0.7929
	Type	1	−2.478	**0.0208**
Normalised generality	PC1	67	1.122	0.2666
	Type	1	−6.531	**<0.0001**
sd(Vulnerability)	PC1	115	−0.395	0.6940
	Type	1	−0.450	0.6570
sd(Generality)	PC1	112	1.952	0.0534
	Type	1	−7.228	**<0.0001**
Redundancy	PC1	91	3.577	**0.0005**
	Type	1	−5.269	**<0.0001**

All models include a random effect of site on the intercept of the linear relationship, and year on the slope of the linear relationship. Bold p values indicate significance at $\alpha = 0.05$.

food webs had lower linkage density than stream food webs (Fig. 5). Normalised consumer generality and normalised vulnerability did not change (Table 2; Fig. 6), nor did their standard deviations, across the acidity gradient. As predicted, however, network efficiency was lower in more acidified conditions (Table 2; Fig. 7), suggesting that more acidified food webs were connected such that the average path length between nodes was greater than for circumneutral food webs. Node redundancy was highest in more acidified food webs (Table 2; Fig. 7), suggesting that circumneutral food webs had more unique feeding pathways, confirming our prediction. Overall, lake food webs had lower network efficiency and higher node redundancy than streams (Table 2; Fig. 7). Fig. 4 indicates more variation in lake food web structure along PCO1, contrasting with our expectation that stream food webs might be more dynamic and variable than lake food webs.

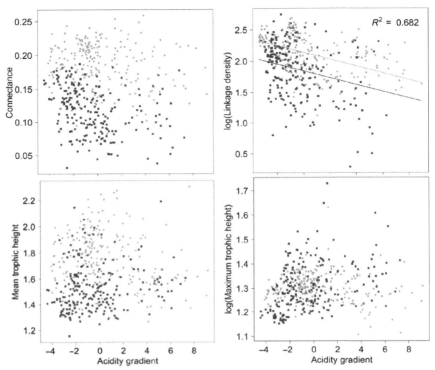

Fig. 5 The relationship between connectance, linkage density, mean and maximum trophic height, and environmental stress. The acidity gradient is PC1 extracted from Fig. 3 and is strongly related to pH, SO_4 and labile aluminium, with increasing environmental stress (acidity) from *left* to *right*. Lines indicate fitted values from GLMM, where $p < 0.05$ and the conditional R^2 as an indication of overall model explanatory power is shown (Johnson, 2014). Where site type (lake or stream) was found to be a significant predictor variable, *separate lines* are given for each site type. Lake food webs are indicated by *dark blue* (*dark grey* in the print version) *squares*, while streams are *light blue* (*light grey* in the print version) *circles*.

3.2 Directional Change in Food Web Structure

There was overall a clear directional trend in chemical recovery: 18 of the 23 sites exhibited a monotonic declining trend in their PC1 axis scores (i.e. decreasing acidity) over time (Figs. 8 and 9). Three of the five sites which showed no trend in their PC1 scores (i.e. no directional change in acidity over time) were located in the north of Scotland, which always experienced less acid deposition and so were not highly acidified at the outset of monitoring (Fig. 1) (Patrick et al., 1995). This suggests that at least partial chemical recovery from acidification has occurred at most sites at which it was expected (Monteith et al., 2014).

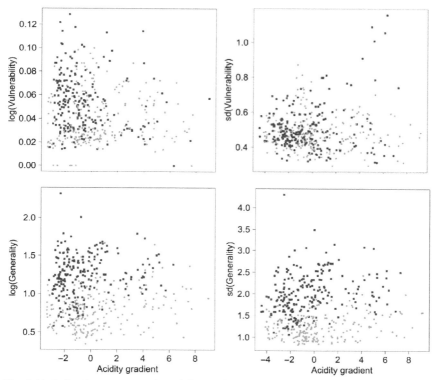

Fig. 6 Variation in resource vulnerability and consumer generality (both normalised to species richness) across the stress (acidity) gradient (greater acidity to the right, see Fig. 5). Lake food webs are indicated by *dark blue* (*dark grey* in the print version) *squares*, while streams are *light blue* (*light grey* in the print version) *circles*.

Directional change in food web structure was also evident across many of the UWMN sites, in line with the chemical trends and with our second main hypothesis, although it was far from ubiquitous. Of the 18 sites showing chemical recovery, around half also showed significant increasing trends in connectance (9 sites; Tables 3 and 4), linkage density (7 sites; Tables 3 and 4), mean trophic height (8 sites; Tables 3 and 4), resource *Vulnerability* (6 sites; Tables 3 and 4), standard deviation in resource *Vulnerability* (5 sites; Figs. 10 and 11) and network efficiency (7 sites; Tables 3 and 4). Of the 18 showing chemical recovery, significant declines were evident in consumer *Generality* (10 sites; Tables 3 and 4), redundancy (10 sites; Table 3) and the standard deviation of *Generality* (10 sites; Figs. 10 and 11). Maximum trophic height increased in one site, decreased in two, and showed no trend in the other 20 (Figs. 10 and 11). See Appendix for more detailed plots of each trend over time.

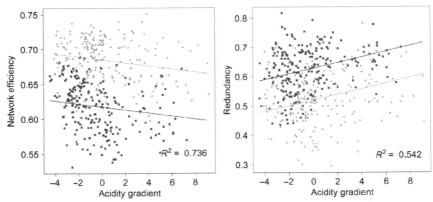

Fig. 7 Network efficiency increases with decreasing environmental stress (acidity, greater acidity to the *right*, see Fig. 5), and node redundancy decreases with decreasing environmental stress. *Lines* indicate fitted values from GLMM, where $p < 0.05$ and the conditional R^2 as an indication of overall model explanatory power is shown (Johnson, 2014). Where site type (lake or stream) was found to be a significant predictor variable, separate lines (*light blue* (*light grey* in the print version) streams; *dark blue* (*dark grey* in the print version) lakes) are given for each site type. Lake food webs are indicated by *dark blue* (*dark grey* in the print version) *squares*, while streams are *light blue* (*light grey* in the print version) *circles*.

There was evidence for a delay in food web recovery after chemical recovery at the UWMN sites; most sites occupied the 'both biological and chemical recovery' or 'chemical but no biological recovery' portions of the conceptual recovery figure (Fig. 2) for each of their food web metrics (Tables 3 and 4). Very few sites exhibited change in their food web structure in the absence of directional change in their acidity (PC1); the food webs of Loch Coire Fionnaraich and Allt na Coire nan Con both showed increasing linkage density and resource vulnerability over time, in the absence of a significant temporal trend in acidity (Figs. 10 and 11). Similarly, the food web of Coneyglen Burn decreased in redundancy over time, despite no significant temporal trend in acidity (Figs. 10 and 11).

3.3 Food Web Recovery from Acidification

Trends in chemistry over time were not linearly related to shifts in food web structure, or at least have not yet related to the latter, strengthening the evidence for inertia in food web recovery. The χ^2 tests revealed that there was little congruence between those sites, exhibiting chemical and biological recovery (Table 5). However, the χ^2 tests did reveal that those sites showing a trend in standard deviation in consumer *Generality* also tended to show a

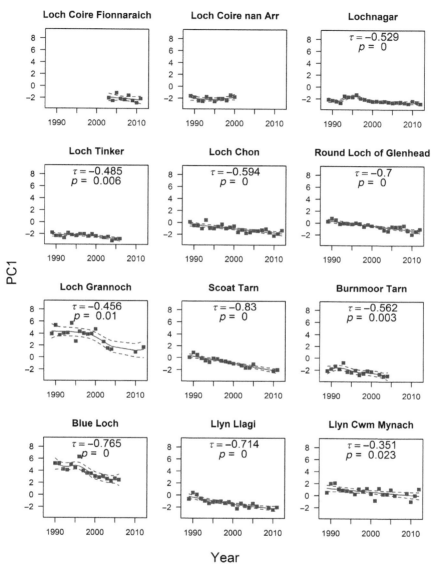

Fig. 8 Trends in overall acidity (PC1 extracted from Fig. 3) at each of the UWMN lake sites. For those sites showing significant monotonic temporal trends (as determined through Mann–Kendall trend tests, see Section 2) the test statistic and associated p values are shown. Sites are arranged in order of their decreasing latitude, which can be used as a proxy for their initial acidified state; more acidified sites were generally in the south (*bottom panels*), while the least acidified sites were more northern (*top panels*).

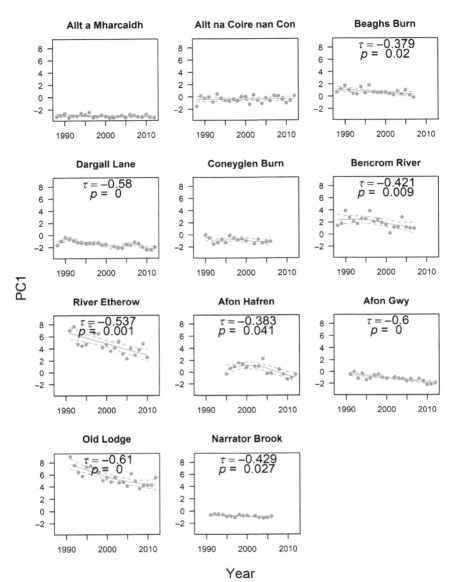

Fig. 9 Trends in overall acidity (PC1 extracted from Fig. 3) at each of the UWMN stream sites. For those sites showing significant monotonic temporal trends (as determined through Mann–Kendall trend tests, see Section 2) the test statistic and associated p values are shown. Sites are arranged in order of their decreasing latitude, which can be used as a proxy for their initial acidified state; more acidified sites were generally in the south (*bottom panels*), while the least acidified sites were more northern (*top panels*).

Table 3 The Number of the 12 Lake Sites Showing a Significant Temporal Trend in Their Food Web Metrics over Time, as Determined by Mann–Kendall Trend Tests (See Section 2)

Biological and Chemical Recovery		Biological but No Chemical Recovery	
C=6	LD=2	C=0	LD=1
meanTH=4	maxTH=0	meanTH=0	maxTH=0
Vul=3	Gen=6	Vul=0	Gen=1
E=4	Red=4	E=0	Red=0
Chemical but No Biological Recovery		**No Biological or Chemical Recovery**	
C=4	LD=8	C=2	LD=1
meanTH=6	maxTH=8	meanTH=2	maxTH=2
Vul=6	Gen=4	Vul=2	Gen=1
E=6	Red=6	E=2	Red=2

Chemical recovery here is indicated by a significant temporal trend in acidity (PC1). *C*, connectance; *LD*, linkage density; *meanTH*, mean trophic height; *maxTH*, maximum trophic height; *Vul*, resource vulnerability; *Gen*, consumer generality; *E*, network efficiency; *Red*, node redundancy.

Table 4 The Number of the 11 Stream Sites Showing a Significant Temporal Trend (in the Direction Predicted in Hypotheses; Recovery) in Their Food Web Metrics over Time, as Determined by Mann–Kendall Trend Tests (see Section 2)

Biological and Chemical Recovery		Biological but No Chemical Recovery	
C=3	LD=5	C=0	LD=1
meanTH=4	maxTH=1	meanTH=0	maxTH=0
Vul=3	Gen=4	Vul=1	Gen=0
E=3	Red=6	E=0	Red=1
Chemical but No Biological Recovery		**No Biological or Chemical Recovery**	
C=5	LD=3	C=3	LD=2
meanTH=4	maxTH=7	meanTH=3	maxTH=3
Vul=5	Gen=4	Vul=2	Gen=3
E=5	Red=2	E=3	Red=2

Chemical recovery here is indicated by a significant temporal trend in acidity (PC1). *C*, connectance; *LD*, linkage density; *meanTH*, mean trophic height; *maxTH*, maximum trophic height; *Vul*, resource vulnerability; *Gen*, consumer generality; *E*, network efficiency; *Red*, node redundancy.

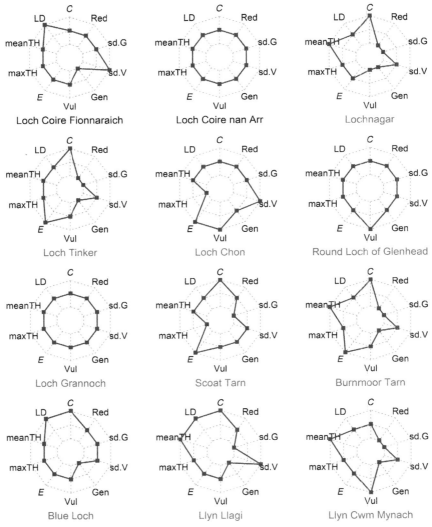

Fig. 10 Trends in food web metrics at each of the UWMN lake sites. Sites are arranged in order of their decreasing latitude (*top left* to *bottom right*; see Fig. 1), which can be used as a proxy for their initial acidified state (generally least acid to the *top left*, more acidified sites at the *bottom*). Site names in *green* (*light grey* in the print version) indicate a monotonic decreasing trend in acidity over time at that site; site names in *black* indicate no trend in acidity. *Points on the inner ring* of each radial plot indicate a negative trend in that variable at that site over time and *points on the middle ring* indicate no significant trend while *points on the outer ring* indicate a positive trend. *C*, connectance; *LD*, linkage density; *meanTH*, mean trophic height; *maxTH*, maximum trophic height; *E*, network efficiency; *Vul*, resource vulnerability; *Gen*, consumer generality; *sd.V*, standard deviation in resource vulnerability; *sd.G*, standard deviation in consumer generality; *Red*, node redundancy. See Appendix for detailed plots of each trend over time.

Food Web Recovery from Acidification 501

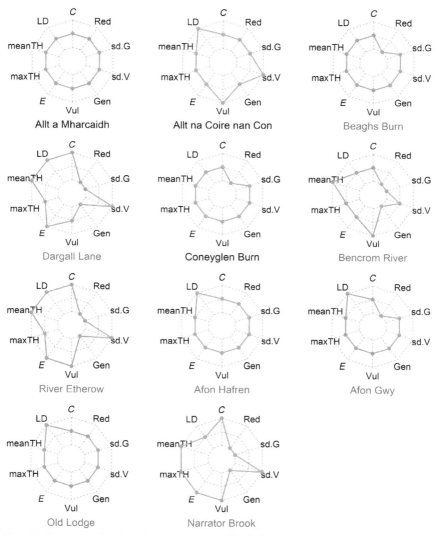

Fig. 11 Trends in food web metrics at each of the UWMN stream sites. Sites are arranged in order of their decreasing latitude (*top left* to *bottom right*; see Fig. 1), which can be used as a proxy for their initial acidified state (generally least acid to the *top left*, more acidified sites at the *bottom*). Site names in *green* (*light grey* in the print version) indicate a monotonic decreasing trend in acidity over time at that site; site names in *black* indicate no trend in acidity. *Points on the inner ring* of each radial plot indicate a negative trend in that variable at that site over time and *points on the middle ring* indicate no significant trend while *points on the outer ring* indicate a positive trend. C, connectance; *LD*, linkage density; *meanTH*, mean trophic height; *maxTH*, maximum trophic height; *E*, network efficiency; *Vul*, resource vulnerability; *Gen*, consumer generality; *sd.V*, standard deviation in resource vulnerability; *sd.G*, standard deviation in consumer generality; *Red*, node redundancy. See Appendix for detailed plots of each trend over time.

Table 5 Results from the χ^2 Contingency Test (See Main Text)

	PC1
C	0.113
LD	1
meanTH	0.123
maxTH	0.574
E	0.146
Vulnerability	1
Generality	1
sd.V	1
sd.G	**0.046**
Redundancy	0.325

Bold p values indicate significance at $\alpha = 0.05$. See the legend of Table 3 for abbreviations.

trend in acidity (Table 5). This generally refutes our third hypothesis and provides more evidence for a lag or inertia in food web recovery; those sites recovering chemically from acidification over time showed little systematic change in their food web structure, suggesting that biological recovery (in terms of food web structure) does not directly track chemical recovery at these sites.

4. DISCUSSION

Our analyses reveals fundamental structural changes occurring in food web structure in response to decreasing acidity over the three decades of the study. These structural changes could have profound implications for the stability of the systems' food webs and could be hindering biological recovery. Confirming our hypotheses, when analysed at the regional (UK) scale, acidified food webs had lower linkage density and network efficiency but had more redundancy within their feeding pathways. Contrary to our other hypotheses, we found no effect on connectance, trophic height, nor on resource *Vulnerability* and consumer *Generality* or the standard deviations of both. When analysed at the site scale some further trends in network metrics over time became clearer, but overall these were mixed and often harder to associate with decreasing acidity per se. There was strong evidence for a lag in biological recovery, as those sites showing a recovery in both their biology and their chemistry tended to be a nested subset of those that only showed a chemical trend.

4.1 Food Web Recovery Across the AWMN Sites and Acidity Gradient

The general increasing mean trophic height of food webs over time at each site (see also Fig. A4) reflected the reverse of the typical responses to acidification, where species are lost throughout the food web, but top predators such as fish (Henriksen et al., 1999) and many predatory macroinvertebrates are especially vulnerable (Layer et al., 2011). The return of these acid-sensitive species over time causes food chain lengths and the trophic height of the web as a whole to increase (Layer et al., 2011; Woodward and Hildrew, 2001)—in our dataset all of the sites which experienced this lengthening of food chains were also decreasing in their acidity. However, not all sites with falling acidity also exhibited increases in trophic height, which again suggests food web inertia. Often site-specific trends were not evident, but were when the data were analysed across the full acidity gradient, also suggesting that other environmental drivers might be modulating the relationship with the food web. For instance DOC, which was closely related to PC2, is known to limit secondary production in lakes (Finstad et al., 2014).

Along the derived acidity gradient, normalised *Vulnerability* and *Generality* were unchanged. Over time at individual sites, however, the latter tended to decrease and the former to increase, but across the 23 sites as a whole only 7 showed increasing *Vulnerability* and most showed complex and nonlinear patterns over time (see Appendix). Decreasing *Generality* and increasing *Vulnerability* with decreasing acidity are consistent with the proposition that specialist consumers and also larger top predators recolonise communities following chemical recovery (Layer et al., 2013; Woodward and Hildrew, 2001). This should result in increased *Vulnerability* (more consumers per resource due to greater consumer species richness) and reduced *Generality* (fewer resources per consumer due to increased specialism). The reappearance of acid-sensitive consumers including both invertebrates (such as species of the mayflies *Baetis* spp. and *Caenis* spp., or the snail *Radix balthica*) and salmonid fish at high pH should lead to both a general elongation and greater compartmentalisation in the web and specialism becomes more prevalent both within and across trophic levels.

The connectivity of the food webs as a whole changed across the derived acidity gradient: network efficiency, which describes how 'reachable' each node is from every other, increased with decreasing acidity. If pockets of species are poorly connected to other species, the average shortest path length between all pairs of nodes will increase. Thus, species within more acidified food webs were less well connected on average across the whole network. The increased species richness and addition of top predators such as salmonid fish to the system (Woodward and Hildrew, 2001) may explain the increased efficiency of these less-acidified food webs. The top predator of these freshwater systems, the brown trout (*S. trutta*), is a highly generalist engulfing predator which will consume anything within a given size range of prey. The addition of these (acid-sensitive) generalist interactions between top predators and those macroinvertebrates within its prey-size range may well increase the reachability between those resource nodes, as well as ultimately linking together different feeding pathways (e.g. the allochthonous vs autochthonous resource base of the food web), even though these may be becoming more compartmentalised horizontally among their increasingly specialist primary consumers.

Acidified food webs contained proportionally more redundant feeding pathways than their circumneutral counterparts, the proportion of 'trophic species', nodes feeding on and being fed on by the same species, is larger in the smaller, more acidified food webs. This is consistent with the increase in specialist consumers as acidity decreases. Additionally, acidified food webs tend to have few species and few links (Layer et al., 2010b), making the scope for unique feeding pathways small.

Contrasting trends emerged when our data were analysed at the site or regional scale. When our data were analysed at the individual site scale, trends were mixed and were not necessarily always related to decreasing acidity, while clearer trends often emerged from the regional-scale analysis. This could arise if communities are highly variable when released from a stressor, and other drivers (e.g. nutrients) that were previously uninfluential start to shape local habitat filtering (e.g. Micheli et al., 1999). Additionally, site-scale sources of variation, such as potentially powerful contingent site characteristics, might have swamped potential underlying trends in food web structure over time. Indeed, site identity was a necessary variable in our models that encompassed a range of site-specific variables, such as latitude. Additionally, weather conditions were uncontrolled and extreme events close to the small sampling window for each site may have caused some sites to lose and regain their significant trends in biotic recovery over time (Kernan et al., 2010; Monteith and Evans, 1998, 2005). Additionally, the portion of the acidity gradient that each site is exposed to is small relative to that of the whole dataset.

4.2 The Recovery of Freshwater Food Webs from Acidification

Although some clear responses were evident, the food web metrics used here might not be the most appropriate for detecting recovery from acidification. There were considerable intersite differences in food web structure, but not all were sensitive to changes in acidity and there was still considerable unexplained variation in the models. It seems likely that, as our understanding grows, more sensitive measures of food web structure will emerge, perhaps through analysis of substructure rather than 'whole network' properties, and that these might be better at capturing ecological responses to environmental change.

That acidified ecosystems might exhibit 'ecological inertia' has increasingly been suggested as a mechanism to explain the delay in biological recovery (Kernan et al., 2010; Layer et al., 2010b; Ledger and Hildrew, 2005; Lundberg et al., 2000). Various lines of evidence increasingly suggest that acidified food webs are dynamically stable and resistant to recolonisation by acid-sensitive species, even as chemical conditions start to improve. Townsend et al. (1987) measured the persistence of 27 stream invertebrate communities across a pH gradient, and found that those communities from the most acidified sites were indeed the most persistent, although data on species interactions and network structure were not available at the time. Later, Layer et al. (2010b) used dynamic modelling to determine the robustness of stream food webs to species extinctions, and found that food webs from more acidified conditions were more robust, but the long-term temporal data were not available to test this prediction empirically. Here we provide the largest scale evidence to test these ideas, which broadly supports the general notion that redundancy is an important component of stability that could confer robustness on the system. In ecosystems redundancy can increase the reliability of process rates and buffer the effects of species loss (Naeem, 1998; Peralta et al., 2014): we found that food webs from acidified waters had higher redundancy, suggesting that they might be more robust, and might therefore provide more stable (albeit often slower) process rates (Naeem, 1998; Peralta et al., 2014). As acidity decreases in fresh waters, decomposition of leaf-litter, which fuels much of the food web, does indeed accelerate (Jenkins et al., 2013), although the extent to which species richness modulates this relationship is still largely unknown (but see Jonsson et al., 2002). Additionally, we found that more acidified food webs had lower global efficiency, which is associated with reduced small-world properties. Ecological networks that are small worlds are often relatively stable (Dunne et al., 2002; Solé and Montoya, 2001), as they offer many alternative pathways of interaction. These apparently contrasting responses to different dimensions of stability warrant further investigation to reveal if acidified food webs are indeed more (or less) stable in some regards and not others (e.g. Donohue et al., 2013).

4.3 Caveats and Future Directions

The use of inferred feeding links in food web studies has been criticised on the basis that they might overestimate diet breadth and fail to detect behavioural differences between sites (Hall and Raffaelli, 1997; Raffaelli, 2007), yet to build complete food webs de novo from replicated natural systems is simply logistically unfeasible, so a trade-off between replication and realism is inevitable. The use of 'summary' food webs, which include the full complement of known possible tropic interactions, can still be a useful tool for understanding community structure, especially as in freshwaters most species are highly generalist and their diets are largely size driven and consistent among systems when presented with the same potential prey species (Gray et al., 2015; Layer et al., 2013; Woodward et al., 2010b). Indeed, given the nature of building summary food webs that they tend to overestimate interactions between species, they are more likely to be insensitive to environmental change rather than reveal erroneous trends (i.e. it is more likely that the structure of summary food webs is conserved given that any changes will be entirely driven by changes in species composition rather than feeding behaviour). Hence, we contend that the trends revealed here are broadly realistic and warrant further examination, especially as the feeding links described in many of our webs had been observed in the same system, albeit only for a snapshot of the full set of sampling occasions. Future work could involve a more formal validation of randomly selected portions of the network via direct analysis of gut contents (as in Layer et al., 2010b; Woodward et al., 2005a) and also the application of new molecular approaches that could potentially capture a more complete picture of the entire food web with a fraction of the current effort required (Gray et al., 2014).

Another potential limitation to the food webs produced here is that they do not include the full freshwater community, in particular the meiofauna and microfauna (e.g. Schmid-Araya et al., 2002) and true apex predators such as the European Dipper (*Cinclus cinclus*) or Otter (*Lutra lutra*). Top predators can have varying effects on food web structure in these systems (Layer et al., 2011; Woodward and Hildrew, 2001), and so their exclusion may be omitting an important source of variation in this data. However, this was unavoidable in this study, as in almost all other food webs described to date, because the presence of these cryptic or very rare species has not been systematically recorded. Additionally, although the fish assemblage of the lakes was sampled from the lake outflows, all these low-productivity upland sites are typically dominated by brown trout (*S. trutta*) and the occasional European eel (*Anguilla anguilla*) in both the running and standing waters across the acidity gradient: of the 434 sampling occasions on which fish were present at a site, the brown trout was always present, reflecting its dominance in these systems. The next most common species, the European eel (*A. anguilla*), was found on 136 occasions and all other species (*Esox lucius, Gasterosteus aculeatus, Lampetra* spp., *Phoxinus phoxinus* and *Salmo salar*) were found on <60 sampling occasions.

5. CONCLUSION

It is clear from this study that both spatial and temporal scales are important considerations when assessing food web responses to environmental change in real time (Chave, 2013). When our data were analysed at the individual site scale, trends were mixed and were not necessarily always related to decreasing acidity. When the data were analysed at the regional (UK) scale, some clear and significant trends emerged, highlighting the need for large, replicated collections of food webs as well as the need for caution when extrapolating from small collections or individual food webs. Identifying the effects of individual chemical drivers was often challenging given the range of potential drivers in a nationwide dataset that also spans several decades.

To the best of our knowledge this is the largest collection of food webs that span both large temporal and spatial gradients: the next largest set of empirical food webs from natural systems of which we are aware is less than half the size (170 soil webs described by Mulder et al., 2011) and the remainder are far smaller still, with most studies being conducted on unreplicated singletons (Ings et al., 2009). Our study is thus one of the first to address macroecological questions relating to the structure of food webs across time and a broad environmental gradient in a (relatively) standardised manner. Our analysis reveals fundamental structural changes occurring in the food webs as they respond to changes in acidity; these structural changes could have profound implications for the stability of the system and may be limiting biological recovery. It would be instructive to investigate further the stability of these food webs, in order to explore more fully whether intrinsic inertia is indeed limiting their recovery, and how that might possibly be manipulated to accelerate the rate of recovery.

APPENDIX

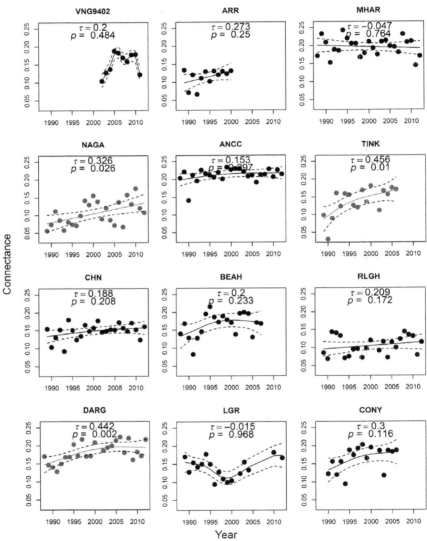

Fig. A1 Trends in connectance at each of the UWMN sites. Sites are arranged in order of their decreasing latitude, which can be used as a proxy for their initial acidified state; more acidified sites were generally in the south (*bottom* of plot), while the least acidified sites were more northern (*top* of plot). See Fig. 1 for site name abbreviations.

Food Web Recovery from Acidification 509

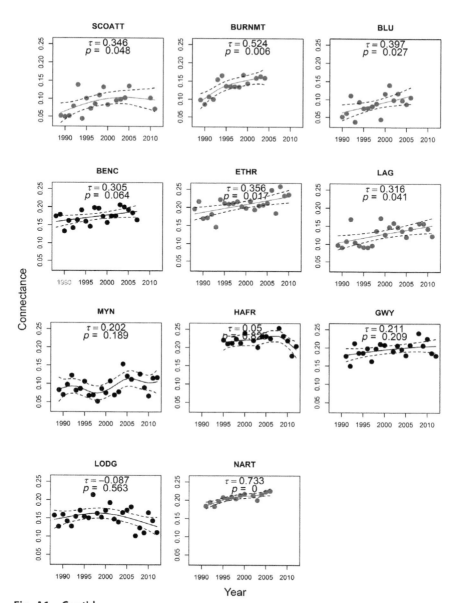

Fig. A1—Cont'd

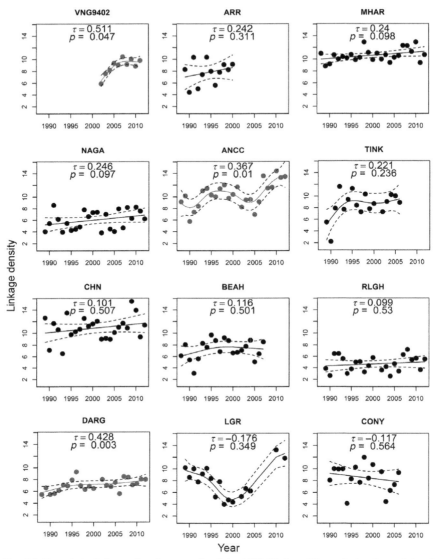

Fig. A2 Trends in linkage density at each of the UWMN sites. Site ordering is explained in the legend of Fig. A1. See Fig. 1 for site name abbreviations.

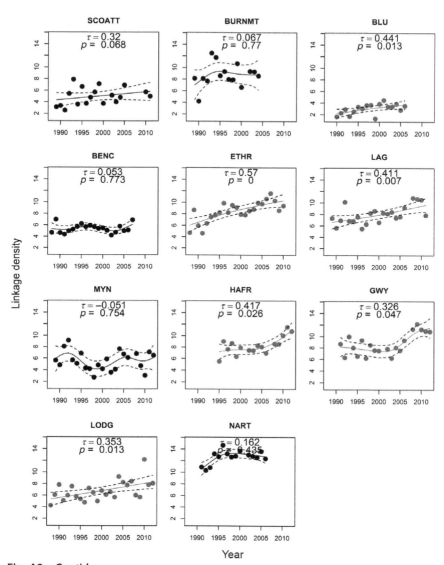

Fig. A2—Cont'd

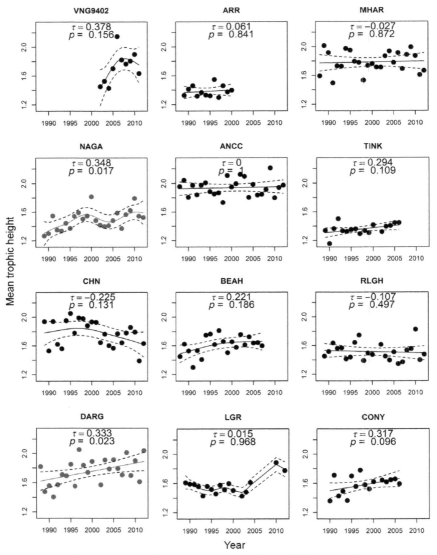

Fig. A3 Trends in mean trophic height at each of the UWMN sites. Site ordering is explained in the legend of Fig. A1. See Fig. 1 for site name abbreviations.

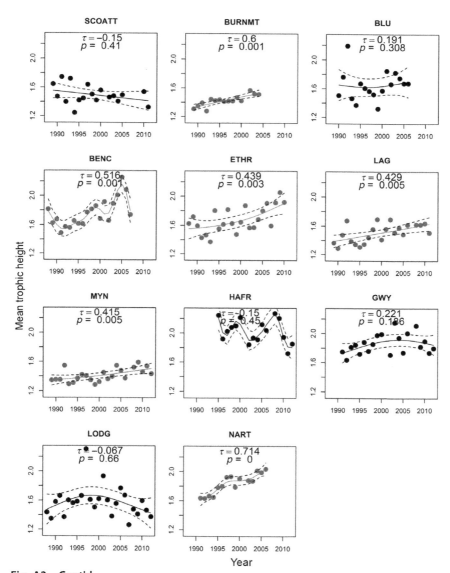

Fig. A3—Cont'd

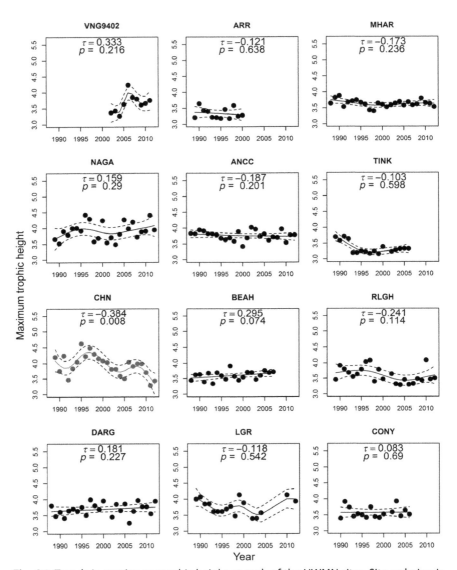

Fig. A4 Trends in maximum trophic height at each of the UWMN sites. Site ordering is explained in the legend of Fig. A1. See Fig. 1 for site name abbreviations.

Food Web Recovery from Acidification 515

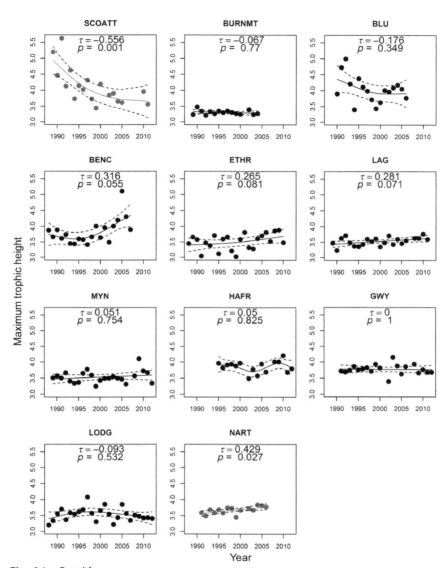

Fig. A4—Cont'd

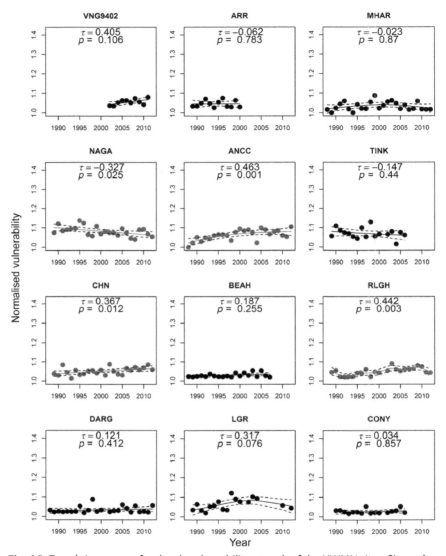

Fig. A5 Trends in average food web vulnerability at each of the UWMN sites. Site ordering is explained in the legend of Fig. A1. See Fig. 1 for site name abbreviations.

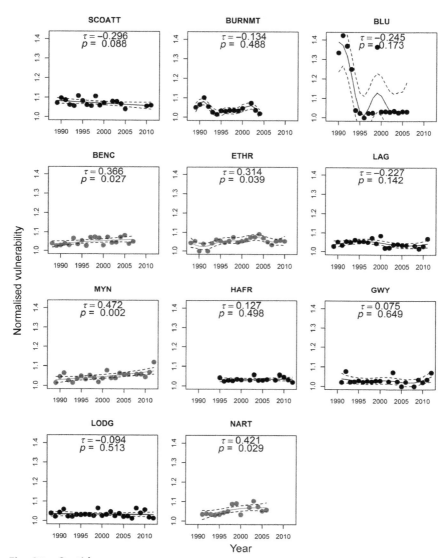

Fig. A5—Cont'd

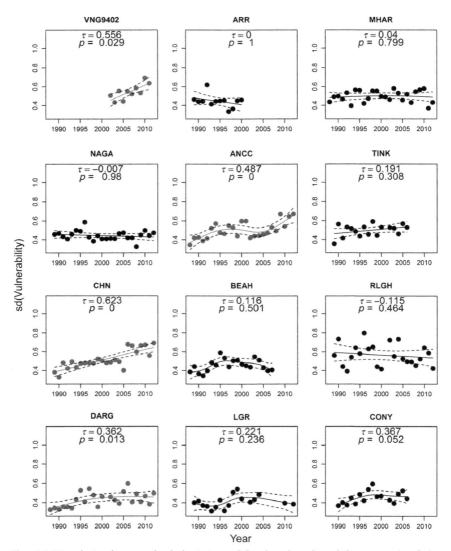

Fig. A6 Trends in the standard deviation of food web vulnerability at each of the UWMN sites. Site ordering is explained in the legend of Fig. A1. See Fig. 1 for site name abbreviations.

Food Web Recovery from Acidification

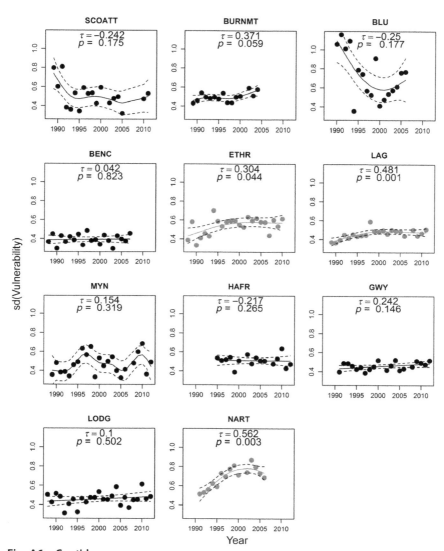

Fig. A6—Cont'd

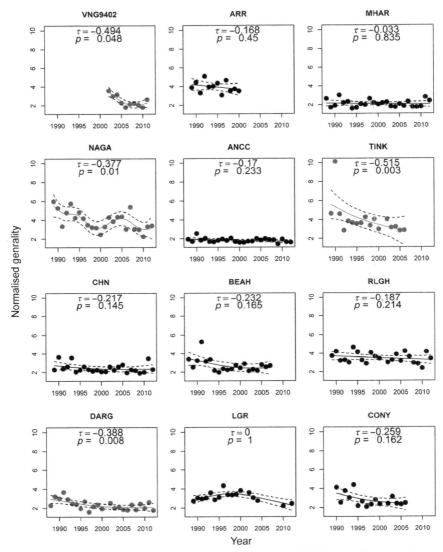

Fig. A7 Trends in food web generality at each of the UWMN sites. Site ordering is explained in the legend of Fig. A1. See Fig. 1 for site name abbreviations.

Food Web Recovery from Acidification

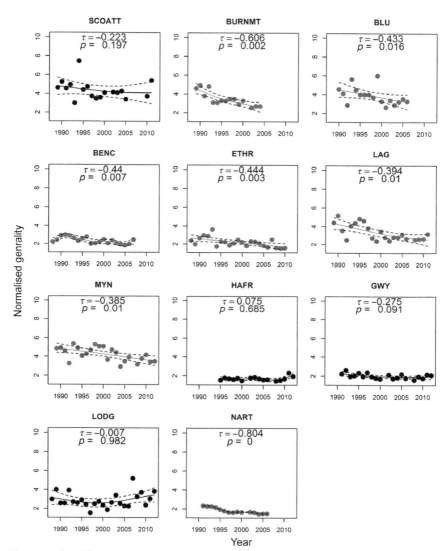

Fig. A7—Cont'd

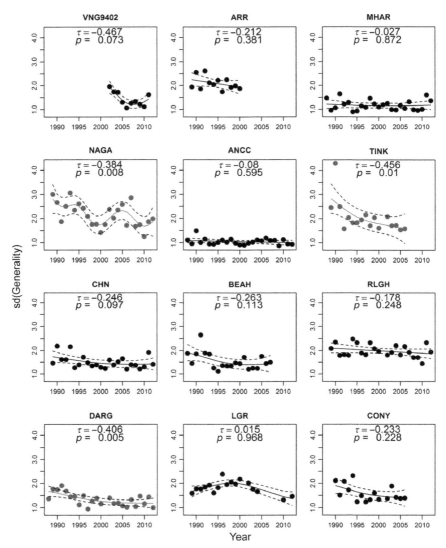

Fig. A8 Trends in the standard deviation of food web generality at each of the UWMN sites. Site ordering is explained in the legend of Fig. A1. See Fig. 1 for site name abbreviations.

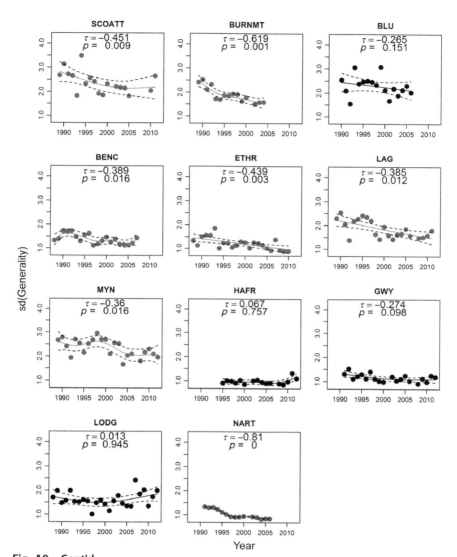

Fig. A8—Cont'd

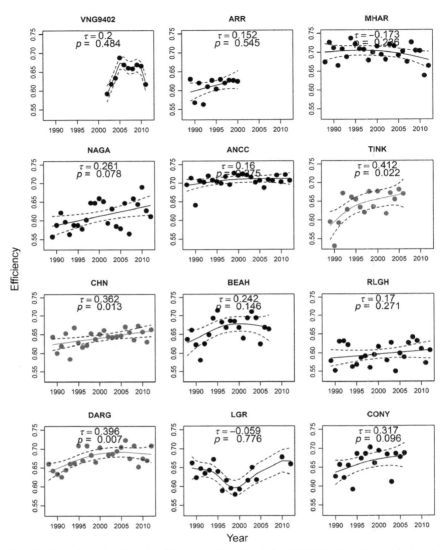

Fig. A9 Trends in food web efficiency at each of the UWMN sites. Site ordering is explained in the legend of Fig. A1. See Fig. 1 for site name abbreviations.

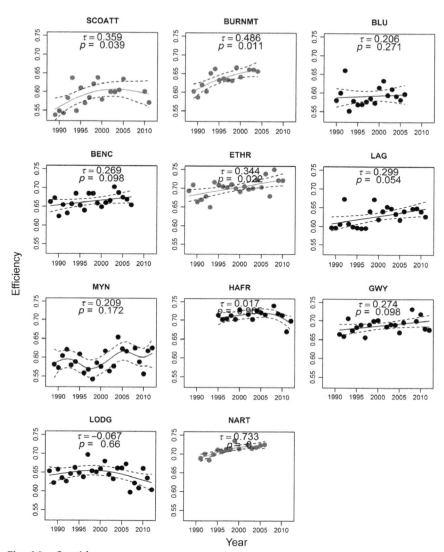

Fig. A9—Cont'd

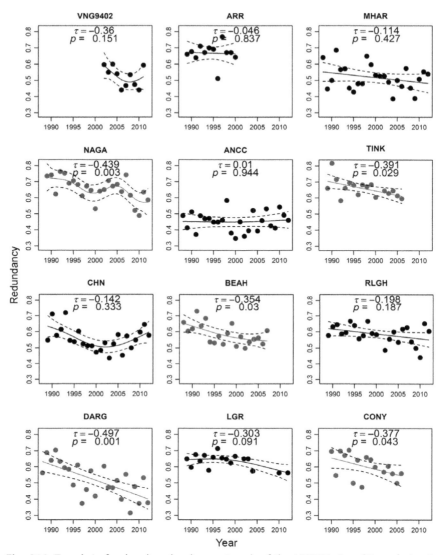

Fig. A10 Trends in food web redundancy at each of the UWMN sites. Site ordering is explained in the legend of Fig. A1. See Fig. 1 for site name abbreviations.

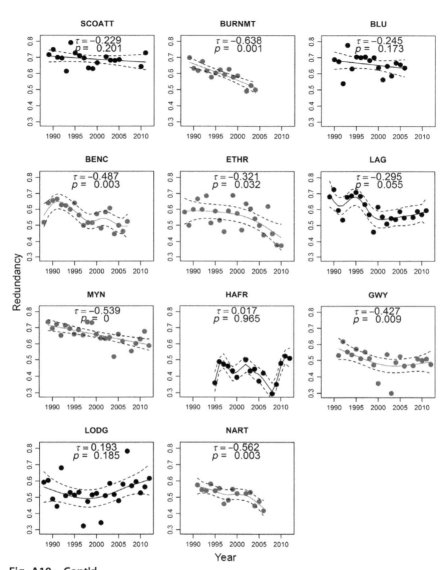

Fig. A10—Cont'd

ACKNOWLEDGEMENTS

This chapter is a contribution to Imperial College's Grand Challenges in Ecosystems and the Environment initiative. C.G. was supported by a Queen Mary University of London studentship and the Freshwater Biology Association. The UK UWMN is supported by the UK Department for Environment Food and Rural Affairs (DEFRA), NERC through the Centre for Ecology & Hydrology (CEH), the Department of the Environment (Northern Ireland), the Environment Agency (EA), the Forestry Commission (FC), Natural Resources Wales (NRW), the Scottish Environmental Protection Agency (SEPA), Scottish Natural Heritage (SNH) and the Welsh Government, the Scottish Government through Marine Scotland Science Pitlochry, Queen Mary University of London and ENSIS Ltd. at the Environmental Change Research Centre, University College London.

REFERENCES

Battarbee, R.W., Anderson, N.J., Appleby, P.G., Flower, R.J., Fritz, S.C., Haworth, E.Y., Higgitt, S., Jones, V.J., Kreiser, A., Munro, M.A.R., Natkanski, J., Oldfield, F., Patrick, S.T., Richardson, N.G., Rippey, B., Stevenson, A.C.L.E., 1988. Lake Acidification in the United Kingdom 1800–1986: Evidence from Analysis of Lake Sediments. ENSIS Ltd., London.

Battarbee, R.W., Jones, V.J., Flower, R.J., Cameron, N.G., Bennion, H., Carvalho, L., Juggins, S., 2001. Diatoms. In: Smol, J.P., Birks, H.J.B., Last, W.M. (Eds.), Tracking Environmental Change Using Lake Sediments. In: Terrestrial, Algal and Siliceous Indicators, vol. 3. Kluwer Academic Publishers, Dordrecht, The Netherlands, pp. 151–202.

Battarbee, R.W., Simpson, G.L., Shilland, E.M., Flower, R.J., Kreiser, A., Yang, H., Clarke, G., 2014. Recovery of UK lakes from acidification: an assessment using combined palaeoecological and contemporary diatom assemblage data. Ecol. Indic. 37 (Part B), 365–380.

Briand, F., Cohen, J.E., 1984. Community food webs have scale-invariant structure. Nature 307, 264–267.

Butts, C., 2013. sna: tools for social network analysis. R package version 2.3-2. http://CRAN.R-project.org/package=sna.

Chave, J., 2013. The problem of pattern and scale in ecology: what have we learned in 20 years? Ecol. Lett. 16, 4–16.

Cohen, J.E., Briand, F., 1984. Trophic links of community food webs. Proc. Natl. Acad. Sci. U.S.A. 81, 4105–4109.

Crawley, M.J., 2013. The R Book, second ed. John Wiley & Sons Ltd., Chichester.

Dillon, P.J., Yan, N.D., Harvey, H.H., Schindler, D.W., 1984. Acidic deposition: effects on aquatic ecosystems. CRC Crit. Rev. Environ. Control 13, 167–194.

Donohue, I., Petchey, O.L., Montoya, J.M., Jackson, A.L., Mcnally, L., Viana, M., Healy, K., Lurgi, M., O'Connor, N.E., Emmerson, M.C., 2013. On the dimensionality of ecological stability. Ecol. Lett. 16, 421–429.

Driscoll, C.T., Lawrence, G.B., Bulger, A.J., Butler, T.J., Cronan, C.S., Eagar, C., Lambert, K.F., Likens, G.E., Stoddard, J.L., Weathers, K.C., 2001. Acidic deposition in the Northeastern United States: sources and inputs, ecosystem effects, and management strategies. BioScience 51, 180.

Dunne, J.A., Williams, R.J., Martinez, N.D., 2002. Food-web structure and network theory: the role of connectance and size. Proc. Natl. Acad. Sci. U.S.A. 99, 12917–12922.

Durance, I., Ormerod, S.J., 2007. Climate change effects on upland stream macroinvertebrates over a 25-year period. Glob. Chang. Biol. 13, 942–957.

Evans, C.D., Monteith, D.T., Harriman, R., 2001. Long-term variability in the deposition of marine ions at west coast sites in the UK Acid Waters Monitoring Network: impacts on surface water chemistry and significance for trend determination. Sci. Total Environ. 265, 115–129.

Evans, C.D., Monteith, D.T., Reynolds, B., Clark, J.M., 2008. Buffering of recovery from acidification by organic acids. Sci. Total Environ. 404, 316–325.

Feld, C.K., Birk, S., Bradley, D.C., Hering, D., Kail, J., Marzin, A., Melcher, A., Nemitz, D., Pedersen, M.L., Pletterbauer, F., Pont, D., Verdonschot, P.F.M., Friberg, N., 2011. From natural to degraded rivers and back again: a test of restoration ecology theory and practice. Adv. Ecol. Res. 44, 119–209.

Findlay, D., Kasian, S., Turner, M., Stainton, M., 1999. Responses of phytoplankton and epilithon during acidification and early recovery of a lake. Freshw. Biol. 42, 159–175.

Finstad, A.G., Helland, I.P., Ugedal, O., Hesthagen, T., Hessen, D.O., 2014. Unimodal response of fish yield to dissolved organic carbon. Ecol. Lett. 17, 36–43.

Friberg, N., Bonada, N., Bradley, D.C., Dunbar, M.J., Edwards, F.K., Grey, J., Hayes, R.B., Hildrew, A.G., Lamouroux, N., Trimmer, M., Woodward, G., 2011. Biomonitoring of human impacts in freshwater ecosystems: the good, the bad and the ugly. Adv. Ecol. Res. 44, 1–68.

Frost, T.M., Montz, P.K., Kratz, T.K., 1998. Zooplankton community responses during recovery from acidification in Little Rock Lake, Wisconsin. Restor. Ecol. 6, 336–342.

Gerson, J.R., Driscoll, C.T., Roy, K.M., 2016. Patterns of nutrient dynamics in Adirondack lakes recovering from acid deposition. Ecol. Appl. 26, 1758–1770.

Grahn, O., Hultberg, H., Landner, L., 1974. Oligotrophication: a self-accelerating process in lakes subjected to excessive supply of acid substances. Ambio 3, 93–94.

Gray, D.K., Arnott, S.E., 2011. Does dispersal limitation impact the recovery of zooplankton communities damaged by a regional stressor? Ecol. Appl. 21, 1241–1256.

Gray, C., Baird, D.J., Baumgartner, S., Jacob, U., Jenkins, G.B., O'Gorman, E.J., Lu, X., Ma, A., Pocock, M.J.O., Schuwirth, N., Thompson, M., Woodward, G., 2014. Ecological networks: the missing links in biomonitoring science. J. Appl. Ecol. 51, 1444–1449.

Gray, C., Figueroa, D.H., Hudson, L.N., Ma, A., Perkins, D., Woodward, G., 2015. Joining the dots: an automated method for constructing food webs from compendia of published interactions. Food Webs 5, 11–20.

Hall, S.J., Raffaelli, D., 1991. Food-web patterns: lessons from a species-rich web. J. Anim. Ecol. 60, 823–841.

Hall, S.J., Raffaelli, D.G., 1997. Food web patterns: what do we really know? In: Gange, A.C., Brown, V.K. (Eds.), Multitrophic Interactions. Blackwells Scientific Publications, Oxford, UK, pp. 395–417.

Hall, R.J., Likens, G.E., Fiance, S.B., Hendrey, G.R., 1980. Experimental acidification of a stream in the Hubbard Brook Experimental Forest, New Hampshire. Ecology 61, 976–989.

Hämäläinen, H., Huttunen, P., 1996. Inferring the minimum pH of streams from macroinvertebrates using weighted averaging regression and calibration. Freshw. Biol. 36, 697–709.

Henriksen, A., Lien, L., Traan, T.S., Rosseland, B.O., Sevaldrud, I.S., 1990. The 1000-lake survey in Norway 1986. In: Mason, B.J. (Ed.), The Surface Waters Acidification Programme. Cambridge University Press, Cambridge, UK, pp. 199–212.

Henriksen, A., Fjeld, E., Hesthagen, T., 1999. Critical load exceedance and damage to fish populations. Ambio 28, 583–586.

Hildrew, A.G., 2009. Sustained research on stream communities: a model system and the comparative approach. Adv. Ecol. Res. 41, 175–312.
Hildrew, A.G., Giller, P.S., 1994. Patchiness, species interactions and disturbance in the stream benthos. In: Giller, P.S., Hildrew, A.G., Raffaelli, D.G. (Eds.), Aquatic Ecology: scale, pattern and process. Symposium of the British Ecological Society. Blackwell Scientific Publications, Oxford, UK, pp. 21–62.
Hildrew, A., Ormerod, S., 1995. Acidification: causes, consequences and solutions. In: Harper, D.M., Ferguson, A.J.D. (Eds.), The Ecological Basis for River Management. John Wiley, Chichester, UK, pp. 147–160.
Hildrew, A.G., Townsend, C.R., Hasham, A., 1985. The predatory chironomidae of an iron-rich stream—feeding ecology and food web structure. Ecol. Entomol. 10, 403–413.
Hudson, L.N., Emerson, R., Jenkins, G.B., Layer, K., Ledger, M.E., Pichler, D.E., Thompson, M.S.A., O'Gorman, E.J., Woodward, G., Reuman, D.C., 2013. Cheddar: analysis and visualisation of ecological communities in R. Methods Ecol. Evol. 4, 99–104.
Ings, T.C., Montoya, J.M., Bascompte, J., Bluthgen, N., Brown, L., Dormann, C.F., Edwards, F., Figueroa, D., Jacob, U., Jones, J.I., Lauridsen, R.B., Ledger, M.E., Lewis, H.M., Olesen, J.M., van Veen, F.J.F., Warren, P.H., Woodward, G., 2009. Ecological networks—beyond food webs. J. Anim. Ecol. 78, 253–269.
Jenkins, A., Whitehead, P.G., Cosby, B.J., Birks, H.J.B., 1990. Modelling long-term acidification: a comparison with diatom reconstructions and the implications for reversibility. Philos. Trans. R. Soc. Lond. B Biol. Sci. 327, 435–440.
Jenkins, G.B., Woodward, G., Hildrew, A.G., 2013. Long-term amelioration of acidity accelerates decomposition in headwater streams. Glob. Chang. Biol. 19, 1100–1106.
Johnson, P.C.D., 2014. Extension of Nakagawa & Schielzeth's R 2 GLMM to random slopes models. Methods Ecol. Evol. 5, 944–946.
Johnson, R.K., Angeler, D.G., 2010. Tracing recovery under changing climate: response of phytoplankton and invertebrate assemblages to decreased acidification. J. North Am. Benthol. Soc. 29, 1472–1490.
Jonsson, M., Dangles, O., Malmqvist, B., Guérold, F., 2002. Simulating species loss following perturbation: assessing the effects on process rates. Proc. Biol. Sci. 269, 1047–1052.
Keller, W., Yan, N.D., Somers, K.M., Heneberry, J.H., 2002. Crustacean zooplankton communities in lakes recovering from acidification. Can. J. Fish. Aquat. Sci. 59, 726–735.
Keller, W., Yan, N.D., Gunn, J.M., Heneberry, J., 2007. Recovery of acidified lakes: lessons from Sudbury, Ontario, Canada. Water Air Soil Pollut. Focus 7, 317–322.
Kernan, M., Battarbee, R.W., Curtis, C., Monteith, D.T., Shillands, E.M., 2010. Recovery of lakes and streams in the UK from acid rain. The United Kingdom Acid Waters Monitoring Network 20 Year Interpretative Report: Report to the Department for Environment, Food and Rural Affairs.
Kowalik, R.A., Cooper, D.M., Evans, C.D., Ormerod, S.J., 2007. Acidic episodes retard the biological recovery of upland British streams from chronic acidification. Glob. Chang. Biol. 13, 2439–2452.
Latora, V., Marchiori, M., 2001. Efficient behavior of small-world networks. Phys. Rev. Lett. 87, 198701.
Lawton, J.H., 1989. Food webs. In: Cherrett, J.M. (Ed.), Ecological Concepts. Blackwell Scientific, Oxford, UK, pp. 43–78.
Layer, K., Hildrew, A., Monteith, D., Woodward, G., 2010a. Long-term variation in the littoral food web of an acidified mountain lake. Glob. Chang. Biol. 16, 3133–3143.
Layer, K., Riede, J.O., Hildrew, A.G., Woodward, G., 2010b. Food web structure and stability in 20 streams across a wide pH gradient. Adv. Ecol. Res. 42, 265–299.

Layer, K., Hildrew, A.G., Jenkins, G.B., Riede, J.O., Rossiter, S.J., Townsend, C.R., Woodward, G., 2011. Long-term dynamics of a well-characterised food web: four decades of acidification and recovery in the Broadstone Stream model system (ed G Woodward). Adv. Ecol. Res. 44, 69–117.

Layer, K., Hildrew, A.G., Woodward, G., 2013. Grazing and detritivory in 20 stream food webs across a broad pH gradient. Oecologia 171, 459–471.

Ledger, M.E., Hildrew, A.G., 2005. The ecology of acidification and recovery: changes in herbivore-algal food web linkages across a stream pH gradient. Environ. Pollut. 137, 103–118.

Ledger, M.E., Brown, L.E., Edwards, F.K., Milner, A.M., Woodward, G., 2012. Drought alters the structure and functioning of complex food webs. Nat. Clim. Chang. 3, 223–227.

Lepori, F., Ormerod, S.J., 2005. Effects of spring acid episodes on macroinvertebrates revealed by population data and in situ toxicity tests. Freshw. Biol. 50, 1568–1577.

Levine, S., 1980. Several measures of trophic structure applicable to complex food webs. J. Theor. Biol. 83, 195–207.

Likens, G.E., Bormann, F.H., 1974. Acid rain: a serious regional environmental problem. Science 184, 1176–1179.

Lundberg, P., Ranta, E., Kaitala, V., 2000. Species loss leads to community closure. Ecol. Lett. 3, 465–468.

Malcolm, I.A., Bacon, P.J., Middlemas, S.J., Fryer, R.J., Shilland, E.M., Cullen, P., 2014. Relationships between hydrochemistry and the presence of juvenile brown trout (*Salmo trutta*) in headwater streams recovering from acidification. Ecol. Indic. 37, 351–364.

Martinez, N.D., 1991. Artifacts or attributes? Effects of resolution on the Little Rock Lake food web. Ecol. Monogr. 61, 367–392.

Masters, Z.O.E., Peteresen, I., Hildrew, A.G., Ormerod, S.J., 2007. Insect dispersal does not limit the biological recovery of streams from acidification. Aquat. Conserv. Mar. Freshw. Ecosyst. 17, 375–383.

McLaughlin, Ó.B., Emmerson, M.C., O'Gorman, E.J., 2013. Habitat Isolation Reduces the Temporal Stability of Island Ecosystems in the Face of Flood Disturbance. Adv. Ecol. Res. 48, 225–284.

Micheli, F., Cottingham, K.L., Bascompte, J., Bjørnstad, O.N., Eckert, G.L., Fischer, J.M., Keitt, T.H., Kendall, B.E., Klug, J.L., Rusak, J.A., 1999. The dual nature of community variability. Oikos 85, 161–169.

Monteith, D.T., Evans, C.D., 1998. United Kingdom Acid Waters Monitoring Network 10 Year Report. Analysis and interpretation of results, April 1988—March 1998: Report to the Department for Environment, Food and Rural Affairs (Contract EPG 1/3/160).

Monteith, D.T., Evans, C.D., 2005. The United Kingdom Acid Waters Monitoring Network: a review of the first 15 years and introduction to the special issue. Environ. Pollut. 137, 3–13.

Monteith, D.T., Hildrew, A.G., Flower, R.J., Raven, P.J., Beaumont, W.R.B., Collen, P., Kreiser, A.M., Shilland, E.M., Winterbottom, J.H., 2005. Biological responses to the chemical recovery of acidified fresh waters in the UK. Environ. Pollut. 137, 83–101.

Monteith, D.T., Evans, C.D., Henrys, P.A., Simpson, G.L., Malcolm, I.A., 2014. Trends in the hydrochemistry of acid-sensitive surface waters in the UK 1988–2008. Ecol. Indic. 37, 287–303.

Montoya, J.M., Solé, R.V., 2002. Small world patterns in food webs. J. Theor. Biol. 214, 405–412.

Montoya, J.M., Pimm, S.L., Solé, R.V., 2006. Ecological networks and their fragility. Nature 442, 259–264.

Mulder, C., Boit, A., Bonkowski, M., De Ruiter, P.C., Mancinelli, G., Van der Heijden, M.G.A., Van Wijnen, H.J., Vonk, J.A., Rutgers, M., 2011. A belowground perspective on Dutch agroecosystems: how soil organisms interact to support ecosystem services. Adv. Ecol. Res. 44, 277–357.
Murphy, J.F., Winterbottom, J.H., Orton, S., Simpson, G.L., Shilland, E.M., Hildrew, A.G., 2014. Evidence of recovery from acidification in the macroinvertebrate assemblages of UK fresh waters: a 20-year time series. Ecol. Indic. 37, 330–340.
Naeem, S., 1998. Species redundancy and ecosystem reliability. Conserv. Biol. 12, 39–45.
Nedbalová, L., Vrba, J., Fott, J., Kohout, L., Kopáček, J., Macek, M., Soldán, T., 2006. Biological recovery of the Bohemian Forest lakes from acidification. Biologia 61, 453–465.
O'Gorman, E.J., Emmerson, M.C., 2010. Manipulating Interaction Strengths and the Consequences for Trivariate Patterns in a Marine Food Web. Adv. Ecol. Res. 42, 302–419.
O'Neill, R.V., 1998. Recovery in complex ecosystems. J. Aquat. Ecosyst. Stress Recovery 6, 181–187.
Patrick, S., Waters, D., Juggins, S., Jenkins, A., 1991. The United Kingdom Acid Waters Monitoring Network: site descriptions and methodology report: Report to the Department of the Environment and Department of the Environment Northern Ireland, London.
Patrick, S., Monteith, D.T., Jenkins, A., 1995. UK Acid Waters Monitoring Network: the first five years. Analysis and interpretation of results, April 1988–March 1993. ENSIS Ltd, London.
Peralta, G., Frost, C.M., Rand, T.A., Didham, R.K., Tylianakis, J.M., 2014. Complementarity and redundancy of interactions enhance attack rates and spatial stability in host-parasitoid food webs. Ecology 95, 1888–1896.
Pimm, S.L., 1980. Properties of food webs. Ecology 61, 219–255.
Pimm, S.L., Russell, G.J., Gittleman, J.L., Brooks, T.M., 1995. The future of biodiversity. Science 269, 347–350.
Pocock, M.J.O., Evans, D.M., Memmott, J., 2012. The robustness and restoration of a network of ecological networks. Science 335, 973–977.
R Core Team, 2013. R: A Language and Environment for Statistical Computing.
Raffaelli, D., 2007. Food webs, body size and the curse of the Latin binomial. In: Rooney, N., McCann, K.S., Noakes, D.L.G. (Eds.), From Energetics to Ecosystems: The Dynamics and Structure of Ecological Systems. Springer, The Netherlands, pp. 53–64.
RoTAP, 2012. Review of transboundary air pollution: acidification, eutrophication, ground level ozone and heavy metals in the UK: Contract Report to the Dept of the Environment, Food and Rural Affairs.
Scheffer, M., Carpenter, S.R., 2003. Catastrophic regime shifts in ecosystems: linking theory to observation. Trends Ecol. Evol. 18, 648–656.
Schindler, D.W., 1988. Effects of acid rain on freshwater ecosystems. Science 239, 149–157.
Schmid-Araya, J.M., Hildrew, A.G., Robertson, A., Schmid, P.E., Winterbottom, J., 2002. The importance of meiofauna in food webs: evidence from an acid stream. Ecology 83, 1271–1285.
Schoener, W., Schoenerz, T.W., 1989. Food webs from the small to the large. Ecology 70, 1559–1589.
Smith, B.D., Zeder, M.A., 2013. The onset of the Anthropocene. Anthropocene 4, 8–13.
Solé, R.V., Montoya, J.M., 2001. Complexity and fragility in ecological networks. Proc. R. Soc. B Biol. Sci. 268, 2039–2045.
Solé, R.V., Ferrer-Cancho, R., Montoya, J.M., Valverde, S., 2003. Selection, tinkering, and emergence in complex networks. Complexity 8, 20–33.
Steffen, W., Grinevald, J., Crutzen, P., McNeill, J., 2011. The Anthropocene: conceptual and historical perspectives. Philos. Trans. R. Soc. A 369, 842–867.

Stoddard, J.L., Jeffries, D.S., Lukewille, A., Clair, T.A., Dillon, P.J., Driscoll, C.T., Forsius, M., Johannessen, M., Kahl, J.S., Kellogg, J.H., Kemp, A., Mannio, J., Monteith, D.T., Murdoch, P.S., Patrick, S., Rebsdorf, A., Skjelkvale, B.L., Stainton, M.P., Traaen, T., van Dam, H., Webster, K.E., Wieting, J., Wilander, A., 1999. Regional trends in aquatic recovery from acidification in North America and Europe. Nature 401, 575–578.

Sutcliffe, D.W., Hildrew, A.G., 1989. Invertebrate communities in acid streams. In: Morris, R., Taylor, E.W., Brown, D.J.A., Brown, J.A. (Eds.), Acid Toxicity and Aquatic Animals. Seminar Series of the Society for Experimental Biology, Cambridge University Press, Cambridge, UK, pp. 13–29.

Sutherland, W.J., Broad, S., Caine, J., Clout, M., Dicks, L.V., Doran, H., Entwistle, A.C., Fleishman, E., Gibbons, D.W., Keim, B., LeAnstey, B., Lickorish, F.A., Markillie, P., Monk, K.A., Mortimer, D., Ockendon, N., Pearce-Higgins, J.W., Peck, L.S., Pretty, J., Rockström, J., Spalding, M.D., Tonneijck, F.H., Wintle, B.C., Wright, K.E., Sutherland, W.J., Broad, S., Caine, J., Clout, M., Dicks, L.V., Doran, H., Entwistle, A.C., Fleishman, E., Gibbons, D.W., Keim, B., LeAnstey, B., Lickorish, F.A., Markillie, P., Monk, K.A., Mortimer, D., Ockendon, N., Pearce-Higgins, J.W., Peck, L.S., Pretty, J., Rockström, J., Spalding, M.D., Tonneijck, F.H., Wintle, B.C., Wright, K.E., 2016. A horizon scan of global conservation issues for 2016. Trends Ecol. Evol. 31, 44–53.

Thébault, E., Fontaine, C., 2010. Stability of ecological communities and the architecture of mutualistic and trophic networks. Science 329, 853–856.

Thompson, R.M., Brose, U., Dunne, J.A., Hall, R.O., Hladyz, S., Kitching, R.L., Martinez, N.D., Rantala, H., Romanuk, T.N., Stouffer, D.B., Tylianakis, J.M., 2012. Food webs: reconciling the structure and function of biodiversity. Trends Ecol. Evol. 27, 689–697.

Thompson M.S.A., Bankier C., Bell T., Dumbrell A.J., Gray C., Ledger M.E., Lehman K., McKew B.A., Sayer C.D., Shelley F., Trimmer M., Warren S.L. and Woodward G., Gene-to-ecosystem impacts of a catastrophic pesticide spill: testing a multilevel bio-assessment approach in a river ecosystem. Freshw. Biol., (in press).

Townsend, C.R., Hildrew, A.G., Francis, J., 1983. Community structure in some southern English streams: the influence of physicochemical factors. Freshw. Biol. 13, 521–544.

Townsend, C.R., Hildrew, A.G., Schofield, K., 1987. Persistence of stream invertebrate communities in relation to environmental variability. J. Anim. Ecol. 56, 597–613.

Tylianakis, J.M., Tscharntke, T., Lewis, O.T., 2007. Habitat modification alters the structure of tropical host-parasitoid food webs. Nature 445, 202–205.

van den Berg, R.A., Hoefsloot, H.C.J., Westerhuis, J.A., Smilde, A.K., van der Werf, M.J., 2006. Centering, scaling, and transformations: improving the biological information content of metabolomics data. BMC Genomics 7, 142.

Warren, P.H., 1994. Making connections in food webs. Trends Ecol. Evol. 9, 136–141.

Watts, D.J., Strogatz, S.H., 1998. Collective dynamics of 'small-world' networks. Nature 393, 440–442.

Webster, K.E., Frost, T.M., Watras, C.J., Swenson, W.A., Gonzalez, M., Garrison, P.J., 1992. Complex biological responses to the experimental acidification of Little Rock Lake, Wisconsin, USA. Environ. Pollut. 78, 73–78.

Williams, R.J., Martinez, N.D., 2000. Simple rules yield complex food webs. Nature 404, 180–183.

Williams, R.J., Martinez, N.D., 2004. Limits to trophic levels and omnivory in complex food webs: theory and data. Am. Nat. 163, 458–468.

Woodward, G., Hildrew, A.G., 2001. Invasion of a stream food web by a new top predator. J. Anim. Ecol. 70, 273–288.

Woodward, G., Hildrew, A.G., 2002. Body-size determinants of niche overlap and intraguild predation within a complex food web. J. Anim. Ecol. 71, 1063–1074.
Woodward, G., Speirs, D.C., Hildrew, A.G., 2005a. Quantification and resolution of a complex, size-structured food web. Adv. Ecol. Res. 36, 85–135.
Woodward, G., Thompson, R., Townsend, C.R., Hildrew, A.G., 2005b. Pattern and process in food webs: evidence from running waters. In: Belgrano, A., Scharler, U.M., Dunne, J., Ulanowicz, R.E. (Eds.), Aquatic Food Webs: An Ecosystem Approach. Oxford University Press, Oxford, UK, pp. 51–66.
Woodward, G., Benstead, J.P., Beveridge, O.S., Blanchard, J., Brey, T., Brown, L.E., Cross, W.F., Friberg, N., Ings, T.C., Jacob, U., Jennings, S., Ledger, M.E., Milner, A.M., Montoya, J.M., O'Gorman, E.J., Olesen, J.M., Petchey, O.L., Pichler, D.E., Reuman, D.C., Thompson, M.S.A., Van Veen, F.J.F., Yvon-Durocher, G., 2010a. Ecological networks in a changing climate. Adv. Ecol. Res. 42, 71–138.
Woodward, G., Blanchard, J., Lauridsen, R.B., Edwards, F.K., Jones, J.I., Figueroa, D., Warren, P.H., Petchey, O.L., 2010b. Individual-based food webs: species identity, body size and sampling effects. Adv. Ecol. Res. 43, 211–266.
Woodward, G., Brown, L.E., Edwards, F.K., Hudson, L.N., Milner, A.M., Reuman, D.C., Ledger, M.E., 2012. Climate change impacts in multispecies systems: drought alters food web size structure in a field experiment. Philos. Trans. R. Soc. Lond. B Biol. Sci. 367, 2990–2997.
Yan, N.D., Leung, B., Keller, W., Arnott, S.E., Gunn, J.M., Raddum, G.G., 2003. Developing conceptual frameworks for the recovery of aquatic biota from acidification. Ambio 32, 165–169.
Yan, N.D., Girard, R., Heneberry, J.H., Keller, W.B., Gunn, J.M., Dillon, P.J., 2004. Recovery of copepod, but not cladoceran, zooplankton from severe and chronic effects of multiple stressors. Ecol. Lett. 7, 452–460.
Yodzis, P., 1989. Introduction to Theoretical Ecology. Harper & Row, Cambridge, UK.

CHAPTER ELEVEN

Effective River Restoration in the 21st Century: From Trial and Error to Novel Evidence-Based Approaches

N. Friberg*,†,1, N.V. Angelopoulos‡, A.D. Buijse§, I.G. Cowx‡, J. Kail¶,
T.F. Moe*, H. Moir∥, M.T. O'Hare#, P.F.M. Verdonschot**, C. Wolter††

*Norwegian Institute for Water Research, Oslo, Norway
†water@leeds, University of Leeds, Leeds, United Kingdom
‡University of Hull, Hull, United Kingdom
§Deltares, Delft, The Netherlands
¶University of Duisburg-Essen, Essen, Germany
∥cbec eco-engineering Ltd., Inverness, Scotland, United Kingdom
#Centre for Ecology & Hydrology, Edinburgh, Scotland, United Kingdom
**Alterra, Wageningen, The Netherlands
††IGB, Berlin, Germany
1Corresponding author: e-mail address: Nikolai.Friberg@niva.no

Contents

1. Introduction 537
   1.1 A Brief Introduction to River Restoration 537
   1.2 The Need for Restoration 538
   1.3 Drivers of River Restoration 540
   1.4 A Short Introduction to the REFORM Project and Scope of this Paper 541
2. Responses of River Biota to Hydrology and Physical Habitats 545
   2.1 Can We Expect the Biota to Respond to River Restoration? 545
   2.2 Importance of Local Physical Habitat Filters in Structuring Stream Biota 549
   2.3 Are We Capable of Detecting Impacts of Hydromorphology on Biodiversity Using Standard Methods? 555
   2.4 Which Are the Best Standard Indicator to Detect HYMO Stress and Recovery Through Restoration? 559
3. The Current Restoration Paradigm 560
   3.1 The Many Ways of Restoring Rivers 560
   3.2 Restoration Measures Are Dependent on River Type 564
   3.3 Using Current Management Plans as Indicators of Restoration Practises 567
   3.4 Limitations of Current and Planned Restoration Approaches 569
4. Effects of Restoration 569
   4.1 General Effect: Does River Restoration Work in General? 570
   4.2 Differences in Responses Among Organism Groups and Species Traits: Which Benefit Most? 571

4.3 Differences Between Restoration Measures: What to Do? 574
4.4 Confounding Factors: Why Do Some Restoration Projects Fail? 576
5. Future Directions 578
  5.1 Future River Restoration Needs Better Planning 578
  5.2 Project Planning at a Catchment Scale: A Necessity 584
  5.3 Exploring the Full Potential of River Restoration 584
  5.4 Project Identification 585
  5.5 Project Planning at a Local Scale 586
  5.6 Project Formulation 586
  5.7 Monitoring, Evaluation and Project Success 587
  5.8 Adjustment and Maintenance 588
  5.9 The Future: Holistic and Process-Oriented Restoration 588
6. Conclusions 593
Acknowledgements 600
References 600

Abstract

This paper is a comprehensive and updated overview of river restoration and covers all relevant aspects from drivers of restoration, linkages between hydromorphology and biota, the current restoration paradigm, effects of restorations to future directions and ways forward in the way we conduct river restoration. A large part of this paper is based on the outcomes of the REFORM (REstoring rivers FOR effective catchment Management, http://reformrivers.eu/) project that was funded by EU's 7th Framework Programme (2011–15). REFORM included the most comprehensive comparison, to date, of existing river restorations across Europe and their effect on biota, both in relation to preintervention state and project size in terms of river length restored. The REFORM project outcomes are supplemented by an extensive literature review and two case studies to illustrate key points. We conclude that river restorations conducted up until now have had highly variable effects with, on balance, more positives than negatives. The largest positive effects have interestingly been in terrestrial and semiaquatic organism groups, in widening projects, while positive effects on truly aquatic organisms groups are only seen when in-stream measures are applied. The positive responses of biota are primarily seen as increased abundance of organisms with very little indication that overall biodiversity has increased: specific traits rather than mere species number or total abundance have benefited from restoration interventions. This modest success rate can partly be attributed to the fact that the catchment filter is largely ignored; large-scale pressures related to catchment land use or the lack of source populations for the recolonisation of the restored habitats are inadequately considered. The key reason for this shortfall is a lack of clear objective setting and planning processes. Furthermore, we suggest that there has been a focus on form rather than processes and functioning in river restoration, which has truncated the evolution of geomorphic features and any dynamic interaction with biota. Finally, monitoring of restoration outcomes is still rare and often uses inadequate statistical designs and inappropriate biological methods which hamper our ability to detect change.

1. INTRODUCTION
1.1 A Brief Introduction to River Restoration

Restoration is a human intervention aimed at improving conditions of an ecosystem to a former, predisturbance state. Inherently, the term 'restoration' elicits confusion as projects are very rarely aimed at restoring ecosystems back to exactly the way they used to be. This is certainly true for rivers, which are often nested within catchments where larger scale impacts such as land use are the rule rather than the exception (Friberg, 2014). In many ways, the definition by the Society of Ecological Restoration (SER, 2004) from 2004 stating that '*Ecological restoration is the process of assisting the recovery of an ecosystem that has been degraded, damaged or destroyed*' was important in focusing a discussion that had become more about semantics than science. Importantly, and in the context of the present paper, the definition is to the point when dealing with the restoration of rivers and fluvial systems, where the main focus has been primarily on smaller reach-scale interventions to improve, longitudinal connectivity, channel planform (the configuration of the river channel in plan view, e.g. meandering) and local habitat conditions (Bernhardt et al., 2005; Feld et al., 2011). Removing dams, weirs and other obstacles for migration of primarily diadromous fish have been, and still are, a very common type of restoration intervention (Kail et al., 2015). Change of planform, in particular from a straight channel to a meandering course, has become popular in river restoration (Feld et al., 2011; Kail et al., 2015); unfortunately, this has been indiscriminate, including instating meanders where they are not historically a naturally occurring geomorphological feature (Walter and Merritts, 2008). Concepts such as physical habitats and mesohabitats are frequently used in river restoration projects as synonyms for improving substrate conditions and increase variability in water velocities for the expected benefit of aquatic species (Friberg, 2010).

River restoration has been high on the global environmental agenda for more than 3 decades with a large number of projects undertaken primarily in Europe and the United States (Feld et al., 2011; Ormerod, 2004; Palmer et al., 2005). Investments have been significant and river restoration has received wide recognition in the public, reflected by the engagement of both stakeholders and scientists. Many previously degraded rivers look different today compared with the recent past and this change in appearance is a very visual outcome of restoration. However, the history of river restoration is also flawed by the lack of a systematic approach in project development and infrequent evaluation of ecosystem responses (Bash and Ryan, 2002; Bernhardt et al., 2005). In the past decade, there has been some improvement with an increasing number of scientific papers that review or report results on effects of river restoration (e.g. Kail et al., 2015; Palmer et al., 2010; Roni et al., 2008; Whiteway et al., 2010).

The capacity of practitioners to implement and achieve ecosystem restoration is often limited by opportunity, time or economic constraints (Borgström et al., 2016). So despite considerable investment, river restoration still depends on trial and error as there is a lack of comprehensive and well-formulated planning, implementation and appraisal techniques. Several guidance documents have been produced over the last few decades to either assist with planning and techniques (Cowx and Welcomme, 1998; FISRWG, 1998; Roni and Beechie, 2013; RRC, 2011; Ward et al., 1994) or provide overviews of key concepts and principles (Brierley and Fryirs, 2008; Clewell and Aronson, 2008). Collectively these publications cover many of the tools, techniques and concepts needed for restoration activities, but not the planning and integration of restoration processes from initial assessment to monitoring of results. With an increased emphasis on restoration has come the need for new techniques and guidance for assessing stream and catchment condition that has the necessary sensitivity to detect postrestoration changes (Rumps et al., 2007). Another weakness of the current river restoration paradigm is that societal needs, such as energy demands, freshwater and food supply, transportation networks, flood protection, etc., have rarely been considered and other potential socioeconomic benefits, e.g. increasing value of private properties, are seldom fully explored, understood or achieved (Ayres et al., 2014; Wortley et al., 2013).

1.2 The Need for Restoration

Ecosystems worldwide face large-scale challenges such as population growth, climate change, land degradation and habitat loss (Foley et al., 2005; Halpern et al., 2008; Parmesan and Yohe, 2003; Sanderson et al., 2002). Freshwater ecosystems, and in particular rivers, have a long history of human pressures and they have been impacted by a number of very deteriorating types of stress relating primarily to sewage influx, land-use intensification and physical degradation. As a result of this sum of stressors, the ecological condition of rivers is still highly impaired and globally these ecosystems rank among those that have seen the greatest loss of biodiversity

(Sala et al., 2000; Vörösmarty et al., 2010). The negative impacts of such pressures on biodiversity and ecosystem services, e.g. the provision of freshwater, are recognised worldwide and have led to a set of international targets including the UN Aichi Biodiversity Target 15 and EU 2020 Biodiversity Strategy Target 2 of restoring at least 15% of degraded ecosystems by 2020 (Convention on Biological Diversity, 2010; European Commission, 2011).

The scope for freshwater restoration is vast with more than 50% of Europe's freshwaters not meeting their environmental quality objectives of good ecological status (GES) or potential (Solheim et al., 2012), despite a costly and largely successful, effort to improve in particular water quality, e.g. lowering concentrations of easy degradable organic matter from sewage (BOD_5) that depletes oxygen in the water. The wide range of services (Millennium Ecosystem Assessment, 2005) that can be delivered by a functioning ecosystem is the target of a multitude of policies and legislations, e.g. the Paris Agreement on Climate Change, the EU Renewable Energy Sources Directive, the UNEP Green Economy Initiative and the US Clean Water Act. Consequently, the focus of river restoration is now predominantly on biodiversity and the delivery of ecosystem services such as carbon sequestration, flood protection and provision of freshwater (Palmer et al., 2014). However, restoring previously degraded physical habitat features in rivers and floodplains affects the interests of multiple stakeholders (Kondolf and Yang, 2008; TEEB, 2011) and implementation of more recent environmental quality targets, such as those specified by the Water Framework Directive (WFD) of the EU or the Aichi Biodiversity Targets, has encountered resistance, slowing down implementation and yielding mixed success (Hart et al., 2012; Jähnig et al., 2011).

The assessment of the first river basin management plans (RBMPs), as part of implementing the EU WFD (see Box 1) indicated that 40% of European rivers are affected by changed hydrological regimes and degraded channel morphology caused predominantly by hydropower, navigation, agriculture, flood protection and urban development. The aim of WFD is that all water bodies should fulfil a requirement of GES by 2027 at the latest. As a consequence, there is increasing emphasis in Europe on river restoration driven by demands of the WFD, where restoring hydrology and morphology, as part of programmes of measures (PoMs), is an important component in achieving GES. Implementation of PoMs requires substantial investment in these measures, but there still remains a great need to better understand and predict the costs and benefits of future river restoration. Furthermore, ecological response to river restoration is complex and poorly understood.

> **BOX 1 Important Policy Drivers for River Restoration**
> There are a number of European Directives to support the ecological health of rivers such as the Water Framework Directive (WFD (2000/60/EC)), Habitats Directive (HD (92/43/EEC)) and Groundwater Directive (GWD (2006/118/EC)), in addition to global initiatives such as Agenda 21 of the Rio Convention and the Convention of Biological Diversity. These have driven the management of inland waters towards rehabilitation of rivers and lakes to improve the aquatic environment for biodiversity and allow for sustainable exploitation of the resources (Hobbs et al., 2011; Pasternack, 2008). Consequently, nature conservation, and in particular river restoration, is increasingly considered as part of a much wider framework of environmental policy and practise (Arlinghaus et al., 2002). In Europe, the main policy driver of restoration is without a doubt WFD that was implemented in 2000, and to understand parts of the present paper it is necessary to introduce some key concepts of the Directive. One of the first steps of implementing the Directive was to identify river basins that could be overall management units and some of these traverse national frontiers. River Basin Management Plans (RBMPs) have to be established and updated every 6 years in three plan periods starting 2009 and ending in 2027. RBMPs are detailed accounts of how objectives set for the river basins (ecological status, quantitative status, chemical status and protected area objectives) are to be reached within the timescale required. The plan will include river basin characteristics, a review of the impact of human activity on the status of waters in the basin, estimation of the effect of existing legislation and the remaining 'gap' to meeting these objectives; as well as a set of measures designed to bridge the gap.

1.3 Drivers of River Restoration

Degradation and loss of physical complexity in river ecosystems have been massive in most parts of Europe through, for instance, channellisation, dredging, wood removal, etc. (Friberg, 2010). In addition, siltation with fine sediments is a major problem in many streams, especially in agricultural catchments (Glendell et al., 2014). The importance of habitat heterogeneity for biota is indisputable but surprisingly few studies have documented clear impacts of habitat degradation (Friberg, 2014). In the following, the collective term *hydromorphology* is used for the physical environment as it is used by the WFD, signifying how important physical features are considered in determining the ecological status of freshwaters. The hydromorphological quality of a river in 'high' status class is defined as follows: '*Channel patterns, width and depth variations, flow velocities, substrate conditions and both structure and condition of the riparian zones correspond totally or nearly totally to undisturbed conditions'*. With regard to the other status classes, hydromorphology is considered a supporting element and there are no normative definitions but can help in explaining why biological quality elements (BQEs) are not achieving GES. It is evident from the RBMPs undertaken thus far that degraded hydromorphology is one the most extensive impacts on river ecosystems in Europe today. Centuries of modification by man to ensure drainage, flood protection, navigation and hydropower have completely altered habitat area, channel form and processes in rivers and floodplains almost everywhere. While the effects of stressors such as low oxygen levels on the river biota are well documented and have been instrumental in reducing sewage loads, specific methods to assess the impact of degraded hydromorphology on ecological status are relatively uncommon. Provisioning and regulating services from development sectors (pressures: Table 1; Fig. 1), such as water resource management, flood protection, inland navigation, hydropower and agriculture, have led to the replacement of naturally occurring and functioning systems with highly modified and human-engineered systems.

Fig. 1, furthermore, exemplifies how complex interactions between the main drivers of hydromorphological degradation, in this case damming of rivers and creating large reservoirs, and a range of associated environmental factors will substantially change living conditions for the biota across scales. Consequently, damming a river will have profound impacts beyond impairing longitudinal connectivity that directly affects anadromous fish species. This complexity in response to hydromorphological degradation signifies the importance of appropriate restoration strategies.

1.4 A Short Introduction to the REFORM Project and Scope of this Paper

A large part of this paper is based on the outcomes of the REFORM (REstoring rivers FOR effective catchment Management, http://reformrivers.eu/) project that was funded by EU's 7 Framework Programme (2011–15) and brought together 26 research institutes and applied partners from 15 European countries (Fig. 2).

The overall aim were to generate tools for cost-effective restoration of river ecosystems, and for improved monitoring of the biological effects of physical change by investigating natural, degradation and restoration processes in a wide range of river types across Europe. In relation to this chapter, the following specific aims are relevant:

Table 1 Identification of Main Pressures on Rivers and their Categorisation in Specific Types of Pressures

Main Pressure	Specific Types of Pressures
Water abstraction	Groundwater abstraction
	Surface water abstraction
Flow regulations	Discharge diversions and returns
	Hydrological regime modification: timing or quantity
	Hydropeaking
	Interbasin flow transfers
	Reservoir flushing
	Sediment discharge from dredging
River fragmentation	Artificial barriers downstream from the site
	Artificial barriers upstream from the site
	Colinear connected reservoir
	Large dams and reservoirs
Morphological alterations	Alteration of in-stream habitat
	Alteration of riparian vegetation
	Channellisation/cross-section alteration
	Embankments, levees or dikes
	Impoundment
	Loss of vertical connectivity
	Sand and gravel extraction
	Sediment input
Water quality	Diffuse source pollution
	Point source pollution

The majority of pressures relates to physical changes of rivers, i.e. hydrological and geomorphological alterations often denoted 'hydromorphology' in accordance with the EU WFD.
After Garcia de Jalón, D., Alonso, C., et al., 2013. Review on pressure effects on hydromorphological variables and ecologically relevant processes. REFORM Deliverable D1.2, Report to the European Union (available at http://reformrivers.eu/).

- Will river biota show a sufficiently strong response to hydromorphological change, either in terms of degradation or restoration that the effects can be singled out from other stressors acting on ecosystem as well as effects of biotic interactions and dispersal mechanisms?
- Which types of river restorations are the most frequently used in Europe and how are these in accordance with the main challenges regarding hydromorphological degradation?
- What are the effects of river restoration on the various components of the river biota and how do spatial and temporal scales influence the outcomes?

Effective River Restoration 543

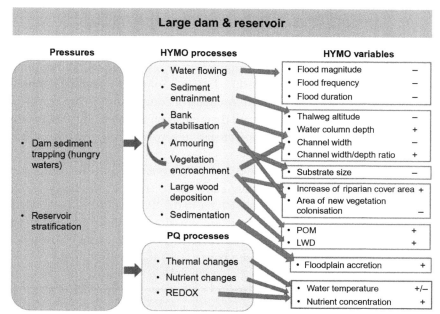

Fig. 1 The link between hydromorphological pressures, the processes they impair and their effect on key variables that is important for river biota here exemplified with large dams and reservoirs as the overall pressure. It is evident that this pressure has a multitude of effects on habitat conditions severely degrading living conditions for a number of species (Garcia de Jalón et al., 2013). *HYMO*: Hydromorphological processes; *PQ*: physical–chemical processes; *LWD*: large woody debris; *POM*: particulate organic matter.

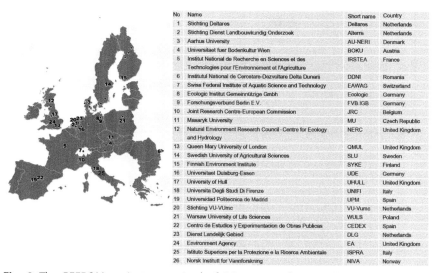

Fig. 2 The REFROM project comprised of 26 partners from 15 European countries covering all major landscapes and river types.

- How can river restoration become more efficient in the future in delivering expected outcomes and secure that environmental objectives are achieved as well as stopping the decline in biodiversity?

These aims are addressed by a combination of reviewing the literature and undertaking new analysis on existing datasets combined with field sampling. REFORM included the most comprehensive comparison to date of existing river restorations across Europe and their effect on the biota, both in relation to preintervention state and project size in terms of river length restored (Fig. 3).

We provide a comprehensive and fully updated status for restoration activities in Europe that hopefully can help to steer future developments in the right direction. In the text we refer to the 'REFORM project' whenever we use primary results if these are not already published. Full details regarding methods and also additional results that we cannot cover here can be found on the project home page.

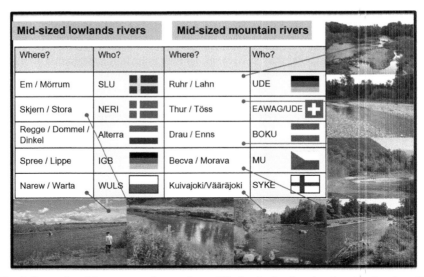

Fig. 3 Ten pairs of river restoration projects differing in size (predominantly river length) were investigated as part of REFORM across Europe. The restoration effect of the large and smaller project size was investigated in a similar length river section and quantified by comparing each of these to a nearby nonrestored section (space-for-time substitution). We sampled the following response variables: habitat composition in the river and its floodplain, three aquatic organism groups (macrophytes, benthic invertebrates and fish), two floodplain-inhabiting organism groups (floodplain vegetation and ground beetles), as well as food web composition and aquatic land interactions as reflected by stable isotopes (Hering et al., 2015).

2. RESPONSES OF RIVER BIOTA TO HYDROLOGY AND PHYSICAL HABITATS

2.1 Can We Expect the Biota to Respond to River Restoration?

Natural rivers depend on catchment-scale structural controls, reach-scale channel pattern differences and microscale variations in channel bed forms, all of which vary over different time scales (Friberg, 2014). In this hierarchical organisation, structure and processes occurring on small spatial and temporal scales are nested within increasingly larger scales, from microhabitat to catchment (Fig. 4).

Naturally, therefore, the effects of restoration will be scale dependent and linked to the spatial and temporal heterogeneity provided by natural stream reaches, which creates a range of biotopes for the biota and is the scale where impact is assessed (Frissell et al., 1986; Wolter et al., 2016). Moreover, lotic ecosystems will be impacted at a range of scales depending on the type of pressure, with larger scale impacts having negative effects on lower levels

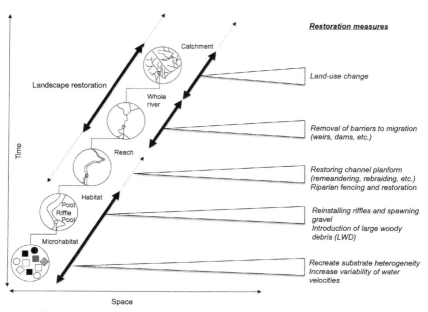

Fig. 4 Effects of restoration is dependent of both spatial and temporal scales. Furthermore, the type of restoration measure is scale dependent.

of organisation. A key challenge to the restoration of rivers is the assumption 'if you build they will come …' which has been the core paradigm in most projects (Palmer et al., 1997). Local physical habitat conditions are without doubt important filters for the biota and species have evolved life histories in response to coarse substrates originating from riverine sediment sorting, supporting the assumption that there is a clear causative link between habitats provided at local scales and biological communities present (Fig. 5).

In effect, other local-scale controls such as biotic interactions and dispersal at larger scales must be of insignificant importance for this assumption to hold and this is clearly not valid in all cases. Local environmental conditions are not the sole determinant of macroinvertebrate or fish community composition as both local biotic interactions (Woodward, 2009) and larger dispersal mechanisms (Heino, 2013; Radinger and Wolter, 2015) play an important structuring role.

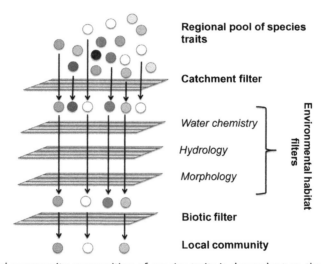

Fig. 5 Local community composition of species traits is dependent on the filtering of the regional pool of species traits by (a) catchment filters; (b) environmental habitat filters and (c) local biotic filters. In the context of river restoration, the focus has been primarily to modify the hydrological and morphology components of the environmental habitat filters. However, it is evident that filtering at the scale of catchments can reduce the potential of recovery even if local habitat conditions are improved (Stoll et al., 2016). Likewise, water chemistry will also be an important filter and poor quality will impair possibilities of recovery if not considered (Violin et al., 2011). The biotic filter is largely ignored in restoration projects although it is important in determining local community composition.

Restoration measures potentially influence river food webs at each trophic level with implications up through the web in terms of, for example, energy transfer (Fig. 6). For instance, increased retention via the restoration of more complex bed forms will increase retention of allochthonous organic matter, which can further be increased by rehabilitating riparian zones so they include a large proportion of woody vegetation. This will increase the detrital food base and increase detritivore biomass, which are important food items for higher trophic levels. In wider rivers with less riparian shading in-stream macrophytes have the ability to increase physical heterogeneity and improve habitat conditions for top predators such as fish. Overall, restoration has the potential to increase the resource base of river ecosystems with knock-on effects throughout the food web. However, while this is at the core of biomanipulation in lakes (Søndergaard et al., 2007), restoration projects that consider the effects at the scale of the food web do not exist for rivers.

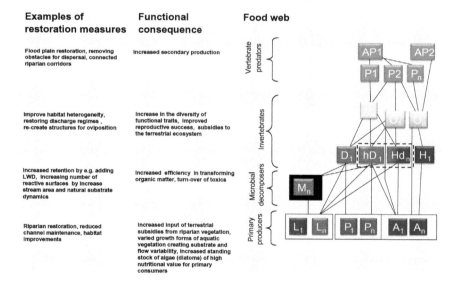

Fig. 6 Different restoration measures will influence both individual trophic levels as well as the entire food web. The figure shows examples of the possible effects on ecosystem functioning of different restoration measures (*left* and *middle panels*) with a schematic freshwater food web mapping possible effects onto trophic position. *AP*, apex vertebrate predator such as fish, amphibians and birds; *C*: carnivorous invertebrate; *O*: omnivorous invertebrate; *D*: detritivore; *hD/Hd*: herbivore–detritivore; *H*: herbivore; *AH*: aquatic hyphomycetes; *L*: leaf-litter; *P*: plant; *A*: algae.

Evaluation of restoration projects, using a space-for-time substitution approach, showed that despite plenty of suitable habitat macroinvertebrate diversity was still lower than in reference streams that had never been significantly modified (Pedersen et al., 2014). Likewise, the influence of other pressures on the system needs to be less important local physical habitat conditions (Fig. 7). Several empirical studies indicated that large-scale pressures related to catchment land use can be more important in shaping macroinvertebrate and fish communities compared to pressures at smaller spatial scales like local hydromorphological alterations (Roth et al., 1996; Stephenson and Morin, 2009; Sundermann et al., 2013), and might even limit macroinvertebrate and fish assemblages (Bryce et al., 2010; Kail et al., 2012; Wang et al., 2007). In addition, Harding et al. (1998) showed a clear land-use legacy, where catchment land use more than 3 decades back determined present day diversity of macroinvertebrates and fish.

Lastly, we have to consider that most studies that have revealed links between biota and local physical conditions have been scientific studies employing very specific sets of methodology. In reality, most assessment of ecological conditions in rivers, and effect studies of restoration interventions, are undertaken as part of monitoring programmes using sampling

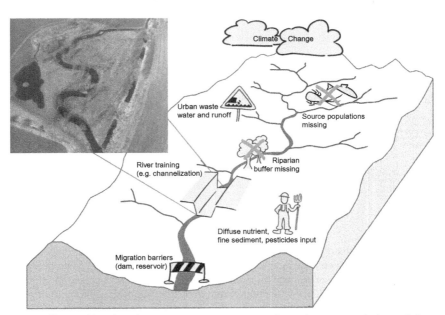

Fig. 7 Effects of local-scale restoration measures such as the remeandering of short river reaches will be dependent on other pressures acting on the catchment. Poor water quality relating to upstream land use will potentially mask any benefits of improved habitat conditions as will dispersal barriers such as dams or lack of source populations.

approaches based on CEN standards and metrics (in Europe) that have been intercalibrated. These may lack the sensitivity required to adequately monitor hydromorphology, reflecting the original development of this methodology was developed to be primarily sensitive to water quality (Friberg, 2014). These observations are important to consider when exploring links between river biota and local physical conditions in the context of restoration.

2.2 Importance of Local Physical Habitat Filters in Structuring Stream Biota

River discharge maintains riverine processes and mediates connectivity and is often the focal point of many restoration projects; however, species do not directly respond to discharge but rather flow velocities and stream power (O'Hare et al., 2011; Statzner et al., 1988). Flow velocity and stream power provide species and size-specific thresholds for habitat utilisation. In addition, the interaction between flowing water and the size and quantity of available sediment leads to diverse substrate calibres emerging from flow-induced sorting. Therefore, the most important components of local physical habitat filter in stream and rivers relate to flow conditions and substrate composition. These are interlinked as flow determines substrate. Far too often, however, substrate is introduced as a restoration measure without considering flow or stream power and this will not provide sufficient habitat conditions (Pedersen et al., 2014). In a comparative metaanalysis of 80 interacting hydromorphological processes and variables aiming to identify the most relevant factors controlling ecological degradation and restoration, Lorenz et al. (2016) identified water flow as the most important process. Accordingly, species with life history traits and ecological characteristics that are directly linked to flowing water and the related processes of sediment erosion, transport and sorting should be diagnostic in their response to hydromorphological degradation and rehabilitation. A challenge, however, is that the biotic response to high flow velocities and shear stresses is not very specific. Common thresholds values and ranges of flow velocities reported in the literature underpin this point (Fig. 8): <0.3 m/s for species rich, diverse macrophyte communities (Janauer et al., 2010), 0.3–1.0 m/s for rheophilic invertebrates (Söhngen et al., 2008; Statzner et al., 1988), and 0.1 and 0.5 m/s for hatchlings and juvenile fish, respectively (Wolter and Arlinghaus, 2003, 2004). However, these thresholds vary widely within taxa and genera, e.g. between <0.8 and >2.0 m/s for various gastropods, selected dipterids and some beetles, respectively (Statzner et al., 1988). While upper flow thresholds selects for few species adapted to a life in high

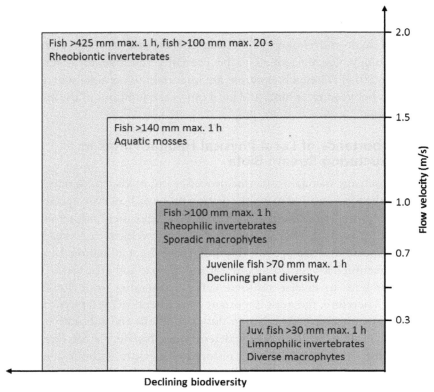

Fig. 8 Physical thresholds for diverse aquatic taxa in response to flow velocity. *Modified from Söhngen, B., Koop, J., Knight, S., Rythönen, J., Beckwith, P., Ferrari, N., Iribarren, J., Kevin, T., Wolter, C., Maynord, S., 2008. Considerations to Reduce Environmental Impacts of Vessels. Brussels: PIANC, PIANC Report 99.*

water velocities; the provision of low flow habitats <0.3 m/s similarly supports nearly all taxa, stressing the importance of these low flow habitats to sustain and improve biodiversity in restoration projects.

Benthic algae are particularly prone to the impact of increased fine sediment loads (Jones et al., 2014) and this is probably the most important type of hydromorphological stress for this organism group. A direct first principle effect is a decrease in light with increased turbidity but the most profound effect of fine sediment is the smothering of substrata to which benthic algae attach. These relatively unstable deposits (compared with larger particles) are not suitable for the attachment of long-lived sedentary species. Hence, non-motile, and particularly chain-forming taxa, cannot establish easily, further pushing the assemblage towards single-celled and motile taxa. A shift in assemblage composition towards motile taxa can be seen even where larger particles are covered with a layer of fines (Dickman et al., 2005). The lack of stability in patches, where easily erodible fine sediments accumulate, tends to result in reduced taxon richness and biomass compared to more stable patches (Biggs et al., 1999; Biggs and Smith, 2002; Matthaei et al., 2003). When comparing across streams, those with stable bed sediments support a higher biomass of diatoms than those that have unstable beds (Biggs, 1996; Biggs et al., 1999; Biggs and Smith, 2002; Jowett and Biggs, 1997). Further negative effects of hydromorphology could be expected through both direct and indirect impacts on the substrate on which benthic algae grow. Reductions in flow velocity, for example caused by impoundments, would tend to increase the deposition of fine sediment altering both bed substrate and the potential for planktonic algae to thrive. Direct modification of in-stream and marginal habitat has the potential to alter the substrate on which benthic algae grow. In restoration context, it is needed not only to provide coarse substrates for diatoms and other benthic algae to grow but to control, in particular, the delivery of fine sediments to the river system.

Water flow and turbulence determines the macrophytes of running waters, governs plant form, dominates the growth-controlling factors and defines the habitats (Biggs, 1996; Dawson, 1988; Folkard, 2011; Schutten et al., 2005). Generally, plants with high drag coefficients and low anchoring strengths are those species most susceptible to high velocities (Biggs, 1996). Therefore, the ability of a plant to tolerate water movement without suffering mechanical damage relies either on minimising the hydrodynamic forces or maximising its breakage and uprooting strengths (Bornette and Puijalon, 2011). Emerged growth is considered favourable under low flow conditions, especially since this growth form is not affected by any reduction of light due to the attenuation by the water column (Bal et al., 2011). Submerged growth is supported by rougher conditions and high flow velocities due to the flattening, compressing and reconfiguration of plants which lower the drag forces (O'Hare et al., 2007; Puijalon et al., 2005; Sand-Jensen and Pedersen, 2008; Sukhodolova, 2008). Slightly enhanced flow velocities promote the growth of aquatic macrophytes due to facilitated diffusion of CO_2 and nutrients (Madsen et al., 2001) with increases in vegetation evident at velocities up to 0.3 m/s with a peak at about 0.3–0.5 m/s depending on the species (Janauer et al., 2010; Riis and Biggs, 2003). At higher velocities, plant biomass and diversity decrease and flow velocities higher than 0.8 m/s dislodge and eliminate most in-stream macrophytes (Bernez et al., 2007; Chambers et al., 1991; Janauer et al., 2010; Madsen et al., 2001). Macrophytes have a large potential as ecoengineers in river restoration due to their ability to changing flow and substrate conditions at the reach scale.

However, macrophytes are important primarily in low energy systems in the lowlands.

Substrate composition influences benthic macroinvertebrate communities, in particular the quality and quantity of organic matter in sediment and the stability of the substrate (Buss et al., 2004; Jowett, 2003; Maxted et al., 2003; Timm et al., 2008). Density and diversity of macroinvertebrates show a general increase with substrate particle size (Beauger et al., 2006; Duan et al., 2009; Reice, 1980) with the exception that high organic content can sustain large numbers of certain macroinvertebrate taxa on fine-grained substrates (Pan et al., 2012). The hydraulic environment and in particular bed shear stress is a good predictor of benthic macroinvertebrates distribution at the microhabitat scale, because it accounts for the turbulences at the bed surface generated by sediment roughness and its associated drag and lift forces (Mérigoux and Dolédec, 2004; Möbes-Hansen and Waringer, 1998). Consequently, Dolédec et al. (2007) and Mérigoux et al. (2009) were able to characterise shear stress preferences of 181 benthic macroinvertebrate taxa using the methodology of Statzner et al. (1991). Only few studies provide velocity dislodgement thresholds for benthic macroinvertebrates (e.g. Holomuzki and Biggs, 2000; Statzner et al., 1988; Wilzbach et al., 1988). The average shear stress necessary to detach and dislodge macroinvertebrates has been estimated for a total of 27 taxa (Borchardt, 1993; Gabel et al., 2012; Hauer et al., 2012). However, responses of macroinvertebrates to hydraulic conditions and shear stress at the microhabitat scale are not independent of larger scale features such as stream size (Dolédec et al., 2007; Hauer et al., 2012; Mérigoux et al., 2009). In addition, critical shear stress thresholds for the same taxa vary between studies (Gibbins et al., 2010; Hauer et al., 2012) and more recently Gibbins et al. (2016) suggested water velocity and Froude numbers are better predictors of macroinvertebrate drift than shear stress.

To identify species whose presence is not primarily determined by substrate preference or hydraulic preference, we plotted preferred substrate fractions reported by Tolkamp (1982) and Singh et al. (2010) in classes from 0 (128–256 mm) to 11 (0.125–0.05 mm) against shear stress calculated from Dolédec et al. (2007) and Mérigoux et al. (2009) (Fig. 9). Outliers above and below the substrate–shear stress relationship obtained are characterised by significantly higher shear force tolerance compared to the preferred substrate calibre and vice versa. When restoring rivers, it is therefore important to acknowledge that macroinvertebrates do respond to shear stress and not just

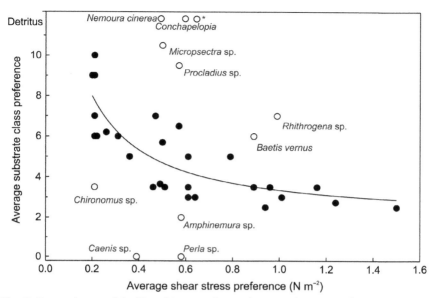

Fig. 9 Regression model of benthic macroinvertebrates substrate preferences in categorical size classes from class 0 (128–256 mm) to 11 (0.125–0.05 mm) plotted against their hydraulic preferences. Outliers (*white circles*) indicate taxa not primarily determined by hydromorphology (* = *genera Brillia, Corynoneura, Diplocladius, Eukiefferiella, Rheocricotopus*).

substrate sizes. The focus must be on restoring natural fluvial processes that creates a range of different shear stress microhabitats both in space and in time. It also evident that range in response to both shear stress and substrate size of most taxa is fairly wide, calling for the creation of a highly variable physical environment.

Fish are comparably long living, mobile organisms with various habitat requirements, habitat shifts during ontogeny, and functional differences between age groups and hence utilising habitats at various spatiotemporal scales (Wolter et al., 2016). Coarse, well-oxygenated and permeable gravel beds are essential for fish species that are depend on such substrates for spawning and gravel spawning are commonly considered as adaptation of fish to faster flowing environmental conditions by protecting eggs and hatchling from becoming washed away (Balon, 1975; DeVries, 1997; Jungwirth et al., 2000). Significant empirical relations between fish length and gravel diameter have been reported for red digging salmonids (Crisp, 1996; Kondolf and Wolman, 1993). Headwater species, such as trout and grayling, lay their eggs into the substrates at depths from a few centimetres and up to 30 cm and, in addition to sediment size, they are dependent on permeable sediment with interstitial flow of oxygen-rich water (DeVries, 1997; Riedl and Peter, 2013; Sternecker et al., 2013). Accumulation of fines <1 mm has been reported causing most significant impacts on hatch and survival of fish larvae even at rather low proportions (Heywood and Walling, 2007; Julien and Bergeron, 2006; Soulsby et al., 2001). As with other aquatic organisms fish are exposed to flow velocity and stream power, which both set physical thresholds for habitat utilisation. Swimming performance of fish has been reviewed and analysed as a proxy for their ability to withstand absolute physical forces set by flow velocity (Wolter and Arlinghaus, 2003, 2004). The individual swimming performance depends on species, swimming mode, size, temperature, ontogenetic stage, photoperiod, oxygen tension, pH, salinity and various pollutants and toxins, with total length as paramount trait (reviewed in Hammer, 1995; Videler, 1993; Wolter and Arlinghaus, 2003). Based on the metaanalysis of 168 swimming performance studies covering 75 freshwater fish species, Wolter and Arlinghaus (2003) derived general models for burst and critical swimming performance of fish. The general models of length-specific burst and critical swimming performance were highly significant. As expected, salmonids exhibited the highest burst swimming performance; however, the differences detected between the small-sized individuals of different taxonomic orders were not significant (F-test, $p = 0.142$). Thus, the swimming performance model applies for all fish up to 60 mm total length, which is important as one would intuitively think that rheophilic fish perform superior to eurytopic and limnophilic fish. Consequently, a 56 mm long fish already maintains a speed of 1.0 m/s for 20 s, but only 0.54 m/s in the critical mode for 1 h (Fig. 10).

In conclusion, the different organism groups found in rivers will respond to different components of the physical environment and some links to restoration are straightforward in substrate preferences of gravel spawning fish as well as critical swimming speeds. In both cases, this knowledge have been used in restoration measures such as instalment of riffles and by adjusting current velocity in bypass streams (at dams and weirs) to allow migration and resting of specific fish species. However, we also show that there is a generally high variability in response and that physical habitat preferences are quite nonspecific in a way that limit our ability to be very prescriptive in the approach to restoration, also considering the complex interaction between hydromorphological degradation and environmental factors.

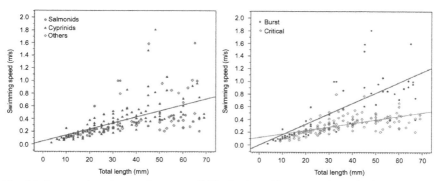

Fig. 10 Burst swimming performance (*left*) of salmonids, cyprinids and other fish species up to 60 mm total length compared to their critical swimming performance (*right*). Regressions did not significantly differ between families (F-test, $p=0.142$) and followed the models $U_{burst} = 0.0068 \, TL^{1.24}$ (df = 84, $R^2 = 0.83$; $p < 0.001$) and $U_{crit} = 0.0067 \, TL^{1.09}$ (df = 155, $R^2 = 0.60$, $p < 0.001$). *Modified from Wolter, C., Arlinghaus, R., 2003. Navigation impacts on freshwater fish assemblages: the ecological relevance of swimming performance. Rev. Fish Biol. Fish. 13, 63–89.*

2.3 Are We Capable of Detecting Impacts of Hydromorphology on Biodiversity Using Standard Methods?

Several recent studies have revealed weak relationships between a standardised measure of habitat quality (River Habitat Survey (RHS)) and a number of macroinvertebrate indices in streams along a hydromorphological degradation gradient (e.g. Vaughan et al., 2009). Existing monitoring datasets collected across Europe, to assess ecological status as stipulated by the WFD, were analysed as part of the REFORM project and we found in general weak statistical relationships between identity-based (taxonomic) ecological indicators and measures of hydromorphological quality as shown in Table 2. In most cases, other impacts such as eutrophication caused by agriculture had a much more pronounced impact on the biological indicators compared with changes to the physical environment, underpinning the importance of considering all pressures together when restoring rivers.

Metrics developed to detect hydrological impairment and hydromorphological degradation were not more discriminative to other types of metrics when tested on a Danish dataset that included a hydrological time series (Table 3). Some of the strongest relationships identified were between macroinvertebrate metrics that measure the presence of taxa with a preference for specific flow or substrate type conditions and between indicators of

Table 2 Spearman Correlation Coefficients of EQRs (Standardised Ecological Quality Ratios) of Diatoms, Benthic Macroinvertebrates and Fish, and the Mean EQR of the Three Groups, Against the PCA-Gradients from Finnish Streams with the Main Environmental Gradients

	EQR Diatoms	EQR Macroinvertebrates	EQR Fish	Mean EQR
Number of sites	132	139	96	91
PCA1 (agriculture)	**−0.550**	**−0.563**	**−0.488**	**−0.670**
PCA2 (morphological alteration)	−0.060	0.023	−0.067	−0.025
PCA3 (urban)	−0.064	−0.086	0.154	0.151
PCA4 (naturalness)	**0.186**	0.097	0.065	0.121

Biological and environmental monitoring data are from 150 river sites across Finland (not all biological groups were present at each site for direct comparison). Significant coefficients are given in bold (for details on results, please consult http://reformrivers.eu/results/effects-of-hydromorphological-changes).

Table 3 The Relationship of Selected Macroinvertebrate Metrics Sensitive to Different Types of Stress and Hydrological Data Shown as Pearson Correlation Matrix for Maroinvertebrates Metrics[a] and Flow Statistics[a]

Sensitive to	MESH HYMO Stress	ASPT Organic Pollution	LIFE Low Flow	EPT General Degradation	SPEAR (%) Pesticide Exposure
Q90	0.61	0.59	0.52	0.44	0.6
Q10	−0.58	−0.52	−0.47	−0.43	−0.55

[a]*MESH*, Maroinvertebrates of Estonia: score of hydromorphology; *ASPT*, average score per taxon; *LIFE*, Lotic-Invertebrate Index for flow evaluation; *EPT*, the total number of taxa belonging to Ephemeroptera, Plecoptera, Trichoptera order; *SPEAR* (%), indicated toxicity of pesticides and organic pollution in water; Q10, flow magnitude exceeded for 10% of the time; Q90, flow magnitude exceeded 90% of the time.
All selected macroinvertebrate metrics revealed a significant relationship ($p < 0.001$) with flow statistics (Q90 and Q10) and they all exhibited a similar direction of response at Q90 (positive) and Q10 (negative).

these specific conditions, which reflects the potential of using trait-based metrics to evaluate hydromorphological conditions. However, a subsequent analysis revealed few traits in macroinvertebrates that potentially could distinguish between hydromorphological and other stressors although relationships were not very significant.

The finding using larger monitoring datasets is somewhat in contrast with the literature reviewed at the beginning of this chapter that showed that both substrate composition and shear forces influenced macroinvertebrate community composition. The reasons for the lack of sensitivity might be most likely attributed to: (1) a number of explanatory variables not being measured as part of routine biological monitoring programmes or (2) the hydromorphological assessment schemes that do not necessarily register elements of importance to the in-stream biota. As an example, riffle habitats that are visually assessed as identical can differ markedly in macroinvertebrate diversity and community composition (Pedersen and Friberg, 2007). This very clearly suggests that human perception is biased, leading to subjective assessments that mask relationships between biota and hydromorphology and drives restoration efforts in a direction which does not optimise ecological recovery.

No effects of hydromorphological degradation on indices based on phytobenthos were detected from analysis on a large-scale dataset with no significant effects detected of channel impairments such as straightening (Fig. 11).

However, indices developed to assess eutrophication stress (e.g. Trophic Diatom Index (TDI) and related indices) appear robust to hydromorphological alteration as there was a significant relationship with $\log_{10}$ orthophosphate for almost all indices. With regard to fine sediment stress specifically, analysis of a spatial dataset across England and Wales showed a strong relationship between fine sediment and the diatom community composition, which suggests that diatoms could be used as an indicator of fine sediment stress. In contrast to this finding, and more in support of general analysis of the response to hydromorphological stress, phytobenthos community composition was primarily impacted by soluble reactive phosphorous concentration in experiments. Fine sediment treatment did not significantly impact on either chlorophyll-a concentrations or ash-free dry weight, suggesting that algal growth was unaffected.

Macrophyte trait characteristics changed significantly in response to hydromorphological degradation in small streams. Several traits could be identified such as species growing from single basal meristems declined and that species with a high overwintering capacity increased. However, with one exception, the trait heterophylly (the ability to have different types of leaves above and below water), impacts of hydromorphological stress could not be separated from eutrophication. More traits were specific to eutrophication, which could aid in the diagnostic of main stressors in multiple stress systems. Although macrophytes are very important in some river types, they have limited applicability as a general indicator for the range of European river types. For lowland streams, which are the type most often

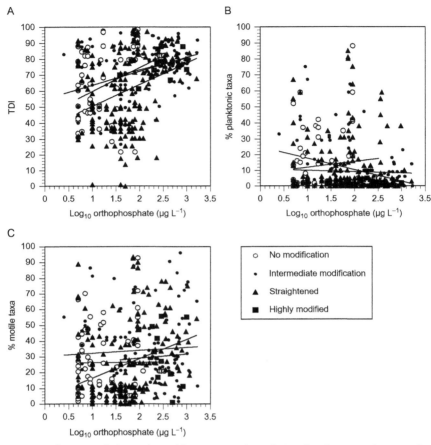

Fig. 11 Influence of channel modification on the relationship between $\log_{10}$ orthophosphate concentration and (a) Trophic Diatom Index (TDI)—an indicator of eutrophication (phosphate) and two ecological species traits: (b) % planktonic taxa and (c) % motile taxa.

physically modified, they could be used if a trait-based metric was developed (Baattrup-Pedersen et al., 2015).

Forty-nine per cent of the studied European freshwater fish species in REFORM showed a significant response to hydromorphological stress. Using conceptual models that link pressures via processes to responses, it was identified that benthic fish like Cobitidae, Balitoridae, Cottidae or Gobiidae showed the most consistent response to HYMO stress. This response can be related to their dependence on substrate dynamics and low mobility. Conceptual models should be viewed as a first step, and although more traditional metric approaches using fish also has significant potential, both approaches require significant development before they can be applied in regular monitoring. The inherent difficulty in fish methods based on number of taxa is that their use is restricted to stream/river types with a minimum of species diversity, leaving out most small streams. In these types more detailed investigations on length frequency distributions can be related to HYMO stress. Many HYMO pressures (but also non-HYMO pressures) generate a clear response in the fish community size structures, particularly inducing changes in the overall shape of the size spectra, which might therefore be useful as a new potential metric for assessing impacts of HYMO stress.

Key to the strength of these relationships was sampling methods and to some degree organism size, with fish showing most promise which is not surprising as they, due to their generally large body size, integrate over larger spatial scales and thereby reduce the variability introduced when assessing small-scale features such as individual microhabitats. Relationships were further improved when using species traits (also using macrophytes) that are more closely linked to the habitat template, including hydromorphology. These findings suggest that scales of sampling are a core issue in assessing the effects of restoration measures and methods using species traits of large(r) organisms are likely to be the most sensitive. However, the focus on in-stream biota that is routinely monitored ignores that many of the pronounced effects of degraded hydromorphology relates to the riparian zones and the wider floodplain. When comparing restored and nonrestored river sections, riparian ground beetles most strongly responded to hydromorphological improvements, followed by fish and floodplain vegetation (Hering et al., 2015).

2.4 Which Are the Best Standard Indicator to Detect HYMO Stress and Recovery Through Restoration?

Our findings leave water managers with a significant challenge when diagnosing the reason for not obtaining GES in a waterbody or detecting positive effects of a restoration intervention, aiming at improving the physical environment. Even for fish, the organism group which showed most promise, there is a significant amount of work to do before sensitive metrics to hydromorphological change can be applied in water management. The issue with the impact of multiple stressors on the biota has been shown to preclude our ability to disentangle hydromorphological change from other stressors. It appears from the analysis undertaken that eutrophication is a stronger driver of community changes. This is, however, most likely also related to the quality of the hydromorphological assessments, which in most of the larger datasets were fairly superficial in the current analysis.

3. THE CURRENT RESTORATION PARADIGM

3.1 The Many Ways of Restoring Rivers

In recent decades, Europe has made significant progress in reducing pollution and water quality has significantly improved, albeit there are still issues relating to diffuse pollution from agriculture and point-source discharges from wastewater treatment plants (EEA, 2012). Concurrently, with the implementation of the WFD, it has been mandatory to set environmental targets focused on ecological status of surface waters as well as good chemical water quality. Hence, despite the significant improvements to water quality and some biological indicators, the continued trend of declining freshwater biodiversity is unacceptable from a legislative point of view (Aarts et al., 2004; SCBD, 2010). Accordingly, a shift in restoration paradigm has occurred from a focus on purely physicochemical water quality to a paradigm encompassing ecological quality, hydromorphological and habitat conditions. Both hydromorphological pressures and altered habitats have been reported as the most common impact for 48.2% and 42.7% of all river water bodies, respectively (Fehér et al., 2012). Hydromophology was considered a larger pressure on water bodies than diffuse pollution across the full range of arable land use, in particular in intensively farmed catchments reflecting the impact of drainage (Fig. 12).

Despite numerous hydromorphological restoration projects implemented since the 1940s, the number of projects has exponentially risen in the last 10–15 years (e.g. Alexander and Allan, 2006; Bernhardt et al., 2005; Feld et al., 2011; Roni et al., 2008). Here we have compiled a database of 813 hydromorphological river restoration projects described in 878 publications, 53% of them implemented after 2003 and published after 2007. Most of the projects (649) were from Europe. The review revealed a range in the number of river hydromorphological elements that have been addressed as well as the employment of a substantial variety of 53 specific measures in total (Table 4). Especially, the improvement of habitat structures in the stream channel itself has been targeted with a number of different measures, most common the removal of artificial embankments, the addition of large wood and the provision of spawning gravel. In-stream measures typically address specific hydromorphologic elements, not the underlying processes. In contrast,

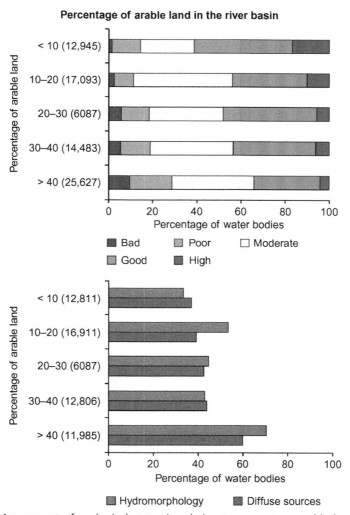

Fig. 12 Assessment of ecological status in relation to percentage arable land in the catchment with the main pressures on aquatic ecosystems listed. Hydromorphology is perceived as a larger pressure than diffuse pollution in most cases. *Figure obtained from the EEA (first published 2012) at: http://www.eea.europa.eu/data-and-maps/figures/ecological-status-potential-and-pollution/ecological-status-potential-and-pollution.*

improvement of the riparian zone is primarily addressed by vegetated buffer strips. Remeandering and hydrological reconnection of floodplain water bodies are the most commonly applied intervention to improve river planform and floodplains, respectively (Table 4). The largest restoration measures, most often river widening, have been implemented in the lower parts of mountain

Table 4 Number of Hydromorphological Restoration Measures Reportedly Implemented in the 813 River Restoration Projects Surveyed

Restoration Measure	Number
Water quantity/Hydrology	*85*
Reduce water surface water abstraction without return	4
Improve water retention (e.g. on floodplain, urban areas)	59
Reduce groundwater abstraction	6
Improve/create water storage (e.g. polders)	7
Increase minimum flow (to generally increase discharge in a reach or to improve flow dynamics)	16
Water diversion and transfer to improve water quantity	3
Recycle used water (off-site measure to reduce water consumption)	4
Reduce water consumption (other measures than recycling used water)	1
Sediment quantity	*47*
Add/feed sediment (e.g. downstream from dam)	21
Reduce undesired sediment input (e.g. from agricultural areas or from bank erosion other than riparian buffer strips!)	8
Prevent sediment accumulation in reservoirs	1
Improve continuity of sediment transport (e.g. manage dams for sediment flow)	9
Trap sediments (e.g. building sediment traps to reduce washload)	16
Reduce impact of dredging	3
Flow dynamics	*60*
Establish environmental flows/naturalise flow regimes (does focus on discharge variability)	36
Modify hydropeaking	7
Increase flood frequency and duration in riparian zones or floodplains	11
Reduce anthropogenic flow peaks	6
Shorten the length of impounded reaches	3
Favour morphogenic flows (could also be considered a measure to improve planform or in-channel habitat conditions)	3
Longitudinal connectivity	*143*

Table 4 Number of Hydromorphological Restoration Measures Reportedly Implemented in the 813 River Restoration Projects Surveyed—cont'd

Restoration Measure	Number
Install fish pass, bypass, side channel for upstream migration	32
Install facilities for downstream migration (including fish friendly turbines)	25
Manage sluice, weir and turbine operation for fish migration	18
Remove barrier (e.g. dam or weir)	103
Modify or remove culverts, syphons, piped streams	13
In-channel habitat conditions	*560*
Remove bed fixation	77
Remove bank fixation	144
Remove sediment (e.g. mud from groyne fields)	30
Add sediment (e.g. gravel)	102
Manage aquatic vegetation (e.g. mowing)	40
Remove or modify in-channel hydraulic structures (e.g. groynes, bridges)	33
Creating shallows near the bank	106
Recruitment or placement of large wood	130
Boulder placement	115
Initiate natural channel dynamics to promote natural regeneration	112
Create artificial gravel bar or riffle	128
Riparian zone	*218*
Develop buffer strips to reduce nutrient input	18
Develop buffer strips to reduce fine sediment input	13
Develop natural vegetation on buffer strips (other reasons than nutrient or sediment input, e.g. shading, organic matter input)	186
River planform	*321*
Remeander water course (actively changing planform)	156
Widening or rebraiding of water course (actively changing planform)	115
Shallow water course (actively increasing level of channel bed)	54
Narrow overwidened water course (actively changing width)	25

Continued

Table 4 Number of Hydromorphological Restoration Measures Reportedly Implemented in the 813 River Restoration Projects Surveyed—cont'd

Restoration Measure	Number
Create low-flow channels in oversized channels	38
Allow/initiate lateral channel migration (e.g. by removing bank fixation and adding large wood)	31
Create secondary floodplain on present low level of channel bed	8
Floodplain	*320*
Reconnect existing backwaters, oxbow lakes, wetlands	158
Create seminatural/artificial backwaters, oxbow lakes, wetlands	107
Lowering embankments, levees or dikes to enlarge inundation and flooding	64
Back-removal of embankments, levees or dikes to enlarge the active floodplain area	23
Remove embankments, levees or dikes or other engineering structures that impede lateral connectivity	32
Remove vegetation	85
Others	*243*

Several measures and measure combinations per project possible; note, connectivity measures were only counted in combination with other hydromorphologic measures. The database excludes fish migration facilities. Fish passes as connectivity measures are prerequisite for fish dispersal and may provide access to restored habitats, but they will rarely improve overall hydromorphological and habitat conditions.

rivers, where typically only one pressure was reported relating to channel narrowing and embankments (Kail and Wolter, 2011).

Generally, evaluations of the outcome of projects are still rare (Bernhardt et al., 2005; Kail et al., 2015; Palmer et al., 2010) despite the increasing number of restoration interventions and an increased societal drive to identify efficient solutions that have economic benefits (Everard, 2012; Reichert et al., 2015; Smith et al., 2014).

3.2 Restoration Measures Are Dependent on River Type

In more detail, we found that restoration effort was dependent on river type across Europe with marked differences between lowland rivers (<200 masl.), lower mountain rivers (<800 masl.), upland and glacial rivers and large rivers (catchment > 10,000 km^2).

Floodplain wide measures in lowland areas, which are predominantly used for agricultural production across Europe, will most often necessitate either buying land off farmers or compensating them for a change into a less profitable land use as increased inundation and ground water tables limits the commercial value of the land. In reality costs of land at the floodplain scale are so high in lowland areas that this type of restoration is rarely executed. Naturally lowland river systems are meandering or have multiple channels configuration (anastomosing) with the active channel covering large parts of the valley (Rinaldi et al., 2016). The most frequently applied measure is lowering the floodplain in combination with a shallow stream bed whereby the stream can shape the floodplain, rewet it and form a single channel with sometimes a secondary channel. Restoring an anastomosing planform, or allowing natural features of lowland systems such swamp forest to develop, often meets resistance from other users of the floodplain. Most of the measures in lowland rivers aim to restore the channel planform (56%), primarily by remeandering previously channellised stretches of river or some intermediate form of channel plan modifications, e.g. digging a two-stage profile. More often these channel planform measures are combined with in-channel measures, like removal of bank fixation and/or adding local structures such as groynes. Probably this is because of the low cost of in-channel measures compared to changes in channel planform that needs adjacent land. In theory, remeandering will affect in-channel habitat conditions but sand-bed rivers have a limited potential for recruiting coarse substrates. They are to a large degree reliant on woody debris and plants to create heterogeneous habitats. Our review of projects carried out in lowland areas furthermore showed that active remeandering of lowland rivers can also decrease microhabitat diversity, i.e. there were cases where remeandering led to a decrease in river velocity resulting in particulate organic material as the main microhabitat, while in the unrestored section more habitats were present. Infrequently, projects will also involve reconnection of former secondary channels or the creation of new where original are lost. Sometimes these 'reconnection' measures are combined with in-channel measures including removal of bank fixation and/or adding local structures such as tree logs. Restoration of the riparian zone in lowland areas is always limited to local areas where rewetting and flooding is a possibility. Often water safety arguments are behind creating areas that can get inundated and store water to reduce risk of downstream flooding. The most frequently applied measure in lowland rivers is reconnecting old meanders and oxbow lakes; removing weirs and restoring river banks. Restoration of the riparian zone is always combined with channel planform and in-channel measures and the extent in term of width restored is often limited.

Most of the restoration projects in single thread, lower mountain rivers applied in-channel measures to increase habitat complexity (75–80%), most frequently by removing bed and bank fixation, adding large wood and boulders and creating shallow slow-flowing areas, while measures to explicitly restore natural sediment dynamics (e.g. by adding sediment, restoring natural sediment transport or limiting fine sediment input) were rarely applied (1–6%). Many projects also aimed to restore a more natural planform (40–54%), e.g. by widening or remeandering the stream channel, or involved developing a riparian buffer strip (~30%) or restored floodplain habitats (~48%).

The majority of measures taken in mountain rivers and glacial rivers with single-thread channels aim to restore the flow alteration as these rivers across Europe have been dammed for energy production, drinking water supply and irrigation of farmland. Most important is the restoration of the natural flow regime, the reestablishment of the natural flow dynamics and the increase of water flow quantity in case of residual water flow not used for hydropower production. Hydropeaking (release of pulses of water), impoundment and water abstraction are relevant topics because of the dynamism of the hydropower sector and the need to mitigate and remediate adverse ecological impacts. Hydrological measures focused on mitigating the flow alteration are often applied at a local/small scale without solving the hydrological dynamics that result from catchment-wide activities. Individual measures at each hydropower plant are usually set without considering the downstream or upstream situation. In addition to restoring flow regimes, natural sediment regime and wood delivery will in some projects be restored as will in-stream habitats to mitigate the negative effects of hydropeaking. Even though the sediment regime in highland river types is usually not compromised, the building of check dams and the subsequent retention of sediment and wood can cause negative effects. These effects (e.g. increased bed and bank erosion, bed incision and negative sediment budget in wide floodplains) are visible far downstream at the lowland rivers. The input of sediment at downstream reaches is a commonly applied even though it is considered an unsustainable countermeasure. Restoring natural processes (e.g. restoration of water and sediment regime by removing blocking debris in the upper catchment) have a better effect on recovery, compared to local-scale interventions (e.g. wood or gravel addition at a lower part of the river catchment), but are rarely undertaken.

Large rivers typically serve inland navigation and have been straightened, regulated and embanked to improve their function as waterways. This designated use sets significant boundaries for river restoration, because it prevents the implementation of all kinds of measures that impact on fairway dimensions, as many of the in-channel measures do. Accordingly, the number of measures implemented is rather low compared to their dimension and the multitude of existing pressures and impacts. Most hydromorphological restoration measures implemented address the floodplain and riparian vegetation; however, by far majority of measures are conceptual. This category sums up investigations, research, pilot studies, but also changes in legislation, stakeholder involvement, information and education.

3.3 Using Current Management Plans as Indicators of Restoration Practises

In parallel to the compilation of the restoration project database, the first RBMPs and PoMs of the EU Member States were accessed for this study through the Water Information System for Europe (http://water.europa.eu). A previous assessment of the German PoMs revealed a reasonable coherent selection of measures in accordance with the analysis of pressures and impacts (Kail and Wolter, 2011). At the same time, this analysis showed a general lack of knowledge on the effectiveness of restoration measures especially those used to enhance the ecological status of lowland rivers and heavily modified water bodies.

We have translated the supplementary measures to enhance ecological status and classified them according to restoration measure groups. For these planned measures, and the preexisting restoration projects, we compared the consistency of measures suggested against identified pressures. Furthermore, we examined for any bias in the selection of measures based on available knowledge and potential ecological effects of the planned measures. In total 49,055 supplementary measures were listed for European river basin districts (17,341 for Continental Europe, i.e. excluding UK which had a more detailed list of measures than other member states). The following refers only to the supplementary measures listed for Continental Europe: conceptual measures (e.g. investigations, stakeholder information and legislation) were the most frequently planned measures, accounting for 56% of all measures followed by measures addressing water quality (18%) and hydromorphology (14%). The share of hydromorphological measures planned was dominated by floodplain rehabilitation (15%), in-stream habitat enhancement (11%), hydrology (16%), connectivity (17%) and riparian buffers (12%)

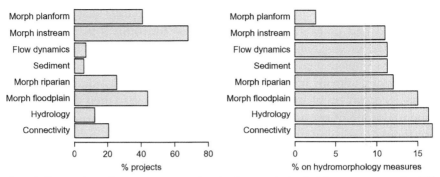

Fig. 13 Proportion of measures targeting hydromorphological improvements in 649 implemented European restoration projects (*left*) and of 2428 supplementary hydromorphological measures planned for Continental Europe river basin districts in the PoMs (*right*).

(Fig. 13, *right panel*). In comparison, in-stream habitat enhancement was the dominating measure (68%) already implemented in European rivers in the past (Fig. 13, *left panel*). Similarly, the proportion of projects addressing river planform already implemented (40%) by far exceeds those planned (<3%).

These findings provide interesting new insights: only 14% of all planned supplementary measures address hydromorphology and this is in spite of the fact that hydromorphological modifications are among the most significant pressures identified throughout Europe (EEA, 2012; Fehér et al., 2012). Furthermore, a surprisingly low percentage (11%) of the planned hydromorphological measures address in-stream habitat enhancement, although this measure retrospectively was very frequently employed (in 68% of European restoration projects in our database), implying that ample experience should be available. This obvious discrepancy between implemented restoration projects and further planning of measures raises questions as to the reasons why. One reason could be that the predominant use of in-stream measures during the first planning period have been assessed as ineffective measures, because they often provide habitat forms while ignoring the habitat forming hydromorphological processes (e.g. Roni et al., 2008; Wolter, 2010). Natural hydromorphological conditions are often only partially reestablished and the natural flooding may not be enabled which prevents the natural links between the stream and the riparian zone. Furthermore, impaired water quality due to land use (agriculture and forestry) in the catchment often prevents achieving the ecological goals of the habitat restorations.

3.4 Limitations of Current and Planned Restoration Approaches

In summary, our review of restoration projects conducted across Europe, as well as planned measures in management plans, revealed a number of limitations that might influence the effectiveness of the interventions:

- Morphological measures are very common in all river types, typically implemented as engineered solutions creating relatively static habitats at small scales with little or no interplay with geomorphic processes.
- Hydrological measures are mostly applied only locally without solving issues related to larger scale catchment-wide activities.
- Most restoration projects do not tackle the underlying processes of natural flow and sediment dynamics that acts out on larger scales.
- Even moderate water pollution and fine sediment input, as well as missing source populations for colonisation and dispersal barriers, may limit restoration effects and these catchment-scale constrains to restoration success are generally not considered.

4. EFFECTS OF RESTORATION

An increasing number of primary research studies have reported results on effects of river restoration which have been summarised in several narrative reviews (Palmer et al., 2010; Roni et al., 2002, 2008), semiquantitative reviews using vote counting (Palmer et al., 2014) and quantitative metaanalyses (Kail et al., 2015; Miller et al., 2010; Stewart et al., 2009; Whiteway et al., 2010). Moreover, there is a growing number of studies on multiple restoration projects which also allow one to draw more general conclusions, either on single organism groups (Jähnig et al., 2010; Lepori et al., 2005; Lorenz et al., 2012; Pretty et al., 2003; Schmutz et al., 2014) or comparing effects on different organism groups (Haase et al., 2013; Jähnig et al., 2009; Januschke et al., 2009). However, a global metaanalysis on restoration effects was missing as well as a comprehensive, harmonised study considering a broad range of response variables and factors affecting restoration outcomes to get an overall picture on restoration effects and influencing factors (including hydromorphology, different organism groups, as well as food web composition and aquatic land interactions as response variables; and catchment and project characteristics—especially project extent—as influencing factors; 20 case study projects). Here we summarise the results of these studies that were conducted in the REFORM project and place them in the broader context of restoration literature as a basis to derive recommendations for future directions in river restoration.

4.1 General Effect: Does River Restoration Work in General?

Studies reported contrasting and highly variable effects of restoration but there is nevertheless evidence for an overall positive outcome of river restoration. Several studies showed that the ecological effect has been small even if local river morphology and habitat conditions have substantially improved (Jähnig et al., 2010; Lepori et al., 2005; Palmer et al., 2010). In contrast, other studies found a significant positive effect of river restoration on specific organism groups (Lorenz et al., 2012; Schmutz et al., 2014). The few narrative reviews that compiled information on a larger number of restoration projects also found highly variable restoration effects (Roni et al., 2002, 2008). Similarly, variability of restoration effects was high in the few quantitative metaanalyses and a substantial part of the projects showed no effect or even a negative effect, but in general, the overall effect of restoration projects on macroinvertebrates and fish were positive in terms of abundance and/or biodiversity (Miller et al., 2010; Whiteway et al., 2010). These results were supported by the global metaanalysis conducted in REFORM, which found a high variability but an overall positive effect on macroinvertebrates, fish and especially macrophytes (Kail et al., 2015, Fig. 14).

The high variability is likely to reflect real differences in effectiveness of restoration measures applied, as well as context-derived factors such as catchment, river and project characteristics, which either enhance or constrain restoration effect. For example, Kail et al. (2015) reported that nearly half of the variance in restoration effect could be related to differences in project characteristics such as time since the restoration intervention, catchment land use, river size and type, organism groups studied and the biological metric considered as well as the restoration measures applied. The substantial unexplained variance might be partly due to missing information on factors enhancing or constraining restoration effect (Roni et al., 2008) but can also be attributed to noise caused by large methodological differences in respect to monitoring design, field sampling and data analysis. The high variability of restoration effects and our inability to predict restoration outcomes stresses a need for postproject appraisals and adaptive management approaches, which has long been recognised (Downs and Kondolf, 2002; Friberg et al., 1994) but still rarely applied in practise (Williams and Brown, 2014).

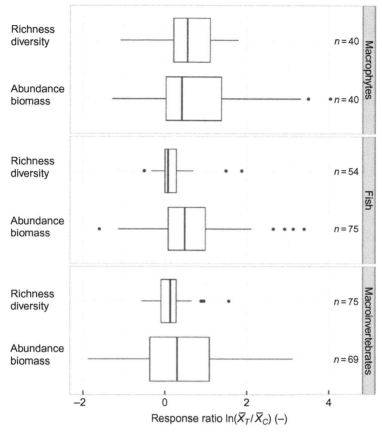

Fig. 14 Overall positive but highly variable effect of restoration on the richness/diversity and abundance/biomass of macrophytes, fish and macroinvertebrates, as reflected by the response ratio of Osenberg et al. (1997) which relates the value of the restored or treatment section X_T to a degraded control section X_C; all mean values were significantly different from zero (t-test, $p < 0.01$). From Kail, J., Brabec, K., Poppe, M., Januschke, K., 2015. *The effect of river restoration on fish, macroinvertebrates and aquatic macrophytes: a meta-analysis. Ecol. Indic. 58, 311–321.*

4.2 Differences in Responses Among Organism Groups and Species Traits: Which Benefit Most?

Not all organism groups benefit from restoration to the same extent. For example, studies reported a significant positive effect on floodplain vegetation (Januschke et al., 2011) and macrophytes (Lorenz et al., 2012), a relatively small effect on fish (Schmutz et al., 2014), and a low or missing effect on macroinvertebrate richness and diversity (Friberg et al., 2013;

Jähnig et al., 2010; Palmer et al., 2010). Comparative studies on several organism groups indicated that in general, restoration effect is the highest for terrestrial and semiaquatic groups such as floodplain vegetation and ground beetles, intermediate for macrophytes, lower for fish and the lowest for macroinvertebrates (Haase et al., 2013; Jähnig et al., 2009; Januschke et al., 2009; Kail et al., 2015). Similarly, the effect of restoration significantly differed between the response variables investigated in REFORM as reflected by the Bray–Curtis dissimilarities of the restored and nearby degraded sections (Hering et al., 2015). Moreover, although the effect on richness did not significantly differ between organism groups (one-way ANOVA, $F_{4/89} = 2.082$, $p = 0.09$), there was a tendency for higher effects on terrestrial and semiaquatic compared to aquatic organism groups (Fig. 15).

In the 20 projects investigated in REFORM, restoration had no or only a small effect on species richness or diversity of macroinvertebrates (Verdonschot et al., 2015) and fish (Schmutz et al., 2016), while restoration had a clear positive effect on richness or diversity of organism groups inhabiting river banks or adjacent shallow shoreline habitats, namely ground

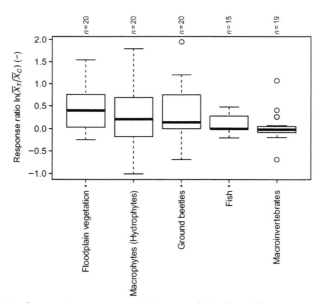

Fig. 15 Effect of restoration on species richness of the five different organism groups investigated in REFORM, as reflected by the response ratio of Osenberg et al. (1997) which relates the value of the restored or treatment section X_T to a degraded control section X_C; mean values significantly different from zero are marked with an asterisk (t-test, $p < 0.05$).

beetles (Januschke and Verdonschot, 2016) and macrophytes (Ecke et al., 2016). While Göthe et al. (2016) found no differences in floodplain vegetation diversity between the 20 restored and unrestored sections, restoration had a significant effect on pure richness (Fig. 15), similar to other studies reporting a significant higher richness in restored compared to degraded sections (Jähnig et al., 2009; Januschke et al., 2011). The different effects on different organism groups stress the need to monitor and assess the effect of river restoration on biota in a holistic way, including semiterrestrial and terrestrial organism groups. Terrestrial (floodplain) and aquatic ecosystems are closely linked and cannot be considered and assessed separately, which is clearly substantiated by findings in the REFORM project.

Within single organism groups, restoration generally has a higher effect on the abundance than on the number of species (richness) with Miller et al. (2010) and Kail et al. (2015) finding larger effects of restoration interventions in terms of increased abundance of macroinvertebrates and fish compared with changes in the number of taxa. This is not surprising as taxa that are already present prior to an intervention can respond rapidly with increased population densities to an improvement in quantitative (i.e. more area created) and qualitative (i.e. more variable living conditions) habitat features that allow coexistence of more individuals. In contrast, dispersal of new taxa into a project area is depending on their presence in the system and number of deterministic and stochastic elements (e.g. Leibold et al., 2004). Higher abundance might reflect increase in reproduction and/or the restored reaches attracting individuals from adjacent areas, the latter being especially true for mobile organism groups like fish (e.g. Hughes, 2007). Large-scale pressures like nutrient and fine sediment input suppressing regional to local biodiversity and absence of source populations nearby will limit the possibility of the introduction of new species into a restoration project area.

Our analysis revealed, however, that specific traits are a more sensitive indicator than both abundance and species richness. An alternative method to evaluate the effects of restoration based on species identity (taxonomy) is the use of species traits, an approach recently advocated in relation to biomonitoring (Beche et al., 2006; Charvet et al., 2000; Costello et al., 2015; Demars et al., 2012; Mcgill et al., 2006; Statzner et al., 1997). Traits are inherited characteristics of species that relate to their biology and the environment they live in and provide a mechanistic understanding of how environmental change, such as a restoration intervention, influences biotic communities through environmental filtering (Dolédec et al., 1999; Menezes et al., 2010; Rabeni et al., 2005). For example, richness especially increased for ground beetle species inhabiting sparsely vegetated river banks (Januschke and Verdonschot, 2016), macrophyte richness did only increase for helophytes (Ecke et al., 2016), abundance of small rheophilic fish increased but not for other flow traits (Schmutz et al., 2016), and effects on floodplain vegetation were the highest for metrics describing community structure like the increase of therophytes and annual floodplain vegetation species (Göthe et al., 2016). Moreover, while there was no effect on macroinvertebrate richness, there was an increase of food source diversity as indicated by stable isotopes (Kupilas et al., 2016). These changes in community structure potentially indicate specific functional changes caused by river restoration and should be used in future to increase our understanding how restoration measures affect aquatic ecosystems as suggested for biomonitoring in general (Menezes et al., 2010).

4.3 Differences Between Restoration Measures: What to Do?

Effects of restoration on different organism groups and traits will also depend on the specific measures applied. Comparing effects of different restoration measures in the 20 REFORM case studies indicated that widening (removing bed and bank fixation, flattening of river banks and considerably widening the cross section, and hence increasing the wetted area) is one of the most effective restoration measures, especially for terrestrial and semiterrestrial organism groups (Fig. 16). The effect of widening was more pronounced compared to other restoration measures on ground beetle richness (Januschke and Verdonschot, 2016) and specific macrophyte traits (Ecke et al., 2016). Moreover, the global metaanalysis of peer-reviewed literature conducted in REFORM indicated that the effect of widening is significantly higher on macrophytes compared to fish and macroinvertebrates, but there is also some evidence that the effect of widening on macrophytes may vanish over time (Kail et al., 2015). In contrast, widening had no effect on macroinvertebrates (Verdonschot et al., 2015), which might be due to the fact that—although mesohabitat diversity was significantly increased— widening often failed to increase microhabitat/substrate diversity relevant for macroinvertebrates (Poppe et al., 2015). The high effect of widening on ground beetles and macrophytes might also explain the overall higher effect of restoration on terrestrial and semiaquatic organism groups because these studies mainly investigated widening/rebraiding and remeandering projects. In contrast to planform measures like widening, pure in-stream measures are primarily beneficial for aquatic species with evidence of

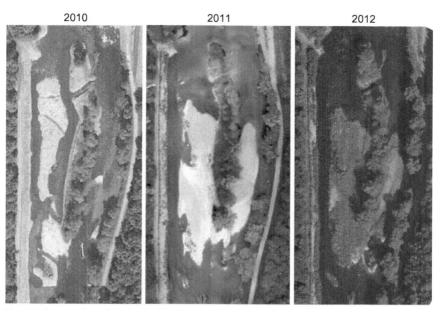

Fig. 16 Succession after river widening in a restored section of the river Ruhr (Germany), restored in 2009.

increasing macroinvertebrate and fish richness/diversity (Kail et al., 2015). In accordance, Miller et al. (2010) reported a significant positive effect of typical in-stream measures (large wood and boulder placement) on macroinvertebrate richness.

These differences between restoration measures and organism groups are intuitively meaningful since widening and other planform measures like remeandering reduce flow velocity and create pioneer habitats like bare riparian areas and bare gravel bars that are often sparsely shaded in the beginning, and hence, favour pioneer species in the riparian area and macrophytes in the aquatic zone in the first years. It is only on a longer time scale that natural processes, if these are given room to evolve and substrate can be mobilised, create more diverse aquatic habitats. However, if natural morphodynamic processes are not restored, aquatic habitat quality will not increase and the lack of disturbances to rejuvenate the pioneer habitats will lead to a succession to less favourable conditions for terrestrial and semi-aquatic organism groups as the riparian vegetation develops and shading increases.

This is consistent with the widely held assumption that restoring natural geomorphological processes at relevant (larger) scales, by removing bed and bank fixation and widening, has a higher effect on hydromorphology and biota compared to other nonprocess-based measures like gravel addition into shorter reaches in physically perturbed streams. Therefore, it is crucial to restore a natural flow and sediment regime (i.e. natural processes) and to restore habitat diversity at relevant spatial scales (including microhabitats) to ensure long-term positive effects on all organism groups.

4.4 Confounding Factors: Why Do Some Restoration Projects Fail?

Confounding factors potentially constraining the effect restoration include (i) characteristics of the typically reach-scale projects like the type of measures applied, as well as project size and age and (ii) factors at larger spatial scales like pressures related to catchment land use and missing source populations for recolonisation.

Project size in terms of river length restored might be of importance since longer reaches might mitigate the influence of large-scale pressures like fine sediment input and provide a minimum area for geomorphological processes to evolve and viable populations to establish, and hence, larger effects of restoration should be expected with increasing extent of projects. However, in the 20 REFORM case studies, the effect of restoration did not differ significantly between large and small restoration projects, although there was a tendency for large restoration projects being more successful (Kail et al., 2014). Moreover, the global metaanalysis of peer-reviewed literature conducted in REFORM revealed that restoration had positive effects even though projects were small and effects did not increase with restored reach length (Kail et al., 2015). The restoration projects investigated in REFORM were probably still too small to detect an effect of project size (most case study projects <2 and <2.6 km in the metaanalysis). Other studies including larger projects indeed found that restoration had higher effects in larger projects (e.g. Schmutz et al., 2014). These contrasting results indicate that even small restoration projects can have a positive effect on some organism groups. Slightly larger projects do not necessarily have larger effects (Stoll et al., 2016). Most probably, restoration projects implemented in the past were simply too small to benefit from possible positive mitigating effects of project size and large projects are needed for extensive effects.

Project age might be of importance due to the time needed for natural channel dynamics to create higher habitat diversity and the lag between habitat creation, colonisation and the establishment of vital populations. However, in the 20 REFORM case studies, project age (time between implementation of the measures and monitoring) only had positive effects on aquatic habitat conditions (Poppe et al., 2015) but not on any of the organism groups investigated. Most projects investigated were just implemented 1–16 years prior to monitoring, which was probably less than what is needed for full community recovery (Hering et al., 2015). In contrast, project age was identified as the most important variable affecting restoration success in the metaanalysis of REFORM (Kail et al., 2015). However, the effect of restoration did not simply increase with time but project age had nonlinear and even negative effects on restoration outcome (e.g. on macrophytes). Other studies also report contrasting results on the effect of project age on restoration outcomes (no effect on macrophytes in Lorenz et al. (2012) and invertebrates in Leps et al. (2015), small positive effect on fish in Haase et al. (2013) but different effects in Schmutz et al. (2014) and nonlinear effects in Whiteway et al. (2010)). These results indicate that the effect of restoration does not simply increase over time but changes nonlinearly and might even decrease.

Besides these project characteristics, the effect of reach-scale river restoration is potentially constrained by large-scale pressures related to catchment land use such as water pollution, high nutrient and fine sediment loads, storm water peak flows or a depleted species pool in the catchment resulting in missing source populations for recolonisation. Projects are prone to failure if large-scale pressures are not adequately considered (Bond and Lake, 2003; Miller et al., 2010; Palmer et al., 2010; Roni et al., 2008). In contrast to the numerous studies on the effect of catchment land use on the biological state (Roth et al., 1996; Stephenson and Morin, 2009; Sundermann et al., 2013), there is limited knowledge if and how the pressures related to catchment land use affect restoration outcomes (but see Miller et al. (2010), who found that projects implemented in forested regions tended to have a higher effect on macroinvertebrate richness and abundance). In the REFORM metaanalysis on peer-reviewed literature, agricultural land use was among the three most important factors affecting restoration outcomes for macrophytes, fish and macroinvertebrates, and restoration effect on fish abundance and biomass was significantly negatively related to agricultural land use (Kail et al., 2015). In contrast, there was no clear negative effect of agricultural land use on restoration outcomes in the 20 restoration projects investigated in REFORM. Possibly the reason for this missing clear constraining effect was due to the relatively high share of agricultural land use in the catchments investigated; even the less intensively used catchments might have already been above a critical threshold limiting biota, with any further increase having only a minor impact. Furthermore, at least for macroinvertebrates richness and diversity, the effect of restoration did not depend on the presence of reaches in the vicinity of the restoration projects being in a high or GES, which was used as a proxy for the species pool available for recolonisation. However, effects on macroinvertebrates were nonsignificant anyhow, resulting in a short gradient in the dataset. Therefore, the REFORM results do not question the findings of other studies indicating that restoration effects on macroinvertebrates and fish are limited by the depleted species pool and sparse source populations for recolonisation (Stoll et al., 2013; Tonkin et al., 2014; Winking et al., 2014). This topic clearly merits further investigation since a limited recolonisation potential would need a completely different restoration strategy compared to habitat improvements or pressures related to catchment land use. As a first rule of thumb, source populations for fish and macroinvertebrates should be located less than 5 and 1 km upstream from the restored reach, respectively (Stoll et al., 2013; Tonkin et al., 2014).

In summary, results of the REFORM project, placed in the context of existing restoration literature, indicate that river restoration (i) generally has a positive but highly variable effect, the highest effects are on (ii) terrestrial and semiaquatic organism groups in widening projects, (iii) on aquatic organism groups if in-stream measures are applied, (iv) on species abundance rather than richness and (v) on specific traits rather than mere species number or total abundance. Moreover, results indicate that restoration projects are prone to failure if (i) large-scale pressures related to catchment land use are not adequately considered, (ii) source populations are missing for the recolonisation of the restored habitats, (iii) relevant and limiting habitats have not been restored (e.g. microhabitats for macroinvertebrates) and (iv) related processes like a natural flow regime and sediment transport have not been restored to rejuvenate the habitats. Finally, long-term monitoring is needed to better understand the trajectories of change induced by restoration measures, and to identify sustainable measures which enhance biota in the long-term (Box 2).

5. FUTURE DIRECTIONS
5.1 Future River Restoration Needs Better Planning

The key to improving the effectiveness of future restoration projects is the use of comprehensive and consistent planning, implementation and appraisal techniques. Today, the state of knowledge is primarily based on experiences

BOX 2 Example of River Restoration Project North of the Polar Circle: Unusual Place but Typical in Term of the Trial-and-Error Approach

Lake Børsvann (68.30825N, 16.72100E) was regulated in 1914, and the water was directed away from the river Børselva, reducing the catchment from 85 to 5.5 km^2. By the end of the last century, low discharge and agricultural activities had made the river highly eutrophic, filled with fine sediments and overgrown with plants (Fig. B.1). The water course was protected as a nature reserve in 1997 due to its importance as a feeding and resting area for migratory water birds. A substantial number of different water birds used to nest in the river and lake system, some of them red listed, but there was a drastic reduction in the number and species present during the last years before restoration was initiated. The river was also earlier the main spawning and recruiting area for the lakes downstream, and had a valuable population of arctic char (*Salvelinus alpinus*) and trout (*Salmo trutta*). Fishing with nets and electro fishing showed no fish in 1998 and 1999.

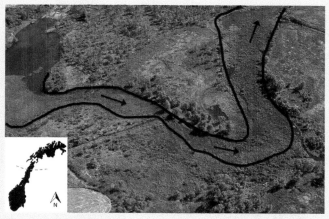

Fig. B.1 The river Børselva with water plants such as *Equisetum fluviatile* and *Carex* spp. covering nearly 70% of the river bed. *Bottom left*: map of Norway showing Børselva (*white cross*) and the arctic circle (*dotted line*).

A 10-year restoration project was initiated in 1997, aiming at: (1) reducing the input of nutrients and fine material from the tributaries. (2) Physically open/alter the river course. (3) Design a new flow regime for the river. Morphology of tributaries was modified to a more natural channel cross section, as they had become very incised by dredging. The more gently sloping banks were covered with coconut mats to prevent erosion until the vegetation was established, and the stream

Continued

BOX 2 Example of River Restoration Project North of the Polar Circle: Unusual Place but Typical in Term of the Trial-and-Error Approach—cont'd

bed was reinforced with geotextiles covered by stones and cobbles to avoid further incision. Furthermore, in the downstream end, a sedimentation pool and two constructed wetlands were created to retain fine sediments and nutrients lost from the fields.

To reestablish the continuum in the system, vegetation had to be cleared to create a new river channel that was dimensioned for the actual discharge regime. Where pools and deeper parts of the river were dug out, an excavator with a long arm was used. The heavy machinery used carpets made for military transport to avoid damaging the natural wetlands (Fig. B.2).

Fig. B.2 Carpets made for military transport over wetlands were used for the excavators to avoid unrepairable damage to the environment, and to create safe passageway for the heavy machinery.

Macrophytes were manually cut, and geotextiles covered with stones were added after clearing, reducing regrowth. During winter, the geotextile and stones were set out on the ice by trucks (Fig. B.3), reducing the need for manpower. This was an easy way of designing bends, side arms and width variation. By varying the thickness of the stone layer and stone size, new habitats for fish and benthic fauna were created. In the end, a new channel was created, connecting the ponds and lakes of the river stretch (Fig. B.4).

BOX 2 Example of River Restoration Project North of the Polar Circle: Unusual Place but Typical in Term of the Trial-and-Error Approach—cont'd

Fig. B.3 Designing a new river with geotextiles and stones during ice cover in winter.

Fig. B.4 Børselva, Norway, post-restoration: a new corridor was created, connecting the different lakes and ponds.

When the new river channel was cleared, the hydrology was restored by changing water release from Lake Børsvatn. This included a minimum residual water discharge and a set number and size of artificial floods in the river Børselva. However, the new water regime was not adjusted to the recreated morphological conditions and caused bank erosion and sedimentation of sand and silt on the newly created geotextile stone habitats. After this, nearly no monitoring or following-up was conducted, and as a result, 2 decades after initiation, the inlet streams were more or less back to prerestoration conditions. Most of the geotextile covered reaches were still open, but there were no published results on substrate structure, flora or fauna in these channels. Furthermore, the river had by itself created new channels, mostly outside the artificial channels.

compiled from a multitude of concepts, tools and techniques used in river restoration over recent decades (Cowx and Welcomme, 1998; Roni and Beechie, 2013). We argue that the way in which knowledge on restoration is available and organised today provides very little concrete guidance to managers wishing to conduct cost-effective projects that deliver set targets. As a part of the REFORM project, we developed an easily applicable project planning framework to support implementation of legislation on surface water by restoration, and to integrate social and economic aspects into decision making. It follows the structure of the well-known project cycle but has been enhanced to assist river restoration in the setting of regional and national objectives by incorporating existing planning strategies (Fig. 17). The framework systematically guides practitioners through two main planning stages of river restoration, from (1) catchment scale to a more (2) project-specific scale, enabling users to put project-specific restoration into a river basin context. It provides detailed information for each of the planning stages and offers tools and guidelines (e.g. Plan–Do–Check–Act (PDCA), Driver–Pressure–State–Impact–Response (DPSIR) (Angelopoulos et al., 2015), Logical Framework (Cowx et al., 2013),

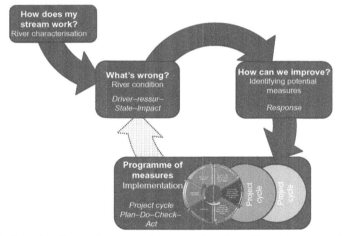

Fig. 17 Project planning cycle at a catchment scale using a five step approach starting at the top left text box: (1) *River characterisation*—at a catchment scale to identify river styles and understand their processes. (2) *River status*—understand the current condition of the aquatic biota or biological quality elements. (3) *River restoration potential*—to understand the level of restoration a river can reach. (4) *Project identification*—to identify specific restoration projects at a reach scale and identify suitable rehabilitation measures and project objectives. (5) *The project cycle*—planning, formulation and implement of projects at a local scale.

Specific Measurable Achievable Realistic Timely (SMART) (Hammond et al., 2011), Before–After Control–Impact (BACI) monitoring, Multi-criteria Decision Analysis (MCDA) (Cowx et al., 2013) and Cost Benefit Analysis (CBA) (Brouwer and Pearce, 2005; Brouwer et al., 2009; Shamier et al., 2013)) to support users (Fig. 18), some of which were developed during the REFORM project. We recommend adopting this integrated project planning framework for river restoration to reduce the uncertainty of management actions by providing five key components:

1. A project planning cycle at a catchment scale, guiding the user through a logical path to design projects linking policy, watershed/catchment assessment, restoration goals, monitoring and evaluation schemes, selection and prioritisation.
2. Concise structured information for each stage of the project cycle, upscaling to river basin to select appropriate restoration measures.
3. Concise structured information for each stage of the project cycle at a project-specific scale and to identify specific measures.
4. Easy access to relevant tools and guidelines that can be used at different stages of the planning process.
5. The choice of more detailed information where needed, giving the option of a more complex planning framework.

These five key components can reduce the uncertainty of management actions and are incorporated in the integrated project planning framework.

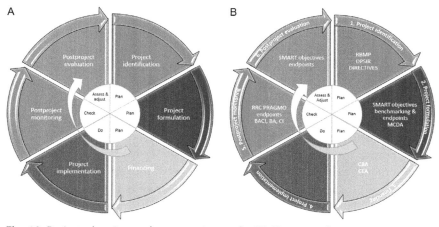

Fig. 18 Project planning cycle at a project scale. (A) Six stages for restoration project planning: (1) project formulation; (2) financing; (3) project implementation; (4) postproject monitoring; (5) postproject evaluation and (6) adjustment or maintenance. (B) The tools needed at each stage.

To aid stakeholders, all information required for the implementation of the framework can be accessed via the interactive REFORM wiki (wiki.reformrivers.eu).

5.2 Project Planning at a Catchment Scale: A Necessity

Applying integrated project planning at a catchment scale takes account of physical, chemical and biological aspects of broad-scale processes of rivers and interfaces between connecting ecosystems, such as the natural habitat continuum from upstream catchments to downstream areas and between the river and its surrounding land use. This ensures river restoration objectives are set to improve ecological status at a river basin level while being defined by institutional, regional and national policies. Subsequent decisions for smaller, local-scale river restoration will still benefit on a larger catchment scale. Here we have categorised project planning at a catchment scale into five main steps composed of *characterising the river* at the catchments scale, assessing its *status* in terms of ecological condition, the *potential* for restoration, project *identification* and implementing the *project cycle* (Fig. 17).

5.3 Exploring the Full Potential of River Restoration

Restoration potential has to be considered across the catchment scale to understand the level of ecological improvement that can be attained. Thus, the current status should be evaluated, including all sector activities that have adverse impacts on hydromorphological characteristics necessary to achieve or support GES, especially as they may further limit the level of restoration that can be achieved. Restoration potential can be assessed when all regional and national policy objectives are considered and constraints in meeting policy objectives are identified. Water body goals and specific objectives identified using all directives/legislations both for and against ecological conservation and finally the current status is compared with objectives to identify constraints.

Synergies between cross-sectoral river ecosystem services and ecological requirements maximise multiple benefits between sectors and enhancing opportunities for restoration of rivers in resource limiting situations (Jackson et al., 2016). Nevertheless, technical barriers to integrated cross-sectoral planning are substantial, related partly to limited data and poor understanding of drivers of each sector, as well as indirect effects of actions (Adams et al., 2014). Thus effective management requires the collaboration between disciplines (e.g. hydrologist, fluvial gemomorphologist, ecologist, economist, sociologist and engineering) and improves the interaction with policy makers and the local stakeholder community to distinguish between the social, economic and environmental requirements of the foreseen project (Letcher and Giupponi, 2005). Understanding these links between ecosystem services and ecological outcomes enables successful integrated planning to produce multiple benefits.

Approaches to achieve multiple benefits are now emerging in river restoration and cross-sectoral interactions, and are supported by various policy documents, for example, synergies between flood risk and river water management or between hydropower development and restoration of longitudinal connectivity for migratory fish (CIS, 2007). The project planning cycle identifies key relationships between society and the environment at both a catchment and project scale to identify and formulate options for synergistic restoration. Here multibenefits can be gained by linking the ecosystem approach, ecosystem services and societal benefits that come from these services to further identify restoration potential and aid decisions for restoration measures.

Agricultural and urban land use, flood protection, inland navigation and hydropower support the key demands of society such as food, water, energy, transport and space to live (Angelopoulos et al., 2015) and these are responsible for pressures that cause biological and abiotic state changes (physical and chemical) and further impacts within the river system. Invasive species and climate change are indirect pressures that can also cause changes in river state and combine with pressures from human activities, intensify impacts on the ecosystem. Disentangling these knock-on effects and identifying mitigation response to the impacts on ecosystem services and ecosystem function through the application of river restoration prevents and improves state changes in the environment.

5.4 Project Identification

For the project identification stage clear objectives are set to improve the health of river ecosystems. Chosen restoration measures are identified at a basin level necessary to meet these environmental objectives. Restoration measures are reviewed and made operational through a number of regulators, therefore, it is important that the planning leading up to the selection of restoration measures is clear and developed in association with regulators and other stakeholders. Catchment-scale planning provides information on river characterisation, river condition and restoration potential, all of which underpin the decisions made to select restoration measures as part of the

PoMs. The end result of project identification is to locate and prioritise reach-scale restoration projects, some of which will combine several rehabilitation measures, in an attempt to ensure smaller scale projects work towards a catchment approach. There is however, no universally accepted approach for prioritising rehabilitation actions and habitat protection (Johnson et al., 2003).

5.5 Project Planning at a Local Scale

A well-designed adaptive planning tool reduces the uncertainty of management actions and establishes the purpose for river restoration to ensure project objectives are set to improve ecological status at local scale, while keeping the project in a river basin/catchment context. The project cycle proposed here is a development of the existing project cycle but includes more detailed planning phases that systematically play a key role when planning river restoration projects (Fig. 18A). The framework has been designed to be transferable to individual restoration projects by drawing on commonalities. Within each stage there are several logical steps to be followed and the option of practical tools and guidelines that can be applied at each specific stage (Fig. 18). A feedback loop provides managers with the ability to account for uncertainty through evaluation of outcomes, and facilitate improved understanding of the efficacy of rehabilitation measures (Fig. 18).

5.6 Project Formulation

In the project identification phase, the project planners should be concerned with the suitability and feasibility of the project, in the formulation phase the emphasis shifts to the acceptability of the project and the desired outcomes. Suitable restoration 'goals' and 'objectives' establish an acceptable state for the system to be restored to, ultimately leading to a self-sustaining river ecosystem (Cowx, 1994; England et al., 2007; Kondolf et al., 2006). However, we still have the recurring limitation to identify restoration success or failure and this is attributable to the absence of adequate aims and objectives besides our limited understanding how ecosystems respond to interventions. Setting benchmarks and endpoints that are linked to clearly defined project goals are a valuable method to help determine the measure of success within river rehabilitation because they provide realistic, quantifiable criteria (Anderson et al., 2005; Buijse et al., 2005). Benchmarks are measurable targets for restoring degraded sections of river within the same river or catchment as representative sites with similar characteristics that have the required ecological status and are relatively undisturbed. Thus attempts to create conditions unrelated to the original ones at the site of interest are avoided and consequently restoration is more likely to result in long-term success (Choi, 2004; Palmer et al., 2004; Suding et al., 2004; Woolsey et al., 2007). Setting benchmarks draws on the assessment of catchment status and identifies restoration needs before selecting appropriate restoration actions to address those needs. Endpoints are target levels of restoration, whether this is an ecological (to restore a level of function/species), social (delivery of services to society) or physical–chemical (hydrology, hydromorphology and water quality) endpoint and are usually linked closely to project objectives. It is important to recognise what is the minimum acceptable achievable level of restoration and what is the desirable level to have as a target endpoint that is still below the benchmark level. Subsequently, what can be compromised for this desired level, will it be cost, ecosystem services or ecological aspects? Natural in-stream habitats consist of complex multidimensional arrays of hydrological and morphological conditions (hydraulic patterns, substrate composition and presence of woody debris) along with the complex life structures and habitat guilds of the biota (Statzner et al., 1988; Strange, 1999) and the environmental conditions (velocity, depth and temperature) and resources (food and space) on which they depend, all of which need to be incorporated in to river rehabilitation. As a result, this level of intricacy needed in river restoration practise is prevented from moving forward as we revisit the reoccurring problem of how to revitalise such a complex systems. The way to move forward is to identify project success of which benchmarking and endpoints will play a vital role in future catchment management.

It is imperative that endpoints accompany benchmarking in the planning process to guarantee the prospect of measuring success because endpoints are feasible targets for river restoration (Buijse et al., 2005). Given that benchmark conditions cannot always be achieved, especially on urban rivers, endpoints will assist in moving restoration effort towards benchmark standards through application of the SMART approach to decide what is specific, measurable, achievable, feasible and realistic in a specific timeframe. There is a need to distinguish endpoints for individual measures, combination of measures, catchment water body measures and river basin district measures. There is thus a need to consider not only the procedures for defining benchmarking and endpoints at the project level but also to integrate the outcomes.

5.7 Monitoring, Evaluation and Project Success

While there is a steady increase of restoration projects each year, the absence of adequate monitoring and evaluation is more frequently a consequence of lack of resources than unwillingness, and this constrains the ability to assess the effectiveness of restoration techniques (FAO, 2008). Monitoring and evaluation plays a key role within the planning framework because it enables identification of river restoration project success by assessing results (outcomes) against objectives (Hammond et al., 2011). Without such analysis it is difficult to assess to what extent the restoration is successful (Possingham, 2012). It is a vital stage in adaptive management as it influences the decisions made to continue, modify or discontinue management actions (Bash and Ryan, 2002). Although the need for monitoring has been acknowledged in recent years (Roni and Beechie, 2013), the majority of river rehabilitation schemes fail to assess outcomes and effectiveness (Cowx et al., 2013). A variety of monitoring techniques are available for detecting environmental impacts of restoration projects whose data collection methods differ spatially and temporally. These monitoring assessment techniques include before/after (BA) contrasts at a single site, BACI sampling sites and repeated BACI and posttreatment design and are well documented in the literature (Ellis and Schneider, 1997; Roni and Beechie, 2013; Sedgwick, 2006). The evaluation phase, for a rehabilitation project which has undergone the initial stages of the project approach, assesses the overall project effects (intentional and unintentional) and the sectoral impact of the project. Evaluation is only possible where a series of measurable indicators or endpoints has been established for the project. The evaluation phase will use measurable indications (in Europe usually WFD compliant BQEs even when not a suitable option as described earlier) to gauge how far the restoration project has developed in relation to the initial objectives and defined endpoints. It should be noted these monitoring tools do not include any CBA, which should be carried out separately.

5.8 Adjustment and Maintenance

Projects may require adjustment when evaluation has demonstrated that the objectives are not reached, i.e. adaptive management. Many regulated rivers have lost much of their natural dynamics lacking rejuvenation of habitats and quite often it is such habitats that improvement projects aim to restore. The altered and regulated conditions, however, may cause such projects to deteriorate and lose their effectiveness, e.g. blocked fish passages by debris or vegetation, pioneer habitats such a bare gravel bar where vegetation settlement has reduced its functionality. They then require maintenance or interventions and to be reset to the initial stage.

5.9 The Future: Holistic and Process-Oriented Restoration

Process-oriented restoration focuses on restoring critical drivers such as hydrological regime and river functions. Process-oriented actions will help to avoid common pitfalls of engineered solutions, such as the creation of localised habitats that cannot be sustained by natural processes (Beechie et al., 2010; Palmer et al., 2014; Palmer and Ruhl, 2015). Restoring natural processes in longer river reaches by letting erosion–sedimentation processes occur by removal of bed and bank fixation; by reprofiling the channel and by reinstalling free water flow has a higher effect on recovery compared to local-scale interventions, such as wood or gravel addition. Overall, the need for a more holistic and process-oriented approach can be condensed into four basic principles in future to make restoration more efficient (Table 5).

Natural flow regimes are the most important driver when restoring rivers (Lorenz et al., 2016) and the success of habitat restoration is likely to be high in rivers with natural flow dynamics, whereas there is a greater likelihood of failure when impaired flow dynamics and processes at larger scales are not addressed (e.g. Jähnig et al., 2010; Palmer et al., 2010). Hydrological measures should therefore focus (1) on groundwater balances and flows at catchment level and (2) on mapping catchment-scale hydrological surface water infrastructure and its functioning. Restoration should involve upscaling of current hydrological measures to reduce discharge dynamics and increase water retention (Richardson et al., 2011; Fig. 19). Other relevant processes such as vegetation encroachment and sediment entrainment are closely linked to water flow.

Holistic measures at the catchment scale and floodplain will have significant hydrological effects also at smaller scales such as individual river reaches. Furthermore, uninterrupted flow and thus lateral connectivity provide continuous potential of exchange of water, substances and propagules. The catchment level is also the appropriate scale to tackle other pressures such as nutrient, organic and toxic load in concert with the hydrological restoration. However, catchment-scale reductions in nutrient emissions are unrealistic in most cases and local impacts of nutrients, organic and toxic substances and sediments can be mitigated at the reach scale by introducing wider or smaller riparian buffers (Figs 20 and 21). There is clear and, in many cases, strong evidence for the role of wooded riparian buffers in controlling nutrient and sediment retention (Hines and Hershey, 2011), water temperature (Kristensen et al., 2013) and improving in-channel habitat structure (Kail et al., 2007).

Small-scale restoration measures should not be ignored in future and profile adaptations at local scales will be necessary in many cases, in particular where stream power is low and there is not sufficient energy to change modified channel profiles back to something more natural (Kondolf et al., 2001;

Table 5 Four Basic Principles to Make Restoration More Efficient in Future

Principle	What to Consider
Target the root causes of river ecosystem change and do this at different scales	Restoration actions that target root causes of degradation rely on knowledge of (1) the processes that drive river ecosystem conditions and (2) effects of human-induced alterations onto those driving processes. This implies that restoration of natural processes with the natural, or near-natural, hydrological, hydromorphological and chemical conditions will have the highest success. In short it means a call for large scale, longer time process-oriented restoration
Tailor restoration measures to the river ecosystem potential in a hierarchical manner	Each river ecosystem is part of a large catchment and the river itself depends strongly on the range of channel and riparian conditions. Both catchment and riparian valley should be or become the logical outcome of the physiographic and climatic setting. Furthermore, understanding the processes controlling restoration outcomes helps to design restoration measures that redirect river valley, river channel and river habitat conditions
Match the scale of restoration to the scale of the problem	When disrupted processes causing degradation are at the reach scale (e.g. channel modification, levees, removal of riparian vegetation), restoration actions at individual reaches can effectively address root causes. When causes of degradation are at the catchment scale (e.g. increased runoff due to impervious surfaces, increased eutrophication), restoration actions need to be taken at catchment scale to restore the root causes
Be explicit about expected outcomes	Process-oriented restoration is a long-term endeavour, and there are often long lag times between implementation and recovery. Ecosystem features will also continuously change through natural dynamics, and biota may not improve dramatically with any single individual action. Hence, quantifying the restoration outcome is critical to setting appropriate expectations for river restoration

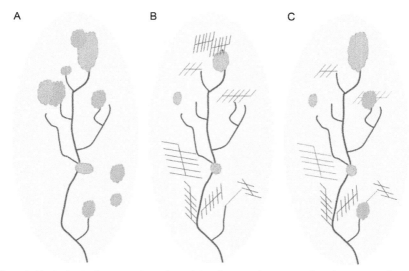

Fig. 19 Hydrological restoration of combined ground water and surface water flows by restoring infiltration capacity and recreating water storage areas at the catchment scale. The image (A) represents a natural catchment with wetlands, bogs and mires that acts as sponges for the water and have a high storage capacity. (B) Catchment degraded by human intervention and a high drainage intensity whereby the natural water stores have been substantially reduced. (C) Restored catchment with water infiltration, reduced drainage intensity, water storage areas (*green* (*dark grey* in the print version)) and water flow retarding by remeandering.

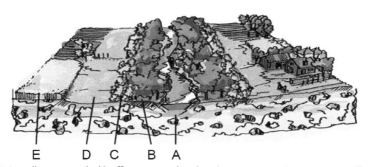

Fig. 20 Installing extended buffer zones at local scale as a restoration measure will control nutrient and sediment run off, cool water temperature and improve in-channel habitat structure (A: river, B: Tree-zone, C: Bush-zone, D: grass-zone, E: adjacent land).

Rinaldi et al., 2016). At local-scale morphological processes (e.g. sorting of bed material, creation of pools, bars and cut-banks) are generally the result of high flows in rich structured beds. By addition of wood or gravel habitat morphology can be improved. Channel incision can alternatively be

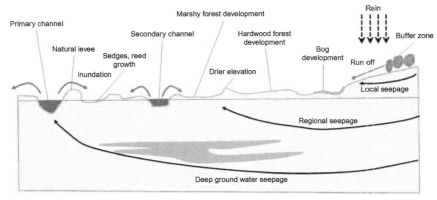

Fig. 21 River reach-scale processes for river floodplain development with natural ground water, rain water and surface water flows that provide the basis for morphological processes. Buffer zone and measures to reduce nutrient levels provide the basis for river restoration.

Fig. 22 The application of large woody debris in combination with the addition of sand in a bare sandy streambed in the Hierden stream (The Netherlands) (*photo left*) results in a strong increase in habitat heterogeneity (*photo right*). Photos: Ralf Verdonschot.

decreased or even reversed by placing a large number of naturally shaped logs randomly in the river. Logs in combination with sand addition to the river will heighten the river bed, increase the flow and flow variability and improve habitat heterogeneity (Fig. 22). Changing channel maintenance, such as dredging, removal of wood and weed cutting, from a negative disturbance to a positive measure is another way of improving local hydromorphological conditions and aid restoration. Soft renaturalisation solutions that are low maintenance and plants, depending on local diversity of growth forms and flow conditions, are bioengineering solutions that create more varied flow and habitat conditions at the river stretch scale (Gurnell, 2014; Iversen et al., 1993).

Table 6 Hierarchical Order of Restoration, Stressors vs Ecological Key Factors and Processes

	Stressors	Order of restoration	Temperature regime	Light regime	Flow regime	Substrate variation	Organic matter	Oxygen regime	Nutrients	Salinity	Toxicity	Connectivity
Catchment	Changed hydrology	1										
	Diffuse sources	2										
	Point sources	3										
Stretch	Current alterations	4										
	Channelisation	5										
	Bank degradation	6										
	Maintenance	7										
Site	Barriers	8										
	Habitat degradation	9										

To summarise, one can hierarchically order in nine steps the measures to restore rivers keeping both stress and key ecological processes into account (Table 6).

6. CONCLUSIONS

Results of the REFORM project, and in the context of existing restoration literature, clearly conclude that river restorations conducted until now has had highly variable effects with, on balance, more positives than negatives. The largest positive effects have interestingly been to terrestrial and semiaquatic organism groups in widening projects, while positive effects on truly aquatic organisms groups are only seen when in-stream measures are applied. The positive responses of biota are primarily to abundance of organisms with very little indication that overall biodiversity has increased: specific traits rather than mere species number or total abundance have benefited from restoration interventions. This modest success rate can partly be attributed to the fact that the catchment filter is largely ignored with large-scale pressures related to catchment land use not adequately considered or the lack of source populations for the recolonisation of the restored habitats. The key reason for this shortfall is a lack of clear objective setting and planning process. Furthermore, we suggest that there has been a focus on form rather than processes and functioning in river restorations to date, which has truncated the evolution of geomorphic features and any dynamic interaction with biota. Finally, monitoring of restoration outcomes is still rarely done and often using inadequate statistical designs and inappropriate biological methods which hamper our ability to detect changes.

We suggest that the first step in any future (river) restoration actions is to acknowledge that they are social constructs made by us to improve a wide variety of aspects relating to the riverine environment (Fig. 23). A range of socioeconomic drivers will determine why restoration is considered an option in the first place and what is feasible in the context of potentially conflicting interests. The next steps are then to implement the full restoration project cycle and optimise the restoration effort by using process-based techniques at appropriate scales to maximise ecosystem resilience and recovery.

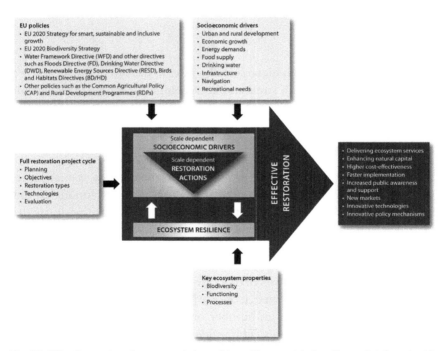

Fig. 23 Effective restoration can only be achieved by considering the societal context in which river restoration projects are set.

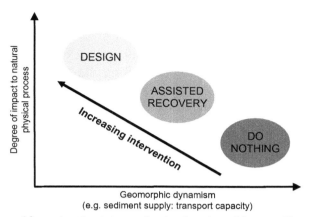

Fig. 24 The need for restoration intervention is a function of degree of impact and geomorphic dynamism of the system at relevant scales.

We advocate the use of a process-based restoration approach based on an initial hydrological and geomorphic assessment of the fluvial conditions across the relevant scales (Fig. 24). Many systems have the potential to recover if they are left to themselves. This is the case in high energy systems with natural geomorphic processes occurring and in rivers that are slightly physically impaired. In contrast, low energy systems with a lack of natural geomorphic processes will need active restoration designs, similar in many cases to what has been done in the past 3 decades. This can be the only option to create a more natural river environment in urban areas or rivers severely impacted by hydropower. In between is the assisted recovery which is exemplified in Box 3. These types of streams have moderate geomorphic dynamism as well as being moderately impacted and are ideal places for soft-engineering approaches involving large woody debris, riparian vegetation and in-stream macrophytes.

The most important catchment-scale stressors that potentially constrain the effects of intermediate and local restoration projects are hydrological changes, disturbed sediments, water pollution and oxygen depletion. The effect of local in-stream and planform measures can potentially be improved by (i) ensuring that catchment-scale pressures do not constrain the effects, (ii) restoring natural sediment dynamics, i.e. processes and (iii) the restored channel pattern corresponds to the channel planform which would develop naturally given the (altered) controls like discharge, sediment load and bank stability. In general, restoring processes should be favoured over restoring form since the risk for failure (created forms being destroyed by channel dynamics) is high.

Better planning acknowledging the social context of all river restoration projects is an essential component in setting clear and realistic objectives (Figs 23 and 25). In future, integration of (river) restoration into the relevant policy agendas could be a strong driver for developing projects and improve excepted outcomes in relation to existing legislation such as WFD and

BOX 3 The Allt Lorgy Restoration in Scotland: An Example of a Process-Based Restoration Approach

Fig. B.5 Assisted recovery by the introduction of wood in the Allt Lorgy changed channel configuration markedly over a 3-year period from a straight single channel to multiple channels with gravel bars.

The Allt Lorgy is a tributary of the River Dulnain in the Cairngorm Mountains, Scotland and part of the Spey catchment. It is a highly hydrologically responsive upland gravel-bed stream that is typical of this region but which was physically managed for agricultural purposes in the late 1980s. This involved significant straightening/canalisation, dredging of the active channel, the construction of artificial embankments (i.e. from the dredged material) and bank revetment/protection, together with in-channel boulder placement grade control structures. The project aimed to restore the physical and ecological conditions of a $\sim$1 km section of the river and its adjoining floodplain and represented a near unique application of the 'process restoration' philosophy (Beechie et al., 2010). This approach aims to reestablish the fundamental physical processes that drive the evolution of a diverse morphology, with the associated benefits for in-stream, riparian and floodplain ecology. Rather than designing a specific channel configuration, the implemented approach of 'assisted recovery' aimed to 'kick-start' natural processes to move the river towards a state of increased physical heterogeneity and improved channel–floodplain connection.

BOX 3 The Allt Lorgy Restoration in Scotland: An Example of a Process-Based Restoration Approach—cont'd

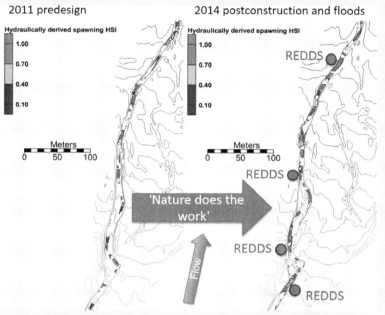

Fig. B.6 Detailed 2D hydraulic modelling of the pre- and postrestoration condition of the site showed a significant increase in hydraulic hetereogeneity. Habitat modelling demonstrated a significant increase in the area of conditions suitable for spawning Atlantic salmon which was subsequently validated by field observations of active REDDS (*filled dots* superimposed on the modelled output map).

The restoration was undertaken in Autumn 2012 and involved a ∼1 km section of the watercourse over an 11 ha site. Five stretches of artificial embankments were removed/lowered, liberating over 900 m³ of material. In-stream boulders were removed and replaced with large wood structures (whole trees with branches and root balls intact). Additional wood structures were introduced in locations identified as key to induce the desired morphological adjustment (e.g. at sites where some degree of meander development had already been initiated). A significant flood event occurred in August 2014, resulting in significant sediment transport, bank erosion and associated morphological adjustment of the channel corridor. A further minor phase of works was conducted in September 2015 with some additional gravel augmentation and minor channel reprofiling.

Continued

BOX 3 The Allt Lorgy Restoration in Scotland: An Example of a Process-Based Restoration Approach—cont'd

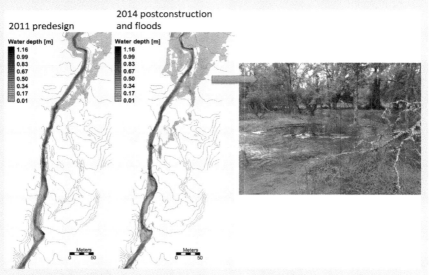

Fig. B.7 Modelling output showing that the restored Allt Lorgy is now better connected to its floodplain, particularly evident in the downstream section of the site.

The site has been extensively monitored to capture physical and ecological/biological evolutions, including repeated high resolution topographic surveys, aerial 'drone' photographic surveys, sediment sampling, REDD mapping, water temperature, electrofishing and invertebrate sampling. Immediately before and after construction, detailed topographic surveys were carried out, with further resurveys undertaken after significant flow events (i.e. 'bank-full' or higher). The monitoring has determined that the restoration measures implemented have significantly reinstated the natural physical and ecological processes that the site would have exhibited under unimpacted (i.e. 'reference') conditions (i.e. an upland 'wandering' gravel-bed river). Three years after design implementation (and a number of high flow events), significant changes in river planform, bedform morphology and sedimentology have been observed. The channel through the site, that had been in a near static 'canalised' condition for ~30 years, has experienced rapid development of three meander bends (~15 m lateral migration, ~2 channel widths) since the restoration measures were implemented. This process has been associated with significant increased sediment storage through the development of complex gravel bar features (i.e. important habitat features), natural tree 'capture' and reconnection of the river with its floodplain (Fig. B.5).

BOX 3 The Allt Lorgy Restoration in Scotland: An Example of a Process-Based Restoration Approach—cont'd

Detailed 2D hydraulic modelling of the pre- and postrestoration conditions of the site showed a significant increase in hydraulic hetereogeneity. This change from the flume-like prerestoration condition greatly improved habitat diversity through the site and there has been a measureable ecological response. For instance, habitat modelling demonstrated a significant increase in the area of conditions suitable for spawning Atlantic salmon, spatially validated by subsequent field-observed REDD distributions (Fig. B.6). The hydraulic modelling also demonstrated greater in-channel shear stress heterogeneity postrestoration, with resultant differential patterns of erosion, transport and deposition of sediments being responsible for the greater physical diversity observed. This has provided enhanced habitat opportunity for a wide range of in-stream and riparian species. Furthermore, the modelling showed that the restored Allt Lorgy is now better connected to its floodplain, particularly evident in the downstream section of the site (Fig. B.7) where the development of a new semipermanent wet woodland habitat has been a direct consequence of the restoration work.

Fish populations are a good indicator of habitat condition and initial monitoring results show significant increases in spawning and juvenile salmon and trout. This provides preliminary evidence that the ecological condition of the channel is responding to physical improvements, although longer-term datasets are required for clear relationships (and the nature of such relationships) to be determined.

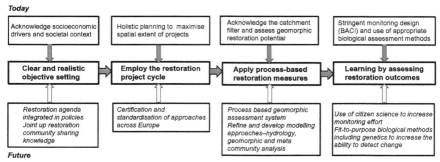

Fig. 25 Summary of the future directions of river restoration-based today's knowledge and future needs for improvement. The future directions can be summarised in four steps that apply to all restorations projects (*central boxes*). *Top boxes* show key elements to be considered for each step from the knowledge we have today. *Lower boxes* illustrate elements that in future need to be implemented to lift river restoration science and application to the next level.

Floods Directive in Europe. Moreover, present knowledge on restoration is scattered among a number actors including policymakers, stakeholders and scientists. A future aspiration should be to promote knowledge sharing by establishing formalised restoration communities. All evidence gathered in the REFORM project, and in the restoration literature, point in the same direction, namely that larger scale projects are more likely to be successful and this take planning (Figs 17 and 25). In future, we need more prescriptive standardised protocols of how to do river restoration that can be certified in a similar manner to analytical approaches. Already today there is sufficient evidence that process-based restoration is likely to deliver better results than restoring solely form (Figs 24 and 25). Hydromorphological assessment methods that record indicators of process and modelling approaches that integrates hydrological, geomorphic and biotic responses are clearly needed in future to aid the development of innovative restoration techniques. Lastly, we need to stop with the current trial-and-error approach to restoration and build an evidence based on how to undertake projects that will deliver the expected outcomes (Fig. 25). We can only do this by monitoring projects in ways in which we can detect change with the necessary degree of certainty. Citizen science and new cost-effective genetic techniques could be some key elements for overcoming the economic barrier than currently hamper a consistent assessment of restoration interventions. The future of river restoration is as bright as we want it to be.

ACKNOWLEDGEMENTS

A large component of this paper results from the EU-funded Integrated Project REFORM (REstoring rivers FOR effective catchment Management), European Union's Seventh Programme for research, technological development and demonstration under Grant Agreement No. 282656. We acknowledge the entire REFORM team for all their efforts that made this paper possible.

REFERENCES

Aarts, B.G.W., van den Brink, F.W.B., Nienhius, P.H., 2004. Habitat loss as the main cause of the slow recovery of fish faunas of regulated large rivers in Europe: the transversal floodplain gradient. River Res. Appl. 20, 3–23.

Adams, V.M., lvarez-Romero, J.G.A., Carwardine, J., Cattarino, L., Hermoso, V., Kennard, M.J., Linke, S., Pressey, R.L., Stoeckl, N., 2014. Planning across freshwater and terrestrial realms: cobenefits and tradeoffs between conservation actions. Conserv. Lett. 7, 425–440.

Alexander, G.G., Allan, J.D., 2006. Stream restoration in the Upper Midwest, U.S.A. Restor. Ecol. 14, 595–604.

Anderson, D.H., Bousquin, S.G., Williams, G.E., Colangelo, D.J. (Eds.), 2005. Defining success: expectations for restoration of the Kissimmee River. South Florida Water Management District, West Palm Beach, FL. Technical Publication ERA #433.

Angelopoulos, N.V., Cowx, I.G., Smith, M., Buijse, T., Smits, S., van Geest, G., Wolter, C., Slawson, D., Nichersu, I., Constantinescu, A., Staras, M., Feodot, I., Bund, W., Brouwer, R., Gonzalez, M., Garcia de Jalon, D., 2015. Effects of climate and land use changes on river ecosystems and restoration practise. Part 1 Main report. REFORM D5.3.

Arlinghaus, R., Engelhardt, C., Sukhodolov, A., Wolter, C., 2002. Fish recruitment in a canal with intensive navigation: implications for ecosystem management. J. Fish Biol. 61, 1386–1402.

Ayres, A., Gerdes, H., Goeller, B., Lago, M., Catalinas, M., Cantón, Á.G., Brouwer, R., Sheremet, O., Vermaat, J., Angelopoulos, N., Cowx, I., 2014. Inventory of the cost of river degradation and the socio-economic aspects and costs and benefits of river restoration. REFORM D1.4.

Baattrup-Pedersen, A., Göthe, E., Riis, T., O'Hare, M.T., 2015. Functional trait composition of aquatic plants can serve to disentangle multiple interacting stressors in lowland streams. Sci. Total Environ. 543, 230–238.

Bal, K.D., Bouma, T.J., Buis, K., Struyf, E., Jonas, S., Backx, H., Meire, P., 2011. Trade-off between drag reduction and light interception of macrophytes: comparing five aquatic plants with contrasting morphology. Funct. Ecol. 25, 1197–1205.

Balon, E.K., 1975. Reproductive guilds of fishes: a proposal and definition. J. Fish. Res. Board Can. 32 (6), 821–864.

Bash, J.S., Ryan, C.M., 2002. Stream restoration and enhancement projects: is anyone monitoring? Environ. Manag. 29, 877–885.

Beauger, A., Lair, N., Reyes-Marchant, P., Peiry, J.L., 2006. The distribution of macroinvertebrate assemblages in a reach of the River Allier (France), in relation to riverbed characteristics. Hydrobiologia 571, 63–76.

Beche, L.A., McElravy, E.P., Resh, V.H., 2006. Long-term seasonal variation in the biological traits of benthic macroinvertebrates in two Mediterranean-climate streams in California, USA. Freshw. Biol. 51, 56–75.

Beechie, T.J., Sear, D.A., Olden, J.D., Pess, G.R., Buffington, J.M., Moir, H., Roni, P., Pollock, M.M., 2010. Process-based principles for restoring river ecosystems. BioScience 60 (3), 209–222.

Bernez, I., Chicouène, D., Haury, J., 2007. Changes of *Potamogeton pectinatus* clumps under variable, artificially flooded river water regimes. Belg. J. Bot. 140, 51–59.

Bernhardt, E.S., Palmer, M.A., Allan, J.D., Alexander, G., Barnas, K., Brooks, S., Carr, J., Clayton, S., Dahm, C., Follstad-Shah, J., Galat, D., Gloss, S., Goodwin, P., Hart, D., Hassett, B., Jenkinson, R., Katz, S., Kondolf, G.M., Lake, P.S., Lave, R., Meyer, J.L., O'Donnell, T.K., Pagano, L., Powell, B., Sudduth, E., 2005. Synthesizing US river restoration efforts. Science 308 (5722), 636–637.

Biggs, B.J.F., 1996. Hydraulic habitat of plants in streams. Regul. Rivers Res. Manag. 12, 131–144.

Biggs, B.J.F., Smith, R.A., 2002. Taxonomic richness of stream benthic algae: effects of flood disturbance and nutrients. Limnol. Oceanogr. 47, 1175–1186.

Biggs, B.J.F., Smith, R.A., Duncan, M.J., 1999. Velocity and sediment disturbance of periphyton in headwater streams: biomass and metabolism. J. N. Am. Benthol. Soc. 18, 222–241.

Bond, N.R., Lake, P.S., 2003. Local habitat restoration in streams: constraints on the effectiveness of restoration for stream biota. Ecol. Manag. Restor. 4, 193–198.

Borchardt, D., 1993. Effects of flow and refugia on drift loss of benthic macroinvertebrates: implications for habitat restoration in lowland streams. Freshw. Biol. 29, 221–227.

Borgström, S., Zachrisson, A., Eckerberg, K., 2016. Funding ecological restoration policy in practice—patterns of short-termism and regional biases. Land Use Policy 52, 439–453.

Bornette, G., Puijalon, S., 2011. Response of aquatic plants to abiotic factors: a review. Aquat. Sci. 73, 1–14.

Brierley, G.J., Fryirs, K.A., 2008. River Futures: An Integrative Scientific Approach to River Repair. Island Press, Washington, DC.

Brouwer, R., Pearce, D., 2005. Cost-Benefit Analysis and Water Resources Management. Edward Elgar Publishing, UK.

Brouwer, R., Barton, D., Bateman, I., Brander, L., Georgiou, S., Martin-Ortega, J., Navrud, S., Pulido-Velazquez, M., Schaafsma, M., Wagtendonk, A., 2009. Economic valuation of environmental and resource costs and benefits in the water framework directive: technical guidelines for practitioners. Project report.

Bryce, S.A., Lomnicky, G.A., Kaufmann, P.R., 2010. Protecting sediment-sensitive aquatic species in mountain streams through the application of biologically based streambed sediment criteria. J. N. Am. Benthol. Soc. 29, 657–672.

Buijse, A.D., Klijn, F., Leuven, R.S.E.W., Middelkoop, H., Schiemer, F., Thorp, J.H., Wolfert, H.P., 2005. Rehabitation of large rivers: references, achievements and integration into river management. Arch. Hydrobiol. Suppl. 155 (Large Rivers 15), 715–738. http://dx.doi.org/10.1127/lr/15/2003/715.

Buss, D., Baptista, D., Nessimian, J., Egler, M., 2004. Substrate specificity, environmental degradation and disturbance structuring macroinvertebrate assemblages in neotropical streams. Hydrobiologia 518, 179–188.

Chambers, P.A., Prepas, E.E., Hamilton, H.R., Bothwell, M.L., 1991. Current velocity and its effect on aquatic macrophytes in flowing waters. Ecol. Appl. 1, 249–257.

Charvet, S., Statzner, B., Usseglio-Polatera, P., Dumont, B., 2000. Traits of benthic macroinvertebrates in semi-natural French streams: an initial application to biomonitoring in Europe. Freshw. Biol. 43, 277–296.

Choi, Y.D., 2004. Theories for ecological restoration in changing environment: towards 'futuristic' restoration. Ecol. Res. 19, 75–81.

CIS, 2007. WFD and Hydro-morphological pressures, Policy paper. Focus on hydropower, navigation and flood defence activities. Recommendation for better policy integration. 44 pp.

Clewell, A., Aronson, J., 2008. Ecological Restoration: Principles, Values, and Structure of an Emerging Profession. Island Press, Washington.

Convention on Biological Diversity, 2010. Strategic Plan for Biodiversity 2011–2020. Convention on Biological Diversity. https://www.cbd.int/sp/default.shtml.

Costello, M.J., Claus, S., Dekeyzer, S., Vandepitte, L., Tuama, E.O., Lear, D., Tyler-Walters, H., 2015. Biological and ecological traits of marine species. PeerJ 3, e1201, 1–29. http://dx.doi.org/10.7717/peerj.1201.

Cowx, I.G., 1994. Strategic approach to fishery rehabilitation. In: Cowx, I.G. (Ed.), Rehabilitation of Freshwater Fisheries. Fishing News Book. Blackwell Scientific Publications Ltd, Oxford, pp. 3–10.

Cowx, I.G., Welcomme, R.L., 1998. Rehabilitation of Rivers for Fish. Fishing News Books, Oxford.

Cowx, I.G., Angelopoulos, N., Noble, R.A., Slawson, D., Buijse, T., Wolter, C., 2013. Measuring success of river restoration actions using end-points and benchmarking. Deliverable REFORM D5.1 available at http://reformrivers.eu.

Crisp, D.T., 1996. Environmental requirements of common riverine European salmonid fish species in fresh water with particular reference to physical and chemical aspects. Hydrobiologia 323, 201–221.

Dawson, E.H., 1988. Water flow and the vegetation of running water. In: Symoens, J.J. (Ed.), Vegetation of Inland Waters. Kluwer Academic Publishers, Dordrecht, pp. 283–309.

Demars, B.O.L., Kemp, J.L., Friberg, N., Usseglio-Polatera, P., Harper, D.M., 2012. Linking biotopes to invertebrates in rivers: biological traits, taxonomic composition and diversity. Ecol. Indic. 23, 301–311.

DeVries, P., 1997. Riverine salmonid egg burial depths: review of published data and implications for scour studies. Can. J. Fish. Aquat. Sci. 54 (8), 1685–1698.

Dickman, M.D., Peart, M.R., Yim, W.W.-S., 2005. Benthic diatoms as indicators of stream sediment concentration in Hong Kong. Int. Rev. Hydrobiol. 90, 412–421.

Dolédec, S., Statzner, B., Bournard, M., 1999. Species traits for future biomonitoring across ecoregions: patterns along a human-impacted river. Freshw. Biol. 42, 737–758.

Dolédec, S., Lamouroux, N., Fuchs, U., Mérigoux, S., 2007. Modelling the hydraulic preferences of benthic macroinvertebrates in small European streams. Freshw. Biol. 52, 145–164.

Downs, P.W., Kondolf, G.M., 2002. Post-project appraisals in adaptive management of river channel restoration. Environ. Manag. 29 (4), 477–496.

Duan, X., Wang, Z., Xu, M., Zhang, K., 2009. Effect of streambed sediment on benthic ecology. Int. J. Sediment Res. 24, 325–338.

Ecke, F., Hellsten, S., Köhler, J., Lorenz, A.W., Rääpysjärvi, J., Scheunig, S., Segersten, J., Baattrup-Pedersen, A., 2016. The response of hydrophyte growth forms and plant strategies to river restoration. Hydrobiologia 769, 41–54.

EEA, 2012. European waters—assessment of status and pressures. EEA Report No 8/2012, European Environmental Agency, Copenhagen.

Ellis, J.I., Schneider, D.C., 1997. Evaluation of a gradient sampling design for environmental impact assessment. Environ. Monit. Assess. 48, 157–172.

England, J., Skinner, K.S., Carter, M.G., 2007. Monitoring, river restoration and the Water Framework Directive. Water Environ. J. 22, 227–234.

European Commission, 2011. Communication from the Commission to the European Parliament, the Council, the Economic and Social Committee and the Committee of the Regions. Our life insurance, our natural capital: an EU biodiversity strategy to 2020. European Commission.

Everard, M., 2012. Why does 'good ecological status' matter? Water Environ. J. 26, 165–174.

FAO, 2008. Inland fisheries. 1. Rehabilitation of inland waters for fisheries. FAO Technical Guidelines for Responsible Fisheries 6 (Suppl. 1), Rome.

Fehér, J., Gáspár, J., Szurdiné-Veres, K., Kiss, A., Kristensen, P., Peterlin, M., Globevnik, L., Kirn, T., Semerádová, S., Künitzer, A., Stein, U., Austnes, K., Spiteri, C., Prins, T., Laukkonen, E., Heiskanen, A.-S., 2012. Hydromorphological alterations and pressures in European rivers, lakes, transitional and coastal waters. Thematic assessment for EEA Water 2012 Report. European Topic Centre on Inland, Coastal and Marine Waters, Prague, ETC/ICM Technical Report 2/2012.

Feld, C.K., Birk, S., Bradley, D.C., Hering, D., Kai, l.J., Marzin, A., Melcher, A., Nemitz, D., Pedersen, M.L., Pletterbauer, F., Pont, D., Verdonschot, P.F.M., Friberg, N., 2011. From natural to degraded rivers and back again: a test of restoration ecology theory and practice. Adv. Ecol. Res. 44, 119–209.

FISRWG (Federal Interagency Stream Restoration Working Group), 1998. Stream Corridor Restoration: Principles, Processes, and Practices. USDA, Washington, DC. GPO Item No. 0120-A.

Foley, J.A., DeFries, R., Asner, G.P., Barford, C., Bonan, G., Carpenter, S.R., Chapin, F.S., Coe, M.T., Daily, G.C., Gibbs, H.K., Helkowski, J.H., Holloway, T., Howard, E.A., Kucharik, C.J., Monfreda, C., Patz, J.A., Prentice, I.C., Ramankutty, N., Snyder, P.K., 2005. Global consequences of land use. Science 309, 570–574.

Folkard, A.M., 2011. Vegetated flows in their environmental context: a review. Proc. Inst. Civil. Eng. 164, 3–24.

Friberg, N., 2010. Ecological consequences of river channel management. In: Ferrier, R.C., Jenkins, A. (Eds.), Handbook of Catchment Management. Wiley-Blackwell, Chichester, UK, pp. 77–106.

Friberg, N., 2014. Impacts and indicators of change in lotic ecosystem. Wiley Interdiscip. Rev.: Water 1, 513–531.

Friberg, N., Kronvang, B., Svendsen, L.M., Hansen, H.O., Nielsen, M.B., 1994. Restoration of channellized reach of the River Gelså, Denmark. Effects on the macroinvertebrate community. Aquat. Conserv. Mar. Freshwat. Ecosyst. 4, 289–297.

Friberg, N., O'Hare, M.T., Poulsen, A.M., 2013. Impacts of hydromorphological degradation and disturbed sediment dynamics on ecological status. Deliverable REFORM D3.1 available at http://reformrivers.eu.

Frissell, C.A., Liss, W.J., Warren, C.E., Hurley, M.D., 1986. A hierarchical framework for stream habitat classification: viewing streams in a watershed context. Environ. Manag. 12, 199–214.

Gabel, F., Garcia, X.F., Schnauder, I., Pusch, M., 2012. Effects of ship-induced waves on littoral benthic invertebrates. Freshw. Biol. 57, 2425–2435.

Garcia de Jalón, D., Alonso, C., et al., 2013. Review on pressure effects on hydromorphological variables and ecologically relevant processes. Deliverable REFORM D1.2 available at http://reformrivers.eu.

Gibbins, C., Batalla, R.J., Vericat, D., 2010. Invertebrate drift and benthic exhaustion during disturbance: response of mayflies (Ephemeroptera) to increasing shear stress and riverbed instability. River Res. Appl. 26, 499–511.

Gibbins, C.N., Vericat, D., Batalla, R.J., Buendia, C., 2016. Which variables should be used to link invertebrate drift to river hydraulic conditions? Fundam. Appl. Limnol. 187, 191–205.

Glendell, M., Extence, C., Chadd, R., Brazier, R.E., 2014. Testing the pressure-specific invertebrate index (PSI) as a tool for determining ecological relevant targets for reducing sedimentation in streams. Freshw. Biol. 59, 353–367.

Göthe, E., Timmermann, A., Januschke, K., Baattrup-Pedersen, A., 2016. Structural and functional response of floodplain vegetation to stream ecosystem restoration. Hydrobiologia 769, 79–92.

Gurnell, A., 2014. Plants as river system engineers. Earth Surf. Process. Landf. 39 (1), 4–25.

Haase, P., Hering, D., Jähnig, S.C., Lorenz, A.W., Sundermann, A., 2013. The impact of hydromorphological restoration on river ecological status: a comparison of fish, benthic invertebrates, and macrophytes. Hydrobiologia 704, 475–488.

Halpern, B.S., Walbridge, S., Selkoe, K.A., Kappel, C.V., Micheli, F., D'Agrosa, C., Bruno, J.F., Casey, K.S., Ebert, C., Fox, H.E., Fujita, R., Heinemann, D., Lenihan, H.S., Madin, E.M.P., Perry, M.T., Selig, E.R., Spalding, M., Steneck, R., Watson, R., 2008. A global map of human impact on marine ecosystems. Science 319, 948–952.

Hammer, C., 1995. Fatigue and exercise tests with fish. Comp. Biochem. Physiol. A Physiol. 112, 1–20.

Hammond, D., Mant, J., Holloway, J., Elbourne, N., Janes, M., 2011. Practical river restoration appraisal guidance for monitoring options (PRAGMO). River Restoration Centre, Cranfield, UK.

Harding, J.S., Benfield, E.F., Bolstad, P.V., Helfman, G.S., Jones, E.B.D., 1998. Stream biodiversity: the ghost of land use past. Proc. Natl. Acad. Sci. U.S.A. 95, 14843–14847.

Hart, K., Allen, B., Linder, M., Keenleyside, C., Burgess, P., Eggers, J., Buckwell, A., 2012. Land as an Environmental Resource. Report Prepared for DG Environment, Contract No ENV.B.1/ETU/2011/0029, Institute for European Environmental Policy, London.

Hauer, C., Unfer, G., Graf, W., Leitner, P., Zeiringer, B., Habersack, H., 2012. Hydromorphologically related variance in benthic drift and its importance for numerical habitat modelling. Hydrobiologia 683, 83–108.

Heino, J., 2013. The importance of metacommunity ecology for environmental assessment research in the freshwater realm. Biol. Rev. 88, 166–178.

Hering, D., Aroviita, J., Baattrup-Pedersen, A., Brabec, K., Buijse, T., Ecke, F., Friberg, N., Gielczewski, M., Januschke, K., Köhler, J., Kupilas, B., Lorenz, A.W., Muhar, S., Paillex, A., Poppe, M., Schmidt, T., Schmutz, S., Vermaat, J., Verdonschot, P., Verdonschot, R., Wolter, C., Kail, J., 2015. Contrasting the roles of section length and instream habitat enhancement for river restoration success: a field study on 20 European restoration projects. J. Appl. Ecol. 52, 1518–1527.

Heywood, M.J.T., Walling, D.E., 2007. The sedimentation of Salmonid spawning gravels in the Hampshire Avon catchment, UK: implications for the dissolved oxygen content of intragravel water and embryo survival. Hydrol. Process. 21, 770–788.

Hines, S.L., Hershey, A.E., 2011. Do channel restoration structures promote ammonium uptake and improve macroinvertebrate-based water quality classification in urban streams? Inland Waters 1, 133–145.

Hobbs, R.J., Hallett, L.M., Ehrlich, P.R., Mooney, H.A., 2011. Intervention ecology: applying ecological science in the twenty-first century. BioScience 61 (6), 442–450.

Holomuzki, J.R., Biggs, B.J.F., 2000. Taxon-specific responses to high-flow disturbance in streams: implications for population persistence. J. N. Am. Benthol. Soc. 19, 670–679.

Hughes, J.M., 2007. Constraints on recovery: using molecular methods to study connectivity of aquatic biota in rivers and streams. Freshw. Biol. 52, 616–631.

Iversen, T.M., Kronvang, B., Madsen, B.L., Markmann, P., Nielsen, M.B., 1993. Re-establishment of Danish streams: restoration and maintenance measures. Aquat. Conserv. Mar. Freshwat. Ecosyst. 3 (2), 73–92.

Jackson, M.C., Weyl, O.L.F., Altermatt, F., Durance, I., Friberg, N., Dumbrell, A.J., Piggott, J.J., Tiegs, S.D., Tockner, K., Krug, C.B., Leadley, P.W., Woodward, G., 2016. Recommendations for the next generation of global freshwater biological monitoring tools. Adv. Ecol. Res. 55, 615–636.

Jähnig, S.C., Brunzel, S., Gacek, S., Lorenz, A.W., Hering, D., 2009. Effects of re-braiding measures on hydromorphology, floodplain vegetation, ground beetles and benthic invertebrates in mountain rivers. J. Appl. Ecol. 46, 406–416.

Jähnig, S.C., Brabec, K., Buffagni, A., Erba, S., Lorenz, A.W., Ofenböck, T., Verdonschot, P.F.M., Hering, D., 2010. A comparative analysis of restoration measures and their effects on hydromorphology and benthic invertebrates in 26 central and southern European rivers. J. Appl. Ecol. 47, 671–680.

Jähnig, S.C., Lorenz, A.W., Hering, D., Antons, C., Sundermann, A., Jedicke, E., Haase, P., 2011. River restoration success: a question of perception. Ecol. Appl. 21, 2007–2015.

Janauer, G.A., Schmidt-Mumm, U., Schmidt, B., 2010. Aquatic macrophytes and water current velocity in the Danube River. Ecol. Eng. 36, 1138–1145.

Januschke, K., Verdonschot, R.C.M., 2016. Effects of river restoration on riparian ground beetles (Coleoptera: Carabidae) in Europe. Hydrobiologia 769, 93–104.

Januschke, K., Sundermann, A., Antons, C., Haase, P., Lorenz, A.W., Hering, D., 2009. Untersuchung und Auswertung von ausgewählten Renaturierungsbeispielen repräsentativer Fließgewässertypen der Flusseinzugsgebiete Deutschlands. Verbesserung der biologischen Vielfalt in Fließgewässern und ihren Auen. Schriftenreihe des Deutschen Rates für Landespflege 82, 3–39.

Januschke, K., Brunzel, S., Haase, P., Hering, D., 2011. Effects of stream restorations on riparian mesohabitats, vegetation and carabid beetles: a synopsis of 24 cases from Germany. Biodivers. Conserv. 20, 3147–3164.

Johnson, G.E., Thom, R.M., Whiting, A.H., Sutherland, G.B., Berquam, T., Ebberts, B.D., Ricci, N.M., Southard, J.A., Wilcox, J.D., 2003. An ecosystem-based approach to habitat restoration projects with emphasis on salmonids in the California River estuary. Report no.PNNL-14412, Battelle Pacific Northwest National Laboratory, Sequim, Washington.

Jones, J.I., Duerdoth, C.P., Collins, A.L., Naden, P.S., Sear, D.A., 2014. Interactions between diatoms and fine sediment. Hydrol. Process. 28, 1226–1237.

Jowett, I.G., 2003. Hydraulic constraints on habitat suitability for benthic invertebrates in gravel-bed rivers. River Res. Appl. 19, 495–507.

Jowett, I.G., Biggs, B.J.F., 1997. Flood and velocity effects on periphyton and silt accumulation in two New Zealand rivers. N. Z. J. Mar. Freshw. Res. 31, 287–300.

Julien, H.P., Bergeron, N.E., 2006. Effect of fine sediment infiltration during the incubation period on Atlantic salmon (*Salmo salar*) embryo survival. Hydrobiologia 563, 61–71.
Jungwirth, M., Muhar, S., Schmutz, S., 2000. Fundamentals of fish ecological integrity and their relation to the extended serial discontinuity concept. Hydrobiologia 422 (423), 85–97.
Kail, J., Wolter, C., 2011. Analysis and evaluation of large-scale river restoration planning in Germany to better link river research and management. River Res. Appl. 27, 985–999.
Kail, J., Hering, D., Muhar, S., Gerhard, M., Preis, S., 2007. The use of large wood in stream restoration: experiences from 50 projects in Germany and Austria. J. Appl. Ecol. 44 (6), 1145–1155.
Kail, J., Arle, J., Jähnig, S.C., 2012. Limiting factors and thresholds for macroinvertebrate assemblages in European rivers: empirical evidence from three datasets on water quality, catchment urbanization, and river restoration. Ecol. Indic. 18, 62–72.
Kail, J., Lorenz, A.W., Hering, D., 2014. Effects of large- and small-scale river restoration on hydromorphology and ecology. Deliverable REFORM D4.3 available at http://reformrivers.eu, http://www.reformrivers.eu/system/files/4.3%20Effects%20of%20large-%20and%20small-scale%20restoration.pdf.
Kail, J., Brabec, K., Poppe, M., Januschke, K., 2015. The effect of river restoration on fish, macroinvertebrates and aquatic macrophytes: a meta-analysis. Ecol. Indic. 58, 311–321.
Kondolf, G.M., Wolman, M.G., 1993. The sizes of salmonid spawning gravels. Water Resour. Res. 29, 2275–2285.
Kondolf, G., Yang, C.N., 2008. Planning river restoration projects: social and cultural dimensions. In: Darby, S., Sear, D. (Eds.), River Restoration: Managing the Uncertainty in Restoring Physical Habitat. John Wiley & Sons, Ltd, Chichester, UK, pp. 43–60.
Kondolf, G.M., Smeltzer, M.W., Railsback, S.F., 2001. Design and performance of a channel reconstruction project in a coastal California gravel-bed stream. Environ. Manag. 28 (6), 761–776.
Kondolf, G.M., Boulton, A.J., O'Daniel, S., Poole, G.C., Rahel, F.J., Stanley, E.H., Wohl, E., Bång, A., Carlstrom, J., Cristoni, C., Huber, H., Koljonen, S., Louhi, P., Nakamura, K., 2006. Process-based ecological river restoration: visualizing three dimensional connectivity and dynamic vectors to recover lost linkages. Ecol. Soc. 11, 5.
Kristensen, P.B., Kristensen, E.A., Riis, T., Baisner, A.J., Larsen, S.E., Verdonschot, P.F.M., Baattrup-Pedersen, A., 2013. Riparian forest as a management tool for moderating future thermal conditions of lowland temperate streams. Hydrol. Earth Syst. Sci. Discuss. 10 (5), 6081–6106.
Kupilas, B., Friberg, N., McKie, B.G., Jochmann, M.A., Lorenz, A.W., Hering, D., 2016. River restoration and the trophic structure of benthic invertebrate communities across 16 European restoration projects. Hydrobiologia 769, 105–120. http://dx.doi.org/10.1007/s10750-015-2569.
Leibold, M.A., Holyoak, M., Mouquet, N., Amarasekare, P., Chase, J.M., Hoopes, M.F., Holt, R.D., Shurin, J.B., Law, R., Tilman, D., Loreau, M., Gonzalez, A., 2004. The metacommunity concept: a framework for multi-scale community ecology. Ecol. Lett. 7, 601–613.
Lepori, F., Palm, D., Brännäs, E., Malmquist, B., 2005. Does restoration of structural heterogeneity in streams enhance fish and macroinvertebrate diversity? Ecol. Appl. 15, 2060–2071.
Leps, M., Tonkin, J.D., Dahm, V., Haase, P., Sundermann, A., 2015. Disentangling environmental drivers of benthic invertebrate assemblages: the role of spatial scale and riverscape heterogeneity in a multiple stressor environment. Sci. Total Environ. 536, 546–556.
Letcher, R.A., Giupponi, C., 2005. Policies and tools for sustainable water management in the European Union. Environ. Model. Software 20, 93–98.
Lorenz, A.W., Korte, T., Sundermann, A., Januschke, K., Haase, P., 2012. Macrophytes respond to reach-scale river restorations. J. Appl. Ecol. 49, 202–212.

Lorenz, S., Martinez-Fernández, V., Alonso, C., Mosselman, E., García de Jalón, D., González del Tánago, M., Belletti, B., Hendriks, D., Wolter, C., 2016. Fuzzy cognitive mapping for predicting hydromorphological responses to multiple pressures in rivers. J. Appl. Ecol. 53, 559–566. http://dx.doi.org/10.1111/1365-2664.12569.

Madsen, J.D., Chambers, P.A., James, W.F., Koch, E.W., Westlake, D.F., 2001. The interaction between water movement, sediment dynamics and submersed macrophytes. Hydrobiologia 444, 71–84.

Matthaei, C.D., Guggelberger, C., Huber, H., 2003. Local disturbance history affects patchiness of benthic river algae. Freshw. Biol. 48, 1514–1526.

Maxted, J.R., Evans, B.F., Scarsbrook, M.R., 2003. Development of standard protocols for macroinvertebrate assessment of soft-bottomed streams in New Zealand. N. Z. J. Mar. Freshw. Res. 37, 793–807.

Mcgill, B.J., Enquist, B.J., Weiher, E., Westoby, M., 2006. Rebuilding community ecology from functional traits. Trends Ecol. Evol. 21, 178–185.

Menezes, S., Baird, D.J., Soares, A., 2010. Beyond taxonomy: a review of macroinvertebrate trait-based community descriptors as tools for freshwater biomonitoring. J. Appl. Ecol. 47, 711–719.

Mérigoux, S., Dolédec, S., 2004. Hydraulic requirements of stream communities: a case study on invertebrates. Freshw. Biol. 49, 600–613.

Mérigoux, S., Lamouroux, N., Olivier, J.M., Dolédec, S., 2009. Invertebrate hydraulic preferences and predicted impacts of changes in discharge in a large river. Freshw. Biol. 54, 1343–1356.

Millennium Ecosystem Assessment, 2005. Ecosystems and Human Well-Being: Synthesis. Island Press, Washington, DC.

Miller, S.W., Budy, P., Schmidt, J.C., 2010. Quantifying macroinvertebrate responses to in-stream habitat restoration: applications of meta-analysis to river restoration. Restor. Ecol. 18, 8–19.

Möbes-Hansen, B., Waringer, J.A., 1998. The influence of hydraulic stress on microdistribution patterns of zoobenthos in a sandstone brook (Weidlingbach, Lower Austria). Int. Rev. Hydrobiol. 83, 381–396.

O'Hare, M.T., Hutchinson, K.A., Clarke, R.T., 2007. The drag and reconfiguration experienced by five macrophytes from a lowland river. Aquat. Bot. 86, 253–259.

O'Hare, J.M., O'Hare, M.T., Gurnell, A.M., Dunbar, M.J., Scarlett, P.M., Laizé, C., 2011. Physical constraints on the distribution of macrophytes linked with flow and sediment dynamics in British rivers. River Res. Appl. 27, 671–683.

Ormerod, S.J., 2004. A golden age of river restoration science? Aquat. Conserv. Mar. Freshwat. Ecosyst. 14, 543–549.

Osenberg, C.W., Sarnelle, O., Cooper, S.D., 1997. Effect size in ecological experiments: the application of biological models in meta-analysis. Am. Nat. 150, 798–812.

Palmer, M.A., Ruhl, J.B., 2015. Aligning restoration science and the law to sustain ecological infrastructure for the future. Front. Ecol. Environ. 13 (9), 512–519.

Palmer, M.A., Ambrose, R.F., Poff, N.L., 1997. Ecological theory and community restoration ecology. Restor. Ecol. 4, 291–300.

Palmer, M.A., Bernhardt, E., Chornesky, E., Collins, S., Dobson, A., Duke, C., Gold, B., Jacobson, R., Kingsland, S., Kranz, R., Mappin, M., Martinez, M.L., Micheli, F., Morse, J., Pace, M., Pascual, M., Palumbi, S., Reichman, O.J., Simons, A., Townsend, A., Turner, M., 2004. Ecology for a crowded planet. Science 304, 1251–1252.

Palmer, M.A., Bernhardt, E.S., Allan, J.D., Lake, P.S., Alexander, G., Brooks, S., Carr, J., Clayton, S., Dahm, C., Follstad Shah, J., Galat, D.J., Gloss, S., Goodwin, P., Hart, D.H., Hassett, B., Jenkinson, R., Kondolf, G.M., Lave, R., Meyer, J.L., O'Donnell, T.K., Pagano, L., Srivastava, P., Sudduth, E., 2005. Standards for ecologically successful river restoration. J. Appl. Ecol. 42, 208–217.

Palmer, M.A., Menninger, H., Bernhardt, E.S., 2010. River restoration, habitat heterogeneity and biodiversity: a failure of theory or practice? Freshw. Biol. 55, 205–222.
Palmer, M.A., Hondula, K.L., Koch, B.J., 2014. Ecological restoration of streams and rivers: shifting strategies and shifting goals. Annu. Rev. Ecol. Evol. Syst. 45, 247–269.
Pan, B.Z., Wang, Z.Y., Xu, M.Z., 2012. Macroinvertebrates in abandoned channels: assemblage characteristics and their indications for channel management. River Res. Appl. 28, 1149–1160.
Parmesan, C., Yohe, G., 2003. A globally coherent fingerprint of climate change impacts across natural systems. Nature 421, 37–42.
Pasternack, G.B., 2008. Spawning habitat rehabilitation: advances in analysis tools. Am. Fish. Soc. Symp. 65, 321–348.
Pedersen, M.L., Friberg, N., 2007. Spatio-temporal variations in substratum stability and macroinvertebrates in lowland stream riffles. Aquat. Ecol. 41, 475–490.
Pedersen, M.L., Kristensen, K.K., Friberg, N., 2014. Re-meandering of lowland streams: will disobeying the laws of geomorphology have ecological consequences? PLoS One 9, e108558. http://dx.plos.org/10.1371/journal.pone.0108558.
Poppe, M., Kail, J., Aroviita, J., Stelmaszczyk, M., Giełczewski, M., Muhar, S., 2015. Assessing restoration effects on hydromorphology in European mid-sized rivers by key hydromorphological parameters. Hydrobiologia 769, 21–40.
Possingham, H., 2012. How can we sell evaluating, analyzing and synthesizing to young scientists. Commentary. Anim. Conserv. 15, 229–230.
Pretty, J.L., Harrison, S.S.C., Shepherd, D.J., Smith, C., Hildrew, A.G., Hey, R.D., 2003. River rehabilitation and fish populations: assessing the benefit of instream structures. J. Appl. Ecol. 40, 251–265.
Puijalon, S., Bornette, G., Sagnes, P., 2005. Adaptations to increasing hydraulic stress: morphology, hydrodynamics and fitness of two higher aquatic plant species. J. Exp. Bot. 56, 777–786.
Rabeni, C.F., Doisy, K.E., Zweig, L.D., 2005. Stream invertebrate community functional responses to deposited sediment. Aquat. Sci. 67, 395–402.
Radinger, J., Wolter, C., 2015. Disentangling the effects of habitat suitability, dispersal and fragmentation on river fish distribution. Ecol. Appl. 25, 914–927.
Reice, S.R., 1980. The role of substratum in benthic macroinvertebrate microdistribution and litter decomposition in a woodland stream. Ecology 61, 580–590.
Reichert, P., Langhans, S.D., Lienert, J., Schuwirth, N., 2015. The conceptual foundation of environmental decision support. J. Environ. Manage. 154, 316–332.
Richardson, C.J., Flanagan, N.E., Ho, M., Pahl, J.W., 2011. Integrated stream and wetland restoration: a watershed approach to improved water quality on the landscape. Ecol. Eng. 37 (1), 25–39.
Riedl, C., Peter, A., 2013. Timing of brown trout spawning in Alpine rivers with special consideration of egg burial depth. Ecol. Freshw. Fish 22, 384–397.
Riis, T.B., Biggs, B.J.F., 2003. Hydrologic and hydraulic control of macrophyte establishment and performance in streams. Limnol. Oceanogr. 48, 1488–1497.
Rinaldi, M., Gurnell, A.M., Gonzalez del Tanago, M., Bussettini, M., Hendriks, D., 2016. Classification of river morphology and hydrology to support management and restoration. Aquat. Sci. 78, 17–33.
Roni, P., Beechie, T., 2013. Stream and Watershed Restoration—A Guide to Restoring Riverine Processes and Habitats. John Wiley & Sons, Ltd, Chichester.
Roni, P., Beechie, T.J., Bilby, R.E., Leonetti, F.E., Pollock, M.M., Pess, G.R., 2002. A review of stream restoration techniques and a hierarchical strategy for prioritizing restoration in Pacific Northwest watersheds. N. Am. J. Fish Manag. 22, 1–20.
Roni, P., Hanson, K., Beechie, T., 2008. Global review of the physical and biological effectiveness of stream habitat rehabilitation techniques. N. Am. J. Fish Manag. 28, 856–890.

Roth, N.E., Allan, J.D., Erickson, D.L., 1996. Landscape influences on stream biotic integrity assessed at multiple spatial scales. Landsc. Ecol. 11, 141–156.
RRC, 2011. Practical River Restoration Appraisal Guidance for Monitoring Options (PRAGMO). Guidance document on sustainable monitoring for river and floodplain restoration projects. http://www.therrc.co.uk/PRAGMO/PRAGMO_2012-01-24.pdf.
Rumps, J.M., Katz, S.L., Morehead, M.D., Jenkinson, R., Goodwin, P., 2007. Stream restoration in the Pacific Northwest: analysis of interviews with project managers. Restor. Ecol. 15, 506–515.
Sala, O.E., Chapin, F.S., Armesto, J.J., Berlow, E., Bloomfield, J., Dirzo, R., Huber-Sanwald, E., Huenneke, L.F., Jackson, R.B., Kinzig, A., Leemans, R., Lodge, D.M., Mooney, H.A., Oesterheid, M., Poff, N.L., Sykes, M.T., Walker, B.H., Walker, M., Wall, D.H., 2000. Global biodiversity scenarios for the year 2100. Science 287, 1770–1774.
Sanderson, E.W., Jaiteh, M., Levy, M.A., Redford, K.H., Wannebo, A.V., Woolmer, G., 2002. The human footprint and the last of the wild. BioScience 52, 14.
Sand-Jensen, K., Pedersen, M.L., 2008. Streamlining of plant patches in streams. Freshw. Biol. 53, 714–726.
SCBD—Secretariat of the Convention on Biological Diversity, 2010. Global Biodiversity Outlook 3. Secretariat of the CBD, Montréal.
Schmutz, S., Kremser, H., Melcher, A., Jungwirth, M., Muhar, S., Waidbacher, H., Zauner, G., 2014. Ecological effects of rehabilitation measures at the Austrian Danube: a meta-analysis of fish assemblages. Hydrobiologia 729, 49–60.
Schmutz, S., Jurajda, P., Kaufmann, S., Lorenz, A.W., Muhar, S., Paillex, A., Poppe, M., Wolter, C., 2016. Response of fish assemblages to hydromorphological restoration in central and northern European rivers. Hydrobiologia 769, 67–78.
Schutten, J., Dainty, J., Davy, A.J., 2005. Root anchorage and its significance for submerged plants in shallow lakes. J. Ecol. 93, 556–571.
Sedgwick, R.W., 2006. Manual of best practice for fisheries impact assessments. Science Report SC020025/SR.
SER, 2004. http://ser.org/resources/resources-detail-view/ser-international-primer-on-ecological-restoration#3.
Shamier, N., Johnstone, C., Whiles, D., Cochrane, D., Moore, K., Lenane, R., Ryder, S., Betts, V., Horton, B., Donovan, C., Harding, E., Bennett, R., Moseley, R., 2013. Water appraisal guidance: assessing costs and benefits for River Basin management planning. pp. 77. http://www.ecrr.org/Portals/27/Publications/Water%20Appraisal%20Guidance.pdf.
Singh, J., Gusain, O.P., Gusain, M.P., 2010. Benthic insect-substratum relationship along an altitudinal gradient in a Himalayan stream, India. Int. J. Ecol. Environ. Sci. 36, 215–231.
Smith, B., Clifford, N.J., Mant, J., 2014. The changing nature of river restoration. Wiley Interdiscip. Rev.: Water 1, 249–261.
Söhngen, B., Koop, J., Knight, S., Rythönen, J., Beckwith, P., Ferrari, N., Iribarren, J., Kevin, T., Wolter, C., Maynord, S., 2008. Considerations to Reduce Environmental Impacts of Vessels. PIANC Report 99, PIANC, Brussels.
Solheim, A.L., Austnes, K., Peterlin, M., Kodeš, V., Filippi, R., Semerádová, S., Prchalová, H., Künitzer, A., Spiteri, C., Prins, T., Collins, R.P., Kristensen, P., 2012. Ecological and chemical status and pressures in European waters. ETC/ICM Technical Report.
Søndergaard, M., Jeppesen, E., Lauridsen, T.L., Skov, C., Van Nes, E.H., Roijackers, R., Lammens, E., Portielje, R.O.B., 2007. Lake restoration: successes, failures and long-term effects. J. Appl. Ecol. 44, 1095–1105.
Soulsby, C., Youngson, A.F., Moir, H.J., Malcolm, I.A., 2001. Fine sediment influence on salmonid spawning habitat in a lowland agricultural stream: a preliminary assessment. Sci. Total Environ. 265, 295–307.
Statzner, B., Gore, J.A., Resh, V.H., 1988. Hydraulic stream ecology: observed patterns and potential applications. J. N. Am. Benthol. Soc. 7, 307–360.

Statzner, B., Kohmann, F., Hildrew, A.G., 1991. Calibration of FST-hemispheres against bottom shear stress in a laboratory flume. Freshw. Biol. 26, 227–231.
Statzner, B., Hoppenhaus, K., Arens, M.F., Richoux, P., 1997. Reproductive traits, habitat use and templet theory: a synthesis of world-wide data on aquatic insects. Freshw. Biol. 38, 109–135.
Stephenson, J.M., Morin, A., 2009. Covariation of stream community structure and biomass of algae, invertebrates and fish with forest cover at multiple spatial scales. Freshw. Biol. 54, 2139–2154.
Sternecker, K., Cowley, D.E., Geist, J., 2013. Factors influencing the success of salmonid egg development in river substratum. Ecol. Freshw. Fish 22, 322–333.
Stewart, G.B., Bayliss, H.R., Showler, D.A., Sutherland, W.J., Pullin, A.S., 2009. Effectiveness of engineered in-stream structure mitigation measures to increase salmonid abundance: a systematic review. Ecol. Appl. 19, 931–941.
Stoll, S., Sundermann, A., Lorenz, A.W., Kail, J., Haase, P., 2013. Small and impoverished regional species pools constrain colonisation of restored river reaches by fishes. Freshw. Biol. 58, 664–674.
Stoll, S., Breyer, P., Tonkin, J.D., Früh, D., Haase, P., 2016. Scale-dependent effects pf river habitat quality on benthic invertebrate communities—implications for stream restoration practice. Sci. Total Environ. 553, 495–503.
Strange, R.M., 1999. Historical biogeography, ecology, and fish distributions: conceptual issues for establishing IBI criteria. In: Simon, T.P. (Ed.), Assessing the Sustainability and Biological Integrity of Water Resources Using Fish Communities. CRC Press, Boca Raton, FL, pp. 65–78.
Suding, K.N., Gross, K.L., Housman, D.R., 2004. Alternative states and positive feedbacks in restoration ecology. Trends Ecol. Evol. 19, 46–53.
Sukhodolova, T., 2008. Studies of turbulent flow in vegetated river reaches with implications for transport and mixing processes. PhD thesis, Humboldt-University, Berlin.
Sundermann, A., Gerhardt, M., Kappes, H., Haase, P., 2013. Stressor prioritisation in riverine ecosystems: which environmental factors shape benthic invertebrate assemblage metrics? Ecol. Indic. 27, 83–96.
TEEB, 2011. The Economics of Ecosystems and Biodiversity in National and International Policy Making. In: ten Brink, P. (Ed.), Earthscan, London and Washington, p. 528.
Timm, H., Mardi, K., Möls, T., 2008. Macroinvertebrates in Estonian streams: the effects of habitat, season, and sampling effort on some common metrics of biological quality. Est. J. Ecol. 57, 37–57.
Tolkamp, H., 1982. Microdistribution of macroinvertebrates in lowland streams. Hydrobiol. Bull. 16, 133–148.
Tonkin, J.D., Stoll, S., Sundermann, A., Haase, P., 2014. Dispersal distance and the pool of taxa, but not barriers, determine the colonisation of restored river reaches by benthic invertebrates. Freshw. Biol. 59 (9), 1843–1855.
Vaughan, I.P., Diamond, M., Gurnell, A.M., Hall, K.A., Jenkins, A., Milner, N.J., Naylor, L.A., Sear, D.A., Woodward, G., Ormerod, S.J., 2009. Integrating ecology with hydromorphology: a priority for river science and management. Aquat. Conserv. Mar. Freshwat. Ecosyst. 19, 113–125.
Verdonschot, R.C.M., Kail, J., McKie, B.G., Verdonschot, P.F.M., 2015. The role of benthic microhabitats in determining the effects of hydromorphological river restoration on macroinvertebrates. Hydrobiologia 769, 55–66.
Videler, J.J., 1993. Fish Swimming. Chapman & Hall, London.
Violin, C.R., Cada, P., Sudduth, E.B., Hassett, B.A., Penrose, D.L., Bernhardt, E.S., 2011. Effects of urbanization and urban stream restoration on the physical and biological structure of stream ecosystems. Ecol. Appl. 21, 1932–1949.

Vörösmarty, C.J., McIntyre, P.B., Gessner, M.O., Dudgeon, D., Prusevich, A., Green, P., Glidden, S., Bunn, S.E., Sullivan, C.A., Lierman, C.R., Davies, P.M., 2010. Global threats to human water security and river biodiversity. Nature 467, 555–561.

Walter, R.C., Merritts, D.J., 2008. Natural streams and the legacy of water-powered mills. Science 319, 299–304.

Wang, L., Robertson, D.M., Garrison, P.J., 2007. Linkages between nutrients and assemblages of macroinvertebrates and fish in wadeable streams: implication to nutrient criteria development. Environ. Manag. 39, 194. http://dx.doi.org/10.1007/s00267-006-0135-8.

Ward, D., Holmes, N., José, P., 1994. The New Rivers & Wildlife Handbook. The Royal Society for the Protection of Birds, Bedfordshire, UK.

Whiteway, S.L., Biron, P.M., Zimmermann, A., Venter, O., Grant, J.W., 2010. Do in-stream restoration structures enhance salmonid abundance? A meta-analysis. Can. J. Fish. Aquat. Sci. 67, 831–841.

Williams, B.K., Brown, E.D., 2014. Adaptive management: from more talk to real action. Environ. Manage. 53, 465–479.

Wilzbach, M.A., Cummins, K.W., Knapp, R.A., 1988. Towards a functional classification of stream invertebrate drift. Verh. Int. Ver. Theor. Angew. Limnol. 23, 1244–1254.

Winking, C., Lorenz, A.W., Sures, B., Hering, D., 2014. Recolonisation patterns of benthic invertebrates: a field investigation of restored former sewage channels. Freshw. Biol. 59, 1932–1944. http://dx.doi.org/10.1111/fwb.12397.

Wolter, C., 2010. Functional vs scenic restoration—challenges to improve fish and fisheries in urban waters. Fish. Manag. Ecol. 17, 176–185.

Wolter, C., Arlinghaus, R., 2003. Navigation impacts on freshwater fish assemblages: the ecological relevance of swimming performance. Rev. Fish Biol. Fish. 13, 63–89.

Wolter, C., Arlinghaus, R., 2004. Burst and critical swimming speeds of fish and their ecological relevance in waterways. Berichte des IGB 20, 77–93.

Wolter, C., Buijse, A.D., Parasiewicz, P., 2016. Temporal and spatial patterns of fish response to hydromorphological processes. River Res. Appl. 32, 190–201.

Woodward, G., 2009. Biodiversity, ecosystem functioning and food webs in fresh waters: assembling the jigsaw puzzle. Freshw. Biol. 54, 2171–2187.

Woolsey, S., Capelli, F., Gonser, T., Hoehn, E., Hostmann, M., Junker, B., Paetzold, A., Roulier, C., Schweizer, S., Tiegs, S.D., Tockner, K., Weber, C., Peter, A., 2007. A strategy to assess river restoration success. Freshw. Biol. 52, 752–769.

Wortley, L., Hero, J.-M., Howes, M., 2013. Evaluating Ecological Restoration Success: A Review of the Literature. Restor. Ecol. 21 (5), 537–543.

PART IV

A Look To the Future

CHAPTER TWELVE

Recommendations for the Next Generation of Global Freshwater Biological Monitoring Tools

M.C. Jackson*,†,1, O.L.F. Weyl‡, F. Altermatt§,¶, I. Durance∥,
N. Friberg#,**, A.J. Dumbrell††, J.J. Piggott‡‡,§§, S.D. Tiegs¶¶,
K. Tockner∥∥,##, C.B. Krug***, P.W. Leadley***, G. Woodward†

*University of Pretoria, Hatfield, Gauteng, South Africa
†Imperial College London, Ascot, Berkshire, United Kingdom
‡South African Institute for Aquatic Biodiversity, Grahamstown, Eastern Cape, South Africa
§Eawag, Swiss Federal Institute of Aquatic Science and Technology, Dübendorf, Switzerland
¶University of Zurich, Zürich, Switzerland
∥Cardiff Water Research Institute, School of Biosciences, Cardiff University, Cardiff, United Kingdom
#Norwegian Institute for Water Research (NIVA), Oslo, Norway
**water@leeds, School of Geography, University of Leeds, Leeds, United Kingdom
††School of Biological Sciences, University of Essex, Colchester, United Kingdom
‡‡University of Otago, Dunedin, New Zealand
§§Center for Ecological Research, Kyoto University, Otsu, Japan
¶¶Oakland University, Rochester, MI, United States
∥∥Leibniz-Institute of Freshwater Ecology and Inland Fisheries, Berlin, Germany
##Freie Universität Berlin, Berlin, Germany
***Laboratoire ESE, Université Paris-Sud, UMR 8079 CNRS, UOS, AgroParisTech, Orsay, France
[1]Corresponding author: e-mail address: m.jackson@imperial.ac.uk

Contents

1. Introduction	616
2. Invertebrates as Indicators of Ecosystem State	618
3. Decomposition-Based Indicators	623
4. Fishery Indicators: Learning from the Marine Realm	624
5. Molecular-Based Indicators	625
6. Indicators of Change Across Space and Time	628
7. Conclusions and Future Directions	630
Acknowledgments	631
References	631

Abstract

Biological monitoring has a long history in freshwaters, where much of the pioneering work in this field was developed over a 100 years ago—but few of the traditional monitoring tools provide the global perspective on biodiversity loss and its consequences for ecosystem functioning that are now needed. Rather than forcing existing monitoring paradigms to respond to questions they were never originally designed to address, we need to take a step back and assess the prospects for novel approaches that could be developed and adopted in the future. To resolve some of the issues with indicators currently used to inform policymakers, we highlight new biological monitoring tools that are being used, or could be developed in the near future, which (1) consider less-studied taxonomic groups, (2) are standardised across regions to allow global comparisons, and (3) measure change over multiple time points. The new tools we suggest make use of some of the key technological and logistical advances seen in recent years—including remote sensing, molecular tools, and local-to-global citizen science networks. We recommend that these new indicators should be considered in future assessments of freshwater ecosystem health and contribute to the evidence base for global to regional (and national) assessments of biodiversity and ecosystem services: for example, within the emerging framework of the Intergovernmental Platform on Biodiversity and Ecosystem Services.

1. INTRODUCTION

The unprecedented rate of decline in global biodiversity has been linked to anthropogenic stressors including habitat loss, pollution, the changing climate, and species invasions (Global Biodiversity Outlook 4, 2014; WWF, 2014). These stressors are expected to become both more widespread and more intense in the future as the growing human population imposes an ever-stronger footprint on natural ecosystems, and temperatures continue to rise (Bellard et al., 2012; Tittensor et al., 2014). In recent years, there has been a call for the use of bioindicators by policymakers to monitor and predict the impacts of these multiple environmental stressors on ecosystems (Hoffmann et al., 2014; Pereira et al., 2013). However, many current indicators for freshwater systems simply quantify drivers of change, such as temperature or the number of nonnative species, and we urgently need new indicators of modifications in ecological state (i.e. loss of biodiversity and associated ecosystem functioning). In particular, these need to enable us to link cause to effect, and structure to function across multiple levels of organisation—from genes to entire ecosystems.

Decision-makers need monitoring that is simple, effective, and which allows them to measure the efficiency of interventions or management actions that were undertaken in response to environmental impacts (Pereira et al., 2013). This has resulted in the recent establishment of collaborative platforms, such as the Intergovernmental Platform on Biodiversity and Ecosystem Services (IPBES) to assess the state of biodiversity and the ecosystem services it provides. However, many current indicators, such as extinction risk, are often based on secondary information of typically rather abstract, derived measures or proxies, rather than on primary data.

Moreover, the predictive power of freshwater indicators is often constrained by measures of biodiversity loss that are taxonomically and trophically biased towards easily sampled and readily identifiable taxa (e.g. fishes and macroinvertebrates), have a patchy global coverage, and/or are incomparable between regions due to the high degree of local contingency in their historical development (Nicholson et al., 2012; Revenga et al., 2005). This may seem surprising because freshwater systems are home to ~10% of the Earths diversity despite only covering <1% of total surface area (Dudgeon et al., 2006; Strayer and Dudgeon, 2010), and the earliest pioneering work in biomonitoring was initiated in these systems well over a century ago (Friberg et al., 2011). Endemism is also unusually high: for example, it is close to 100% in the diverse fish communities of Lake Malawi (~1000 species) and Lake Tanganyika (~200 species; Salzburger et al., 2005). This has resulted in a focus on uniqueness, rarity, and species richness, rather than aiming to develop globally applicable monitoring schemes and those focused on ecosystem properties. Freshwater ecosystems are also increasingly recognised for the essential 'goods and services' that they provide, such as flood mitigation, fish protein, and in particular, drinking water, yet many of the underlying links between biodiversity and the delivery of these outputs are still poorly understood (Durance et al., 2016; Raffaelli et al., 2014). Even the link between the precursors to service delivery—ecosystem functioning, and biodiversity is still poorly characterised, especially in terms of how its relationship to biodiversity is shaped by environmental conditions (but see Perkins et al., 2015).

The relatively few larger-scale surveys that have been conducted in freshwater systems show that diversity is declining much faster than in terrestrial and marine realms (Strayer and Dudgeon, 2010), underlining how vulnerable they are to environmental change. For example, The Living Planet report estimated alarming average population declines of 76% between 1970 and 2010 in freshwater species (WWF, 2014). However, this indicator only focuses on vertebrates and therefore does not consider some of the most diverse groups in freshwaters (e.g. invertebrates, microbes, parasites, and plants), highlighting the need for new monitoring approaches which consider other important taxonomic groups. These surveys are also often patchy and scarce in both time and space and are therefore highly aggregated into coarse metrics at large scales: thus, there is a pressing need for new indicators that quantify biodiversity loss across wide spatial and temporal scales. A wide global coverage with multiple time points would allow policy makers to compare diversity change between regions, and identify drivers of diversity loss over time. However, some form of taxon-free or functional approach is likely to be needed to circumvent some of the problems arising from biogeographical differences among functionally similar systems (Friberg et al., 2011), which is still largely nonexistent beyond local to regional scales. Here we gauge the potential of some indicators which could meet some or all of these requirements, and which we argue should be urgently considered for adoption in future global monitoring programmes.

2. INVERTEBRATES AS INDICATORS OF ECOSYSTEM STATE

Invertebrates are one of the most diverse groups in freshwater ecosystems and are essential for the delivery of many ecosystem processes, such as decomposition (Chauvet et al., 2016, this volume), and they link plant primary production to higher trophic levels, such as fish and the ecosystem services they underpin. Despite this, indicators of macroinvertebrate abundance or diversity are rarely considered by policymakers beyond local to national scales. Their many shortcomings have been reviewed extensively elsewhere in recent years (Baird and Hajibabaei, 2012; Friberg et al., 2011; Gray et al., 2014), so we will not revisit these issues in any great detail here beyond a cursory overview.

One approach to providing regional and global perspectives would be to develop a uniform measure of change in invertebrate biodiversity (and potentially any other group of taxa) by comparing disturbed sites to relatively pristine sites as a baseline (e.g. Scholes and Biggs, 2005). There are already many existing regional indicators of river health based on macroinvertebrate community composition. For instance, the South African Scoring System (SASS) assigns invertebrate families a sensitivity score (according to the water quality conditions they are known to tolerate) which is then used to calculate a score of overall river health based on the community assemblage (Dickens and Graham, 2002). Similar indicators, which are founded on typologies or comparisons to reference conditions, include RIVPACS (River Invertebrate Prediction and Classification System; Wright et al., 1998) in the United Kingdom, PERLA in the Czech Republic (Kokeš et al., 2006), AUSRIVAS (Australian River Assessment System; Smith et al., 1999) in Australia, MCI (Macroinvertebrates Community Index; Stark, 1985) in New Zealand, and the IBCH index in Switzerland (Altermatt et al., 2013; Stucki, 2010). Differences between each approach make it difficult to compare absolute biodiversity; however, it is possible to use the data to determine important trends. Since the scores of river health are typically based on natural reference conditions of relatively unimpacted invertebrate diversity, the data will indicate if a location has experienced biodiversity change. This information can then be synthesised at a larger scale to make more meaningful comparisons of macroinvertebrate community health. Such indices, however, are not designed to disentangle the impacts of multiple stressors (Box 1), and most have focused on either a single stressor such as organic pollution or acidification, or provide 'black box' measures of overall stress of an undefined and unknown suite of stressors.

Citizen science groups are increasingly expanding the spatial and temporal coverage of invertebrate-based approaches. For instance, the Riverfly Partnership (http://www.riverflies.org/) in the United Kingdom is a network of citizen scientists who are trained in a simple invertebrate monitoring technique to classify river health (Box 2). miniSASS (http://www.minisass.org/) in South Africa uses a similar approach and both have large and growing online data repositories (Box 2). We suggest that combining these river health indicators across a global scale with the aid of citizen science networks will increase our knowledge on the extent of biodiversity change in macroinvertebrate communities across the globe. There is increasing evidence that these ideas are being taken up at an accelerating rate, often from a combination of top-down and bottom-up grassroots coordination (Huddart et al., 2016). The key challenge to overcome in the coming years is to find an effective means to maintain continuity and quality control in these largely self-organising and otherwise unregulated bodies (Follett and Strezov, 2015; Huddart et al., 2016).

The use of invertebrate species functional traits (i.e. inherited characteristics of species, such as the presence of external gills) to monitor diversity change has recently gathered momentum because it allows comparison across spatial and temporal scales, avoiding problems associated with biogeographical uniqueness and endemism (Beche et al., 2006; Costello et al., 2015; Mcgill et al., 2006; Statzner et al., 1997). They are also rooted in ecological and evolutionary concepts related to ecological equivalence and convergent evolution, as well as redundancy, which resonate with the body of theory built up around research into the resilience of biodiversity–ecosystem functioning relationships under environmental change. Trait-based indicators can thus be standardised across areas with different species diversity to predict changes in ecosystem functioning, as well as providing complementary and unique information that goes beyond simple species richness-based indices (Seymour et al., 2016). For instance, the morphological traits of some invertebrate species allow the consumption of leaf litter at different rates

BOX 1 Multiple Indicators for Multiple Stressors

While ecological indicators provide a useful tool to track and compare ecosystem states or trends, disentangling indicator drivers can be particularly challenging due to multiple interacting stressors. Freshwater indicator responses frequently demonstrate complex synergistic, antagonistic, or reversal of effects to multiple stressors that cannot be predicted from knowledge of their single effects acting in isolation (Jackson et al., 2016). Additionally, some response indicators which have traditionally been used to assess ecosystem state may become increasingly unreliable in the future, e.g. under a warming climate (Piggott et al., 2015a). Coupled with this challenge are the shifting baselines and emergent novel ecosystem properties that the multiple drivers of global change may manifest. Responses to multiple stressors are also highly context dependent, with the same taxonomic group often responding to the same set of stressors in different ways in different places (Jackson et al., 2016).

Integrative indices may be less susceptible to multiple stressor effects that dampen and diffuse effects at lower levels of organisation (Crain et al., 2008); however, evidence for this is inconclusive. A possible solution might involve the use of multiple indicators in combination to help disentangle stressor effects when they are in opposite directions (i.e. one stressor causes diversity loss, and the other an increase; Piggott et al., 2015b). In this case, a complementary approach which employs both indicators of drivers of change (i.e. the number of nonnative species and temperature) and indicators of ecosystem state (i.e. diversity measures) could be used to disaggregate stressor effects. A similar problem occurs when different taxonomic groups respond in different ways (e.g. fish and plants experience a decline and increase in diversity, respectively).

Indicators of whole ecosystem change at the network level (e.g. food webs) will also be valuable when unravelling the multiple responses of ecosystems to multiple stressors. This approach will provide a whole ecosystem perspective which is consistent and comparable between regions (Gray et al., 2014; QUINTESSENCE Consortium, 2016). Network metrics such as connectance (i.e. the number of food web links) will permit taxonomic free whole food web responses and, in some cases, the impacts of multiple stressors may only manifest at this level of organisation (Gray et al., 2014; O'Gorman et al., 2012). For example, a network approach was successfully employed to disentangle the roles of flood disturbance and urbanisation in structuring stream biodiversity in Italy (Calizza et al., 2015), and to quantify the impact of inorganic nutrients and organic matter on intertidal food web in Ireland (O'Gorman et al., 2012; Fig. B1). At present though, these more holistic and information-rich approaches are still in the early stages of development as novel monitoring tools (e.g. Gray et al., 2016, this volume; Thompson et al., 2016) and still some way behind the other measures we have highlighted here as approaches that could realistically be applied globally in the near future.

BOX 1 Multiple Indicators for Multiple Stressors—cont'd

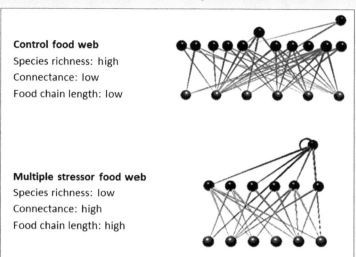

Control food web
Species richness: high
Connectance: low
Food chain length: low

Multiple stressor food web
Species richness: low
Connectance: high
Food chain length: high

Fig. B1 Conceptual figure showing an intertidal food web in its natural state (control food web) and affected by nutrient enrichment and organic matter supplements (multiple stressor food web) where the nodes and lines represent species and trophic links, respectively. The food web under the influence of multiple stressors supported larger and more generalist top predators, had higher connectance between species, and had lower biodiversity. *Reproduced from O'Gorman, E.J., Fitch, J.E., Crowe, T.P., 2012. Multiple anthropogenic stressors and the structural properties of food webs. Ecology 93, 441–448.*

(Chauvet et al., 2016, this volume). Therefore, leaf-litter decomposition, a key ecosystem function, depends on the diversity of this trait within the ecosystem (Chauvet et al., 2016, this volume; Crowl et al., 2001). Additionally, because traits relate directly to environmental characteristics such as temperature and water flow, they provide a good mechanistic understanding of how environmental change influences biotic communities, and are therefore a step towards diagnosing cause and effect (Doledec et al., 1999; Rabeni et al., 2005).

Numerous studies have shown that macroinvertebrate trait metrics can be sensitive to human stressors (Boersma et al., 2016; Charvet et al., 2000; Doledec et al., 1999; Menezes et al., 2010), but there is still a need to gather additional evidence concerning the overall effectiveness of a species trait

BOX 2 Citizen Science Networks

A citizen scientist is a volunteer who collects or processes data, usually with the backing of a scientist or research project. Ecological projects involving citizen scientists are growing (Silvertown, 2016) with the general public's mounting interest in protecting the environment. Two examples from the freshwater realm are miniSASS in South Africa and the Riverfly Partnership in the United Kingdom, both use simple macroinvertebrate monitoring to categorise river health (Fig. B2). In South Africa, 462 sites are monitored by miniSASS volunteers and in the last assessment (data obtained in March 2016), 46% were categorised as very poor, 12% as poor, 16% as fair, 17% as good, and 9% as very good (Fig. B2A). In the United Kingdom, the Riverfly Partnership uses a slightly different scoring technique (Huddart et al., 2016), but we were able to use the information to assign comparable health categories (Fig. B2). A total of 449 sites are monitored in the United Kingdom and in the last assessment (data obtained in March 2016) 8% were categorised as very poor, 31% as poor, 39% as fair, 20% as good, and 2% as very good (Fig. B2B). If such schemes could be expanded to other countries, the data could be added to a global database of invertebrate community health and, therefore, overall river health.

Fig. B2 Locations of citizen science monitoring sites in (A) South Africa (miniSASS) and (B) United Kingdom (Riverfly Partnership). Sites are categorised as very poor (*filled triangle*), poor (*filled circle*), fair (*grey circle*), good (*open circle*), or very good (*open triangle*) health (and therefore from low to high invertebrate diversity).

approach at global scales (Demars et al., 2012). Trait-based indicators are also still subject to many of the same constraints as taxonomic-based macroinvertebrates indices, in that both need to be clearly linked to the stressor in question, and also to provide sufficient accuracy and precision to detect changing conditions. However, there is undoubtedly real scope for using species traits as global indicators of biodiversity change because they are spatially robust across biogeographically distinct regions, but additional research is needed to make these indicators operative at such scales. In particular, careful selection of traits would allow differentiation between response to environmental variability (response traits) and effects on ecosystem functioning (effect traits) (Sterk et al, 2013).

3. DECOMPOSITION-BASED INDICATORS

Despite not always relating directly to biodiversity, an alternative to an invertebrate-based approach is the direct measurement of the important ecosystem services that invertebrates and microorganisms perform. For instance, decomposition, the processing of organic matter, is a universal ecosystem process with many attributes that make it an obvious and logical choice for use as an indicator of human impacts at local to global scales (Chauvet et al., 2016, this volume). Foremost among these is its sensitivity to some of the most pressing threats in freshwaters, including chemical pollution, warming, and land-use change (e.g. Griffiths and Tiegs, 2016; Hladyz et al., 2011; Perkins et al., 2015; Woodward et al., 2012). Additionally, because of the significance of decomposition to the functioning of freshwater ecosystems, given that it underpins the bulk of the trophic basis of production in many systems around the globe, a large body of literature has been devoted to its study. This body of work can now be used to conduct formal large-scale analyses or meta-analyses to interpret data in the context of a wide range of human impacts (Chauvet et al., 2016, this volume). Most decomposition assays are 'taxon-free' and therefore do not require any taxonomic expertise; in most cases they simply measure mass loss (or an associated proxy) of a substrate (usually leaf litter) over time. They are also simple and inexpensive, and some can be highly standardised. For example, cotton strip assays (CSAs) provide an easily repeatable way to quantify organic matter decomposition by microbes in aquatic habitats. Unlike most traditional decomposition assays that rely on the quantification of mass loss of terrestrial leaf litter, the CSA operates by determining the loss of tensile strength of standardised cotton fabric, a process that equates to the degradation of cellulose—an important source of carbon in its globally most abundant forms as a food resource, in many ecosystems (Tiegs et al., 2013). Given its pivotal role at the base of the food web, changes in the processing rates of cellulose decomposition provide a universal measure of one of the most common ecosystem processes on the planet, and one that is sensitive to environmental change (e.g. Jenkins et al., 2013). Since this approach is comparable at any scale and simple to use (and therefore could also be deployed by citizen science networks), it has the potential to become one of the first bioindicators available to ecologists at a spatio-temporal resolution previously only known for physio-chemical indicators, such as temperature or water chemistry (Chauvet et al., 2016, this volume).

4. FISHERY INDICATORS: LEARNING FROM THE MARINE REALM

At the opposite end of aquatic food chains, indicators related to fisheries are widely used in the marine realm (e.g. Coll et al., 2016; http://www.indiseas.org/) although their inland counterparts are still surprisingly rarely considered by policymakers despite the importance of freshwater fisheries, especially in many developing countries (Brummett et al., 2013). This is exemplified by the Food and Agriculture Organisation of the United Nations (FAO, 2015; http://www.fao.org/fishery) fishery catch database, which found in 2008 alone that >10 million metric tonnes of fish were produced by freshwater fisheries (Brummett et al., 2013). These productivity data are freely available and, in marine ecosystems, are employed as useful indicators of both fishery health and overall ecosystem condition. *Indicators for the Seas* (http://www.indiseas.org/), a programme that evaluates effects of fishing on marine ecosystems, suggests using multiple fish-based indicators simultaneously, such as average fish size (which depends on the impact of harvests), the average trophic level of landed catch (i.e. catch composition), and total biomass of surveyed species (i.e. total production; Coll et al., 2016) as a fundamental set of indicators. Analogous indicators can be derived from freshwater ecosystems, and where fishery yields are influenced by both human and environmental drivers, including harvesting, nutrient loading, climate, and pollution; clear links can be established to these stressors that could be rolled out as a basic global bio-assessment scheme (Allan et al., 2005; Pauly and Froese, 2012; Welcomme et al., 2010). Catch data allow for the assessment of not only changes in total landings but also for changes in species composition, trophic level, and abundance, which are all recognised as good indicators of the impacts of exploitation (Allan et al., 2005). Additionally, recent evidence suggests that yield is directly correlated with fish biodiversity (Brooks et al., 2016).

Inland fisheries catches are reported via the FAO for 156 countries (Bartley et al., 2015; Garibaldi, 2012) and more disaggregated catch and effort data (e.g. number of fishers, boats, and fishing gear) are also often available. However, data quality is variable and while some countries have a >40 year time series of catch and effort data (e.g. Malawi; Weyl et al., 2010), data are still lacking for others (Welcomme, 2011). Additionally, some lake systems are not fished, and in others there are considerable challenges regarding the accuracy of reported data (Bartley et al., 2015; Garibaldi, 2012; Welcomme et al., 2010). The social and economic value of inland fisheries is gaining increased recognition (Brummett et al., 2013; Welcomme et al., 2010), and we contend that they have additional value as global indicators for ecosystem change, which could be adopted relatively quickly by building on the approaches already developed in marine systems.

5. MOLECULAR-BASED INDICATORS

Classic indicators of freshwater diversity focus only on a small subset of taxonomic groups embedded within the wider food web, such as fish. Furthermore, these indices are often not standardised across regions, so comparisons of ecological state are difficult. Trait-based approaches or measures of functional diversity provide one avenue to standardise indicator assessments. Alternatively, assessments based on molecular markers, particularly nucleic acids (DNA or RNA), have the potential to provide both taxonomic and functional data, across all three domains of life, and at a fraction of the cost and time associated with traditional approaches. Recent reductions in the cost and capacity of DNA sequencing technologies (i.e. Next-Generation Sequencing; NGS), alongside more effective methods of extracting DNA from environmental samples and other novel DNA-based techniques (e.g. metabarcoding of pooled invertebrate samples), have highlighted the potential for these molecular methods to become a unifying and globally comparable indicator of freshwater diversity (Bohmann et al., 2014; Lodge et al., 2012; Mächler et al., 2016).

Environmental DNA (eDNA) is ubiquitous in ecosystems, originating from organismal excretions and loose cells (including microbes), among many other sources. This ubiquity of eDNA, coupled with the assumption that all taxa produce it and that sampling methods can be standardised, has led to an emergent research field, promoting eDNA as the solution for rapid, noninvasive monitoring of species from any environment (Bohmann et al., 2014). eDNA approaches typically focus on the isolation of free DNA directly from the environment followed by some form of taxon-specific quantification based around PCR amplification of the eDNA (e.g. for NGS community-level analysis, or to quantify specific populations via qPCR). Thus, these approaches are still intrinsically linked to providing taxonomic-based assessments of biodiversity, albeit at a range of taxonomic breadths decided by the practitioner. This link to taxonomy in itself is not a major problem, as eDNA approaches can be designed to simultaneously target the majority of major taxonomic or functional groups, moving freshwater monitoring away from a narrow focus around a few specific species (e.g. fish). This combined with the speed and ever-decreasing cost of eDNA approaches makes them a very attractive monitoring option. Furthermore, by combining eDNA-based data targeting maximum taxonomic coverage with machine-learning-based approaches, the ecological networks (e.g. food webs and/or species interaction networks) present in the system can be resolved (Vacher et al., 2016). Subsequently this information can generate a range of network properties (e.g. connectance, linkage strength; see Box 1) that can provide a universal set of measures for monitoring that do not require an a priori knowledge of the ecosystem or a specific composition of species to be present. A major advantage of these more holistic measures over traditional trait-based approaches is that they can take into account whole system properties and also indirect effects of species interactions—which often give rise to 'ecological surprises' or hystereses, such as trophic cascades and regime shifts, in biomonitoring schemes. In essence, they represent a next step beyond autecological species traits to more synecological attributes that arise from those traits at the whole-system level (Gray et al., 2014).

Despite its huge potential and appeal, our current ability to detect eDNA within aquatic systems is still primarily a function of time from cellular release, as exposure to UV light and extracellular enzymatic activity from microbes cleaves eDNA into shorter fragments until it is impractical to investigate or is completely degraded (Rees et al., 2014). These constraints need to be far better characterised and understood as part of the quality control process we now need to develop if the field is to mature sufficiently to deliver fully on its promise. Spatial limits to detection are determined by how far eDNA diffuses or actively travels before it is fully degraded, which can be over 10 km from its source within a few days or weeks (Deiner and

Altermatt, 2014). Alternatively, these methodological constraints could be considered strengths depending on the context and scale of assessment in which eDNA approaches are deployed—especially for global monitoring that aims to use a relatively small number of widely dispersed spot measurements to characterise ecological systems over much larger scales. If monitoring is to move away from point measures of biodiversity to catchment-scale evaluations, then eDNA holds massive potential as it has the ability to integrate measure of biodiversity over large spatial, but short temporal, scales. Furthermore, if more precise predictions of eDNA sources are required (e.g. for fine-scale evaluation of an organisms range), then it is feasible that DNA degradation rates from controlled sources may be used to evaluate how long eDNA persists within the environment. In turn this generates additional data for monitoring, as it provides information on UV penetration and microbial activity within aquatic systems, which could contribute towards a metric of aquatic ecosystem health and potentially of ecosystem services, such as water purification by microbes.

Arguably, one of the main strengths of nucleic acid (DNA or RNA)-based approaches for monitoring is broad-scale evaluations of functionally important groups, notably microbes, via either nontargeted shotgun methods (i.e. NGS metagenomics and metatranscriptomics) or targeted approaches (NGS amplicon techniques or qPCR). Microbial communities underpin almost all ecosystem functions and support the key biogeochemical processes and macronutrient cycles. In addition, microbial populations and the associated functional genes that regulate macronutrient cycling are reasonably well known (Nedwell et al., 2016, this volume), and changes in these across environmental gradients are increasingly being characterised in the field (e.g. Lansdown et al., 2016). Yet despite the functional importance of these groups, and established NGS and qPCR approaches for quantifying them (e.g. Lansdown et al., 2016; Li et al., 2015; Papaspyrou et al., 2014; Thompson et al., 2016), they are still rarely included in monitoring of diversity change. Combining information on functionally important groups of microbes with eDNA-based assessments of biodiversity could provide a new paradigm in monitoring, as it provides direct insights into the functional capacity and health of ecosystems. Furthermore, a nontargeted metagenomic approach will also capture microbial responses to novel energy sources used for their metabolism (e.g. organophosphates, hydrocarbons, lipids), providing a rapid indicator of potential pollution events. For example, a small spill of the organophosphate chlorpyrifos, which wiped out invertebrate life over a 15-km stretch of a UK river, triggered a greater than three orders of magnitude increase in the abundance of microbial populations containing the organophosphate hydrolase gene (*opd*; Thompson et al., 2016). In addition, collection of the samples required for all molecular approaches is straightforward and therefore there is potential for convergence with citizen science approaches (Box 2) and global coverage is feasible.

6. INDICATORS OF CHANGE ACROSS SPACE AND TIME

Macroinvertebrate, fishery, molecular-based, and decomposition-based indicators all have the potential to be standardised across space and time, but current coverage is limited and, in many instances, data from developing countries are absent or estimated. There is, therefore, a need for indicators which already have a standardised methodology across a global scale and long time span. For instance, The Global Lake Ecological Observatory Network (GLEON; http://gleon.org/) has a worldwide network of data gathering buoys in lakes to document changes that occur in response to different stressors. Data collected include water temperature, dissolved oxygen, and phytoplankton chlorophyll fluorescence. Similarly, International Rivers (www.internationalrivers.org/worldsrivers) has coverage of river health across 50 major river basins. Many indicators are associated with drivers of change (e.g. number of dams and percentage of nonnative fish) but indicators of species richness are also available, although they are taxonomically biased with a focus on mammals and birds. These global observatories clearly offer potential for building on existing infrastructure or sampling programmes to incorporate some of the new measures advocated here, and at relatively minimal cost when compared with the creation of new projects from scratch (Nedwell et al., 2016, this issue).

Remote sensing technology also has the advantage of offering global coverage data that can be standardised and repeatable over multiple scales in time and space. Currently, despite initially high development and deployment costs, sensors on board satellites offer arguably the most cost-effective long-term global monitoring capability, and recent developments have opened new possibilities for freshwater ecosystems. Earth observation platforms are now providing increasingly higher spatial and spectral resolution data that can be adapted to gain meaningful measurements as biological indicators of change, for instance aquatic vegetation coverage, phytoplankton functional types, and water column chlorophyll-*a* concentration (Devred et al., 2013; Legleiter and Overstreet, 2014). Applications to lake ecosystems are widespread (Matthews, 2011), and global monitoring programmes are already being implemented, including Diversity II (www.diversity2.info)

and Globolakes (Fig. 1, www.globolakes.ac.uk) which, even in these early stages, already cover 340 and 971 lakes each, respectively. A promising avenue for freshwater ecosystem applications lies in hyperspectral technologies: the hundreds of narrow spectral bands provide more information on water column constituents such as phycocyanin, chlorophyll-*a*, suspended mater, phytoplankton functional types, and benthic composition (Devred et al., 2013). However, there are still high costs of transmitting, storing, and manipulating these very large datasets, although the rapidly moving developments in infrastructure and computational capacity are likely to resolve these issues in the near future (Hestir et al., 2015), and several new hyperspectral global mapping programmes are due to launch soon (e.g. HyspIRI, https://hyspiri.jpl.nasa.gov and EnMap, http://www.enmap.org). Moreover, lightweight hyperspectral cameras on board small airborne devices, such as drones, open significant possibilities for monitoring key freshwater ecosystems such as headwater streams that are currently out of reach from satellites (Lee et al., 2011; Pajares, 2015).

Remote sensing can also be used to evaluate past environmental stressors by using repositories of historical aerial imagery and maps that can be digitised and compared with more recent images. For example, historical aerial photographs have been used to quantify the long-term geomorphic

Legend
- UK NERC GloboLakes sites. Source: Politi et al. (2016)
- Country Boundaries. Source: Digital Chart of the World v1.0.0, Harvard University (c) 2015

Fig. 1 Locations of 1000 lakes monitored by the Globolakes project using MERIS data between 2002 and 2012. Data cover 15 spectral bands across the globe every 3 days with 260 × 300 m resolution. Indicators measured include physical variables such as temperature and turbidity, and biotic indicators such as chlorophyll-*a*. E. Politi, S. MacCallum, M.E.J. Cutler, C.J. Merchant, J.S. Rowan and T.P. Dawson (2016). Selection of a network of large lakes and reservoirs suitable for global environmental change analysis using Earth Observation, *International Journal of Remote Sensing*, 37 (13), 3042–3060, http://dx.doi.org/10.1080/01431161.2016.1192702.

response of river channels to dam construction and water abstraction (e.g. Gurnell et al., 1994; Tiegs and Pohl, 2005). Another option involves palaeo techniques, where hindcasting of past conditions and diversity is achieved by examining organisms preserved in lake sediment profiles (e.g. Smol and Cumming, 2008). This approach could benefit from taxonomy-free DNA techniques to reconstruct past community responses to stressors, and evaluate ecosystem and diversity change through time. For example, ancient DNA (aDNA), an emerging tool which is used in a similar way to eDNA (Bohmann et al., 2014; Kidwell, 2015; Rawlence et al., 2014), could be employed to identify long-term shifts in community structure due to past climate change or environmental perturbations.

7. CONCLUSIONS AND FUTURE DIRECTIONS

From even our cursory 'horizon scanning' overview presented here, it is clear that all of the indicators of ecosystem state suggested are potentially relevant at a global scale, where currently there is a huge lack of standardised and suitable biological monitoring data. However, further research or development is needed either to synthesise regional methods (e.g. macroinvertebrate and fishery indices), to expand global coverage (e.g. decomposition), or in other cases to carry out additional cross-validation and quality control checks (e.g. molecular techniques, remote sensing). Given that many key environmental drivers of freshwater biodiversity and ecosystem processes and services are now available at a global scale with relatively fine resolution (e.g. Domisch et al., 2015), these data can be used to parameterize high-level models in order to extrapolate diversity to the complete landscape level. For example, Kaelin and Altermatt (2016) used a set of 39 environmental variables to predict diversity at different taxonomic resolution in >20,000 catchments, giving fine-scale insights into the spatial patterns of aquatic diversity across a vast area in ways that would have been inconceivable even just a few years ago.

We suggest that these indices should be used with a high temporal consistency where methods are equivalent among years, so that diversity or ecosystem functioning measures collected today can be directly compared to those in the future. This consistency will be challenging to achieve as, for example, the methods for simple measurements of ambient temperatures have changed across the 19th or 20th century, and the emerging technologies of the 21st century are developing at such a pace that the baselines are continually shifting, so a 'lowest common denominator' needs to be established as soon as possible, while also having sufficient redundancy and scope to also enrich the information further in the future as the technology matures. Another challenge is that indices often only provide metrics which are loosely connected to ecosystem state and diversity (e.g. GLEON and remote sensing) or are in the early stages of development, so expenses are often still high (e.g. molecular techniques) and quality control/assurance is also relatively embryonic compared with the long-established but localised monitoring schemes that have already been in place in some parts of the world for many decades.

Furthermore, the future of freshwater diversity monitoring should also involve indicators to disentangle the multiple impacts of various cooccurring environmental stressors (Box 1). In particular, this is where new molecular tools and a focus on the huge functional and phylogenetic diversity contained within the members of the most neglected of freshwater taxa—those in the microbial 'black box'—hold particular promise, as the potential number and scope of gene-to-ecosystem response variables is so vast that they are arguably the best placed as future bioindicators for disentangling multiple stressor effects at global scales (Jackson et al., 2016). In summary, we recommend that future monitoring in freshwaters should use a combination of the above molecular, remote sensing, and citizen science approaches to tackle the big issues facing our planet.

ACKNOWLEDGMENTS

We are grateful to Future Earth, The Global Mountain Biodiversity Assessment and IPBES for the organisation of a workshop: 'Global Biodiversity Assessment and Monitoring—Science, Data and Infrastructure Needs for IPBES and Beyond' in March 2016, where the idea for this paper was developed.

REFERENCES

Allan, J.D., Abell, R., Hogan, Z.E.B., Revenga, C., Taylor, B.W., Welcomme, R.L., Winemiller, K., 2005. Overfishing of inland waters. Bioscience 55, 1041–1051.

Altermatt, F., Seymour, M., Martinez, N., 2013. River network properties shape α-diversity and community similarity patterns of aquatic insect communities across major drainage basins. J. Biogeogr. 40, 2249–2260.

Baird, D.J., Hajibabaei, M., 2012. Biomonitoring 2.0: a new paradigm in ecosystem assessment made possible by next-generation DNA sequencing. Mol. Ecol. 21, 2039–2044.

Bartley, D.M., De Graaf, G.J., Valbo-Jørgensen, J., Marmulla, G., 2015. Inland capture fisheries: status and data issues. Fish. Manag. Ecol. 22, 71–77.

Beche, L.A., McElravy, E.P., Resh, V.H., 2006. Long-term seasonal variation in the biological traits of benthic macroinvertebrates in two Mediterranean-climate streams in California, USA. Freshw. Biol. 51, 56–75.

Bellard, C., Bertelsmeier, C., Leadley, P., Thuiller, W., Courchamp, F., 2012. Impacts of climate change on the future of biodiversity. Ecol. Lett. 15, 365–377.

Boersma, K.S., Dee, L.E., Miller, S.J., Bogan, M.T., Lytle, D.A., Gitelman, A.I., 2016. Linking multidimensional functional diversity to quantitative methods: a graphical hypothesis-evaluation framework. Ecology 97, 583–593.

Bohmann, K., Evans, A., Gilbert, M.T.P., Carvalho, G.R., Creer, S., Knapp, M., Yu, D.W., de Bruyn, M., 2014. Environmental DNA for wildlife biology and biodiversity monitoring. Trends Ecol. Evol. 29, 358–367.

Brooks, E.G.E., Holland, R.A., Darwall, W.R.T., Eigenbrod, F., 2016. Global evidence of positive impacts of freshwater biodiversity on fishery yields. Glob. Ecol. Biogeogr. 25, 553–562.

Brummett, R.E., Beveridge, M., Cowx, I.G., 2013. Functional aquatic ecosystems, inland fisheries and the Millennium Development Goals. Fish Fish. 14, 312–324.

Calizza, E., Costantini, M.L., Rossi, L., 2015. Effect of multiple disturbances on food web vulnerability to biodiversity loss in detritus-based systems. Ecosphere 6, 1–20.

Charvet, S., Statzner, B., Usseglio-Polatera, P., Dumont, B., 2000. Traits of benthic macroinvertebrates in semi-natural French streams: an initial application to biomonitoring in Europe. Freshw. Biol. 43, 277–296.

Chauvet, E., Ferreira, V., Giller, P.S., Mckie, B.G., Tiegs, S.D., Woodward, G., et al., 2016. Litter decomposition as an indicator of stream ecosystem functioning at local-to-continental scales: insights from the European *RivFunction* Project. Adv. Ecol. Res. 55, 99–182.

Coll, M., Shannon, L.J., Kleisner, K.M., Juan-Jordá, M.J., Bundy, A., Akoglu, A.G., Banaru, D., Boldt, J.L., Borges, M.F., Cook, A., Diallo, I., 2016. Ecological indicators to capture the effects of fishing on biodiversity and conservation status of marine ecosystems. Ecol. Indic. 60, 947–962.

Costello, M.J., Claus, S., Dekeyzer, S., Vandepitte, L., Tuama, E.O., Lear, D., Tyler-Walters, H., 2015. Biological and ecological traits of marine species. PeerJ 3, e1201. http://dx.doi.org/10.7717/peerj.1201.

Crain, C.M., Kroeker, K., Halpern, B.S., 2008. Interactive and cumulative effects of multiple human stressors in marine systems. Ecol. Lett. 11, 1304–1315.

Crowl, T.A., McDowell, W.H., Covich, A.P., Johnson, S.L., 2001. Freshwater shrimp effects on detrital processing and nutrients in a tropical headwater stream. Ecology 82, 775–783.

Deiner, K., Altermatt, F., 2014. Transport distance of invertebrate environmental DNA in a natural river. PLoS One 9, e88786.

Demars, B.O.L., Kemp, J.L., Friberg, N., Usseglio-Polatera, P., Harper, D.M., 2012. Linking biotopes to invertebrates in rivers: biological traits, taxonomic composition and diversity. Ecol. Indic. 23, 301–311.

Devred, E., Turpie, K., Moses, W., Klemas, V., Moisan, T., Babin, M., et al., 2013. Future retrievals of water column bio-optical properties using the hyperspectral infrared imager (HyspIRI). Remote Sens. 5, 6812–6837.

Dickens, C.W., Graham, P.M., 2002. The South African Scoring System (SASS) version 5 rapid bioassessment method for rivers. Afr. J. Aquat. Sci. 27, 1–10.

Doledec, S., Statzner, B., Bournard, M., 1999. Species traits for future biomonitoring across ecoregions: patterns along a human-impacted river. Freshw. Biol. 42, 737–758.

Domisch, S., Amatulli, G., Jetz, W., 2015. Near-global freshwater-specific environmental variables for biodiversity analyses in 1 km resolution. Sci. Data 2, 150073.

Dudgeon, D., Arthington, A.H., Gessner, M.O., Kawabata, Z., Knowler, D.J., Lévêque, C., Naiman, R.J., et al., 2006. Freshwater biodiversity: importance, threats, status and conservation challenges. Biol. Rev. 81, 163–182.

Durance, I., Bruford, M.W., Chalmers, R., Chappell, N.A., Christie, M., Cosby, B.J., Noble, D., Ormerod, S.J., Prosser, H., Weightman, A., Woodward, G., 2016. The challenges of linking ecosystem services to biodiversity: lessons from a large-scale freshwater study. Adv. Ecol. Res. 54, 87–134.

FAO, 2015. FishStat Plus: universal software for fishery statistical time series. Data and Statistics Unit, FAO, Rome. http://www.fao.org/fishery/statistics/software/fishstat/en.
Follett, R., Strezov, V., 2015. An analysis of citizen science based research: usage and publication patterns. PLoS One 10, e0143687.
Friberg, N., Bonada, N., Bradley, D.C., Dunbar, M.J., Edwards, F.K., Grey, J., 2011. Biomonitoring of human impacts in freshwater ecosystems: the good, the bad and the ugly. Adv. Ecol. Res. 44, 1–68.
Garibaldi, L., 2012. The FAO global capture production database: a six-decade effort to catch the trend. Mar. Policy 36, 760–768.
Global Biodiversity Outlook 4, 2014. A mid-term assessment of progress towards the implementation of the strategic plan for biodiversity 2011–2020. Secretariat of the CBD, Montréal, Canada. Available at https://www.cbd.int/gbo/gbo4/publication/gbo4-en.pdf.
Gray, C., Baird, D.J., Baumgartner, S., Jacob, U., Jenkins, G.B., O'Gorman, E.J., Lu, X., Ma, A., Pocock, M.J., Schuwirth, N., Thompson, M., 2014. FORUM: ecological networks: the missing links in biomonitoring science. J. Appl. Ecol. 51, 1444–1449.
Gray, C., Hildrew, A., Lu, X., Ma, A., McElroy, D., Monteith, D., et al., 2016. Recovery and nonrecovery of freshwater food webs from the effects of acidification. Adv. Ecol. Res. 55, 475–534.
Griffiths, N.A., Tiegs, S.D., 2016. Organic-matter decomposition along a temperature gradient in a forested headwater stream. Freshw. Sci. 35, 518–533.
Gurnell, A.M., Downward, S.R., Jones, R., 1994. Channel planform change on the River Dee meanders, 1876–1992. Regul. Rivers Res. Manag. 9, 187–204.
Hestir, E., Brando, V.E., Bresciani, M., Giardino, C., Matta, E., Villa, P., Dekker, A., 2015. Measuring freshwater aquatic ecosystems: the need for a hyperspectral global mapping satellite mission. Remote Sens. Environ. 167, 181–195.
Hladyz, S., Åbjörnsson, K., Giller, P.S., Woodward, G., 2011. Impacts of an aggressive riparian invader on community structure and ecosystem functioning in stream food webs. J. Appl. Ecol. 48, 443–452.
Hoffmann, A., Penner, J., Vohland, K., Cramer, W., Doubleday, R., Henle, K., Kõljalg, U., Kühn, I., Kunin, W., Negro, J.J., Penev, L., 2014. The need for an integrated biodiversity policy support process—building the European contribution to a global Biodiversity Observation Network (EU BON). Nat. Conserv. 6, 49.
Huddart, J.E., Thompson, M.S., Woodward, G., Brooks, S.J., 2016. Citizen science: from detecting pollution to evaluating ecological restoration. WIREs Water 3, 287–300.
Jackson, M.C., Loewen, C.J., Vinebrooke, R.D., Chimimba, C.T., 2016. Net effects of multiple stressors in freshwater ecosystems: a meta-analysis. Glob. Chang. Biol. 22, 180–189.
Jenkins, G.B., Woodward, G., Hildrew, A.G., 2013. Long-term amelioration of acidity accelerates decomposition in headwater streams. Glob. Chang. Biol. 19, 1100–1106.
Kaelin, K., Altermatt, F., 2016. Landscape-level predictions of diversity in river networks reveal opposing patterns for different groups of macroinvertebrates. Aquat. Ecol. 50, 283–295.
Kidwell, S.M., 2015. Biology in the anthropocene: challenges and insights from young fossil records. Proc. Natl. Acad. Sci. U.S.A. 112, 4922–4929.
Kokeš, J., Zahrádková, S., Němejcová, D., Hodovský, J., Jarkovský, J., Soldán, T., 2006. The PERLA system in the Czech Republic: a multivariate approach for assessing the ecological status of running waters. In: The Ecological Status of European Rivers: Evaluation and Intercalibration of Assessment Methods. Springer, Netherlands, pp. 343–354.
Lansdown, K., McKew, B.A., Whitby, C., Heppell, C.M., Dumbrell, A.J., Binley, A., Olde, L., Trimmer, M., 2016. Importance and controls of anaerobic ammonium oxidation influenced by riverbed geology. Nat. Geosci. 9, 357–360.
Lee, B.S., McGwire, K.C., Fritsen, C.H., 2011. Identification and quantification of aquatic vegetation with hyperspectral remote sensing in western Nevada rivers, USA. Int. J. Remote Sens. 32, 9093–9117.

Legleiter, C.J., Overstreet, B.T., 2014. Retrieving river attributes from remotely sensed data: an experimental evaluation based on field spectroscopy at the outdoor stream lab. River Res. Appl. 30, 671–684.

Li, J., Nedwell, D.B., Beddow, J., Dumbrell, A.J., McKew, B.A., Thorpe, E.L., Whitby, C., 2015. amoA gene abundances and nitrification potential rates suggest that benthic ammonia-oxidizing bacteria (AOB) not archaea (AOA) dominate N cycling in the Colne estuary, UK. Appl. Environ. Microbiol. 81, 159–165.

Lodge, D.M., Turner, C.R., Jerde, C.L., Barnes, M.A., Chadderton, L., Egan, S.P., Feder, J.L., Mahon, A.R., Pfrender, M.E., 2012. Conservation in a cup of water: estimating biodiversity and population abundance from environmental DNA. Mol. Ecol. 21, 2555–2558.

Mächler, E., Deiner, K., Spahn, F., Altermatt, F., 2016. Fishing in the water: effect of sampled water volume on environmental DNA-based detection of macroinvertebrates. Environ. Sci. Technol. 50, 305–312.

Matthews, M.W., 2011. A current review of empirical procedures of remote sensing in inland and near-coastal transitional waters. Int. J. Remote Sens. 32, 6855–6899.

Mcgill, B.J., Enquist, B.J., Weiher, E., Westoby, M., 2006. Rebuilding community ecology from functional traits. Trends Ecol. Evol. 21, 178–185.

Menezes, S., Baird, D.J., Soares, A., 2010. Beyond taxonomy: a review of macroinvertebrate trait-based community descriptors as tools for freshwater biomonitoring. J. Appl. Ecol. 47, 711–719.

Nedwell, D.B., Underwood, G.J.C., McGenity, T.J., Whitby, C., Dumbrell, A.J., 2016. The Colne Estuary: a long-term microbial ecology observatory. Adv. Ecol. Res. 55, 225–281.

Nicholson, E., Collen, B., Barausse, A., Blanchard, J.L., Costelloe, B.T., Sullivan, K.M., Underwood, F.M., Burn, R.W., Fritz, S., Jones, J.P., McRae, L., 2012. Making robust policy decisions using global biodiversity indicators. PLoS One 7, e41128.

O'Gorman, E.J., Fitch, J.E., Crowe, T.P., 2012. Multiple anthropogenic stressors and the structural properties of food webs. Ecology 93, 441–448.

Pajares, G., 2015. Overview and current status of remote sensing applications based on unmanned aerial vehicles (UAVs). Photogramm. Eng. Remote. Sens. 81, 281–330.

Papaspyrou, S., Smith, C.J., Dong, L.F., Whitby, C., Dumbrell, A.J., Nedwell, D.B., 2014. Nitrate reduction functional genes and nitrate reduction potentials persist in deeper estuarine sediments. Why? PLoS One 9, e94111.

Pauly, D., Froese, R., 2012. Comments on FAO's state of fisheries and aquaculture, or 'SOFIA 2010'. Mar. Policy 36, 746–752.

Pereira, H.M., Ferrier, S., Walters, M., Geller, G.N., Jongman, R.H.G., Scholes, R.J., Bruford, M.W., Brummitt, N., Butchart, S.H.M., Cardoso, A.C., Coops, N.C., 2013. Essential biodiversity variables. Science 339, 277–278.

Perkins, D.M., Bailey, R.A., Dossena, M., Gamfeldt, L., Reiss, J., Trimmer, M., Woodward, G., 2015. Higher biodiversity is required to sustain multiple ecosystem processes across temperature regimes. Glob. Chang. Biol. 21, 396–406.

Piggott, J.J., Salis, R.K., Lear, G., Townsend, C.R., Matthaei, C.D., 2015a. Climate warming and agricultural stressors interact to determine stream periphyton community composition. Glob. Chang. Biol. 21, 206–222.

Piggott, J.J., Townsend, C.R., Matthaei, C.D., 2015b. Reconceptualizing synergism and antagonism among multiple stressors. Ecol. Evol. 5, 1538–1547.

QUINTESSENCE Consortium, 2016. Networking our way to better ecosystem service provision. Trends Ecol. Evol. 31, 105–115.

Rabeni, C.F., Doisy, K.E., Zweig, L.D., 2005. Stream invertebrate community functional responses to deposited sediment. Aquat. Sci. 67, 395–402.

Raffaelli, D., Bullock, J.M., Cinderby, S., Durance, I., Emmett, B., Harris, J., Hicks, K., Oliver, T.H., Patersonk, D., White, P.C., 2014. Big data and ecosystem research programmes. Adv. Ecol. Res. 51, 41–77.

Rawlence, N.J., Lowe, D.J., Wood, J.R., Young, J.M., Churchman, G., Huang, Y.T., Cooper, A., 2014. Using palaeoenvironmental DNA to reconstruct past environments: progress and prospects. J. Quat. Sci. 29, 610–626.

Rees, H.C., Maddison, B.C., Middleditch, D.J., Patmore, J.R., Gough, K.C., 2014. REVIEW: the detection of aquatic animal species using environmental DNA—a review of eDNA as a survey tool in ecology. J. Appl. Ecol. 51, 1450–1459.

Revenga, C., Campbell, I., Abell, R., De Villiers, P., Bryer, M., 2005. Prospects for monitoring freshwater ecosystems towards the 2010 targets. Philos. Trans. R. Soc. Lond. B Biol. Sci. 360, 397–413.

Salzburger, W., Mack, T., Verheyen, E., Meyer, A., 2005. Out of Tanganyika: genesis, explosive speciation, key-innovations and phylogeography of the haplochromine cichlid fishes. BMC Evol. Biol. 5, 17.

Scholes, R.J., Biggs, R., 2005. A biodiversity intactness index. Nature 434, 45–49.

Seymour, M., Deiner, K., Altermatt, F., 2016. Scale and scope matter when explaining varying patterns of community diversity in riverine metacommunities. Basic Appl. Ecol. 17, 134–144.

Silvertown, J., 2016. A new dawn for citizen science. Trends Ecol. Evol. 24, 467–471.

Smith, M.J., Kay, W.R., Edward, D.H.D., Papas, P.J., Richardson, K.S.J., Simpson, J.C., Pinder, A.M., Cale, D.J., Horwitz, P.H.J., Davis, J.A., Yung, F.H., 1999. AusRivAS: using macroinvertebrates to assess ecological condition of rivers in Western Australia. Freshw. Biol. 41, 269–282.

Smol, J.P., Cumming, B.F., 2008. Tracking long-term changes in climate using algal indicators in lake sediments. J. Phycol. 36, 986–1011.

Stark, J.D., 1985. A macroinvertebrates community index of water quality for stony streams. National Water and Soil Conservation Authority Wellington, Wellington, New Zealand.

Statzner, B., Hoppenhaus, K., Arens, M.F., Richoux, P., 1997. Reproductive traits, habitat use and templet theory: a synthesis of world-wide data on aquatic insects. Freshw. Biol. 38, 109–135.

Sterk, M., Gort, G., Klimkowska, A., van Ruijven, J., van Teeffelen, A.J.A., Wamelink, G.W.W., 2013. Assess ecosystem resilience: linking response and effects traits to environmental variability. Ecol. Indic. 30, 21–27.

Strayer, D.L., Dudgeon, D., 2010. Freshwater biodiversity conservation: recent progress and future challenges. J. N. Am. Benthol. Soc. 29, 344–358.

Stucki, P., 2010. Methoden zur Untersuchung und Beurteilung der Fliessgewässer: Makrozoobenthos Stufe F. Bundesamt für Umwelt, Bern. Umwelt-Vollzug, 1026, 61.

Thompson, M.S.A., Bankier, C., Bell, T., Dumbrell, A.J., Gray, C., Ledger, M.E., Lehmann, K., McKew, B.A., Sayer, C.D., Shelley, F., Trimmer, M., Warren, S.L., Woodward, G., 2016. Gene-to-ecosystem impacts of a catastrophic pesticide spill: testing a multilevel bioassessment approach in a river ecosystem. Freshw. Biol., in press.

Tiegs, S.D., Pohl, M.M., 2005. Planform channel dynamics of the lower Colorado river: 1976–1999. Geomorphology 69, 14–27.

Tiegs, S.D., Clapcott, J.E., Griffiths, N.A., Boulton, A.J., 2013. A standardized cotton-strip assay for measuring organic-matter decomposition in streams. Ecol. Indic. 32, 131–139.

Tittensor, D.P., Walpole, M., Hill, S.L., Boyce, D.G., Britten, G.L., Burgess, N.D., Butchart, S.H., Leadley, P.W., Regan, E.C., Alkemade, R., Baumung, R., 2014. A mid-term analysis of progress toward international biodiversity targets. Science 346, 241–244.

Vacher, C., Tamaddoni-Nezhad, A., Kamenova, S., Peyrard, N., Moalic, Y., Sabbadin, R., Schwaller, L., Chiquet, J., Smith, M.A., Vallance, J., Fievet, V., Jakuschkin, B., Bohan, D.A., 2016. Chapter one-learning ecological networks from next-generation sequencing data. Adv. Ecol. Res. 54, 1–39.

Welcomme, R.L., 2011. An overview of global catch statistics for inland fish. ICES J. Mar. Sci. 68, 1751–1756.

Welcomme, R.L., Cowx, I.G., Coates, D., Béné, C., Funge-Smith, S., Halls, A., Lorenzen, K., 2010. Inland capture fisheries. Philos. Trans. R. Soc. Lond. B Biol. Sci. 365, 2881–2896.
Weyl, O.L., Ribbink, A.J., Tweddle, D., 2010. Lake Malawi: fishes, fisheries, biodiversity, health and habitat. Aquat. Ecosyst. Health Manag. 13 (3), 241–254.
Woodward, G., Gessner, M.O., Giller, P.S., Gulis, V., Hladyz, S., Lecerf, A., Malmqvist, B., McKie, B.G., Tiegs, S.D., Cariss, H., Dobson, M., Elosegi, A., Ferreira, V., Graça, M.A., Fleituch, T., Lacoursière, J.O., Nistorescu, M., Pozo, J., Risnoveanu, G., Schindler, M., Vadineanu, A., Vought, L.B., Chauvet, E., 2012. Continental-scale effects of nutrient pollution on stream ecosystem functioning. Science 336 (6087), 1438–1440.
Wright, J.F., Furse, M.T., Moss, D., 1998. River classification using invertebrates: RIVPACS applications. Aquat. Conserv. Mar. Freshwat. Ecosyst. 8, 617–631.
WWF, 2014. Living planet report 2014: species and spaces, people and places. In: McLellan, R., Iyengar, L., Jeffries, B., Oerlemans, N. (Eds.), World Wide Fund for Nature, Gland, Switzerland.

INDEX

Note: Page numbers followed by "*f*" indicate figures, "*t*" indicate tables, and "*b*" indicate boxes.

A

Abiotic factors
  gaseous regime, 295–296, 296*f*
  mineral nutrients, 298
  soil pH, 296–297
Aboveground biomass, 441–442
  short-term *vs.* long-term responses, 443–444
Acidified food webs, 481–484, 504
  food web recovery, 496–502, 502*t*
  metrics in, 483*b*
  small-world properties, 484
Acid Neutralising Capacity, 479–480
Acid-tolerant predators, 482
Aerobic autotrophic ammonia oxidisers, 260–261
Aerobic metabolism, of organic matter, 247
Agricultural ecosystems, 47, 53
  taxonomic and functional diversity (FD), 48
Agricultural intensification (AI), 53–54
  AGRIPOPES project, 54–59
  biodiversity and ecosystem services
    general model, 83
    structural equation modelling (SEM), 84–87
    study area, 84
  biodiversity loss, 45
  CAP and, 46*f*, 49–59
  local-level and landscape-level effects, 60–79
    farm-level components, 76–77
    landscape-level components, 74–76
    local (field)-level components, 60–74
  multiple spatial scales, 45–46
  objective and goals, 48–49
  organic-conventional comparisons, 79–83
    functional and taxon-specific responses, 82–83
    landscape context, 80–81

physical environment, 47–48
AGRIcultural POlicy-Induced landscaPe changes: effects on biodiversity and Ecosystem Services (AGRIPOPES), 48–49, 54–59, 55*f*
  general methodology, 55–59
    biodiversity sampling, 56–57
    biological control potential, 57–58
    farms and fields, 55–56
    field-level intensification variables, 58, 62*f*
    landscape-level intensification variables, 58–59, 63*f*
    pesticides, 67–68
Agri-Environmental Schemes (AESs), 51–52, 69
AGRIPOPES. *See* AGRIcultural POlicy-Induced landscaPe changes: effects on biodiversity and Ecosystem Services (AGRIPOPES)
Agroecosystems, 48
AI. *See* Agricultural intensification (AI)
Air-dried litter, 157–158
Allochthonous trophic pathway, 118–119
Allt Lorgy Restoration, 596–599*b*
Allyl thiourea (ATU), 260
Ammonium-oxidising archaea (AOA), 260, 262
Ammonium-oxidising bacteria (AOB), 260, 262
Anaerobic ammonium oxidation, 259
Anammox (AN) pathway, 259
Ancient DNA (aDNA), 629–630
Annual net primary production (ANPP), 439
  plant biomass, 461
  responses to drought, 457–458
Anthropogenic acidification, 477–478
Anthropogenic climate change, 287–288, 344–345
Anthropogenic stressors, 476

Index

Aquatic ecosystems, 107–109, 188
  large-scale ecology and human impacts, 184–193
    in micropollutants (MPs), 188–193
    real-world and research-led experiments in large-scale ecology, 212–215
    urban infrastructure, 211–212
  water management
    EcoImpact project, 196–211
    real-world experiments, 193–196
Aquatic hyphomycetes, 139–141
Arbuscular mycorrhizal (AM) fungi, 293–295, 310f, 311
Archaea, 262–263
Ash-free dry mass (AFDM), 104–106, 120–121
Atlantic Multidecadal Oscillation, 326–327
Autecology to synecology, 401–407
  dispersal importance, 401–404
  species interactions, 404–407
Autochthonous trophic pathways, 118–119
Autotrophic microorganisms, 260

B

Belowground biodiversity, 76
Benthic algae, 550–551
Benthic denitrification, 253
Benthic diatoms, 487–488
Benthic ecosystems
  European kelp forests, 343
  long-term time-series, 341–344
  marine biodiversity and climate change (MarClim), 341–342
  North Sea soft sediment benthos, 343–344
Benthic macroinvertebrate communities, 552
  regression model, 552–553, 553f
Betaproteobacteria nitrifiers, 260–261
Bioclimatic indicators, 390–391
Biodiversity, 185, 373
  ecosystem functioning, 107–109, 116–117, 140–141f
  ecosystem services
    agricultural intensification (AI), 48
    agricultural management, 48
    AGRIPOPES project, 54–59
    CAP affect, 53–54
    cereal agroecosystems, 54
    European-wide scale, 53
    general model, 83
    local-$vs.$ landscape-level components, 77–79
    structural equation modelling (SEM), 84–87
    study area, 84
  functional and taxon-specific responses, 82–83
  hydromorphology, 555–559
  litter decomposition in, 137–153
    decomposition rates, 150–152
    use and interpretation, 152–153
  sampling, 56–57
Biofilm productivity, 244–245
Biogeochemistry, 250, 260–261
Biogeographic distribution, 402
Biogeographic modellers, 373
Biogeographic range limits
  biological processes, 337–338
  biotic $vs.$ abiotic factors, 338
  ecological factors, 338
  environmental conditions, 336–337
  habitat, 338–341
Biological control potential, 57–58
Biological forecasting system
  energetics and cumulative stress, 398–401
  multiple stressors, 397–398
  performance curves, 394–397
  weather, climate and climate indices, 388–394
Biological network exploration
  mofette fields, 313–314
Biota to hydrology, 545–549, 545–547f
  hydromorphology on biodiversity, 555–559
  local physical habitat filters, 549–554
  standard indicator, 559–560
Biotic carbon sequestration, 461
Black box approach, 380
Boreal–Lusitanian biogeographic breakpoint, 332–336
Bray–Curtis dissimilarities, 571–572
Broadbalk and Geescroft Wildernesses, 10–11
Broadbalk wheat experiment, 8–9, 8f
  microbial communities on, 34–35

C

CAP. See Common Agricultural Policy (CAP)
Capital-intensive operations, 49
Carbon capture and storage (CCS) systems, 284–285, 312–313
CEllulose Decomposition EXperiment (CELLDEX), 158–160, 159f
Cereal agroecosystems, 54
Citizen science networks, 622b, 622f
Climate change, 388–394, 439–444
  common pitfalls and consequences, 383–407
    autecology to synecology, 401–407
    biological details, 388–401
    scale and data resolution, 384–388
  data analysis methods, 448–449
    accumulated approach across sites, 449
    field site experiments, 444–447
    piecewise approach within sites, 449–450
    treatment effect size and certainty, 447–448
  ecological niche, 376–379
  forecasting biogeographic responses, 372–376
  on intertidal benthic species, 347–352
  limitations, 407–413
    ensemble approaches, 411–413
    hybrid models, 408–410
    physiologically and trait-based indicators, 410–411
  mechanism/correlation, 379–383
  in physical environment, 344–347
  quantifying and modelling distributional responses, 352–356
  response pattern types, 456–457
    drought, 457–458
    elevated CO_2, 459–460
    plant biomass, 461
    warming, 458–459
  results
    accumulated patterns across sites, 450–451
    piecewise regression within sites, 454–455
    sites grouped by climate parameters, 452–454

Climate-change velocity, 392–393
Climate indices, 388–394
  vs. local weather data, 393–394
Climate manipulation factors, 439, 450
ClimMani, 462
Coastal Biodiversity and Ecosystem Service Sustainability (CBESS), 272–273
Coldest MAT, 452–454, 459
Cold water species, 351–352
Colne-Blackwater Estuary Ecosystem, 231–233
Colne estuary, 230f
  ecological importance, 228–229
  estuarine microbes, functional ecology in
    archaea, 262–263
    muddy estuarine systems, MPB in, 240–246
    nitrate respiration, 251–259
    nitrification in, 260–262
    organic matter breakdown and recycling, 247–251
    primary production, 238–240
  gas production and estuaries
    isoprene cycling, 267–268
    sulphur gases, 266–267
    trace gas, 265–266
  model estuary-microbial ecology, 229–231, 230f
  saltmarshes, 263–265
  stressors and pollution
    crude oil degradation, 268–269
    engineered nanoparticles, 269–270
  study site description
    catchment and estuary, 231–233, 232f
    nitrogen inputs, 233–234
    phosphate inputs and N:P ratios, 234–236
    physical factors, 236–238
Common Agricultural Policy (CAP)
  agricultural intensification (AI), 48–59
  in Europe, 45, 49–52
Community ecology, 17–24, 303–312
  advances in, 311–312
  archaea and bacteria, 310–311
  fauna, 305–307, 306f
  flora, 305
  fungi, 307–310, 308t
  insect abundance and distribution, 23–24

Community ecology (*Continued*)
  plant community ecology, 18–20
  soil microalgae, 307
  trophic interactions, 21–23
Community-level thermal sensitivity metrics, 407
Conifer plantations, 124–126, 125t
Consumer Stress Models, 405
Correlative approaches, 379–383
Critical mixing ratio (CMR), 240
Crop production, 44
Crude oil degradation, 268–269

D

Dampening response, 457
Data analysis methods, 448–449
  accumulated approach across sites, 449
  break point analysis, 448
  field site experiments, 444–447, 444f
  piecewise approach within sites, 449–450
  treatment effect size and certainty, 447–448
Deciduous broadleaf plantations, 122–124, 123t
Decomposition-based indicators, 623–624
Decomposition rates, 150–152. *See also* Leaf litter decomposition
DecoTabs, 158
Denitrification (DN), 252–255, 254f
Density-dependent diversity effects, 146–147
Dimethyl sulphide (DMS), 266–267
Dispersal potential, 401–404
  functional connectivity, 403–404
  structural connectivity, 403–404
Dissimilatory nitrate reduction to ammonium (DNRA), 252–253, 252f, 256, 259
Dissolved inorganic nitrogen (DIN), 104
Dissolved organic carbon (DOC), 479–480
Dissolved organic nitrogen (DON), 233, 235t, 239, 263–265
Diverse empirical approaches, 193–194
Diversity begets diversity, 185
Drought, 450, 451f, 455t, 457–458
Dynamic energy budget (DEB), 399

E

Earth system models, 461
EcoImpact project, 196–211, 208–210b
  diatom data, 209–210
  field survey, 197–205, 198f, 199t
    biological endpoints, 200
    first insights, 200–205
    quantifying environmental drivers, 199–200
  inferring causality, 205–211
  macroinvertebrate survey, 208–209
  spiking experiments, 209
  water chemistry, 208
Ecological focus areas, 52
Ecological forecast horizon, 184, 376–379
  hybrid models, 408–410
  macrophysiological studies, 378–379
  physiological stress, 376
Ecological hysteresis, 476–477
Ecological niche, 376–379, 377b
Ecological quality ratios (EQRs), 555, 556t
Ecological restoration, 537
Ecological surprises, 476–477
Ecological theory, 345
Ecology
  biological communities, adaptation of, 284–285
  locally extreme environments, as long-term experiments, 284–290
    space for space, 288–289
    space-for-time substitutions, 287–288
    tractable natural model systems, 289–290
  mofettes, 290–314
    carbon capture and storage (CCS), 312–313
    community ecology, 303–312
    food webs and biological network exploration, 313–314
    free air carbon-dioxide enrichment (FACE) experiments, 298–299
    plant ecophysiology, 299–303
    specific abiotic factors, 295–298
Economic and global climatic changes, 44
Ecosystem
  functions, 101–102
    biodiversity, 140–141f, 142–149
    litter decomposition, 163–164

national adoption, 164–166
proposed metrics, 166–168
in stream management, 161–168
services
  agricultural intensification (AI), 48
  agricultural management, 48
  AGRIPOPES project, 54–59
  CAP affect, 53–54
  cereal agroecosystems, 54
  European-wide scale, 53
  general model, 83
  local-*vs*. landscape-level components, 77–79
  structural equation modelling (SEM), 84–87
  study area, 84
stability
  plant community stability, 24–26
  resilience of, 26–28
Ecotoxicological assays, 200
Ectothermic plants and animals, 385–386
eDNA. *See* Environmental DNA (eDNA)
Effective restoration, 594, 594*f*
Electronic Rothamsted Archive (e-RA), 6
Elevated CO_2 (eCO_2), 284–285, 299–300, 440, 449–450, 451*f*, 454, 455*t*, 459–460
El Niño-Southern Oscillation (ENSO), 156–157, 326–327, 380
Ems Dollard estuary, 244–245
Ems estuary, 243–244
Energetics models, 398–401, 399*b*
  dynamic energy budget (DEB), 399
  metabolic theory of ecology (MTE), 399
  niche mapper, 399
  scope for growth (SFG), 399
Engineered nanoparticles, 269–270
Ensemble forecasting approaches, 411–413
Environmental change monitoring, 13–17
  phenological change and trophic asynchrony, 17
  plant communities, 13–15
  plant pathogens, 15–16
Environmental Change Network (ECN), 13
Environmental DNA (eDNA), 270–272, 625–627

Environmental impact assessments (EIA), 284–285, 291*f*
Environmental Stress Models (ESMs), 405
Estuaries
  ecological importance, 228–229
  inorganic N and P ratio, 234–236
  microphytobenthos (MPB), 240–241
Estuarine microbes, functional ecology in
  archaea, 262–263
  muddy estuarine systems, MPB in, 240–246
  nitrate respiration, 251–259
  nitrification, 260–262
  organic matter breakdown and recycling, 247–251
  primary production, 238–240
Estuarine saltmarshes, 263–265
Eucalyptus plantations, 126–129, 127–128*t*
European Academy of Sciences (EASAC), 70
European Agricultural Fund for Rural Development, 51–52
European agroecosystems, 47, 51–52, 54
European cereal-dominated agroecosystems, 54, 55*f*
European kelp forests, 343
European Science Foundation, 54
Evolutionary ecology, 28–32
  beyond Snaydon and Davies, 29–30
  pathogens and weeds, 30–32
Exotic woody species
  in riparian areas, 130, 131*t*, 132*f*
Extracellular polymeric substances (EPS), 246, 248–249
Extreme environments, 285–287

F

FACE. *See* Free air carbon-dioxide enrichment (FACE)
Farmlands, 44–45
Farm-level components, of AI, 76–77
Farms and fields, 55–56
Fertilization effects, 71–72
Field-level AI, 60–61
  intensification variables, 58, 78–79
Fishery indicators, 624–625
Food web construction, 488–489
Food web recovery research, 477

Food web recovery research (*Continued*)
  from acidification, 496–502
  acidified food webs, 481–484
  caveats and future directions, 506–507
  freshwater acidification, 477–481, 505–506
  methods, 487–491
    biota, 487–488
    chemistry, 487
    food web construction, 488–489
    network metrics, 489–490
    sites, 487
    statistical analyses, 489–490
  structure
    acidity effects, 491–493
    directional change in, 494–496
  upland waters monitoring network (UWMN), 485–486, 503–504
Food web structure
  acidity effects, 491–493, 492f, 493t
  directional change in, 494–496
Forecasting models, 401–402, 413f
Forest clear cutting, 131–134, 133t
Fosters Ley-Arable experiment, 9–10, 27–28
Free air carbon-dioxide enrichment (FACE), 298–299, 440–443
Freshwater acidification, 477–481
Freshwater ecosystems, 107, 538–539, 617–618
  analogous indicators, 624–625
  invertebrates, 618
Freshwater flushing time (FWFT), 236–238, 253
Freshwater food webs
  from acidification, 505–506
Freshwater restoration, 539
Functional diversity (FD), 48, 78

G

Gas production and estuaries
  isoprene cycling, 267–268
  sulphur gases, 266–267
  trace gas, 265–266
Gene-area relationships (GARs), 34
Geographic mosaics, 387
Geothermal freshwater systems, 287–288
Global biodiversity, 616

Global freshwater biological monitoring tools, 616–618
  indicators
    decomposition-based, 623–624
    fishery, 624–625
    invertebrates, 618–623
    molecular-based, 625–628
    space and time change, 628–630
Global Lake Ecological Observatory Network (GLEON), 628
Global metaanalysis
  on restoration effects, 569–570
Global warming, 288
Good ecological status (GES), 539
  restoration projects, 577–578
Greening measurement, 52

H

Habitat degradation, 540–541
Highfield Ley-Arable experiment, 9–10, 27–28
Hydrocarbon degradation, 268–270
Hydrological restoration, 589, 591f
Hydromorphology, 550–551, 560–564, 561f
  on biodiversity, 555–559
  degradation, 549–550, 557, 558f
  restoration measures, 560–564, 562–564t
  river restoration projects, 560–564
HYMO stress, 558–560
Hypernutrified Colne estuary, 233, 234f

I

Indicators
  decomposition-based, 623–624
  fishery, 624–625
  invertebrates, 618–623
  molecular-based, 625–628
  space and time change, 628–630
Intergovernmental Platform on Biodiversity and Ecosystem Services (IPBES), 616–617
Intermediate MAT, 452–454
International tundra experiment (ITEX) network, 442
International Union for Conservation of Nature (IUCN), 353
Intertidal benthic species, 347–352
  biogeographic range shifts, 348–349

biological mechanisms, 350–352
population dynamics, 349–350
Invertebrate-mediated decomposition, 115–116
Invertebrates, as indicators, 618–623
In vitro ecotoxicological assays.
    See Ecotoxicological assays
Isoprene cycling, 267–268

L

Landscape diversity, 52–53
Landscape homogenization, 53–54
Landscape-level effects, 60–79
  AI and biodiversity effects, 74–76
  farm-level components, 76–77
  intensification variables, 58–59, 59–60t, 61
  $vs.$ local (field)-level components, 77–79
Large-scale ecology and human impacts, 194
  on ecosystems, 184–193
  micropollutants (MPs), 188–193
  real-world and research-led experiments, 212–215
Large-scale field-level intensification, 45–46
Leaf-associated spore production, 166–168
Leaf litter decomposition, 102
  in bioassessment, 152–153
  ecosystem function, 619–621
  nutrient enrichment effects on, 107–118, 111f
  riparian forest modifications, 118–137
    conifer plantations, 124–126
    deciduous broadleaf plantations, 122–124
    eucalyptus plantations, 126–129
    exotic woody species, 130
    forest clear cutting, 131–134
    pasture, 134–137
Leaf-shredding invertebrates, 145
Litter-consuming macroinvertebrates, 155–156
Litter decomposition
  biodiversity-related mechanisms, 137–153
    decomposition rates, 150–152
    ecosystem functioning, 142–149
    use and interpretation, 152–153
  brown pathways, 101–102

  for functional assessment, 163–164
  stream assessment, 153–161
    extrinsic factors, 153–154
    intrinsic factors, 157–161
    temporal variability, 154–157
Local (field)-level components, 60–74.
    See also Agricultural intensification (AI)
  fertilization, 71–72
  $vs.$ landscape-level effects, 77–79
  pesticides, 67–71
  relationships with yield, 61–66
  sowing density, 73–74
  tillage, 72–73
Locally extreme environments
  in ecology, 284–285
  geographic distribution, 286f
  as long-term experiments, 284–290
  space for space, 288–289
  space-for-time substitutions, 287–288
  tractable natural model systems, 289–290
Local-scale morphological processes, 589–592
Local-scale restoration measures, 548, 548f
Log ratio response (LRR), 447–450
Long-term experiments (LTEs), 4–13
  Broadbalk and Geescroft Wildernesses, 10–11
  classical experiments, 5–9, 5f
  highfield, fosters and Woburn Ley-Arable experiments, 9–10
  insect survey, 11–13
Long-term time-series, benthic ecosystems, 341–344
  European kelp forests, 343
  marine biodiversity and climate change (MarClim), 341–342
  North Sea soft sediment benthos, 343–344
LTEs. See Long-term experiments (LTEs)

M

Macroinvertebrate communities, 203
Macroinvertebrate-driven leaf litter decomposition, 106, 109–110, 134–136
Macroinvertebrates, 487–488

Macrophytes, 479–480, 557–558
  river restoration, 550–551
MacSharry reform, 51
Maiandros flume system, 206*b*
Mann-Kendall trend tests, 494, 497–498*f*, 499*t*
Marine biodiversity and climate change (MarClim), 341–342
Marine biogeographic transition zone, 326–328
  biogeographic range limits
    biological processes, 337–338
    ecological factors, 338
    environmental conditions, 336–337
    habitat, 338–341
  climate change
    in physical environment, 344–347
    quantifying and modelling distributional responses, 352–356
  history, 328–330
  intertidal benthic species, 347–352
    biogeographic range shifts, 348–349
    biological mechanisms, 350–352
    population dynamics, 349–350
  Northeast Atlantic
    benthic ecosystems, 341–344
    biogeographic research, 330–332
    Boreal–Lusitanian biogeographic breakpoint, 332–336
Marine Biological Association, 329, 332
Mean annual temperature (MAT), 447, 452–454, 452*f*
Mechanistic models, 382–383
Metabolic theory of ecology (MTE), 399, 405–406
Methanogenesis, 248–249
Methanogenic Archaea (MA), 248–249
Methanotrophic bacteria, 248–249
Microbial decomposition, 114
Microbial ecology, 310–312
Microhabitats, 386–388
Microphytobenthos (MPB)
  Colne estuary, 228–229
  in muddy estuarine systems
    benthic invertebrates, 241
    Chl *a* concentrations, 241–243
    normalised differential vegetation index (NVDI), 243–244
    photosynthesis and biomass accumulation, 245
    photosynthetic organisms, 240–241
    primary production, 240–241
    sediment Chl *a*, 244
Micropollutants (MPs), 187–188*f*, 187–188*b*
  aquatic ecosystems, 189*t*
  biological organization, 188–193
  definition of, 187–188
  ecotoxicological tests, 190–191
  large-scale ecology and human impacts, 188–193
  stress ecology, 191
  wastewater-borne, 185–188
  wastewater treatment plants (WWTPs), 186*b*, 195*b*
Model estuarine ecosystem, 230–231
Mofette fields, 290–314, 291*f*
  in antiquity, 292*b*, 292*f*
  carbon capture and storage (CCS), 312–313
  community ecology, 303–312
  food webs and biological network exploration, 313–314
  free air carbon-dioxide enrichment (FACE) experiments, 298–299
  plant ecophysiology, 299–303
  specific abiotic factors, 295–298
Molecular-based indicators, 625–628
Molecular microbiology, 270–272
MPB. *See* Microphytobenthos (MPB)
Muddy estuarine systems
  Colne estuary, 228–229
  microphytobenthos (MPB)
    benthic invertebrates, 241
    Chl *a* concentrations, 241–243
    estuarine primary production, 240–241
    normalised differential vegetation index (NVDI), 243–244
    photosynthesis and biomass accumulation, 245
    photosynthetic organisms, 240–241
    sediment Chl *a*, 244
Multiple indicators
  for multiple stressors, 620–621*f*, 620–621*b*
Multiple regression models, 113–114
Multiple stressors, 397–398

N

National Biodiversity Indices, 47
National water agencies, 164–165
Natural ecosystems, 161–162, 476
Net primary production (NPP), 240
Network metrics, 489–490
Next-generation sequencing (NGS), 262, 310–312
  eDNA-based assessments, 627–628
  food-web ecology, 314
  molecular-based indicators, 625–628
Nitrate-accumulating microorganisms, 259
Nitrate respiration, 251–259, 252f
Nitrification, in Colne estuary, 260–262
Nitrite reductase (NirK and NirS), 256, 258f
Nonanalogue climatic conditions, 380–381
Normalised differential vegetation index (NVDI), 243–244
North Atlantic Oscillation (NAO), 17, 243–244, 326–327, 393–394
Northeast Atlantic region
  benthic ecosystems, 341–344
  biogeographic research, 330–332
  Boreal–Lusitanian biogeographic breakpoint, 332–336, 333f
North Sea soft sediment benthos, 343–344
Nutrient enrichment effects
  leaf litter decomposition, 107–118, 111f

O

Oil degraders
  biodiversity and ecology, 287–288
Operational taxonomic units (OTUs), 250
Organic-conventional comparisons, 76, 79–83
  functional and taxon-specific responses, 82–83
  landscape context, 80–81
Organic matter breakdown
  and recycling, 247–251
Oxygen and capacity limitation of thermal tolerance (OCLTT) hypothesis, 395–396

P

Pacific Decadal Oscillation (PDO), 326–327, 380
Paired isotope technique, 253–255
Park Grass experiment (PGE), 6–7, 7f, 16f, 29
  evolutionary ecology, 30–31
  microbial communities on, 33–34
  plant community stability, 25–26
Particulate organic nitrogen (PON), 263–265
Pasture, on leaf litter decomposition, 134–137, 135t
Pelagic and benthic primary production, 247
Periphyton biomass and community tolerance assays, 207
Pesticides, 67–71
PGE. *See* Park Grass experiment (PGE)
Phenological change and trophic asynchrony, 17
Physiological temperature ange (PTR), 394–395
Phytoplankton primary production, 238–240
  critical mixing ratio (CMR), 240
  microphytobenthos (MPB), 246
Plant biomass, 443–444, 461
Plant community, 13–15
  composition of, 440
  reconciling resource ratio and C-S-R theories, 19–20
  Rothamsted experiments, 18–20
  stability, 24–26
Plant ecophysiology, in mofette areas, 299–303
  aerenchyma formation, 301–303
  facilitation, 302–303
  mineral nutrition, 301–303
  photosynthesis and transpiration, 299–301
  root respiration, 301–303
Plant pathogens, 15–16
Point-intercept method, 447
Pollution-induced community tolerance (PICT), 191
Precipitation of wettest month (PWM), 447, 452–454, 453f, 460

Process-based restoration approach, 588–593, 596–599b
Programmes of measures (PoMs), 539, 567, 568f

Q
Quantitative polymerase chain reaction (qPCR), 15–16, 257, 261–262

R
Reach-scale river restoration, 577–578
Real-world and research-led experiments, 185–188
  comparison of, 212, 213t
  in large-scale ecology, 212–215
  urban infrastructure, 211–212
Redundancy analysis (RDA), 21f
REFORM. See REstoring rivers FOR effective catchment Management (REFORM)
Response pattern types, 441–442, 443f, 456–457
  drought, 457–458
  elevated CO_2, 459–460
  plant biomass, 461
  response curve, 441–442
  short-term vs. long-term, 443–444
  warming, 458–459
Restoration effects, 569–578
  confounding factors, 576–578
  differences in measure, 574–576, 575f
  general effect, 570, 571f
  organism groups, 571–574
  stressors vs. ecological key factors, 593t
Restoration intervention, 595, 595f
Restoration measures, 564–567
Restoration paradigm
  current and planned approaches, 569
  indicators, 567–568
  measures, 564–567
  restoring rivers, ways of, 560–564
Restoration potential, 584–585
REstoring rivers FOR effective catchment Management (REFORM), 541–544, 543–544f, 555, 569–570, 578
  global metaanalysis, 576
  hydromorphological stress, 558–559
  macroinvertebrates, 573–574
  organism groups, 572–573, 572f
  river widening, 574–575, 575f
Riparian forest modifications, 118–137, 121t
  conifer plantations, 124–126
  deciduous broadleaf plantations, 122–124
  eucalyptus plantations, 126–129
  exotic woody species, 130
  forest clear cutting, 131–134
  pasture, 134–137
River basin management plans (RBMPs), 539, 567
  policy drivers for, 540b
River floodplain development, 589, 592f
River Habitat Survey (RHS), 555
Riverine ecosystems, 101–102, 162–163
River Invertebrate Prediction and Classification System (RIVPACS), 618–619
River management and restoration, 161–162
River reach-scale processes, 589, 592f
River restoration, in 21st century
  adjustment and maintenance, 588
  biota to hydrology, 545–549
    hydromorphology on biodiversity, 555–559
    local physical habitat filters, 549–554
    standard indicator, 559–560
  definition, 537–538
  effects, 569–578
  need for, 538–539
  policy drivers, 540–541, 540b, 542t
  principles, 590t
  project formulation, 586–587
  project identification, 585–586
  project planning, 587–588
    at catchment scale, 582f, 584
    at local scale, 583f, 586
  REFORM project, 541–544
  restoration paradigm
    current and planned approaches, 569
    indicators, 567–568
    measures, 564–567
    restoring rivers, ways of, 560–564
  trial-and-error approach, 579–581b
RivFunction project, 102, 118, 156f
  biodiversity, 138–139, 142

CELLDEX, 158–160, 159f
  leaf decomposition, 137–138
  litter-bag approach, 165
  for pan-European comparisons, 103, 105f
  riparian forest modifications, 119
Rothamsted classical experiments, 5–9, 5f
  Broadbalk wheat, 8–9, 8f
  Park Grass experiment (PGE), 6–7, 7f
Rothamsted Electronic Archive, 13
Rothamsted Estate
  community ecology, 17–24
    insect abundance and distribution, 23–24
    plant community ecology, 18–20
    trophic interactions, 21–23
  ecosystem stability
    plant community stability, 24–26
    resilience of, 26–28
  environmental change monitoring, 13–17
    phenological change and trophic asynchrony, 17
    plant communities, 13–15
    plant pathogens, 15–16
  evolutionary ecology, 28–32
    beyond Snaydon and Davies, 29–30
    pathogens and weeds, 30–32
  long-term research, 4–13
    Broadbalk and Geescroft Wildernesses, 10–11
    classical experiments, 5–9, 5f
    highfield, fosters and Woburn Ley-Arable experiments, 9–10
    insect survey, 11–13
  soil microbial ecology, 32–35
    on Broadbalk, 34–35
    on PGE, 33–34
Rothamsted Insect Survey' (RIS), 11–13, 12f, 21–23
  insect abundance and distribution, 23–24
Rothamsted Sample Archive (RSA), 6
RuBisCO, 300

S

Saltmarshes, 263–265. *See also* Estuarine saltmarshes
Scale and data resolution, 384–388
  microhabitats, 386–388
  site *vs.* body temperature, 385–386

Scope for growth (SFG), 399
Sea surface temperatures (SSTs), 378–379
Sediment-water interface, 253–255
Sewage fungus, 109–110
Sewage treatment work (STW), 233, 235t
Single-nucleotide polymorphisms (SNPs), 30–31
Site *vs.* body temperature, 385–386, 386f
Slow-decomposing litter, 155
Sluggish biological recovery, 479–480
Small-scale restoration measures, 589–592
Society of Ecological Restoration (SER), 537
Soil hypoxia, 290–291, 301–302
Soil microbial ecology, 32–35
  on Broadbalk, 34–35
  on PGE, 33–34
Soil O_2 concentration, 293–295, 294–295f, 297f
Soil organic matter (SOM), 9–10, 28, 298
  mofette food webs and biological networks, 313–314
  Woburn Ley-Arable experiment, 10
Soluble reactive phosphorus (SRP), 104
South African Scoring System (SASS), 618–619
Species-area relationships (SARs), 34
Species distribution models (SDMs), 340–341, 383, 390–391, 409
Species interactions, 404–407
Statistical analyses, food web structure, 490–491
  acidity effects, 490–491
  directional change in, 491
Stavešinci mofette area, 293–295
Storage effect hypothesis, 25–26
Stream assessment, in litter decomposition, 153–161
  extrinsic factors, 153–154
  intrinsic factors, 157–161
  temporal variability, 154–157
Stressors and pollution
  crude oil degradation, 268–269
  engineered nanoparticles, 269–270
Structural equation modelling (SEM), 83–87, 83f, 85f
Sulphate reduction, 248–249, 249t
  and methanogenesis, 250

Sulphate respiring bacteria (SRB), 248–249, 262–263
Supporting ecosystem service, 246

T
Taxonomic and functional diversity (FD), 48–49, 54
Temperature and rainfall parameters, 447
Temperature-corrected litter decomposition, 121–122, 157
Temperature-dependent enzyme reactions, 346–347
Terrestrial and semiaquatic groups, 572–573
 restoration effect, 571–572
Terrestrial mofettes, 290–291
Thermal performance curves (TPCs), 394–397, 395f
Thermophilization, 442
Tillage effects, 72–73
Total oxidised nitrogen (TOxN), 253, 255
Trace gas production
 in estuary, 265–266
Tractable natural model systems, 289–290
Trait-based indicators, 410–411, 619–621
Trait-litter lignin concentration, 144–145
Trophic Diatom Index (TDI), 557, 558f
Trophic interactions, 21–23, 191, 192f, 193

U
UK Acid Waters Monitoring Network (UKAWMN), 477–478
Upland Waters Monitoring Network (UWMN), 478, 479f, 480–481, 485–487, 486f, 503–504
 acidity and food web structure, 485, 486t
 and acidity gradient, 503–504
 annual electrofishing surveys, 488
 benthic diatoms, 487–488
 biological and chemical recovery, 496
 connectance, 508–509f
 food web efficiency, 524–525f
 food web generality, 520–523f
 food web metrics, 496, 500–501f
 food web redundancy, 526–527f
 food web vulnerability, 516–519f
 linkage density, 510–511f
 maximum trophic height, 514–515f
 mean trophic height, 512–513f
US Clean Water Act, 107

V
Van't Hoff's rule, 151–152

W
Warmest MAT, 452–454
Warming, 444, 450, 451f, 455t, 458–459
Wastewater (WW), 185–188, 186b
 micropollutants (MPs) in, 186b
 spatial field comparisons, 197
Wastewater-born MPs, 185–188, 190–191, 194–195, 212
Wastewater treatment plants (WWTPs), 186b, 188–190, 560
 in Switzerland, 194, 195b
 temporal field comparisons, 197
Water Framework Directive (WFD), 107, 161–162, 539–540
 hydromorphology, 540–541
 water quality, 560
Water Information System for Europe, 567
Water management
 EcoImpact project, 196–211
 field survey, 197–205
 inferring causality, 205–211
 real-world experiments, 193–196
Water quality, 185, 194
Weather, 388–394. *See also* Climate change
 latitudinal gradients, 390–391
 trends *vs.* predictive relationships, 390–391b
Weed traits syndrome, 19
WorldClim database, 447

ADVANCES IN ECOLOGICAL RESEARCH VOLUME 1–55

 CUMULATIVE LIST OF TITLES

Aerial heavy metal pollution and terrestrial ecosystems, **11**, 218
Age determination and growth of Baikal seals (*Phoca sibirica*), **31**, 449
Age-related decline in forest productivity: pattern and process, **27**, 213
Allometry of body size and abundance in 166 food webs, **41**, 1
Analysis and interpretation of long-term studies investigating responses to climate change, **35**, 111
Analysis of processes involved in the natural control of insects, **2**, 1
Ancient Lake Pennon and its endemic molluscan faun (Central Europe; Mio-Pliocene), **31**, 463
Ant-plant-homopteran interactions, **16**, 53
Anthropogenic impacts on litter decomposition and soil organic matter, **38**, 263
Arctic climate and climate change with a focus on Greenland, **40**, 13
Arrival and departure dates, **35**, 1
Assessing the contribution of micro-organisms and macrofauna to biodiversity-ecosystem functioning relationships in freshwater microcosms, **43**, 151
A belowground perspective on Dutch agroecosystems: how soil organisms interact to support ecosystem services, **44**, 277
The benthic invertebrates of Lake Khubsugul, Mongolia, **31**, 97
Big data and ecosystem research programmes, **51**, 41
Biodiversity, species interactions and ecological networks in a fragmented world **46**, 89
Biogeography and species diversity of diatoms in the northern basin of Lake Tanganyika, **31**, 115
Biological strategies of nutrient cycling in soil systems, **13**, 1
Biomanipulation as a restoration tool to combat eutrophication: recent advances and future challenges, **47**, 411
Biomonitoring of human impacts in freshwater ecosystems: the good, the bad and the ugly, **44**, 1
Bray-Curtis ordination: an effective strategy for analysis of multivariate ecological data, **14**, 1

Body size, life history and the structure of host-parasitoid networks, **45**, 135
Breeding dates and reproductive performance, **35**, 69
Can a general hypothesis explain population cycles of forest Lepidoptera? **18**, 179
Carbon allocation in trees; a review of concepts for modeling, **25**, 60
Catchment properties and the transport of major elements to estuaries, **29**, 1
A century of evolution in *Spartina anglica*, **21**, 1
Changes in substrate composition and rate-regulating factors during decomposition, **38**, 101
The challenge of future research on climate change and avian biology, **35**, 237
The challenges of linking ecosystem services to biodiversity: lessons from a large-scale freshwater study, **54**, 87
The colne estuary: a long-term microbial ecology observatory, **55**, 227
Climate change and eco-evolutionary dynamics in food webs, **47**, 1
Climate change impacts on community resilience: evidence from a drought disturbance experiment **46**, 211
Climate change influences on species interrelationships and distributions in high-Arctic Greenland, **40**, 81
Climate-driven range shifts within benthic habitats across a marine biogeographic transition zone, **55**, 325
Climate influences on avian population dynamics, **35**, 185
Climatic and geographic patterns in decomposition, **38**, 227
Climatic background to past and future floods in Australia, **39**, 13
The climatic response to greenhouse gases, **22**, 1
Coevolution of mycorrhizal symbionts and their hosts to metal-contaminated environment, **30**, 69
Community genetic and competition effects in a model pea aphid system, **50**, 239
Communities of parasitoids associated with leafhoppers and planthoppers in Europe, **17**, 282
Community structure and interaction webs in shallow marine hardbottom communities: tests of an environmental stress model, **19**, 189
A complete analytic theory for structure and dynamics of populations and communities spanning wide ranges in body size, **46**, 427
Complexity, evolution, and persistence in host-parasitoid experimental systems with *Callosobruchus* beetles as the host, **37**, 37
Connecting the green and brown worlds: Allometric and stoichiometric predictability of above- and below-ground networks, **49**, 69

Conservation of the endemic cichlid fishes of Lake Tanganyika; implications from population-level studies based on mitochondrial DNA, **31**, 539

Constructing nature: laboratory models as necessary tools for investigating complex ecological communities, **37**, 333

Construction and validation of food webs using logic-based machine learning and text mining, **49**, 225

The contribution of laboratory experiments on protists to understanding population and metapopulation dynamics, **37**, 245

The cost of living: field metabolic rates of small mammals, **30**, 177

Cross-scale approaches to forecasting biogeographic responses to climate change, **55**, 371

Decomposers: soil microorganisms and animals, **38**, 73

The decomposition of emergent macrophytes in fresh water, **14**, 115

Delays, demography and cycles; a forensic study, **28**, 127

Dendroecology; a tool for evaluating variations in past and present forest environments, **19**, 111

Determinants of density-body size scaling within food webs and tools for their detection, **45**, 1

Detrital dynamics and cascading effects on supporting ecosystem services, **53**, 97

The development of regional climate scenarios and the ecological impact of green-house gas warming, **22**, 33

Developments in ecophysiological research on soil invertebrates, **16**, 175

The direct effects of increase in the global atmospheric CO_2 concentration on natural and commercial temperate trees and forests, **19**, 2; **34**, 1

Disentangling the pathways and effects of ecosystem service co-production, **54**, 245

Distributional (In)congruence of biodiversity—ecosystem functioning, **46**, 1

The distribution and abundance of lake dwelling Triclads-towards a hypothesis, **3**, 1

DNA metabarcoding meets experimental ecotoxicology: Advancing knowledge on the ecological effects of Copper in freshwater ecosystems, **51**, 79

Do eco-evo feedbacks help us understand nature? Answers from studies of the trinidadian guppy, **50**, 1

The dynamics of aquatic ecosystems, **6**, 1

The dynamics of endemic diversification: molecular phylogeny suggests an explosive origin of the Thiarid Gastropods of Lake Tanganyika, **31**, 331

The dynamics of field population of the pine looper, *Bupalis piniarius* L. (Lep, Geom.), **3**, 207
Earthworm biotechnology and global biogeochemistry, **15**, 369
Ecological aspects of fishery research, **7**, 114
Eco-evolutionary dynamics of agricultural networks: implications for sustainable management, **49**, 339
Eco-evolutionary dynamics: experiments in a model system, **50**, 167
Eco-evolutionary dynamics of individual-based food webs, **45**, 225
Eco-evolutionary dynamics in a three-species food web with intraguild predation: intriguingly complex, **50**, 41
Eco-evolutionary dynamics of plant–insect communities facing disturbances: implications for community maintenance and agricultural management, **52**, 91
Eco-evolutionary interactions as a consequence of selection on a secondary sexual trait, **50**, 143
Eco-evolutionary spatial dynamics: rapid evolution and isolation explain food web persistence, **50**, 75
Ecological conditions affecting the production of wild herbivorous mammals on grasslands, **6**, 137
Ecological networks in a changing climate, **42**, 71
Ecological and evolutionary dynamics of experimental plankton communities, **37**, 221
Ecological implications of dividing plants into groups with distinct photosynthetic production capabilities, **7**, 87
Ecological implications of specificity between plants and rhizosphere microorganisms, **31**, 122
Ecological interactions among an Orestiid (Pisces: Cyprinodontidae) species flock in the littoral zone of Lake Titicaca, **31**, 399
Ecological studies at Lough Ine, **4**, 198
Ecological studies at Lough Hyne, **17**, 115
Ecology of mushroom-feeding Drosophilidae, **20**, 225
The ecology of the Cinnabar moth, **12**, 1
Ecology of coarse woody debris in temperate ecosystems, **15**, 133; **34**, 59
Ecology of estuarine macrobenthos, **29**, 195
Ecology, evolution and energetics: a study in metabolic adaptation, **10**, 1
Ecology of fire in grasslands, **5**, 209
The ecology of pierid butterflies: dynamics and interactions, **15**, 51
The ecology of root lifespan, **27**, 1
The ecology of serpentine soils, **9**, 225

Ecology, systematics and evolution of Australian frogs, **5**, 37
Ecophysiology of trees of seasonally dry Tropics: comparison among phonologies, **32**, 113
Ecosystems and their services in a changing world: an ecological perspective, **48**, 1
Effect of flooding on the occurrence of infectious disease, **39**, 107
Effects of food availability, snow, and predation on breeding performance of waders at Zackenberg, **40**, 325
Effect of hydrological cycles on planktonic primary production in Lake Malawi Niassa, **31**, 421
Effects of climatic change on the population dynamics of crop pests, **22**, 117
Effects of floods on distribution and reproduction of aquatic birds, **39**, 63
The effects of modern agriculture nest predation and game management on the population ecology of partridges (*Perdix perdix* and *Alectoris rufa*), **11**, 2
Effective river restoration in the 21st century: From trial and error to novel evidence-based approaches, **55**, 529
El Niño effects on Southern California kelp forest communities, **17**, 243
Empirically characterising trophic networks: What emerging DNA-based methods, stable isotope and fatty acid analyses can offer, **49**, 177
Empirical evidences of density-dependence in populations of large herbivores, **41**, 313
Endemism in the Ponto-Caspian fauna, with special emphasis on the Oncychopoda (Crustacea), **31**, 179
Energetics, terrestrial field studies and animal productivity, **3**, 73
Energy in animal ecology, **1**, 69
Environmental warming in shallow lakes: a review of potential changes in community structure as evidenced from space-for-time substitution approaches, **46**, 259
Environmental warming and biodiversity-ecosystem functioning in freshwater microcosms: partitioning the effects of species identity, richness and metabolism, **43**, 177
Estimates of the annual net carbon and water exchange of forests: the EUROFLUX methodology, **30**, 113
Estimating forest growth and efficiency in relation to canopy leaf area, **13**, 327
Estimating relative energy fluxes using the food web, species abundance, and body size, **36**, 137

Evolution and endemism in Lake Biwa, with special reference to its gastropod mollusc fauna, **31**, 149

Evolutionary and ecophysiological responses of mountain plants to the growing season environment, **20**, 60

The evolutionary ecology of carnivorous plants, **33**, 1

Evolutionary inferences from the scale morphology of Malawian Cichlid fishes, **31**, 377

Explosive speciation rates and unusual species richness in haplochromine cichlid fishes: effects of sexual selection, **31**, 235

Extreme climatic events alter aquatic food webs: a synthesis of evidence from a mesocosm drought experiment, **48**, 343

The evolutionary consequences of interspecific competition, **12**, 127

The exchange of ammonia between the atmosphere and plant communities, **26**, 302

Faster, higher and stronger? The Pros and Cons of molecular faunal data for assessing ecosystem condition, **51**, 1

Faunal activities and processes: adaptive strategies that determine ecosystem function, **27**, 92

Fire frequency models, methods and interpretations, **25**, 239

Floods down rivers: from damaging to replenishing forces, **39**, 41

Food webs, body size, and species abundance in ecological community description, **36**, 1

Food webs: theory and reality, **26**, 187

Food web structure and stability in 20 streams across a wide pH gradient, **42**, 267

Forty years of genecology, **2**, 159

Foraging in plants: the role of morphological plasticity in resource acquisitions, **25**, 160

Fossil pollen analysis and the reconstruction of plant invasions, **26**, 67

Fractal properties of habitat and patch structure in benthic ecosystems, **30**, 339

Free air carbon dioxide enrichment (FACE) in global change research: a review, **28**, 1

From Broadstone to Zackenberg: space, time and hierarchies in ecological networks, **42**, 1

From natural to degraded rivers and back again: a test of restoration ecology theory and practice, **44**, 119

Functional traits and trait-mediated interactions: connecting community-level interactions with ecosystem functioning, **52**, 319

The general biology and thermal balance of penguins, **4**, 131
General ecological principles which are illustrated by population studies of Uropodid mites, **19**, 304
Generalist predators, interactions strength and food web stability, **28**, 93
Genetic correlations in multi-species plant/herbivore interactions at multiple genetic scales: implications for eco-evolutionary dynamics, **50**, 263
Genetic and phenotypic aspects of life-history evolution in animals, **21**, 63
Geochemical monitoring of atmospheric heavy metal pollution: theory and applications, **18**, 65
Global climate change leads to mistimed avian reproduction, **35**, 89
Global persistence despite local extinction in acarine predator-prey systems: lessons from experimental and mathematical exercises, **37**, 183
Habitat isolation reduces the temporal stability of island ecosystems in the face of flood disturbance, **48**, 225
Heavy metal tolerance in plants, **7**, 2
Herbivores and plant tannins, **19**, 263
High-Arctic plant–herbivore interactions under climate influence, **40**, 275
High-Arctic soil CO_2 and CH_4 production controlled by temperature, water, freezing, and snow, **40**, 441
Historical changes in environment of Lake Titicaca: evidence from Ostracod ecology and evolution, **31**, 497
How agricultural intensification affects biodiversity and ecosystem services **55**, 43
How well known is the ichthyodiversity of the large East African lakes? **31**, 17
Human and environmental factors influence soil faunal abundance-mass allometry and structure, **41**, 45
Human ecology is an interdisciplinary concept: a critical inquiry, **8**, 2
Hutchinson reversed, or why there need to be so many species, **43**, 1
Hydrology and transport of sediment and solutes at Zackenberg, **40**, 197
The Ichthyofauna of Lake Baikal, with special reference to its zoogeographical relations, **31**, 81
Impact of climate change on fishes in complex Antarctic ecosystems, **46**, 351
Impacts of warming on the structure and functioning of aquatic communities: individual- to ecosystem-level responses, **47**, 81
Implications of phylogeny reconstruction for Ostracod speciation modes in Lake Tanganyika, **31**, 301

Importance of climate change for the ranges, communities and conservation of birds, **35**, 211
Increased stream productivity with warming supports higher trophic levels, **48**, 285
Individual-based food webs: species identity, body size and sampling effects, **43**, 211
Individual variability: the missing component to our understanding of predator–prey interactions, **52**, 19
Individual variation decreases interference competition but increases species persistence, **52**, 45
Industrial melanism and the urban environment, **11**, 373
Individual trait variation and diversity in food webs, **50**, 203
Inherent variation in growth rate between higher plants: a search for physiological causes and ecological consequences, **23**, 188; **34**, 283
Insect herbivory below ground, **20**, 1
Insights into the mechanism of speciation in Gammarid crustaceans of Lake Baikal using a population-genetic approach, **31**, 219
Interaction networks in agricultural landscape mosaics, **49**, 291
Integrated coastal management: sustaining estuarine natural resources, **29**, 241
Integration, identity and stability in the plant association, **6**, 84
Intrinsic and extrinsic factors driving match–mismatch dynamics during the early life history of marine fishes, **47**, 177
Inter-annual variability and controls of plant phenology and productivity at Zackenberg, **40**, 249
Introduction, **38**, 1
Introduction, **39**, 1
Introduction, **40**, 1
Isopods and their terrestrial environment, **17**, 188
Lake Biwa as a topical ancient lake, **31**, 571
Lake flora and fauna in relation to ice-melt, water temperature, and chemistry at Zackenberg, **40**, 371
The landscape context of flooding in the Murray–Darling basin, **39**, 85
Landscape ecology as an emerging branch of human ecosystem science, **12**, 189
Late quaternary environmental and cultural changes in the Wollaston Forland region, Northeast Greenland, **40**, 45
Learning ecological networks from next-generation sequencing data, **54**, 1
Linking biodiversity, ecosystem functioning and services, and ecological resilience: Towards an integrative framework for improved management, **53**, 55

Linking spatial and temporal change in the diversity structure of ancient lakes: examples from the ecology and palaeoecology of the Tanganyikan Ostracods, **31**, 521

Litter decomposition as an indicator of stream ecosystem functioning at local-to-continental scales: insights from the European *RivFunction* project, **55**, 99

Litter fall, **38**, 19

Litter production in forests of the world, **2**, 101

Locally extreme environments as natural long-term experiments in ecology, **55**, 283

Long-term changes in Lake Balaton and its fish populations, **31**, 601

Long-term dynamics of a well-characterised food web: four decades of acidification and recovery in the broadstone stream model system, **44**, 69

Macrodistribution, swarming behaviour and production estimates of the lakefly *Chaoborus edulis* (Diptera: Chaoboridae) in Lake Malawi, **31**, 431

Making waves: the repeated colonization of fresh water by Copepod crustaceans, **31**, 61

Manipulating interaction strengths and the consequences for trivariate patterns in a marine food web, **42**, 303

Manipulative field experiments in animal ecology: do they promise more than they can deliver? **30**, 299

Marine ecosystem regime shifts induced by climate and overfishing: a review for the Northern Hemisphere, **47**, 303

Mathematical model building with an application to determine the distribution of Durshan® insecticide added to a simulated ecosystem, **9**, 133

Mechanisms of microthropod-microbial interactions in soil, **23**, 1

Mechanisms of primary succession: insights resulting from the eruption of Mount St Helens, **26**, 1

Mesocosm experiments as a tool for ecological climate-change research, **48**, 71

Methods in studies of organic matter decay, **38**, 291

The method of successive approximation in descriptive ecology, **1**, 35

Meta-analysis in ecology, **32**, 199

Microbial experimental systems in ecology, **37**, 273

Microevolutionary response to climatic change, **35**, 151

Migratory fuelling and global climate change, **35**, 33

The mineral nutrition of wild plants revisited: a re-evaluation of processes and patterns, **30**, 1

Modelling interaction networks for enhanced ecosystem services in agroecosystems, **49**, 437

Modelling terrestrial carbon exchange and storage: evidence and implications of functional convergence in light-use efficiency, **28**, 57

Modelling the potential response of vegetation to global climate change, **22**, 93

Module and metamer dynamics and virtual plants, **25**, 105

Modeling individual animal histories with multistate capture–recapture models, **41**, 87

Mutualistic interactions in freshwater modular systems with molluscan components, **20**, 126

Mycorrhizal links between plants: their functioning and ecological significances, **18**, 243

Mycorrhizas in natural ecosystems, **21**, 171

The nature of species in ancient lakes: perspectives from the fishes of Lake Malawi, **31**, 39

Networking agroecology: Integrating the diversity of agroecosystem interactions, **49**, 1

A network-based method to detect patterns of local crop biodiversity: Validation at the species and infra-species levels, **53**, 259

Nitrogen dynamics in decomposing litter, **38**, 157

Nocturnal insect migration: effects of local winds, **27**, 61

Nonlinear stochastic population dynamics: the flour beetle *Tribolium* as an effective tool of discovery, **37**, 101

Nutrient cycles and H^+ budgets of forest ecosystems, **16**, 1

Nutrients in estuaries, **29**, 43

On the evolutionary pathways resulting in C_4 photosynthesis and crassulacean acid metabolism (CAM), **19**, 58

Origin and structure of secondary organic matter and sequestration of C and N, **38**, 185

Oxygen availability as an ecological limit to plant distribution, **23**, 93

Parasitism between co-infecting bacteriophages, **37**, 309

Persistence of plants and pollinators in the face of habitat loss: Insights from trait-based metacommunity models, **53**, 201

Scaling-up trait variation from individuals to ecosystems, **52**, 1

Temporal variability in predator–prey relationships of a forest floor food web, **42**, 173

The past as a key to the future: the use of palaeoenvironmental understanding to predict the effects of man on the biosphere, **22**, 257

Towards an integration of biodiversity–ecosystem functioning and food web theory to evaluate relationships between multiple ecosystem services, **53**, 161

Pattern and process of competition, **4**, 11

Permafrost and periglacial geomorphology at Zackenberg, **40**, 151

Perturbing a marine food web: consequences for food web structure and trivariate patterns, **47**, 349

Phenetic analysis, tropic specialization and habitat partitioning in the Baikal Amphipod genus *Eulimnogammarus* (Crustacea), **31**, 355

Photoperiodic response and the adaptability of avian life cycles to environmental change, **35**, 131

Phylogeny of a gastropod species flock: exploring speciation in Lake Tanganyika in a molecular framework, **31**, 273

Phenology of high-Arctic arthropods: effects of climate on spatial, seasonal, and inter-annual variation, **40**, 299

Phytophages of xylem and phloem: a comparison of animal and plant sapfeeders, **13**, 135

Population and community body size structure across a complex environmental gradient, **52**, 115

The population biology and Turbellaria with special reference to the freshwater triclads of the British Isles, **13**, 235

Population cycles in birds of the Grouse family (Tetraonidae), **32**, 53

Population cycles in small mammals, **8**, 268

Population dynamical responses to climate change, **40**, 391

Population dynamics, life history, and demography: lessons from *Drosophila*, **37**, 77

Population dynamics in a noisy world: lessons from a mite experimental system, **37**, 143

Population regulation in animals with complex life-histories: formulation and analysis of damselfly model, **17**, 1

Positive-feedback switches in plant communities, **23**, 264

The potential effect of climatic changes on agriculture and land use, **22**, 63

Predation and population stability, **9**, 1

Predicted effects of behavioural movement and passive transport on individual growth and community size structure in marine ecosystems, **45**, 41

Predicting the responses of the coastal zone to global change, **22**, 212

Predictors of individual variation in movement in a natural population of threespine stickleback (*Gasterosteus aculeatus*), **52**, 65

Present-day climate at Zackenberg, **40**, 111

The pressure chamber as an instrument for ecological research, **9**, 165
Primary production by phytoplankton and microphytobenthos in estuaries, **29**, 93
Principles of predator-prey interaction in theoretical experimental and natural population systems, **16**, 249
The production of marine plankton, **3**, 117
Production, turnover, and nutrient dynamics of above and below ground detritus of world forests, **15**, 303
Protecting an ecosystem service: approaches to understanding and mitigating threats to wild insect pollinators, **54**, 135
Quantification and resolution of a complex, size-structured food web, **36**, 85
Quantifying the biodiversity value of repeatedly logged rainforests: gradient and comparative approaches from borneo, **48**, 183
Quantitative ecology and the woodland ecosystem concept, **1**, 103
Realistic models in population ecology, **8**, 200
Recovery and nonrecovery of freshwater food webs from the effects of acidification, **55**, 469
Recommendations for the next generation of global freshwater biological monitoring tools, **55**, 609
References, **38**, 377
The relationship between animal abundance and body size: a review of the mechanisms, **28**, 181
Relative risks of microbial rot for fleshy fruits: significance with respect to dispersal and selection for secondary defence, **23**, 35
Renewable energy from plants: bypassing fossilization, **14**, 57
Responses of soils to climate change, **22**, 163
Rodent long distance orientation ("homing"), **10**, 63
The role of body size in complex food webs: a cold case, **45**, 181
The role of body size variation in community assembly, **52**, 201
Scale effects and extrapolation in ecological experiments, **33**, 161
Scale dependence of predator-prey mass ratio: determinants and applications, **45**, 269
Scaling of food-web properties with diversity and complexity across ecosystems, **42**, 141
Scaling from traits to ecosystems: developing a general trait driver theory via integrating trait-based and metabolic scaling theories, **52**, 249
Secondary production in inland waters, **10**, 91
Seeing double: size-based and taxonomic views of food web structure, **45**, 67
The self-thinning rule, **14**, 167

Shifting impacts of climate change: Long-term patterns of plant response to elevated CO_2, drought, and warming across ecosystems, **55**, 437

Shifts in the Diversity and Composition of Consumer Traits Constrain the Effects of Land Use on Stream Ecosystem Functioning, **52**, 169

A simulation model of animal movement patterns, **6**, 185

Snow and snow-cover in central Northeast Greenland, **40**, 175

Soil and plant community characteristics and dynamics at Zackenberg, **40**, 223

Soil arthropod sampling, **1**, 1

Soil diversity in the Tropics, **21**, 316

Soil fertility and nature conservation in Europe: theoretical considerations and practical management solutions, **26**, 242

Solar ultraviolet-b radiation at Zackenberg: the impact on higher plants and soil microbial communities, **40**, 421

Some economics of floods, **39**, 125

Spatial and inter-annual variability of trace gas fluxes in a heterogeneous high-Arctic landscape, **40**, 473

Spatial root segregation: are plants territorials? **28**, 145

Species abundance patterns and community structure, **26**, 112

Stochastic demography and conservation of an endangered perennial plant (*Lomatium bradshawii*) in a dynamic fire regime, **32**, 1

Stomatal control of transpiration: scaling up from leaf to regions, **15**, 1

Stream ecosystem functioning in an agricultural landscape: the importance of terrestrial–aquatic linkages, **44**, 211

Structure and function of microphytic soil crusts in wildland ecosystems of arid to semiarid regions, **20**, 180

Studies on the cereal ecosystems, **8**, 108

Studies on grassland leafhoppers (Auchenorrhbyncha, Homoptera) and their natural enemies, **11**, 82

Studies on the insect fauna on Scotch Broom *Sarothamnus scoparius* (L.) Wimmer, **5**, 88

Sustained research on stream communities: a model system and the comparative approach, **41**, 175

Systems biology for ecology: from molecules to ecosystems, **43**, 87

The study area at Zackenberg, **40**, 101

Sunflecks and their importance to forest understorey plants, **18**, 1

A synopsis of the pesticide problem, **4**, 75

The temperature dependence of the carbon cycle in aquatic ecosystems, **43**, 267

Temperature and organism size – a biological law for ecotherms? **25**, 1

Terrestrial plant ecology and ^{15}N natural abundance: the present limits to interpretation for uncultivated systems with original data from a Scottish old field, **27**, 133

Theories dealing with the ecology of landbirds on islands, **11**, 329

A theory of gradient analysis, **18**, 271; **34**, 235

Throughfall and stemflow in the forest nutrient cycle, **13**, 57

Tiddalik's travels: the making and remaking of an aboriginal flood myth, **39**, 139

Towards understanding ecosystems, **5**, 1

Tradeoffs and compatibilities among ecosystem services: biological, physical and economic drivers of multifunctionality, **54**, 207

Trends in the evolution of Baikal amphipods and evolutionary parallels with some marine Malacostracan faunas, **31**, 195

Trophic interactions in population cycles of voles and lemmings: a model-based synthesis **33**, 75

The use of perturbation as a natural experiment: effects of predator introduction on the community structure of zooplanktivorous fish in Lake Victoria, **31**, 553

The unique contribution of rothamsted to ecological research at large temporal scales, **55**, 3

The use of statistics in phytosociology, **2**, 59

Unanticipated diversity: the discovery and biological exploration of Africa's ancient lakes, **31**, 1

Understanding ecological concepts: the role of laboratory systems, **37**, 1

Understanding the social impacts of floods in Southeastern Australia, **39**, 159

Unravelling the impacts of micropollutants in aquatic ecosystems: interdisciplinary studies at the interface of large-scale ecology, **55**, 183

Using fish taphonomy to reconstruct the environment of ancient Lake Shanwang, **31**, 483

Using large-scale data from ringed birds for the investigation of effects of climate change on migrating birds: pitfalls and prospects, **35**, 49

Vegetation, fire and herbivore interactions in heathland, **16**, 87

Vegetational distribution, tree growth and crop success in relation to recent climate change, **7**, 177

Vertebrate predator–prey interactions in a seasonal environment, **40**, 345

The visualisation of ecological networks, and their use as a tool for engagement, advocacy and management, **54**, 41

Water flow, sediment dynamics and benthic biology, **29**, 155

When ranges collide: evolutionary history, phylogenetic community interactions, global change factors, and range size differentially affect plant productivity, **50**, 293

When microscopic organisms inform general ecological theory, **43**, 45

10 years later: Revisiting priorities for science and society a decade after the millennium ecosystem assessment, **53**, 1

Zackenberg in a circumpolar context, **40**, 499

The zonation of plants in freshwater lakes, **12**, 37.

Edwards Brothers Malloy
Ann Arbor MI. USA
October 6, 2016